Stressors in the Marine Environment

Stressors in the Marine Environment

Physiological and ecological responses; societal implications

EDITED BY

Martin Solan
University of Southampton, UK

Nia M. Whiteley
Bangor University, UK

OXFORD

UNIVERSITY PRESS

Great Clarendon Street, Oxford, OX2 6DP,
United Kingdom

Oxford University Press is a department of the University of Oxford.
It furthers the University's objective of excellence in research, scholarship,
and education by publishing worldwide. Oxford is a registered trade mark of
Oxford University Press in the UK and in certain other countries

Published in the United States of America by Oxford University Press
198 Madison Avenue, New York, NY 10016, United States of America

British Library Cataloguing in Publication Data

Data available

Library of Congress Control Number: 2015955002

ISBN 978–0–19–871883–3

Preface

The biological composition and richness of most of the Earth's major ecosystems are being dramatically transformed—to a significant extent irreversibly—by anthropogenic activity. The oceans form a considerable sink for heat and carbon dioxide, and ozone-related increases in UV-B radiation can negatively influence many aquatic species and ecosystems. At the same time, terrestrial flooding and polar meltwaters contribute to the freshening of many coastal regions, while land run-off brings chemical pollutants and nutrients that can lead to eutrophication and the development of harmful algal blooms and hypoxic 'dead zones'. Human activities, such as offshore construction and the transport of cargo using ships, also generate novel sound fields that can affect species behaviour. Marine environments are particularly vulnerable to such changes because approximately 40% of the world's population live within 100 km of the coast, yet a significant proportion of these inhabitants also depend on the ocean for food, economic prosperity, and well-being. Consequently, the cumulative effect of multiple stressors on ecosystems is now a major source of concern to society and is fast becoming a prominent research goal, attracting interest from those tasked with managing the environment or developing environmental policy.

Understanding and predicting the combined impacts of single and multiple stressors is, however, particularly challenging because observed ecological responses are underpinned by a number of physiological and behavioural responses that are affected by the type, severity, and timing of stressors, yet integration between the traditional domains of physiology and ecology is fragmented and often focused towards a specific set of circumstances. Environmental or comparative animal physiology texts either treat the subject area system by system, moving, for example, from cardiovascular responses to nerve and muscle function, or by considering the challenges posed by specific environments, such as polar or estuarine habitats. Similarly, most general ecology-based textbooks, across a range of systems, provide only superficial considerations of environmental stressors, tending (in a marine context) to deal mainly with the adjustments required to survive or tolerate changes in temperature, salinity, oxygen, and/or pressure. Hence, the purpose of this book is to provide in a single volume an overview of the physiological and ecological responses of marine species to a wide range of potential stressors resulting from contemporary anthropogenic activity, while referencing the effects that this may have for the oceans, other systems, and for human well-being.

Recent syntheses of the available literature continue to present strong evidence that the cumulative effect of multiple stressors on a variety of marine species and habitats tends to be variable (additive, antagonistic, and synergistic effects), but they also reveal that the suite of effects brought about by individual anthropogenic forcing has been poorly represented in experimental manipulations. Indeed, it is clear from the literature that individual studies tend to refrain from considering physiological and behavioural mechanisms alongside the ecological and societal significance of their findings, and that studies tend to be either ecologically or physiologically themed. Our vision in compiling this volume was to provide a gateway to targeted and authoritative information within a discipline, while presenting alternative perspectives in relevant sister contributions. In doing so, we focus on eight stressors (salinity, hypoxia, ocean acidification, temperature, chemical pollution, nitrogen deposition, ultraviolet radiation, and noise) that are particularly prevalent in coastal and shelf sea environments, before offering perspectives on the concepts of thresholds and tipping points with just social foundations, economic valuation of stressor-mediated change, and, considering the many sectors of human society that utilize the oceans, how best to manage multiple stressors for society as a whole.

In order to cover both physiological and ecological aspects as well as the societal implications, we have brought together a range of expertise from within the marine community, including early career researchers and established leaders in the field, and combined this with internationally recognized researchers from the fields of physiological processes and ecological systems to broaden the scope and generic value of the volume. Collectively, the authors have broad interests across many different marine species (animals/plants/microbes) and habitats, and cover a range of technical expertise including environmental physiology and ecological theory, mathematical modelling, statistics, empirical and field research, social science, economics, and policy. Throughout, there is an emphasis on the species and communities that are the most affected, such as marine calcifiers in the case of ocean acidification and marine mammals in the case of noise pollution. Consideration is given to the damage caused by each pollutant, whether this is morphological or biochemical, and the means by which some species can recover, as well as the energetic implications that have far-reaching consequences by influencing the function and survival of populations and hence marine communities and ecosystems. Common themes include differences in sensitivity to specific stressors among taxa, species, and life stages, and the logistical challenges represented by introducing additional stressors to experiments, along with requirements to provide local, regional, and global perspectives. The requirement for laboratory experiments to be carried out alongside, and be informed by, field and 'natural' experiments wherever possible is also apparent. Overall, we envisage that this book will be a valuable reference for students, researchers, and those tasked with conserving or managing marine systems in both the natural and social sciences by outlining our current but patchy understanding of the complex interactions between various stressors faced by species, populations, and communities, and informing on future avenues for experimentation and observation using a combination of both laboratory and field-based studies.

Martin Solan and Nia M. Whiteley

Acknowledgement

We are particularly grateful, in no small measure, to the administrative support provided by Natalie Tulett, University of Southampton, in the compilation of this book.

Contents

Part II Ecological Responses   159

9 Effects of changing salinity on the ecology of the marine environment   161
Katie Smyth and Mike Elliott

10 The ecological consequences of marine hypoxia: from behavioural to ecosystem responses   175
Bettina Riedel, Robert Diaz, Rutger Rosenberg, and Michael Stachowitsch

11 Ecological effects of ocean acidification   195
M. Débora Iglesias-Rodriguez, Katharina E. Fabricius, and Paul McElhany

Part III Societal Implications   **299**

17 Managing complex systems to enhance sustainability   **301**
Simon Willcock, Sarwar Hossain, and Guy M. Poppy

18 Using the Ecosystem Approach to manage multiple stressors in marine environments   **313**
Zoë Austin and Piran C.L. White

List of Contributors

Editors

Martin Solan Ocean and Earth Science, National Oceanography Centre Southampton, University of Southampton Waterfront Campus, European Way, Southampton SO14 3ZH, United Kingdom

Nia M. Whiteley School of Biological Sciences, College of Natural Sciences, Bangor University, Bangor, Gwynedd LL57 2UW, United Kingdom

Authors

Natacha Aguilar de Soto BIOECOMAC, University of la Laguna, La Laguna 38206, Tenerife, Canary Islands, Spain, Sea Mammal Research Unit, Scottish Oceans Institute, University of St Andrews, St Andrews, Fife KY16 8LB, United Kingdom

Sabine E. Apitz SEA Environmental Decisions Ltd, 1 South Cottages, The Ford, Little Hadham, Hertfordshire SG11 2AT, United Kingdom

Zoë Austin Environment Department, University of York, Heslington, York YO10 5DD, United Kingdom

Elia Beniash University of Pittsburgh, School of Dental Medicine, Pittsburgh, PA 15261, USA

John A. Berges Department of Biological Sciences and School of Freshwater Sciences, University of Wisconsin-Milwa, Milwaukee, WI 53211, USA

Alastair Brown Ocean and Earth Science, University of Southampton, National Oceanography Centre Southampton, European Way, Southampton SO14 3ZH, United Kingdom

David J. Burritt Department of Botany, University of Otago, Dunedin, New Zealand

Zanna Chase Institute for Marine and Antarctic Studies, University of Tasmania, Private Bag 49, Hobart, Tasmania 7001, Australia

Benjamin J. Ciotti Ocean and Earth Science, University of Southampton, National Oceanography Centre Southampton, European Way, Southampton SO14 3ZH, United Kingdom

Robert Diaz Virginia Institute of Marine Science, College of William and Mary, Gloucester Point, VA 23062, USA

Gary H. Dickinson Biology Department, The College of New Jersey, Ewing, NJ 08628, USA

Mike Elliott Institute of Coastal and Estuarine Studies, School of Biological Biomedical and Environmental Sciences, University of Hull, Hull HU6 7RX, United Kingdom

Ruth S. Eriksen Institute for Marine and Antarctic Studies, University of Tasmania, Private Bag 49, Hobart, Tasmania 7001, Australia

Katharina E. Fabricius Australian Institute of Marine Science, PMB 3, Townsville Q4810, Queensland, Australia

Gustavo Ferreyra Institut des sciences de la mer de Rimouski (ISMER), Université du Québec à Rimouski (UQAR), 310 allée des Ursulines, Rimouski, QC G5L 3A1, Canada

Nick Hanley Department of Geography and Sustainable Development, Irvine Building, University of St Andrews, St Andrews, KY16 9AL Fife Scotland, United Kingdom

Chris Hauton Ocean and Earth Science, National Oceanography Centre Southampton, University of Southampton Waterfront Campus, European Way, Southampton SO14 3ZH, United Kingdom

Sarwar Hossain Geography and Environment, University of Southampton, Southampton SO17 1BJ, United Kingdom

M. Débora Iglesias-Rodriguez Department of Ecology, Evolution and Marine Biology, University of California Santa Barbara, Santa Barbara, CA 93106, USA

Andrew G. Jeffs Leigh Marine Laboratory, Institute of Marine Science, University of Auckland, Auckland, New Zealand

Caitlin Kight Centre for Ecology and Conservation Biosciences, College of Life and Environmental Sciences, University of Exeter, Penryn Campus, Penryn, Cornwall TR10 9FE, United Kingdom

Miles D. Lamare Department of Marine Science, University of Otago, Dunedin 9054, New Zealand

Ceri Lewis Biosciences, College of Life and Environmental Sciences, University of Exeter, Exeter EX4 4QD, United Kingdom

Clara L. Mackenzie Centre for Marine Biodiversity and Biotechnology, School of Life Sciences, Heriot-Watt University, Edinburgh EH14 4AS, United Kingdom

Catriona K. Macleod Institute for Marine and Antarctic Studies, University of Tasmania, Hobart, Tasmania 7001, Australia

Omera B. Matoo Department of Biological Sciences, University of North Carolina at Charlotte, Charlotte, NC 28223, USA

Paul McElhany NOAA Fisheries, Northwest Fisheries Science Center, East Seattle, WA 98112, USA

Sébastien Moreau Georges Lemaître Centre for Earth and Climate Research, Earth and Life Institute, Universite catholique de Louvain, Louvain-La-Neuve, Belgium

Elizabeth A. Morgan Ocean and Earth Science, University of Southampton, National Oceanography Centre Southampton, European Way, Southampton SO14 3ZH, United Kingdom

Behzad Mostajir Center of Marine Biodiversity Exploitation and Conservation (MARBEC), UMR 9190: CNRS-Université de Montpellier-IRD-Ifremer, Place E. Bataillon Université de Montpellier Case 93, Montpellier Cedex 05, France

Anouska Panton Ocean and Earth Science, University of Southampton, National Oceanography Centre Southampton, European Way, Southampton SO14 3ZH, United Kingdom

Guy M. Poppy Centre for Biological Sciences, Faculty of Natural and Environmental Sciences, Life Sciences, University of Southampton, Highfield Campus, Southampton SO17 1BJ, United Kingdom

Bettina Riedel Department of Limnology and Bio-Oceanography, University of Vienna, 1090 Vienna, Austria, Laboratoire LPG-BIAF Bio-Indicateurs Actuels et Fossiles, UMR CNRS 6112, Université d'Angers, 2 Bd Lavoisier, Angers 49045 CEDEX, France

Rutger Rosenberg Marine Monitoring AB, Strandvägen 9, SE-453 30 Lysekil, Sweden, Department of Biology and Environmental Science—Kristineberg, University of Gothenburg, SE-451 78 Fiskebäckskil, Sweden

Eduarda M. Santos Biosciences, College of Life and Environmental Sciences, University of Exeter, Exeter EX4 4QD, United Kingdom

Katie Smyth Institute of Coastal and Estuarine Studies, School of Biological Biomedical and Environmental Sciences, University of Hull, Hull HU6 7RX, United Kingdom

Inna M. Sokolova Department of Biological Sciences, University of North Carolina at Charlotte, Charlotte, NC 28223, USA

Martin Solan Ocean and Earth Science, National Oceanography Centre Southampton, University of Southampton Waterfront Campus, European Way, Southampton SO14 3ZH, United Kingdom

John I. Spicer Marine Biology and Ecology Research Centre, School of Marine Science and Engineering, University of Plymouth, Plymouth Devon PL4 8AA, United Kingdom

Michael Stachowitsch Department of Limnology and Bio-Oceanography, University of Vienna, Althanstrasse 14, 1090 Vienna, Austria

Jenni A. Stanley Leigh Marine Laboratory, Institute of Marine Science, University of Auckland, Auckland, New Zealand

Francesca Vidussi Center of Marine Biodiversity Exploitation and Conservation (MARBEC), UMR 9190: CNRS-Université de Montpellier-IRD-Ifremer, Place E. Bataillon Université de Montpellier Case 93, Montpellier Cedex 05, France

Piran C.L. White Environment Department, University of York, Heslington, York YO10 5DD, United Kingdom

Nia M. Whiteley School of Biological Sciences, College of Natural Sciences, Bangor University, Bangor, Gwynedd LL57 2UW, United Kingdom

Simon Willcock Centre for Biological Sciences, Faculty of Natural and Environmental Sciences, Life Sciences, University of Southampton, Highfield Campus, Southampton SO17 1BJ, United Kingdom

Erica B. Young Department of Biological Sciences and School of Freshwater Sciences, University of Wisconsin-Milwaukee, Milwaukee, WI 53211, USA

PART I

Physiological Responses

CHAPTER 1

Effects of salinity as a stressor to aquatic invertebrates

Chris Hauton

1.1 Introduction

The homeostases of ionic composition and cell volume regulation are fundamentally important prerequisites for successful persistence, growth, and development of aquatic species. Ion regulation is directly necessary to maintain optimal electrostatic interactions of enzymes and substrates and receptors and their ligands (Dubyak, 2004; Fernandez-Reiriz et al., 2005), as well as to maintain ion gradients across membranes, protein phosphorylation, and genomic integrity (Kültz, 2005) and the transduction of impulses in nerve cells (Silver et al., 1997). Cellular ionic homeostasis is also essential for the maintenance of cellular osmotic potential, as cell cytoplasm ion concentrations can be altered to permit the uptake of necessary organic molecules that have their own osmotic potential. Maintenance of cell and tissue volume within an organism is essential as dramatic changes in volume can disrupt cell membrane integrity and cell structure. In addition, excess water in cells can have a fundamental impact on protein function and performance within the cell (Lang, 2007).

Deviations from the maintenance of cell ionic concentration or volume can lead to stress. At a cellular level, stress can be defined as the impact of: 'environmental force(s) on macromolecules,' (Kültz, 2005). If not corrected, this can result in the manifestation of stress response at the level of the organism. Organismal stress has been variously defined in the literature previously, but remains a challenging concept. Barton (2002) has defined organismal stress as: 'a non-specific response of a body to any demand placed upon it such that it cases an extension of a physiological stated beyond a normal resting state,' which reflects the integrated and longer–term response of the organism to perturbation. Both definitions are of value to this chapter, which considers both cellular and whole organism integrated responses to changes in environmental salinity that have been reported in the literature.

For open ocean species these demands are limited as the salinity of the world's oceans is generally within the range of 34.6–34.8 (Worthington, 1981). However, for neritic and estuarine species deviations from this mean can be extreme, extending from salinities of 0 to above 40 and subject to change on a tidal or even hourly basis (McAllen and Taylor, 2001). Nonetheless, these environments are some of the most productive across the world and support shellfish production for many of the world's nations (Field et al., 1998). The importance of marine invertebrate shellfish fisheries and aquaculture within these challenging and changing aquatic environments has been the motivation for a considerable body of research on the direct and indirect impacts of salinity on marine invertebrates. This chapter will focus on the impacts of changes in salinity as a primary stressor of aquatic invertebrates, ranging from freshwater crustaceans to fully marine corals and echinoderms.

For coastal marine and freshwater aquatic species homeostasis is complex as their body surfaces are not completely impermeable to the external environment, an environment which often has a different osmotic potential to the organism's internal (cellular/tissue) environment. In considering osmoregulatory capacity, species can be defined along a continuum from stenohaline—having a narrow tolerance range for environment salinity—to euryhaline—tolerating a wide range of environment salinity (Schmidt-Nielsen, 1997). Species can also be considered along a spectrum from complete osmoconformers, where the body fluid osmolarity matches the external environment, to complete osmoregulators in which the organism actively controls the osmolarity of the body fluids irrespective of the external osmolarity. In general terms the majority of marine invertebrates are stenohaline in habit

Stressors in the Marine Environment. Edited by Martin Solan and Nia M. Whiteley
© Oxford University Press 2016. Published in 2016 by Oxford University Press.

and a minority are euryhaline, although coastal and estuarine environments are dominated by euryhaline species. Within the euryhaline group most are osmoconformers that can only control their osmolarity at a cellular level or by behavioural modification. Euryhaline osmoregulators are mainly comprised of the Crustacea, and this group actively regulate the osmolarity of their body fluids (Davenport, 1985; see also Henry, 2001).

1.2 An overview of the mechanisms for osmotic control

Davenport (1985) and Lang (2007) have provided comprehensive reviews of the mechanisms for osmotic control in cells and, specifically, in marine fauna. Dubyak (2004), Henry et al. (2012), and McNamara and Faria (2012) have also provided excellent detailed accounts of cellular molecular mechanisms of ion homeostasis. Complete details of mechanisms for cellular osmotic control are beyond the scope of this chapter to review. In brief, however, maintenance of the cell osmolarity and cell volume through regulation of the free amino acid pool (FAAP) and ion exchange are features of all cells adjusting to a new extracellular osmotic environment, whether that be the osmoconforming extracellular fluid of a bivalve such as the mussel *Mytilus edulis*, or the more regulated extracellular fluid of a crustacean osmoregulator, such as the European shore crab *Carcinus maenas*. Both intracellular ion regulation and regulation of the FAAP can be used to regulate cell volume (Fig. 1.1). However, a major constraint on varying ionic concentrations intracellularly is that very quickly this can have significant detrimental impacts on enzyme interactions and metabolic pathways. As a result, intracellular osmotic pressure is substantially created (< 60–70%) and regulated using organic molecules (Yancey et al., 1982; Davenport, 1985; see also Deaton and Pearce, 1994, and other papers in that special issue). Control of cell volume by FAAP regulation is considered in further detail in Section 1.4, which considers the cellular homeostatic response to salinity stress.

In addition to the maintenance of cell volume, osmoregulating organisms regulate the ionic composition and osmolarity of the extracellular fluids relative to the changing environmental conditions. As described in Davenport (1985), most aquatic invertebrates are hyperosmotic regulators which maintain their osmolarity above that of the environment at low salinities, becoming iso-osmotic at high salinity. Ionic regulation in osmoregulating crustaceans is achieved via membrane-bound ion exchange pumps that are concentrated within the gill epithelial, and in many species predominantly within the posterior gills (Neufeld et al., 1980; Henry and Cameron, 1982; Boettcher et al., 1995).

A large suite of ion regulatory pumps and channels have been identified, mainly from crustaceans, including Na^+/K^+-ATPases, V-type proton- ATPases, bicarbonate-ATPases, K^+ and Cl^- channels, Na^+ channels, Cl^-/bicarbonate exchangers, $Na^+/K^+/2Cl^-$ co-transporter, Ca^{2+}-pumps, Na^+/Ca^{2+} exchangers, and carbonic anhydrases (CA) (Henry, 1984; Henry et al., 2012; and McNamara and Faria, 2012). Increased activity of membrane-associated Na^+/K^+-ATPases and CAs, as well as the gene transcription of new Na^+/K^+-ATPases (e.g. Towle et al., 2001) and CAs (e.g. Henry et al., 2003) have been generally associated with acclimation to low salinity environments in crustaceans (e.g. Pacific white shrimp *Litopenaeus vannamei*, Palacios et al., 2004 and Sun et al., 2011; shore crab *Pachygrapsus marmoratus*, Jayasundara et al., 2007; tiger shrimp *Penaeus monodon*, Pongsomboon et al., 2009; and shore crab *Carcinus maenas*, Towle et al., 2011).

Changes in gene transcription and protein expression do not necessarily have the same temporal profile however. A detailed study of *Carcinus maenas* by Jillette et al. (2011) demonstrated that although there were rapid (<1 week) changes in gene expression of two isoforms of carbonic anhydrase in the posterior gills in response to hypo- and hyperosmotic acclimation, changes in gill carbonic anhydrase enzyme activity had a different time course. On exposure to low salinity (to a salinity of 15, from a control salinity of 32) there was a significant (approximately fourfold) increase in enzyme activity in the posterior gills within one week. However, it took a period of four weeks for that enzyme activity to return to baseline levels once the crabs were returned to the control conditions (salinity 32). These differential time courses identify the importance of establishing changes in physiology at multiple levels of biological organization and not relying on a single measure, for example gene transcription alone.

Further, longer term, acclimation to chronic salinity change has been shown to require the significant synthesis of new protein pumps, in addition to the increased activity of existing pumps (e.g. Towle et al., 2001; Henry et al., 2003). Lovett et al. (2006b) determined the expression profile of Na^+/K^+-ATPase during acute and chronic hypoosmotic stress in the blue crab *Callinectes sapidus*. These authors reported that

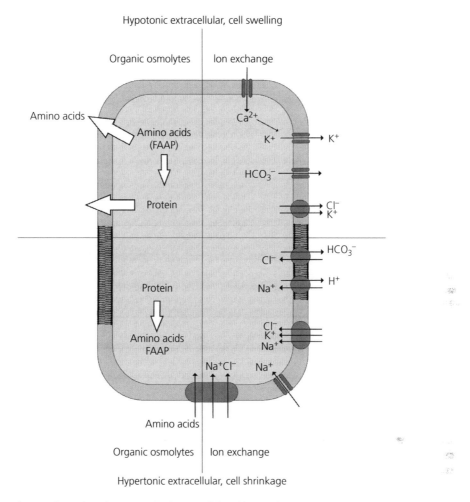

Hypotonic extracellular, cell swelling

Organic osmolytes | Ion exchange

Amino acids

Amino acids
(FAAP)

Ca²⁺

K⁺ → K⁺

HCO₃⁻ →

Protein

Cl⁻
K⁺

HCO₃⁻

Cl⁻ ←

H⁺

Protein

Na⁺ ←

Amino acids
FAAP

Cl⁻
K⁺
Na⁺

Na⁺Cl⁻   Na⁺

Amino acids

Organic osmolytes | Ion exchange

Hypertonic extracellular, cell shrinkage

Figure 1.1 Overview of ionic and organic mechanisms involved in intracellular volume regulation in animal cells, showing an increase in ion uptake and the FAAP in response to hyperosmotic stress, and a reduction in the FAAP and excretion of ions in response to hypoosmotic stress. Adapted from Davenport (1985) and Lang (2007).

crabs exposed to dilute seawater for over 18 days showed a 300% increase in Na⁺/K⁺-ATPase specific activity and a 200% increase in Na⁺/K⁺-ATPase protein levels; chronic exposure to low salinity resulted in the synthesis of new enzyme (Lovett et al., 2006b). Of significance to subsequent discussion is the fact that the sustained activity of membrane-bound ATPases and the transcription and translation of new enzymes represents a significant energetic commitment from the individual in the form of ATP. Na⁺/K⁺-ATPase activity is a major demand on maintenance metabolism (3–40%; Leong and Manahan, 1997) and so osmoregulatory adjustments clearly represent a significant energetic cost to the individual.

Further evidence for the energetic cost of chronic osmoregulation in crustaceans is provided by data showing that, within a species, gill ultrastructure is plastic and can be modified in response to salinity change. Changes in the apical border of gill epithelial cells, mitochondria, and cytoplasmic lacunae have been recorded in amphipods *Gammarus duebeni* grown in high salinity environments (Shires et al., 1994), and proliferation of mitochondria-rich cells in the gills of *C. maenas* transferred to low salinity environments for 4–7 days (Compère et al., 1989). Such tissue reorganization in response to environmental challenge will present a significant energetic cost to the individual. The implications of increased metabolic demands of osmoregulation for whole organism respiration rates are considered further in Section 1.5.

It should not be overlooked that osmoconformers, and some osmoregulators, can also employ behavioural modifications to control tissue osmolarity; this is especially the case for mobile species. Behavioural avoidance is widespread in mobile species (Davenport, 1985) such as the homarid lobsters (Charmantier et al., 2001) and the sandhopper *Talitrus saltator* (Fanini et al., 2012), which preferentially burrows into high salinity sediments. Nonetheless, behavioural control of osmolarity is also common among the bivalve and gastropod molluscs. These groups can either close their valves, clamp down on to the substrate, or seal themselves inside the shell behind the opercular plate. For example, bivalves such as the mussel *Mytilus edulis* close their valves in response to low salinity water (below a salinity 20; Davenport, 1985). Suspension feeding in the gastropod *Crepipatella dilatata* is also arrested at salinities below 20 and is associated with the isolation of the mantle cavity from the environment, a mechanism that also can be employed to create a brood chamber for developing embryos (discussed in Section 1.7; Chapparro et al. (2008)). Alternately, deep burrowing species such as the clam *Mya arenaria* (Davenport, 1985; Deaton, 1992) and the lugworm *Arenicola marina* (Shumway and Davenport, 1977) can retract deeper into their burrow or retreat behind a mucus boundary to avoid exposure to salinity stress from surficial waters. While effective, prolonged exposure to low salinity environments through valve closure can lead to anaerobic metabolism, reduced feeding rates, and, subsequently, impacts to growth (Poulain et al., 2011). Behavioural regulation of osmotic pressure is reviewed more fully in Davenport (1985).

In summary, neritic, estuarine, and freshwater invertebrates are presented with significant challenges to osmotic control that are acute, but temporary, as well as chronic. To meet these challenges all species have evolved mechanisms for behavioural avoidance and intracellular volume regulation. In addition, and mainly a characteristic of the Crustacea, some groups have evolved ion regulatory mechanisms which permit them to also regulate the osmolarity and ionic composition of their extracellular fluids, relative to a changing external osmolarity. All of these mechanisms: (1) behavioural, (2) cell volume control via intracellular amendment of the FAAP and ionic regulation, and (3) regulation of the extracellular fluid in osmoregulators, require the provision of energy in the form of ATP. This requirement underlies the majority of the 'stress responses' which are developed through this chapter.

1.3 The cellular stress response (CSR) to salinity perturbation

At a cellular level responses to environmental perturbation can be divided into two stages. Early responses to stress or insult focus on the counteraction and repair of stress-induced damaged, increased tolerance against further stress damage, and apoptosis or maintenance of the cell cycle (Kültz, 2005). This evolutionarily conserved response, the 'cell stress response (CSR)' is ubiquitous across the different Kingdoms of Life and is triggered as a result of non-specific macromolecular damage. The CSR is a transient state and is followed by the cellular homeostatic response (CHR), which is a semi-permanent state that remains until the environmental conditions of the cell change again (Kültz, 2005). While the CSR is a conserved response, the CHR has the potential to be species-, cell- and even stressor-specific. The homeostases of ion balance and cell volume described in Section 1.2 are components of the CHR.

Of the 44 proteins with known functions in the CSR, a number include members of different molecular weight (MW) families of heat shock proteins (HSPs), including the 60 kDa, 70 kDa, and 90 kDa families, which variously: (a) assist with the refolding of denatured proteins, or (b) chaperone irreversibly damaged proteins for polyubiquitination and degradation at the proteasome (Hochstrasser, 1996). Up-regulation of HSP gene expression in response to high salinity stress has been reported in many aquatic invertebrates (e.g. the ascidian *Styela plicata*, Carmen Pineda et al., 2012; the Chinese mitten crab *Eriocheir sinensis*, Sun et al., 2012; and the estuarine copepod *Eurytemora affinis*, Xuereb et al., 2012). In contrast, in the osmoconforming echinoderm *Apostichopus japonicus* both hyper- and hypoosmotic stress have been shown to increase the expression of 70 kDa HSP (HSP70) proteins (Dong et al., 2008), although the temporal profile of expression differed at different salinities.

However, notable differences do exist, for example the euryhaline osmoregulating European shore crab *Carcinus maenas* did not show elevated HSP70 gene expression in response to salinity stress (Towle et al., 2011) and Werner and Hinton (2000) reported data from field collections and laboratory experiments which identified a *decrease* in the expression of HSP70 proteins in the Asian clam *Potamocorbula amurensis* at extremely low salinities. One requirement for the action of heat shock proteins is the supply of energy in the form of ATP, which drives the conformational changes required in the molecule to support function

(Mayer, 2010); this again represents a cost to the individual which can translate into longer term impacts identified later in this chapter.

While the measurement of HSP expression in response to salinity stress—either in terms of gene transcription or protein concentration by Western blot—is common within the literature, Morris et al. (2013) have recently questioned its widespread continued use, in light of many uncertainties over what is being measured (heat shock protein versus heat shock cognates) and inconsistent responses reported from different populations of the same species, especially in field studies. Morris et al. (2013) have instead argued in favour of the monitoring of stressor specific responses, triggered as part of the cellular homeostatic response (CHR—described in Section 1.4), as being more informative.

A second component of the CSR is the expression of proteins and enzymes of various antioxidant pathways. Intracellular reactive oxygen species (ROS)—including superoxide anions and hydrogen peroxide—are produced continuously as a by-product of routine respiration within the mitochondria. However, cellular ROS production can be increased in response to environment perturbations, such as salinity stress (Paital and Chainy, 2012) or uptake of pollutants, when cellular homeostasis cannot be restored (reviewed by: Luschak, 2011; Galluzzi et al., 2012). Extracellular ROS production can also be increased in response to pathogen infection, accompanying phagocytosis to break down the cell membranes of invading pathogens (described in Section 1.8). Unregulated production of ROS causes harm to the host, which was initially considered in the 'free radical theory of aging' (Harman, 1956), although the significance of this idea has since been questioned, especially for the case of marine invertebrates (Buttemer et al., 2010).

The damaging effects of an excess production of ROS are controlled through the expression of diverse antioxidant protective pathways, including the redox enzymes catalase (CAT), superoxide dismutase (SOD), and glutathione (GSH), as well as the peroxiredoxins (Prxs). The rapid action of all of these proteins protects cells and tissues from the damage induced by environmental insult and also inappropriate immune responses. Copper and zinc conjugated SODs are widespread in the cytoplasm of many eukaryotic cells while manganese-conjugated SODs are found within the mitochondria. SOD enzymes catalyse the dismutation of superoxide anions ($O_2 \bullet^-$) into oxygen and hydrogen peroxide, while catalases further breakdown hydrogen peroxide to water and oxygen.

Peroxiredoxins, which contain a redox-active cysteine residue, also detoxify hydrogen peroxide to water. Glutathione (GSH) functions by providing reducing equivalents for key antioxidant defence enzymes and can also scavenge hydroxyl radicals directly. High levels of glutathione disulphide (GSSG), the oxidized disulphide form of GSH, accumulate during the ROS detoxification processes and it is therefore necessary to recycle GSSG back to the reduced glutathione form, requiring the enzyme glutathione reductase (GR).

Antioxidant responses to the production of free radicals associated with salinity exposure have been reported in marine invertebrates, but as with the HSPs, responses have been shown to vary as a function of species, population, and tissues studied. For example, Paital and Chainy (2010) have reported extensive changes in the antioxidant pathway in the mud crab *Scylla serrata*. In response to an increase in environmental salinity from 10 to 35, they reported a decline in activity of superoxide dismutase (SOD) in abdominal muscle, contrasting with an increase in activity of the catalase enzyme and no change in the activity of glutathione peroxidase (GPx). In the hepatopancreas, however, SOD and CAT activities eventually decreased, while GPx and GR activities consistently decreased with increasing salinity. In gill tissue, SOD activity initially fell before increasing as salinity increased, while decreases in CAT and GPx activities were noted after 21 days (Fig. 1.2). Freire et al. (2011) compared the antioxidant response of the swimming crabs *Callinectes danae* (euryhaline) and *C. ornatus* (stenohaline) in response to hyper-and hypoosmotic stress. *C. danae* displayed higher baseline activities of GPx (in hepatopancreas and muscle) and CAT (in hepatopancreas, muscle, anterior and posterior gills) than *C. ornatus*, which only demonstrated activation of these enzymes when exposed to hypersalinity (40). Rodrigues et al. (2012) also reported little perturbation in the antioxidant response of the euryhaline European shore crab *Carcinus maenas* exposed for seven days in the salinity range of 4–45. Differential responses have also been reported between species of bivalves. Zanette et al. (2011) reported that in the Pacific oyster *Crassostrea gigas*, salinity perturbations of between 35 and 9 did not produce major changes in the gill CAT or GST activity.

From field studies, Philipp et al. (2012) reported population level differences in the expression of genes coding for antioxidant enzymes between populations of the oceanic quahog *Arctica islandica*, a species which can have an extremely long lifespan (> 500 years; Treaster et al., 2014). Philipp et al. (2012) compared

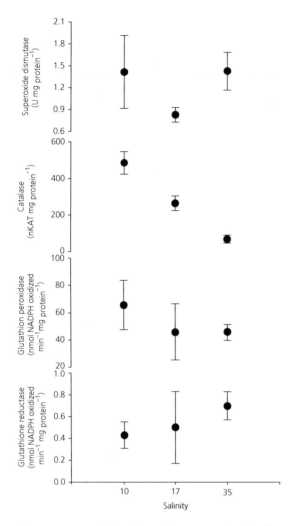

Figure 1.2 Activity of selected antioxidant enzymes within the gill tissue of the mud crab *Scylla serrata* after 21 days acclimation to different salinity, showing the mean ± SD. Figure plotted from data presented in Paital and Chainy (2010) (their Table 3).

lifespan of German Bight clams might be a function of an initial low rate of ROS formation, resulting from a low metabolic rate combined with a high damage repair (antioxidant) capacity.

Antioxidant responses to salinity perturbation have also been shown to vary as a function of the nature of the salinity change, whether acute or chronic. Cailleaud et al. (2007) measured glutathione-S-transferase GST activity in the calanoid copepod *Eurytemora affinis* sampled from the Seine Estuary in France and reported that activity was maximal during acute exposures to salinities within the range of 5–15, while long-term exposure resulted in maximal GST activity at a salinity of 5.

A complication in all of these data is that, as with HSP expression, the temporal component of antioxidant response, and potentially the magnitude of response, can vary as a function the type of measurement taken. The immediate antioxidant responses to perturbations in salinity rely on the activity of existing mature proteins, while longer-term exposures are likely to require increases in gene transcription to maintain or increase the capacity of the antioxidant pathways. As an example, Seo et al. (2006) have reported that expression of the gene encoding for glutathione reductase was significantly increased and sustained from 6 h after exposure to high salinity stress (24 and 40, controls acclimated to 18) in the copepod *Tigriopus japonicus*, while expression to low salt stress (0 and 12) resulted in a down-regulation in the expression. Van Horn et al. (2010) have reported also that in the flatback mud crab *Eurypanopeus depressus* the gene coding for peroxiredoxin was transcribed initially at low levels in the gill, hypodermis, and hepatopancreas of crabs under non-stressed conditions and was only elevated about threefold in gills after 48 h exposure to hypoosmotic stress (acclimation salinity 30, exposure salinity of 10).

1.4 The cellular homeostatic response (CHR) and maintenance of cell volume

Once the initial damage from osmotic stress has been contained the cellular homeostatic response (CHR) regulates cell processes to achieve acclimation to the new extracellular environment. This includes the maintenance of cell volume and hydration by the regulation of cell osmolarity. Osmolarity, and therefore cell volume, in both osmoregulators and osmoconformers can be achieved intracellularly through adjustments to ionic regulation (in part) and the free amino acid

the expression of genes for antioxidant enzymes in German Bight quahogs (maximum lifespan (MSLP) of ~150 years; typical environmental salinity 33) with those of the Baltic Sea (MSLP ~40 years; typical salinity 20–25). These authors concluded that the existence of populations of shorter lifespan quahogs within the variable environment (including salinity) of the Baltic was not a result of metabolic rate depression but a consequence of 'stress-hardening'; an increase in their ability to up-regulate the expression of genes coding for antioxidant enzymes at times of stress. However, Basova et al. (2012) have also argued that the long pool (FAAP). Intracellular amino acids contribute to the intracellular osmotic pressure and, by adjusting the concentration of the FAAP, changes in this osmotic pressure can be achieved to regulate the osmotic water flux between the intracellular and extracellular compartments. Adjustments to the FAAP (Fig. 1.1) occur via the catabolism and anabolism of intracellular protein (e.g. Gaspar Martins and Bianchini, 2009), by the *de novo* synthesis of amino acids in high salinity environments, or by the expulsion of FAA from the cells for deamination and excretion (e.g. Rosas et al., 1999). Ultimately the osmoregulatory CHR can significantly affect organism excretion rates (e.g. Tirard et al., 1997; Shinji and Wilder, 2012).

While decreases in the FAAP are qualitatively associated with acclimation to low salinity, considerable variation has been reported in the dominant amino acids involved in different species, and also in the temporal profile of different amino acids. Glycine, proline, and taurine were reported as quantitatively the most important amino acid osmolytes in crustaceans (Bishop and Burton, 1993) but decreases in glycine, taurine, proline, *and* alanine were reported as responsible for a reduction in the FAAP in *Palaemon elegans* following acclimation from a salinity of 40 to 10 (Dalla Via, 1989).

In bivalves, taurine and glycine have been reported to be two major contributors to the FAAP, at least in gill tissues. In the oyster *Crassostrea gigas,* decreases in taurine concentration were a major contributor to the overall decrease in FAAP during acclimation from a salinity of 30 to 7 (Lee et al., 2004). In the osmoconforming blood worm *Glycera dibranchiata* red coelomocytes have been identified to regulate cell volume during hypoosmotic stress (498 milliosmoles (mOsm) compared to an acclimation osmolality of 996 mOsm) by reducing the intracellular concentration of the FAAP, principally by reductions in the amino acids proline, asparagine, and also again taurine (Costa and Pierce, 1983).

In hyperosmotic environments amino acids are either synthesized *de novo* or are produced by the catabolism of cellular protein, thus producing an increase in the FAAP and intracellular osmotic pressure (Fig. 1.1). Proline and alanine were the primary contributors to increases in the FAAP pool in the copepod *Tigriopus californicus* (Burton, 1991) while the red swamp crayfish *Procambarus clarkii* largely accumulated D- and L-alanine together with glycine, L-glutamine, and L-proline in both muscle and hepatopancreas on transfer from fresh water to salinities of 17 and 25 (Fujimori and Abe, 2002).

Similar responses have also been reported in osmoconforming bivalves. In the brackish water bivalve *Rangia cuneata*, glutamic acid, glycine, alanine, and arginine constituted 70–80% of the total muscle FAAP, and concentrations were elevated by as much as 300% during acclimation to seawater (Henry et al., 1980; Otto and Pierce, 1981); however, they also demonstrated that not all amino acids were regulated in response to salinity stress. Amino acids such as serine varied little after 42 days exposure to increasing salinity (Fig. 1.3). In the clam *Meretrix lusoria* hyperosmotic conditions (150% sea water) led to the accumulation of alanine in adductor muscle, gills, and midgut gland (Okuma et al., 1988), while in the commercially important Pacific oyster *Crassostrea gigas* the accumulation of taurine and glycine drove increases in the FAAP within 48 h of being exposed to increased salinities from 30 to 39 (Lee et al., 2004).

Conflicting data have raised questions over whether the accumulation of amino acids and other organic osmolytes in response to hyperosmotic conditions is a rapid process that results only from the activity of existing enzymes or changes in cell permeability, or rather if amino acid synthesizing enzymes must first be translated from mRNA. The fourfold accumulation of proline in response to hyperosmotic stress in the copepod *Tigriopus californicus* could be inhibited by the protein synthesis inhibitor cycloheximide, leading to the conclusion that this required the synthesis of one or more enzymes of the proline biosynthetic pathway (Burton, 1991). However, Deaton (2001) concluded that

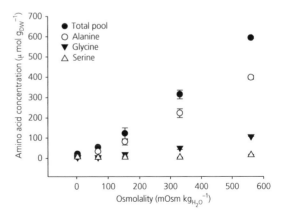

Figure 1.3 Changes in the intracellular free amino acid concentration of selected amino acids in the foot muscle of the brackish water bivalve *Rangia cuneata* acclimated to different salinities for 42 days, showing the mean ± SEM. Figure plotted from data presented in Otto and Pierce (1981) (their Table 1).

gene transcription and translation were *not* responsible for the increase from 45 to 150 µmol g dry weight of betaine⁻¹ in the gills of the ribbed mussel *Geukensia demissa* within 12 h of transfer from 250 to 1000 mOsm (Deaton, 2001).

It is of relevance to this book that while trends in the FAAP established from laboratory manipulations broadly appear consistent across diverse invertebrate phyla, outcomes of laboratory manipulations of salinity are not necessarily replicated in field studies, where multiple stressors may interact in complex ways. Kube et al. (2006) identified a complex pattern of cell volume regulation via the FAAP in different populations of the bivalves *Macoma balthica* and *Mytilus* spp. along their European distribution. They classified these patterns into a northern Baltic type, a southern Baltic type, and an Atlantic/Mediterranean type. These three types differed in the relative importance of two amino acids: alanine and taurine. Kube et al. (2007) further developed this study and concluded that salinity was not the main factor in determining FAAP concentration; the seasonal patterns of FAAP components varied as a complex function of environmental conditions (salinity and temperature) and physiological state of the bivalve (glycogen content and reproductive stage).

1.5 The energetic and metabolic requirements of osmotic control and the consequences for organism fitness

From the overview provided in Sections 1.3 and 1.4 it is apparent that, in many cases, the initial CSR and subsequent regulation of cell volume by the FAAP, as well as the regulation of cellular and extracellular ionic composition using membrane-bound ATPases, require an energy source which is obtained by the hydrolysis of the phosphoanhydride bonds in adenosine triphosphate (ATP) and adenosine diphosphate (ADP). As such, this requirement can be directly measured as alterations in the adenylate energy charge (AEC) of a tissue (Atkinson and Walton, 1967). The AEC is proportional to the mole fraction of ATP plus half the mole fraction of ADP (as ATP contains two high energy phosphoanhydride bonds whereas ADP contains one) and is given by the equation:

$$\frac{[ATP]+0.5\,[ADP]}{[ATP]+[ADP]+[AMP]}$$

Rainer et al. (1979) reported a 17% reduction in mean AEC following a reduction in salinity from 35 to

approximately 10 for the gastropod *Pyrazus ebeninus* and the bivalves *Anadara trapezia* and *Saccostrea commercialis*. Nevertheless, while the AEC has been applied as an index of salinity stress in experimental studies on invertebrates (e.g. Matsushima et al., 1984), its validity in the study of stress in organisms sampled directly from the field has been questioned (Veldhuizentsoerkan et al., 1991), as the gross requirement for ATP is an integrated function of responding to multiple environmental conditions simultaneously, again reflecting the challenge of interpreting stress indices in a multi-stressor environment. A further practical limitation to the use of AEC as a measure of stress in field populations is that it is very difficult, if not impossible, to determine *in situ*. Tissues for AEC determination must be flash frozen using a tissue clamp as the relative concentrations of the different adenosine pools can be altered over timescales of seconds. The handling stress imposed on an organism as it is removed from the field, returned to the laboratory, and then sampled will, inevitably, confound any measurement of relative abundance of ATP, ADP, and AMP.

To provision ATP for ion and volume regulation, osmoregulation places increased demands on oxidative metabolism that increases the requirement for oxygen, the ultimate electron acceptor within the mitochondrial electron transport system, as well as substrates of glycolysis and the tricarboxylic acid cycle. These demands are reflected in the mobilization of carbohydrate and lipid (e.g. Telahigue et al., 2010; Martins et al., 2011) and even protein reserves. These requirements can be summarized by the determination of the Cellular Energy Allocation (De Coen and Janssen, 1997). The aim of the CEA is to quantify the available energy reserves and consumption within a cell to produce a single integrated measure of metabolic status. The CEA is calculated from the following equations:

$$CEA = \frac{E_a}{E_c}$$

where: E_a (available energy) $= E_{Carbohydrate} + E_{Lipid} + E_{Protein}(mJ\,mg^{-1}WW)$

and: E_c(energy consumption) $=$ ETS activity (mJ mg⁻¹ WW h⁻¹)

A decrease in the CEA indicates either a reduction in available energy or a higher energy expenditure, both of which will reduce the energy available for growth, reproduction, or other processes such as the maintenance of an immune system (see Section 1.8). The CEA has been used to determine the stress of osmoregulation, in isolation and in combination with other factors, in the digestive gland tissue of the mussel *Mytilus galloprovincialis* (Erk et al., 2011) and also in tissues of the estuarine mysid shrimp *Neomysis integer* (Verslycke and Janssen, 2002; Erk et al., 2008). Mussels *Mytilus galloprovincialis* collected from an estuarine site with a high salinity (and temperature) variation exhibited generally lower CEA (~200) in both summer and winter 2008 than those collected from a coastal region (CEA ~400) that had a smaller temperature and salinity range (Erk et al., 2011), as a complex function of changes in the available energy fractions $E_{(Carbohydrate)}$, $E_{(Protein)}$ and $E_{(Lipid)}$ (Fig. 1.4).

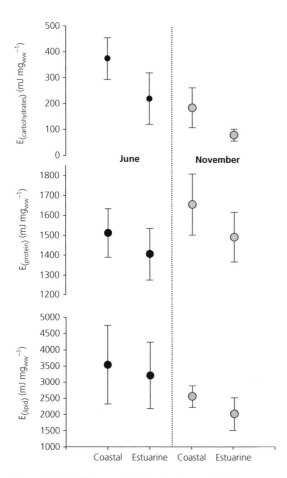

Figure 1.4 Available energy fractions for the mussel *Mytilus galloprovincialis* between summer and winter (June and November 2008) for a variable estuarine site (Martinska) and a less variable open coastal site (Zablac'e), showing the mean ± SD. Figure plotted from data presented in Erk et al. (2011) (their Table 1).

Alternatively, the net effect of any trade-offs in energy expenditure can be established at the level of the whole animal through Scope for Growth (SfG) determination (Widdows and Johnson, 1988). This represents the balance between energy input from food, energy expenditure via metabolism (maintenance and specific dynamic), and energy loss via excretion. The net result gives an indication of the energy available for somatic and gonadal growth (SfG). SfG can be determined from the following:

$$C = P + R + U + F$$

where C = total consumption of food energy; P = production of both somatic tissue and gametes (SfG); R = respiratory energy expenditure; U = energy lost as excreta; and F = faecal energy loss. Changes in SfG have been reported in oysters *Ostrea edulis* (Hutchinson and Hawkins, 1992); the white shrimp *Litopenaeus vannamei* (Valdez et al., 2008); the green-lipped mussel *Perna viridis* (Wang et al., 2011); and the Mexican oyster *Crassostrea corteziensis* (Eduardo Guzman-Agueero et al., 2013) in response to salinity change. Generally, in the above examples SfG decreased as salinity was adjusted away from the acclimation salinity and most often changes in SfG were driven by changes in feeding rates and respiration rates, with assimilation being constant across a range of salinities (Urbina et al., 2010). However, SfG again represents an integrated index of a series of physiological processes acting simultaneously. As an overall index, SfG can lack useful resolution of these processes, especially in environments that are not resource (food) limited. For example SfG approaches were used to study the effects of salinity exposure on the estuarine crab *Hemigrapsus crenulatus* (Urbina et al., 2010). Ingestion, excretion, and respiration rates increased at low salinity (5) but overall the SfG remained constant, suggesting that for this species the increased energy losses at low salinity due to respiration and excretion were compensated by an increase in the ingestion rate in an environment in which food was in excess. Over-reliance on a single stress index can mask significant detail of the fine-scale impacts of hyper- and hypoosmotic stress.

As well as influencing tissue or organism energetics, changes in environment salinity (both hyper- and hypoosmotic) affect tissue biochemistry. Perturbations in environmental salinity have been shown to increase the expression of the crustacean hyperglycaemic hormone (CHH) which, in part, regulates haemolymph glucose in crustaceans (Chang et al., 1998; Chang, 2005). These authors have reported an increase in the expression of CHH in the American lobster *Homarus americanus* and the langoustine *Nephrops norvegicus*, causing an increase in circulating glucose (hyperglycaemia) in response to salinity perturbation. Similarly, Shinji et al. (2012) have reported increases in the concentration of sinus gland peptide-G (SGP-G), one of the CHH family of peptides, coincident with tissue hyperglycaemia in the shrimp *Litopenaeus vannamei* exposed to low salinity stress for 3 h (0, compared to an acclimation salinity of 28). With reference to other metabolites, Luvizotto-Santos et al. (2003), working with starved individuals of the euryhaline crab *Chasmagnathus granulata*, identified a depletion of muscle and gill lipid when transferred from an acclimation salinity of 20 to 0 for a period of seven days. The authors concluded that that lipid mobilization was taking place in the absence of other energy reserves to drive processes of hypoosmotic acclimation, although this mobilization was not recorded in response to hyperosmotic exposure of 40. Similar results have been reported by Verslycke and Janssen (2002) who identified that salinity was the most important environmental constraint on sugar reserves in the mysid shrimp *Neomysis integer*.

In summary, it can be concluded that in response to hyper- or hypoosmotic stress, especially in resource-limited situations, there is a depletion of multiple different energy reserves to provide substrates to support oxidative metabolism and the production of ATP required to drive energetically demanding mechanisms for cellular osmotic control (Fig. 1.1). In chronic scenarios of salinity perturbation this can have implications for physiological performance (Section 1.6), growth, ontogeny and reproduction (Section 1.7), and other processes such as the maintenance of the immune system (Section 1.8).

1.6 Physiological and endocrine responses to salinity stress

To supply tissues and cells with sufficient metabolites to support the increased demands summarized in the previous sections of this chapter, salinity stress has been shown to increase cardiac output, scaphognathite beating in crustaceans, and whole organism oxygen consumption—all indicators of an increase in metabolic rate. However, while acclimation to hypo- and hyperosmotic stress can produce an increase in oxygen consumption rate (e.g. in speckled shrimp *Metapenaeus monoceros*, Pillai and Diwan, 2002; the estuarine shrimps *Palaemonetes pugio* and *P. vulgaris*, Rowe, 2002; and the white shrimp *Litopenaeus vannamei*, Li et al., 2007), in other studies species acclimation to hyperosmotic stress produced no measureable change in

oxygen consumption rate (e.g. in the freshwater prawn *Macrobrachium rosenbergii*, Diaz Herrera and Ramirez, 1993; and the opossum shrimp *Neomysis integer*, Vilas et al., 2006). Changes in the rates of oxygen consumption in response to acute low salinity stress have also been recorded in hermatypic corals. Moberg et al. (1997) have reported that the hourly ratio of gross production (Pg) to respiration (R) were significantly reduced in *Porites lutea* and *Pocillopora damicornis* from the Inner Gulf of Thailand when the salinity was reduced from 30 (Pg/R ratios 2.34 for *P. lutea*, 2.92 for *P. damicornis*) to a salinity of 20 (1.75 and 1.75, respectively) and a salinity of 10 (1.12 and 0.77, respectively).

Associated with increases in the oxygen consumption rate in response to salinity stress, particularly in crustaceans, researchers have recorded increases in cardiac output (e.g. the amphipod *Talitrus saltator*, Calosi et al., 2005; the blue crab *Portunus pelagicus*, Ketpadung and Nongnud Tangkrock-olan, 2006; and the freshwater shrimp *Palaemonetes antennarius*, Ungherese et al., 2008). Early reports (e.g. Larimer and Riggs, 1964; Mason et al., 1983) concluded that salinity impacts on rates of oxygen consumption were a function of changes in haemolymph ion concentrations which altered the affinity of haemocyanin for oxygen, necessitating an increase in cardiac output to meet demand. However, Taylor et al. (1985) reported that both the haemocyanin oxygen affinity and the Bohr Effect were not affected by exposure to low salinities (10–30) in the intertidal prawn *Palaemon elegans*, and were actually increased in response to exposure to elevated salinity of 40. Nevertheless, the effects of exposure salinity stress on oxygen consumption rate might be the result of both the direct effects of extracellular ion concentration on haemocyanin affinity and also indirect effects of a requirement for increased oxidative metabolism to meet demand.

Changes in cardiac output have also been reported for bivalves in response to salinity stress; however, the nature of the response reported was more variable than in some of the above examples. Sara and de Pirro (2011) established the heartbeat of two intertidal Mediterranean bivalves, the euryhaline invasive Lessepsian species *Brachidontes pharaonis* and the stenohaline native *Mytilaster minimus*. In response to salinity changes from brackish (20) to extreme hypersaline (75) the two species displayed different responses to varying salinity. *M. minimus* showed signs of stress (tachycardia) at salinities above 37 while *B. pharaonis* did not display tachycardia until the environment salinity exceeded 45. Furthermore the native and stenohaline bivalve entered cardiac arrest at a salinity of 75, while *Brachidontes pharaonis* retained a cardiac output above a salinity of 75. Bakhmet et al. (2012) have also recently reported the cardiac response of the White Sea bivalves *Hiatella arctica* and *Modiolus modiolus* exposed to different salinities. Changes in cardiac activity were monitored for nine days of the animals' acclimation to salinities of 15, 20, 30, and 35, and for four days of re-acclimation (return to the initial salinity of 25). The initial response to salinity change was a significant bradycardia although cardiac activity in *M. modiolus* was intensified at salinities of 30 and 35. Re-acclimation induced different cardiac responses—increases and decreases depending on the species considered and the salinity applied in the experiment.

As a consequence of the increased energy demand imposed by salinity stress, chronic exposure can also produce alterations to the endocrine system of aquatic invertebrates. Lovett et al. (2006a) reported that exposure to hypoosmotic environments (salinity of 5, compared to an acclimation salinity of 27) produced a tenfold increase in the level of haemolymph methyl farnesoate (MF) in the crab *Carcinus maenas*, which was not seen on exposure to a hyperosmotic medium. Their data suggested that the change in methyl farnesoate was triggered internally as a specific response to low ion concentrations, rather than just hypoosmotic conditions. They also suggested that the changes in MF might be a rather indirect effect of low ionic conditions, as not all crabs subjected to a salinity of 5 showed the same response. Methyl farnesoate has been suggested to have a similar role to juvenile hormone in insects in relation to ovarian development and metamorphosis (Nagaraju, 2007), but from their study Lovett et al. (2006a) also suggested that it may be expressed in response to low ion concentrations in this species. The equivocal role of MF in ion regulation in crustaceans is, however, further evidenced by comparing the work of Ogan et al. (1997), who reported increased MF concentrations in the stenohaline crab *Libinia emarginata* exposed to dilute seawater, with that of Henry and Borst (2006), who found no change in MF concentrations in the euryhaline crab *Callinectes sapidus* in low salinity environments. The later work of Lee et al. (2008) has shown that hyperosmotic salinity stress (control salinity 32, stress salinity 42) in the copepod *Tigriopus japonicus* results in up-regulation of the expression of corticotropin-releasing hormone (CRH)-binding protein (CRH-BP). This hormone is considered as having a chaperoning role in response to stress (Lee et al., 2008).

Salinity perturbation can also result in increases in a number of catecholamines in bivalves, with downstream effects of growth, reproduction, and immune function (considered in Sections 1.7 and 1.8). Lacoste et al. (2001) reported that exposure to salinity decreases from 34 to 24 in the oyster *Crassostrea gigas* produced long lasting (at least 72 h) increases in the concentration of circulating catecholamines, including noradrenaline (NA) and dopamine (DA), that peaked within 12 h of the start of the exposure at concentrations of approximately 22 and 2 ng DA ml^{-1} (controls 1 and 0.5 ng DA ml^{-1}). These catecholamines have been shown to modulate the ciliary activity of gills (Aiello, 1990) and the authors concluded that the increases seen in response to salinity stress contributed to the maintenance of ionic regulation in this oyster. Chen et al. (2008) have reported similar responses in adrenaline and NA in the haemolymph of the zhikong scallop *Chlamys farreri*, again in the first 72 h after transfer from a salinity of 31 to 20.

1.7 Impacts of salinity exposure on ontogeny, larval settlement, and growth

For aquatic invertebrates the impacts of salinity stress can occur from the point at which eggs are released to the environment or when the larvae hatch. Impacts of salinity stress on larval development in echinoderms (Cowart et al., 2009), molluscs (see below), and crustaceans have been widely reported. For crustaceans the impacts of hyper- and hypoosmotic stress have been reviewed extensively by Anger (2003) and will not be repeated here; the examples developed below consider the results of studies on molluscs.

To an extent, bivalve and gastropod molluscs have developed mechanisms to protect developing embryos from salinity stress during early critical phases. Brooding molluscs, including examples of gastropods and bivalves, can either close their valves or seal their opercula to avoid hyper- or hyposaline waters and create a 'brood chamber' in which the embryos are protected and can develop. For example, ostreid bivalves and the gastropod *Crepipatella dilatata* incubate their embryos in the pallial cavity for at least four weeks before the young emerge as larvae (oyster) or juveniles (gastropod). Chaparro et al. (2009) determined the changing conditions within the brood chamber during salinity stress in these two species to establish the limits of this mechanism as a means of avoiding salinity stress. Both species were exposed to water of reduced salinity (<22 to 24) and, although salinity within the pallial cavity remained high for both species, proving that the mechanism was an effective means of avoiding salinity stress, the oxygen available to the adult tissues was reduced to hypoxic levels (<1.5 mg O_2 l^{-1}) in *C. dilatata* and *Ostrea chilensis* within 12 h and 20 min, respectively. The subsequent respiratory acidosis reduced the pH of the intrapallial fluid to as low as pH 6.4 in *Crepipatella dilatata*, during extended isolation, leading to the conclusion that this might impact on the ability of brooded veligers to form nascent calcified shells.

The creation of an external egg mass is more widespread within the Gastropoda. These egg masses, which consist of many capsules each containing multiple developing eggs (e.g. a mean of 1094 eggs per capsule reported in the whelk *Buccinum undatum*, Smith and Thatje, 2013), are also susceptible to salinity stress during development in shallow coastal environments and in the intertidal. As reviewed by Przeslawski (2004), extremes of salinity exposure can lead to a protracted embryo or larval development and an increase in mortality. Again, however, in the field the outcome of interactions between salinity and other abiotic and biotic factors can complicate interpretation. Nevertheless, there is evidence that the vitelline capsule and the gelatinous matrix of the capsule produced by the whelks *Nucella lamellosa*, *N. lima*, and *Ilyanassa obsoleta* afford a degree of protection for the developing embryos, particularly with reference to salinity stress. The survival of encapsulated embryos remained approximately 100% after 48 h of exposure to dilute media, while quickly falling to 0% for embryos that were removed from their capsule (Fig. 1.5; Pechenik, 1982). This protection is, however, perhaps

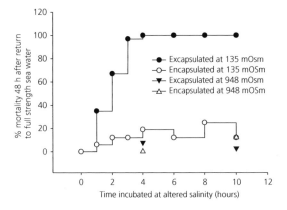

Figure 1.5 Percent mortality of encapsulated (16 capsules of embryos) and excapsulated embryos (60 individual embryos) of the whelk *Ilyanassa obsoleta* exposed to low salinity (135 mOsm) for increasing periods. Mortality was recorded 48 h after the capsules and embryos had been returned to full strength sea water. Figure produced from data presented in Pechenik (1982) (Table 1).

not universal. Deschaseaux et al. (2011) established the stress response induced by salinity changes on hatched larvae of three marine gastropodss: *Bembicium nanum*, *Siphonaria denticulata*, and *Dolabrifera brazieri*. Encapsulated embryos were maintained at three different salinities (25, 25, and 45) at two temperatures until the larvae hatched. The time to hatching was significantly affected by the salinity of incubation; in the case of *Bembicium nanum* hatching time at 22 °C doubled from a mean of 1.5 days at a salinity of 25 to a mean of 3 days at a salinity of 45.

For bivalves, however, and even brooding bivalves, eventually the larvae are released to develop in the water column before settlement. Rodstrom and Jonsson (2000) have examined survival and feeding activity of spat of the European flat oyster (*Ostrea edulis* L) as a function of temperature and salinity on the Swedish west coast. They determined that the spat feeding rate started to decline at salinities below 18 and stopped completely at 16. Moreover, spat that were exposed to salinities below 16 did not recover feeding activity when they were returned to full salinity, which indicated that permanent physiological damage had taken place.

Undoubtedly, early mollusc developmental stages are sensitive to salinity perturbation but there is evidence to suggest that once a key stage has been reached the surviving larvae can compensate and develop successfully to adulthood. In the case of the calyptraeid gastropods *Crepidula fornicata*, *Crepidula onyx*, and *Crepipatella fecunda*, larval development was slowed and mortality increased as the salinity was reduced from 30 to 10, 15, or 20 for less than 48 h. However, all surviving larvae that reached and survived metamorphosis grew successfully into juveniles and no 'latent effects' of the low salinity exposure were recorded (Diederich et al., 2011).

In itself, however, growth post metamorphosis just represents one aspect of physiological performance necessary for an individual long-term success. At the time of spat settlement and into adulthood the production of byssus becomes essential for bivalves to remain in secure contact with the substrate. Rupp and Parsons (2004) established the salinity tolerance window and also the environmental limits for byssus production for spat, juveniles, and adults of the stenohaline scallop *Nodipecten nodosus*. Lethal salinities (LC50) for a 48-h exposure at a temperature of 23.5 °C were identified as 23.2, 23.6, and 20.1 for adults, juveniles, and spat, respectively. However, the percent attachment by spat with byssus was significantly reduced to ~ 40% at salinities of 29 and dropped to ~ 6% at a salinity of 21.

Clearly, even within the salinity tolerance window of a species and life stage, physiological performance may become disrupted to an extent that the successful completion of the full life cycle is impacted.

Growth in invertebrate juveniles and adults can effectively represent the integrated outcome of chronic salinity stress at the level of the individual. As described in Section 1.5, low salinity stress results in disturbance to protein and amino acid synthesis/degradation and, through the requirement for ATP, places additional demands on the energy balance and biochemistry of tissues and hence individuals. It is therefore to be expected that exposure to periods of hypoosmotic stress may compromise the growth rate of post-larval aquatic invertebrates. Consequently, growth responses to salinity stress have similar profiles to SfG responses or measures of AEC or CEA, although critical thresholds of salinity for growth will inevitably differ to other measured indices. In general terms extremes of high and low salinity impact the growth rate of aquatic invertebrates including molluscs (e.g. Schone et al., 2003a; Parada et al., 2012) and echinoderms (e.g. Zhang et al., 2012). In bivalves, impacts to growth (accumulation of live wet weight) can be a function of the rate of calcification of the shell. Malone and Dodd (1967) reported that shell calcification in the mussel *Mytilus edulis* varied as a function of salinity in the range of 20–37 at 18.5 °C, and hypoosmotic environments have been shown to impact the shell increment of the ocean quahog *Arctica islandica* as well (Hiebenthal et al., 2012).

In contrast, and as outlined in Section 1.4, it has been known for a long time that growth rates of crustaceans can be increased at low salinity (hypoosmotic stress), within the salinity tolerance window of the species (e.g. in the penaeid shrimps *Parapenaeopsis hardwickii* and *Parapenaeopsis stylifera*, Kulkarni et al., 1978). During ecdysis at these low salinities the osmotic gradient favours the net inward movement of water immediately post moult. This allows the new exoskeleton to swell to a larger size, resulting in a larger growth increment at each moult stage. Inevitably though, as the lower limit of salinity tolerance is reached, growth becomes compromised (McKenney and Celestial, 1995; Anger et al., 1998) as internal perturbation of homeostasis and tissue damage become more pronounced. Interestingly, Anger et al. (1998) identified that, in larvae of the shore crab *Carcinus maenas* at salinities below 20, the decrease in growth rate was, in part, the result of a reduction in food assimilation efficiency due to salinity impacts on the kinetics of digestive enzymes within the larval gut, emphasizing that extreme hypoosmotic environments can impact both the internal and external (gut) environment of individuals. As stated above, growth in crustaceans is not continuous and takes place immediately post moult when the carapace is soft and so permits expansion. As such, a reduction in growth rate in response to extreme hyposaline or hypersaline environments in crustaceans can be realized through a reduction in the frequency of moulting, particularly in resource-limited environments. Romano and Zeng (2006) have demonstrated that the intermoult period of juvenile blue swimming crabs *Portunus pelagicus* was significantly longer when they were held at a salinity of 10 and also at a salinity of 40.

As before, however, species and population differences do exist and measurements made from laboratory exposures do not always transfer simply to field populations where multiple stressors interact. For example, while growth rates of the blue *Mytilus edulis* are strongly impacted by the low salinity waters of the Baltic (Kautsky, 1982), populations of *Macoma balthica* transplanted from across Europe and into the Baltic (above a salinity of 3) exhibited no impacts on growth (Jansen et al., 2009). Groener et al. (2011) compared resistance to salinity stress in two species of colonial ascidians in the Irish Sea, the invasive species *Didemnum vexillum* that is relatively tolerant of low salinity stress, and the native *Diplosoma listerianum*. In laboratory experiments which mimicked chronic exposure to low and varying salinity sea water (treatments from a salinity of 27 down to 10) they determined that the invasive *D. vexillum* had higher growth and survival under low salinities than the cosmopolitan *D. listerianum*, suggesting that the invasive species was better in adapting to fluctuating and low salinity conditions. Growth rates for the bivalve *Phacosoma japonicum* established in the field have also been shown to vary only slightly with salinity and were better explained by variations in nutrient availability (Schone et al., 2003b).

1.8 Impacts of salinity stress on immune function, the incidence of disease, and tissue pathology

Maintenance of an effective immune system has conventionally been regarded as an energetically expensive constraint on an organism's energy budget (Folstad and Karter, 1992). A functionally competent immune system is comprised of two arms, the receptor and effector arms. The ability of organisms to protect themselves from infection relies both on the detection of invading pathogens (the receptor arm) and on the triggering of an appropriate response (the effector arm). As such, a functional immune system requires the synthesis and release of a range of proteins, including pathogen receptor proteins (PRPs), as well as effector molecules including antimicrobial peptides, agglutinins, opsonins, and proteins required for the production of reactive oxygen, in addition to cell-based responses such as phagocytosis.

Impacts of salinity exposure on immune function can be mediated through the direct suppression of enzyme activity or through the indirect suppression of protein synthesis, resulting in a decrease in synthesis of key immune enzymes and peptides. Here again, however, consistent impacts of exposure to changes in salinity are often difficult to identify because of differences in exposure period (acute versus chronic exposures) and relative differences in the salinity stress imposed—a function of the acclimation and exposure salinity.

In general terms, low salinity exposure, i.e. lower than the organism's acclimation salinity, has a negative impact on immune performance in the Crustacea and causes an increased incidence of disease. Exposure to low salinities has been shown to result in a decrease in circulating haemocyte number (total haemocyte count, THC) in penaeids—for example in the pink shrimp *Farfantepenaeus paulensis* (Perazzolo et al., 2002), tiger shrimp *Penaeus monodon* (Joseph and Philip, 2007) and the white shrimp *Litopenaeus vannamei* (Pan et al., 2010). Low salinity exposure of tiger shrimp *P. monodon* also decreased phenoloxidase enzyme activity and intracellular respiratory burst, as evidenced by the NBT reduction assay, as well as alkaline phosphatase (ALP) activity and acid phosphatase (ACP) activity (Joseph and Philip, 2007). These changes correlated with an increase in susceptibility to infection with White Spot Syndrome Virus (WSSV), one of the most economically significant economic pathogens of the global shrimp farming industry. Transfer of the white shrimp *Litopenaeus schmitti* from 35 to 18 and 8 for a period of 48 h also reduced the expression of phenoloxidase activity, and similar effects have been reported for the white shrimp *L. vannamei* in the first 24 h after exposure to reduced salinity, which also had a direct impact on disease resistance when infected with the Gram-negative bacteria *Vibrio harveyi* (Li et al., 2010). As a contrasting example, five-day exposures to increased (29–33) or reduced (21–9) salinity, relative to acclimation salinity (25), have been shown to reduce THC while increasing the total phenoloxidase enzyme activity in the kuruma shrimp *Marsupenaeus japonicus* (Yu et al., 2003). Active phenoloxidase is produced *in vivo* following the

degranulation of circulating haemocytes and so cell counts and phenoloxidase enzyme activity are often, although not exclusively, negatively correlated (Hauton et al., 1997).

Similar responses to low salinity have been reported in bivalves, including the European flat oyster *Ostrea edulis*. In this species, compromised cellular function at low salinities was reported from measurements of lipid membrane stability at low salinity (Hauton et al., 1998). These data were indicative of lipid membrane disruption associated with cell volume changes that might have resulted from disruption to the cellular osmotic pressure. At salinities of 16 and 19 retention of the supra-vital cationic dye neutral red within the lysosomal compartment of the phagocytically active cells was reduced to approximately 40 minutes, compared to a retention time of approximately 80 minutes at salinities of 25 and 32 (Fig. 1.6). For bivalves there is also evidence that compromised immune function translates into an increased susceptibility to infection. The Manila clam *Ruditapes philippinarum* showed reduced haemocyte counts at a salinity of 20, which coincided with an increased susceptibility to Gram-negative bacteria *Vibrio tapetis*, the causative agent of Brown Ring Disease (BRD) (Reid et al., 2003). Rearing of clams at elevated salinity (40) produced a significant reduction in the incidence of BRD and this correlated with increases in THC, phenoloxidase, and phagocytic activity in the clams.

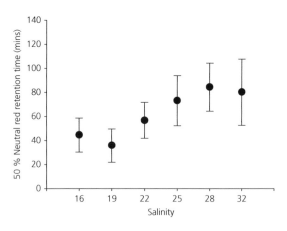

Figure 1.6 Mean neutral red retention time (NRR) times in haemocyte lysosomes from oysters *Ostrea edulis* acclimated for 7 days to different salinities at 15 °C, showing the mean ± SD. NRR reflects lipid membrane stability and gives an indication of cellular performance in response to stress. Figure replotted from original data but first published by Hauton et al. (1998).

Studies of changes in immune gene transcription have been reported in response to changes in incubation salinity. However, transcriptional responses show a different aspect of immune response and so are not directly comparable with the above studies of enzyme activity. An increase in the transcription of the prophenoloxidase gene represents an increase in the production of the inactive zymogen, perhaps to replace that which is released on granulocyte degranulation. Terwilliger (2007) reported little effect of low salinity exposure on the expression of genes coding for prophenoloxidase in the osmoregulating shore crab, *Carcinus maenas*. Conversely, Vatanavicharn et al. (2009) have reported short-term (at 6 h) increased gene expression of two classes of antimicrobial peptide, *crustinPm1* and *crustinPm5*, in epipodites and eyestalks of the tiger shrimp *Penaeus monodon* in response to transfer from an acclimation salinity of 25 to 40. The crustin antimicrobial peptides are a diverse group of antimicrobial peptides (reviewed by Smith et al., 2008) with reported activity against Gram-positive bacteria, but also with some reports of activity against Gram-negative species. However, the response to environmental stress (temperature and salinity) and the presence of crustin-like peptides in eyestalks and epipodites suggests that these peptides might have a wider role within the Crustacea, other than simply immune function.

Predictably, changes in circulating haemocyte populations or the activity of key immune enzymes have often been associated with the increased incidence of disease in specific host pathogen models. However, this relationship is not always direct. Indeed, outcomes of infection studies, especially from field experiments, are often too complex to interpret fully. Insufficient consideration has been paid to the effects of salinity perturbation on pathogen virulence. As originally proposed by Snieszko (1974), the occurrence of a disease depends upon the favourable conjunction of three factors within the 'disease triumverate' in which the effects of the environment on both host and pathogen physiology must be simultaneously considered. In the case of infection with the protozoan *Perkinsus marinus* in the Eastern oyster *Crassostrea virginica*, exposure to low salinity has been shown to have both a positive effect on the host immune system, in terms of the expression of plasma lysozyme (Chu and Volety, 1997), and to simultaneously reduce the virulence of the pathogen (Hoffman et al., 1995; Powell et al., 1996), resulting in a reduced incidence of disease outbreaks at low salinity.

Charismatic invertebrates widely known to suffer from pathology and mortality events are the scleractinian reef-forming corals of tropical surface waters. Coral bleaching is a worldwide phenomenon, but the mechanism by which it occurs is disputed. While the majority of bleaching events have been linked to extremes of water temperature, or combinations of high water temperature and low salinity (Chavanich et al., 2009), evidence is available which also has identified low salinity as a causative factor. Fang et al. (1995) reported that salinity stress reduced both the number of zooxanthellae in each polyp and the amount of total pigment in each zooxanthella and that this coincided with a reduction in the porphyrin pigments. Meehan and Ostrander (1997) have argued that coral bleaching could be used as an indicator of environmental stress, including salinity stress. In an experimental study Downs et al. (2009) reported several tissue pathologies associated with exposure to decreasing salinity from 39 to 24, including tissue oedema, degradation, plus loss of zooxanthellae and tissue necrosis, as evidence for increased oxidative damage. However, with current limitations in our knowledge of the triggers of bleaching events it cannot be regarded as a salinity-specific indicator, perhaps only representing the integrated impact of multiple stressors in a field environment.

Outbreaks of infectious disease in invertebrates are often evidenced by individual mortality or the development of tissue pathology, including tissue oedema, cavitation, hyperplasy, necrosis, and disintegration. However, it should be highlighted that tissue pathology has also been reported as a result of extreme salinity stress in isolation. Santos-Gouvea and Freire (2007) have reported tissue pathology in the peristomial gill ultrastructure of the intertidal and subtidal echinoderm *Echinometra lucunter* in response to exposure to hyperosmotic environments (salinity 45). Tissue pathology is also not necessarily limited to the gills. For example, Cristina Diaz et al. (2010) have reported that the hepatopancreas of fresh water prawn *Palaemonetes argentinus* becomes significantly modified 30 days after transfer from fresh water to salinities of 16 and 24. Hepatopancreatic tissue from saline-exposed individuals showed enlarged tubular lumens with degenerated and infolded basal lamina. Ultrastructurally, nuclear retraction, cytoplasmolysis, and rough endoplasmic reticulum (RER) membrane separation were observed, with no apparent recovery in cell ultrastructure on return to fresh water. Clearly, the diagnosis of pathology arising from disease via a compromised immune system must be compared against tissue pathology associated with salinity exposure alone.

1.9 Summary

This chapter has provided a review of the acute and chronic impacts associated with changing environmental salinity on aquatic invertebrates. Direct impacts of salinity change to the intracellular osmotic potential result in the induction of various pathways of the cellular stress response (CSR) and the cellular homeostatic response (CHR). The purpose of the CSR is to limit the damage caused to the cell through oxidative damage and protein degradation. In the case of osmotic stress the CHR, which may be induced at a slightly later stage and persist for a longer interval, specifically restores ionic balance and cell volume. In the main, pathways within the CSR and CHR are energy consuming and many of the organism-level responses to salinity perturbation, responses that have been variously advocated as indices of salinity stress, reflect the consequences of an increased energy demand. These indirect energetic impacts can be manifest through changes in physiology and biochemistry to growth, development, and immune function.

While certain organismal responses to salinity challenge are consistent throughout the invertebrate phyla, notable taxon-, species-, and even population-specific differences are known to occur. These differences make the identification of consistent responses difficult. Furthermore, in many instances, clear responses to hyper- or hypoosmotic environments have been determined from laboratory exposures in which only the salinity is varied. It is often the case that when studied in the field, a multiple stressor environment, the consistent laboratory response disappears or interpretation becomes more complex.

References

Aiello, E. (1990). Nervous control of gill ciliary activity in *Mytilus edulis*. In: Stefano, G.B. (ed.), *Neurobiology of Mytilus edulis*, pp. 189–208. Manchester University Press, Manchester.

Anger, K. (2003). Salinity as a key parameter in the larval biology of decapod crustaceans. *Invertebrate Reproduction and Development*, **43**, 29–45.

Anger, K., Spivak, E., & Luppi, T. (1998). Effects of reduced salinities on development and bioenergetics of early larval shore crab, *Carcinus maenas*. *Journal of Experimental Marine Biology and Ecology*, **220**, 287–304.

Atkinson, D.E. & Walton, G.M. (1967). Adenosine triphosphate conservation in metabolic regulation. Rat liver cleavage enzyme. *Journal of Biological Chemistry*, **242**, 3239–3241.

Bakhmet, I.N., Komendantov, A.J., & Smurov, A.O. (2012). Effect of salinity change on cardiac activity in *Hiatella arctica* and *Modiolus modiolus*, in the White Sea. *Polar Biology*, **35**, 143–148.

Barton, B.A. (2002). Stress in fishes: a diversity with particular reference to changes in circulating corticosteroids. *Integrative and Comparative Biology*, **42**, 517–525.

Basova, L., Begum, S., Strahl, J., et al. (2012). Age-dependent patterns of antioxidants in *Arctica islandica* from six regionally separate populations with different lifespans. *Aquatic Biology*, **14**, 141–152.

Bishop, J.S. & Burton, R.S. (1993). Amino-acid synthesis during hyperosmotic stress in *Penaeus aztecus* postlarvae. *Comparative Biochemistry and Physiology Part A*, **106**, 49–56.

Boettcher, K., Siebers, D., & Sender, S. (1995). Carbonic anhydrase, a respiratory enzyme in the gills of the shore crab *Carcinus maenas*. *Helgolaender Meeresuntersuchungen*, **49**, 737–745.

Burton, R.S. (1991). Regulation of proline synthesis in osmotic response—effects of protein-synthesis inhibitors. *Journal of Experimental Zoology*, **259**, 272–277.

Buttemer, W.A., Abele, D., & Costantini, D. (2010). The ecology of antioxidants and oxidative stress in animals. From bivalves to birds: oxidative stress and longevity. *Functional Ecology*, **24**, 971–983.

Cailleaud, K., Maillet, G., Budzinski, H., et al. (2007). Effects of salinity and temperature on the expression of enzymatic biomarkers in *Eurytemora affinis* (Calanoida, Copepoda). *Comparative Biochemistry and Physiology Part A*, **147**, 841–849.

Calosi, P., Ugolini, A., & Morritt D. (2005). Physiological responses to hypoosmotic stress in the supralittoral amphipod *Talitrus saltator* (Crustacea: Amphipoda). *Comparative Biochemistry and Physiology Part A*, **142**, 267–275.

Carmen Pineda, M., Turon, X., & Lopez-Legentil, S. (2012). Stress levels over time in the introduced ascidian *Styela plicata*: the effects of temperature and salinity variations on hsp70 gene expression. *Cell Stress and Chaperones*, **17**, 435–444.

Chang, E.S. (2005). Stressed-out lobsters: crustacean hyperglycemic hormone and stress proteins. *Integrative and Comparative Biology*, **45**, 43–50.

Chang, E.S., Keller, R., & Chang, S.A. (1998). Quantification of crustacean hyperglycemic hormone by ELISA in hemolymph of the lobster, *Homarus americanus*, following various stresses. *General and Comparative Endocrinology*, **111**, 359–366.

Chaparro, O.R., Montiel, Y.A., Segura, C.J., et al. (2008). The effect of salinity on clearance rate in the suspension-feeding estuarine gastropod *Crepipatella dilatata* under natural and controlled conditions. *Estuarine, Coastal and Shelf Science*, **76**, 861–868.

Chaparro, O.R., Segura, C.J., Montory, J.A., et al. (2009). Brood chamber isolation during salinity stress in two estuarine mollusk species: from a protective nursery to a dangerous prison. *Marine Ecology Progress Series*, **374**, 145–155.

Charmantier, G., Haond, C., Lignot, J.H., & Charmantier-Daures, M. (2001). Ecophysiological adaptation to salinity throughout a life cycle: a review in homarid lobsters. *Journal of Experimental Biology*, **204**, 967–977.

Chavanich, S., Viyakarn, V., Loyjiw, T., et al. (2009). Mass bleaching of soft coral, *Sarcophyton* spp. in Thailand and the role of temperature and salinity stress. *ICES Journal of Marine Science*, **66**, 1515–1519.

Chen, M., Yang, H.S., Xu, B., et al. (2008). Catecholaminergic responses to environmental stress in the hemolymph of zhikong scallop *Chlamys farreri*. *Journal of Experimental Zoology Part A*, **309**, 289–296.

Chu, F.L.E., Volety, A.K. 1997. The interaction of the oyster protozoan parasite, *Perkinsus marinus* and its host, the eastern oyster *Crassostrea virginica*: a progress report. Journal of Shellfish Research 16:261 (Abstract).

Compère, P., Wanson, S., Pequeux, A., et al. (1989). Ultrastructural changes in the gill epithelium of the green crab *Carcinus maenas* in relation to external salinity. *Tissue and Cell*, **21**, 299–318.

Costa, C.J. & Pierce, S.K. (1983). Volume regulation in the red coelomcytes of *Glycera dibranchiata*—an interaction of amino acid and K + effluxes. *Journal of Comparative Physiology Part B*, **151**, 133–144.

Cowart, D.A., Ulrich, P.N., Miller, D.C., & Marsh, A.G. (2009). Salinity sensitivity of early embryos of the Antarctic sea urchin, *Sterechinus neumayeri*. *Polar Biology*, **32**, 435–441.

Cristina Diaz, A., Graciela Sousa, L., & Maria Petriella, A. (2010). Functional cytology of the hepatopancreas of *Palaemonetes argentinus* (Crustacea, Decapoda, Caridea) under osmotic stress. *Brazilian Archives of Biology and Technology*, **53**, 599–608.

Dalla Via, G.J. (1989). The effect of salinity on free amino acids in the prawn *Palaemon elegans* Rathke. *Archiv fuer Hydrobiologie*, **115**, 125–136.

Davenport, J. (1985). Osmotic control in marine animals. *Symposia of the Society for Experimental Biology*, **39**, 207–244.

Deaton, L.E. (1992). Osmoregulation and epithelial permeability in 2 euryhaline bivalve mollusks—*Mya arenaria* and *Geukensia demissa*. *Journal of Experimental Marine Biology and Ecology*, **158**, 167–177.

Deaton, L.E. (2001). Hyperosmotic volume regulation in the gills of the ribbed mussel, *Geukensia demissa*: rapid accumulation of betaine and alanine. *Journal of Experimental Marine Biology and Ecology*, **260**, 185–197.

Deaton, L.E. & Pierce, S.K. (1994). Introduction: cellular volume regulation—mechanisms and control. *Journal of Experimental Zoology*, **268**, 77–79.

De Coen, W.M. & Janssen, C.R. (1997). The use of biomarkers in *Daphnia magna* toxicity testing: IV. Cellular energy allocation: a new methodology to assess the energy budget of toxicant-stressed *Daphnia* populations. *Journal of Aquatic Ecosystem Stress and Recovery*, **6**, 43–55.

Deschaseaux, E., Taylor, A., & Maher, W. (2011). Measure of stress response induced by temperature and salinity changes on hatched larvae of three marine gastropod species. *Journal of Experimental Marine Biology and Ecology*, **397**, 121–128.

Diaz Herrera, F. & Ramirez, L.F.B. (1993). Effect of salinity on oxygen consumption and ammonium excretion in *Macrobrachium rosenbergii* (Crustacea: Palaemonidae). *Revista De Biologia Tropical*, **41**, 239–243.

Diederich, C.M., Jarrett, J.N., Chaparro, O.R., et al. (2011). Low salinity stress experienced by larvae does not affect post-metamorphic growth or survival in three calyptraeid gastropods. *Journal of Experimental Marine Biology and Ecology*, **397**, 94–105.

Dong, Y., Dong, S., & Meng, X. (2008). Effects of thermal and osmotic stress on growth, osmoregulation and Hsp70 in sea cucumber (*Apostichopus japonicus* Selenka). *Aquaculture*, **276**, 179–186.

Downs, C.A., Kramarsky-Winter, E., Woodley, C.M., et al. (2009). Cellular pathology and histopathology of hyposalinity exposure on the coral *Stylophora pistillata*. *Science of the Total Environment*, **407**, 4838–4851.

Dubyak, G.R. (2004). Ion homeostasis, channels, and transporters: an update on cellular mechanisms. *Advances in Physiological Education*, **28**, 143–154.

Eduardo Guzmán-Agüero, J., Nieves-Soto, M., Ángel Hurtado, M., et al. (2013). Feeding physiology and scope for growth of the oyster *Crassostrea corteziensis* (Hertlein, 1951) acclimated to different conditions of temperature and salinity. *Aquaculture International*, **21**, 283–297.

Erk, M., Ivanković, D., & Strižak Ž. (2011). Cellular energy allocation in mussels (*Mytilus galloprovincialis*) from the stratified estuary as a physiological biomarker. *Marine Pollution Bulletin*, **62**, 1124–1129.

Erk, M., Muyssen, B.T.A., Ghekiere, A., & Janssen, C.R. (2008). Metallothionein and cellular energy allocation in the estuarine mysid shrimp *Neomysis integer* exposed to cadmium at different salinities. *Journal of Experimental Marine Biology and Ecology*, **357**, 172–180.

Fang, L.-S., Liao, C.-W., & Liu, M.-C. (1995). Pigment composition in different-colored scleractinian corals before and during the bleaching process. *Zoological Studies*, **34**, 10–17.

Fanini, L., Gecchele, L.V., Gambineri, S., et al. (2012). Behavioural similarities in different species of sandhoppers inhabiting transient environments. *Journal of Experimental Marine Biology and Ecology*, **420**, 8–15.

Fernandez-Reiriz, M.J., Navarro, J.M., & Labarta, U. (2005). Enzymatic and feeding behaviour of *Argopecten purpuratus* under variation in salinity and food supply. *Comparative Biochemistry and Physiology Part A*, **141**, 153–163.

Field, C.B., Behrenfeld, M.J., Randerson, J.T., & Falkowski, P. (1998). Primary production of the biosphere: integrating terrestrial and oceanic components. *Science*, **281**, 237–240.

Folstad, I. & Karter, A.J. (1992). Parasites, bright males, and the immunocompetence handicap. *American Naturalist*, **139**, 603–622.

Freire, C.A., Togni, V.G., & Hermes-Lima, M. (2011). Responses of free radical metabolism to air exposure or salinity stress, in crabs (*Callinectes danae* and *C. ornatus*) with different estuarine distributions. *Comparative Biochemistry and Physiology Part A*, **160**, 291–300.

Fujimori, T. & Abe, H. (2002). Physiological roles of free D- and L-alanine in the crayfish *Procambarus clarkii* with special reference to osmotic and anoxic stress responses. *Comparative Biochemistry and Physiology Part A*, **131**, 893–900.

Galluzzi, L., Kepp, O., Trojel-Hansen, C., & Kroemer, G. (2012). Mitochondrial control of cellular life, stress, and death. *Circulation Research*, **111**, 1198–1207.

Gaspar Martins, C.D.M. & Bianchini, A. (2009). Metallothionein-like proteins in the blue crab *Callinectes sapidus*: effect of water salinity and ions. *Comparative Biochemistry and Physiology Part A*, **152**, 366–371.

Groener, F., Lenz, M., Wahl, M., & Jenkins, S.R. (2011). Stress resistance in two colonial ascidians from the Irish Sea: the recent invader *Didemnum vexillum* is more tolerant to low salinity than the cosmopolitan *Diplosoma listerianum*. *Journal of Experimental Marine Biology and Ecology*, **409**, 48–52.

Harman, D. (1956). Aging: a theory based on free radical and radiation chemistry. *Journal of Gerontology*, **11**, 298–300

Hauton, C., Hawkins, L.E., & Hutchinson, S. (1998). The use of neutral red to examine the effects of temperature and salinity on haemocytes of the European flat oyster *Ostrea edulis* (L.). *Comparative Biochemistry and Physiology Part B*, **119**, 619–623.

Hauton, C., Hawkins, L.E., & Williams, J.A. (1997). *In situ* variability in phenoloxidase activity in the shore crab, *Carcinus maenas* (L). *Comparative Biochemistry and Physiology Part B*, **117**, 267–271.

Henry, R.P. (1984). The function of invertebrate carbonic anhydrase in ion transport. *Annals of the New York Academy of Sciences*, **429**, 544–546.

Henry, R.P. (2001). Environmentally mediated carbonic anhydrase induction in the gills of euryhaline crustaceans. *Journal of Experimental Biology*, **204**, 991–1002.

Henry, R.P. & Borst, D.W. (2006). Effects of eyestalk ablation on carbonic anhydrase activity in the euryhaline blue crab *Callinectes sapidus*: neuroendocrine control of enzyme expression. *Journal of Experimental Zoology Part A*, **305**, 23–31.

Henry, R.P. & Cameron, J.N. (1982). Acid-base balance in *Callinectes sapidus* during acclimation from high to low salinity. *Journal of Experimental Biology*, **101**, 255–264.

Henry, R.P., Gehnrich, S., Weihrauch, D., & Towle, D.W. (2003). Salinity-mediated carbonic anhydrase induction in the gills of the euryhaline green crab, *Carcinus maenas*. *Comparative Biochemistry and Physiology Part A*, **136**, 243–258.

Henry, R.P., Lucu, Č., Onken, H., & Weihrauch, D. (2012). Multiple functions of the crustacean gill: osmotic/ionic regulation, acid-base balance, ammonia excretion, and bioaccumulation of toxic metals. *Frontiers in Physiology*, **3**(431), 1–33.

Henry, R.P., Mangum, C.P., & Webb, K.L. (1980). Salt and water balance in the oligohaline clam, *Rangia cuneata*. 2. Accumulation of intracellular free amino-acids during high salinity adaptation. *Journal of Experimental Zoology Part A*, **211**, 11–24.

Hiebenthal, C., Philipp, E.E.R., Eisenhauer, A., & Wahl, M. (2012). Interactive effects of temperature and salinity on shell formation and general condition in Baltic Sea *Mytilus edulis* and *Arctica islandica*. *Aquatic Biology*, **14**, 289–298.

Hochstrasser, M. (1996). Ubiquitin-dependent protein degradation. *Annual Review Genetics*, **30**, 405–439.

Hoffmann, E.E., Powell, E.N., Klinck, J.M., Saunders, G. (1995). Modelling diseased oyster populations: 1 modelling *Perkinsus marinus* infection in oysters. *Journal of Shellfish Research*, **14**, 121–151.

Hutchinson, S. & Hawkins, L.E. (1992). Quantification of the physiological responses of the European flat oyster *Ostrea edulis* L. to temperature and salinity. *Journal of Molluscan Studies*, **58**, 215–226.

Jansen, J.M., Koutstaal, A., Bonga, S.W., & Hummel, H. (2009). Salinity-related growth rates in populations of the European clam *Macoma balthica* and in field transplant experiments along the Baltic Sea salinity gradient. *Marine and Freshwater Behaviour and Physiology*, **42**, 157–166.

Jayasundara, N., Towle, D.W., Weihrauch, D., & Spanings-Pierrot, C. (2007). Gill-specific transcriptional regulation of Na^+/K^+-ATPase alpha-subunit in the euryhaline shore crab *Pachygrapsus marmoratus*: sequence variants and promoter structure. *Journal of Experimental Biology*, **12**, 2070–2081.

Jillette, N., Cammack, L., Lowenstein, M., & Henry, R.P. (2011). Down-regulation of activity and expression of three transport-related proteins in the gills of the euryhaline green crab, Carcinus maenas, in response to high salinity acclimation. *Comparative Biochemistry and Physiology Part A*, **158**, 189–193.

Joseph, A. & Philip, R. (2007). Acute salinity stress alters the haemolymph metabolic profile of *Penaeus monodon* and reduces immunocompetence to white spot syndrome virus infection. *Aquaculture*, **272**, 87–97.

Kautsky, N. (1982). Growth and size structure in a Baltic *Mytilus edulis* population. *Marine Biology*, **68**, 117–133.

Ketpadung, R. & Tangkrock-olan, N. (2006). Changes in oxygen consumption and heart rate of the blue swimming crab, *Portunus pelagicus* (Linnaeus, 1766) following exposure to sublethal concentrations of copper. *Journal of Environmental Biology*, **27**, 7–12.

Kube, S., Gerber, A., Jansen, J.M., & Schiedek, D. (2006). Patterns of organic osmolytes in two marine bivalves, *Macoma balthica*, and *Mytilus* spp., along their European distribution. *Marine Biology*, **149**, 1387–1396.

Kube, S., Sokolowski, A., Jansen, J.M., & Schiedek, D. (2007). Seasonal variability of free amino acids in two marine bivalves, *Macoma balthica* and *Mytilus* spp., in relation to environmental and physiological factors. *Comparative Biochemistry and Physiology Part A*, **147**, 1015–1027.

Kulkarni, G.K., Joshi, P.K., & Nagabhushanam, R. (1978). Osmoregulation in the marine penaeid prawns *Parapenaeopsis hardwickii* and *Parapenaeopsis stylifera* Crustacea Decapoda Penaeidae. *Bioresearch (Ujjain)*, **2**, 89–94.

Kültz, D. (2005). Molecular and evolutionary basis of the cellular stress response. *Annual Review of Physiology*, **67**, 225–257.

Lacoste, A., Malham, S.K., Cueff, A., & Poulet, S.A. (2001). Stress-induced catecholamine changes in the hemolymph of the oyster *Crassostrea gigas*. *General and Comparative Endocrinology*, **122**, 181–188.

Lang, F. (2007). Mechanisms and significance of cell volume regulation. *Journal of the American College of Nutrition*, **26**, 613S–623S.

Larimer, J.L. & Riggs, A.F. (1964). Properties of hemocyanins. 1. The effect of calcium ions on the oxygen equilibrium of crayfish hemocyanin. *Comparative Biochemistry and Physiology*, **13**, 35–46.

Lee, K.-W., Rhee, J.-S., Raisuddin, S., et al. (2008). A corticotropin-releasing hormone binding protein (CRH-BP) gene from the intertidal copepod, *Tigriopus japonicus*. *General and Comparative Endocrinology*, **158**, 54–60.

Lee, N.H., Han, K.N., & Choi, K.S. (2004). Effects of salinity and turbidity on the free amino acid composition in gill tissue of the Pacific oyster, *Crassostrea gigas*. *Journal of Shellfish Research*, **23**, 129–133.

Leong, P.K.K. & Manahan, D.T. (1997). Metabolic importance of Na$^+$/K$^+$-ATPase activity during sea urchin development. *Journal of Experimental Biology*, **200**, 2881–2892.

Li, C.-C., Yeh, S.-T., & Chen, J.-C. (2010). Innate immunity of the white shrimp *Litopenaeus vannamei* weakened by the combination of a *Vibrio alginolyticus* injection and low-salinity stress. *Fish and Shellfish Immunology*, **28**, 121–127.

Li, E., Chen, L.Q., Zeng, C., et al. (2007). Growth, body composition, respiration and ambient ammonia nitrogen tolerance of the juvenile white shrimp, *Litopenaeus vannamei*, at different salinities. *Aquaculture*, **265**, 385–390.

Lovett, D.L., Tanner, C.A., Glomski, K., et al. (2006a). The effect of seawater composition and osmolality on hemolymph levels of methyl farnesoate in the green crab *Carcinus maenas*. *Comparative Biochemistry and Physiology Part A*, **143**, 67–73.

Lovett, D.L., Verzi, M.P., Burgents, J.E., et al. (2006b). Expression profiles of Na$^+$, K$^+$-ATPase during acute and chronic hypoosmotic stress in the blue crab *Callinectes sapidus*. *Biological Bulletin*, **211**, 58–65.

Lushchak, V.I. (2011). Environmentally induced oxidative stress in aquatic animals. *Aquatic Toxicology*, **101**, 13–30.

Luvizotto-Santos, R., Lee, J.T., Branco, Z.P., et al. (2003). Lipids as energy source during salinity acclimation in the euryhaline crab *Chasmagnathus granulata* Dana, 1851 (Crustacea-Grapsidae). *Journal of Experimental Zoology Part A*, **295**, 200–205.

Malone, P.G. & Dodd, J.R. (1967). Temperature and salinity effects on calcification rate in *Mytilus edulis* and its paleoecological implications. *Limnology and Oceanography*, **12**, 432–436.

Martins, T.L., Chitto, A.L.F., Rossetti, C.L., et al. (2011). Effects of hypo- or hyperosmotic stress on lipid synthesis and gluconeogenic activity in tissues of the crab *Neohelice granulata*. *Comparative Biochemistry and Physiology Part A*, **158**, 400–405.

Mason, R.P., Mangul, C.P., & Godette, G. (1983). The influence of inorganic ions and acclimation salinity on hemocyanin-oxygen binding in the blue crab *Callinectes sapidus*. *Biological Bulletin*, **164**, 104–123.

Matsushima, O., Katayama, H., Yamada, K., & Kado, Y. (1984). Effect of external salinity change on the adenylate energy-charge in the brackish bivalve *Corbicula japonica*. *Comparative Biochemistry and Physiology Part A*, **77**, 57–61.

Mayer, M.P. (2010). Gymnastics of molecular chaperones. *Molecular Cell*, **39**, 321–331.

McAllen, R. & Taylor, A.C. (2001). The effect of salinity change on the oxygen consumption and swimming activity of the high-shore rockpool copepod *Tigriopus brevicornis*. *Journal of Experimental Marine Biology and Ecology*, **263**, 227–240.

McKenney, C.L. & Celestial, D.M. (1995). Interactions among salinity, temperature, and age on growth of the estuarine mysid *Mysidopsis bahia* reared in the laboratory through the complete life cycle. 1. Body-mass and age-specific growth rate. *Journal of Crustacean Biology*, **15**, 169–178.

McNamara, J.C. & Faria, S.C. (2012). Evolution of osmoregulatory patterns and gill ion transport mechanisms in the decapod Crustacea: a review. *Journal of Comparative Physiology Part B*, **182**, 997–1014.

Meehan, W.J. & Ostrander, G.K. (1997). Coral bleaching: a potential biomarker of environmental stress. *Journal of Toxicology and Environmental Health*, **50**, 529–552.

Moberg, F., Nystrom, M., Kautsky, N., et al. (1997). Effects of reduced salinity on the rates of photosynthesis and respiration in the hermatypic corals *Porites lutea* and *Pocillopora damicornis*. *Marine Ecology Progress Series*, **157**, 53–59.

Morris, J.P., Thatje, S., & Hauton, C. (2013). The use of stress-70 proteins in physiology: a re-appraisal. *Molecular Ecology*, **22**, 1494–1502.

Nagaraju, G.P.C. (2007). Is methyl farnesoate a crustacean hormone? *Aquaculture*, **272**, 39–54.

Neufeld, G.J., Holliday, C.W., & Pritchard, J.B. (1980). Salinity adaption of gill Na,K-ATPase in the Blue crab, *Callinectes sapidus*. *Journal of Experimental Zoology*, **211**, 215–224.

Ogan, J., Shaub, A., Lovett, D.L., & Borst, D.W. (1997). Relationship of methyl transferase activity and methyl farnesoate levels in the spider crab *Libinia emarginata*. *Biological Bulletin*, **193**, 267–268.

Okuma, E., Watanabe, K., & Abe, H. (1988). Distribution of free D-amino acids in bivalve mollusks and the effects of physiological conditions on the levels of D- and L-alanine in the tissues of the hard clam, *Meretrix lusoria*. *Fisheries Science*, **64**, 606–611.

Otto, J. & Pierce, S.K. (1981). Water balance systems of *Rangia cuneata*: ionic and amino acid regulation in changing salinities. *Marine Biology*, **61**, 185–192.

Paital, B. & Chainy, G.B.N. (2010). Antioxidant defenses and oxidative stress parameters in tissues of mud crab (*Scylla serrata*) with reference to changing salinity. *Comparative Biochemistry and Physiology Part C*, **151**, 142–151.

Paital, B. & Chainy, G.B.N. (2012). Effects of salinity on O$_2$ consumption, ROS generation and oxidative stress status of gill mitochondria of the mud crab *Scylla serrata*. *Comparative Biochemistry and Physiology Part C*, **155**, 228–237.

Palacios, E., Bonilla, A., Luna, D., & Racotta, I.S. (2004). Survival, Na$^+$/K$^+$-ATPase and lipid responses to salinity challenge in fed and starved white pacific shrimp (*Litopenaeus vannamei*) postlarvae. *Aquaculture*, **234**, 497–511.

Pan, L.Q., Xie, P., & Hu, F.W. (2010). Responses of prophenoloxidase system and related defence parameters of *Litopenaeus vannamei* to low salinity. *Journal of Ocean University of China*, **9**, 273–278.

Parada, J.M., Molares, J., & Otero, X. (2012). Multispecies mortality patterns of commercial bivalves in relation to estuarine salinity fluctuation. *Estuaries and Coasts*, **35**, 132–142.

Pechenik, J.A. (1982). Ability of some gastropod egg capsules to protect against low-salinity stress. *Journal of Experimental Marine Biology and Ecology*, **63**, 195–208.

Perazzolo, L.M., Gargioni, R., Ogliari, P., & Barracco, M.A.A. (2002). Evaluation of some hemato-immunological parameters in the shrimp *Farfantepenaeus paulensis* submitted to environmental and physiological stress. *Aquaculture*, **214**, 19–33.

Philipp, E.E.R., Wessels, W., Gruber, H., et al. (2012). Gene expression and physiological changes of different populations of the long-lived bivalve *Arctica islandica* under low oxygen conditions. *PLoS ONE*, **7**, e44621.

Pillai, B.R. & Diwan, A.D. (2002). Effects of acute salinity stress on oxygen consumption and ammonia excretion rates of the marine shrimp *Metapenaeus monoceros*. *Journal of Crustacean Biology*, **22**, 45–52.

Pongsomboon, S., Udomlertpreecha, S., Amparyup, P., et al. (2009). Gene expression and activity of carbonic anhydrase in salinity stressed *Penaeus monodon*. *Comparative Biochemistry and Physiology Part A*, **152**, 225–233.

Poulain, C., Lorrain, A., Flye-Sainte-Marie, J., et al. (2011). Environmentally induced tidal periodicity of microgrowth increment formation in subtidal populations of the clam *Ruditapes philippinarum*. *Journal of Experimental Marine Biology and Ecology*, **397**, 58–64.

Powell, E.N., Klinck, J.M., Hofman, E.E. (1996). Modelling diseased oyster populations, 2 triggering mechanisms for *Perkinsus marinus* epizootics. *Journal of Shellfish Research*, **15**, 141–165.

Przeslawski, R. (2004). A review of the effects of environmental stress on embryonic development within intertidal gastropod egg masses. *Molluscan Research*, **24**, 43–63.

Rainer, S.F., Ivanovici, A.M., & Wadley, V.A. (1979). Effect of reduced salinity on adenylate energy charge in 3 estuarine mollusks. *Marine Biology*, **54**, 91–99.

Reid, H.I., Soudant, P., Lambert, C., et al. (2003). Salinity effects on immune parameters of *Ruditapes philippinarum* challenged with *Vibrio tapetis*. *Diseases of Aquatic Organisms*, **56**, 249–258.

Rodrigues, A.P., Oliveira, P.C., Guilhermino, L., & Guimaraes, L. (2012). Effects of salinity stress on neurotransmission, energy metabolism, and anti-oxidant biomarkers of *Carcinus maenas* from two estuaries of the NW Iberian Peninsula. *Marine Biology*, **159**, 2061–2074.

Rodstrom, E.M. & Jonsson, P.R. (2000). Survival and feeding activity of oyster spat (*Ostrea edulis* L) as a function of temperature and salinity with implications for culture

policies on the Swedish west coast. *Journal of Shellfish Research*, **19**, 799–808.

Romano, N. & Zeng, C. (2006). The effects of salinity on the survival, growth and haemolymph osmolality of early juvenile blue swimmer crabs, *Portunus pelagicus*. *Aquaculture*, **260**, 151–162.

Rosas, C., Martinez, E., Gaxiola, G., et al. (1999). The effect of dissolved oxygen and salinity on oxygen consumption, ammonia excretion and osmotic pressure of *Penaeus setiferus* (Linnaeus) juveniles. *Journal of Experimental Marine Biology and Ecology*, **234**, 41–57.

Rowe, C.L. (2002). Differences in maintenance energy expenditure by two estuarine shrimp (*Palaemonetes pugio* and *P. vulgaris*) that may permit partitioning of habitats by salinity. *Comparative Biochemistry and Physiology Part A*, **132**, 341–351.

Rupp, G.S. & Parsons, G.J. (2004). Effects of salinity and temperature on the survival and byssal attachment of the lion's paw scallop *Nodipecten nodosus* at its southern distribution limit. *Journal of Experimental Marine Biology and Ecology*, **309**, 173–198.

Santos-Gouvea, I.A. & Freire, C.A. (2007). Effects of hypo- and hypersaline seawater on the microanatomy and ultrastructure of epithelial tissues of *Echinometra lucunter* (Echinodermata: Echinoidea) of intertidal and subtidal populations. *Zoological Studies*, **46**, 203–215.

Sara, G. & de Pirro, M. (2011). Heart beat rate adaptations to varying salinity of two intertidal Mediterranean bivalves: the invasive *Brachidontes pharaonis* and the native *Mytilaster minimus*. *Italian Journal of Zoology*, **78**, 193–197.

Schmidt-Nielsen, K. (1997). Water and osmotic regulation. In: Schmidt-Nielsen K. (ed.), *Animal Physiology Adaptation and Environment*, 5th edition, pp. 301–354. Cambridge University Press India Pvt. Ltd, New Delhi, India.

Schone, B.R., Flessa, K.W., Dettman, D.L., & Goodwin, D.H. (2003a). Upstream dams and downstream clams: growth rates of bivalve mollusks unveil impact of river management on estuarine ecosystems (Colorado River Delta, Mexico). *Estuarine Coastal and Shelf Science*, **58**, 715–726.

Schone, B.R., Tanabe, K., Dettman, D.L., & Sato, S. (2003b). Environmental controls on shell growth rates and delta O-18 of the shallow-marine bivalve mollusk *Phacosoma japonicum* in Japan. *Marine Biology*, **142**, 473–485.

Seo, J.S., Lee, K.-W., Rhee, J.-S., et al. (2006). Environmental stressors (salinity, heavy metals, H_2O_2) modulate expression of glutathione reductase (GR) gene from the intertidal copepod *Tigriopus japonicus*. *Aquatic Toxicology*, **80**, 281–289.

Shinji, J., Kang, B.J., Okutsu, T., et al. (2012). Changes in crustacean hyperglycemic hormones in Pacific whiteleg shrimp *Litopenaeus vannamei* subjected to air-exposure and low-salinity stresses. *Fisheries Science*, **78**, 833–840.

Shinji, J. & Wilder, M.N. (2012). Dynamics of free amino acids in the hemolymph of Pacific whiteleg shrimp *Litopenaeus vannamei* exposed to different types of stress. *Fisheries Science*, **78**, 1187–1194.

Shires, R., Lane, N.J., Inman, C.B.E., & Lockwood, A.P.M. (1994). Structural changes in the gill cells of *Gammarus duebeni* (Crustacea, Amphipoda) under osmotic stress—with notes on microtubules in association with the septate junctions. *Tissue and Cell*, **26**, 767–778.

Shumway, S.E. & Davenport, J. (1977). Some aspects of the physiology of *Arenicola marina* (Polychaeta) exposed to fluctuating salinities. *Journal of the Marine Biological Association of the United Kingdom*, **57**, 907–924.

Silver, I.A., Deas, J., & Erecińska, M. (1997). Ion homeostasis in brain cells: differences in intracellular ion responses to energy limitation between cultured neurons and glial cells. *Neuroscience*, **78**, 589–601.

Smith, K.E. & Thatje, S. (2013). Nurse egg consumption and intracapsular development in the common whelk *Buccinum undatum* (Linnaeus 1758). *Helgoland Marine Research*, **67**, 109–120.

Smith, V.J., Fernandes, J.M.O., Kemp, G.D., & Hauton, C. (2008). Crustins: enigmatic WAP domain-containing antibacterial proteins from crustaceans. *Developmental and Comparative Immunology*, **32**, 758–772.

Snieszko, S.F. (1974). The effects of environmental stress on outbreaks of diseases in fishes. *Journal of Fish Biology*, **6**, 197–208.

Sun, H., Zhang, L., Ren, C., et al. (2011). The expression of Na, K-ATPase in *Litopenaeus vannamei* under salinity stress. *Marine Biology Research*, **7**, 623–628.

Sun, M., Jiang, K., Zhang, F., et al. (2012). Effects of various salinities on Na+-K+-ATPase, Hsp70 and Hsp90 expression profiles in juvenile mitten crabs, *Eriocheir sinensis*. *Genetics and Molecular Research*, **11**, 978–986.

Taylor, A.C., Morris, S., & Bridges, A.C. (1985). Modulation of haemocyanin oxygen affinity in the prawn, *Palaemon elegans* (Rathke) under environmental salinity stress. *Journal of Experimental Marine Biology and Ecology*, **94**, 167–180

Telahigue, K., Rabeh, I., Chetoui, I., et al. (2010). Effects of decreasing salinity on total lipids and fatty acids of mantle and gills in the bivalve *Flexopecten glaber* (Linnaeus, 1758) under starvation. *Cahiers De Biologie Marine*, **51**, 301–309.

Terwilliger, N.B. (2007). Hemocyanins and the immune response: defense against the dark arts. *Integrative and Comparative Biology*, **47**, 662–665.

Tirard, C.T., Grossfeld, R.M., Levine, J.F., & Kennedy Stoskopf, S. (1997). Effect of osmotic shock on protein synthesis of oyster hemocytes *in vitro*. *Comparative Biochemistry and Physiology Part A*, **116**, 43–49.

Towle, D.W., Henry, R.P., & Terwilliger, N.B. (2011). Microarray-detected changes in gene expression in gills of green crabs (*Carcinus maenas*) upon dilution of environmental salinity. *Comparative Biochemistry and Physiology Part D*, **6**, 115–125.

Towle, D.W., Paulsen, R.S., Weihrauch, D., et al. (2001). Na+ + K+-ATPase in gills of the blue crab *Callinectes sapidus*: cDNA sequencing and salinity-related expression of a-subunit mRNA and protein. *Journal of Experimental Biology*, **204**, 4005–4012.

Treaster, S.B., Ridgway, I.D., Richardson, C.A., et al. (2014). Superior proteome stability in the longest lived animal. *Age*, **36**, 1009–1017.

Ungherese, G., Boddi, V., & Ugolini, A. (2008). Ecophysiology of *Palaemonetes antennarius* (Crustacea, Decapoda): the influence of temperature and salinity on cardiac frequency. *Physiological Entomology*, **33**, 155–161.

Urbina, M., Paschke, K., Gebauer, P., & Chaparro, O.R. (2010). Physiological energetics of the estuarine crab *Hemigrapsus crenulatus* (Crustacea: Decapoda: Varunidae): responses to different salinity levels. *Journal of the Marine Biological Association of the United Kingdom*, **90**, 267–273.

Valdez, G., Diaz, F., Denisse Re, A., & Sierra, E. (2008). Effect of salinity on physiological energetics of white shrimp Litopenaeus vannamei (Boone). *Hydrobiologica*, **18**, 105–115.

Van Horn, J., Malhoe, V., Delvina, M., et al. (2010). Molecular cloning and expression of a 2-Cys peroxiredoxin gene in the crustacean *Eurypanopeus depressus* induced by acute hypoosmotic stress. *Comparative Biochemistry and Physiology Part B*, **155**, 309–315.

Vatanavicharn, T., Supungul, P., Puanglarp, N., et al. (2009). Genomic structure, expression pattern and functional characterization of crustin Pm5, a unique isoform of crustin from *Penaeus monodon*. *Comparative Biochemistry and Physiology Part B*, **153**, 244–252.

Veldhuizentsoerkan, M.B., Holwerda, D.A., Debont, A.M.T., et al. (1991). A field-study on stress indexes in the sea mussel, *Mytilus edulis*—application of the stress approach to biomonitoring. *Archives of Environmental Contamination and Toxicology*, **21**, 497–504.

Verslycke, T. & Janssen, C.R. (2002). Effects of a changing abiotic environment on the energy metabolism in the estuarine mysid shrimp *Neomysis integer* (Crustacea: Mysidacea). *Journal of Experimental Marine Biology and Ecology*, **279**, 61–72.

Vilas, C., Drake, P., & Pascual, E. (2006). Oxygen consumption and osmoregulatory capacity in *Neomysis integer* reduce competition for resources among mysid shrimp in a temperate estuary. *Physiological and Biochemical Zoology*, **79**, 866–877.

Wang, Y.J., Hu, M.H., Wong, W.H., et al. (2011). Combined effects of dissolved oxygen and salinity on growth and body composition of the juvenile green-lipped mussel. *Journal of Shellfish Research*, **30**, 851–857.

Werner, I. & Hinton, D.E. (2000). Spatial profiles of hsp70 proteins in Asian clam (*Potamocorbula amurensis*) in northern San Francisco Bay may be linked to natural rather than anthropogenic stressors. *Marine Environmental Research*, **50**, 379–384.

Widdows, J. & Johnson D. (1988). Physiological energetics of *Mytilus edulis*: scope for growth. *Marine Ecology Progress Series*, **46**, 113–121.

Worthington, L.V. (1981). The water masses of the World Ocean: some results of a fine-scale census. In: Warren, B.A. & Wunsch, C. (eds), *Evolution of Physical Oceanography: Scientific Surveys in Honor of Henry Stommel*, pp. 42–69. Massachusetts Institute of Technology, Cambridge, MA.

Xuereb, B., Forget-Leray, J., Souissi, S., et al. (2012). Molecular characterization and mRNA expression of grp78 and hsp90A in the estuarine copepod *Eurytemora affinis*. *Cell Stress and Chaperones*, **17**, 457–472.

Yancey, P.H., Clark, M.E., Hand, S.C., et al. (1982). Living with water stress: evolution of osmolyte systems. *Science*, **217**, 1214–1222.

Yu, Z.M., Li, C.W., & Guan, Y.Q. (2003). Effect of salinity on the immune responses and outbreak of white spot syndrome in the shrimp *Marsupenaeus japonicus*. *Ophelia*, **57**, 99–106.

Zanette, J., de Almeida, E.A., da Silva, A.Z., et al. (2011). Salinity influences glutathione S-transferase activity and lipid peroxidation responses in the *Crassostrea gigas* oyster exposed to diesel oil. *Science of the Total Environment*, **409**, 1976–1983.

Zhang, P., Dong, S.L., Wang F., et al. (2012). Effect of salinity on growth and energy budget of red and green colour variant sea cucumber *Apostichopus japonicus* (Selenca). *Aquaculture Research*, **43**, 1611–1619.

CHAPTER 2

Respiratory responses of marine animals to environmental hypoxia

John I. Spicer

2.1 Introduction

Low O_2 (hypoxia) is a natural feature of many marine environments, such as tide pools, within soft sediments, midwater O_2 minimum zones (OMZ), and (semi-) enclosed areas with reduced water circulation (Spicer, 2014). In addition, anthropogenic disturbance, global climate change, and regional eutrophication are increasing the incidence and severity of hypoxia in marine environments with marked ecological effects (this volume). Hypoxia is lethal to a large proportion of marine species, although how lethal depends on species identity, exposure time, and intensity (Vaquer-Sunyer and Duarte, 2008). Many phyla have species that maintain aerobic metabolism in hypoxic waters, or have other adaptations to survive hypoxia. It is likely that in the future marine animals that have never encountered hypoxia will begin to do so, and those currently inhabiting periodically or chronically hypoxic environments will experience more severe hypoxia in terms of duration and intensity. Therefore there is a pressing need to inform those attempting to manage the ecological and biodiversity consequences of that hypoxia of possible physiological responses, chief among which is our current understanding of the respiratory responses of marine animals to environmental hypoxia.

It has proved contentious to define quantitatively what is meant by hypoxia. A value of less than 2 mgO_2.l^{-1} is frequently taken as the threshold for defining hypoxia (Vaquer-Sunyer and Duarte, 2008). However, arguably hypoxic exposure for an animal normally living in the open ocean is different from one living in a burrow where in the latter 'normal O_2' would be classed as severely hypoxic for an open water species or a human being. Thus Seibel (2011) argued convincingly that hypoxic habitats should be defined by the biological response to hypoxia and not by some arbitrary O_2 concentration, and so this is the approach adopted here.

There are two different, but interdependent, ways of approaching a scientific question. Both have merit and limitations. An *a posteriori* approach is where observations are made as part of a biological comparison and then an adaptive hypothesis is formulated to explain those differences. An *a priori* approach is where predictions are made about how the biological world should work, and those predictions are then tested through observation and experiment. The aim of this chapter is to take an *a priori*, or predictive, approach to understanding how, in broad terms, the respiratory biology of marine invertebrates responds to hypoxia. This is done by listing as many possible responses to hypoxia we can imagine a marine animal (water breathers only, i.e. not marine reptiles, birds, or mammals) might make, before testing our predictions against our current knowledge of respiratory responses to hypoxia. Firstly we consider the different ways the metabolism of an animal could change when exposed to hypoxia, before asking how the physiological components responsible for 'fuelling' that metabolism might be modified. The piece finishes with a perspective posing future challenges to knowledge of how marine animals respond to environmental hypoxia, and suggests some reviews for those who want to go further.

2.2 Oxyregulators and oxyconformers

Because of the technical difficulties in quantifying total metabolism, measurement of aerobic metabolism (rate of O_2 uptake) is often used as a proxy. While this seems sufficient for most purposes, we must be careful for anaerobic metabolism may play quite a substantial role in the metabolism of some taxa, e.g. molluscs and

Stressors in the Marine Environment. Edited by Martin Solan and Nia M. Whiteley

annelids (see under Mechanism 9 in Section 2.3). When exposed to declining O_2 tensions (pO_2), the rate of O_2 uptake of an aquatic animal could theoretically (i) increase, (ii) stay the same, (iii) decrease, or (iv) respond idiosyncratically. In reality only two of these responses are common.

Those species that maintain rate of O_2 uptake over a wide range of environmental pO_2s are termed oxyregulators (Fig. 2.1, 2.2A). They seem to be comprised of species drawn from numerous taxa which encounter periodic or chronic hypoxia in their natural environment, although, in truth, evidence for this generalization is not strong. However, no species is able to regulate O_2 uptake rates across the entire range (Lutz and Storey, 1997; Burnett and Stickle, 2001). The point at which oxyregulation breaks down is termed the P_c, or critical O_2 tension. Below this point O_2 uptake rate decreases precipitously with decreasing environmental pO_2, and there is often the possibility of a shift from aerobic to anaerobic energy generation pathways. However, being a poor oxyregulator does not always preclude a species from living in a hypoxic environment. The lugworm, *Arenicola marina*, has a high P_c (16.00 kPa where 1 kPa~7.5 torr or mmHg). Below the P_c O_2 uptake decreases with declining pO_2. However, at about $pO_2 = 6.00$ kPa, anaerobic metabolism

is switched on allowing them to survive quite severe hypoxia (Schottler and Bennet, 1991).

Species in which O_2 uptake rate decreases with increasing hypoxia are termed oxyconformers (Fig. 2.1) and are present in a wide range of taxa. The assumption that these species do not normally encounter hypoxia, and if they do they do not fare well, seems too simplistic. A sipunculid worm, though classed as an oxyconformer, lives in chronically hypoxic sediments. Pörtner et al. (1985) found that it reduced its O_2 uptake rate with progressive hypoxia, but this reduction was not indicative of compromised O_2 supply. Under severely hypoxic conditions there was a switch from aerobic to anaerobic metabolism. In this case an oxyconformer, and not just an oxyregulator, has a P_c or critical O_2 tension, although in this case it is defined as the pO_2 at which there is a shift from aerobic to anaerobic metabolism (Fig. 2.1).

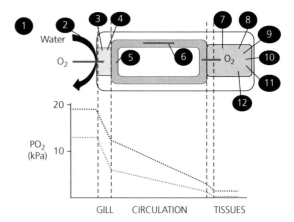

Figure 2.2 Hypothetical marine animal depicting the mechanisms that could *potentially* respond to environmental hypoxia. 1. Animals could move—avoid hypoxia or emerse themselves into air. 2. Anything that causes more water to pass across potential areas of gas exchange could improve O_2 uptake rates. 3. Increase surface area and reduced thickness of gas exchange surfaces. 4. Actively take up O_2 from the environment. 5. Increase perfusion of gas exchange organs and tissues. 6. Increase the concentration of, and affinity for, O_2 in circulatory fluid. 7. Increase the amount of stored O_2 for use during hypoxic exposure. 8. Reduce overall metabolic rate. 9. Increase capacity for anaerobic metabolism and improve energetic efficiency (i.e. maximize ATP production). 10. Fuel metabolism without O_2. 11. Use a simple, all-embracing, mechanism to control and co-ordinate the respiratory response to hypoxia. 12. Reduce damage caused by production of reactive O_2 species. The grey arrows depict O_2 movements from the environment to the tissue. The black arrow depicts water movement across the gas exchange surface. The graph below depicts the pO_2 gradient from the environment to the tissues in hypoxia (grey dotted line) and normoxia (black dotted line) conditions.

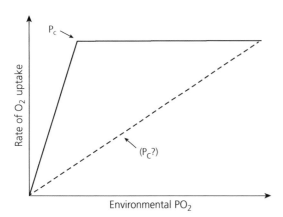

Figure 2.1 Relationship between O_2 uptake rate and declining environmental pO_2 in marine animals. Oxyregulators (solid line) are able to maintain O_2 uptake rates over a wide range of environmental pO_2 until a critical point, the P_c, where respiratory regulation breaks down. This P_c point may also coincide with a shift from aerobic to anaerobic metabolism. Oxyconformers (dashed line) show a steady reduction in O_2 uptake rate with declining environmental pO_2. While in this case there is no respiratory regulation to break down there still may be a critical O_2 tension (marked as P_c?) where there is a shift from aerobic to anaerobic metabolism.

The efficacy of using terms like oxyregulator and ox-yconformer is still hotly debated. In reality they are, at best, two extremes of a broad continuum, and there can be variable responses not just between species, but between individuals in a population, and even between the same individual in different physiological states (Lutz and Storey, 1997; Burnett and Stickle, 2001; Seibel, 2011). Broadly speaking oxyregulatory ability tends to be better developed in (a) species that permanently inhabit, or regularly migrate into, chronically hypoxic areas, such as the OMZ, (b) species and individuals with lower metabolic rates, (c) taxa inhabiting colder temperature environments, and (d) more 'settled' individuals. The pO_2 difference between the environment and the mitochondria drives O_2 uptake and delivery (Fig. 2.2). When environmental pO_2 is reduced, that difference is reduced. We now suggest potential mechanisms (Fig. 2.2) which could either maintain rates of O_2 uptake or provide an alternative to aerobic metabolism in the face of a reduction in the pO_2 difference between environment and mitochondria.

2.3 Potential respiratory responses to environmental hypoxia

Mechanism 1. *Animals could move—avoid hypoxia or emerse themselves into comparatively O_2-rich air.* Avoidance behaviour is a common response of many marine animals to hypoxia (Burnett and Stickle, 2001). A number of crustacean, mollusc, and fish species exhibit a hypoxia-induced emersion response (Hsia et al., 2013). However, emersion, for all but semi-terrestrial species, is only an effective strategy in the short term (hours to days) if there are adaptations facilitating good O_2 exchange in air. Thus we find that those, mainly intertidal, species tend to have structural support within the gills that prevent their collapse in air, an enclosed space where water can be retained for gas exchange, and the development of extrabranchial gas exchange surfaces ('lungs' and vascularization of the mouth in fish, vascularized mantle cavity in molluscs, and vascularized gill chambers in crabs) (Hsia et al., 2013). However, an increasing number of marine (even deep water) animals with no such adaptations still emerse themselves in response to widespread seasonal hypoxia, e.g. crawfish off the US Gulf Coast or rock lobsters off the coasts of Namibia and South Africa. This emersion response is lethal and internal hypoxia is produced even though environmental O_2 is plentiful. Some species such as the intertidal shrimp *Palaemon elegans* exhibit a hypoxia-induced, *partial* emersion response, which allows them to aerate water in one of its partially emersed gill chambers, increasing the water pO_2, while ensuring the shrimp does not dry out (Spicer, 2014). However, the partial-emersion response is a short-term (minutes to hours) strategy.

Many species found in the OMZ have well-developed mechanisms for coping with hypoxia but some are transients, migrating in and out of this hypoxic water body as a way of reducing long-term hypoxic stress (Seibel, 2011).

Mechanism 2. *Anything that causes more water to pass across potential areas of gas exchange could improve rates of O_2 uptake.* This may simply be a small and/or highly permeable individual increasing locomotory activity. Increased activity may act as a hypoxia-avoidance mechanism, enabling mobile animals such as fish, crustaceans, echinoderms, and molluscs to escape hypoxia either periodically or completely (Herreid, 1980). The small marine flatworm *Macrostomum lignano* seems to maintain O_2 uptake rates under conditions of declining pO_2 by increasing the beating of cilia that cover the body surface (Rivera-Ingraham et al., 2013). For species with localized gas exchange organs such as gills (e.g. fish, crustaceans, and annelids) hypoxia-related hyperventilation of those organs is an effective oxyregulation strategy, particularly if the animal inhabits a burrow, or the gills are enclosed (Burnett and Stickle, 2001). However, powering ventilatory structures (e.g. scaphognathite, of lobsters and crabs) is energy demanding (Herreid, 1980) and so is only viable as a short-term (minutes to hours) hypoxic response, although there are exceptions, e.g. a midwater mysid which maintains comparatively high ventilation rates in the OMZ (Seibel, 2011). In some oxyregulators ventilation rate increases with exposure to declining pO_2 until the P_c point after which there is a marked decrease (Fig. 2.3C). Both hypoxia-tolerant (e.g. Atlantic cod) and hypoxia-intolerant (e.g. tuna) fish show a hypoxia-related hyperventilation but the former shows a higher P_c than the latter (Lutz and Storey, 1997). Many taxa that inhabit periodically hypoxic environments tend to show hypoxia-induced hyperventilation down to very low P_cs, e.g. the intertidal prawn *Palaemon elegans*. Interestingly there are oxyregulating and oxyconforming jellyfish species (Rutherford and Thuesen, 2005). Neither increase the rate of bell pulsation (ventilation) in response to declining pO_2 but the oxyconformers tend to have relatively narrow velar apertures which restricts seawater exchange between the underside of the bell and the outside water, suggesting that 'normal' ventilation in oxyregulators is sufficient to meet metabolic needs down to low environmental pO_2s.

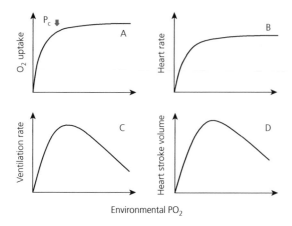

Figure 2.3 Respiratory response to hypoxia of a crustacean oxyregulator. The animal is able to maintain O_2 uptake down to a critical level, the P_c point (A), by increasing the ventilation rate (C) and total cardiac output. Cardiac output increases not as a result of increasing heart rate (B) but by increasing stroke volume (D).

Mechanism 3. *Increase surface area and reduce thickness of gas exchange surfaces.* In the short term some sea anemones and sea cucumbers may increase the surface area available for gas exchange when challenged with acute hypoxia. In the medium term (weeks to months) chronically hypoxic fish exhibit a proliferation and re-modelling of gill tissue (Mitrovic et al., 2009). While evidence for plasticity in other groups is sparse, there is some evidence that over evolutionary timescales species that encounter chronic hypoxia develop gills with greater surface areas and thinner cuticles than those that do not, although this is not invariant (Decelle et al., 2010; Seibel, 2011). The flat, leaf-like structure of the cnidarian-like animals of the 600 my old Ediacaran fauna has been interpreted as respiratory adaptations to low O_2. Furthermore cutaneous gas exchange may even occur in animals with well-developed circulatory systems (see Mechanism 5), as in the case for cephalopods where the surface of the mantle contains mitochondria-rich red muscle.

Mechanism 4. *Actively take up O_2 from the environment.* O_2 uptake is driven by diffusion alone. There is no evidence for active transport of O_2 across gas exchange surfaces.

Mechanism 5. *Increase perfusion of gas exchange organs and tissues.* This could be achieved by means of a circulatory system with at least one pump (heart) that is responsive to hypoxia. We see the development of circulatory systems in numerous taxa above a critical body size. Although 'closed' systems (i.e. circulatory

fluid never leaves the circulatory system—annelids, cephalopods, and vertebrates) were thought to be superior in function to 'open' systems (i.e. circulatory fluid bathes the organs—found in the remaining invertebrate groups) this assessment is too simplistic (McMahon, 2012). Both systems may respond to hypoxia. A reduction in heart rate (bradycardia) is not an uncommon hypoxic response of hypoxia-intolerant species, and even some that are hypoxia-tolerant. For crustaceans the heart rate of oxyregulators tends to be insensitive to hypoxia (and does not always decrease as one might have predicted), at least until the P_c after which it decreases (Fig. 2.3B). During embryonic development of the intertidal snail *Littorina obtusata* there was a shift from heart rate (adult and another transient structure that appears earlier in embryonic development termed the 'larval heart') declining with declining pO_2 to a pattern of heart rate being maintained over a wide range of pO_2s (Bitterli et al., 2012). However, in some oxyregulators stroke volume (unit haemolymph/blood pumped per heart beat) increases with declining pO_2. Thus cardiac output, the total amount of haemolymph/blood delivered to the tissues per unit time, in some crustaceans and fish species increases under hypoxic conditions, due to an increase in stroke volume, not heart rate (Fig. 2.3B,D). In hypoxia-intolerant species (e.g. tuna fish) hypoxia reduces heart rate (bradycardia) and there is no compensatory increase in stroke volume, resulting in a reduction in cardiac output (Bushnell and Brill, 1992, cited in Lutz and Storey, 1997). As with hyperventilation increasing cardiac output is a highly energy demanding response and so, when it occurs, may only viable in the short-to-medium term. There is also evidence that in crustaceans and fish there is redirection of circulatory fluid towards some organs and away from others during hypoxic exposure (McMahon, 2001). In hypoxic crabs, *Cancer magister*, haemolymph is directed away from the viscera and towards the limbs, via the gills, which increases the gill's capacity for O_2 uptake.

Mechanism 6. *Increase the concentration of, and affinity for, O_2 in circulatory fluid.* While changes in cardiac output can increase the amount of circulatory fluid perfusing the gas exchange organs and the tissues, respiratory gas transport could be further improved by carrying more O_2 per unit volume of circulatory fluid, and by loading and unloading that O_2 in the most efficient way possible. For many marine invertebrates there is enough O_2 carried in physical solution in the body fluids to meet resting metabolism. However, larger, more active animals, and taxa that encounter hypoxia or are hypoxia-tolerant tend to have a respiratory pigment (e.g. haemoglobin or haemocyanin) in their blood/haemolymph which binds O_2 reversibly and cooperatively. The more hypoxic the water, the greater the role of the respiratory pigment in transporting O_2 (Wells, 2009).

During short-term hypoxia, hyperventilation will blow off CO_2 resulting in an increase in blood/haemolymph pH. This increase in pH will increase the O_2 affinity of the respiratory pigment (i.e. push the dissociation curve to the left), making it easier to pick up O_2 from the gas exchange surfaces at low pO_2 (Fig. 2.4). If anaerobic end products are produced and released into blood/haemolymph, and/or hypoxia induces erythrocyte organic phosphate production, these metabolites are allosteric modulators of O_2 affinity and will further increase the O_2 affinity of the respiratory pigment (Wells, 2009; Fig. 2.4). During medium- to long-term hypoxia, hyperventilation will cease (or at least reduce) and the pH of the blood/haemolymph will fall. However, an increase in the O_2 affinity of the respiratory pigment may be sustained, or even enhanced, by the hypoxia-related production of a modified respiratory pigment with an intrinsically higher O_2 affinity (Wells, 2009; Fig. 2.4). Not only does hypoxia increase the production of a respiratory pigment that binds O_2 differently but it also produces greater concentrations of that pigment, enabling a unit of blood/haemolymph to carry more O_2. Species from chronically hypoxic environments, such as the OMZ or burrows, tend to possess high concentrations of high O_2 affinity respiratory pigments, but this is not invariant.

Hypoxia-tolerant species also tend to have respiratory pigments with pronounced Bohr effects (i.e. small increases in CO_2 effect changes in pH which in turn elicit large reductions in O_2 affinity). In some species a large Bohr effect could reduce the O_2 affinity of the respiratory pigment, making venous blood/haemolymph give up its bound O_2 to hypoxic tissues, thus saving energy elsewhere (i.e. this is a 'low-cost' option compared to increasing cardiac output). However, there is still the possibility that for some species a large Bohr effect is maladaptive and that the allosteric effect on O_2 affinity of anaerobic end products, such as L-lactate, evolved to counteract and so negate the Bohr effect (Mangum, 1997). In many teleost fish species haemoglobin shows a Root effect (pH-related reduction in both O_2 affinity and O_2 carrying capacity), which in the presence of carbonic anhydrase and acidified blood can greatly increase O_2 delivery to the tissues with the cost of modifying other aspects of their physiology, e.g. perfusion (Rummer et al., 2013).

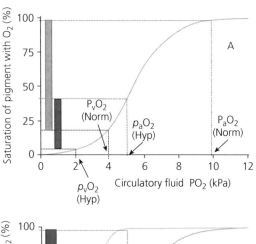

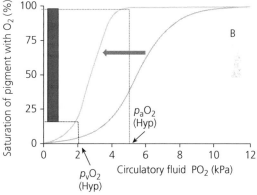

Figure 2.4 Hypothetical O_2 binding curve for a respiratory pigment, either in cells or dissolved in the extracellular fluid (ECF), that binds O_2 reversibly and cooperatively. Shifting the curve to the left means the pigment performs better under hypoxic conditions when there is a reduction in arterialized and venous ECF values. A. Under normoxic conditions arterialized ECF (P_aO_{2Norm}) is almost 100% saturated with O_2, and the venous ECF is about 20% saturated ($P_vO_{2\ Norm}$), meaning that 80% of the bound O_2 is transported to and unloaded at the tissues (light shaded bar). Consider first, if under hypoxic conditions both the arterialized and venous ECF are reduced each by about a half, *but the position of the curve does not change*. Then arterialized ECF (P_aO_{2Hyp}) leaves the gas exchange surfaces less than 40% saturated with O_2, and as the venous ECF is less than 10% saturated (P_vO_{2Hyp}) only about 30% of the bound O_2 is delivered to, and unloaded at the tissues (dark shaded bar). B. When animals experience long-term hypoxia the O_2 binding curve may be shifted to the left (i.e. it has a higher affinity for O_2) either through intrinsic changes to the molecule or extrinsic effectors introduced into the ECF, such as L-lactate. Before the curve shifted arterialized ECF (P_aO_{2Hyp}) was only 40% saturated with O_2. Now, even although P_aO_{2Hyp} is the same low value, because the curve has shifted to the left the ECF leaves the gas exchange surfaces 100% saturated. Instead of 30% of the bound O_2 being delivered to the tissues it is now nearer 80% again (dark shaded bar).

Mechanism 7. *Increase the amount of O_2 stored in an animal that can be used during hypoxic exposure.* There are few examples of marine water breathers that utilize O_2 stores when they encounter hypoxia unless the exposure period is very brief (minutes to hours). To be viable for any longer, there must be a large O_2 reservoir and/or a comparatively low O_2 demand from the tissues. There are some examples where these requirements are present.

Cephalopods are considered a hypoxia-sensitive group. However, the shelled *Nautilus* is hypoxia-tolerant. It maintains ventilation down to $pO_2 = 5.33$ kPa, but below 3.33 kPa it becomes inactive. Ventilation ceases, heart rate is greatly reduced, the large amount of high affinity pigment in the haemolymph slowly releases bound O_2 from the large venous haemolymph reserve, and O_2 gas within buoyancy chambers in the shell acts like an aqualung (Boutilier et al., 1996). This aqualung can last approx. 40 min at 19 °C (Lutz and Storey, 1997).

Gelatinous animals, such as jellyfish, passively absorb O_2 from the environment into their mesogleal gel. If they enter hypoxic water, this absorbed O_2 may be sufficient to maintain aerobic metabolism for a time (Thuesen et al., 2005). Some tissues contain respiratory pigments (e.g. myoglobin) that could act as O_2 stores. The recent discovery of a membrane-bound haemoglobin (the first in eukaryotes) in crab gills with a very high O_2 affinity ($P_{50} = 0.07$ kPa) at first looks like a good candidate. However, it does not act as an O_2 carrier. Instead it has an enzymatic function, protecting membranes from reactive O_2 species (Ertas et al., 2010). Despite evidence for hypoxia increasing the concentration of respiratory pigments in the blood/haemolymph and tissues, their use for storage and slow release (rather than transport) of O_2 in aquatic breathers is still a relatively short-term strategy.

Mechanism 8. *Reduce overall metabolic rate as much as possible.* When ATP demand exceeds supply many cellular processes will be disrupted, some with few consequences in the short term, others with more immediate serious consequences, e.g. use of ATP-fuelled membrane ion pumps which maintain differences in membrane potential essential for cell viability. For most taxa that can respond to hypoxia, those responses centre on energy conservation rather than a greater ability of hypoxia-tolerant cells to withstand major ion or energy imbalances (Hochachka and Somero, 2002). Metabolism is reduced (termed hypometabolism, a reduction in energy demand) to a level which, if the individual is to persist in that hypoxic environment, can be sustained solely by anaerobic ATP production

(2–30% in marine molluscs and polychaetes (Lutz and Storey, 1997)). Hypoxia causes body temperature to fall in many endotherms, but in ectothermic marine animals hypoxia induces behavioural selection of a lower environmental temperature, thus reducing O_2 demand (Morris, 2004). In terms of physiological responses, reversible hypometabolism tends to involve (a) dramatic reductions in protein and glucose synthesis and protein degradation, (b) lesser reductions in maintaining electrochemical gradients, with ion exchangers the main energy consumers, and (c) changing the activities of existing enzymes in the cell rather than constructing new ones. In marine animals the latter has best been studied in marine bivalve molluscs that live in hypoxic and even anoxic sediments, e.g. alanine, an anaerobic end product in some molluscs, can inhibit key glycolytic enzymes. In the Gulf killifish, *Fundulus grandis*, activities of enzymes associated with glycolysis and glycogen metabolism decreased as a result of hypoxia in some tissues (e.g. muscle) but not all (e.g. liver, brain), and in fact the activities of some gluconeogenic enzymes associated with carbohydrate metabolism actually increased in the long term (Martinez et al., 2006).

Seibel (2011) has argued convincingly that metabolic depression (and a good anaerobic capacity) may be essential for enabling short but regular excursions of transients into the oxygen minimum zone (Fig. 2.5).

Mechanism 9. *Increase capacity for anaerobic metabolism and improve energetic efficiency (i.e. maximize the amount of ATP produced per unit substrate).* If generation of ATP using aerobic respiration is compromised by hypoxia it may be advantageous to maximize ATP anaerobic production. This should be by means of efficient pathways of supplementing ATP production from fermentation reactions and provisioning of substrates to fuel those reactions, preferably ones where metabolic by-products (a) are not toxic, (b) are excretable, and (c) will not disturb the chemical balance of the extracellular fluid, e.g. pH. And this is broadly what we see across a wide range of taxa (Greishaber et al., 1994). Most marine animals can resort to anaerobic metabolism to some extent. However, the critical pO_2 at which anaerobic metabolism is co-opted, the metabolic pathways involved, and the efficiency of those pathways differ widely between taxa (Fig. 2.6). The aspartate–succinate and opine pathways require both carbohydrates and amino acids as substrates and are considered to have evolved before the glucose–succinate and L-lactate pathways which use only carbohydrates as their substrates. Livingstone (1991) suggested that there are two main types of anaerobic metabolic pathway.

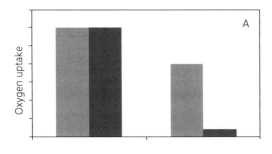

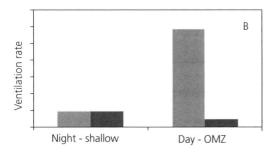

Figure 2.5 The importance of metabolic suppression for transients in the Oxygen Minimum Zone. Metabolism and ventilation of squid *Doscidiscus gigas* migrating down into the OMZ during the day (250 m depth, T = 10 °C, pO_2 = 0.8 kPa) and returning to shallower layers (70 m depth, T = 20 °C, pO_2 = 21 kPa) during the day, in the Guaymas Basin, Gulf of California. The lighter shaded bars represent what would happen if there was no metabolic depression. O_2 uptake rate would decrease as the squid entered the OMZ but mainly because of the lower temperature at this depth and there would be a high energetic cost of increasing ventilation to maintain O_2 under hypoxic conditions. The darker shaded bars represent what happens because there is metabolic depression. There is a dramatic decrease in O_2 uptake rate (only a small proportion of it explained by the reduction in temperature) as the squid enter the OMZ, with a further saving because of reduced ventilation, with some of overall reduced metabolism being fuelled anaerobically (modified from Seibel 2011). Without metabolic depression these squid would not be able to pump enough water across the gills to meet metabolic demands when they were in the O_2 poor waters of the OMZ.

(i) Those with high rates of energy production, but poor efficiency, e.g. various opine pathways (tauropine, strombine, and opine) and the L-lactate pathways (both generate two ATPs per glucose molecule). Such pathways can only be short-term solutions to hypoxic exposure. Hypoxic crustaceans and fish would normally accumulate and 'poison' their tissues with L-lactate, but in some very hypoxic-tolerant species soluble ethanol is produced instead and excreted into the external medium. L-lactate is also the metabolic end

product of echinoderms. Opines are used by sponges, cnidarians, molluscs, and annelids; they generally are less acidic end products than L-lactate and, because they consume an amino acid in their production, tend to be osmotically neutral.

(ii) Those that, while still producing considerably less ATP than would be produced by aerobic pathways, are more efficient (four to six ATPs per glucose molecule) but as a result have low rates of energy production per unit time. These are the glucose–succinate and aspartate–succinate pathways which characterize some bivalves, polychaetes, and other 'worm' phyla, such as sipunculids (Müller et al., 2012). However, in the last three taxa referred to, although it is true that succinate, proponiate, and acetate are the main metabolic end products generated in the mitochondria, something different occurs in the cell cytosol outside. When mussels first switch to anaerobic metabolism, opines (octopine, strombine, and alanine) are produced in the cytosol before the mitochondria-based pathways take over. The same opines are produced in the cytosol during environmental hypoxia in the sipunculid worm and the lugworm (except that octopine is replaced by alanopine). In active taxa, with relatively high overall metabolic rates, generally the first type dominates, whereas in more sedentary taxa, with relatively low overall metabolic rates, the second type dominates.

The extent to which anaerobic energy output compares with aerobic output under normoxic conditions across a range of taxa is still not clear. This is because how total caloric or heat output changes with declining pO_2 has not, because of logistical constraints in using calorimetry, received the attention it should.

Mechanism 10. *Fuel metabolism without any O_2 . . . ever.* While there are many metazoans which can survive protracted periods of no O_2 (anoxia), until recently it was thought that only a few unicellular organisms could survive anoxia indefinitely. However, sediments in deep anoxic hypersaline environments in the Mediterranean Sea are inhabited by three species of meiofauna belonging to the phylum Lorocifera. These species do not possess mitochondria, but instead hydrogenosome-like organelles, and individuals complete their life cycles in anoxic conditions, the first metazoans to be known to do so (Danovaro et al., 2010).

Mechanism 11. *Use a simple, all-embracing mechanism to control and co-ordinate the respiratory response to hypoxia.*

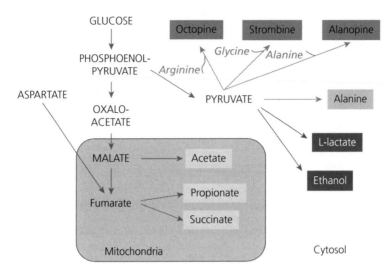

Figure 2.6 Summary of metabolic pathways that may be used to sustain anaerobic metabolism in the tissues (mitochondria and cell cytosol) of marine animals. Derived from various sources.

Cells of many organisms have a system for switching on a suite of genes that help the cell cope with hypoxia. The 'master switch' can involve a family of related proteins, transcription factors termed hypoxia-inducible factors (HIFs). HIFs are considered essential for maintaining homeostasis in metazoans and can be induced by exposure to hypoxia (Semenza, 2008). However, HIF involvement in regulating gene expression is much wider than just responding to hypoxia: HIF-1 controls the expression of hundreds of genes. Because these genes vary from one cell type to another, it is probable that the HIF-1 transcriptome includes literally thousands of genes with roles in development, physiology, and disease. HIF-1 is formed from two subunits (α and β) expressed constitutively, with the α-subunit post-translationally modified in response to changes in O_2. The production of HIF-1 during hypoxia controls the expression of genes responsible for producing glucose transporters and glycolytic enzymes involved in the anaerobic production of ATP (Hochachka and Lutz, 2001). In fish HIF-1α/β is also involved in enhancing O_2 delivery through increased angiogenesis and erythropoiesis and in crustaceans HIF initiates the synthesis of respiratory pigment under hypoxic (×15–20 normoxic levels) but not normoxic conditions (Gorr et al., 2004). Silencing the α and β subunits of HIF-1 by intramuscular injection of double-stranded RNA (dsRNA) affected both glucose (energy source) and L-lactate (anaerobic end product) in the shrimp *Litopenaeus vannamei* exposed to 24 h environmental hypoxia (Soñanez-Organis et al., 2010). Although HIF is a transcription factor it also displays some non-transcriptional roles during hypoxic stress, such as slowing down DNA replication and reducing cell proliferation (Huang, 2013). We now consider hypoxia-induced transcription changes likely to be controlled by HIF.

Gracey et al. (2001) found that under hypoxia the intertidal fish *Gillichthys mirabilis* showed tissue specific patterns of gene regulation, but generally up-regulated genes associated with anaerobic ATP production and gluconeogenesis, and down-regulated genes associated with protein synthesis and cell growth. Hypoxic fish also conserved energy by reducing movement, and redirected energy to essential cellular processes. Subsequent transcriptomic studies of marine teleosts, e.g. three-spined stickleback, *Gasterosteus aculeatus* (Leveelahti et al., 2011), killifish, *Fundulus grandis* (Everett et al., 2012), and Japanese rice fish, *Oryzias latipes* (Zhenlin et al., 2007), indicate that some of the responses observed are broadly similar across species, although they can differ between tissues even in the same individual. This is supported by the work on model species, which are predominantly freshwater forms (zebrafish), although the overall story is more complicated and involved than originally thought, largely because many of the hypoxic responses are not necessarily hypoxia-specific (Richards, 2009; Zhang et al., 2009). Understanding the effects of hypoxia on the transcriptome of marine invertebrates is poorly developed compared with our understanding in fish, but again, though complex, with marked differences between laboratory and field studies, very broadly some of the same themes seem to be emerging as shown by the fish (Spicer, 2014).

Mechanism 12. *Reduce damage caused by production of reactive O_2 species.* It is recognized that too much O_2 is toxic to animal life, because of the production of reactive O_2 species (ROS). While their production is a normal feature of aerobic metabolism, it can be influenced by environmental factors. ROS oxidize nucleic acids and proteins resulting in cellular damage, dysfunction, and even death. Within strict limits electrons transferred through the mitochondrial respiratory chain will, in the presence of the enzyme cytochrome-c-oxidase, react with O_2 to produce water (H_2O). However, seemingly small increases *or sometimes decreases* in cellular O_2 can result in some electrons escaping the respiratory chain and combining with O_2 to produce ROS, i.e. the superoxide ion, which is then converted into hydrogen peroxide (H_2O_2) (Abele et al., 2007). Thus ROS may also be a potential problem during exposure to environmental hypoxia (Burnett and Stickle, 2001). It is probable that post-hypoxic re-oxygenation is likely to result in an increase in ROS formation. To reduce ROS formation marine invertebrates could potentially (a) maintain tissue pO_2 within a very narrow range, (b) within the mitochondria control the membrane potential and leakage of hydrogen ions across the membrane (both of which influence ROS formation), or (c) use nitric oxide to lower the affinity of cytochrome oxidase for O_2, thus controlling the reduction state of electron transport. Currently there is some evidence for (a) and (b) but (c) has so far only been observed in mammals (Abele et al., 2007).

2.4 Perspectives

Broadly speaking, in response to environmental hypoxia, in the short term many marine animals increase ventilation of respiratory surfaces and/or perfusion of tissues. In the longer term many modify respiratory gas transport (O_2 carrying capacity and intrinsic O_2 affinity) and aspects of the intensity and efficiency of tissue respiration, and have recourse to anaerobic metabolism. If we are to understand 'how animals work in the wild' (Spicer, 2014), and generate information which will help understand environmental hypoxia in its larger context of global climate change, it is suggested that future studies on the effects of hypoxia on respiratory performance of marine animals should investigate:

(i) what pO_2 conditions animals actually encounter *in situ*, and how they change spatially and temporally, at different scales. Taking small-scale spatial variation into account has been critical, for example, to understand the effects of realistic environmental temperatures on intertidal animals (Helmuth et al., 2010). While severe hypoxia has been a useful tool to probe animal function, and its effects on the respiratory system are comparatively well documented, this is not so for moderate hypoxia, which is not uncommon in ecological systems. Also exposure to hypoxia is often cyclical, or periodic, and there is some evidence that cyclical and chronic exposure have different physiological effects (Spicer and El-Gamal, 1999; Burnett and Stickle, 2001; Flück et al., 2007).

(ii) respiratory performance under 'realistic' conditions, i.e. taking account of the fact that in nature a number of factors alter at the same time, and sometimes in concert, e.g. hypoxia is often accompanied by elevated CO_2 (hypercapnia) (Burnett and Stickle, 2001). It is true that we need to understand multi-stressor effects on marine organisms, e.g. current ocean acidification research is incorporating the effects of elevated temperature and reduced O_2 into its experimental designs (Pörtner, 2012; although see Burnett and Stickle (2001) who were studying hypercapnic hypoxia more than a decade before this). However, adding additional stressors to our experiments is logistically challenging, and still may not bring us any closer to the ecological realism which drives such multi-stressor experimental designs. Such laboratory experiments need to be carried out alongside, and be informed by, field and 'natural' experiments wherever possible, if our aim is to understand how animals work in the wild (Spicer, 2014).

(iii) the respiratory responses to hypoxia of animals, particularly oceanic and pelagic species that (as far as we know) never encounter environmental hypoxia naturally. This is important in understanding potential effects of spreading dead zones on animal physiology and ecology.

(iv) how the respiratory system responds to hypoxia at different times in one generation and in subsequent generations, and how and when those responses appear during development. Embryos and larvae are not just large animals writ small, and often have their own, complicated physiological responses to hypoxia. Furthermore, we know little of transgenerational effects of hypoxia, and the costs, in terms of Darwinian fitness, of hypoxic exposure (Spicer and El-Gamal, 1999).

(v) how we can produce a truly integrative approach to understanding respiratory responses to hypoxia, learning how best to interpret and combine -omics approaches (genomic, transcriptomic, proteomic, and metabolomics) with more traditional physiological approaches. This has started but there is still a long way to go (Spicer, 2014).

For more detail on the respiratory responses of marine animals to hypoxia, Farrell's (2011) edited book on fish physiology contains excellent sections on anaerobic metabolism, metabolic suppression, and respiratory and circulatory responses of elasmobranches and bony fish to hypoxia (see also Richards, 2009; Wells, 2009), and there are a number of reviews (some old but still worthwhile) that cover physiological and metabolic responses of marine invertebrates to hypoxia (Herreid, 1980; Livingstone, 1991; Grieshaber et al., 1994; Lutz and Storey, 1997; Childress and Seibel, 1998; Burnett and Stickle, 2001; Hochachka and Lutz, 2001; McMahon, 2001; Hochachka and Somero, 2002; Larade and Storey, 2002; Wu, 2002; Flück et al., 2007; Hourdez and Lallier, 2007; Gorr et al., 2010; Seibel, 2011; Müller et al., 2012).

References

Abele, D., Philipp, E., Gonzalez, P.M., & Puntarulo, S. (2007). Marine invertebrate mitochondria and oxidative stress. *Frontiers in Bioscience*, **12**, 933–946.

Bitterli, T.S., Rundle, S.D., & Spicer, J.I. (2012). Development of cardiovascular function in the marine gastropod *Littorina obtusata* (Linnaeus). *Journal of Experimental Biology*, **215**, 2327–2333.

Boutilier, R.G., West, T.G., Pogson, G.H., et al., (1996). *Nautilus* and the art of metabolic maintenance. *Nature*, **382**, 534–536.

Burnett, L.E. & Stickle, W.B. (2001). Physiological responses to hypoxia. In: Rabalais, N.B. & Turner, R.E. (eds), *Coastal Hypoxia: Consequences for Living Resources and Ecosystems*. Coastal and Estuarine Studies 58, pp. 101–114. American Geophysical Union, Washington, DC.

Childress, J.J. & Seibel, B.A. (1998). Life at stable low oxygen: adaptations of animals to oceanic oxygen minimum layers. *Journal of Experimental Biology*, **201**, 1223–1232.

Danovaro, R., Dell'Anno, A., Pusceddu, A., et al. (2010). The first metazoan living in permanently anoxic conditions. *BMC Biology*, **8**, 30.

Decelle, J., Anderson, A., & Hourdez, S. (2010). Morphological adaptations to chronic hypoxia in deep-sea decapods crustaceans from hydrothermal vents and cold seeps. *Marine Biology*, **157**, 1259–1269.

Ertas, B., Kiger, L., Blank, M., et al. (2010). A membrane bound haemoglobin from gills of the green shore crab *Carcinus maenas*. *Journal of Biological Chemistry*, **286**, 3185–3193.

Everett, M.A., Antal, C.E., & Crawford, D.L. (2012). The effect of short-term hypoxic exposure on metabolic gene expression. *Journal of Experimental Zoology Part A*, **317**, 9–23.

Farrell, A.P. (ed.) (2011). *Encyclopedia of Fish Physiology. From Genome to Environment. Volume 3: Energetics, Interactions with the Environment, Lifestyles and Applications*, pp.1221–1228. Academic Press, New York.

Flück, M., Webster, K.A., Graham, J., et al. (2007). Coping with cyclic oxygen availability: evolutionary aspects. *Integrative and Comparative Biology*, **47**, 524–531.

Gorr, T.A., Cahn, J.D., Yamagata, H., & Bunn, H.F. (2004). Hypoxia-induced synthesis of haemoglobin in the crustacean *Daphnia magna* is hypoxia-inducible factor dependent. *Journal of Biological Chemistry*, **279**, 36038–36047.

Gorr, T.A., Wichmann, D., Hu, J., et al. (2010). Hypoxia tolerance in animals: biology and applications. *Physiological and Biochemical Zoology*, **83**, 733–752.

Gracey, A.Y., Troll, J.V., & Somero, G.N. (2001). Hypoxia-induced gene expression profiling in the euryoxic fish, *Gillichthys mirabilis*. *Proceedings of the National Academy of Sciences of the USA*, **98**, 1993–1998.

Greishaber, M.K., Hardewig, I., Kreutzer, U., & Pörtner, H.-O. (1994). Physiological and metabolic responses to hypoxia in invertebrates. *Reviews of Physiology, Biochemistry and Pharmacology*, **125**, 43–147.

Helmuth, B., Broitman, B.R., Yamane, L., et al. (2010). Organismal climatology: analyzing environmental variability at scales relevant to physiological stress. *Journal of Experimental Biology*, **213**, 995–1003.

Herreid, C.F. (1980). Hypoxia in invertebrates. *Comparative Biochemistry and Physiology Part A*, **67**, 311–320.

Hochachka, P.W. & Lutz, P.L. (2001). Mechanism, origin and evolution of anoxia tolerance in animals. *Comparative Biochemistry and Physiology Part B*, **130**, 435–459.

Hochachka, P.W. & Somero, G.N. (2002). *Biochemical Adaptation: Mechanism and Process in Physiological Evolution*. Oxford University Press, Oxford.

Hourdez, S. & Lallier, F.H. (2007). Adaptations to hypoxia in hydrothermal-vent and cold-seep invertebrates. *Review of Environmental Science and Biotechnology*, **6**, 143–159.

Hsia, C.C., Schmitz, A., Lambertz, M., et al. (2013). Evolution of air breathing: oxygen homeostasis and the transitions from water to land and sky. *Comprehensive Physiology*, **3**, 8499–8915.

Huang, L.E. (2013). How HIF-1α handles stress. *Science*, **339**, 1285–1286.

Larade, K. & Storey, K.B. (2002). A profile of the metabolic responses to anoxia in marine invertebrates. In: Storey, K.B. & Storey, J.M. (eds), *Cellular and Molecular Responses to Stress. Volume 3: Sensing, Signalling and Cell adaptation*, pp. 27–46. Elsevier Press, Amsterdam.

Leveelahti, L., Leskinen, P., Leder, E.H., et al. (2011). Responses of threespine stickleback (*Gasterosteus aculeatus*, L) transcriptome to hypoxia. *Comparative Biochemistry and Physiology Part D*, **6**, 370–381.

Livingstone, D.R. (1991). Origins and evolution of pathways of anaerobic metabolism in the animal kingdom. *American Zoologist*, **31**, 522–534.

Lutz, P.L. & Storey, K.B. (1997). Adaptations to variations in oxygen tension by vertebrates and invertebrates. In: Dantzler, W.H. (ed.), *Handbook of Physiology, Section 13: Comparative Physiology*, Vol. 2, pp.1479–1522. Oxford University Press, Oxford.

Mangum, C.P. (1997). Invertebrate blood oxygen carriers. In: Dantzler, W.H. (ed.), *Handbook of Physiology, Section 13: Comparative Physiology*, Vol. 2, pp. 1097–1135. Oxford University Press, Oxford.

Martinez, M.L., Landry, C., Boehm, R., et al. (2006). Effects of long-term hypoxia on enzymes of carbohydrate metabolism in the Gulf killifish *Fundulus grandis*. *Journal of Experimental Biology*, **209**, 3851–3861.

McMahon, B.R. (2001). Respiratory and circulatory compensation to hypoxia in crustaceans. *Respiration Physiology*, **128**, 349–364.

McMahon, B.R. (2012). Comparative evolution and design in non-vertebrate cardiovascular systems. In: Sedmera, D. & Wang, T. (eds), *Ontogeny and Phylogeny of the Vertebrate Heart*, pp.1–33. Springer, New York.

Mitrovic, D., Dymowska, A., Nilsson, G.E., & Perry, S.F. (2009). Physiological consequences of gill remodelling in goldfish (*Carassius auratus*) during exposure to long term hypoxia. *American Journal of Physiology*, **297**, R224–R234.

Morris, S. (2004). HIF and anapyrexia. *International Congress Series*, **1275**, 79–88.

Müller, M., Mentel, M., van Hellemond, J.J., et al. (2012). Biochemistry and evolution of anaerobic energy metabolism in eukaryotes. *Microbiological and Molecular Biology Reviews*, **76**, 444.

Pörtner, H.-O. (2012). Integrating climate-related stressor effects on marine organisms: unifying principles linking molecule to ecosystem-level changes. *Marine Ecology Progress Series*, **470**, 273–290.

Pörtner, H.-O., Heisler, N., & Grieshaber, M.K. (1985). Oxygen consumption and mode of energy production in the intertidal worm *Sipunculus nudus* L: definition and characterisation of the critical PO$_2$ for an oxyconformer. *Respiration Physiology*, **59**, 361–377.

Richards, J.G. (2009). Metabolic and molecular responses of fish to hypoxia. *Fish Physiology*, **27**, 443–485.

Rivera-Ingraham, G.A., Bickmeyer, U., & Abele, D. (2013). The physiological response of the marine platyhelminth *Macrostomum lignano* to different environmental oxygen concentrations. *Journal of Experimental Biology*, **216**, 2741–2751.

Rummer, J.L., McKenzie, D.J., Innocenti, A, Superan, C.T., & Brauner, C.J. (2013). Root effect haemoglobin may have evolved to enhance general tissue oxygen delivery. *Science*, **340**, 1327–1329.

Rutherford, L.D. & Thuesen, E.V. (2005). Metabolic performance and survival of medusa in estuarine hypoxia. *Marine Ecology Progress Series*, **294**, 189–200.

Schottler, U. & Bennet E.M. (1991). Annelids. In: Bryant, C. (ed.), *Metazoan Life without Oxygen*. Chapman and Hall,London.

Seibel, B.A. (2011). Critical oxygen levels and metabolic suppression in oceanic oxygen minimum zones. *Journal of Experimental Biology*, **214**, 326–336.

Semenza, G.L. (2008). Regulation of oxygen homeostasis by hypoxia-inducible factor 1. *Physiology*, **24**, 97–106.

Soñanez-Organis, J.G., Racotta, I.S., & Yepiz-Plascencia, G. (2010). Silencing of the hypoxia inducible factor 1—HIF-1—obliterates the effects of hypoxia on glucose and lactate concentrations in a tissue specific manner in the shrimp *Litopenaeus vannamei*. *Journal of Experimental Marine Biology and Ecology*, **393**, 51–58.

Spicer, J.I. (2014). What can an ecophysiological approach tell us about the physiological responses of marine invertebrates to hypoxia? *Journal of Experimental Biology*, **217**, 46–56.

Spicer, J.I. & El-Gamal, M.M. (1999). Hypoxia accelerates the development of respiratory respiration in brine shrimp—but at a cost. *Journal of Experimental Biology*, **202**, 3637–3646.

Thuesen, E.V., Rutherford, L.D., Brommer, P.L., et al. (2005). Intragel oxygen promotes hypoxia tolerance of scyphomedusae. *Journal of Experimental Biology*, **208**, 2475–2482.

Vaquer-Sunyer, R. & Duarte, C.M. (2008). Thresholds of hypoxia for marine biodiversity. *Proceedings of the National Academy of Sciences of the USA*, **105**, 15452–15457.

Wells, R.M.G. (2009). Blood-gas transport and hemoglobin function: adaptations for functional and environmental hypoxia. *Fish Physiology*, **27**, 255–299.

Wu, R.S.S. (2002). Hypoxia: from molecular responses to ecosystem responses. *Marine Pollution Bulletin*, **45**, 35–45.

Zhang, Z., Ju, Z., Wells, M.C., & Walter, R.B. (2009). Genomic approaches in the identification of hypoxia biomarkers in model fish species. *Journal of Experimental Marine Biology and Ecology*, **381**, S180–S187.

Zhenlin, J., Wells, M.C., Heater, S.J., & Walter, R.B. (2007). Multiple tissue gene expression analyses in Japanese medaka (*Oryzias latipes*) exposed to hypoxia. *Comparative Biochemistry and Physiology Part C*, **145**, 134–144.

CHAPTER 3

Physiological effects of ocean acidification on animal calcifiers

Inna M. Sokolova, Omera B. Matoo, Gary H. Dickinson, and Elia Beniash

3.1 Introduction

An increase in CO_2 concentrations in the atmosphere over the last 250 years has led to increased absorption of CO_2 by surface ocean waters, causing changes in seawater carbonate chemistry due to an increase in partial pressure of CO_2 (pCO_2) and a consequent drop in seawater pH. This phenomenon, called ocean acidification (OA), has been recently recognized as a major environmental factor which can negatively affect marine ecosystems and ocean-related economic activities on the global scale (Cooley and Doney, 2009; Doney et al., 2009, 2012). The current rate of OA is unprecedented and in the next four decades pH of the ocean is projected to hit the lowest point in the past 21 million years. Although the effects of OA are likely to be felt by all marine organisms, marine calcifiers (the organisms that build their shells and skeletons of calcium carbonate) are most severely affected. The CO_2-driven changes in carbonate chemistry and pH lower the degree of calcium carbonate saturation of seawater thereby impacting the ability of marine calcifiers to build skeletal structures. Furthermore, the effects of OA extend beyond biomineralization and impact other important fitness-related traits such as growth, survival, reproduction, and development (Dupont et al., 2010b; Kroeker et al., 2010; Harvey et al., 2013).

Marine calcifiers play key ecological roles and constitute a large fraction of biodiversity in the ocean. As early as the 1990s, a growing recognition of the negative effects of elevated CO_2 levels on marine calcifiers raised concerns about the potential implications of OA for marine ecosystems and spurred intensive research on the biological effects of OA. Since the early 2000s, when OA was generally recognized by the scientific community as a major problem, thousands of studies on the effects of OA on marine organisms and ecosystems

have been published. It is impossible to cover all these data in a short chapter; instead, we chose to focus on marine calcifiers from four ecologically important and evolutionary distant phyla—corals, molluscs, crustaceans, and echinoderms. Our aim is to provide an overview of the current state of knowledge about the diversity of physiological effects of OA in these key groups of marine calcifiers and identify commonalities in their responses to OA. We specifically focus on the effects of OA on energy metabolism, acid–base balance, and biomineralization. These physiological functions directly affect fitness-related traits such as growth, survival, reproduction, and development and thus are important for understanding and predicting the impacts of OA on affected populations. A recent study by Wittmann and Pörtner conducted a meta-analysis of the sensitivities of physiological processes (including metabolic rate, regulation of acid–base balance, and calcification) in five animal taxa (corals, molluscs, echinoderms, crustaceans, and fishes) over a range of pCO_2 concentrations (Wittmann and Portner, 2013). The study demonstrated that all the considered taxa were impacted negatively, but differentially, even by moderate scenarios of OA. Among the invertebrates, the pCO_2 at which 50% of the species are negatively affected (P_{50}) was significantly lower (632–1003 µatm) for corals, echinoderms, and molluscs compared to crustaceans (2086 µatm). In general, species with higher metabolic rates or levels of activity, a higher capacity to adjust body fluid pH, and less calcified structures (i.e. crustaceans) cope better with elevated CO_2 levels than more inactive, sessile groups with heavier skeletons and a lower capacity to regulate pH (i.e. corals, echinoderms, and molluscs). We also analyse the available (and currently rather limited) data on the interactions between acidification and other environmental factors such as temperature, salinity, hypoxia,

Stressors in the Marine Environment. Edited by Martin Solan and Nia M. Whiteley
© Oxford University Press 2016. Published in 2016 by Oxford University Press.

and trace metals on physiological responses of animal calcifiers to OA. For further reference on general and taxon-specific responses to OA, interested readers are encouraged to see recently published reviews: for marine calcifiers in general (Kroeker et al., 2013a, b; Secretariat of the Convention on Biological Diversity, 2014); for corals (Albright, 2011; Doo et al., 2014; Lough and Cantin, 2014; Ordoñez et al., 2014); for marine shelled molluscs (Gazeau et al., 2013); for crustaceans (Whiteley, 2011); and for echinoderms (Byrne et al., 2013a, b; Evans and Watson-Wynn, 2014).

3.2 Corals

Reef-building corals are one of the keystone groups of marine calcifiers. The ability of corals to deposit large amounts of calcium carbonate allowed them over the millennia to build structures on a geological scale such as the Great Barrier Reef, atoll islands, and other reef structures. Corals and other inhabitants of coral reefs such as calcareous algae, foraminifera, and molluscs account for a staggering 50% of the total biogenic calcium carbonate precipitation in the ocean (Erez et al., 2011). The coral reefs support unique and very diverse

ecosystems, with thousands of species of fish, invertebrates, plants, algae, and protozoa depending on the reef structure for their survival. Over the last decades the growth of coral reefs has slowed down globally, and in some areas the coral reefs are disappearing. These declines have been linked to global climate change, specifically rising temperatures, that result in frequent and widespread bleaching events when coral reefs lose their algal symbionts, stop growing, and eventually die (Hoegh-Guldberg, 1999; Knowlton, 2001). In recent years, the negative effects of OA on corals have also come into focus. OA slows down calcification and staggers growth of the reefs (Schneider and Erez, 2006; Anthony et al., 2008; de Putron et al., 2011; Crook et al., 2013; Lough and Cantin, 2014). If continued, this trend is projected to result in a precipitous decline in the global footprint of coral reefs (Fig. 3.1) (Kleypas et al., 1999; Hoegh-Guldberg et al., 2007; Silverman et al., 2009; Lough and Cantin, 2014).

Negative effects of OA on the rates of coral calcification are well documented, although the exact mechanisms are not yet fully understood and likely differ among taxonomic groups. Most calcifying corals belong to the order Scleractinia (subclass Hexacorralia)

Figure 3.1 Ecological structures predicted to form in place of the coral reefs for three different scenarios of global climate change, the Coral Reef Scenario (CRS)-A, CRS-B, and CRS-C. The typical anticipated ecological structures are illustrated using extant examples of reefs from the Great Barrier Reef. The atmospheric CO_2 concentration and temperature increases are shown for each Coral Reef Scenario (note that these conditions do not refer to the values measured at the photographed locations). CRS-A scenario assumes that the atmospheric CO_2 concentrations have stabilized at ~380 ppmv (note that as of September 2013, the atmospheric CO_2 levels have already passed that point, reaching ~395 ppmv). CRS-B scenario assumes an increase in CO_2 levels to approximately 500 ppmv, which is slightly below the predictions of a conservative IPCC B1 scenario for the year 2100, at ~550 ppmv. CRS-C scenario assumes an increase of CO_2 to levels above 500 ppmv. For comparison, a moderate IPCC A2 emission scenario predicts atmospheric CO_2 levels of ~800 ppmv by the year 2100, and the current trajectory of CO_2 increase indicates that it is a conservative estimate likely to be exceeded. (A) Reef slope communities at Heron Island. (B) Mixed algal and coral communities associated with inshore reefs around St. Bees Island near Mackay. (C) Reefs not dominated by corals illustrated by an inshore reef slope around the Low Isles near Port Douglas. Reprinted from *Science*, Vol. 318, by Hoegh-Guldberg et al. 'Coral reefs under rapid climate change and ocean acidification', pp. 1737–1742, Copyright 2007, with permission of the American Association for the Advancement of Science. [PLATE 1]

and produce an aragonitic exoskeleton secreted by specialized cells of aboral ectoderm called calicoblasts. Among different physiological functions of corals, calcification appears to be the most sensitive to OA (Moya et al., 2012). In scleractinian corals, the rate of calcification is positively correlated with the degree of aragonite saturation (Ω_{arag}) (Gattuso et al., 1998; Langdon et al., 2000; Schneider and Erez, 2006; Anthony et al., 2008; Cohen et al., 2009; Holcomb et al., 2010; Chan and Connolly, 2013; Comeau et al., 2013; Crook et al., 2013; Gagnon et al., 2013). Although the calcification rates slow down with decreasing Ω_{arag}, corals have the ability to compensate and can produce skeletons even at $\Omega_{arag} < 1$, as was shown for both larvae and adults (Cohen et al., 2009; Maier et al., 2009; Ries et al., 2010; de Putron et al., 2011). However, at $\Omega_{arag} \leq 1$ mineral deposition is significantly delayed and the corallites are generally small and deformed (Kurihara, 2008; Cohen et al., 2009; de Putron et al., 2011) indicating that this compensation comes at a cost. Moreover, a net carbonate accretion only occurs in the coral reefs when $\Omega_{arag} >$ 3.3 indicating that at Ω_{arag} below 3.3 the dissolution of $CaCO_3$ outstrips calcification rates (Hoegh-Guldberg et al., 2007).

A unique feature of corals, distinguishing them from other animal calcifiers, is the presence of symbiotic algae (zooxanthellae) in their endoderm. These photosynthetic symbionts are a source of food and energy for the animal host, while the corals provide CO_2 and shelter for the algae. In corals, the calcification rates are much higher under light than in the dark and this phenomenon is called light enhanced calcification (LEC) (Goreau, 1959; Chalker and Taylor, 1975; Gattuso et al., 1999; Moya et al., 2006; Schneider and Erez, 2006). The exact mechanisms of LEC are still debated (Erez et al., 2011). The proposed mechanisms of LEC involve scrubbing of CO_2 from the mesoglea and coelenteron by zooxanthellae during photosynthesis, which leads to an increase in Ω_{arag}, increased energy supply to the coral tissues, and increased production of matrix molecules (Goreau, 1959; Chalker and Taylor, 1975; Gattuso et al., 1999; Moya et al., 2006). Importantly, the rates of both dark and light calcification decrease with decreasing Ω_{arag} and the offsets between calcification rates in light and dark are more or less equal in the range of Ω_{arag} between 1 and 7 (Schneider and Erez, 2006). Several reports show dissolution of mineral in the dark at Ω_{arag} as high as 4.0 (Ohde and Hossain, 2004; Schneider and Erez, 2006), which might suggest that the pCO_2 in these systems is increased due to respiration.

The relationship between calcification rates of corals and Ω_{arag} of the seawater is not simple. While some

studies show a progressive decrease in calcification rates with decreasing Ω_{arag} (Langdon et al., 2000; Ohde and Hossain, 2004; Schneider and Erez, 2006), others demonstrate a nonlinear response, with calcification rates holding steady or showing only a slight decline at $\Omega_{arag} \geq 2$ and dropping sharply at $\Omega_{arag} < 2$ (Gattuso et al., 1998; Ries et al., 2010; de Putron et al., 2011). These differences might be related to the way the experiments were performed. Typically, the data acquired in long-term (weeks to months) experiments are a composite of light and dark calcification while in short-term (hours to days) experiments the dark and light calcification are measured separately. Another possibility is that there are major interspecific differences in the response to OA. The general trend of decreased calcification with decreasing Ω_{arag} has been confirmed for most studied coral species (Erez et al., 2011; Chan and Connolly, 2013) but important exceptions were found when calcification rates are not affected by Ω_{arag} (Maier et al., 2009; Edmunds, 2011; Comeau et al., 2013; Maier et al., 2013; Takahashi and Kurihara, 2013). Interestingly, in some species elevated pCO_2 also leads to an increase in soft tissue mass of polyps (Fine and Tchernov, 2007; Comeau et al., 2013). At the current level of our understanding of coral physiology, it is impossible to determine what drives these species-specific differences in sensitivity of calcification to Ω_{arag}. OA can potentially affect multiple life-history stages of calcifying corals and current studies raise an intriguing possibility that resistant species will predominate in the acidified ocean with the consequent shifts in population dynamics and community structure of the reefs (Albright, 2011; Ordoñez et al., 2014).

Corals can regulate intracellular pH and Ω_{arag} in the mineralization compartment located beneath the aboral calicoblastic layer. Several studies using microelectrodes implanted into the mineralization compartment showed that both pH and Ω_{arag} in this compartment are several times higher than in the outside medium (Al-Horani et al., 2003; Ries, 2011; Venn et al., 2013). Elevated pH at the mineralization site was also detected over a wide range of external pH values in several coral species using the boron isotope ratio as a pH proxy (McCulloch et al., 2012). The mechanisms by which corals create high supersaturation levels in their mineralization compartment are not yet fully characterized but likely involve active proton pumping. Studies with pharmacological inhibitors and *in situ* hybridization indicate that Ca^{2+} ATPase, a proton pump which exchanges one Ca^{2+} per 2 H^+, plays an essential role in biomineralization of corals (Al-Horani et al., 2003; Zoccola et al., 2004). Carbonic anhydrase (CA), an enzyme that converts CO_2 to bicarbonate, also plays an important role (Goreau, 1959; Tambutte et al., 1996; Furla et al., 2000; Bertucci et al., 2013). CA is highly expressed in calicoblasts (Moya et al., 2008) and its inhibition leads to decrease in calcification (Goreau, 1959). Presently, the effects of OA on the acid–base regulation of intra- and extracellular fluids of corals are not fully understood. The expression of CA was suppressed at elevated pCO_2 levels (Moya et al., 2012) but the implications of this change to pH regulation are unknown. Given the ATP dependence of proton pumping, a decline in seawater pH and Ω_{arag} will likely result in a higher load on the systems that regulate pH and an increase in the energy cost of acid–base homeostasis.

Another important gap in our knowledge of the effects of OA on coral concerns the potential impact of multiple environmental stressors than can co-occur with OA such as warming, changes in light intensity, and other factors. The combination of high CO_2 and warming was shown to dramatically increase bleaching of corals (Anthony et al., 2008; Doo et al., 2014). These studies also revealed that while acidification decreases the rate of calcification in all studied coral species, the effect of temperature is species-specific. In Mediterranean corals, the effects of OA on mortality and calcification rates are temperature-dependent, with negative impacts observed only at elevated temperatures (Rodolfo-Metalpa et al., 2011).

3.3 Molluscs

Molluscs are a large and diverse group of marine calcifiers encompassing over 30 000 species, which are found in various environments from rocky intertidal shores to deep sea terrains and hydrothermal vents (Gosling, 2003). Molluscs provide essential ecosystem services, regulating water quality, creating habitat for other species, contributing to global carbon flux and serving as a food source to other organisms (Gutiérrez et al., 2003; Jackson et al., 2008). The phylum also includes many economically important species. At 12–16% of total seafood consumption (FAO, 2008), annual global shellfish aquaculture has an estimated worth of 10–13 billion USD (Cooley and Doney, 2009). Marine molluscs deposit large quantities of calcium carbonate in their shells, and shell mineralogy differs among species and life stages. In most molluscs, larval shells are made of aragonite while adult shells can be made from aragonite, low Mg calcite, high Mg calcite, or a mixture of different calcium carbonate isoforms (Weiss et al., 2002). The shells also contain embedded organic material and in some species are covered with

a protective proteinaceous layer called periostracum. This diversity in shell composition and structure, as well as in species' physiology is reflected in the diversity of responses of molluscs' biomineralization processes to OA (Gazeau et al., 2013).

The rates of biomineralization of molluscs are dependent on pH and the degree of calcium carbonate saturation of seawater (Ries et al., 2009; Gazeau et al., 2013). However, like in corals, the relationship between the calcium carbonate saturation of seawater and calcification rate differs between species and is not always linear. In many intertidal molluscs, including oysters, mussels, and conchs, a decrease in shell mass, growth, and/or calcification rates has been reported in response to elevated pCO_2 (Michaelidis et al., 2005; Shirayama and Thorton, 2005; Gazeau et al., 2007; Beniash et al., 2010; Dickinson et al., 2012; Parker et al., 2012). Out of the nine shelled molluscs studied by Reis and co-workers (Ries et al., 2009), seven showed decreased calcification at elevated pCO_2, one (a gastropod *Crepidula fornicata*) increased calcification, and one (a blue mussel *Mytilus edulis*) no response to elevated pCO_2. A relative insensitivity of calcification of blue mussels (*M. edulis* and *M. galloprovincialis*) to elevated CO_2 was also confirmed in other studies (Thomsen and Melzner, 2010; Fernández-Reiriz et al., 2012). Similarly, shell deposition of an intertidal clam, *Ruditapes decussatus*, did not change at elevated CO_2 levels predicted by OA scenarios (Range et al., 2011), while in the hard clam, *Mercenaria mercenaria*, shell growth was highest at an intermediate pCO_2 (~800 μatm) and decreased at higher and lower CO_2 levels (Dickinson et al., 2013; Ivanina et al., 2013b). Resistance of calcification to elevated CO_2 levels in intertidal molluscs (such as hard clams, carpet clams, and mussels) appears associated with the presence of a thick protective periostracum that shields the mineral layers of the shell from direct contact with corrosive water. Interestingly, in cephalopods that have internal shells protected from contact with outside water and a strong capacity to regulate pH of extracellular fluids, elevated pCO_2 (6000 μatm, a value far greater than the end-of-century predicted value of 800–1000 μatm) led to an increase of the shell (cuttlebone) mass by up to 55% after six weeks of exposure (Gutowska et al., 2010b). In contrast, biomineralization of larvae and species whose post-metamorphic shells lack a periostracum (such as oysters) is more sensitive to the negative effects of OA (Ries et al., 2009). Planktonic pteropods that have aragonitic shells covered with a thin periostracum might be especially susceptible to the negative impacts of OA (Orr et al., 2005; Fabry et al., 2008). Pteropods are found in high-latitude waters that have a naturally low degree of calcium carbonate saturation. In both Arctic and temperate pteropods, elevated pCO_2 results in decreased calcification, increased shell dissolution, and appearance of perforated corroded shells (Comeau et al., 2009; Lischka et al., 2011; Manno et al., 2012).

As external CO_2 equilibrates with the extra- and intracellular compartments by diffusion, a reduction of pH of the body fluids (hypercapnic acidosis) results, which can affect metabolism, development, growth, and reproduction (Pörtner, 2008; Hammer et al., 2011). Molluscs show diverse abilities to maintain acid–base balance in the face of OA that can be linked to their habitat and life styles (Fabry et al., 2008; Gutowska et al., 2010c; Range et al., 2011; Parker et al., 2013). Benthic bivalves and gastropods that lead hypometabolic sessile life styles and lack well-developed ventilation and circulation systems are weak ion and acid–base regulators with low non-bicarbonate buffering capacity of the body fluids (Michaelidis et al., 2005; Melzner et al., 2009). In these groups, the extracellular pH (pH_e) typically decreases with increasing ambient pCO_2 (Michaelidis et al., 2005; Lannig et al., 2010; Thomsen et al., 2010; Schalkhausser et al., 2013; Thomsen et al., 2013). In some species, passive buffering mechanisms are invoked, such as shell dissolution that provides HCO_3^- ions to partially compensate pH_e (Michaelidis et al., 2005; Lannig et al., 2010; Marchant et al., 2010; Thomsen et al., 2010; Schalkhausser et al., 2013). Active pelagic species such as the cuttlefish *Sepia officinalis* are capable of active compensation of pH_e through the action of ion-regulatory epithelia of the basolateral gill membranes; however, this active acid–base regulation only leads to partial compensation of pH_e (Gutowska et al., 2010a; Hu et al., 2011). Intracellular pH (pH_i) of molluscs is generally better regulated than pH_e and can be fully compensated within a certain range of environmentally relevant pH/ pCO_2 levels through passive buffering and active ion transport mechanisms (Sokolova et al., 2000; Thomsen et al., 2010; Hammer et al., 2011; Ivanina et al., 2013a). However, the ability to regulate pH_i is limited in molluscs and extreme acidification, due to environmental or respiratory hypercapnia and/or anaerobiosis, results in a drop of pH_i and may lead to metabolic rate depression (Sokolova et al., 2000; Michaelidis et al., 2005).

Shifts in the extracellular acid–base balance and the necessity to maintain pH_i in the face of changing pH_e and CO_2 levels can affect metabolism and increase the energy costs to maintain homeostasis in molluscs (Pörtner et al., 2004b; Fabry et al., 2008). Metabolic responses to elevated CO_2 vary both between and within species of molluscs depending on physiological status, developmental stage, and other factors (Parker et al., 2013). Additionally, regional water chemistry parameters as well as their temporal fluctuations in the organisms' habitat can strongly modulate the metabolic responses of organisms to OA (Thomsen et al., 2010; Fernández-Reiriz et al., 2012). An increase in standard metabolic rate (SMR) during exposure to elevated CO_2 has been reported in several species of marine bivalves, including temperate oysters and mussels and an Antarctic clam *Laternula elliptica* (Beniash et al., 2010; Lannig et al., 2010; Thomsen and Melzner, 2010; Cummings et al., 2011). In oysters, the increase in SMR was accompanied by reduced survival and decreased shell and somatic growth (Beniash et al., 2010), decreased tissue ATP levels, and reduced condition index (Lannig et al., 2010), indicating that elevated energy demand cannot be fully met by energy supply. In contrast, in several subtropical and tropical species of bivalves and gastropods, no change in SMR in response to OA was found (Marchant et al., 2010; Fernández-Reiriz et al., 2012; Liu and He, 2012). This has been attributed to increased absorption efficiency of food (Fernández-Reiriz et al., 2012) and/or near-complete compensation of extracellular acidosis by HCO_3^- (Marchant et al., 2010). No change in SMR in response to elevated pCO_2 (6000 µatm, a value far greater than the end-of-century predicted value of 800–1000 µatm) was also found in an active nektobenthic cephalopod *S. officinalis* (Gutowska et al., 2010a). A third type of metabolic response of molluscs to OA involves metabolic rate depression (a decrease in SMR) as seen in some bivalves and gastropods and a jumbo squid (Michaelidis et al., 2005; Rosa and Seibel, 2008; Fernández-Reiriz et al., 2011; Hammer et al., 2011; Melatunan et al., 2011; Liu and He, 2012; Dickinson et al., 2013; Navarro et al., 2013). Reduced SMR is often accompanied by a net degradation in proteins (Fernández-Reiriz et al., 2011), reduced clearance rate and ingestion (Fernández-Reiriz et al., 2011; Liu and He, 2012; Navarro et al., 2013), lower feeding and absorption efficiency (Navarro et al., 2013), and in the long term can lead to reduced growth and survival (Pörtner et al., 2004a; Michaelidis et al., 2005; Navarro et al., 2013; Schalkhausser et al., 2013).

It is worth noting that early life stages of molluscs (embryos and especially larvae) are typically more sensitive to OA than juveniles and adults (Kurihara, 2008; Kroeker et al., 2010). The higher sensitivity of early life stages to OA may reflect the underdeveloped mechanisms of ion and acid–base regulation (Melzner et al., 2009), higher metabolic rates, and thus energy costs of basal maintenance (Sokolova et al., 2011, 2012) as well as the presence of a more soluble polymorph of $CaCO_3$ (amorphous calcium carbonate) in larval shells (Weiss et al., 2002; Ross et al., 2011).

Effects of OA on physiology and biomineralization of molluscs can be modulated by other environmental factors, such as temperature, salinity, pollution, and hypoxia. Elevated temperatures and reduced salinities exacerbate the negative effects of OA on growth,

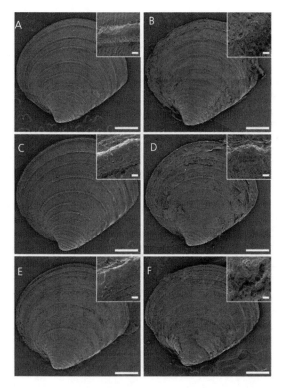

Figure 3.2 SEM micrographs of the exterior of the shells of hard shell clams *Mercenaria mercenaria* after 16 weeks exposure to different salinities and CO_2 levels. The inset on each panel is a high magnification image of a growth ridge near the periphery of the shell. (A) 395 µatm pCO_2, salinity 32 practical salinity units; (B) 395 µatm, salinity 16; (C) 800 µatm, salinity 32; (D) 800 µatm, salinity 16; (E) 1500 µatm, salinity 32; (F) 1500 µatm, salinity 16. Scale bars, 500 µm main panels; 5 µm insets. In high salinity exposures (A–C) Ω_{Arg} remained above the saturation level and only minor differences in the structure of shell exterior were observed. At low salinity (D–F), less distinct growth ridges and flaking of the periostracum was observed at elevated CO_2 levels (800 and 1500 µatm; $\Omega_{Arg} < 1$), with a nearly complete loss of periostracum at ~1500 µatm pCO_2. Reprinted from the *Journal of Experimental Biology*, Vol. 216, by Dickinson et al. 'Environmental salinity modulates the effects of elevated CO_2 levels on juvenile hard-shell clams, *Mercenaria mercenaria*', pp. 2607–2618, Copyright 2013, with permission of the Society of Biologists.

survival, and metabolism (Lannig et al., 2010; Lischka et al., 2011; Waldbusser et al., 2011a, b; Dickinson et al., 2012, 2013; Gazeau et al., 2013; Ivanina et al., 2013b; Matoo et al., 2013; Melatunan et al., 2013; Parker et al., 2013; Schalkhausser et al., 2013). Furthermore, elevated temperatures and reduced salinity may increase the negative impact of elevated pCO_2 on biomineralization resulting in weakened shells with altered structure, and reduced hardness and fracture toughness (Fig. 3.2) (Dickinson et al., 2012; Manno et al., 2012; Dickinson et al., 2013). On the other hand, negative changes in biomineralization and physiology of molluscs can be partially offset by increased food supply or changes in seawater chemistry, particularly elevated alkalinity (Green et al., 2009; Range et al., 2011; Thomsen et al., 2013). Some recent studies have interestingly shown planktonic larvae of some species of *Mytilus* spp. to be resistant to effects of low O_2 in pH variable regions of coastal zones (Eerkes-Medrano et al., 2013; Frieder et al., 2014). Given that OA co-occurs with other environmental changes, including elevated temperature, intensification of precipitation and salinity fluctuations, coastal hypoxia, and potential changes in algal production, understanding of these interactions will be crucial for our ability to predict the potential effects of OA on marine molluscs especially in complex environments such as estuaries and coastal shelf zones.

3.4 Crustaceans

Crustaceans are foundational species in many marine communities and support valuable fisheries worldwide. While research on crustacean responses to OA has historically lagged behind other major groups of calcifiers, the body of research on the topic has steadily risen over the past ten years with the majority of work done on decapods (e.g. shrimp, crabs, and lobsters) and cirripeds (i.e. barnacles). The effects of OA vary considerably among and within species based on differences in metabolic rate, activity level, and/or iono-/osmoregulatory capacity (Melzner et al., 2009; Whiteley, 2011; Carter et al., 2013; Ceballos-Osuna et al., 2013; Pansch et al., 2013). Most crustaceans appear to be relatively tolerant to OA, at least in short-term (days to weeks) exposures, and this generalization holds for both adult and larval stages (Melzner et al., 2009; Whiteley, 2011).

The mineralized exoskeleton of crustaceans differs from that of other calcifiers in that it is covered by a relatively thick, waxy epicuticle and is moulted periodically in most species (Cameron and Wood, 1985). The effects of OA on biomineralization of the crustacean exoskeleton varies among species, but in most cases the effect is either negligible (Kurihara and Ishimatsu, 2008; Hauton et al., 2009; Small et al., 2010; Carter et al., 2013) or increased mineralization (sometimes associated with faster growth; Fig. 3.3) is observed (McDonald et al., 2009; Ries et al., 2009). Only few crustacean species show decreased mineralization in response to OA (Kurihara et al., 2008; Long et al., 2013b). A similar trend was found for Ca^{2+} content of the exoskeleton; most studies found no change (Findlay et al., 2009;

Small et al., 2010; Donohue et al., 2012; Long et al., 2013b) or increased Ca^{2+} content with increasing pCO_2 (Long et al., 2013a) with only one report of a decreased Ca^{2+} content in response to elevated pCO_2 (Long et al., 2013b). If crustaceans are able to use HCO_3^- in the mineralization process via a proton pumping mechanism that converts HCO_3^- to CO_3^{2-} (Cameron and Wood, 1985), they may be able to maintain or increase calcification rates as HCO_3^- becomes more available in CO_3^{2-}-enriched seawater (Wickins, 1984) as long as the rate of dissolution is controlled (Ries et al., 2009).

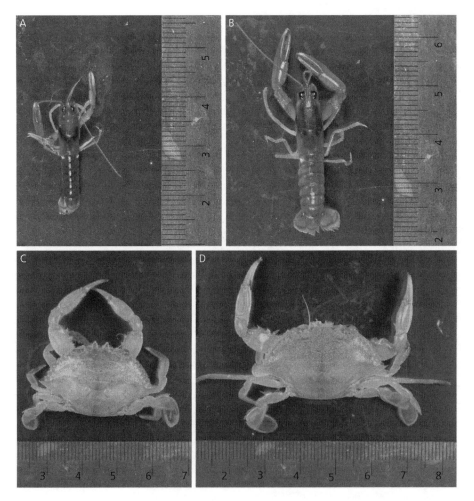

Figure 3.3 Effects of elevated CO_2 levels on growth of decapod crustaceans. A, B—the American lobster, *Homarus americanus*, raised under normocapnia (400 ppmv CO_2; A) and elevated CO_2 levels (2850 ppmv; B). C, D—the blue crab, *Callinectes sapidus*, raised under normocapnia (400 ppmv CO_2; C) and elevated CO_2 levels (2850 ppmv; D). Higher biomineralization rates of the decapod crustaceans observed at elevated CO_2 levels were associated with faster growth as shown by larger sizes of the representative crustaceans shown on the photo. Photo credit: Justin B. Ries (Northeastern University, USA). Reproduced with permission from J. B. Ries. [PLATE 2]

The implications of CO_2-induced changes in biomineralization for the structural and mechanical properties of crustacean exoskeletons are not well understood. In the American lobster, the blue crab, and a shrimp (*Penaeus plebejus*), OA exposure did not result in changes in the polymorph of $CaCO_3$ deposited (percent calcite versus aragonite or $Mg^{2+}:Ca^{2+}$ ratio; Ries, 2011). In a barnacle, *Amphibalanus amphitrite*, elevated CO_2 had opposite effects on the mechanical properties of the base plate and the parietal plates due to differential dissolution and calcium carbonate deposition in different parts of the exoskeleton (McDonald et al., 2009). These findings clearly indicate that the effects of OA on the crustacean exoskeleton are complex and likely taxonomic group-specific, but the currently available information does not permit any broad generalizations.

Crustaceans possess a well-developed ability to regulate the pH of their extra- and intracellular fluids. It is important to note that many of the studies assessing mechanisms of acid–base regulation described in this section employed a pCO_2 far greater than the end-of-century predicted value of 800–1000 µatm. The mechanisms, and the capacity of acid–base regulation, differ between species and depend on the species' habitat and lifestyle (Melzner et al., 2009; Whiteley, 2011). In an intertidal crab, *Necora puber*, blood pH was maintained constant in the range of pCO_2 from ~3200 to 12 000 µatm, mostly due to increased HCO_3^- buffering of the blood (Spicer et al., 2007; Small et al., 2010). Similarly, intertidal/shallow water crabs *Callinectes sapidus* and *Cancer magister* were able to restore a near-control blood pH within 48 h of exposure to extreme hypercapnia (~10 000 µatm pCO_2) using bicarbonate buffering (Cameron, 1978; Pane and Barry, 2007). The same trend of increased HCO_3^- concentration to maintain blood pH was found in an intertidal crab, *Carcinus maenas*, during exposure to air, a condition that limits elimination of CO_2 leading to increased blood pCO_2 (Truchot, 1975). HCO_3^- that serves as a buffer against rising [H^+] is taken up primarily within the gills in electroneutral ion exchangers, with HCO_3^- exchanged for Cl^- and H^+ for Na^+, although a small portion of blood HCO_3^- may be mobilized from the exoskeleton (Cameron, 1978, 1985; Melzner et al., 2009). This effective bicarbonate buffering to maintain constant blood pH may help crabs to survive broad fluctuations of pCO_2 due to both environmental and respiratory hypercapnia in the intertidal zone. In contrast, in the burrowing shrimp *Upogebia deltaura* HCO_3^- buffering was not involved in pH compensation of the blood (Donohue et al., 2012). Blood pH was maintained at the control levels in *U. deltaura* exposed to ~1400 µatm,

but dropped significantly at 2700 µatm. In a deep-sea tanner crab, *Chionoecetes tanneri*, exposure to elevated pCO_2 (~10 000 µatm, a value far greater than the end-of-century predicted value of 800–1000 µatm) led to a steady drop in blood pH and no change in HCO_3^- (Pane and Barry, 2007). This may be related to the reduced ion exchange in the gills and lower metabolic activity of deep-sea crustaceans (Seibel and Walsh, 2001).

Ion exchange mechanisms to maintain acid–base homeostasis in the face of elevated pCO_2 incur energetic costs that must be covered by higher ATP production and can in the long term lead to depletion of energy stores and/or divert energy from other physiological processes (e.g. growth and reproduction). However, additional energy costs to maintain acid–base homeostasis at elevated pCO_2 levels are small in most adult crustaceans, at least in the range of pCO_2 predicted by OA scenarios. In most studied crustaceans, exposure to elevated pCO_2 did not affect the rates of basal metabolism of adults (Kurihara et al., 2008; Donohue et al., 2012; Carter et al., 2013). In the isopod *Paradella dianae*, a decrease in metabolic rate was observed when animals were exposed for 21 days to end-of-century CO_2 levels (Munguia and Alenius, 2013). If, however, *P. dianae* was gradually acclimated to high pCO_2 (low pH) over the course of 21 days, a change in metabolic rate was not observed, suggesting scope for metabolic adaptation in this species. Decreased metabolic rate was also observed in the intertidal crab *Necora puber* when exposed to very high pCO_2 (~12 000 µatm, a value far greater than the end-of-century predicted value of 800–1000 µatm), but not at a more moderate pCO_2 of ~3200 µatm (Small et al., 2010). Reduced metabolic rate at very high pCO_2 may enable short-term ATP conservation (Pörtner et al., 1998).

Unlike adults, bioenergetics of the early life stages (embryos, larvae, and juveniles) of crustaceans appears to be more sensitive to the effects of OA. In embryos, exposure to elevated pCO_2 led to decreased metabolic rate (Carter et al., 2013), heart rate (Ceballos-Osuna et al., 2013), and yolk size (Long et al., 2013a). In larvae, a decreased heart rate and a shift from lipid to protein metabolism were observed in response to OA (Carter et al., 2013). These changes in embryonic and larval metabolism may negatively affect development by reducing hatching and settlement success and increasing the risk of predation due to the longer time spent in the water column (Keppel et al., 2012). Lower hatching success in response to OA has been reported in many (Mayor et al., 2007; Findlay et al., 2009; Kawaguchi et al., 2011) but not all species of crustaceans

(Kurihara and Ishimatsu, 2008; Ceballos-Osuna et al., 2013). The effects of OA on the larval stages is more variable, with extended larval development, reduced larval mass, and/or reduced carapace length observed in certain species (Arnold et al., 2009; Bechmann et al., 2011; Keppel et al., 2012), and no effects found in others (Kurihara and Ishimatsu, 2008; McDonald et al., 2009; Wong et al., 2011; Carter et al., 2013; Pansch et al., 2013). Notably, metabolic responses to elevated pCO_2 may vary among larval broods of crustaceans (Carter et al., 2013; Ceballos-Osuna et al., 2013; Pansch et al., 2013) suggesting a genetic basis for individual responses to OA. In juveniles, reduced condition index (Long et al., 2013b), decreased total protein (Carter et al., 2013), and diminished somatic growth and altered cuticle composition (Fitzer et al., 2012) have all been documented during exposure to elevated pCO_2, suggesting that OA can result in energy deficiency affecting survival and maturation of juveniles.

Similar to other groups of marine calcifiers, the effects of OA on crustaceans can be modulated by other stressors. The existing data indicate that both low dissolved O_2 (Pane and Barry, 2007) and salinity (Egilsdottir et al., 2009; Miller et al., 2014) act synergistically with OA. In the deep-sea tanner crab *Chionoecetes tanneri* hypoxic conditions led to slightly lower blood pH levels and serum protein levels than exposure to highly elevated pCO_2 (~10 000 μatm) alone (Pane and Barry, 2007). Low salinity exerts a dominant influence on the amphipod *Echinogammarus marinus*, with a trend towards delayed larval development in individuals exposed to OA and low salinity (Egilsdottir et al., 2009). Sequential exposure of porcelain crab (*Petrolisthes cinctipes*) larvae to elevated pCO_2 (at an end-of-century predicted level of 1000 μatm), followed by exposure to salinity stress (increased from 34 to 40 or decreased from 34 to 22), led to a significant increase in metabolic rate, whereas metabolic rate in larvae exposed to elevated pCO_2 or salinity stress alone did not differ significantly different from those held under ambient conditions (Miller et al., 2014).

The combined effects of high pCO_2 and temperature stress vary among species and depend on where animals were collected within the species' geographic range (Findlay et al., 2010b; Walther et al., 2010). A significant reduction in thermal tolerance upon exposure to high pCO_2 has been shown in several decapods (Metzger et al., 2007; Walther et al., 2009), although thermal tolerance was unaffected by exposure to pCO_2 as high as ~12 000 μatm in *Necora puber* (Small et al., 2010). Elevated pCO_2 and temperature independently reduced survival of the amphipod

Peramphithoe parmerong; however, the interaction temperature × pCO_2 was not significant (Poore et al., 2013). Barnacles exhibit a complex and species-specific response to the combination of temperature and OA (Findlay et al., 2009, 2010a, b, c; Pansch et al., 2012). In *Amphibalanus improvisus* a significant temperature × pCO_2 interaction was observed for the extent and duration of larval development but not for final larval settlement or survival (Pansch et al., 2012). Temperature × pCO_2 interactions were also not significant for juvenile survival in *Semibalanus balanoides* or *Elminius modestus*, although the combination of high pCO_2 and high temperature resulted in reduced shell Ca^{2+} content in *S. balanoides* and diminished growth in *E. modestus* (Findlay et al., 2010b). Given that a large portion of crustacean species live in habitats characterized by fluctuating environmental conditions at some point in their life cycle, further assessments of the combined effects of OA and other stressors is critical.

3.5 Echinoderms

The phylum Echinodermata encompasses about 7000 species that are found exclusively in the oceans, from shallow rocky shores to deep sea and from poles to tropical areas. Echinoderms play a key role in marine food chains and provide important ecosystem services, modifying habitat structure, stabilizing communities as keystone species, and affecting sediment quality as bioturbators. Echinoderms build calcified rods, shells (tests), spines, and other skeletal structures from high magnesium calcite, which is the most soluble of all biogenic isoforms of calcium carbonate (Donnay and Pawson, 1969; Okazaki and Inoue, 1976), through a transient amorphous calcium carbonate phase (Beniash et al., 1997; Politi et al., 2004). Calcification of echinoderms occurs within a syncytial compartment created by primary mesenchymal cells (in larvae) or sclerocytes (in adults) (Markel et al., 1989; Beniash et al., 1999). Recent studies indicate that echinoderms are resilient to OA in the range predicted for end-of-the-century scenarios (Dupont et al., 2010a, b). This conclusion has an important caveat, however, because most studies to date used short-term exposures (hours to days) and may not be fully representative of the chronic exposures to elevated pCO_2 expected during OA. Furthermore, responses to OA are variable among different species and life stages of echinoderms (Byrne, 2012; Byrne and Przeslawski, 2013). Overall, early life stages (larvae and juveniles) are more susceptible to OA than adults (Kurihara, 2008; Clark et al., 2009; Stumpp et al., 2012a). Echinoderm embryos are generally robust to OA, but negative effects appear in the gastrulation stage (Byrne et al., 2009; Byrne and Przeslawski, 2013), with calcifying (e.g. echinoplutei) larvae generally more sensitive than non-calcifying ones (Byrne and Przeslawski, 2013). In calcifying larvae and juveniles of echinoderms, exposure to elevated CO_2 can result in delayed larval skeletogenesis, reduced size and abnormal morphology (Fig. 3.4), lower regenerative capacity, reduced survival, and

impaired swimming performance (Kurihara and Shirayama, 2004; Dupont et al., 2008; Clark et al., 2009; Donnell et al., 2010; Chan et al., 2011; Stumpp et al., 2011; Catarino et al., 2012b; Gonzalez-Bernat et al., 2013). In both larval and adult echinoderms exposed to elevated CO_2 levels, a decrease (Miles et al., 2007; Clark et al., 2009; Gooding et al., 2009; Donnell et al., 2010; Sheppard Brennand et al., 2010; Wood et al., 2011), increase (Wood et al., 2008), or no change in calcification

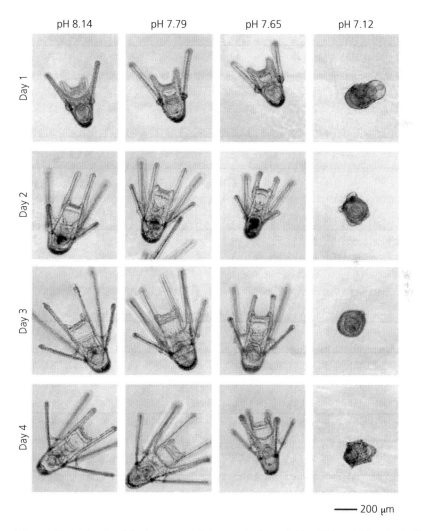

Figure 3.4 Effects of elevated CO_2 levels on larval development and skeletogenesis of a tropical intertidal sand dollar, *Arachnoides placenta*. Larvae were reared under four seawater pH levels (pH 8.14, pH 7.79, pH 7.65, and pH 7.12) from fertilization until four days of age. Extreme acidification (pH 7.12) led to arrested embryonic development. Larvae reared at pH 7.65 were smaller and developed at slower rates compared with those reared at pH 7.79 and 8.14. Scale bar = 200 μm. Reprinted from *Marine Biology*, Vol. 160, by Gonzalez-Bernat et al., 'Fertilisation, embryogenesis and larval development in the tropical intertidal sand dollar *Arachnoides placenta* in response to reduced seawater pH', pp. 1927–1941, Copyright 2013, with permission from Springer-Verlag.

(Kurihara and Shirayama, 2004; Wood et al., 2010; Schram et al., 2011) was observed. This species-specific variability of responses has been attributed to the presence of external protective covering (Ries et al., 2009), differences in acid–base regulation capabilities, and in the capacity to up-regulate enzymes and structural proteins involved in biomineralization (Todgham and Hofmann, 2009; Evans et al., 2013; Evans and Watson-Wynn, 2014), and may reflect adaptations to different habitats.

Echinoderms are weak ion regulators and therefore pH of their extracellular coelomic fluids decreases with increases in pCO_2 (Miles et al., 2007; Catarino et al., 2012a; Dupont and Thorndyke, 2012; Stumpp et al., 2012a, b; Collard et al., 2013b). Partial compensation of acidosis may be achieved by increased extracellular concentrations of HCO_3^- either due to shell dissolution (as indicated by a concomitant increase in Mg^{2+} ions) or by up-regulation of the expression of bicarbonate transporters (Donnell et al., 2010). Among echinoderms, echinoids (sea urchins) can better compensate for extracellular acidosis than sea stars when exposed to elevated pCO_2 (Dupont and Thorndyke, 2012). Some species of sea urchins, such as a eurybiont *Strongylocentrotus droebachiensis*, can fully compensate for hypercapnic extracellular acidosis within ten days of exposure to moderate (but not high) levels of pCO_2 (Stumpp et al., 2012a). Sea urchins also have a higher bicarbonate buffering capacity of their coelomic fluid than starfishes and holothurians, and can supplement bicarbonate buffering with protein-based buffering mechanisms and possibly use lactate ions in acid–base regulation (Collard et al., 2013b). The higher capacity for acid–base regulation correlates with higher skeletal organization in sea urchins compared to sea stars and holothurians and suggests that more complex biomineralization may require more efficient acid–base regulation (Dupont and Thorndyke, 2012).

Similar to biomineralization, metabolic responses to OA are highly variable among echinoderms. In most studied species, an increase in metabolic rate was found at elevated pCO_2 (Wood et al., 2010; Stumpp et al., 2011; Collard et al., 2013a; Dorey et al., 2013), but several studies also reported a decrease (McElroy et al., 2012), no change (Stumpp et al., 2012b), or a nonlinear response with reduced metabolic rates at intermediate pCO_2 and an increase at higher pCO_2 levels (Christensen et al., 2011; Uthicke et al., 2013). In many species, exposure to elevated pCO_2 led to reduced feeding rates (Donnell et al., 2010; Stumpp et al., 2012b), lower rates of protein degradation, reduced scope for growth (Stumpp et al., 2012b), and down-regulation

of mRNA expression of proteins involved in protein synthesis, tricarboxylic acid cycle, electron transport chain, and ion regulation (Todgham and Hofmann, 2009; Donnell et al., 2010). This indicates that metabolic rate depression may help echinoderms conserve energy in the face of elevated pCO_2 stress. However, this strategy may come at a cost during long-term exposure to elevated CO_2 levels, causing reduced growth and reproduction and/or developmental delays, and thus negatively affecting survival.

Elevated temperature can strongly modulate responses of echinoderms to OA. In line with the species-specific variability of physiological responses to OA, the interactions between temperature and pCO_2 also vary between different species of echinoderms (for a comprehensive review see Byrne, 2011). In some species, elevated temperature exacerbates the negative effects of OA on metabolism and calcification (Gooding et al., 2009; Wood et al., 2010, 2011) leading to reduced calcification, suppressed ability to regenerate, elevated metabolic rates, and delayed development. In other species, OA and warming act antagonistically so that temperature-induced increases in growth, feeding rates, and metabolism compensate for the negative effects of OA on larval development and growth (Gooding et al., 2009; Sheppard Brennand et al., 2010; Catarino et al., 2012a; Byrne et al., 2013a).

3.6 Conclusions and perspectives

OA is an ongoing and urgent problem that can, alone and in combination with other factors such as pollution, overfishing, temperature stress, and hypoxia, strongly affect marine ecosystems. The present brief review of four major taxonomic groups of animal calcifiers demonstrates that in general calcifiers are sensitive to OA and will experience a reduction in fitness due to delayed or disrupted development and calcification, reduced survival, growth, and reproduction. Several meta-analyses of literature on OA arrived at the same conclusions (Dupont et al., 2010b; Kroeker et al., 2010; Chan and Connolly, 2013; Harvey et al., 2013; Kroeker et al., 2013b), but also highlighted variations in sensitivity to OA between different phyla of marine calcifiers. Specifically, Kroeker et al. (2013b) in a meta-analysis of 155 original studies demonstrated that survival is negatively affected in corals, molluscs, and echinoderms, but not in crustaceans, while calcification is affected to a much greater degree in corals and molluscs than in echinoderms and crustaceans. This analysis of the literature also showed profound differences in sensitivity to OA between different life stages (Kroeker et al., 2013b). In many species, the early developmental stages (especially larvae that have high metabolic rates, and have not yet fully developed mechanisms of acid–base regulation) may represent the bottleneck for the survival and persistence of populations of marine calcifiers. However, adults of many species are also sensitive to elevated pCO_2 and reduced pH; moreover, transgenerational, carry-over effects of OA may exist that affect the offspring of the adults exposed to OA that are as yet poorly understood (Hettinger et al., 2012; Parker et al., 2012). Furthermore, there is great variability in the responses of different species of the same phylum to OA. Some of these differences may be related to differences in experimental design, such as the length of experimental exposure, the way the pH was adjusted, or laboratory versus field experiments, as well as other factors (Chan and Connolly, 2013). However, many publications report significant differences in the responses of different species studied under the same experimental conditions (Clark et al., 2009; Ries et al., 2009; Collard et al., 2013b; Comeau et al., 2013; Ivanina et al., 2013b; Matoo et al., 2013). The inter-phylum and inter-species variability in sensitivity to OA suggests that changes in compositions of marine communities will likely occur in the future, with more resistant species gaining a larger share and more sensitive species significantly decreasing in their global footprint.

The diversity of responses among calcifiers underscores the difficulty in making specific predictions for ecosystem shifts during OA and emphasizes the need for mechanistic approaches that can help in discerning the common physiological patterns associated with sensitivity or tolerance to OA in different species. Furthermore, it will never be possible to document responses of all marine species to OA and future studies should continue to focus on key ecosystem-forming species such as corals, reef-forming bivalves, keystone predators, and others that have a large impact in the structure and dynamics of the ecosystems in which they reside. Evidence from natural acidified environments (such as CO_2 vents) shows that while biodiversity of calcifiers drops most strongly with decreasing pH, the effects also propagate to non-calcifying taxa (Fig. 3.5; Hall-Spencer et al., 2008; Bijma et al., 2013; Kroeker et al., 2013a). Evolutionary potential in the face of OA also needs to be evaluated to predict realistic trajectories of biological response in the ecosystem-forming species, as well as in species that are important for humans, to determine the likelihood that resilience to OA can evolve fast enough to prevent extinctions (for an evolutionary perspective of OA see reviews by

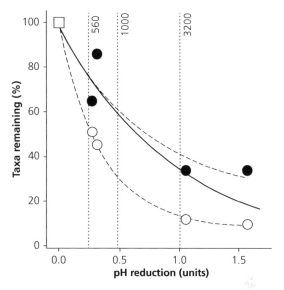

Figure 3.5 A natural experiment in ocean acidification: changes in species diversity as a function of pH near the Ischia CO_2 vents. The Ischia vents are located in the Mediterranean Sea near the coast of Italy, where CO_2 seeps through the vents in the seabed as result of volcanic activity around Mount Vesuvius. The biodiversity depending on the degree of reduction of pH is expressed as percent of taxa compared to that found in areas with the normal pH (pH 8.1–8.2) (open square). Calcifying taxa (51 total) are shown in white circles and non-calcifying taxa (71 total) in black circles. Atmospheric pCO_2 levels (ppmv) that would be required to cause corresponding pH changes are indicated by dotted vertical lines. At seawater pH expected by 2100, the loss of biodiversity predicted by the exponential regression lines shown in the figure is ~40% for non-calcifiers and ~70% for calcifiers. Reprinted from *Marine Pollution Bulletin*, in press, by Bijma et al. Climate change and the oceans – What does the future hold?, Copyright 2013, with permission from Elsevier.

Sunday et al., 2011; Munday et al., 2013). Given that the ongoing rate of carbon release and associated OA is unprecedented in the past 55 million years, and possibly 300 million years, and is at least ten times faster than preceded the last major species extinction in the geological record (Kump et al., 2009; Greene et al., 2012; Honisch et al., 2012), the current rate of OA may surpass the capacity of many marine species, especially those that have long generation time, to adapt to the change.

Another important aspect of OA that urgently requires further investigation is the interaction of OA with other environmental stressors that co-occur during global change, especially the so called 'deadly trio' of OA, warming, and hypoxia (Gruber, 2011), as well as changes in the salinity of the surface waters and increasing pressure from pollution, fishing, and habitat loss (Jackson et al., 2001; Durack et al., 2012; Pierce et al., 2012; Nikinmaa, 2013). This requires that the studies of OA are placed in a realistic environmental context and involve long-term exposures to major environmental factors (and their variation) relevant to the organism's ecology. Integrative approaches such as recently proposed concepts of energy-limited stress tolerance and oxygen- and capacity-limited thermal tolerance can be useful in integrating the physiological effects of multiple stressors and predicting their fitness- and population-level consequences (Pörtner, 2010, 2012; Sokolova et al., 2012; Sokolova, 2013).

OA and the associated changes in marine ecosystems are inevitable and it is important that humankind finds ways to mitigate the changes and adapt to the ones that are already occurring and will intensify in the near future. Therefore, future research should move beyond documenting the impacts towards finding potential solutions. Of course, the ultimate solution to OA will involve curbing carbon emissions into the atmosphere; however, research-based temporary solutions can buy some critical time and preclude the worst damage to marine ecosystems. Such solutions may involve identifying the most vulnerable ecosystems as a focus for conservation efforts, reducing the impact of other stressors such as pollution and overfishing to reduce the pressure on the ecosystems, and aquaculture and selective breeding efforts to improve resilience of existing stocks of economically important marine organisms to OA (Bijma et al., 2013). An understanding of the physiological mechanisms of the impacts of OA (alone or combined with other stressors) on marine calcifiers will play a major role in these efforts.

References

Albright, R. (2011). Reviewing the effects of ocean acidification on sexual reproduction and early life history stages of reef-building corals. *Journal of Marine Biology*, **2011**, 14.

Al-Horani, F.A., Al-Moghrabi, S.M., & de Beer, D. (2003). The mechanism of calcification and its relation to photosynthesis and respiration in the scleractinian coral *Galaxea fascicularis*. *Marine Biology*, **142**, 419–426.

Anthony, K.R.N., Kline, D.I., Diaz-Pulido, G., et al. (2008). Ocean acidification causes bleaching and productivity loss in coral reef builders. *Proceedings of the National Academy of Sciences of the USA*, **105**, 17442–17446.

Arnold, K.E., Findlay, H.S., Spicer, J.I., et al. (2009). Effect of CO_2-related acidification on aspects of the larval development of the European lobster, *Homarus gammarus*. *Biogeosciences*, **6**, 3087–3107.

Bechmann, R.K., Taban, I.C., Westerlund, S., et al. (2011). Effects of ocean acidification on early life stages of shrimp (*Pandalus borealis*) and mussel (*Mytilus edulis*). *Journal of Toxicology and Environmental Health Part A*, **74**, 424–438.

Beniash, E., Addadi, L., & Weiner, S. (1999). Cellular control over spicule formation in sea urchin embryos: a structural approach. *Journal of Structural Biology*, **125**, 50–62.

Beniash, E., Aizenberg, J., Addadi, L., & Weiner, S. (1997). Amorphous calcium carbonate transforms into calcite during sea urchin larval spicule growth. *Proceedings of the Royal Society of London Series B*, **264**, 461–465.

Beniash, E., Ivanina, A., Lieb, N.S., et al. (2010). Elevated level of carbon dioxide affects metabolism and shell formation in oysters *Crassostrea virginica*. *Marine Ecology Progress Series*, **419**, 95–108.

Bertucci, A., Moya, A., Tambutte, S., et al. (2013). Carbonic anhydrases in anthozoan corals: a review. *Bioorganic and Medicinal Chemistry*, **21**, 1437–1450.

Bijma, J., Pörtner, H.-O., Yesson, C., & Rogers, A.D. (2013). Climate change and the oceans—what does the future hold? *Marine Pollution Bulletin*, **79**(2), 495–505.

Byrne, M. (2011). Impact of ocean warming and ocean acidification on marine invertebrate life history stages: vulnerabilities and potential for persistence in a changing ocean. In: Gibson, R.N., Atkinson, R.J.A., Gordon, J.D.M., et al. (eds), *Oceanography and Marine Biology: An Annual Review*, Vol. 49, pp. 1–42. CRC Press, Boca Raton.

Byrne, M. (2012). Global change ecotoxicology: identification of early life history bottlenecks in marine invertebrates, variable species responses and variable experimental approaches. *Marine Environmental Research*, **76**, 3–15.

Byrne, M., Foo, S., Soars, N.A., et al. (2013a). Ocean warming will mitigate the effects of acidification on calcifying sea urchin larvae (*Heliocidaris tuberculata*) from the Australian global warming hot spot. *Journal of Experimental Marine Biology and Ecology*, **448**, 250–257.

Byrne, M., Ho, M., Selvakumaraswamy, P., et al. (2009). Temperature, but not pH, compromises sea urchin fertilization and early development under near-future climate change scenarios. *Proceedings of the Royal Society of London Series B*, **276**, 1883–1888.

Byrne, M., Lamare, M., Winter, D., et al. (2013b). The stunting effect of a high CO_2 ocean on calcification and development in sea urchin larvae, a synthesis from the tropics to the poles. *Philosophical Transactions of the Royal Society Series B*, **368**, 20120439.

Byrne, M. & Przeslawski, R. (2013). Multistressor impacts of warming and acidification of the ocean on marine invertebrates' life histories. *Integrative and Comparative Biology*, 53(4), 582–596.

Cameron, J.N. (1978). Effects of hypercapnia on blood acid-base status, NaCl fluxes, and trans-gill potential in freshwater blue crabs, *Callinectes sapidus*. *Journal of Comparative Physiology*, **123**, 137–141.

Cameron, J.N. (1985). Compensation of hypercapnic acidosis in the aquatic blue crab, *Callinectes sapidus*: the predominance of external sea water over carapace carbonate as the proton sink. *Journal of Experimental Biology*, **114**, 197–206.

Cameron, J.N. & Wood, C.M. (1985). Apparent H⁺ excretion and CO_2 dynamics accompanying carapace mineralization in the blue crab (*Callinectes sapidus*) following moulting. *Journal of Experimental Biology*, **114**, 181–196.

Carter, H.A., Ceballos-Osuna, L., Miller, N.A., & Stillman, J.H. (2013). Impact of ocean acidification on metabolism and energetics during early life stages of the intertidal porcelain crab *Petrolisthes cinctipes*. *Journal of Experimental Biology*, **216**, 1412–1422.

Catarino, A.I., Bauwens, M., &Dubois, P. (2012a). Acid-base balance and metabolic response of the sea urchin Paracentrotus lividus to different seawater pH and temperatures. *Environmental Science and Pollution Research International*, **19**, 2344–2353.

Catarino, A., Ridder, C., Gonzalez, M., et al. (2012b). Sea urchin *Arbacia dufresnei* (Blainville 1825) larvae response to ocean acidification. *Polar Biology*, **35**, 455–461.

Ceballos-Osuna, L., Carter, H.A., Miller, N.A., &Stillman, J.H. (2013). Effects of ocean acidification on early life-history stages of the intertidal porcelain crab *Petrolisthes cinctipes*. *Journal of Experimental Biology*, **216**, 1405–1411.

Chalker, B.E. & Taylor, D.L. (1975). Light-enhanced calcification, and the role of oxidative phosphorylation in calcification of the coral *Acropora cervicornis*. *Proceedings of the Royal Society of London Series B*, **190**, 323–331.

Chan, N.C.S. & Connolly, S.R. (2013). Sensitivity of coral calcification to ocean acidification: a meta-analysis. *Global Change Biology*, **19**, 282–290.

Chan, K.Y.K., Grünbaum, D., & O'Donnell, M.J. (2011). Effects of ocean-acidification-induced morphological changes on larval swimming and feeding. *Journal of Experimental Biology*, **214**, 3857–3867.

Christensen, A.B., Nguyen, H.D., & Byrne, M. (2011). Thermotolerance and the effects of hypercapnia on the metabolic rate of the ophiuroid *Ophionereis schayeri*: inferences for survivorship in a changing ocean. *Journal of Experimental Marine Biology and Ecology*, **403**, 31–38.

Clark, D., Lamare, M., & Barker, M. (2009). Response of sea urchin pluteus larvae (Echinodermata: Echinoidea) to reduced seawater pH: a comparison among a tropical, temperate, and a polar species. *Marine Biology*, **156**, 1125–1137.

Cohen, A.L., McCorkle, D.C., de Putron, S., et al. (2009). Morphological and compositional changes in the skeletons of new coral recruits reared in acidified seawater: insights into the biomineralization response to ocean acidification. *Geochemistry Geophysics Geosystems*, **10** Q07005, doi:10.1029/2009GC002411.

Collard, M., Catarino, A.I., Bonnet, S., et al. (2013a). Effects of CO_2-induced ocean acidification on physiological and mechanical properties of the starfish Asterias rubens. *Journal of Experimental Marine Biology and Ecology*, **446**, 355–362.

Collard, M., Laitat, K., Moulin, L., et al. (2013b). Buffer capacity of the coelomic fluid in echinoderms. *Comparative Biochemistry and Physiology Part A*, **166**, 199–206.

Comeau, S., Edmunds, P.J., Spindel, N.B., & Carpenter, R.C. (2013). The responses of eight coral reef calcifiers to increasing partial pressure of CO_2 do not exhibit a tipping point. *Limnology and Oceanography*, **58**, 388–398.

Comeau, S., Gorsky, G., Jeffree, R., et al. (2009). Impact of ocean acidification on a key Arctic pelagic mollusc (*Limacina helicina*). *Biogeosciences*, **6**, 1877–1882.

Cooley, S.R. & Doney, S.C. (2009). Anticipating ocean acidification's economic consequences for commercial fisheries. *Environmental Research Letters*, **4**, 1–8.

Crook, E.D., Cohen, A.L., Rebolledo-Vieyra, M., et al. (2013). Reduced calcification and lack of acclimatization by coral colonies growing in areas of persistent natural acidification. *Proceedings of the National Academy of Sciences of the USA*, **110**, 11044–11049.

Cummings, V., Hewitt, J., Van Rooyen, A., et al.(2011). Ocean acidification at high latitudes: potential effects on functioning of the Antarctic bivalve Laternula elliptica. *PLoS ONE*, **6**, e16069.

de Putron, S.J., McCorkle, D.C., Cohen, A.L., & Dillon, A.B. (2011). The impact of seawater saturation state and bicarbonate ion concentration on calcification by new recruits of two Atlantic corals. *Coral Reefs*, **30**, 321–328.

Dickinson, G.H., Ivanina, A.V., Matoo, O.B., et al. (2012). Interactive effects of salinity and elevated CO_2 levels on juvenile eastern oysters, *Crassostrea virginica*. *Journal of Experimental Biology*, **215**, 29–43.

Dickinson, G.H., Matoo, O.B., Tourek, R.T., et al. (2013). Environmental salinity modulates the effects of elevated CO_2 levels on juvenile hard-shell clams, *Mercenaria mercenaria*. *Journal of Experimental Biology*, **216**, 2607–2618.

Doney, S.C., Fabry, V.J., Feely, R.A., & Kleypas, J.A. (2009). Ocean acidification: the other CO_2 problem. *Annual Review of Marine Science*, **1**, 169–192.

Doney, S.C., Ruckelshaus, M., Duffy, J.E., et al. (2012). Climate change impacts on marine ecosystems., *Annual Review of Marine Science*, **4**, 11–37.

Donnay, G. & Pawson, D.L. (1969). X-ray diffraction studies of echinoderm plates. *Science*, **166**, 1147.

Donnell, M., Todgham, A., Sewell, M., et al. (2010). Ocean acidification alters skeletogenesis and gene expression in larval sea urchins. *Marine Ecology Progress Series*, **398**, 157–171.

Donohue, P., Calosi, P., Bates, A.H., et al. (2012). Impact of exposure to elevated pCO_2 on the physiology and behaviour of an important ecosystem engineer, the burrowing shrimp *Upogebia deltaura*. *Aquatic Biology*, **15**, 73–86.

Doo, S.S., Fujita, K., Byrne, M., & Uthicke, S. (2014). Fate of calcifying tropical symbiont-bearing large benthic foraminifera: living sands in a changing ocean. *Biological Bulletin*, **226**, 169–186.

Dorey, N., Lancon, P. Thorndyke, M., & Dupont, S. (2013). Assessing physiological tipping point of sea urchin larvae exposed to a broad range of pH. *Global Change Biology*, **19**, 3355–3367.

Dupont, S., Havenhand, J., Thorndyke, W., et al. (2008). Near-future level of CO_2-driven ocean acidification radically affects larval survival and development in the brittlestar *Ophiothrix fragilis*. *Marine Ecology Progress Series*, **373**, 285–294.

Dupont, S., Lundve, B. & Thorndyke, M. (2010a). Near future ocean acidification increases growth rate of the lec-
ithotrophic larvae and juveniles of the sea star *Crossaster
papposus*. *Journal of Experimental Zoology Part B*, **314**, 382–
389.

Dupont, S., Ortega-Martínez, O., & Thorndyke, M. (2010b).
Impact of near-future ocean acidification on echinoderms.
Ecotoxicology, **19**, 449–462.

Dupont, S. & Thorndyke, M. (2012). Relationship between
CO_2-driven changes in extracellular acid–base balance
and cellular immune response in two polar echinoderm
species. *Journal of Experimental Marine Biology and Ecology*,
424–425, 32–37.

Durack, P.J., Wijffels, S.E., & Matear, R.J. (2012). Ocean salinities reveal strong global water cycle intensification
during 1950 to 2000. *Science*, **336**, 455–458.

Edmunds, P.J. (2011). Zooplanktivory ameliorates the effects
of ocean acidification on the reef coral *Porites spp. Limnology and Oceanography*, **56**, 2402–2410.

Eerkes-Medrano, D., Menge, B., Sislak, C., & Langdon, C.
(2013). Contrasting effects of hypoxic conditions on survivorship of planktonic larvae of rocky intertidal invertebrates. *Marine Ecology Progress Series*, **478**, 139–151.

Egilsdottir, H., Spicer, J.I., & Rundle, S.D. (2009). The effect
of CO_2 acidified sea water and reduced salinity on aspects
of the embryonic development of the amphipod *Echinogammarus marinus* (Leach). *Marine Pollution Bulletin*, **58**,
1187–1191.

Erez, J., Reynaud, S., Silverman, J., et al. (2011). Coral calcification under ocean acidification and global change. In:
Dubinsky, Z. & Stambler, N. (eds), *Coral Reefs: An Ecosystem in Transition*, pp. 151–176. Springer, Dordrecht.

Evans, T.G., Chan, F., Menge, B.A., & Hofmann, G.E. (2013).
Transcriptomic responses to ocean acidification in larval
sea urchins from a naturally variable pH environment.
Molecular Ecology, **22**, 1609–1625.

Evans, T.G. & Watson-Wynn, P. (2014). Effects of seawater
acidification on gene expression: resolving broader-scale
trends in sea urchins. *Biological Bulletin*, **226**, 237–254.

Fabry, V.J., Seibel, B.A., Feely, R.A., & Orr, J.C. (2008). Impacts of ocean acidification on marine fauna and ecosystem processes. *ICES Journal of Marine Science*, **65**,
414–432.

FAO (2008). *The State of World Fisheries and Aquaculture*.
Technical report of the Fisheries Aquaculture Depart-
ment, Food and Agriculture Organization of the United
Nations: Rome, Italy, 2010.

Fernández-Reiriz,M.J., Labarta, U., Range, P., & Álvarez-
Salgado, X.A. (2011). Physiological energetics of juvenile
clams *Ruditapes decussatus* in a high CO_2 coastal ocean.
Marine Ecology Progress Series, **105**, 97–105.

Fernández-Reiriz, M., Range, P., Álvarez-Salgado, X., et al.
(2012). Tolerance of juvenile *Mytilus galloprovincialis* to experimental seawater acidification. *Marine Ecology Progress
Series*,**454**, 65–74.

Findlay, H.S., Burrows, M.T., Kendall, M.A., et al. (2010a).
Can ocean acidification affect population dynamics of

the barnacle *Semibalanus balanoides* at its southern range
edge? *Ecology*, **91**, 2931–2940.

Findlay, H.S., Kendall, M.A., Spicer, J.I., & Widdicombe, S.
(2009). Future high CO_2 in the intertidal may compromise
adult barnacle *Semibalanus balanoides* survival and embryonic development rate. *Marine Ecology Progress Series*, **389**,
193–202.

Findlay, H.S., Kendall, M.A., Spicer, J.I., & Widdicombe, S.
(2010b). Post-larval development of two intertidal barnacles at elevated CO_2 and temperature. *Marine Biology*,
157, 725–735.

Findlay, H.S., Kendall, M.A., Spicer, J.I., & Widdicombe,
S. (2010c). Relative influences of ocean acidification and
temperature on intertidal barnacle post-larvae at the
northern edge of their geographic distribution. *Estuarine,
Coastal and Shelf Science*, **86**, 675–682.

Fine, M. & Tchernov, D. (2007). Scleractinian coral species
survive and recover from decalcification. *Science*, **315**, 1811.

Fitzer, S.C., Caldwell, G.S., Close, A.J., et al. (2012). Ocean
acidification induces multi-generational decline in copepod naupliar production with possible conflict for reproductive resource allocation. *Journal of Experimental Marine
Biology and Ecology*, **418**, 30–36.

Frieder, C.A., Gonzalez, J.P., Bockmon, E.E., et al. (2014).
Can variable pH and low oxygen moderate ocean acidification outcomes for mussel larvae? *Global Change Biology*,
20, 754–764.

Furla, P., Galgani, I., Durand, I., & Allemand, D. (2000).
Sources and mechanisms of inorganic carbon transport
for coral calcification and photosynthesis. *Journal of Experimental Biology*, **203**, 3445–3457.

Gagnon, A.C., Adkins, J.F., Erez, J., et al. (2013). Sr/Ca sensitivity to aragonite saturation state in cultured subsamples
from a single colony of coral: mechanism of biomineralization during ocean acidification. *Geochimica et Cosmochimica Acta*, **105**, 240–254.

Gattuso, J.P., Allemand, D., & Frankignoulle, M. (1999).
Photosynthesis and calcification at cellular, organismal
and community levels in coral reefs: a review of interactions and control by carbonate chemistry. *American Zoologist*, **39**, 160–183.

Gattuso, J.P., Frankignoulle, M., Bourge, I., et al. (1998). Effect of calcium carbonate saturation of seawater on coral
calcification. *Global and Planetary Change*, **18**, 37–46.

Gazeau, F., Parker, L., Comeau, S., et al. (2013). Impacts of
ocean acidification on marine shelled molluscs. *Marine
Biology*, **160**, 2207–2245.

Gazeau, F., Quiblier, C., Jansen, J.M., et al. (2007). Impact
of elevated CO_2 on shellfish calcification. *Geophysical Research Letters*, **34** L07603, doi:10.1029/2006GL028554.

Gonzalez-Bernat, M., Lamare, M., Uthicke, S., & Byrne, M.
(2013). Fertilisation, embryogenesis and larval development in the tropical intertidal sand dollar *Arachnoides placenta* in response to reduced seawater pH. *Marine Biology*,
160, 1927–1941.

Gooding, R.A., Harley, C.D.G., & Tang, E. (2009). Elevated
water temperature and carbon dioxide concentration

increase the growth of a keystone echinoderm. *Proceedings of the National Academy of Sciences of the USA*, **106**(23), 9316–9321.

Goreau, T.F. (1959). The physiology of skeleton formation in corals. I. A method for measuring the rate of calcium deposition by corals under different conditions. *Biological Bulletin*, **116**, 59–75.

Gosling, E. (2003). *Bivalve Molluscs: Biology, Ecology and Culture*. Wiley-Blackwell, Oxford.

Green, M.A., Waldbusser, G.G., Reilly, S.L., et al. (2009). Death by dissolution: sediment saturation state as a mortality factor for juvenile bivalves. *Limnology and Oceanography*, **54**, 1037–1047.

Greene, S.E., Martindale, R.C., Ritterbush, K.A., et al. (2012). Recognising ocean acidification in deep time: an evaluation of the evidence for acidification across the Triassic-Jurassic boundary. *Earth Science Reviews*, **113**, 72–93.

Gruber, N. (2011). Warming up, turning sour, losing breath: ocean biogeochemistry under global change. *Philosophical Transactions of the Royal Society Series A*, **369**, 1980–1996.

Gutiérrez, J.L., Jones, C.G., Strayer, D.L., & Iribarne, O.O. (2003). Mollusks as ecosystem engineers: the role of shell production in aquatic habitats. *Oikos*, **101**, 79–90.

Gutowska, M., Melzner, F., Pörtner, H., & Meier, S. (2010b). Cuttlebone calcification increases during exposure to elevated seawater pCO_2 in the cephalopod *Sepia officinalis*. *Marine Biology*, **157**, 1653–1663.

Gutowska, M., Melzner, F., Langenbuch, M., et al. (2010a). Acid–base regulatory ability of the cephalopod (*Sepia officinalis*) in response to environmental hypercapnia. *Journal of Comparative Physiology Part B*, **180**, 323–335.

Gutowska, M.A., Melzner, F., Langenbuch, M., et al. (2010c). Acid-base regulatory ability of the cephalopod (*Sepia officinalis*) in response to environmental hypercapnia. *Journal of Comparative Physiology Part B*, **180**, 323–335.

Hall-Spencer, J.M., Rodolfo-Metalpa, R., Martin, S., et al. (2008). Volcanic carbon dioxide vents show ecosystem effects of ocean acidification. *Nature*, **454**, 96–99.

Hammer, K.M., Kristiansen, E., & Zachariassen, K.E. (2011). Physiological effects of hypercapnia in the deep-sea bivalve *Acesta excavata* (Fabricius, 1779) (Bivalvia; Limidae). *Marine Environmental Research*, **72**, 135–142.

Harvey, B.P., Gwynn-Jones, D., & Moore, P.J. (2013). Meta-analysis reveals complex marine biological responses to the interactive effects of ocean acidification and warming. *Ecology and Evolution*, **3**, 1016–1030.

Hauton, C., Tyrrell, T., & Williams, J. (2009). The subtle effects of sea water acidification on the amphipod *Gammarus locusta*. *Biogeosciences*, **6**, 1479–1489.

Hettinger, A., Sanford, E., Gaylord, B., et al. (2012). Extended larval carry-over effects: synergisms from a stressful benthic existence in juvenile Olympia oysters. *Journal of Shellfish Research*, **31**, 296–296.

Hoegh-Guldberg, O. (1999). Climate change, coral bleaching and the future of the world's coral reefs. *Marine and Freshwater Research*, **50**, 839–866.

Hoegh-Guldberg, O., Mumby, P.J., Hooten, A.J., et al. (2007). Coral reefs under rapid climate change and ocean acidification. *Science*, **318**, 1737–1742.

Holcomb, M., McCorkle, D.C., & Cohen, A.L. (2010). Long-term effects of nutrient and CO_2 enrichment on the temperate coral *Astrangia poculata* (Ellis and Solander, 1786). *Journal of Experimental Marine Biology and Ecology*, **386**, 27–33.

Honisch, B., Ridgwell, A., Schmidt, D.N., et al. (2012). The geological record of ocean acidification. *Science*, **335**, 1058–1063.

Hu, M.Y., Tseng, Y.-C., Stumpp, M., et al. (2011). Elevated seawater pCO_2 differentially affects branchial acid-base transporters over the course of development in the cephalopod *Sepia officinalis*. *American Journal of Physiology*, **300**, R1100–R1114.

Ivanina, A.V., Beniash, E., Etzkorn, M., et al. (2013a). Short-term acute hypercapnia affects cellular responses to trace metals in the hard clams *Mercenaria mercenaria*. *Aquatic Toxicology*, **140–141**, 123–133.

Ivanina, A.V., Dickinson, G.H., Matoo, O.B., et al. (2013b). Interactive effects of elevated temperature and CO_2 levels on energy metabolism and biomineralization of marine bivalves *Crassostrea virginica* and *Mercenaria mercenaria*. *Comparative Biochemistry and Physiology Part A*, **166**, 101–111.

Jackson, A., Chapman, M., & Underwood, A. (2008). Ecological interactions in the provision of habitat by urban development: whelks and engineering by oysters on artificial seawalls. *Austral Ecology*, **33**, 307–316.

Jackson, J.B.C., Kirby, M.X., Berger, W.H., et al. (2001). Historical overfishing and the recent collapse of coastal ecosystems. *Science*, **293**, 629–638.

Kawaguchi, S., Kurihara, H., King, R., et al. (2011). Will krill fare well under Southern Ocean acidification? *Biology Letters*, **7**, 288–291.

Keppel, E.A., Scrosati, R.A., & Courtenay, S.C. (2012). Ocean acidification decreases growth and development in American lobster (*Homarus americanus*) larvae. *Journal of Northwest Atlantic Fishery Science*, **44**, 61–66.

Kleypas, J.A., Buddemeier, R.W., Archer, D., et al. (1999). Geochemical consequences of increased atmospheric carbon dioxide on coral reefs. *Science*, **284**, 118–120.

Knowlton, N. (2001). The future of coral reefs. *Proceedings of the National Academy of Sciences of the USA*, **98**, 5419–5425.

Kroeker, K.J., Gambi, M.C., & Micheli, F. (2013a). Community dynamics and ecosystem simplification in a high-CO_2 ocean. *Proceedings of the National Academy of Sciences of the USA*, **110**, 12721–12726.

Kroeker, K.J., Kordas, R.L., Crim, R., et al. (2013b). Impacts of ocean acidification on marine organisms: quantifying sensitivities and interaction with warming. *Global Change Biology*, **19**, 1884–1896.

Kroeker, K.J., Kordas, R.L., Crim, R., & Singh, G.G. (2010). Meta-analysis reveals negative yet variable effects of ocean acidification on marine organisms. *Ecology Letters*, **13**, 1419–1434.

Kump, L.R., Bralower, T.J., & Ridgwell, A. (2009). Ocean acidification in deep time. *Oceanography*, **22**, 94–107.

Kurihara, H. (2008). Effects of CO_2-driven ocean acidification on the early developmental stages of invertebrates. *Marine Ecology Progress Series*, **373**, 275–284.

Kurihara, H. & Ishimatsu, A. (2008). Effects of high CO_2 seawater on the copepod (*Acartia tsuensis*) through all life stages and subsequent generations. *Marine Pollution Bulletin*, **56**, 1086–1090.

Kurihara, H., Matsui, M., Furukawa, H., et al. (2008). Long-term effects of predicted future seawater CO_2 conditions on the survival and growth of the marine shrimp *Palaemon pacificus*. *Journal of Experimental Marine Biology and Ecology*, **367**, 41–46.

Kurihara, H. & Shirayama, Y. (2004). Effects of increased atmospheric CO_2 on sea urchin early development. *Marine Ecology Progress Series*, **274**, 161–169.

Langdon, C., Takahashi, T., Sweeney, C., et al. (2000). Effect of calcium carbonate saturation state on the calcification rate of an experimental coral reef. *Global Biogeochemical Cycles*, **14**, 639–654.

Lannig, G., Eilers, S., Pörtner, H.-O., et al. (2010). Impact of ocean acidification on energy metabolism of oyster, *Crassostrea gigas*—changes in metabolic pathways and thermal response. *Marine Drugs*, **8**, 2318–2339.

Lischka, S., Büdenbender, J., Boxhammer, T., & Riebesell, U. (2011). Impact of ocean acidification and elevated temperatures on early juveniles of the polar shelled pteropod *Limacina helicina*: mortality, shell degradation, and shell growth. *Biogeosciences*, **8**, 919–932.

Liu, W. & He, M. (2012). Effects of ocean acidification on the metabolic rates of three species of bivalve from southern coast of China. *Chinese Journal of Oceanology and Limnology*, **30**, 206–211.

Long, W.C., Swiney, K.M., & Foy, R.J. (2013a). Effects of ocean acidification on the embryos and larvae of red king crab, *Paralithodes camtschaticus*. *Marine Pollution Bulletin*, **69**, 38–47.

Long, W.C., Swiney, K.M., Harris, C., et al. (2013b). Effects of ocean acidification on juvenile red king crab (*Paralithodes camtschaticus*) and tanner crab (*Chionoecetes bairdi*) growth, condition, calcification, and survival. *PLoS ONE*, **8**, e60959.

Lough, J.M. & Cantin, N.E. (2014). Perspectives on massive coral growth rates in a changing ocean. *Biological Bulletin*, **226**, 187–202.

Maier, C., Hegeman, J., Weinbauer, M.G., & Gattuso, J.P. (2009). Calcification of the cold-water coral *Lophelia pertusa* under ambient and reduced pH. *Biogeosciences*, **6**, 1671–1680.

Maier, C., Schubert, A., Berzunza Sanchez, M.M., et al. (2013). End of the century pCO_2 levels do not impact calcification in Mediterranean cold-water corals. *PLoS ONE*, **8**, e62655–e62655.

Manno, C., Morata, N., & Primicerio, R. (2012). *Limacina retroversa*'s response to combined effects of ocean acidification and sea water freshening. *Estuarine, Coastal and Shelf Science*, **113**, 163–171.

Marchant, H.K., Calosi, P., & Spicer, J.I. (2010). Short-term exposure to hypercapnia does not compromise feeding, acid–base balance or respiration of *Patella vulgata* but surprisingly is accompanied by radula damage. *Journal of the Marine Biological Association of the United Kingdom*, **90**, 1379–1384.

Markel, K., Roser, U., & Stauber, M. (1989). On the ultrastructure and the supposed function of the mineralizing matrix coat of sea-urchins (Echinodermata, Echinoida). *Zoomorphology*, **109**, 79–87.

Matoo, O.B., Ivanina, A.V., Ullstad, C., et al. (2013). Interactive effects of elevated temperature and CO_2 levels on metabolism and oxidative stress in two common marine bivalves (*Crassostrea virginica* and *Mercenaria mercenaria*). *Comparative Biochemistry and Physiology Part A*, **164**, 545–553.

Mayor, D.J., Matthews, C., Cook, K.B., et al. (2007). CO_2-induced acidification affects hatching success in *Calanus finmarchicus*. *Marine Ecology Progress Series*, **350**, 91–97.

McCulloch, M., Trotter, J., Montagna, P., et al. (2012). Resilience of cold-water scleractinian corals to ocean acidification: boron isotopic systematics of pH and saturation state up-regulation. *Geochimica et Cosmochimica Acta*, **87**, 21–34.

McDonald, M.R., McClintock, J.B., Amsler, C.D., et al. (2009). Effects of ocean acidification over the life history of the barnacle *Amphibalanus amphitrite*. *Marine Ecology Progress Series*, **385**, 179–187.

McElroy, D.J., Nguyen, H.D., & Byrne, M. (2012). Respiratory response of the intertidal seastar *Parvulastra exigua* to contemporary and near-future pulses of warming and hypercapnia. *Journal of Experimental Marine Biology and Ecology*, **416–417**, 1–7.

Melatunan, S., Calosi, P., Rundle, S., et al. (2013). Effects of ocean acidification and elevated temperature on shell plasticity and its energetic basis in an intertidal gastropod. *Marine Ecology Progress Series*, **472**, 155–168.

Melatunan, S., Calosi, P., Rundle, S., et al. (2011). Exposure to elevated temperature and Pco(2) reduces respiration rate and energy status in the periwinkle *Littorina littorea*. *Physiological and Biochemical Zoology*, **84**, 583–594.

Melzner, F., Gutowska, M.A., Langenbuch, M., et al. (2009). Physiological basis for high CO_2 tolerance in marine ectothermic animals: pre-adaptation through lifestyle and ontogeny? *Biogeosciences*, **6**, 4693–4738.

Metzger, R., Sartoris, F.J., Langenbuch, M., & Pörtner, H.-O. (2007). Influence of elevated CO_2 concentrations on thermal tolerance of the edible crab *Cancer pagurus*. *Journal of Thermal Biology*, **32**, 144–151.

Michaelidis, B., Ouzounis, C., Paleras, A., & Pörtner, H.-O. (2005). Effects of long-term moderate hypercapnia on acid-base balance and growth rate in marine mussels *Mytilus galloprovincialis*. *Marine Ecology Progress Series*, **293**, 109–118.

Miles, H., Widdicombe, S., Spicer, J.I., & Hall-Spencer, J. (2007). Effects of anthropogenic seawater acidification on acid–base balance in the sea urchin *Psammechinus miliaris*. *Marine Pollution Bulletin*, **54**, 89–96.

Miller, S.H., Zarate, S., Smith, E.H., et al. (2014). Effect of elevated pCO_2 on metabolic responses of porcelain crab (*Petrolisthes cinctipes*) larvae exposed to subsequent salinity stress. *PLoS ONE*, **9**, e109167.

Moya, A., Huisman, L., Ball, E.E., et al. (2012). Whole transcriptome analysis of the coral *Acropora millepora* reveals complex responses to CO_2-driven acidification during the initiation of calcification. *Molecular Ecology*, **21**, 2440–2454.

Moya, A., Tambutté, S., Bertucci, A., et al. (2008). Carbonic anhydrase in the scleractinian coral *Stylophora pistillata*: characterization, localization, and role in biomineralization. *Journal of Biological Chemistry*, **283**, 25475–25484.

Moya, A., Tambutté, S., Tambutté, E., et al. (2006). Study of calcification during a daily cycle of the coral *Stylophora pistillata*: implications for 'light-enhanced calcification'. *Journal of Experimental Biology*, **209**, 3413–3419.

Munday, P.L., Warner, R.R., Monro, K., et al. (2013). Predicting evolutionary responses to climate change in the sea. *Ecology Letters*, **16**, 1488–1500.

Munguia, P. & Alenius, B. (2013). The role of preconditioning in ocean acidification experiments: a test with the intertidal isopod *Paradella dianae*. *Marine and Freshwater Behaviour and Physiology*, **46**, 33–44.

Navarro, J.M., Torres, R., Acuña, K., et al. (2013). Impact of medium-term exposure to elevated pCO_2 levels on the physiological energetics of the mussel *Mytilus chilensis*. *Chemosphere*, **90**, 1242–1248.

Nikinmaa, M. (2013). Climate change and ocean acidification interactions with aquatic toxicology. *Aquatic Toxicology*, **126**, 365–372.

Ohde, S. & Hossain, M.M.M. (2004). Effect of $CaCO_3$ (aragonite) saturation state of seawater on calcification of *Porites* coral. *Geochemical Journal*, **38**, 613–621.

Okazaki, K. & Inoue, S. (1976). Crystal property of the larval sea urchin spicule. *Development, Growth and Differentiation*, **18**, 413–434.

Ordoñez, A., Doropoulos, C., & Diaz-Pulido, G. (2014). Effects of ocean acidification on population dynamics and community structure of crustose coralline algae. *Biological Bulletin*, **226**, 255–268.

Orr, J.C., Fabry, V.J., Aumont, O., et al. (2005). Anthropogenic ocean acidification over the twenty-first century and its impact on calcifying organisms. *Nature*, **437**, 681–686.

Pane, E.F. & Barry, J.P. (2007). Extracellular acid-base regulation during short-term hypercapnia is effective in a shallow-water crab, but ineffective in a deep-sea crab. *Marine Ecology Progress Series*, **334**, 1–9.

Pansch, C., Nasrolahi, A., Appelhans, Y.S., & Wahl, M. (2012). Impacts of ocean warming and acidification on the larval development of the barnacle *Amphibalanus improvisus*. *Journal of Experimental Marine Biology and Ecology*, **420**, 48–55.

Pansch, C., Schlegel, P., & Havenhand, J. (2013). Larval development of the barnacle *Amphibalanus improvisus* responds variably but robustly to near-future ocean acidification. *ICES Journal of Marine Science*, **70**, 805–811.

Parker, L., Ross, P., Connor, W., et al. (2013). Predicting the response of molluscs to the impact of ocean acidification. *Biology*, **2**, 651–692.

Parker, L.M., Ross, P.M., O'Connor, W.A., et al. (2012). Adult exposure influences offspring response to ocean acidification in oysters. *Global Change Biology*, **18**, 82–92.

Pierce, D.W., Gleckler, P.J., Barnett, T.P., et al. (2012). The fingerprint of human-induced changes in the ocean's salinity and temperature fields. *Geophysical Research Letters*, **39**, L21704.

Politi, Y., Arad, T., Klein, E., et al. (2004). Sea urchin spine calcite forms via a transient amorphous calcium carbonate phase. *Science*, **306**, 1161–1164.

Poore, A.G.B., Graba-Landry, A., Favret, M., et al. (2013). Direct and indirect effects of ocean acidification and warming on a marine plant–herbivore interaction. *Oecologia*, **173**, 1113–1124.

Pörtner, H., Langenbuch, M., & Reipschläger, A. (2004a). Biological impact of elevated ocean CO_2 concentrations: lessons from animal physiology and earth history. *Journal of Oceanography*, **60**, 705–718.

Pörtner, H.-O. (2008). Ecosystem effects of ocean acidification in times of ocean warming: a physiologist's view. *Marine Ecology Progress Series*, **373**, 203–217.

Pörtner, H.-O. (2010). Oxygen- and capacity-limitation of thermal tolerance: a matrix for integrating climate-related stressor effects in marine ecosystems. *Journal of Experimental Biology*, **213**, 881–893.

Pörtner, H.-O. (2012). Integrating climate-related stressor effects on marine organisms: unifying principles linking molecule to ecosystem-level changes. *Marine Ecology Progress Series*, **470**, 273–290.

Pörtner, H.-O., Reipschlager, A., & Heisler, N. (1998). Acid-base regulation, metabolism and energetics in *Sipunculus nudus* as a function of ambient carbon dioxide level. *Journal of Experimental Biology*, **201**, 43–55.

Pörtner, H.-O., Langenbuch, M., & Reipschlager, A. (2004b). Biological impact of elevated ocean CO_2 concentrations: lessons from animal physiology and earth history. *Journal of Oceanography*, **60**, 705–718.

Range, P., Chicharo, M.A., Ben-Hamadou, R., et al. (2011). Calcification, growth and mortality of juvenile clams *Ruditapes decussatus* under increased pCO_2 and reduced pH: variable responses to ocean acidification at local scales? *Journal of Experimental Marine Biology and Ecology*, **396**, 177–184.

Ries, J.B. (2011). Skeletal mineralogy in a high-CO_2 world. *Journal of Experimental Marine Biology and Ecology*, **403**, 54–64.

Ries, J.B., Cohen, A.L., & McCorkle, D.C. (2009). Marine calcifiers exhibit mixed responses to CO_2-induced ocean acidification. *Geology*, **37**, 1131–1134.

Ries, J.B., Cohen, A.L., & McCorkle, D.C. (2010). A nonlinear calcification response to CO_2-induced ocean acidification by the coral *Oculina arbuscula*. *Coral Reefs*, **29**, 661–674.

Rodolfo-Metalpa, R., Houlbreque, F., Tambutté, E., et al. (2011). Coral and mollusc resistance to ocean acidification adversely affected by warming. *Nature Climate Change*, **1**, 308–312.

Rosa, R. & Seibel, B.A. (2008). Synergistic effects of climate-related variables suggest future physiological impairment in a top oceanic predator. *Proceedings of the National Academy of Sciences of the USA*, **105**, 20776–20780.

Ross, P.M., Parker, L., O'Connor, W.A., & Bailey, E.A. (2011). The impact of ocean acidification on reproduction, early development and settlement of marine organisms. *Water*, **3**, 1005–1030.

Schalkhausser, B., Bock, C., Stemmer, K., et al. (2013). Impact of ocean acidification on escape performance of the king scallop, *Pecten maximus*, from Norway. *Marine Biology*, **160**, 1995–2006.

Schneider, K. & Erez, J. (2006). The effect of carbonate chemistry on calcification and photosynthesis in the hermatypic coral *Acropora eurystoma*. *Limnology and Oceanography*, **51**, 1284–1293.

Schram, J.B., McClintock, J.B., Angus, R.A., & Lawrence, J.M. (2011). Regenerative capacity and biochemical composition of the sea star *Luidia clathrata* (Say) (Echinodermata: Asteroidea) under conditions of near-future ocean acidification. *Journal of Experimental Marine Biology and Ecology*, **407**, 266–274.

Secretariat of the Convention on Biological Diversity (2014). *An Updated Synthesis of the Impacts of Ocean Acidification on Marine Biodiversity*. In: Hennige, S., Roberts, J.M., & Williamson, P. (eds), CBD Technical Series No. 75. Secretariat of the Convention on Biological Diversity, Montreal.

Seibel, B.A. & Walsh, P.J. (2001). Potential impacts of CO_2 injection on deep-sea biota. *Science*, **294**, 319–320.

Sheppard Brennand, H., Soars, N., Dworjanyn, S.A., et al. (2010). Impact of ocean warming and ocean acidification on larval development and calcification in the sea urchin *Tripneustes gratilla*. *PLoS ONE*, **5**, e11372.

Shirayama, Y. & Thorton, H. (2005). Effect of increased atmospheric CO_2 on shallow water marine benthos. *Journal of Geophysical Research*, **110**, 1–7.

Silverman, J., Lazar, B., Cao, L., et al. (2009). Coral reefs may start dissolving when atmospheric CO_2 doubles. *Geophysical Research Letters*, **36** L05606, doi:10.1029/2008GL036282.

Small, D., Calosi, P., White, D., et al. (2010). Impact of medium-term exposure to CO_2 enriched seawater on the physiological functions of the velvet swimming crab *Necora puber*. *Aquatic Biology*, **10**, 11–21.

Sokolova, I.M. (2013). Energy-limited tolerance to stress as a conceptual framework to integrate the effects of multiple stressors. *Integrative and Comparative Biology*, **53**, 597–608.

Sokolova, I.M., Bock, C., & Pörtner, H.-O. (2000). Resistance to freshwater exposure in White Sea *Littorina* spp. II: Acid-base regulation. *Journal of Comparative Physiology Part B*, **170**, 105–115.

Sokolova, I.M., Frederich, M., Bagwe, R., et al. (2012). Energy homeostasis as an integrative tool for assessing limits of environmental stress tolerance in aquatic invertebrates. *Marine Environmental Research*, **79**, 1–15.

Sokolova, I.M., Sukhotin, A.A., & Lannig, G. (2011). Stress effects on metabolism and energy budgets in mollusks. In: Abele, D., Vazquez-Medina, J.P., & Zenteno-Savín, T. (eds), *Oxidative Stress in Aquatic Ecosystems*, pp. 263–280. John Wiley & Sons, Ltd, Chichester.

Spicer, J.I., Raffo, A., & Widdicombe, S. (2007). Influence of CO_2-related seawater acidification on extracellular acid-base balance in the velvet swimming crab *Necora puber*. *Marine Biology*, **151**, 1117–1125.

Stumpp, M., Dupont, S., Thorndyke, M.C., & Melzner, F. (2011). CO_2-induced seawater acidification impacts sea urchin larval development II: Gene expression patterns in *Pluteus* larvae. *Comparative Biochemistry and Physiology Part A*, **160**, 320–330.

Stumpp, M., Hu, M.Y., Melzner, F., et al. (2012a). Acidified seawater impacts sea urchin larvae pH regulatory systems relevant for calcification. *Proceedings of the National Academy of Sciences of the USA*, **109**, 18192–18197.

Stumpp, M., Trubenbach, K., Brennecke, D., et al. (2012b). Resource allocation and extracellular acid-base status in the sea urchin *Strongylocentrotus droebachiensis* in response to CO_2-induced seawater acidification. *Aquatic Toxicology*, **110**, 194–207.

Sunday, J.M., Crim, R.N., Harley, C.D.G., & Hart, M.W. (2011). Quantifying rates of evolutionary adaptation in response to ocean acidification. *PLoS ONE*, **6**, e22881.

Takahashi, A. & Kurihara, H. (2013). Ocean acidification does not affect the physiology of the tropical coral *Acropora digitifera* during a 5-week experiment. *Coral Reefs*, **32**, 305–314.

Tambutté, E., Allemand, D., Mueller, E., & Jaubert, J. (1996). A compartmental approach to the mechanism of calcification in hermatypic corals. *Journal of Experimental Biology*, **199**, 1029–1041.

Thomsen, J., Casties, I., Pansch, C., et al. (2013). Food availability outweighs ocean acidification effects in juvenile *Mytilus edulis*: laboratory and field experiments. *Global Change Biology*, **19**, 1017–1027.

Thomsen, J., Gutowska, M.A., Saphörster, J., et al. (2010). Calcifying invertebrates succeed in a naturally CO_2-rich coastal habitat but are threatened by high levels of future acidification. *Biogeosciences*, **7**, 3879–3891.

Thomsen, J. & Melzner, F. (2010). Moderate seawater acidification does not elicit long-term metabolic depression in the blue mussel *Mytilus edulis*. *Marine Biology*, **157**, 2667–2676.

Todgham, A.E. & Hofmann, G.E. (2009). Transcriptomic response of sea urchin larvae *Strongylocentrotus purpuratus* to CO_2-driven seawater acidification. *Journal of Experimental Biology*, **212**, 2579–2594.

Truchot, J.P. (1975). Blood acid-base changes during experimental emersion and reimmersion of the intertidal crab *Carcinus maenas*. *Respiration Physiology*, **23**, 351–360.

Uthicke, S., Soars, N., Foo, S., & Byrne, M. (2013). Effects of elevated pCO_2 and the effect of parent acclimation on development in the tropical pacific sea urchin *Echinometra mathaei*. *Marine Biology*, **160**, 1913–1926.

Venn, A.A., Tambutté, E., Holcomb, M., et al. (2013). Impact of seawater acidification on pH at the tissue-skeleton interface and calcification in reef corals. *Proceedings of the National Academy of Sciences of the USA*, **110**, 1634–1639.

Waldbusser, G.G., Steenson, R.A., & Green, M.A. (2011b). Oyster shell dissolution rates in estuarine waters: effects of pH and shell legacy. *Journal of Shellfish Research*, **30**, 659–669.

Waldbusser, G., Voigt, E., Bergschneider, H., et al. (2011a). Biocalcification in the Eastern Oyster (*Crassostrea virginica*) in relation to long-term trends in Chesapeake Bay pH. *Estuaries and Coasts*, **34**, 221–231.

Walther, K., Anger, K., & Pörtner, H.-O. (2010). Effects of ocean acidification and warming on the larval development of the spider crab *Hyas araneus* from different latitudes (54 vs. 79 N). *Marine Ecology Progress Series*, **417**, 159–170.

Walther, K., Sartoris, F.-J., Bock, C., & Pörtner, H.-O. (2009). Impact of anthropogenic ocean acidification on thermal tolerance of the spider crab *Hyas araneus. Biogeosciences*, **6**, 2837–2861.

Weiss, I.M., Tuross, N., Addadi, L., & Weiner, S. (2002). Mollusc larval shell formation: amorphous calcium carbonate is a precursor phase for aragonite. *Journal of Experimental Zoology Part A*, **293**, 478–491.

Whiteley, N.M. (2011). Physiological and ecological responses of crustaceans to ocean acidification. *Marine Ecology Progress Series*, **430**, 257–271.

Wickins, J.F. (1984). The effect of hypercapnic sea water on growth and mineralization in penaied prawns. *Aquaculture*, **41**, 37–48.

Wittmann, A.C. & Portner, H.-O. (2013). Sensitivities of extant animal taxa to ocean acidification. *Nature Climate Change*, **3**, 995–1001.

Wong, K.K.W., Lane, A.C., Leung, P.T.Y., & Thiyagarajan, V. (2011). Response of larval barnacle proteome to CO_2-driven seawater acidification. *Comparative Biochemistry and Physiology Part D*, **6**, 310–321.

Wood, H., Spicer, J.I., Lowe, D.M., & Widdicombe, S. (2010). Interaction of ocean acidification and temperature; the high cost of survival in the brittlestar *Ophiura ophiura. Marine Biology*, **157**, 2001–2013.

Wood, H., Spicer, J.I., Kendall, M.A., et al. (2011). Ocean warming and acidification; implications for the Arctic brittlestar *Ophiocten sericeum. Polar Biology*, **34**, 1033–1044.

Wood, H.L., Spicer, J.I., & Widdicombe, S. (2008). Ocean acidification may increase calcification rates, but at a cost. *Proceedings of the Royal Society of London Series B*, **275**, 1767–1773.

Zoccola, D., Tambutté, E., Kulhanek, E., et al. (2004). Molecular cloning and localization of a PMCA P-type calcium ATPase from the coral *Stylophora pistillata. Biochimica et Biophysica Acta*, **1663**, 117–126.

CHAPTER 4

Physiological responses of marine invertebrates to thermal stress

Nia M. Whiteley and Clara L. Mackenzie

4.1 Introduction

Marine invertebrates occupy a broad range of thermal habitats, from those that are permanently cold, such as the continental shelf region of the Southern Ocean, to those where substrate temperatures can reach > 40 °C during emersion in the littoral zone. As marine invertebrates are ectothermic and unable to regulate body temperatures independently from ambient, survival at such extreme temperatures is a remarkable achievement. Many species are adapted to their particular thermal environment and may fail to survive outside of their own thermal niche (Somero, 2002). As a result, they show tremendous variability in thermal sensitivities, thermal optima, and tolerance limits, and while mobile marine invertebrates are able to avoid unfavourable temperatures to some extent by moving to more equitable microclimates, sessile invertebrates are at a distinct disadvantage, especially in the littoral zone (Helmuth and Hofmann, 2001; Somero, 2002).

Physiological responses to temperature are thought to determine thermal tolerance limits in marine invertebrates. Although heat death is ultimately caused by the degradation of proteins, tolerance limits occur at lower temperatures as whole animal physiological responses are more sensitive to warming and are more relevant in defining optimal temperatures for growth and reproduction (Pörtner, 2002, 2010). Physiology can therefore be linked to biogeographic patterning and the determination of species distribution patterns (Somero, 2002; Tomanek, 2008). Variations in physiological capacities to compensate for change occur among and within species, and can vary over time in response to predictable environmental fluctuations (i.e. oscillating circadian and seasonal change) and over spatial scales in response to environmental gradients (Sanford and Kelly, 2011). Physiological differences among species

or populations are influenced by thermal histories and rates of temperature change (Peck et al., 2014). They can also occur as a result of transgenerational phenotypic plasticity where mothers alter the phenotype of their offspring in response to the local environment, and/or in response to the conditions experienced by early life stages (Marshall et al., 2008). Thermal adaption to variable versus constant temperatures may also have an effect (Sinclair et al., 2006; Rock et al., 2009) with those species living in stable environments possessing reduced acclimatory capacities for environmental change (Peck et al., 2014).

The purpose of this chapter is to review the physiological responses shown by marine invertebrates to thermal stress, where thermal stress is taken to be exposure to elevated temperatures which can reduce performance and cause potential damage (Morris et al., 2013; Schulte, 2014). If subjected to short bouts of thermal stress, it is possible for organisms to survive by resorting to anaerobic respiration, and to recover by producing a series of protective molecules, such as heat shock proteins and antioxidants. Longer-term exposures at temperatures that match or exceed optimal temperatures are likely to affect survival as such temperatures are thought to decrease phenotypic plasticity and exceed the temperatures required for reproduction and growth (Pörtner, 2002, 2010). Many marine invertebrates are likely to be exposed to the warming effects associated with climate change, and if they are unable to migrate into cooler waters, they will depend in the first instance on physiological adjustments followed in the longer term by adaptive change. This chapter concentrates on the spatial variations in physiological rate processes and repair mechanisms shown by marine invertebrates in response to thermal stress. We will examine changes in cardiorespiratory physiology and nerve function, as well as alterations to oxygen

Stressors in the Marine Environment. Edited by Martin Solan and Nia M. Whiteley
© Oxford University Press 2016. Published in 2016 by Oxford University Press.

transport and mitochondrial function. At the cellular level we will review the mechanisms involved in protein repair and protection, as well as those involved in counteracting oxidative stress. As temperature rise will not be occurring on its own in a changing climate, we will also review our current understanding of the effects of thermal stress in combination with other environmental factors in order to more fully appreciate the physiological capacities of marine invertebrates to survive contemporary environmental change. This is not intended to be an exhaustive review. Instead it provides an introduction to this very broad topic and refers to some pertinent studies which have formed the basis for our current understanding of the physiological effects of thermal stress on marine invertebrates. Throughout the chapter, examples will be used to demonstrate the variability in responses observed among populations and species to emphasize that 'one size does not fit all'. The chapter will not consider temperature-related changes in enzyme quality and quantity, or the recent advances in 'omic' research, because these topics are covered elsewhere by several excellent reviews (Stillman and Tagmount, 2009; Somero, 2010; Tomanek, 2014; Stillman and Hurt, 2015).

4.2 Thermal dependency of physiological responses: underlying principles and concepts

Temperature has a profound effect on the physiology of ectothermic organisms by influencing the rate at which biochemical reactions proceed and by influencing the structural and functional properties of macromolecules (Hochacka and Somero, 2002). A convenient term for expressing the extent of the increase in biological rate processes with rise in temperature is the temperature coefficient (Q_{10}), which can be quantified via the van't Hoff equation and used to define thermal dependence. Most biochemical reactions increase with temperature in a predictable fashion to give a Q_{10} of 2–3, but physiological functions, such as heart and ventilation rate, are complex interactions comprising different physical and biochemical processes that have slightly different thermal sensitivities and less predictable Q_{10} values. The relationship between rates of physiological processes and temperature is further complicated by a reduction in Q_{10} values with increase in temperature range, and the fact that Q_{10} can acclimate as a result of thermal experiences, acclimatize as a result of thermal change in combination with other environmental parameters, or be under strong selective

pressure in order to show thermal independence (i.e. $Q_{10} = 1$) (Moyes and Schulte, 2008). To avoid inconsistencies introduced by changes in Q_{10} with temperature, physiological rate processes can be plotted against thermodynamic temperature (absolute temperature in degrees Kelvin) and the Arrhenius equation used to calculate the apparent activation energy of the reaction. Discontinuities in this relationship (termed Arrhenius break points) indicate a change in biochemical pathways, mechanisms, and/or other physical and chemical processes.

Limits for thermal tolerance ranges are determined by genetics but within this framework they are restricted by physiological capabilities (Hochachka and Somero, 2002). Over an organism's normal thermal range, physiological parameters increase with temperature at a specific Q_{10} value until a critical temperature (CT_{max}) is reached when physiological rate processes rapidly decline (e.g. Whiteley and Fraser, 2009; Schulte et al., 2011; Giomi and Pörtner, 2013). In water breathing invertebrates, the capacity to maintain aerobic scope and supply oxygen to the tissues is limited by the functional capacity of the circulatory and ventilatory systems, as explained by the concept of oxygen- and capacity-limited thermal tolerance (OCLTT, Pörtner, 2002, 2010). This concept maintains that the ability of ectotherms to survive temperature change is ultimately influenced by whole animal responses, which tend to be more sensitive to changes in environmental temperature than biochemical reactions and the denaturation of macromolecules (Pörtner, 2010; Schulte et al., 2011). As optimal temperatures are reached, oxygen supply and aerobic scope is maximal and the body fluids are characterized by high oxygen levels. On further warming, there is a point when the increase in heart rate and ventilation rate with temperature are unable to match increasing oxygen demand, even though oxygen is available. This temperature, known as the pejus temperature, defines the beginning of a reduction in oxygen supply to the tissues, and indicates early limitations to thermal tolerance. As temperatures continue to increase, the organism experiences mildly hypoxic conditions and a progressive loss in aerobic scope as the circulatory and ventilatory systems begin to fail as they have reached the limits of their capacity and the organism is unable to take up sufficient oxygen (Frederich and Pörtner, 2000). Once CT_{max} is reached, the decrease in systemic oxygen levels (hypoxaemia) finally leads to anaerobic respiration and a reduction in protein synthesis rates, and marks the temperature at which there is insufficient energy production to cover the high energy demands associated with elevated temperatures (Sommer et al., 1997; Melzner et al., 2006). Beyond CT_{max}, mechanisms for protection and repair of molecular structures are required as denaturation temperatures are reached to prolong survival over limited periods (Pörtner, 2010; Sokolova, 2013). Longer-term exposure to temperatures above CT_{max} is typically fatal due to heat coma and loss of nervous integration, as well as the loss of cardiac and skeletal muscle function. Continuous recordings of haemolymph oxygen levels as partial pressure of oxygen in the haemolymph of the spider crab, *Maja squinado*, during warming and cooling helped to reveal that high oxygen levels and maximum oxygen supply only occur within an optimal temperature window defined by two sets of threshold temperatures (Frederich and Pörtner, 2000; Pörtner, 2002).

Alternative metabolic responses to elevated temperatures also occur with some high intertidal species, such as the gastropod *Echinolittorina malaccana*, undergoing metabolic depression at temperatures close to CT_{max} (Marshall et al., 2011). Pörtner (2012) and Sokolova (2013) argue that such responses help to extend the range between pejus and critical temperatures where survival is possible but the prevailing hypoxaemia prevents growth and reproduction. In the limpet *Lottia austrodigitalis* metabolic depression has the added advantage of reducing energy demand at a time when food is in short supply as the elevated summer temperatures desiccate the algal food source (Parry, 1978). Temperature can also have an indirect effect on oxygen delivery to the tissues in marine invertebrates. An increase in temperature reduces haemolymph oxygen levels in both molluscs and crustaceans via a temperature-related decrease in dissolved oxygen levels and a reduction in the oxygen affinity of the respiratory pigment, haemocyanin (Seibel, 2013; Whiteley and Taylor, 2015). The latter is caused by direct allosteric effects on haemocyanin and by the indirect effects of temperature on haemolymph pH, which falls as temperature rises. Temperature-related alterations in haemocyanin oxygen affinity can facilitate haemolymph oxygen delivery in marine invertebrates during moderate hypoxia, but can compromise oxygen delivery during extreme hypoxia further limiting survival during thermal stress (Seibel, 2013). A reduction in haemocyanin oxygen affinity can also increase thermal tolerances in the eurythermal green crab, *Carcinus maenas*, during warming (Giomi and Pörtner, 2013). As temperatures rise, the species becomes increasingly dependent on oxygen delivery to the tissues via oxygen bound to haemocyanin assisted by the low oxygen partial pressures present in the venous haemolymph.

The thermal dependency of physiological processes in marine invertebrates can be subject to compensatory adjustments to infer some independence from elevated temperature and to ensure optimal function despite thermal fluctuations. The capacity for physiological compensation, whether this occurs as a result of plasticity within the lifespan of an individual (acclimation and acclimatization) or by adaptation across the generations in response to natural selection, has also been linked to changes in thermal tolerances (Pörtner, 2010). Moreover, the passive tolerance of marine invertebrates to thermal extremes is influenced by the ability to activate protective and damage-repair mechanisms. Such abilities can vary among species and populations and is subject to acclimation in the laboratory (e.g. Tomanek and Somero, 1999; Anestis et al., 2008) and acclimatization on the shore (e.g. Hofmann and Somero, 1995). The ability to tolerate increases in temperature can also be quantified and related to the period of exposure, as demonstrated by Antarctic marine invertebrates where long-term moderate warming leads to lower thermal limits than shorter-term exposure to higher temperatures (Peck et al., 2009, 2014). Physiological compensation for temperature change along with the ability to activate protective mechanisms enables marine invertebrates to elevate critical and lethal temperatures as long as they are not living towards their thermal maxima (Stillman, 2003; Pörtner, 2010). Moreover, it has become increasingly apparent that physiological compensation in response to thermal stress is influenced by the nature of any interactions with other environmental factors, which can be synergistic, antagonistic, or additive (Kroeker et al., 2013; Todgham and Stillman, 2013).

4.3 Repair mechanisms during short-term thermal challenges

4.3.1 Protection against protein denaturation

In order to survive, in the short term, acute increases in temperature, mechanisms are necessary to protect and repair any resulting effects of temperature on metabolism and macromolecular structure. Proteins are particularly vulnerable to temperature rise as they function over narrow ranges of temperature and are susceptible to denaturation by unfolding or misfolding into a non-functional conformation. The accumulation of denatured proteins within the cell is also fatal. In order to protect proteins from denaturation and to prevent incompatible protein aggregation, many marine invertebrates synthesize molecular chaperones, a large group of proteins which include the heat shock proteins (Kültz, 2005; Morris et al., 2013). The heat shock proteins most often studied in marine invertebrates in response to thermal stress are those belonging to the hsp70 family. An elevation in temperature increases hsp70 levels in echinoderms (e.g. the sea cucumber *Apostichopus japonicus*, Dong et al., 2011), gastropod molluscs (e.g. *Tegula* species, Tomanek and Somero, 1999; *Cellana* species, Dong and Williams, 2011), bivalve molluscs (*Mytilus* species, Hofmann and Somero, 1995; Anestis et al., 2010), and crustaceans (crabs and shrimps, Madeira et al., 2012, 2014). Collectively, heat shock studies have demonstrated that species or populations with little experience of thermal stress have a reduced heat shock response and lower threshold temperatures for inducing hsps (Tomanek and Somero, 1999; Somero, 2002; Madeira et al., 2012). Patterns of hsp70 responses to thermal stress, however, are not consistent across species, taxa, and habitat type, and are likely to involve a suite of hsps, as well as constituent heat shock proteins (hsc) which are present during normal cell functioning. The latter may play an important role in those species or populations that experience extreme temperatures, such as those inhabiting deep sea and Antarctic marine environments, or those inhabiting the high intertidal (Berger and Young, 2006; Dong et al., 2008; Morris et al., 2013). Although heat shock protein production is ubiquitous, hsp responses are predicted to be energetically demanding, and only synthesized when absolutely necessary (Feder and Hofmann, 1999; Somero, 2002).

4.3.2 Protection against oxidative stress

Increases in seawater temperature towards high pejus temperatures and CT_{max} initiate oxidative stress in ectothermic species via temperature-related increases in metabolic rate leading to a rise in reactive oxygen species (ROS) production as a side product (Ahmad, 1995; Lesser, 2006; Lushchak, 2011). Oxygen limitation under thermal stress also contributes to oxidative stress (Pörtner, 2002; Box 4.1) and has been associated with significant rises in the production of ROS in ectothermic species with resulting structural, and therefore functional, damage to macromolecules. For example, thermal stress causes a significant rise in ROS (H_2O_2) and increased lipid peroxidation in marine bivalves such as *Scapharca broughtonii* (An and Choi, 2010), and leads to significant DNA strand breakage in the mussels *Mytilus galloprovincialis* and *Mytilus californianus* with the degree of damage dependent on time of exposure, temperature, and species (Yao and Somero,

---

Box 4.1 Oxidative stress

Oxidative stress arises due to the reduction of oxygen free radicals into water, resulting in the production of a number of reactive and highly toxic intermediate products termed reactive oxygen species (ROS) (Ahmad, 1995; Davies, 1995; Abele and Puntarulo, 2004). The production of ROS is directly and positively related to the concentration of oxygen within biological systems. All respiring cells produce ROS and the production and accumulation of ROS beyond the capacity of an organism to reduce these reactive species can lead to lipid, protein, carbohydrate, and DNA damage (Cadenas, 1995; Lesser, 2006). Marine organisms are highly susceptible to ROS, particularly hydrogen peroxide (H_2O_2) and the hydroxyl radical (HO•). The former can readily diffuse across cellular membranes, has an extended lifetime in seawater, and can cause direct damage to DNA and enzymes (Lesser, 2006). HO•, the most destructive of the ROS, attacks biological molecules in a diffusion-controlled manner and thus has huge potential for causing biological damage. The high reactivity of HO• also allows it to be relatively non-specific in its targets for oxidation (Davies, 2005; Sheehan and McDonagh, 2008; Winterbourn, 2008). For instance, it can set off free radical chain reactions, oxidize membrane lipids, and cause proteins and nucleic acids to denature (Ahmad, 1995; Lesser, 2006; Dhawan et al., 2009).

2012). Increased ROS damage may also occur during thermal stress due to increases in the growth and reproduction of marine pathogens and parasites, as opposed to a direct effect of temperature as observed in the scallop *Chlamys farreri* (Wang et al., 2012).

To combat ROS accumulation during thermal stress and to prevent subsequent cellular damage, marine invertebrates produce antioxidant enzymes, including superoxide dismutase (SOD), catalase, and glutathione peroxidase, to inactivate ROS by reduction to water (Ahmad, 1995; Lesser, 2006). Antioxidant production under thermal stress has been observed in a range of invertebrate taxa including cnidarians, bivalves, and crustaceans (Griffin and Bhagooli, 2004; Lesser, 2006; An and Choi, 2010; Matozzo et al., 2013), but as with other metabolic pathways, antioxidant activities vary largely over environmental gradients, across taxa and according to life-history stage. An example of the latter can be observed in the squid species *Loligo vulgaris*, where antioxidant capabilities of late embryos differ considerably from those of hatchlings reared under identical thermal stress conditions (Rosa et al., 2012). Thermal stress can also lead to higher antioxidant output but decreased antioxidant capabilities as observed in juvenile abalone, *Haliotis midae*, suggesting that the organism is overwhelmed by ROS when approaching thermal maxima (Vosloo et al., 2013). Additionally, antioxidant activities have been shown to vary across age-size classes and with sex (Mourente and Diaz-Salvago, 1999), be dependent on levels of aerobic metabolism (Lesser, 2006) and change in relation to feeding behaviour (Regoli et al., 1997). Antioxidant activity under thermal stress has also been correlated with disease resistance in tropical coral species with *Porites astreoides*, a disease-resistant species, showing increased antioxidant enzyme activities under thermal stress compared with its counterpart *Montastraea faveolata*, a disease-susceptible species (Palmer et al., 2011).

4.4 Variation in physiological responses to thermal stress across marine environments

In marine environments, various physical factors including temperature, salinity, current speed, and turbidity, vary enormously across the geographical range of a species. Consequently, adaptive variation tends to emerge between populations situated across environmental gradients, or between populations that have limited connectivity giving rise to differences in acclimatization potential between biogeographically distinct populations or closely related species (Osovitz and Hofmann, 2007; Kuo and Sanford, 2009; Whitehead, 2012). Consequently, regular stress events may potentially restructure ecosystems as the distribution patterns of species are influenced by population-based responses to stressors. This is particularly notable for temperature, which is well acknowledged as a physical driver for setting a species' abundance and distribution limits (Somero, 2002; Hofmann and Todgman, 2010). Furthermore, physiological acclimatization of populations to a particular set of environmental conditions is correlated with increasing genetic differentiation between those populations, which can occur over a range of spatial scales (from metres to hundreds of kilometres) (Sanford and Kelly, 2011). In marine invertebrates, intra- and interspecific variation in thermal tolerance limits, thermal tolerance ranges, and physiological responses have been well illustrated along spatial gradients including latitudinal clines and vertical zones in the intertidal.

4.4.1 Physiological variations with latitude

Some of the best-known naturally occurring thermal gradients in the marine environment occur along latitudinal clines. As latitude increases from the equator through temperate to boreal regions, there is, among other factors, a decrease in mean sea surface temperature (Somero, 2002). There is also an increase in the variation in temperature extremes in the shallow marine environment with the temperate regions being the most variable and seasonal fluctuations the most extreme (can be as great as 20 °C) (Stillman and Somero, 1996). Other factors, however, such as aspect and local currents give rise to 'hot spots' showing that localized temperatures on a finer scale are not directly related to latitude (Helmuth et al., 2006). In the case of high-latitude polar marine systems, seawater temperatures are constantly low and there is a general lack of variation with depth. Studies on physiological variation among populations of the same species living over a wide range of latitudinal temperatures provide researchers with a convenient system in which to explore comparative adaptation with respect to natural changes in temperature.

In general, upper thermal tolerances (CT_{max}) vary with latitude along with associated changes in habitat temperature. At high Antarctic latitudes, stenothermal polar marine invertebrates typically have CT_{max} values of 6 °C when warmed at a rate of 1 °C per month (Peck et al., 2009). At low latitudes, CT_{max} values of 34 species of stenothermal tropical marine ectotherms from seven phyla were close to the maximum habitat temperature (MHT) of 32 °C (Nguyen et al., 2011). A similar response was reported for a tropical species of the porcelain crab, which had a higher CT_{max} when compared with a cold-temperate species (Stillman and Somero, 1996; Stillman, 2003). In temperate regions, thermal tolerances are broader and can also vary with season with a reduction in thermal tolerance range and CT_{max} in the winter as observed in the polychaete *Arenicola marina* (Wittmann et al., 2008). In general, upper thermal limits also decrease with an increase in latitude between populations of the same species when comparisons are made between two populations (Spicer and Gaston, 1999), and among multiple populations (Logan et al., 2012). Overall it appears that polar and tropical marine invertebrate species are most vulnerable to warming when compared with species living at intermediate latitudes, due to a combination of living in less variable thermal habitats and having limited physiological capacities for acclimation, and in the case of tropical species, living close to thermal maxima (Stillman, 2003;

Somero, 2010; Peck et al., 2014). A recent study, however, suggests that the capacity of Arctic benthic marine invertebrates to acclimate to temperatures 3–5 °C above their normal summer temperatures are higher than those reported for Antarctic and tropical species indicating differences between the poles (Richard et al., 2012).

Physiological capacities for change can be assessed by determining a number of rate processes at different levels of biological organization. Of all of the physiological parameters studied to date, whole-organism metabolic rate and its relationship to temperature has received the most attention. The ability to compensate metabolic rates for changes in temperature indicates greater phenotypic plasticity and an ability to cope with further change. Metabolic compensation for changes in habitat temperature with latitude has been reported among populations of the same species (e.g. the polychaete *Arenicola marina* and the gammarid amphipod *Gammarus locusta*) occupying more eurythermal habitats as long as food is available and extreme temperatures are avoided (Sommer and Pörtner, 2002; Rastrick and Whiteley, 2011). The species that do not show metabolic compensation with latitude tend to be less eurythermal but also experience cold temperatures (boreal/cold temperate) such as the gastropod *Littorina saxatilis*, the bivalves *Mytilus* spp. and *Macoma balthica*, and the gammarid amphipod *Gammarus setosus* (Sokolova and Pörtner, 2003; Jansen et al., 2007; Rastrick and Whiteley, 2011). Stenothermal Antarctic marine invertebrates are characterized by low metabolic rates and are metabolically limited with respect to further environmental change which helps to explain their limited ability to survive thermal stress (Clarke, 1993; Pörtner et al., 2007; Rastrick and Whiteley, 2011).

Metabolic rates have also been associated with rates of protein synthesis as the latter may account for 11–42% of basal metabolic rate (Houlihan et al., 1995). Although the relationship between habitat temperature and protein synthesis rates is less well known than that for metabolic rate, determination of whole-organism fractional rates of protein synthesis in marine crustaceans from a range of thermal habitats (tropical, temperate, and polar) increases exponentially with acclimation temperature (Whiteley and Taylor, 2015). This relationship is only observed when rates of protein synthesis have been standardized for body mass and the measurements taken from animals fed *ad libitum*. When fractional rates of protein synthesis are examined in acclimatized animals at the summer temperatures they normally experience, fractional rates of protein synthesis remain unchanged with fall in habitat temperature across populations of the temperate gammarid species *Gammarus locusta*, but decrease in the polar population of the subarctic/boreal species *G. oceanicus* (Rastrick and Whiteley, 2013). Consequently both rates of metabolism and protein synthesis remain uncompensated in the extreme cold, with the lowest whole-organism rates of protein synthesis measured to date occurring in adult polar marine invertebrates at 0 ± 2 °C (Whiteley and Fraser, 2009). The relatively slow rates of protein synthesis in association with the increase in protein degradation rates observed at < 5 °C could further limit recovery from thermal stress in these cold-water invertebrates (Fraser et al., 2007).

Thermal adaption of the nervous system has mainly been demonstrated by a limited number of studies on marine crustaceans (Macdonald, 1981; Young et al., 2006). When the conduction velocities of the peripheral nerves of an Antarctic (*Glyptonotus antarcticus*) and a temperate (*Ligia oceanica*) isopod species were compared at the temperatures they normally experience, conduction velocities in the Antarctic species were faster than those in the temperate species suggesting full compensation for the effects of living in stable but cold temperatures (Young et al., 2006). Conduction velocities, however, had a reduced thermal dependency in both isopod species showing little change with increasing temperatures when compared with the temperate crab *Carcinus maenas*, demonstrating that factors other than thermal habitat modify nerve conduction velocities with temperature. In addition, the upper limit for nerve conduction of around 20 °C was the same regardless of thermal habitats illustrating that nerve conduction on its own cannot account for whole-organism thermal tolerance limits of 6–8 °C in Antarctic marine invertebrates (Pörtner et al., 2007). Instead, it is more likely that the processes involved in neuromuscular activation, which fail at 16–22 °C, have an influence (Macdonald, 1981). A different response, however, was observed in two congeneric species of porcelain crabs (genus *Petrolisthes*) from different geographic regions but acclimated to the same temperatures as the upper thermal limit of sensory nerve function and heart function were related to habitat temperature (Miller and Stillman, 2012). More specifically, CT_{max} for sensory nerve function was higher in the warm-adapted southern species *Petrolisthes gracilis*, which experiences a maximal microhabitat temperature (MHT) of 37–43 °C, compared with the temperate species *P. cinctipes* experiencing a MHT of 31 °C (Stillman, 2003; Miller and Stillman, 2012). Sensory nerve CT_{max} was more responsive to temperature acclimation than heart rate CT_{max}, although these observations require further investigation because experimental differences (rate of temperature change and measurements of performance) among studies may have influenced these observations (Miller and Stillman, 2012).

Mitochondria, the organelles responsible for energy metabolism and the determinants of aerobic capacity, are capable of changing thermal tolerance windows via alterations in their numbers and functional properties (Pörtner, 2002). Low temperatures are well known to have an effect on mitochondrial function, as a lowering of temperature increases proton leak across the inner mitochondrial membrane and reduces ATP production. Marine invertebrates show a range of responses to compensate for the loss of mitochondrial efficiency in their muscles in the cold: (1) abalone, *Haliotis* spp., show compensatory adjustments in mitochondrial enzyme activities (Dalhoff and Somero, 1993); (2) a subpolar population of the polychaete *Arenicola marina* shows an increase in mitochondrial volume density as well as an increase in mitochondrial enzyme activities after acclimation to 4 °C (Sommer and Pörtner, 2002); and (3) two Antarctic molluscs, *Nacella concinna* and *Laternula elliptica*, show no increases in mitochondrial volume densities but do show an increase in the surface density of the inner mitochondrial membrane (cristae) (Morley et al., 2009). Recent comparisons among Antarctic and temperate species of brachiopod (genus *Liothyrella*) demonstrate higher mitochondrial capacities (densities) in the polar species but lower cristae surface density (Lurman et al., 2010), and in the actively swimming pteropods (genus *Clione*), an increase in both mitochondrial and cristae densities in the Antarctic versus the temperate species (Dymowska et al., 2012). Clearly there are taxa-related differences but the overall effect is an up-regulation of aerobic capacity in the locomotory muscles in order to maintain swimming activities in the cold. Compensation for temperature change can also be observed in the oxygen uptake rates of isolated mitochondria from a wide range of marine invertebrates, because Arrhenius break temperatures (ABTs) for mitochondrial oxygen uptake rates increase among species with increase in habitat temperature, and increase within species during thermal acclimation (Somero, 2002). Such relationships were demonstrated in five species of eastern Pacific abalone where the relatively warm-adapted southern species had higher ABTs than the more cold-adapted, northern species (Dalhoff and Somero, 1993). Acclimation of mitochondrial respiration occurred in four of the five species but only over the temperature ranges normally experienced by each species/

population, demonstrating a relationship between compensatory capacity and biogeographical distribution. Unfortunately, very little is known about the relationship between warm acclimation temperatures and mitochondrial plasticity, but studies on two subspecies of killifish with different thermal niches found they had similar mitochondrial properties despite differences in thermal tolerances and whole-organism rates of oxygen uptake (Fangue et al., 2009). Mitochondrial plasticity may not be involved in improving aerobic capacities as temperatures increase, but further studies are required before this can be verified.

The relationship between thermal stress and hsp responses across latitudinal gradients is complex and involves heat-inducible and constitutive heat shock proteins. In marine snails (*Tegula* species), the subtropical southern species (*T. rugosa*) has higher hsp70 induction temperatures than the temperate northern species making it more able to cope with thermal stress (Tomanek and Somero, 1999). However, studies on heat shock responses in *Mytilus californianus* and *Nucella ostrina* distributed along the Pacific coast from Baja, California to Vancouver, Canada, failed to find a relationship between hsp70 levels and latitude (Sagarin and Somero, 2006). The southern species, *Nucella emarginata*, had the lowest hsc70 levels, but the highest level of hsps when compared with its northern, high intertidal counterpart, *Nucella lima* (Sorte and Hofmann, 2005). These variations in hsp response may be related to non-linear changes in environmental gradients known to exist along the Pacific coast as intertidal sites in northern Oregon are disproportionately exposed to daytime emersion (Helmuth et al., 2002). Moreover, hsp responses are lacking in some, but not all, Antarctic marine invertebrate species studied to date. For example, heat shock responses are not observed in the Antarctic sea star *Odontaster validus* or the gammarid amphipod *Paraceradocus gibber*, which normally experience temperatures of –1.86 to + 1.0 °C (Clark and Peck, 2008). This may be related to the increase in protein degradation rates in the cold at temperatures < 5 °C and the relatively slow rates of protein synthesis (Fraser et al., 2007). Not all Antarctic species, however, lack a heat shock response, because intertidal species such as the bivalve *Laternula elliptica* and the Antarctic limpet *Nacella concinna* live in an environment where thermal relationships are more complex and exposure to other stressors is likely (Clark and Peck, 2008). Heat shock responses are also compromised in species living at depth as the cold-seep mussel *Bathiomodiolus childressi* does not express an inducible form of hsp70 but does express high levels of constitutive hsp70

(Berger and Young, 2006). The latter may explain how this species is able to survive for 6 h after exposure to a 20 °C increase in temperature.

Polar ectothermic species are likely to be under increased threat of oxidative stress because of the low temperatures, high levels of dissolved oxygen, and seasonal changes in ice cover and light intensity (Regoli et al., 2012). The associated increase in oxygen solubility in the body fluids and lowered rates of oxygen uptake typical of polar ectotherms combine to increase oxygen concentrations in the tissues and enhance the risk of oxidative damage (Abele and Punturalo, 2004). The polar bivalve *Laternula elliptica*, for instance, shows substantially higher rates of lipid hydroperoxidate formation compared with its temperate counterpart *Mya arenaria* (Abele and Punturalo, 2004). Regular exposure to environmental conditions that are conducive to oxidative stress is likely to account for the higher antioxidant defences observed in some polar species compared with temperate or tropical counterparts. Enzymes of the polar bivalve species *Adamussium colbecki*, for example, were shown to be largely more active in comparison with Mediterranean bivalve species *Mytilus galloprovincialis* and *Pecten jacabaeus* (Regoli et al., 1997). While increases in antioxidant activities in polar species may be a consequence of other environmental conditions (e.g. low food availability, reduced levels of pollution, or parasite loads in polar areas), these observations illustrate the innate defences of polar species to such high oxygen conditions (Regoli et al., 1997; Abele and Punturalo, 2004; Camus et al., 2005). However, it should also be noted that this assumption has been proved to be inconsistent across all polar species and tissue types (Viarengo et al., 1991; Regoli et al., 1997).

As most tropical marine invertebrates are already living towards their upper thermal limits (Nguyen et al., 2011), the potential for ROS accumulation is high for such organisms and antioxidant defences are critically required to ensure a low but steady-state concentration of ROS (Lesser, 2012). Thermal stress can increase ROS production in tropical coral species, but the nature of oxidative stress response is complex due to their symbiosis with zooxanthellae (i.e. photosynthetic algae living within the coral surface) (Mydlarz and Jacobs, 2006). Under increased temperature, parallel increases in ROS occur in the chloroplasts of the zooxanthellae and these algal-generated ROS may diffuse into the cells of the coral. Ultimately, coral bleaching can be viewed as a final defence strategy for corals under oxidative stress as expulsion of the zooxanthellae reduces ROS levels for the host (Downs et al., 2002).

Both cnidarian hosts and zooxanthellae express antioxidants (Lesser, 2006).

4.4.2 Physiological variations with vertical zonation

Intertidal invertebrates are more tolerant of thermal extremes than almost any other animals because they are exposed by the tides and removed from the buffering effects of the seawater (Helmuth and Hofmann, 2001; Somero, 2002). During periods of emersion they experience higher air temperatures, increased solar radiation, and changes in humidity and wind speed. In general, the intensity of thermal and desiccation stress is dependent on the height up the shore, as well as the timing of low tides, climate, aspect, angle of substratum, and the occurrence of microclimates (reviewed by Helmuth et al., 2006). Marine invertebrates higher up on the shore will be emersed for longer and experience longer periods of thermal stress, especially if the species are sessile and unable to move into more favourable microclimates. During emersion, changes in body temperature can be rapid, for example, up to 3 °C per hour, and exceed air temperatures due to solar radiation despite some loss of heat by evaporative cooling (Helmuth et al., 2011). Marine invertebrates can also experience diurnal fluctuations in temperature in the intertidal even when submerged. This can occur when they become stranded in rock pools higher up on the shore, and isolated from the ebb and flow of the tides (Morris and Taylor, 1983).

Thermal tolerances are reported to be highly variable in the intertidal region and differ in relation to vertical zonation and thermal conditions (Somero, 2002). Mean upper critical temperatures in gastropods and porcelain crabs, for instance, are related to the maximum height of distribution (Southward, 1958; Newell, 1979; Stillman, 2003; Davenport and Davenport, 2005), but work on sessile intertidal invertebrates has revealed that thermal relationships on the shore are complex mainly because of the localized variations in physical factors mentioned above (Helmuth and Hofmann, 2001; Helmuth et al., 2002; Davenport and Davenport, 2005). Species living in the upper intertidal are thought to be more vulnerable to climate change because they are already living close to their physiological limits (Stillman, 2003; Somero, 2010). Environmental conditions are considered to be even more stressful on tropical rocky shores, where substrate temperatures can reach > 40 °C during emersion and intertidal invertebrates are also exposed to desiccation stress (Dong and Williams, 2011).

Differences in CT_{max} have been associated with ABTs for heart rate and nerve function in two sympatric species of porcelain crabs (genus *Petrolisthes*) occupying different vertical zones on the shore (Stillman and Somero, 1996, 2000; Miller and Stillman, 2012). The ABT for heart rate is higher in the upper shore species, *Petrolisthes cinctipes*, at 31.5 °C, than in the low intertidal species, *P. eriomerus*, at 26.6 °C. Moreover, the upper shore species can survive at temperatures not tolerated by the lower shore species, but is operating at temperatures close to its MHT. The mechanisms responsible for the differences in thermal sensitivity of heart rate could be related to the failure of the cardiac ganglia and/or to the heart muscle (Stillman, 2003). However, CT_{max} of sensory afferent nerves is higher in *P. cinctipes* compared with *P. eriomerus*, and has the greater ability to acclimate (Miller and Stillman, 2012). This contrasts with observations on CT_{max} of heart rate where acclimatory capacities are reduced in *P. cinctipes* (Stillman, 2003). One explanation for these observed differences is that *ex vivo* preparations of sensory afferents are more likely to show plasticity than complex integrated processes, such as heart rate, which are typically examined *in vivo* (Macdonald, 1981). It is also possible that nerves innervating the heart are able to tolerate temperature extremes but have less potential to acclimate to temperature change (Miller and Stillman, 2012). Higher ABTs for heart rate have also been reported in high intertidal versus mid-intertidal tropical limpets (genus *Cellana*, Dong and Williams, 2011), and in emersed versus immersed limpets (*Lottia digitalis*, Bjelde and Todgham, 2013).

ROS production and the ability to mount an antioxidant defence also vary with vertical zonation, as does the ability to show a heat shock response. Intertidal invertebrate species, largely adapted to regular exposure events, show reduced ROS production under thermal stress conditions compared with subtidal species, but can still produce antioxidant enzymes which tend to be more stable under comparable stress conditions (Pörtner, 2002; Lesser, 2006). Pre-exposure to low intensity oxidative stress appears to enhance tolerance to more intense stress events so that animals exposed to regular stress may be better able to cope with further change (Lushchak, 2011). Within populations, individuals living higher up on the shore and experiencing longer bouts of thermal stress have higher levels of hsp70 isoforms than individuals lower down on the shore (mussels, Roberts et al., 1997; amphipod gammarids, Bedulina et al., 2010). However, patterns of hsp induction can also vary, as observed in turban snails (genus *Tegula*), as species on the upper shore are characterized by higher threshold temperatures for heat shock protein induction than subtidal to low intertidal species (Tomanek and Somero, 1999). Moreover, on tropical rocky shoes, heat shock responses differ among species, with a high intertidal limpet species, *Cellana gata*, up-regulating one inducible hsp (hsp75) and the mid-intertidal species, *C. toreuma*, relying on two constituent hsps during thermal stress (Dong and Williams, 2011). Mussels (*Mytilus trossulus*) in the high intertidal also contain higher levels of heat damaged proteins indicating that even though heat shock responses become more pronounced up the shore and involve inducible hsps synthesized at higher temperatures, they are not fully effective (Hofmann and Somero, 1995).

4.5 Future threats: physiological responses to multiple stressors

The threat of increasing sea surface and inshore water temperatures due to climate change has greatly increased our interest in the physiological and biochemical mechanisms that limit thermal tolerances in marine ectotherms. To date, studies have indicated that global warming could alter rates of metabolism, protein synthesis, acid–base status, and immunological responses, as well as calcification rates and growth (Pörtner et al., 2004; Malagoli et al., 2007; Somero, 2010; Ellis et al., 2011). Warming can also affect cellular and molecular based responses including changes in cellular energetics, oxidative stress responses, and gene expression (Cherkasov et al., 2006; Lesser, 2006; Stillman and Tagmount, 2009; Erk et al., 2011). However, global warming of the oceans will not occur in isolation and it is becoming increasingly evident that the emergence and increasing intensity of various additional climate-induced stressors in marine environments, including changes in pH (i.e. ocean acidification), oxygen content (i.e. hypoxia), and salinity, will interact with the effects of warming to further influence physiological responses and their effect on thermal tolerances (Hofmann and Todgman, 2010; Pörtner, 2010; Gazeau et al., 2013). While it is thought that multiple abiotic or biotic (e.g. food availability) stressors may narrow the thermal niche and lower the maximal performance of a given species (Pörtner, 2010), the cumulative effect of multiple stressors remains largely unknown (Crain et al., 2008). Nevertheless, recent investigations into the effect of climate multi-stressors highlight some potential impacts and demonstrate the complexity of responses.

4.5.1 Physiological responses to warming and ocean acidification

Several investigations have demonstrated that concurring ocean warming and acidification may have a synergistic effect on marine invertebrates, i.e. physiological compensation in response to one stressor may increase sensitivity to another (Hofmann and Todgham, 2010). For example, an increase in seawater pCO_2 (hypercapnia) caused enhanced sensitivity to thermal stress in the crabs *Cancer pagurus* and *Hyas araneus* (Metzger et al., 2007; Walther et al., 2010) and reduced the ability of echinoderm larvae to respond to 1 h of acute temperature stress in high CO_2 conditions (O'Donnell et al., 2009). Both studies suggest that increased CO_2 conditions narrow the thermal tolerances of marine invertebrates even at early life stages. The effects of ocean acidification and warming on metabolic rate vary among species. In highly active pelagic species, such as the jumbo squid *Dosidicus gigas*, elevated pCO_2 levels reduce metabolic rates and activity levels especially at high temperature (Rosa and Seibel, 2008), but combined exposure to thermal stress and ocean acidification reduces metabolic rates in marine bivalves, although there is no subsequent effect on oxidative damage (Mattoo et al., 2013). Invertebrate species reared under acidified conditions, including a cnidarian (*Acropora sp.*) and echinoid (*Strongylocentrotus franciscanus*), show an increased sensitivity to thermal events demonstrated by reduced ability to up-regulate heat shock proteins (Reynaud et al., 2003; O'Donnell et al., 2009).

Co-stressors of temperature and acidification do not necessarily lead to synergistic impacts. One stressor may be the sole driver of observed effects, and in some instances concurring conditions may even lead to an antagonistic effect (Todgham and Stillman, 2013). For example, in the sea star *Pisaster ochraceus* a doubling of pCO_2 concentrations (780 ppm) increased growth rates in both the presence and absence of increased temperatures (Gooding et al., 2009). In the blue mussel, *M. edulis*, however, temperature was the predominant driver with regard to effects on shell strength with a decrease in strength resulting from elevated temperature rather than elevated pCO_2 (Mackenzie et al., 2014). These differences, however, may be explained by changes in condition brought about by holding the mussels in the laboratory with limited food supply. Such conditions are known to negatively affect the ability of invertebrates to compensate for ocean acidification as demonstrated by mussels studied in the wild (Thomsen et al., 2013) and copepods reared

in the laboratory (Pedersen et al., 2014). In calcifying marine invertebrates, concurrent ocean acidification reduces calcite availability and is thereby not conducive to increased growth rates even though warming conditions tend to promote growth via increased metabolic rates (Sheppard-Brennand et al., 2010). Collectively these studies demonstrate the highly variable nature of the physiological responses of marine invertebrates to the combined effects of warming and ocean acidification, as interactions tend to be species-specific, variable across life stages, and dependent on holding conditions.

4.5.2 Physiological responses to warming and hypoxia

A serious consequence of global warming is the deoxygenation of the world's oceans due to the reduced solubility of oxygen in warmer water and corresponding increases in upper ocean stratification (Keeling and Garcia, 2002; Keeling et al., 2010). There is evidence that oxygen concentrations in the thermoclines of most ocean basins have decreased since the 1970s (Bindoff et al., 2007) and climate change projections to the end of the century predict a four to seven percent decline in dissolved oxygen in the oceans (Matear and Hirst, 2003). Likewise, modelling studies have predicted that oxygen content will decrease with ocean warming as a result of changes in ocean circulation and biology (Keeling et al., 2010). Ocean warming is also expected to increase the frequency, duration, intensity, and extent of hypoxia/anoxia in certain zones (Keeling et al., 2010; Gruber, 2011), pushing coastal ecosystems towards tipping points (Conley et al., 2009; Rabalais et al., 2010).

Benthic marine invertebrates are particularly susceptible to hypoxia because they live farthest from contact with the atmospheric oxygen supply and because coastal sediments tend to be depleted in oxygen relative to the overlying water column (Vaquer-Sunyer and Duarte, 2008). In marine invertebrates, physiological responses to hypoxia include adjustments to the cardiorespiratory system and changes in the functional properties of the respiratory pigment in order to improve oxygen transport to the tissues, and, during severe hypoxia, a shift to anaerobic metabolism (Pörtner, 2010; Whiteley and Taylor, 2015). However, responses to hypoxia are affected by elevated temperature because of an associated increase in oxygen demand, higher ventilation rates and cardiac activity, and a temperature-related decrease in dissolved oxygen and oxygen affinity of the respiratory pigment

(Metzger et al., 2007; Seibel, 2013; Whiteley and Taylor, 2015). For those species that are able to regulate rates of oxygen uptake down to a critical value of ambient oxygen partial pressure (P_{crit}), temperature increases P_{crit} and therefore affects the oxygen levels at which a species is likely to survive (Keeling et al., 2010; Whiteley and Taylor, 2015). For example, under warming conditions, median lethal thresholds were shown to increase by an average of 16% in marine benthic organisms and survival times during hypoxic events decreased by as much 74% (Vaquer-Sunyer and Duarte, 2011). Invertebrates can also depress metabolic rates under conditions of hypoxia in order to prolong survival when energy supply is limited (Seibel and Walsh, 2003).

4.5.3 Physiological responses to warming and salinity stress

Ocean salinity is an important indicator of change in the climate system (Bindoff and Hobbs, 2013). There is growing evidence for climate-induced changes in ocean salinity and an overall acceleration of the global hydrological cycle. Observed increases in precipitation at high latitudes result in a freshening effect (i.e. decreased salinity) while decreases in subtropical areas are having a concentrating effect (Helm et al., 2010). Concurring ocean warming and changes in salinity may lead to a range of potential physiological effects on marine invertebrates which are mainly iso-osmotic with full strength seawater (30–35 ppt). For instance, comparison of estuarine and coastal bivalve populations suggests that decreased salinity and increased temperature may lead to increases in cellular energy expenditure to maintain osmoregulation and cell volume control, as well as the temperature-related increase in metabolic rates. Decreased energy stores may further affect the ecologically relevant processes of growth, reproduction, or immune defence (Erk et al., 2011). Additionally, increases in temperature and decreases in salinity have been observed to cause increased mortality of bivalve haemocyte cells, a key component of the bivalve immune system (Fisher et al., 1989; Gagnaire et al., 2006) and therefore such conditions may have repercussions for the ability of marine ectotherms to mount an immune response. Thermal and osmotic stress also caused increased oxidative stress in the bivalve species *Scapharca broughtonii* with increased water temperature (30 °C) and decreased salinity (25ppt) causing a significant rise in ROS production (An and Choi, 2010).

4.6 Conclusions

In general, stenothermal high-latitude and low-latitude marine invertebrate species are characterized by limited physiological capacities to compensate for thermal stress compared with mid-latitude species. Responses vary between physiological traits as invertebrates living in cold polar habitats can compensate for nerve function but not for rates of oxygen uptake and protein synthesis. At the other thermal extreme of the tropics, there is a reduced capacity for the CT_{max} for heart rate to acclimate in those species living close to their thermal optima, but a greater capacity for nerve function to acclimate. Damage repair and protective mechanisms also vary among species with geographical location and position on the shore, with exposure to highly variable and extreme temperatures increasing the ability to mount a heat shock response and to produce anti-oxidative enzymes to protect against temperature-related increases in ROS. Hsp responses among species and populations over spatial scales, however, are complex and constitutive heat shock proteins appear to be more important than previously appreciated, especially in those species that normally inhabit stable environments. Even though an up-regulation of aerobic capacity is possible in polar marine invertebrates, the contribution of any changes in mitochondrial properties to aerobic scope in tropical species is unknown, but some species increase their thermal tolerances by undergoing metabolic depression, and some can even survive heat coma (Wong et al., 2014). Those species with reduced physiological capacities to adjust to change are more likely to be under threat from thermal stress (Hofmann and Todgman, 2010; Pörtner, 2010). These are also the species that tend to be exposed to greater oxidative stress, although they appear to have the mechanisms required to deal with these problems.

Thermal stress may intensify the effects of seawater acidification, hypoxia, and salinity changes via altering the organism's sensitivity (i.e. tolerance and performance) and depleting (via increased metabolism and higher maintenance costs) important energy stores that the animal might draw upon to deal with a particular stressor (Hofmann and Todgman, 2010). However, from the limited number of studies carried out to date, physiological responses to a thermal stress plus another environmental factor are highly variable. They tend to be species or habitat specific and vary in response to life stage, food availability, and also immune and reproductive status (Guderley and Pörtner, 2010; Todgham and Stillman, 2013). In their review of anthropogenic stressors in marine environments, Crain et al. (2008) report that, to date, responses from multi-stressor exposures have been additive (26%) and synergistic (36%), but also largely antagonistic (38%), highlighting the variability of responses across species.

In this chapter, we have concentrated on the main physiological responses known to vary in marine invertebrates with an elevation in environmental temperature. Physiological responses, however, occur across different levels of biological organization and involve both biochemical and molecular adjustments which may have different sensitivities to thermal stress (Schulte, 2014). This highlights the complexity of the physiological responses on exposure to the warming conditions expected as a result of climate change. Future challenges involve the integration of molecular, cellular, and whole-animal studies, as well as investigations across life stages and generations, and a better appreciation of the spatial scales over which physiological divergence can occur. By increasing our understanding of the variability in physiological responses across marine invertebrate species, we can better predict the possible restrictions for acclimatization and identify evolutionary adaptation of specific traits under conditions of thermal stress and in response to multiple stressors (Pörtner, 2010; Mattoo et al., 2013; Schulte, 2014). Ultimately this information will allow us to use physiological knowledge to help assess the degree of vulnerability for a particular species or population in response to climate change (Hofmann and Todgman, 2010).

References

Abele, D. & Puntarulo, S. (2004). Formation of reactive species and induction of antioxidant defence systems in polar and temperate marine invertebrates and fish. *Comparative Biochemistry and Physiology Part A*, **138**(4), 405–415.

Ahmad, S. (1995). Oxidative stress from environmental pollutants. *Archives of Insect Biochemistry and Physiology*, **29**(2), 135–157.

An, M.I. & Choi, C.Y. (2010). Activity of antioxidant enzymes and physiological responses in ark shell, *Scapharca broughtonii*, exposed to thermal and osmotic stress: effects on hemolymph and biochemical parameters. *Comparative Biochemistry and Physiology Part B*, **155**(1), 34–42.

Anestis, A.., Pörtner, H.-O., Karagiannis, D., et al. (2010). Responses of *Mytilus galloprovincialis* (L.) to increasing seawater temperature and to marteliosis: metabolic and physiological parameters. *Comparative Biochemistry and Physiology Part A*, **156**, 57–66.

Anestis, A., Pörtner, H.O., Lazou, A., & Michaelidis, B. (2008). Metabolic and molecular stress responses of sublittoral bearded horse mussel *Modiolus barbaratus* to warming seawater. Implications for vertical zonation. *Journal of Experimental Biology*, **211**, 2889–2898.

Bedulina, D.S., Zimmer, M., & Timofeyev, M.A. (2010). Sublittoral and supra-littoral amphipods respond differently to acute thermal stress. *Comparative Biochemistry and Physiology Part B*, **155**, 413–418.

Berger, M.S. & Young, C.M. (2006). Physiological response of the cold-seep mussel *Bathymodiolus childressi* to acutely elevated temperature. *Marine Biology*, **149**, 1397–1402.

Bindoff, N.L. & Hobbs, W.R. (2013). Deep ocean freshening. *Nature Climate Change*, **3**, 864–865.

Bindoff, N.L., Willebrand, J., Artale, V., et al. (2007). Observations: oceanic climate change and sea level. In: Solomon, S., Qin, D., Manning, M., et al. (eds), *Climate Change 2007: The Physical Science Basis. Contribution of Working Group I to the Fourth Assessment Report of the Intergovernmental Panel on Climate Change*. Cambridge University Press, Cambridge, UK and New York, NY.

Bjelde, B.E. & Todgham, A.E. (2013). Thermal physiology of the fingered limpet *Lottia digitalis* under emersion and immersion. *Journal of Experimental Biology*, **216**, 2858–2869.

Cadenas, E. (1995), Mechanisms of oxygen activation and reactive oxygen species detoxification. In: Ahmad, S. (ed.), *Oxidative Stress and Antioxidant Defenses in Biology*, pp. 1–61. Springer, New York.

Camus, L., Gulliksen, B., Depledge, M.H., & Jones, M.B. (2005). Polar bivalves are characterized by high antioxidant defences. *Polar Research*, **24**, 111–118.

Cherkasov, A.S., Biswas, P.K., Ridings, D.M., et al. (2006). Effects of acclimation temperature and cadmium exposure on cellular energy budgets in the marine mollusk *Crassostrea virginica*: linking cellular and mitochondrial responses. *Journal of Experimental Biology*, **209**(Pt 7), 1274–1284.

Clark, M.S. & Peck, L.S. (2008). HSP70 heat shock proteins and environmental stress in Antarctic marine organisms: a mini-review. *Marine Genomics*, **2**, 11–18.

Clarke, A. (1993). Seasonal acclimatization and latitudinal compensation in metabolism—do they exist? *Functional Ecology*, **7**,139–149.

Conley, D.J., Carstensen, J., Vaquer-Sunyer, R., & Duarte, C.M. (2009). Ecosystem thresholds with hypoxia. *Hydrobiologia*, **629**, 21–29.

Crain, C.M., Kroeker, K., & Halpern, B.S. (2008). Interactive and cumulative effects of multiple human stressors on marine systems. *Ecology Letters*, **11**, 1304–1315.

Dalhoff, E. & Somero, G.N. (1993). Effects of temperature on mitochondria from abalone (Genus *Haliotis*)—adaptive plasticity and its limits. *Journal of Experimental Biology*, **185**, 151–168.

Davenport, J. & Davenport, J.L. (2005). Effects of shore height, wave exposure and geographical distance on thermal niche width of intertidal fauna. *Marine Ecology Progress Series*, **292**, 41–50.

Davies, K.J. (1995). Oxidative stress: the paradox of aerobic life. In: Rice-Evans, C. & Halliwell, B. (eds), *Free Radicals and Oxidative Stress: Environment, Drugs and Food Additives*

(Biochemical Society Symposia), Vol. 61, pp. 1–32. Portland Press, London.

Davies, M.J. (2005). The oxidative environment and protein damage. *Biochimica et Biophysica Acta*, **1703**(2), 93–109.

Dhawan, A., Bajpayee, M., & Parmar, D. (2009). Comet assay: a reliable tool for the assessment of DNA damage in different models. *Cell Biology and Toxicology*, **25**(1), 5–32.

Dong, Y.-W., Miller, L.P., Sanders, J.G., & Somero, G.N. (2008). Heat shock protein 70 (Hsp70) expression in four limpets of the Genus *Lottia*: interspecific variation in constitutive and inducible synthesis correlates with an *in situ* exposure to heat stress. *Biological Bulletin*, **215**, 173–181.

Dong, Y.-W. & Williams, G.A. (2011). Variations in cardiac performance and heat shock protein expression to thermal stress in two differently zoned limpets on a tropical rocky shore. *Marine Biology*, **158**, 1223–1231.

Dong, Y.-W., Yu, S.-S., Wang, Q.-L., & Dong, S.-L. (2011). Physiological responses in a variable environment: relationships between metabolism, hsp and thermotolerance in an intertidal-subtidal species. *PLoS ONE*, **6**, e26446.

Downs, C., Fauth, J.E., Halas, J.C., et al. (2002). Oxidative stress and seasonal coral bleaching. *Free Radical Biology and Medicine*, **33**(4), 533–543.

Dymowska, A.K., Manfredi, T., Rosenthal, J.J.C., & Seibel, B.A. (2012). Temperature compensation of aerobic capacity and performance in the Antarctic pteropod, *Clione antarctica*, compared with its northern congener, *C. limaci*. *Journal of Experimental Biology*, **215**, 3370–3378.

Ellis, R.P., Parry, H., Spicer, J.I., et al. (2011). Immunological function in marine invertebrates: responses to environmental perturbation. *Fish and Shellfish Immunology*, **30**(6), 1209–1222.

Erk, M., Ivanković, D., & Strižak, Ž. (2011). Cellular energy allocation in mussels (*Mytilus galloprovincialis*) from the stratified estuary as a physiological biomarker. *Marine Pollution Bulletin*, **62**(5), 1124–1129.

Fangue, N.A, Podrabsky, J.E., Crawshaw, L.I., & Schulte, P.M. (2009). Countergradient variation in gradient in temperature preference in populations of killifish *Fundulus heteroclitus*. *Physiological and Biochemical Zoology*, **82**, 776–786.

Feder, M.E. & Hofmann, G.E. (1999). Heat-shock proteins, molecular chaperones, and the stress response: evolutionary and ecological physiology. *Annual Review of Physiology*, **61**, 243–282.

Fisher, W.S., Chidtala, M.M., & Mark, A. (1989). Annual variation of estuarine and oceanic oyster *Crassostrea virginica* Gmelin hemocyte capacity. *Journal of Experimental Marine Biology and Ecology*, **127**(2), 105–120.

Fraser, K.P.P., Clarke, A., & Peck, L.S. (2007). Growth in the slow lane: protein metabolism in the Antarctic limpet *Nacella concinna* (Strebel, 1908). *Journal of Experimental Biology*, **210**, 2691–2699.

Frederich, M. & Pörtner, H.-O. (2000). Oxygen limitation of thermal tolerance defines by cardiac and ventilator performance in the spider crab *Maja squinado*. *American Journal of Physiology*, **279**, R1531–R1538.

Gagnaire, B., Frouin, H., Moreau, K., et al. (2006). Effects of temperature and salinity on haemocyte activities of the Pacific oyster, *Crassostrea gigas* (Thunberg). *Fish and Shellfish Immunology*, **20**(4), 536–547.

Gazeau, F., Parker, L.M., Comeau, S., et al. (2013). Impacts of ocean acidification on marine shelled molluscs. *Marine Biology*, **160**(8), 2207–2245.

Giomi, F. & Pörtner, H.-O. (2013). A role for haemolymph carrying capacity in heat tolerance of eurythermal crabs. *Frontiers in Physiology*, **4**, 100.

Gooding, R.A, Harley, C.D.G., & Tang E. (2009). Elevated water temperature and carbon dioxide concentration increase the growth of a keystone echinoderm. *Proceedings of the National Academy of Sciences of the USA*, **106**(23), 9316–9321.

Griffin, S.P. & Bhagooli, R. (2004). Measuring antioxidant potential in corals using the FRAP assay. *Journal of Experimental Marine Biology and Ecology*, **302**(2), 201–211.

Gruber, N. (2011). Warming up, turning sour, losing breath: ocean biogeochemistry under global change. *Philosophical Transactions of the Royal Society Series A*, **369**(1943), 1980–1996.

Guderley, H. & Pörtner, H.-O. (2010). Metabolic power budgeting and adaptive strategies in zoology: examples from scallops and fish. *Canadian Journal of Zoology*, **88**, 753–763.

Helm, K.P., Bindoff, N.L., & Church, J.A. (2010). Changes in the global hydrological-cycle inferred from ocean salinity. *Geophysical Research Letters*, **37**(18), L18701.

Helmuth, B.S.T., Harley, C.D.G., Halpin, P.M., et al. (2002). Climate change and latitudinal patterns of intertidal thermal stress. *Science*, **298**, 1015–1017.

Helmuth, B.S.T. & Hofmann, G.E. (2001). Microhabitats, thermal heterogeneity, and patterns of physiological stress in the intertidal zone. *Biological Bulletin*, **201**, 374–384.

Helmuth, B.S.T., Mieszkowska, N., Moore, P., & Hawkins, S.J. (2006). Living on the edge of two changing worlds: forecasting the responses of rocky intertidal ecosystems to climate change. *Annual Review of Ecology Evolution and Systematics*, **37**, 373–404.

Helmuth, B.S.T., Yamane, L., Lalwani, S., et al. (2011).Hidden signals of climate change in intertidal ecosystems: what (not) to expect when you are expecting. *Journal of Experimental Marine Biology and Ecology*, **400**, 191–199.

Hochachka, P.W. & Somero, G.N. (2002). *Biochemical Adaptation: Mechanisms and Processes in Physiological Evolution*. Oxford University Press, New York.

Hofmann, G.E. & Somero, G.N. (1995). Evidence for protein damage at environmental temperatures—seasonal changes in levels of ubiquitin conjugates and hsp70 in the intertidal mussel *Mytilus trossulus*. *Journal of Experimental Biology*, **198**, 1509–1518.

Hofmann, G.E. & Todgham, A.E. (2010). Living in the now: physiological mechanisms to tolerate a rapidly changing environment. *Annual Review of Physiology*, **72**, 127–145.

Houlihan, D.F., Carter, C.G., & McCarthy, I.D. (1995). Protein turnover in animals. In: Walsh, P. & Wright, P. (eds),

Nitrogen Metabolism and Excretion, pp. 1–32. CRC Press, Boca Raton, Florida.

Jansen, J.M., Pronker, A.E., Kube, S., et al. (2007). Geographic and seasonal patterns and limits on the adaptive response to temperature of European *Mytilus* spp. and *Macoma balthica* populations. *Oecologia*, **154**, 23–34.

Keeling, R.F. & Garcia, H.E. (2002). The change in oceanic O_2 inventory associated with recent global warming. *Proceedings of the National Academy of Sciences of the USA*, **99**, 7848–7853.

Keeling, R.F., Körtzinger, A., & Gruber, N. (2010). Ocean deoxygenation in a warming world. *Annual Review of Marine Science*, **2**(1), 199–229.

Kroeker, K.J., Kordas, R.L., Crim, R., et al. (2013). Impacts of ocean acidification on marine organisms: quantifying sensitivities and interaction with warming. *Global Change Biology*, **19**, 1884–1896.

Kültz, D. (2005). Molecular and evolutionary basis of the cellular stress response. *Annual Review of Physiology*, **67**(1), 225–257.

Kuo, E.S. & Sanford, E. (2009). Geographic variation in the upper thermal limits of an intertidal snail: implications for climate envelope models. *Marine Ecology Progress Series*, **388**, 137–146.

Lesser, M.P. (2006). Oxidative stress in marine environments: biochemistry and physiological ecology. *Annual Review of Physiology*, **68**(3), 253–278.

Lesser, M.P. (2012). Oxidative stress in tropical marine ecosystems. In: Abele, D., Vazquez-Medina, J.P., & Zenteno-Savin, T. (eds), *Oxidative Stress in Aquatic Ecosystems*, pp. 9–19. John Wiley & Sons Ltd, Chichester, UK.

Logan, C.A., Kost, L.E., & Somero, G.N. (2012). Latitudinal differences in *Mytilus californianus* thermal physiology. *Marine Ecology Progress Series*, **450**, 93–105.

Lurman, G.L., Blaser, T., Lamare, M., et al. (2010). Mitochondrial plasticity in brachiopod (*Liothyrella* spp.) smooth adductor muscle as a result of season and latitude. *Marine Biology*, 157, 907–913.

Lushchak, V.I. (2011). Environmentally induced oxidative stress in aquatic animals. *Aquatic Toxicology*, **101**(1), 13–30.

Macdonald, J.A. (1981). Temperature compensation in the peripheral nervous system: Antarctic vs temperate poikilotherms. *Journal of Comparative Physiology*, **142**, 411–418.

Mackenzie, C.L., Ormondroyd, G.A, Curling, S.F., et al. (2014). Ocean warming, more than acidification, reduces shell strength in a commercial shellfish species during food limitation. *PLoS ONE*, **9**(1), e86764.

Madeira, D., Narciso, L., Cabral, H.N., et al. (2014). Role of thermal niche in the cellular response to thermal stress: lipid peroxidation and HSP70 expression in coastal crabs. *Ecological Indicators*, **36**, 601–606.

Madeira, D., Narciso, L., Cabral, H.N., et al. (2012). HSP70 production patterns in coastal and estuarine organisms facing increasing temperatures. *Journal of Sea Research*, 73, 137–147.

Malagoli, D., Casarini, L., Sacchi, S., & Ottaviani, E. (2007). Stress and immune response in the mussel *Mytilus galloprovincialis*. *Fish and Shellfish Immunology*, **23**(1), 171–177.

Marshall, D.J., Allen, R.M., & Crean, A.J. (2008). The ecological and evolutionary importance of maternal effects in the sea. In: Gibson, R.N., Atkinson, R.J.A., & Gordon, J.D.M. (eds), *Oceanography and Marine Biology: An Annual Review*, Vol. 46, pp. 203–250. Taylor & Francis, Abingdon.

Marshall, D.J., Dong, Y.-W., McQuaid, C.D., & Williams, G.A. (2011). Thermal adaptation in the intertidal snail *Echinolittorina malaccana* contradicts current theory by revealing the crucial roles of resting metabolism. *Journal of Experimental Biology*, **214**, 3649–3657.

Matear, R.J. & Hirst, A.C. (2003). Long-term changes in dissolved oxygen concentrations in the ocean caused by protracted global warming. *Global Biogeochemical Cycles*, **17**(4), 1125.

Matozzo, V., Chinellato, A., Munari, M., et al. (2013). Can the combination of decreased pH and increased temperature values induce oxidative stress in the clam *Chamelea gallina* and the mussel *Mytilus galloprovincialis*? *Marine Pollution Bulletin*, **72**(1), 34–40.

Mattoo, O.B., Ivanina, A.V, Ullstad, C., et al. (2013). Interactive effects of elevated temperature and CO_2 levels on metabolism and oxidative stress in two common marine bivalves (*Crassostrea virginica* and *Mercenaria mercenaria*). *Comparative Biochemistry and Physiology Part A*, **164**(4), 545–553.

Melzner, F., Bock, C., & Portner, H.-O. (2006). Critical temperature on the cephalopod *Sepia officinalis* investigated using in vivo $^{31}PNMR$ spectroscopy. *Journal of Experimental Biology*, **209**, 891–906.

Metzger, R., Sartoris, F., Langenbuch, M., & Pörtner, H.-O. (2007). Influence of elevated CO_2 concentrations on thermal tolerance of the edible crab *Cancer pagurus*. *Journal of Thermal Biology*, **32**(3), 144–151.

Miller, N.A. & Stillman, J.H. (2012). Neural thermal performance in Porcelain crabs, Genus *Petrolisthes*. *Physiological and Biochemical Zoology*, **85**, 29–39.

Morley, S.A., Lurman, G.R., Skepper, J.M., et al. (2009). Thermal plasticity of mitochondria: a latitudinal comparison between Southern Ocean molluscs. *Comparative Biochemistry and Physiology Part A*, **152**, 423–430.

Morris, S. & Taylor, A.C. (1983). Diurnal and seasonal-variation in physicochemical conditions within intertidal rock pools. *Estuarine, Coastal and Shelf Science*, **17**, 339–355

Morris, J.P., Thatje, S., & Hauton, C. (2013). The use of stress-70 proteins in physiology: a re-appraisal. *Molecular Ecology*, **22**(6), 1494–1502.

Mourente, G. & Díaz-Salvago, E. (1999). Characterization of antioxidant systems, oxidation status and lipids in brain of wild-caught size-class distributed *Aristeus antennatus* (Risso, 1816) Crustacea, Decapoda. *Comparative Biochemistry and Physiology Part B*, **124**(4), 405–416.

Moyes, C.D. & Schulte, P.M. (2008). *Principles of Animal Physiology*. Benjamin Cummings, San Francisco.

Mydlarz, L.D. & Jacobs, R.S. (2006). An inducible release of reactive oxygen radicals in four species of gorgonian corals. *Marine and Freshwater Behaviour and Physiology*, **39**(2), 143–152.

Newell, R.C. (1979). *Biology of Intertidal Animals*. Marine Ecological Surveys Ltd, Faversham, UK.

Nguyen, K.D.T., Morley, S.A., Lai, C-H., et al. (2011). Upper temperature limits of tropical marine ectotherms: global warming implications. *PLoS ONE*, **6**(12), e29340.

O'Donnell, M.J., Hammond, L.M., & Hofmann, G.E. (2009). Predicted impact of ocean acidification on a marine invertebrate: elevated CO_2 alters response to thermal stress in sea urchin larvae. *Marine Biology*, **156**(3), 439–446.

Osovitz, C.J. & Hofmann, G.E. (2007). Marine macrophysiology: studying physiological variation across large spatial scales in marine systems. *Comparative Biochemistry and Physiology Part A*, **147**(4), 821–827.

Palmer, C.V, McGinty, E.S., Cummings, D.J., et al. (2011). Patterns of coral ecological immunology: variation in the responses of Caribbean corals to elevated temperature and a pathogen elicitor. *Journal of Experimental Biology*, **214**(Pt 24), 4240–4249.

Parry, G.D. (1978). Effects of growth and temperature acclimation on metabolic rate in limpet, *Cellana tamosercia* (Gastropoda Patellidae). *Journal of Animal Ecology*, **47**, 351–368.

Peck, L.S., Clark, M.S., Morley, S. A., et al. (2009). Animal temperature limits and ecological relevance: effects of size, activity and rates of change. *Functional Ecology*, **23**, 248–256.

Peck, L.S., Morley, S.A., Richard, J., & Clark, M.S. (2014). Acclimation and thermal tolerance in Antarctic marine ectotherms. *Journal of Experimental Biology*, **217**, 16–22.

Pedersen, S.A., Hakedal, O.J., Seleberria, L., et al. (2014). Multigenerational exposure to ocean acidification during food limitation reveals consequences for copepod scope for growth and vital rates. *Environmental Science and Technology*, **48**, 12275–12284.

Pörtner, H.-O. (2002). Climate variations and the physiological basis of temperature dependent biogeography: systemic to molecular hierarchy of thermal tolerance in animals. *Comparative Biochemistry and Physiology Part A*, **132**(4), 739–761.

Pörtner, H.-O. (2010). Oxygen- and capacity-limitation of thermal tolerance: a matrix for integrating climate-related stressor effects in marine ecosystems. *Journal of Experimental Biology*, **213**(6), 881–893.

Pörtner, H.-O. (2012). Integrating climate-related stressor effects on marine organisms: unifying principles linking molecule to ecosystem-level changes. *Marine Ecology Progress Series*, **470**, 273–290.

Pörtner, H.-O., Langenbuch, M., & Reipschläger, A. (2004). Biological impact of elevated ocean CO_2 concentrations: lessons from animal physiology and earth history. *Journal of Oceanography*, **60**(4), 705–718.

Pörtner, H.-O., Peck, L.S., & Somero, G.N. (2007). Thermal limits and adaptation in marine Antarctic ectotherms: an integrative view. *Philosophical Transactions of the Royal Society Series B*, **362**, 2233–2258.

Rabalais, N.N., Díaz, R.J., Levin, L.A., et al. (2010). Dynamics and distribution of natural and human-caused hypoxia. *Biogeosciences*, **7**, 585–619.

Rastrick, S.P.S. & Whiteley, N.M. (2011). Congeneric amphipods show differing abilities to maintain metabolic rates with latitude. *Physiological and Biochemical Zoology*, **84**, 154–165.

Rastrick, S.P.S. & Whiteley, N.M. (2013). Influence of natural thermal gradients on whole animal rates of protein synthesis in marine gammarid amphipods. *PLoS ONE*, **8**(3) e60050.

Regoli, F., Benedetti, M., Krell, A., & Abele, D. (2012). Oxidative challenges in polar seas. In: Abele, D., Vazquez-Medina, J.P., & Zenteno-Savin, T. (eds), *Oxidative Stress in Aquatic Ecosystems*, pp. 20–40. John Wiley & Sons Ltd, Chichester, UK.

Regoli, F., Principato, G., Bertoli, E., et al. (1997). Biochemical characterization of the antioxidant system in the scallop *Adamussium colbecki*, a sentinel organism for monitoring the Antarctic environment. *Polar Biology*, **17**, 251–258.

Reynaud, S., Leclercq, N., Romaine-Lioud, S., et al. (2003). Interacting effects of CO_2 partial pressure and temperature on photosynthesis and calcification in a scleractinian coral. *Global Change Biology*, **9**(11), 1660–1668.

Richard, J., Morley, S.A., Deloffre, J., & Peck, L.S. (2012). Thermal acclimation capacity for four Arctic marine benthic species. *Journal of Experimental Marine Biology and Ecology*, **424**, 38–43.

Roberts, D.A., Hofmann, G.E., & Somero, G.N. (1997). Heat-shock protein expression in *Mytilus californianus*: acclimatization (seasonal and tidal-height comparisons) and acclimation effects. *Biological Bulletin*, **192**, 309–320.

Rock, J., Magnay, J.L., Beech, S., et al. (2009). Linking functional molecular variation with environmental gradients: myosin gene diversity in a crustacean broadly distributed across variable thermal environments. *Gene*, **437**, 60–70.

Rosa, R., Pimentel, M.S., Boavida-Portugal, J., et al. (2012). Ocean warming enhances malformations, premature hatching, metabolic suppression and oxidative stress in the early life stages of a keystone squid. *PLoS ONE*, **7**(6), e38282.

Rosa, R. & Siebel, B. (2008). Synergistic effects of climate-related variables suggests future physiological impairment in a top oceanic predator. *Proceedings of the National Academy of Sciences of the USA*, **105**(52), 20776–20780.

Sagarin, R.D. & Somero, G.N. (2006). Complex patterns of expression of heat-shock protein 70 across the southern biogeographical ranges of the intertidal mussel *Mytilus californianus* and snail *Nucella ostrina*. *Journal of Biogeography*, **33**, 622–630.

Sanford, E. & Kelly, M.W. (2011). Local adaptation in marine invertebrates. *Annual Review of Marine Science*, **3**(1), 509–535.

Schulte, P.M. (2014). What is environmental stress? Insights from fish living in a variable environment. *Journal of Experimental Biology*, **217**, 23–34.

Schulte, P.M., Healy, T.M., & Fangue, N.A. (2011). Thermal performance curves, phenotypic plasticity, and the time scale of temperature exposure. *Integrative and Comparative Biology*, **51**, 691–702.

Seibel, B.A. (2013). The jumbo squid, *Dosidicus gigas* (Ommastrephidae), living in oxygen minimum zones II: blood-oxygen binding. *Deep Sea Research Part II*, **95**, 139–144.

Seibel, B.A. & Walsh, P. J. (2003). Biological impacts of deep-sea carbon dioxide injection inferred from indices of physiological performance. *Journal of Experimental Biology*, **206**(4), 641–650.

Sheehan, D. & McDonagh, B. (2008). Oxidative stress and bivalves: a proteomic approach. *Invertebrate Survival Journal*, **5**, 110–123.

Sheppard-Brennand, H., Soars, N., Dworjanyn, S.A., et al. (2010). Impact of ocean warming and ocean acidification on larval development and calcification in the sea urchin *Tripneustes gratilla*. *PLoS ONE*, **5**(6), e11372.

Sinclair, E.L.E., Thompson, M.B., & Seebacher, F. (2006). Phenotypic flexibility in the metabolic response of the limpet *Cellana tramoserica* to thermally different microhabitats. *Journal of Experimental Marine Biology and Ecology*, **335**, 131–141.

Sokolova, I.M. (2013). Energy-limited tolerance to stress as a conceptual framework to integrate the effects of multiple stressors. *Integrative and Comparative Biology*, **53**, 597–608.

Sokolova, I.M. & Pörtner, H.-O. (2003). Metabolic plasticity and critical temperatures for aerobic scope in a eurythermal marine invertebrate (*Littorina saxitalis*, Gastropoda: Littorinidae) from different latitudes. *Journal of Experimental Biology*, **206**, 195–207.

Somero, G.N. (2002). Thermal physiology and vertical zonation of intertidal animals: optima, limits, and costs of living. *Integrative and Comparative Biology*, **42**, 780–789.

Somero, G.N. (2010). The physiology of climate change: how potentials for acclimatization and genetic adaptation will determine 'winners' and 'losers'. *Journal of Experimental Biology*, **213**(6), 912–920.

Sommer, A., Klein, B., & Pörtner, H.-O. (1997). Temperature induced anaerobiosis in two populations of the polychaete worm *Arenicola marina* (L). *Journal of Comparative Physiology Part B*, **167**, 25–35.

Sommer, A.M. & Pörtner, H.-O. (2002). Metabolic cold adaptation in the lugworm *Arenicola marina*: comparison of a North Sea and a White Sea population. *Marine Ecology Progress Series*, **240**, 171–182.

Sorte, C.J.B. & Hofmann, G.E. (2005). Thermotolerance and heat-shock protein expression in Northeastern Pacific *Nucella* species with different biogeographical ranges. *Marine Biology*, **146**, 985–993.

Southward, A.J. (1958). Note on the temperature tolerances of some intertidal animals in relation to environmental temperatures and geographical distribution. *Journal of the Marine Biological Association of the United Kingdom*, **37**, 49–66.

Spicer, J.I. & Gaston, K.J. (1999). *Physiological Diversity and Its Ecological Implications*. Blackwell Science, Oxford.

Stillman, J.H. (2003). Acclimation capacity underlies susceptibility to climate change. *Science*, **301**, 65.

Stillman, J.H. & Hurt, D.A. (2015). Crustacean genomics and functional genomic response to environmental stress and infection. In: Chang, E.S. & Thiel, M. (eds), *The Natural History of the Crustacea*, Vol. 4, p. 420. Oxford University Press, Oxford, UK.

Stillman, J.H. & Somero, G.N. (1996). Adaptation to temperature stress and aerial exposure in congeneric species of intertidal porcelain crabs (genus *Petrolisthes*): correlation of physiology, biochemistry and morphology with vertical distribution. *Journal of Experimental Biology*, **199**, 1845–1855.

Stillman, J.H. & Somero, G.N. (2000). A comparative analysis of the upper thermal tolerance limits of eastern Pacific porcelain crabs, genus *Petrolisthes*: influences of latitude, vertical zonation, acclimation, and phylogeny. *Physiological and Biochemical Zoology*, **73**, 200–208.

Stillman, J.H. & Tagmount, A. (2009). Seasonal and latitudinal acclimatization of cardiac transcriptome responses to thermal stress in porcelain crabs, *Petrolisthes cinctipes*. *Molecular Ecology*, **18**(20), 4206–4226.

Thomsen, J., Casties, I., Pansch, C., et al. (2013). Food availability outweighs ocean acidification effects in juvenile *Mytilus edulis*: laboratory and field experiments. *Global Change Biology*, **19**(4), 1017–1027.

Todgham, A.E. & Stillman, J.H. (2013). Physiological responses to shifts in multiple environmental stressors: relevance in a changing world. *Integrative and Comparative Biology*, **53**(4), 539–544.

Tomanek, L. (2008). The importance of physiological limits in determining biogeographical range shifts due to global climate change: the heat-shock response. *Physiological and Biochemical Zoology*, **81**, 709–717.

Tomanek, L. (2014). Proteomics to study adaptations in marine organisms to environmental stress. *Journal of Proteomics*, **105**, 92–106.

Tomanek, L. & Somero, G.N. (1999). Evolutionary and acclimation-induced variation in the heat-shock responses of congeneric marine snails (Genus *Tegula*) from different thermal habitats: implications for limits of thermotolerance and biogeography. *Journal of Experimental Biology*, **202**, 2925–2936.

Vaquer-Sunyer, R. & Duarte, C.M. (2008). Thresholds of hypoxia for marine biodiversity. *Proceedings of the National Academy of Sciences of the United States of America*, **105**(40), 15452–15457.

Vaquer-Sunyer, R. & Duarte, C.M. (2011). Temperature effects on oxygen thresholds for hypoxia in marine benthic organisms. *Global Change Biology*, **17**(5), 1788–1797.

Viarengo, A., Canesi, L., Pertica, M., & Livingstone, D.R. (1991). Seasonal variations in the antioxidant defence systems and lipid peroxidation of the digestive gland of mussels. *Comparative Biochemistry and Physiology Part C*, **100**(1), 187–190.

Vosloo, D., van Rensburg, L., & Vosloo, A. (2013). Oxidative stress in abalone: the role of temperature, oxygen and L-proline supplementation. *Aquaculture*, **416**, 265–271.

Walther, K., Anger, K., & Pörtner, H.-O. (2010). Effects of ocean acidification and warming on the larval development of the spider crab *Hyas araneus* from different latitudes (54 vs. 79 N). *Marine Ecological Progress Series*, **417**, 159–170.

Wang, X., Wang, L., Zhang, H., et al. (2012). Immune response and energy metabolism of *Chlamys farreri* under *Vibrio anguillarum* challenge and high temperature exposure. *Fish and Shellfish Immunology*, **33**(4), 1016–1026.

Whitehead, A. (2012). Comparative genomics in ecological physiology: toward a more nuanced understanding of acclimation and adaptation. *Journal of Experimental Biology*, **215**(Pt 6), 884–891.

Whiteley, N.M. & Fraser, K.P.P. (2009). The effects of temperature on ectotherm protein metabolism. In: Esterhouse, T.E. & Petrinos, L.B. (eds), *Protein Biosynthesis*, pp. 249–265. Nova Science Publishers, New York.

Whiteley, N.M. & Taylor, E.W. (2015). Responses to environmental stresses: oxygen, temperature and pH. In: Chang, E.S. & Thiel, M. (eds), *The Natural History of the Crustacea, Growth, Moulting and Physiology*. Oxford University Press, Oxford, UK.

Winterbourn, C.C. (2008). Reconciling the chemistry and biology of reactive oxygen species. *Nature Chemical Biology*, **4**(5), 278–286.

Wittmann, A.C., Schröer, M., Bock, C., et al. (2008). Indicators of oxygen- and capacity-limited thermal tolerance in the lugworm *Arenicola marina*. *Climate Research*, **37**, 227–240.

Wong, K.K.W., Tsang, L.M., Cartwright, S.R., et al. (2014). Physiological responses of two acorn barnacles, *Tetraclita japonica* and *Megabalanus volcano*, to summer heat stress on a tropical shore. *Journal of Experimental Marine Biology and Ecology*, **461**, 243–249.

Yao, C.-L. & Somero, G.N. (2012). The impact of acute temperature stress on hemocytes of invasive and native mussels (*Mytilus galloprovincialis* and *Mytilus californianus*): DNA damage, membrane integrity, apoptosis and signaling pathways. *Journal of Experimental Biology*, **215**(Pt 24), 4267–4277.

Young, J.S., Peck, L.S., & Matheson, T. (2006). The effects of temperature on peripheral neuronal function in eurythermal and stenothermal crustaceans. *Journal of Experimental Biology*, **209**, 1976–1987.

CHAPTER 5

Physiological impacts of chemical pollutants in marine animals

Ceri Lewis and Eduarda M. Santos

5.1 Introduction

The marine environment is fundamental to the sustainability of the global ecosystem and is frequently under threat from a number of stressors involving changes in a number of biotic and abiotic conditions. These may include changes in temperature, noise, radiation, pH, salinity, dissolved gases, etc., and the physiological responses of marine organisms to these stressors are discussed elsewhere in this book. A key stressor affecting many marine ecosystems is the presence of chemical contaminants and/or the alteration of the concentrations of existing substances able to cause adverse physiological effects in organisms inhabiting the affected waters. The marine environment is often the ultimate recipient of a wide and increasing range of natural and anthropogenically derived chemicals. For example, in 1994 the OECD estimated that ~1500 new chemicals are being added annually to the 100 000 already present in the natural environment (Steinberg et al., 1994). In addition, chemical stressors rarely occur alone, but instead, organisms are often exposed to complex mixtures of chemicals simultaneously. Furthermore, we are now starting to observe evidence that climate change and ocean acidification have the potential of modifying the toxicity of the chemicals present in the water.

Chemical stressors in the marine environment can be grouped according to their chemical properties and according to the biological effects they cause to exposed organisms. These properties determine the uptake routes, transport, and metabolism within organisms. Importantly, they also determine the interaction of chemicals or their metabolites with target molecular systems, leading to adverse effects on organisms or populations. The adverse outcome pathway framework has recently been proposed to describe this process in its entirety and to allow for mathematical models to be developed to predict adverse effects at higher levels of biological organization following exposure to individual chemicals or mixtures.

Uptake can occur via a number of mechanisms, including via ingestion of contaminated food or water and absorption through the digestive system, and via the gills or skin, where some chemicals (including metals) are transported across the gill epithelia, while others diffuse across the lipophilic cell membranes. Once within the body, chemicals are transported to target tissues usually via the blood, either alone or in association with transporter proteins (for example for sex steroids and their mimics). Biological responses to the presence of contaminants in target tissues include the induction of metabolizing and detoxification processes, leading to the excretion of toxic chemicals. These include the activation of P450 enzyme systems following exposure to organic compounds (including polycyclic aromatic hydrocarbons (PAHs)), which metabolize target compounds leading to their detoxification and excretion. However, for some chemicals, metabolism can lead to the formation of toxic metabolites that are sometimes more toxic than the parent compound, before they can be further metabolized and excreted. For metals, response to increased concentrations of toxic metals typically includes induction of metal binding proteins, including metallothionein, which bind, store, and detoxify toxic metals by reducing their bioavailability within the cells. Once toxic chemicals exceed the detoxification and excretion ability of a given organism, toxic effects start to occur, which may lead to damage and adverse health effects to the individuals affected. Populations may recover from exposures and damage when the phenotypic plasticity of their individuals is sufficient to elicit an effective response to the toxic insult, or, in extreme cases, by selection of the most resistant genotypes and genetic adaptation. However, recovery can incur a cost in affected populations, which may lose resilience to respond to other stressors in their environment.

It is, therefore, important to document the presence of chemical stressors in the marine environment, to understand their adverse effects to organisms within those systems and to adopt strategies in order to reduce and/or remediate their presence and impact in marine environments.

Given the increase in human population and land use, in particular in coastal areas, it is expected that the incidence, severity, and persistence of chemical pollution will continue to increase, with potential adverse effects to organisms inhabiting receiving waters. This chapter will discuss the main mechanisms by which chemical stressors in the marine environment can negatively impact the health and physiology of marine organisms, including oxidative damage, reproductive, immunotoxic, genotoxic, and endocrine disruption. In addition, the effects that other stressors, such as ocean acidification and hypoxia, may have on chemical toxicity will also be discussed.

5.2 Oxidative stress

Oxidative stress is an important component of the stress response in marine organisms exposed to a variety of insults as a result of changes in environmental conditions such as thermal stress, exposure to ultraviolet radiation, or exposure to pollution. Many of the chemical contaminants found in marine and estuarine environments, such as metals, nanoparticles, polychlorinated biphenyls (PCBs), and polycyclic aromatic hydrocarbons (PAHs), cause damage to exposed aquatic organisms via oxidative stress mechanisms. Oxidative stress is caused by an imbalance between the production of reactive oxygen species (ROS) and an organism's ability to detoxify the reactive intermediates or easily repair the resulting oxidative damage. Reactive oxygen species, such as free radicals and peroxides, form as a natural by-product of the normal metabolism of oxygen and have important roles in cell signalling and homeostasis. All respiring and photosynthetic cells produce ROS, and it is estimated that 1–3% of oxygen consumed by an organism will be converted to ROS. When the rate of ROS production exceeds the rate of its decomposition by antioxidant defences and repair systems, oxidative stress can be established. Some of the less reactive of reactive oxygen species (such as superoxide) can be converted by oxidoreduction reactions with transition metals or other redox cycling compounds (including quinones) into more aggressive radical species that can cause extensive cellular damage. The most significant consequence of oxidative stress in the body of any organism is via oxidative damage to important biological molecules such as lipids and DNA, as well as disrupting normal cellular signalling pathways.

A suite of cellular defences that include low molecular weight antioxidants, antioxidant enzymes, and DNA repair enzymes, protect against and repair this constant attack. The best studied cellular antioxidants are the enzymes superoxide dismutase (SOD), catalase, and glutathione peroxidase. SOD and catalase appear to be important antioxidant enzymes in most aquatic invertebrates, and their activities are often measured as a 'biomarker' of oxidative stress. SOD is a metalloprotein, and most marine invertebrates and fish have a copper- and zinc-based SOD cytosolic enzyme. However, marine arthropods lack this Cu/Zn SOD and instead have an unusual manganese-based cytosolic Mn SOD (Lesser, 2006). Small molecule antioxidants such as ascorbic acid (vitamin C), tocopherol (vitamin E), uric acid, and glutathione also play important roles as cellular antioxidants. However, during times of environmental stress (e.g. UV or heat exposure) or pollution exposure, ROS levels can increase dramatically and these defences can become overwhelmed, leading to oxidative stress and damage.

A number of environmental chemicals present in marine and coastal waters, such as polycyclic aromatic hydrocarbons (PAHs) and polychlorinated biphenyls (PBCs), generate oxidative stress in exposed biota as a result of metabolism and detoxification pathways, with the subsequent production of electrons that can be transferred to molecular oxygen producing the superoxide ($\bullet O_2^-$) radical. Transition metals such as copper and zinc, which are often present in elevated levels in coastal environments, produce ROS via redox cycling in which a single electron may be accepted or donated by the metal which catalyses reactions that produce reactive hydroxyl (HO$\bullet$) radicals and produce reactive oxygen species via Fenton chemistry. ROS may also be generated by ionizing or ultraviolet radiation. Rates and amounts of ROS produced can be increased by the presence of a large range of man-made xenobiotics, such as redox cycling compounds (quinones, nitroaromatics, bipyridyl herbicides), polycyclic aromatic hydrocarbons (PAHs), dioxins, and toxic metals (e.g. aluminium, copper, cadmium, arsenic, and mercury). Contaminants can stimulate ROS production via a number of direct or indirect mechanisms, e.g. redox reactions with O_2, autoxidation (cytochromes p450 and PCBs), enzyme induction, disruption of membrane bound electron transport, and depletion of antioxidant enzyme defences (glutathione).

Many marine invertebrates and fish produce ROS in response to both varying environmental conditions and chemical pollutants, and we can measure oxidative stress as an indicator of general health and as a biomarker for exposure to pollution using a range of techniques that include measuring the transcription levels or activities of antioxidant enzymes (such as SOD), the total oxyradical scavenging capacity, or the products of oxidative damage. These measures of oxidative stress are therefore often used as an indicator of pollution levels for a population living in a contaminated habitat. Bivalve molluscs, a number of echinoderms, and polychaete worms have all been shown to produce ROS in response to a range of xenobiotics (Pellerin-Massicotte, 1994) and changes in temperature, especially heat stress. For example, one study looked at levels of oxidative stress in the mussel *Mytilus galloprovincialis* collected from the Saronikos Gulf, Greece from a number of polluted and unpolluted sites and found that metal pollution in the Elefsis Bay (the most polluted coastal area) caused elevated levels of oxidative stress in tissues of the mussels (Vlahogianni et al., 2007). Another study looking at fish from rural and industrial areas in Western Ukraine found elevated levels of superoxide anion radical ($^\bullet O_2^-$) production in the liver and gills of the common carp *Cyprinus carpio* in fish from the polluted industrial site (Falfushynska and Stolyar, 2009).

Oxidative damage of important biological molecules such as DNA (see Section 5.3), proteins, and lipids (such as those in cell membranes) is one of the main consequences of oxidative stress in aquatic organisms and can result in cell and tissue damage. In oxidative damage to lipids (lipid peroxidation) reactive oxygen species readily attack the polyunsaturated fatty acids of the fatty acid membrane, initiating a self-propagating chain reaction. The destruction of membrane lipids and the end-products of such lipid peroxidation reactions are especially dangerous for the viability of cells. Since lipid peroxidation is a self-propagating chain-reaction, the initial oxidation of only a few lipid molecules can result in significant tissue damage. A cell can recover from certain levels of oxidative stress; however, more severe oxidative stress can lead to cell death and necrosis, and even moderate oxidative stress can trigger apoptosis (programmed cell death). Lipid peroxidation, generally measured as the level of thiobarbituric acid reactive substances (TBARS), is often used to analyse the effect of pollutants that exert oxidative stress

on marine animals (Livingstone, 2002). For example a population of the clam *Scrobicularia plana* inhabiting a shallow tidal creek affected by fish farm effluents (Rio San Pedro, SW Spain) was found to have elevated lipid peroxidation, measured using a TBARS assay (Silva et al., 2012).

5.3 Genotoxicity in marine organisms

DNA damage occurs naturally in all cells as a result of normal oxidative stress from metabolism. For example, in human cells normal metabolic activities and environmental factors such as UV light can induce as many as one million individual molecular lesions per cell per day. A comprehensive suite of DNA repair enzymes are present in all cells in order to repair these natural levels of DNA damage and maintain normal cell functioning. These DNA repair processes can become overwhelmed, however, when additional DNA damage occurs through exposure to xenobiotic chemicals. A significant proportion of the chemicals entering our marine environments have the potential to induce DNA damage or interfere with the processes involved in cell division (Depledge, 1998; Livingstone et al., 2000). These include, among others, persistent organics such as polycyclic aromatic hydrocarbons (PAHs) and metals, which can damage DNA either directly by forming DNA adducts or indirectly via the production of free radicals or after metabolic activation. Chemical pollutants or environmental radiation that induce various types of DNA damage are referred to as being 'genotoxic'. In a study by Steinart et al. (1998) mussels were deployed in mesh bags at sites around San Diego bay and then recovered 12 and 32 days later. Sites with the most contamination, including high levels of metals such as mercury and copper, were easily identified by increased levels of DNA damage measured in mussels from those sites.

Ultraviolet and other types of radiation or oxidative stress can damage DNA by inducing DNA strand breaks. This involves a cut in one or both DNA strands. Double-strand breaks are particularly indicative of radiation-induced damage and are especially dangerous since they can be mutagenic and can potentially affect the expression of multiple genes. UV-induced damage can also result in the production of pyrimidine dimers, where covalent cross-links occur in cytosine and thymine residues. Pyrimidine dimers can disrupt DNA polymerases and hence prevent proper replication of DNA. Oxidative DNA damage can be a direct effect of exposure to reactive oxygen species generated from reactions with contaminants such as metals, but also the secondary effect of many other pollutants including nitroaromatic compounds and polycyclic aromatic hydrocarbons (PAHs) that may produce reactive oxygen species through redox cycling of metabolites. The impact of such damage will be influenced by an organism's antioxidant status and DNA repair capacity.

Polycyclic aromatic hydrocarbons (PAHs) are potent, ubiquitous pollutants commonly associated with oil, coal, cigarette smoke, and automobile exhaust fumes, and are widespread in coastal habitats. Most PAHs are quickly metabolized and it is their metabolites that will covalently bind to DNA, forming 'adducts'. A suite of cytochrome P450-dependent monooxygenase enzymes (CYPs) play a pivotal role in this PAH metabolism. A common marker for DNA damage due to PAHs is benzo(a)pyrene diol epoxide (BPDE). BPDE is found to be very reactive, and known to bind covalently to proteins, lipids, and guanine residues of DNA to produce BPDE adducts. If left unrepaired, BPDE–DNA adducts may lead to permanent mutations resulting in cell transformation and ultimately tumour development. PAH–DNA adducts trigger nucleotide excision repair (NER) and various DNA damage responses that might include apoptosis. Further, PAHs can also confer DNA damage via an alternative route of metabolic activation, which leads to the generation of PAH semi-quinone radicals and reactive oxygen species (ROS). PAHs are hydrophobic chemicals so tend to accumulate in marine sediments, with bioaccumulation also then occurring in the lipids of exposed benthic marine organisms. Oil spills have been found to cause long-term DNA damage in affected marine biota, such as that measured in Mediterranean mussels (*Mytilus galloprovincialis*) following the Prestige Oil spill of 2002 on the coast of Galicia, Spain (Laffon et al., 2006).

Another source of genetic toxicity in marine and aquatic environments that is an area of current concern comes from radiation. Ionizing radiation is generated through nuclear reactions and is composed of particles that individually carry enough energy to liberate an electron from an atom or molecule, ionizing it. Ionizing radiation is ubiquitous in the environment, and its presence in marine and estuarine ecosystems can be a result of naturally occurring radionuclides and both the controlled and accidental release of radioactive materials stemming from anthropogenic activities such as nuclear power generation, medical treatments, mining, and military operations. When ionizing radiation is emitted by or absorbed by an atom, it can liberate a particle (usually an electron, but sometimes an entire nucleus) from the atom. Such an event can alter chemical bonds and produce ions, usually in ion-pairs,

that are especially chemically reactive. This greatly magnifies the chemical and biological damage per unit energy of radiation. Ionizing radiation is genotoxic, causing DNA damage in exposed biota. In radiosensitive tissue, this genetic damage may result in stochastic (e.g. lesions and tumours resulting from cellular mutations) or deterministic effects (e.g. tissue necrosis resulting from cell death). Exposure to ionizing radiation can ultimately result in organ failure, reduced fecundity, mutations, cancer, and even death.

As energy demands rise, it is estimated there will be global investment of $300 billion in the nuclear industry over the next twenty years. Most new nuclear power generation facilities will be coastal in order to utilize seawater for cooling and dilution of discharges, making estuaries and coastal habitats a likely repository for any resulting radionuclide discharges. Radionuclides can adsorb to the fine, organic-rich sediment common to these ecosystems leading to their accumulation in sediments in much the same way as other chemical contaminants. Contamination of the marine environment following the accident in the Fukushima Dai-ichi nuclear power plant represented the most important artificial radioactive release flux into the sea ever known (du Bois et al., 2012). The radioactive marine pollution came from atmospheric fallout onto the ocean, direct release of contaminated water from the plant, and transport of radioactive pollution from leaching through contaminated soil. Caesium-137 is essentially soluble in seawater; it will be carried over very long distances by marine currents and dissipated throughout the ocean water masses. The caesium-137 from Fukushima will remain detectable for several years throughout the North Pacific.

A commonly used technique for assessing DNA damage in cells is the single-cell electrophoresis technique, otherwise known as the comet assay (example comet images for the common mussel *Mytilus edulis* are shown in Fig. 5.1). The comet assay is based on the ability of negatively charged fragments of DNA to be drawn through an agarose gel in response to an electric field. The extent of DNA migration depends directly on the DNA damage present in the cells. A suspension of cells is mixed with low melting point agarose and spread onto a microscope glass slide. Following lysis of cells with detergent, DNA unwinding and electrophoresis is carried out under either alkaline or neutral conditions (depending on the type of DNA damage to be measured). When subjected to an electric field, the DNA migrates out of the cell, in the direction of the anode, appearing like a 'comet' which can be visualized using fluorescent microscopy and quantified

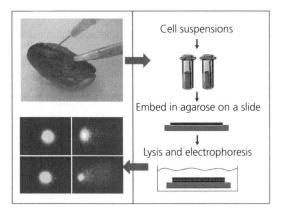

Figure 5.1 Flow diagram for performing the comet assay for measuring DNA damage in mussel haemocytes.

using image analysis software. The size and shape of the comet and the distribution of DNA within the comet correlate with the extent of DNA damage (Fairbairn et al., 1995).

Another well-established method for assessing genetic damage in aquatic invertebrates is the micronucleus frequency assay. A micronucleus is a third nucleus that is formed during the anaphase of mitosis or meiosis. Micronuclei are produced in dividing cells that contain chromosome breaks lacking centromeres (acentric fragments) and/or whole chromosomes that are not carried to the spindle poles during the anaphase of mitosis. At telophase, a nuclear envelope forms around the lagging chromosomes and fragments, which then uncoil and gradually assume the morphology of an interphase nucleus with the exception that they are smaller than the main nuclei in the cell, hence the term 'micronucleus'. These are measurements of permanent genotoxic damage, measured as chromosomal damage or loss, as opposed to the repairable damage measured by the comet assay. The micronucleus assay has been extensively used in haemocytes and gill cells of many molluscan species, including the oyster *Crassostrea gigas*, the Mediterranean mussel *Mytilus galloprovinciallis*, and the common blue mussel *Mytilus edulis*, to measure pollutant-induced genotoxic damage (Klobucar et al., 2008; Bolognesi and Hayashi, 2011). The incidence of micronuclei has also been linked to the induction of leukaemia cells in the clam *Mya arenaria*, suggesting that the micronucleus test is a very good indicator of the potentially life-threatening consequences of genotoxic exposure.

Carcinogenesis, i.e. the generation of cancers and tumours, is a process involving multiple genetic alterations ultimately leading to uncontrolled cell proliferation (neoplasia). Compared to mammals, there have been limited studies pertaining to identification and expression of oncogenes (genes with the potential to cause cancer) in fish and marine invertebrates. Many species of molluscs, including the clam *Mya arenaria* and the mussel *Mytilus edulis*, are susceptible to haemocytic leukaemia (St Jean et al., 2005), a disease with similarities to leukaemia in humans (Elston et al., 1992). A range of neoplasmic damage has been documented in marine invertebrates that have been exposed to genotoxins. A study in Prince Edward Island, Canada, has linked high levels of leukaemia in the clam *Mya arenaria* with high levels of pesticides in the seawater originating from farmland used to grow potatoes (Muttray et al., 2012). In one area of the USA where herbicide use had increased, elevated incidences of gonadal tumours were reported in both bivalves and humans (Van Beneden, 1994). Other studies in dab and flounder found high incidence of liver tumours in areas receiving contaminated waters (Vethaak et al., 2009; Small et al., 2010; Mirbahai et al., 2011).

5.4 Immunotoxicity in marine organisms

Most marine organisms have some form of immune system that works to inactivate or eliminate foreign invaders, including pathogenic microorganisms and their products by either adaptive (memory) or innate (no memory) strategies. Marine invertebrates rely mainly on cellular-based innate immune defences based on phagocytosis, encapsulation, and the production of soluble effector molecules with antimicrobial activity (Galloway and Handy, 2003). The immune systems of fish are more complex and very similar to most vertebrates, with innate and adaptive immune responses. Their innate systems comprise physical barriers (e.g. gills and epithelia) and humoral and cellular receptor molecules that are soluble in plasma (Uribe et al., 2011). Humoral factors include lysozymes, cytokines, and chemokines, together with enzymes that act to promote inflammation, assist in the lysis of foreign cells, and stimulate the adaptive immune system (Segner et al., 2012). The adaptive immune responses of fish include the capacity to produce antibody molecules, although their innate immune systems play an important role in particular during their early stages of development. The lymphoid organs (i.e. those involved in immune function) found in fish include the thymus, spleen, and kidney. Immunoglobulins are the principal components of the immune response against pathogenic organisms. Immunotoxicity is defined as an adverse effect on the immune system or its component parts, whether or not the net result is an alteration of host resistance (Galloway and Handy, 2003).

A number of environmental pollutants have been demonstrated to have immunotoxicological impacts on exposed marine animals, including toxic metals, benzene, pentachlorophenol, polycyclic aromatic hydrocarbons, dioxins, and pesticides (Galloway and Depledge, 2001) as well as emerging pharmaceuticals (Segner et al., 2012). Measuring immunotoxicity can be difficult due to the complexity of the immune system, however; widely used assays in both vertebrates and invertebrates involve the determination of phagocytotic activity of immune cells, or the inhibition of immune cell proliferation (Galloway and Depledge, 2001). Several studies in the mussel *Mytilus edulis* have found that low levels of metal contamination can elicit immunosuppressive effects (e.g. Pipe and Coles, 1995). Short-term metal exposure has also been shown to reduce the number of circulating haemocytes (a condition called haemocytopenia) in the shrimp *Palaemon elegans*, which is indicative of impaired immune function and reduced ability to fight infection (Lorenzon et al., 2001). PAHs can also act as immunosuppressants reducing the phagocytotic ability of haemocytes, as demonstrated by studies looking at the Arctic scallop *Chlamys islandica* (Hannam et al., 2010b) and temperate scallops *Pecten maximus* (Hannam et al., 2010a).

Increased disease susceptibility and incidence in fishes have been demonstrated from the polycyclic aromatic hydrocarbon (PAH) contaminated Puget Sound (Arkoosh et al., 2001). Immunotoxicity responses have also been documented in marine mammals. For example, persistent, lipophilic polyhalogenated aromatic hydrocarbons (PHAHs), which accumulate readily in the aquatic food chain and are found in high concentrations in seals and other marine mammals, have been shown to suppress immune responses in harbour seals. An impairment of T-lymphocyte function, antigen-specific lymphocyte responses, and antibody responses was observed in the seals fed herring containing high levels of PCBs (Ross et al., 1996). Because the immune system plays an important role in host resistance to disease as well as normal homeostasis, any animal suffering from chemical-induced immunotoxicity becomes more susceptible to infectious disease and negative health impacts.

5.5 Endocrine disruption

The endocrine system regulates many of the functions essential for life, such as development, growth, and reproduction. In vertebrates, this system is comprised of a number of endocrine organs that produce signalling molecules (hormones) capable of activating physiological responses in target cells. Structurally, hormones may be grouped into four categories including polypeptides, steroids, amines, and fatty acid derivates, and they activate molecular responses in target cells via activation of specific hormone receptors. In recent years, a wide range of evidence demonstrated that many natural and man-made chemicals have the potential to interfere with the endocrine system of vertebrates and invertebrates, resulting in adverse health effects to the affected organisms and/or their progeny. These chemicals are generally designated as endocrine disrupting chemicals (EDCs). Disruption of the endocrine system can occur via multiple mechanisms, including via mimicking of natural hormones, resulting in binding and activation of their receptors (hormone agonists), blocking of the natural hormone receptors (hormone antagonists), or by modification of the synthesis, transport, or metabolism of the natural hormones. Some of the most well-studied examples of endocrine disruption focus on chemicals that interfere with sex steroid hormone signalling, causing disruption of reproduction in vertebrates (reviewed in Tyler et al., 1998). Sex steroids are produced mainly in the gonads and regulate not only gonadal development and reproduction, but also a wide variety of other processes in the body, including growth, development, metabolism, and immune function. They act in target cells through binding and activating specific receptors, including nuclear receptors that act as transcription factors and elicit the regulation of extensive molecular pathways, or receptors located at the cell membrane, generating rapid responses following hormone stimulation (Fig. 5.2). Environmental chemicals that mimic endogenous oestrogens and androgens are generally designated as oestrogenic and androgenic chemicals, respectively, and those blocking the action of these hormones are generally designated as anti-oestrogenic and anti-androgenic chemicals, respectively.

In the last three decades, extensive evidence has been documented for the presence of oestrogenic chemicals in aquatic systems worldwide. Examples of oestrogenic chemicals include natural (e.g. oestradiol and oestrone) and synthetic oestrogens (e.g. ethinyl oestradiol), plasticizers (e.g. bisphenol A), detergents (e.g. nonylphenol), organochloride pesticides, brominated flame retardants, and others (Tyler et al., 1998). Exposure of vertebrates to these chemicals causes alterations in oestrogenic signalling pathways and results in substantial alterations in gonadal development and reproductive function in both males and females.

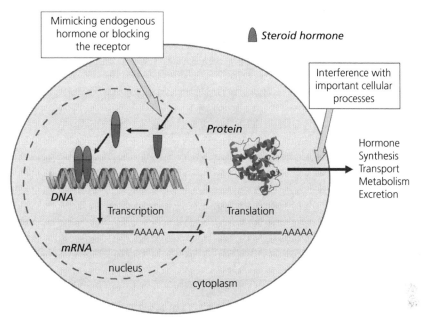

Figure 5.2 Schematic illustration of the mechanisms of hormone disruption, using the sex steroid hormone nuclear receptor pathway as an example.

These effects have been extensively documented to occur in fish both in the laboratory and in field studies. Most of the data currently available are focused on freshwater species including the zebrafish *Danio rerio* (Nash et al., 2004; Santos et al., 2007; Scholz and Mayer, 2008), the fathead minnow *Pimephales promelas* (Filby et al., 2007; Kidd et al., 2007; Scholz and Mayer, 2008; Watanabe et al., 2009), the roach *Rutilus rutilus* (Jobling et al., 2002; Lange et al., 2009; Hamilton et al., 2014), the medaka *Oryzias latipes* (Urushitani et al., 2007), and the three-spined stickleback *Gasterosteus aculeatus* (Katsiadaki et al., 2007; Santos et al., 2013), with a gap in data sets for marine species. The range of effects documented include alterations in gonadal development (e.g. the formation of an ovarian cavity and/or the production of female gametes (eggs) in the testis) (van Aerle et al., 2001, 2002; Jobling et al., 2006), alterations in the concentrations of sex steroid hormones (e.g. testosterone), gonadotropins, and vitellogenin (VTG; an egg yolk precursor protein, normally only expressed in females) in the blood (Tyler et al., 1996; Harris et al., 2001; van Aerle et al., 2001), and alterations in reproductive behaviour (Martinovic et al., 2007). EDCs enter the aquatic environment via sewage treatment works effluent and surface/agricultural run-off, and oestrogenic activity has been reported to be widespread in freshwater systems, in particular in areas associated with sewage treatment works discharges (Tyler et al., 1998). In the marine environment, the concentrations of oestrogenic chemicals are generally lower than those recorded for rivers and estuaries, but in some cases concentrations well above the thresholds for biological effects have been recorded. For example, in the Ria Formosa, Portugal, concentrations of ethinyl oestradiol reached up to 24.3 ng/l (Rocha et al., 2013). Comparatively fewer studies have been performed to determine the effects of model chemicals or environmentally relevant mixtures on the health of marine fish species, but the data obtained to date align well with those obtained for freshwater species. For example, exposure of Atlantic cod (*Gadus morhua*) to alkylphenols resulted in reproductive impairment in both males and females (Meier et al., 2007), with acceleration of the onset of puberty in juvenile fish, and delayed gonadal development in mature females (Meier et al., 2011). In addition, exposure to produced water (waste water generated as a by-product of the oil-production process) disrupted development in Atlantic cod (Meier et al., 2010), and reproductive function in polar cod (*Boreogadus saida*) in both males and females (Geraudie et al., 2014). Furthermore, in the Japanese sea bass (*Lateolabrax japonicus*), high concentrations of oestradiol disrupted the endocrine and biotransformation system during early life stages (Thilagam et al., 2014).

Studies have shown that endocrine disruption in wild marine fish may be occurring on a broad scale, in particular in shallow waters that are near highly populated areas or are affected by industrial discharges. Surveys of flounder (*Platichthys flesus*) in UK estuaries conducted between 1996 and 2001 demonstrated that male fish had been exposed to EDCs, evidenced by the presence of both male and female tissues in the gonad (intersex) and increased levels of VTG. Follow-up studies showed less pronounced disruption of reproductive development in flounder caught in some areas, indicating that a potential recovery had occurred, likely related to decreased discharges of chemical pollutants (Kirby et al., 2004a, b). In addition, studies in Atlantic cod and in dab (*Limanda limanda*) captured from UK offshore waters revealed inappropriate production of VTG in some males, suggesting that these fish had been exposed to oestrogenic chemicals (Scott et al., 2007). A more recent study on adult three-spined stickleback (*Gasterosteus aculeatus*) derived from areas impacted by sewage effluent discharges in the Finnish Baltic Sea coast also showed evidence of exposure to oestrogenic compounds (Bjorkblom et al., 2013).

Androgenic activity has been reported in many freshwater and marine systems, and in particular for waters impacted by pulp and paper mill effluents, where masculinization of fish has been shown to occur (Ellis et al., 2003; Orlando et al., 2007). In marine environments, androgenic activity has also been detected as a result of water treatment effluent inputs, with measurable effects on fish populations. For example, in San Francisco Bay sites were identified where oestrogenic and androgenic activities were present. Mississippi silverside (*Menidia audens*) males from a location impacted with both oestrogenic and androgenic activity had smaller gonads relative to body weight and higher incidence of severe testicular necrosis, and the sex ratio was significantly skewed towards males compared to locations impacted by oestrogens alone. These findings demonstrate that androgens in marine environments, in this case when acting in combination with oestrogens, can cause changes in organismal and population level end-points with the potential to result in adverse effects at the population level (Brander et al., 2013).

More recently, concern has risen about the presence of anti-androgenic chemicals in the aquatic environment, potentially contributing to the feminization of fish populations (Jobling et al., 2009). This anti-androgenic activity has been detected in the marine environment, including in effluents from oil production platforms (Tollefsen et al., 2007), and in marine sediments in Norway (Grung et al., 2011), but generally the specific compounds responsible for this activity have not been identified.

It is important to note that sex steroid receptors from different species can have different binding affinities to various environmental chemicals, with the potential for species-specific effects to occur (Miyagawa et al., 2014). These differences have seldom been explored but it is essential that they are considered for the appropriate risk assessment of environmental chemicals, where often model species are used as surrogates to determine the safety thresholds for chemical contamination in the aquatic environment.

Less is known about EDCs that affect other hormonal systems in vertebrates, despite their importance for the physiology of organisms. Of note, data are available on the thyroid disrupting effects of environmental chemicals including polychlorinated biphenyls (PCBs), a group of persistent organochlorine compounds suspected to disrupt the homeostasis of thyroid hormones (THs). Studies in Japanese flounder (*Paralichthys olivaceus*) showed that exposure to Aroclor 1254 (a mixture of PCBs) for 50 days induced changes in thyroid morphology, including increased follicular cell height, colloid depletion, and hyperplasia (Dong et al., 2014). In the wild, disruption of thyroid function has also been observed in areas impacted by PCB congeners. In San Francisco Bay, studies reported significant reductions in the plasma concentrations of thyroxine (T4) in fish sampled from highly impacted locations compared with fish from locations with relatively lower human impact. The changes in thyroid endocrine parameters were correlated with hepatic concentrations of a number of environmental contaminants including PCBs, indicating that thyroid disruption is occurring in wild populations of fish inhabiting areas contaminated with PCBs, as predicted based on data from laboratory studies (Brar et al., 2010).

For marine mammals, studies investigating the body burden of chemical contaminants have documented high concentrations of a range of contaminants, including those with endocrine activity. For example, for bottlenose dolphins (*Tursiops truncatus*) in Florida, the presence of contaminants including PCBs in blubber samples and the potential for maternal transfer were documented (Wells et al., 2005). The presence of high concentrations of contaminants has also been recorded in toothed whales off the Northwest Iberian Peninsula (NWIP), with concentrations of certain PCBs exciding the toxicity threshold in some species (Mendez-Fernandez et al., 2014). Recently, analysis of ear wax in an individual male blue whale (*Balaenoptera musculus*) allowed the reconstitution of the uptake of pollutants over its life history, including the significant maternal transfer of certain PCBs (Trumble et al., 2013). The use of this technique proved to be highly successful not only in identifying the compounds that this individual whale was exposed to, but also the approximate timing and associated life stage when exposures occurred, and promises to be a viable technique for studying the impacts of environmental chemicals in endangered marine mammals.

Many marine mammals act as top predators in their ecosystems, therefore many organic chemicals biomagnify over the food chain, placing them at particular risk (Diaz and Rosenberg, 2008). Given the scarcity of studies on marine mammals and the difficulties in conducting experimental work in many of the species of interest, it is difficult to predict to what extent the concentrations of chemicals found in blubber and other tissues are related to population-level impacts, but this possibility cannot be excluded. Together, evidence suggests that marine mammals are exposed to, and accumulate, a large range of marine contaminants, with the potential to impact their health and ability to reproduce.

In addition to chemical exposures, hypoxia can also disrupt reproductive processes in marine fish species. Effects observed range from the masculinization of females to inhibition of gonadal development and maturation, via a range of mechanisms including disruption of progestogen signalling and aromatase inhibition (Wu et al., 2003; Thomas et al., 2007; Thomas and Rahman, 2009, 2012). Given the extensive areas affected by hypoxic events annually, and the increased incidence and persistence of those events, this is a cause for significant concern because of its potential contribution to the population and ecosystem level consequences associated with hypoxic zones in marine environments.

The endocrinology of marine invertebrates remains poorly understood. Many of the hormone systems present in vertebrates are absent in invertebrates and, conversely, hormone signalling pathways have evolved independently in invertebrates, and regulate important functions without homologue pathways in vertebrates. Overall, our lack of understanding of the endocrine system for the majority of invertebrate species is compromising our ability to identify the potential effects of EDCs in these organisms, or the population-level consequences of exposure. Considering the essential role that invertebrates play in marine ecosystems, this is a critical area for further research.

Among invertebrates, the effects of EDCs are best studied for molluscs and crustaceans. The endocrine system of crustaceans comprises a range of hormones including peptides, ecdysteroids, and terpenoid hormones that control aspects of development, growth, moulting, and reproduction. Studies investigating for the presence of vertebrate-like sex steroid pathways failed to conclusively demonstrate that such pathways exist and are functional, despite showing that vertebrate hormones cause biological effects in crustaceans, including alterations in reproductive function (reviewed in LeBlanc, 2007).

Disruption of the reproductive system has been reported for a number of species including several species of crabs, shrimps, lobsters, and crayfish (reviewed in LeBlanc, 2007) and effects range from decreased gonadal development and gamete quality, intersexuality, and induction of female-specific proteins such as vitellogenins. The use of intersex as an indicator for endocrine disruption in crustaceans may be inappropriate, as intersex is known to occur naturally in many species. However, increased incidence of intersex has been reported in polluted areas, for example for *Echinogammarus marinus* living on the coast of Scotland, together with reduced sperm quality (Yang et al., 2008). The causation and the mechanistic pathways associated with this higher incidence of intersex have not been established in most studies. In addition, the use of biomarkers for reproductive disruption in crustaceans has been difficult to establish. Many attempts have been made to develop vitellogenin-like proteins as biomarkers for feminization effects in a range of species, with inconsistent results. This has been partly due to the fact that vitellogenins may not be associated only with reproduction for crustaceans but may also be involved in a range of other processes including in immune defence (Short et al., 2014). The lack of knowledge on the fundamental physiology of these organisms, as well as the lack of genomic resources for crustacean species, has constituted a significant hurdle for the progression of this research.

Some of the best evidence for endocrine disruption in crustaceans involves disruption of the ecdysteroid signalling pathway. Ecdysteroids play fundamental roles in moulting and are targets for the development of pesticides for pests affecting many crops across the world. When those pesticides enter the aquatic environment, they can target homologous hormonal systems in aquatic crustaceans, causing significant adverse effects. In addition, many of the chemicals that act as oestrogen mimics in vertebrates have been found to act as antagonists of ecdysteroids, but often only when present at very high concentrations (LeBlanc, 2007; Rodriguez et al., 2007).

In molluscs, peptide and steroid hormones have been described and are thought to control aspects of development, growth, and reproduction. Steroid hormones have been reported to be present in some species, who also are capable of converting precursors such as cholesterol into steroid hormones, similarly to that found in vertebrates (Fernandes et al., 2011). Despite advances for some species, our understanding of the endocrine system is very limited for the majority of species and the hormones and their receptors involved in regulating reproduction are still subject to controversy. The best described case of population collapses due to EDCs is that of the effects of the anti-fouling compound, tributyltin (TBT) on marine gastropods. TBT was generally used on boats and other marine equipment to prevent settlement of aquatic organisms. Due to its generalized use, in the proximity of areas of heavy boat traffic and ports the presence of this compound in the water and associated with organic materials and living organisms became sufficiently high to be a cause for concern. In affected areas, very significant decreases in the populations of a number of molluscan species were observed, and closer investigation revealed that snails in contaminated areas displayed a condition designated as imposex, whereby females developed male-like reproductive structures (Smith, 1981a, b, c). These and other observations resulted in international measures to reduce the use of TBT, and as a consequence, recoveries were observed in affected populations (Matthiessen et al., 1995). The physiological mechanisms linking TBT to the imposex condition remained under debate for a number of years, and initial hypotheses suggested that this was due to an increase in the concentrations of androgens and suppression of oestrogens in exposed molluscs (Morcillo and Porte, 1999). More recent studies demonstrated that retinoic acid signalling pathways play a key role in organotin-induced imposex in molluscs (Nishikawa et al., 2004; Castro et al., 2007; Stange et al., 2012).

The case of TBT illustrates the critical importance of investigating the endocrinology and physiology of invertebrate species in order to allow for investigations on how the fundamental processes such as development, growth, and reproduction are disrupted by environmental chemicals. In many cases, researchers used the knowledge on vertebrate endocrinology as initial hypotheses to investigate the same processes in invertebrates, given the potential for these hypotheses to speed up research. However, because of the evolutionary distance between vertebrates and many groups of invertebrates, this strategy is fraught with problems and has led to misleading conclusions at times (discussed in Scott, 2012, 2013).

Significant knowledge gaps exist on what the concentrations of EDCs in the marine environment are, and how species and ecosystems are affected by exposure to those chemicals, alone or combined with other stressors in their environment. In addition, most of the evidence for endocrine disruption in the marine environment refers to the disruption of steroid hormone function in fish or other vertebrates. It is important to note that much less is known about the effects of environmental chemicals in invertebrates, despite the wide range of species and fundamental role that they play in marine ecosystems. Therefore, this is a critical knowledge gap that needs to be addressed in the future.

5.6 Spermiotoxicity

Successful fertilization for any species depends on the production of high quality sperm, maintenance of DNA integrity and fertilization capacity, and appropriate motility responses to oocytes and spawning medium (in this case water). All of these aspects of fertilization have the potential to be disrupted by environmental perturbances or exposure to xenobiotics. This is particularly relevant in the marine environment where many species reproduce by directly spawning their sperm into the water column, thereby directly exposing their sperm to any contaminants that are present (Lewis and Ford, 2012). While somatic cells and oocytes contain a variety of proteins, antioxidants, and DNA repair enzymes that repair and protect against environmentally induced damage, sperm are generally considered to have little or no capacity for DNA repair or antioxidant defence (Aitken et al., 2004). Human health research has clearly demonstrated that the male reproductive system is a major target of environmental chemicals and that sperm DNA integrity can be adversely affected by exposure to ubiquitous pollutants such as polycyclic aromatic hydrocarbons (PAHs) and phthalates. Sperm are thought to be particularly susceptible to oxidative damage due to the abundance of polyunsaturated fatty acids acting as substrates for reactive oxygen species.

DNA damage has now been measured in the sperm of a number of marine invertebrate species, both in natural populations, such as in the king ragworm, *Nereis virens*, living in contaminated sites (Lewis and Galloway, 2008) and from laboratory-based exposures to a number of chemicals. The comet assay has been used in a number of studies into spermiotoxicity in marine species, such as that caused by exposures to zero-valent iron nanoparticles in sperm of the mussel *Mytilus galloprovincialis* (Kadar et al., 2011). DNA damage in the sperm of the polychaetes *Nereis virens* and *Arenicola marina* (Lewis and Galloway, 2008, 2009; Caldwell et al., 2011) and the mussel *Mytilus edulis* (Lewis and Galloway, 2009) have also been measured following laboratory exposures of males to the common coastal contaminants benzo(a)pyrene and copper prior to spawning. DNA damage in sperm potentially has greater implications for populations than that measured in adult organisms. Sperm DNA damage alone does not necessarily affect sperm motility nor the fertilization reaction; however, it does lead to severe developmental abnormalities of the resulting embryos and larvae (Lewis and Galloway, 2009).

Fertilization for most marine species also depends on the ability of sperm to swim to the egg. Even in a turbulent wave-swept environment, the last few micrometres of distance to an egg requires sperm motility to penetrate the egg boundary layer. Many environmental contaminants have been demonstrated to impair sperm motility. Sperm motility is therefore regularly used as a proxy for 'sperm quality' and can be measured using computer assisted sperm analysis (CASA) similar to that used in human infertility clinics. Individual sperm can be tracked for various time intervals providing data on percent motility and a number of swimming speed and directional parameters (e.g. curvilinear velocity) which have been demonstrated to be positively correlated with capacity for fertilization for a number of species. This technique has been used to demonstrate that a number of environmental contaminants, such as copper (Fitzpatrick et al., 2008) and nonylphenol, significantly reduce sperm motility in blue mussels. Recent data also suggest that the change in seawater pH associated with CO_2 induced acidification of the world's oceans can act to reduce sperm swimming speeds, and hence fertilization success in some marine invertebrates (Havenhand et al., 2008; Morita et al., 2010; Lewis et al., 2013), although other species do not appear to be so sensitive (Havenhand and Schlegel, 2009).

5.7 Mechanical harm—marine plastics

Since the mass production of plastics began in the 1940s plastic contamination of the marine environment has become a growing problem. Plastics are synthetic polymers derived from the polymerization of monomers from oil and gas and plastic manufacturing accounts for about 8% of global oil production. It is estimated that 280 million tonnes of plastics are produced each year globally (Plastics Europe, 2012), approximately 10% of which is thought to end up entering the world's oceans (Thompson et al., 2004), where it may take centuries to break down. While the societal benefits of plastics as a commodity have been far reaching, there has long been concern for the potential for long-lasting marine plastic debris to cause harm to wildlife.

The impacts of large plastic debris, termed 'macroplastics', have long been recognized by environmental scientists, and can range from strangling and entanglement to internal damage cause by ingestion in a range of fish, marine mammals, and seabirds. Over 250 marine species are believed to be impacted by plastic ingestion (Laist, 1987). Other documented impacts of macroplastics include smothering of the seabed to prevent gaseous exchange (Moore, 2008) and the transport of invasive species of floating plastic debris (Barnes, 2002). A more recent area of concern is the fate of much smaller plastic particles, termed 'microplastics'. Defined as less than 5 mm in size by the National Oceanic and Atmospheric Administration (NOAA), microplastics can be purposefully manufactured to be small (such as the scrubbers added to cosmetics), or can be the product of degradation of larger plastic debris or fibrous polymer material released from fishing ropes and the washing of synthetic clothes. These microplastics can either float or sink depending on the type and density of the plastic and the level of biofouling that occurs on the plastic by microorganisms; hence plastics can be found in both the water column and benthic sediments. They are known to accumulate in 'plastics hotspots', such as the oceanic gyres, as a result of their movement in ocean currents; however, they can be found throughout the world's oceans, including in the most remote locations. As a result of the continued widespread use of plastics, microplastics have now accumulated in oceans and sediments worldwide, with maximum concentrations reaching 100 000 particles m^{-3} (Norén and Naustvoll, 2011), and are continuing to increase in prevalence (Wright et al., 2013b). Fig. 5.3 shows Neuston net trawling for microplastics.

Owing to their small size and presence in both pelagic and benthic ecosystems, microplastics are considered to be bioavailable to biota and have the potential to be ingested by an array of marine organisms (Betts, 2008; Thompson et al., 2009). While observing microplastic ingestion in the wild has proven to be methodologically challenging, an increasing number of studies have now reported microplastic ingestion throughout the food chain in both laboratory-based (i.e. manipulated microplastics exposures) and field-based studies.

Figure 5.3 Sampling for marine microplastics; (a) Neuston nets towed on the surface; (b) sampling the strandline by hand; (c) a Ponar sediment grab used for sampling marine sediments. Photos by Ceri Lewis. [PLATE 3]

Lower-trophic level marine organisms, such as the filter feeders, suspension feeders, deposit feeders, and detritivores, are susceptible to ingesting microplastics as they are often indiscriminate feeders that are unable to differentiate between microplastics and their natural food. Ingestion of microplastic particles has now been demonstrated in a range of zooplankton, invertebrates, and echinoderm larvae (Browne et al., 2008; Graham and Thompson, 2009; Murray and Cowie, 2011; Cole et al., 2013). For example, gut content analysis of the crustacean *Nephrops norvegicus* found that 83% of animals collected from the Clyde Sea contained plastic, most of which took the form of nylon-strand balls most probably from fishing rope (Murray and Cowie, 2011).

The potential for microplastic ingestion to cause damage to the organisms ingesting them is less clear. Evidence is starting to mount that microplastics might clog and block the feeding appendages of filter or suspension feeding marine invertebrates (Derraik, 2002), reducing feeding efficiency. Accumulation of microplastic particles in the guts of marine invertebrates could potentially cause blockages throughout the digestive system, suppressing feeding due to satiation, which might ultimately lead to reduced growth rates and reproductive output. For example in the polychaete *Arenicola marina* microplastic ingestion made the worms less able to acquire energy from their diet

and therefore lose weight (Wright et al., 2013a). A few studies have also shown microplastics can be translocated across the gut epithelium into the circulatory system (Browne et al., 2008) and taken up into haemocytes (invertebrate blood cells). If ingested, microplastics are accumulated within the tissues of organisms at these lower trophic levels, so there is also the potential for them to be transferred to any higher trophic organisms consuming them.

Toxicity from microplastic ingestion could also arise from the desorption in the gut of contaminants adsorbed onto the plastic materials, including plastic additives used in production processes, capable of causing carcinogenesis and endocrine disruption (Teuten et al., 2007; Oehlmann et al., 2009) or other hydrophobic persistent organic pollutants (POPs) from seawater, which have a greater affinity for the hydrophobic surface of plastic compared to seawater.

5.8 Combined effects of multiple stressors

In the environment, pollutants rarely occur in isolation, and generally aquatic systems are affected by a range of stressors simultaneously. In addition, the biotic and abiotic characteristics of water systems fluctuate continuously due to natural causes, such as diurnal or seasonal patterns. Organisms in a given environment, therefore, are rarely exposed to a single stressor in the absence of any variation in natural conditions. In order to appropriately consider how stressors affect the physiology of aquatic organisms, it is imperative that we consider how stressors may be combined to affect target organisms, and what happens to the ability of organisms to respond to stressors when other biotic and abiotic factors are fluctuating. This area of research is a vast and complex one, and there is a striking lack of knowledge on how stressors may interact to cause harm to individuals and populations. This is particularly relevant in a world where global changes are occurring and where marine systems are increasingly affected by a wide range of stressors acting in combination. In the following sections we provide examples of combinations of stressors that affect marine ecosystems globally, and we discuss how ocean acidification and hypoxia affect the responses of aquatic organisms to chemical toxicity.

5.8.1 Ocean acidification and contaminant interactions

The change in seawater pH associated with ocean acidification (OA) (described in Chapter 3) has the potential to alter the speciation and behaviour, and therefore the bioavailability and toxicity, of a number of marine contaminants. Many environmental chemicals form complexes with the ions and organic materials found in seawater, and these complexes can be very pH sensitive. Metals are one of the most common types of coastal contaminant and are found in high concentrations in the waters and sediments of many coastal and estuarine systems (Bryan and Langston, 1992). OA is expected to alter the bioavailability of waterborne metals (Millero et al., 2009), either increasing or decreasing their free ion concentration depending on the metal. For example for both copper and nickel, inorganic speciation is dominated by complexation to the bicarbonate $[HCO_3^-]$ ion which will reduce under OA conditions. The toxic free-ion concentration of metals such as copper (Cu) may therefore increase by as much as 115% in coastal waters in the next 100 years due to reduced pH (Pascal et al., 2010; Richards et al., 2011), while the free-ion concentration of other metals including cadmium (Cd) may decrease or be unaffected (Lacoue-Labarthe et al., 2009, 2011, 2012; Pascal et al., 2010). Altered bioavailability of metals would potentially lead to altered uptake and bioaccumulation of these metals in any exposed marine organism. For example, bioaccumulations of some trace

metals have been shown to be altered by $p CO_2$, e.g. in the eggs of the squid *Loligo vulgaris* and the cuttlefish *Sepia officinalis*, such that accumulation of some metals increased under high $p CO_2$ while others decreased.

There is now evidence from a few marine invertebrate species that the lethal and sublethal toxicity effects of low-level metal contamination are significantly increased when organisms are exposed to these metals under near future (i.e. higher) $p CO_2$ levels. One study looked at metal fluxes from contaminated sediments under future $p CO_2$ scenarios and the toxicity impacts of this combined exposure on the sediment dwelling crustacean *Corophium volutator* (Roberts et al., 2013). No evidence of substantial changes in metal fluxes between the sediment and the seawater was found (although Ni flux was altered slightly); however, increased mortality and DNA damage in *Corophium volutator* in the contaminated sediments was observed under high CO_2 exposures. DNA damage in body tissues increased 1.7-fold in the clean reference sediment at 1140 µatm $p CO_2$ and 2.7-fold in the contaminated sediment at just 750 µatm $p CO_2$. In another study using the intertidal tubeworm *Pomatoceros lamarckii*, larvae were found to have much lower survival under ocean acidification conditions when there was also a low (environmentally realistic) level of the metal copper also present in the seawater (Lewis et al., 2013). The increased metal ion concentration under higher $p CO_2$ is just one of a number of potential explanations for these early findings. An energetic trade-off between maintaining acid–base balance in the face of elevated CO_2 and maintaining detoxification and repair processes has been suggested as a possible mechanism to explain these findings. Alternatively, the mechanism could be more direct: if metals disrupt osmoregulation this will directly impact the capacity to regulate acid–base balance, and, combined with OA (which also impacts acid–base balance), could exacerbate internal acidosis. Carbonic anhydrase, a key enzyme in acid–base regulation and calcification, has been shown to be inhibited by copper, suggesting another potential direct toxicity mechanism for this interaction. There are a number of contaminants for which behaviour and speciation in seawater will change within the pH change predicted with ocean acidification by the year 2100 and the research on contaminant–ocean acidification interactions is very much in its infancy.

5.8.2 Hypoxia and chemical stress

Hypoxia occurs naturally in marine systems and is associated with areas where mixing of the water column is limited, for example due to the formation of thermoclines, and areas of upwelling where, due to the influx of nutrients supporting the food chain, the oxygen consumption by aerobic organisms exceeds oxygen production by photosynthetic organisms and diffusion from the atmosphere. As a result, large areas of the marine environment are characterized by low oxygen concentrations. Climate change and increased nutrient input from agricultural run-off and sewage effluent discharges have contributed to the increase in the incidence, severity, and prevalence of hypoxia over time. Currently, hypoxia is considered to be one of the fastest increasing stressors in marine environments and during the last forty years the number of hypoxic zones has increased exponentially (Diaz and Rosenberg, 2008). As a result, marine communities have sustained the negative impacts of exposure to hypoxia and in certain areas, including the Gulf of Mexico, this has led to the collapse of populations of fish and associated fisheries (Diaz and Rosenberg, 2008). One of the principal causes of the recent increases in marine hypoxia is the increased nutrient input originating from agriculture or sewage effluent discharges. These nutrient-rich waters may also contain a large number of chemical contaminants including natural and synthetic hormones, toxic metals, alkylphenols, PAHs, pesticides, plastics, etc. Therefore, marine organisms in hypoxic waters are likely to be exposed simultaneously to a mixture of chemical contaminants. Despite this, data on the interactions between hypoxia and chemical toxicity are scarce (Nikinmaa, 2013).

A number of studies have investigated this question using model organisms in order to better understand the physiological and molecular interactions between chemical toxicity and hypoxia. Prasch and colleagues (Prasch et al., 2004) investigated the effects of dioxin (TCDD) and hypoxia on zebrafish embryos and observed a reduction in toxicity when exposures occurred simultaneously, but were unable to determine the exact mechanisms responsible for this unexpected effect. Co-exposure to hypoxia and a range of PAHs (that act via activation of the aryl hydrocarbon receptor; AhR) resulted in a range of outcomes, including an increase (for fluoranthene; Matson et al., 2008), a decrease (for pyrene; Fleming and Di Giulio, 2011), or a lack of change (Fleming and Di Giulio, 2011) in chemical toxicity. These effects were attributed to the fact that the signalling molecule, hypoxia inducible factor (HIF), a key factor regulating the cellular response to hypoxia, shares a dimerization partner (aryl hydrocarbon receptor nuclear translocator protein; ARNT) with the AhR pathway and therefore combined exposure to

both stressors is likely to result in attenuation of one or both signalling pathways, due to competition for ARNT (Fig. 5.4; reviewed in Hooper et al., 2013). The AhR signalling pathway is activated following exposure to various chemicals and results in the expression of Cyp1a, a key enzyme for xenobiotic metabolism. The changes in chemical toxicity observed when exposures occur under hypoxia, therefore, may be associated with the suppression in the chemical-induced activation of the AhR pathway in the presence of hypoxia. For chemicals for which the parental compound is more toxic than its metabolites, an increase in chemical toxicity is likely to occur when exposures occur under hypoxia and, conversely, for chemicals where the metabolites are more toxic than the parent compound, suppression of chemical toxicity in the presence of hypoxia may result from combined exposures to both stressors. However, it is also important to consider the impacts of chemical exposures on the organism's ability to cope with decreased oxygen. If competition for the HIF dimerization partner occurs, the resulting suppression in the activation of the HIF pathway may result in an increased sensitivity to hypoxia in organisms simultaneously exposed to combinations of hypoxia and AhR activators. Furthermore, chemicals classified as AhR activators often act through a variety of other signalling pathways, which may respond differently to interactions with the activation of hypoxia signalling pathways. These complex modes of action may further modify the outcomes of interactions between hypoxia and chemical toxicity for this class of environmental chemicals.

For metals, chemical toxicity has been shown to increase when exposures occur in the presence of hypoxia, for a number of fish species (Garcia Sampaio et al., 2008; Mustafa et al., 2012). However, published data sets have focused on adult stages and much less attention has been given to embryonic and early life stages. Recent data from our laboratory indicate that during early embryogenesis the toxicity of copper to both zebrafish and three-spined stickleback is significantly reduced under hypoxia, with increased toxicity observed for exposures occurring after hatching (Fitzgerald et al., in preparation). For some metals, including manganese, hypoxia increases its bioavailability by favouring the solubility of the reduced ions in marine sediments, potentially increasing its toxicity, in particular for burrowing organisms exposed to increased concentrations of manganese in pore waters, as has been reported for *Nephrops norvegicus* (Eriksson et al., 2013). Generally, knowledge about the effects of hypoxia on metal toxicity in the aquatic environment is

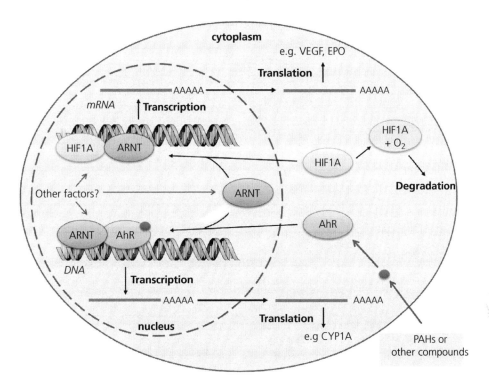

Figure 5.4 Schematic illustration of potential molecular mechanisms of interaction between hypoxia and aryl hydrocarbon receptor (AhR) agonists. Abbreviations: HIF1A—hypoxia-inducible factor 1-alpha; ARNT—aryl hydrocarbon receptor nuclear translocator protein; VEGF—vascular endothelial growth factor; EPO—erythropoietin; PAH—polycyclic aromatic hydrocarbon; CYP1A—cytochrome P4501A.

limited, highlighting a significant knowledge gap that requires further investigation.

Results from recent studies illustrate the knowledge gap in the interactions between hypoxia and chemical toxicity. These are likely to be highly dependent on the mode of action of the chemicals being considered, the severity of hypoxia and its prevalence, the species affected and their life stages, and other environmental variables (such as pH, temperature, and salinity) that may influence the response of organisms to xenobiotics. Further research is needed for these relationships to be understood, in order to provide a framework for predictions of the global consequences of the increase in incidence of chemical pollution overlapping with hypoxic events.

5.9 Conclusion

Oceans cover 70% of the Earth's surface, or 99% of the three-dimensional living space, and act as sinks for anthropogenic chemicals across the globe. Much of the ocean remains under-studied, with new species being frequently discovered, which makes the assessment of the consequences of anthropogenic pollution for marine biodiversity particularly difficult to predict. Factors contributing to this are the lack of data on the fundamental biology of the majority of marine species, and on the effects of chemical exposures on populations and communities. The most well-studied communities are those inhabiting the shoreline, due to accessibility for study and human interests, and conversely some of the least accessible environments, such as the deep sea, remain largely unknown. This is further complicated by the fact that chemicals rarely occur in otherwise pristine environments and their toxicity is likely to be modified by changing conditions in a species-specific and life stage-specific manner.

Substantial efforts have been made in recent decades to evaluate the risk that pollutants pose to the marine environment and to implement appropriate legislation and management measures, such as the recent Marine Strategy Framework, to protect threatened species and communities. These measures are often based on data for sentinel species, chosen based not just on their sensitivity to environmental pollutants (as the relative sensitivity is often unknown due to the lack of information available for the majority of marine species), but for practical reasons, related to their amenability to experimentation. A continuous effort of research has been directed to generate more in-depth and reliable data, and to establish guidelines for chemical testing in marine species, and this must continue in the future to allow for the improvement of the management measures put in place to promote the sustainability of marine ecosystems.

The marine environment has continuously changed over geological time but never as rapidly as in the last century. Challenges that xenobiotics pose to marine organisms can be modified by these changing conditions, and the lack of data on what happens to the physiology of organisms when various environmental variables are altered make predictions of global change impossible to perform. To do this, interdisciplinary approaches combining information and expertise from various disciplines will be required in order to provide in-depth understanding of how biological systems adapt, or incur long-lasting harm, in a changing climate. In the future, considerations of all variables impacting on marine systems must be taken into account when legislating and managing the health of the world's oceans.

References

Aitken, R.J., Koopman, P., & Lewis, S.E.M. (2004). Seeds of concern. *Nature*, **432**, 48–52.

Arkoosh, M.R., Clemons, E., Huffman, P., & Kagley, A.N. (2001). Increased susceptibility of juvenile Chinook salmon to vibriosis after exposure to chlorinated and aromatic compounds found in contaminated urban estuaries. *Journal of Aquatic Animal Health*, **13**, 257–268.

Barnes, D.K.A. (2002). Human rubbish assists alien invasions of seas. *Scientific World Journal*, **2**, 107–112.

Betts, K. (2008). Why small plastic particles may pose a big problem in the oceans. *Environmental Science and Technology*, 42, 8995–8995.

Bjorkblom, C., Mustamaki, N., Olsson, P.E., et al. (2013). Assessment of reproductive biomarkers in three-spined stickleback (*Gasterosteus aculeatus*) from sewage effluent recipients. *Environmental Toxicology*, **28**, 229–237.

Bolognesi, C. & Hayashi, M. (2011). Micronucleus assay in aquatic animals. *Mutagenesis*, **26**, 205–213.

Brander, S.M., Connon, R.E., He, G., et al. (2013). From 'omics to otoliths: responses of an estuarine fish to endocrine disrupting compounds across biological scales. *PLoS ONE*, **8**, e74251.

Brar, N.K., Waggoner, C., Reyes, J.A., et al. (2010). Evidence for thyroid endocrine disruption in wild fish in San Francisco Bay, California, USA. Relationships to contaminant exposures. *Aquatic Toxicology*, **96**, 203–215.

Browne, M.A., Dissanayake, A., Galloway, T.S., et al.(2008). Ingested microscopic plastic translocates to the circulatory system of the mussel, *Mytilus edulis (L.)*. *Environmental Science and Technology*, **42**, 5026–5031.

Bryan, G.W. & Langston, W.J. (1992). Bioavailability, accumulation and effects of heavy-metals in sediments with special reference to United Kingdom estuaries—a review. *Environmental Pollution*, **76**, 89–131.

Caldwell, G.S., Lewis, C., Pickavance, G., et al. (2011). Exposure to copper and a cytotoxic polyunsaturated aldehyde induces reproductive failure in the marine polychaete *Nereis virens* (Sars). *Aquatic Toxicology*, **104**, 126–134.

Castro, L.F., Lima, D., Machado, A., et al. (2007). Imposex induction is mediated through the Retinoid X Receptor signalling pathway in the neogastropod *Nucella lapillus*. *Aquatic Toxicology*, **85**, 57–66.

Cole, M., Lindeque, P., Fileman, E., et al. (2013). Microplastic Ingestion by Zooplankton. *Environmental Science and Technology*, **47**, 6646–6655.

Depledge, M.H. (1998). The ecotoxicological significance of genotoxicity in marine invertebrates. *Mutation Research*, **399**, 109–122.

Derraik, J.G.B. (2002). The pollution of the marine environment by plastic debris: a review. *Marine Pollution Bulletin*, **44**, 842–852.

Diaz, R.J. & Rosenberg, R. (2008). Spreading dead zones and consequences for marine ecosystems. *Science*, **321**, 926–929.

Dong, Y., Tian, H., Wang, W., et al. (2014). Disruption of the thyroid system by the thyroid-disrupting compound Aroclor 1254 in juvenile Japanese flounder (*Paralichthys olivaceus*). *PLoS ONE*, **9**, e104196.

Du Bois PB, Laguionie P, Boust D, Korsakissok I, Didier D, Fievet B (2012) Estimation of marine source-term following Fukushima Dai-ichi accident. *Journal of Environmental Radioactivity*, **114**, 2–9.

Ellis, R.J., Van Den Heuvel, M.R., Bandelj, E., et al. (2003). In vivo and in vitro assessment of the androgenic potential of a pulp and paper mill effluent. *Environmental Toxicology Chemistry*, **22**, 1448–1456.

Elston, R.A., Moore, J.D., & Brooks, K. (1992). Disseminated neoplasia of bivalve mollusks. *Reviews in Aquatic Sciences*, **6**, 405–466.

Eriksson, S.P., Hernroth, B., & Baden, S.P. (2013). Stress biology and immunology in *Nephrops norvegicus*. *Advances in Marine Biology*, **64**, 149–200.

Fairbairn, D.W., Olive, P.L., & O'Neill, K.L. (1995). The comet assay—a comprehensive review. *Mutation Research*, **339**, 37–59.

Falfushynska, H.I. & Stolyar, O.B. (2009). Responses of biochemical markers in carp *Cyprinus carpio* from two field sites in Western Ukraine. *Ecotoxicology and Environmental Safety*, **72**, 729–736.

Fernandes, D., Loi, B., & Porte, C. (2011). Biosynthesis and metabolism of steroids in molluscs. *Journal of Steroid Biochemistry and Molecular Biology*, **127**, 189–195.

Filby, A.L., Neuparth, T., Thorpe, K.L., et al. (2007). Health impacts of estrogens in the environment, considering complex mixture effects. *Environmental Health Perspectives*, **115**, 1704–1710.

Fitzpatrick, J.L., Nadella, S., Bucking, C., et al. (2008). The relative sensitivity of sperm, eggs and embryos to copper in the blue mussel (*Mytilus trossulus*). *Comparative Biochemistry and Physiology Part C*, **147**, 441–449.

Fleming, C.R. & Di Giulio, R.T. (2011). The role of CYP1A inhibition in the embryotoxic interactions between hypoxia and polycyclic aromatic hydrocarbons (PAHs) and PAH mixtures in zebrafish (*Danio rerio*). *Ecotoxicology*, **20**, 1300–1314.

Galloway, T.S. & Depledge, M.H. (2001). Immunotoxicity in invertebrates: measurement and ecotoxicological relevance. *Ecotoxicology*, **10**, 5–23.

Galloway, T. & Handy, R. (2003). Immunotoxicity of organophosphorus pesticides. *Ecotoxicology*, **12**, 345–363.

Garcia Sampaio, F., De Lima Boijink, C., Tie Oba, E., et al. (2008). Antioxidant defenses and biochemical changes in pacu (*Piaractus mesopotamicus*) in response to single and combined copper and hypoxia exposure. *Comparative Biochemistry and Physiology Part C*, **147**, 43–51.

Geraudie, P., Nahrgang, J., Forget-Leray, J., et al. (2014). *In vivo* effects of environmental concentrations of produced water on the reproductive function of polar cod (*Boreogadus saida*). *Journal of Toxicology and Environmental Health Part A*, **77**, 557–573.

Graham, E.R. & Thompson, J.T. (2009). Deposit- and suspension-feeding sea cucumbers (Echinodermata) ingest plastic fragments. *Journal of Experimental Marine Biology and Ecology*, **368**, 22–29.

Grung, M., Naes, K., Fogelberg, O., et al. (2011). Effects-directed analysis of sediments from polluted marine sites in Norway. *Journal of Toxicology and Environmental Health Part A*, **74**, 439–454.

Hamilton, P.B., Nicol, E., De-Bastos, E.S., et al. (2014). Populations of a cyprinid fish are self-sustaining despite widespread feminization of males. *BMC Biology*, **12**, 1.

Hannam, M.L., Bamber, S.D., Galloway, T.S., et al. (2010a). Effects of the model PAH phenanthrene on immune function and oxidative stress in the haemolymph of the temperate scallop *Pecten maximus*. *Chemosphere*, **78**, 779–784.

Hannam, M.L., Bamber, S.D., Moody, A.J., et al. (2010b). Immunotoxicity and oxidative stress in the Arctic scallop *Chlamys islandica*: effects of acute oil exposure. *Ecotoxicology and Environmental Safety*, **73**, 1440–1448.

Harris, C.A., Santos, E.M., Janbakhsh, A., et al. (2001). Nonylphenol affects gonadotropin levels in the pituitary gland and plasma of female rainbow trout. *Environmental Science and Technology*, **35**, 2909–2916.

Havenhand, J.N., Buttler, F.R., Thorndyke, M.C., & Williamson, J.E. (2008). Near-future levels of ocean acidification reduce fertilization success in a sea urchin. *Current Biology*, **18**, R651–R652.

Havenhand, J.N. & Schlegel, P. (2009). Near-future levels of ocean acidification do not affect sperm motility and fertilization kinetics in the oyster *Crassostrea gigas*. *Biogeosciences*, **6**, 3009–3015.

Hooper, M.J., Ankley, G.T., Cristol, D.A., et al. (2013). Interactions between chemical and climate stressors: a role for mechanistic toxicology in assessing climate change risks. *Environmental Toxicology and Chemistry*, **32**, 32–48.

Jobling, S., Burn, R.W., Thorpe, K., et al. (2009). Statistical modeling suggests that antiandrogens in effluents from wastewater treatment works contribute to widespread sexual disruption in fish living in English rivers. *Environmental Health Perspectives*, **117**, 797–802.

Jobling, S., Coey, S., Whitmore, J.G., et al. (2002). Wild intersex roach (*Rutilus rutilus*) have reduced fertility. *Biology of Reproduction*, **67**, 515–524.

Jobling, S., Williams, R., Johnson, A., et al. (2006). Predicted exposures to steroid estrogens in U.K. rivers correlate with widespread sexual disruption in wild fish populations. *Environmental Health Perspectives*, **114**(Suppl. 1), 32–39.

Kadar, E., Tarran, G.A., Jha, A.N., & Al-Subiai, S.N. (2011). Stabilization of engineered zero-valent nanoiron with Na-acrylic copolymer enhances spermiotoxicity. *Environmental Science and Technology*, **45**, 3245–3251.

Katsiadaki, I., Sanders, M., Sebire, M., et al. (2007). Three-spined stickleback: an emerging model in environmental endocrine disruption. *Environmental Science*, **14**, 263–283.

Kidd, K.A., Blanchfield, P.J., Mills, K.H., et al. (2007). Collapse of a fish population after exposure to a synthetic estrogen. *Proceedings of the National Academy of Sciences of the USA*, **104**, 8897–901.

Kirby, M.F., Allen, Y.T., Dyer, R.A., et al. (2004a). Surveys of plasma vitellogenin and intersex in male flounder (*Platichthys flesus*) as measures of endocrine disruption by estrogenic contamination in United Kingdom estuaries: temporal trends, 1996 to 2001. *Environmental Toxicology and Chemistry*, **23**, 748–758.

Kirby, M.F., Neall, P., Bateman, T.A., & Thain, J.E. (2004b). Hepatic ethoxyresorufin O-deethylase (EROD) activity in flounder (*Platichthys flesus*) from contaminant impacted estuaries of the United Kingdom: continued monitoring 1999–2001. *Marine Pollution Bulletin*, **49**, 71–78.

Klobucar, G.I.V., Stambuk, A., Hylland, K., & Pavlica, M. (2008). Detection of DNA damage in haemocytes of *Mytilus galloprovincialis* in the coastal ecosystems of Kastela and Trogir bays, Croatia. *Science of the Total Environment*, **405**, 330–337.

Lacoue-Labarthe, T., Martin, S., Oberhänsli, F., et al. (2012). Temperature and pCO(2) effect on the bioaccumulation of radionuclides and trace elements in the eggs of the common cuttlefish, *Sepia officinalis*. *Journal of Experimental Marine Biology and Ecology*, **413**, 45–49.

Lacoue-Labarthe, T., Martin, S., Oberhänsli, F., et al. (2009). Effects of increased pCO(2) and temperature on trace element (Ag, Cd and Zn) bioaccumulation in the eggs of the common cuttlefish, *Sepia officinalis*. *Biogeosciences*, **6**, 2561–2573.

Lacoue-Labarthe, T., Reveillac, E., Oberhänsli, F., et al. (2011). Effects of ocean acidification on trace element accumulation in the early-life stages of squid *Loligo vulgaris*. *Aquatic Toxicology*, **105**, 166–176.

Laffon, B., Rabade, T., Pasaro, E., & Mendez, J. (2006). Monitoring of the impact of Prestige oil spill on *Mytilus galloprovincialis* from Galician coast. *Environment International*, **32**, 342–348.

Laist, D.W. (1987). Overview of the biological effects of lost and discarded plastic debris in the marine-environment. *Marine Pollution Bulletin*, **18**, 319–326.

Lange, A., Paull, G.C., Coe, T.S., et al. (2009). Sexual reprogramming and estrogenic sensitization in wild fish exposed to ethinylestradiol. *Environmental Science and Technology*, **43**, 1219–1225.

Leblanc, G.A. (2007). Crustacean endocrine toxicology: a review. *Ecotoxicology*, **16**, 61–81.

Lesser, M.P. (2006). Oxidative stress in marine environments: biochemistry and physiological ecology. *Annual Review of Physiology*, **68**, 253–278.

Lewis, C., Clemow, K., & Holt, W.V. (2013). Metal contamination increases the sensitivity of larvae but not gametes to ocean acidification in the polychaete *Pomatoceros lamarckii* (Quatrefages). *Marine Biology*, **160**, 2089–2101.

Lewis, C. & Galloway, T. (2008). Genotoxic damage in polychaetes: a study of species and cell-type sensitivities. *Mutation Research*, **654**, 69–75.

Lewis, C. & Galloway, T. (2009). Reproductive consequences of paternal genotoxin exposure in marine invertebrates. *Environmental Science and Technology*, **43**, 928–933.

Lewis C, Ford AT (2012) Infertility in male aquatic invertebrates: A review. *Aquatic Toxicology*, **120**, 79–89.

Livingstone, D.R. (2002). Pollution and oxidative stress in aquatic organisms. *Revue De Medecine Veterinaire*, **153**, 522–522.

Livingstone, D.R., Chipman, J.K., Lowe, D.M., et al. (2000). Development of biomarkers to detect the effects of organic pollution on aquatic invertebrates: recent molecular, genotoxic, cellular and immunological studies on the common mussel (*Mytilus edulis L.*) and other mytilids. *International Journal of Environment and Pollution*, **13**, 56–91.

Lorenzon, S., Francese, M., Smith, V.J., & Ferrero, E.A. (2001). Heavy metals affect the circulating haemocyte number in the shrimp *Palaemon elegans*. *Fish and Shellfish Immunology*, **11**, 459–472.

Martinovic, D., Hogarth, W.T., Jones, R.E., & Sorensen, P.W. (2007). Environmental estrogens suppress hormones, behavior, and reproductive fitness in male fathead minnows. *Environmental Toxicology and Chemistry*, **26**, 271–278.

Matson, C.W., Timme-Laragy, A.R., & Di Giulio, R.T. (2008). Fluoranthene, but not benzo[a]pyrene, interacts with hypoxia resulting in pericardial effusion and lordosis in developing zebrafish. *Chemosphere*, **74**, 149–154.

Matthiessen, P., Waldock, R., Thain, J.E., et al. (1995). Changes in periwinkle (*Littorina littorea*) populations following the ban on TBT-based antifoulings on small boats in the United Kingdom. *Ecotoxicology and Environmental Safety*, **30**, 180–194.

Meier, S., Andersen, T.E., Norberg, B., et al. (2007). Effects of alkylphenols on the reproductive system of Atlantic cod (*Gadus morhua*). *Aquatic Toxicology*, **81**, 207–218.

Meier, S., Craig Morton, H., Nyhammer, G., et al. (2010). Development of Atlantic cod (*Gadus morhua*) exposed to produced water during early life stages: effects on embryos, larvae, and juvenile fish. *Marine Environmental Research*, **70**, 383–394.

Meier, S., Morton, H.C., Andersson, E., et al. (2011). Low-dose exposure to alkylphenols adversely affects the sexual development of Atlantic cod (*Gadus morhua*): acceleration of the onset of puberty and delayed seasonal gonad development in mature female cod. *Aquatic Toxicology*, **105**, 136–150.

Mendez-Fernandez, P., Webster, L., Chouvelon, T., et al. (2014). An assessment of contaminant concentrations in toothed whale species of the NW Iberian Peninsula: part I. Persistent organic pollutants. *Science of the Total Environment*, **484**, 196–205.

Millero, F.J., Woosley, R., Ditrolio, B., & Waters, J. (2009). Effect of ocean acidification on the speciation of metals in seawater. *Oceanography*, **22**, 72–85.

Mirbahai, L., Yin, G., Bignell, J.P., et al. (2011). DNA methylation in liver tumorigenesis in fish from the environment. *Epigenetics*, **6**, 1319–1333.

Miyagawa, S., Lange, A., Hirakawa, I., et al. (2014). Differing species responsiveness of estrogenic contaminants in fish is conferred by the ligand binding domain of the estrogen receptor. *Environmental Science and Technology*, **48**, 5254–5263.

Moore, C.J. (2008). Synthetic polymers in the marine environment: a rapidly increasing, long-term threat. *Environmental Research*, **108**, 131–139.

Morcillo, Y. & Porte, C. (1999). Evidence of endocrine disruption in the imposex-affected gastropod *Bolinus brandaris*. *Environmental Research*, **81**, 349–354.

Morita, M., Suwa, R., Iguchi, A., et al. (2010). Ocean acidification reduces sperm flagellar motility in broadcast spawning reef invertebrates. *Zygote*, **18**, 103–107.

Murray, F. & Cowie, P.R. (2011). Plastic contamination in the decapod crustacean *Nephrops norvegicus* (Linnaeus, 1758). *Marine Pollution Bulletin*, **62**, 1207–1217.

Mustafa, S.A., Davies, S.J., & Jha, A.N. (2012). Determination of hypoxia and dietary copper mediated sub-lethal toxicity in carp, *Cyprinus carpio*, at different levels of biological organisation. *Chemosphere*, **87**, 413–422.

Muttray, A., Reinisch, C., Miller, J., et al. (2012). Haemocytic leukemia in Prince Edward Island (PEI) soft shell clam (*Mya arenaria*): spatial distribution in agriculturally impacted estuaries. *Science of the Total Environment*, **424**, 130–142.

Nash, J.P., Kime, D.E., Van Der Ven, L.T., et al. (2004). Long-term exposure to environmental concentrations of the pharmaceutical ethynylestradiol causes reproductive failure in fish. *Environmental Health Perspectives*, **112**, 1725–1733.

Nikinmaa, M. (2013). Climate change and ocean acidification-interactions with aquatic toxicology. *Aquatic Toxicology*, **126**, 365–372.

Nishikawa, J., Mamiya, S., Kanayama, T., et al. (2004). Involvement of the retinoid X receptor in the development of imposex caused by organotins in gastropods. *Environmental Science and Technology*, **38**, 6271–6276.

Norén, F. & Naustvoll, L.J. (2011). *Survey of Microscopic Anthropogenic Particles in Skagerrak*. Institute of Marine Research, Flødevigen, Norway.

Oehlmann, J., Schulte-Oehlmann, U., Kloas, W., et al. (2009). A critical analysis of the biological impacts of plasticizers on wildlife. *Philosophical Transactions of the Royal Society Series B*, **364**, 2047–2062.

Orlando, E.F., Bass, D.E., Caltabiano, L.M., et al. (2007). Altered development and reproduction in mosquitofish exposed to pulp and paper mill effluent in the Fenholloway River, Florida, USA. *Aquatic Toxicology*, **84**, 399–405.

Pascal, P.Y., Fleeger, J.W., Galvez, F., & Carman, K.R. (2010). The toxicological interaction between ocean acidity and metals in coastal meiobenthic copepods. *Marine Pollution Bulletin*, **60**, 2201–2208.

Pellerin-Massicotte, J. (1994). Oxidative processes as indicators of chemical stress in marine bivalves. *Journal of Aquatic Ecosystem Health*, **3**, 101–111.

Pipe, R.K. & Coles, J.A. (1995). Environmental contaminants influencing immune function in marine bivalve mollusks. *Fish and Shellfish Immunology*, **5**, 581–595.

Plastics Europe (2012) 'Plastics – the Facts 2014/2015: An analysis of European plastics production, demand and waste data' Published by the Association of Plastics Manufacturers.

Prasch, A.L., Andreasen, E.A., Peterson, R.E., & Heideman, W. (2004). Interactions between 2,3,7,8-tetrachlorodibenzo-p-dioxin (TCDD) and hypoxia signaling pathways in zebrafish: hypoxia decreases responses to TCDD in zebrafish embryos. *Toxicological Sciences*, **78**, 68–77.

Richards R, Chaloupka M, Sano M, Tomlinson R (2011) Modelling the effects of 'coastal' acidification on copper speciation. *Ecological Modelling*, **222**, 3559–3567.

Roberts, D.A., Birchenough, S.N.R., Lewis, C., et al. (2013). Ocean acidification increases the toxicity of contaminated sediments. *Global Change Biology*, **19**, 340–351.

Rocha, M.J., Cruzeiro, C., Reis, M., et al. (2013). Determination of seventeen endocrine disruptor compounds and their spatial and seasonal distribution in Ria Formosa Lagoon (Portugal). *Environmental Monitoring and Assessment*, **185**, 8215–8226.

Rodriguez, E.M., Medesani, D.A., & Fingerman, M. (2007). Endocrine disruption in crustaceans due to pollutants: a review. *Comparative Biochemistry and Physiology Part A*, **146**, 661–671.

Ross, P., Deswart, R., Addison, R., et al. (1996). Contaminant-induced immunotoxicity in harbour seals: wildlife at risk? *Toxicology*, **112**, 157–169.

Santos, E.M., Hamilton, P.B., Coe, T.S., et al. (2013). Population bottlenecks, genetic diversity and breeding ability of the three-spined stickleback (*Gasterosteus aculeatus*) from three polluted English Rivers. *Aquatic Toxicology*, **142–143**, 264–271.

Santos, E.M., Paull, G.C., Van Look, K.J., et al. (2007). Gonadal transcriptome responses and physiological consequences of exposure to oestrogen in breeding zebrafish (*Danio rerio*). *Aquatic Toxicology*, **83**, 134–142.

Scholz, S. & Mayer, I. (2008). Molecular biomarkers of endocrine disruption in small model fish. *Molecular and Cellular Endocrinology*, **293**, 57–70.

Scott, A.P. (2012). Do mollusks use vertebrate sex steroids as reproductive hormones? Part I: critical appraisal of the evidence for the presence, biosynthesis and uptake of steroids. *Steroids*, **77**, 1450–1468.

Scott, A.P. (2013). Do mollusks use vertebrate sex steroids as reproductive hormones? II. Critical review of the evidence that steroids have biological effects. *Steroids*, **78**, 268–281.

Scott, A.P., Sanders, M., Stentiford, G.D., et al. (2007). Evidence for estrogenic endocrine disruption in an offshore flatfish, the dab (*Limanda limanda* L.). *Marine Environmental Research*, **64**, 128–148.

Segner, H., Wenger, M., Moller, A.M., et al. (2012). Immunotoxic effects of environmental toxicants in fish—how to assess them? *Environmental Science and Pollution Research*, **19**, 2465–2476.

Short, S., Yang, G., Kille, P., & Ford, A.T. (2014). Vitellogenin is not an appropriate biomarker of feminisation in a crustacean. *Aquatic Toxicology*, **153**, 89–97.

Silva, C., Mattioli, M., Fabbri, E., et al. (2012). Benthic community structure and biomarker responses of the clam *Scrobicularia plana* in a shallow tidal creek affected by fish farm effluents (Rio San Pedro, SW Spain). *Environment International*, **47**, 86–98.

Small, H.J., Williams, T.D., Sturve, J., et al. (2010). Gene expression analyses of hepatocellular adenoma and hepatocellular carcinoma from the marine flatfish *Limanda limanda*. *Diseases in Aquatic Organisms*, **88**, 127–141.

Smith, B.S. (1981a). Male characteristics on female mud snails caused by antifouling bottom paints. *Journal of Applied Toxicology*, **1**, 22–25.

Smith, B.S. (1981b). Reproductive anomalies in stenoglossan snails related to pollution from marinas. *Journal of Applied Toxicology*, **1**, 15–21.

Smith, B.S. (1981c). Tributyltin compounds induce male characteristics on female mud snails *Nassarius obsoletus = Ilyanassa obsoleta*. *Journal of Applied Toxicology*, **1**, 141–144.

St Jean, S.D., Bishay, F., & Reinisch, C.L. (2005). Leukemia incidence in two mussels species, *Mytilus edulis* and *M. trossulus* caged in Burrard Inlet, Vancouver, BC in 2004. *Canadian Technical Report of Fisheries and Aquatic Sciences*, **2617**, 47.

Stange, D., Sieratowicz, A., & Oehlmann, J. (2012). Imposex development in *Nucella lapillus*—evidence for the involvement of retinoid X receptor and androgen signalling pathways in vivo. *Aquatic Toxicology*, **106–107**, 20–24.

Steinberg CEW, Geyer HJ, Kettrup AaF (1994) Evaluation xenobiotic effects by ecological techniques. *Chemosphere*, **28**, 357–374.

Steinert, S.A., Streib-Montee, R., Leather, J.M., & Chadwick, D.B. (1998). DNA damage in mussels at sites in San Diego Bay. *Mutation Research*, **399**, 65–85.

Teuten, E.L., Rowland, S.J., Galloway, T.S., & Thompson, R.C. (2007). Potential for plastics to transport hydrophobic contaminants. *Environmental Science and Technology*, **41**, 7759–7764.

Thilagam, H., Gopalakrishnan, S., Bo, J., & Wang, K.J. (2014). Comparative study of 17 beta-estradiol on endocrine disruption and biotransformation in fingerlings and juveniles of Japanese sea bass *Lateolabrax japonicus. Marine Pollution Bulletin*, **85**, 332–327.

Thomas, P. & Rahman, M.S. (2009). Chronic hypoxia impairs gamete maturation in Atlantic croaker induced by progestins through nongenomic mechanisms resulting in reduced reproductive success. *Environmental Science and Technology*, **43**, 4175–4180.

Thomas, P. & Rahman, M.S. (2012). Extensive reproductive disruption, ovarian masculinization and aromatase suppression in Atlantic croaker in the northern Gulf of Mexico hypoxic zone. *Proceedings of the Royal Society of London Series B*, **279**, 28–38.

Thomas, P., Rahman, M.S., Khan, I.A., & Kummer, J.A. (2007). Widespread endocrine disruption and reproductive impairment in an estuarine fish population exposed to seasonal hypoxia. *Proceedings of the Royal Society of London Series B*, **274**, 2693–2701.

Thompson, R.C., Olsen, Y., Mitchell, R.P., et al. (2004). Lost at sea: where is all the plastic? *Science*, **304**, 838–838.

Thompson, R.C., Swan, S.H., Moore, C.J., & Vom Saal, F.S. (2009). Our plastic age. *Philosophical Transactions of the Royal Society Series B*, **364**, 1973–1976.

Tollefsen, K.E., Harman, C., Smith, A., & Thomas, K.V. (2007). Estrogen receptor (ER) agonists and androgen receptor (AR) antagonists in effluents from Norwegian North Sea oil production platforms. *Marine Pollution Bulletin*, **54**, 277–283.

Trumble, S.J., Robinson, E.M., Berman-Kowalewski, M., et al. (2013). Blue whale earplug reveals lifetime contaminant exposure and hormone profiles. *Proceedings of the National Academy of Sciences of the USA*, **110**, 16922–16926.

Tyler, C.R., Jobling, S., & Sumpter, J.P. (1998). Endocrine disruption in wildlife: a critical review of the evidence. *Critical Reviews in Toxicology*, **28**, 319–361.

Tyler, C.R., Vandereerden, B., Jobling, S., et al. (1996). Measurement of vitellogenin, a biomarker for exposure to oestrogenic chemicals, in a wide variety of cyprinid fish. *Journal of Comparative Physiology Part B*, **166**, 418–426.

Uribe, C., Folch, H., Enriquez, R., & Moran, G. (2011). Innate and adaptive immunity in teleost fish: a review. *Veterinarni Medicina*, **56**, 486–503.

Urushitani, H., Katsu, Y., Kato, Y., et al. (2007). Medaka (*Oryzias latipes*) for use in evaluating developmental effects of endocrine active chemicals with special reference to gonadal intersex (testis-ova). *Environmental Sciences*, **14**, 211–233.

Van Aerle, R., Nolan, T.M., Jobling, S., et al. (2001). Sexual disruption in a second species of wild cyprinid fish (the gudgeon, *Gobio gobio*) in United Kingdom freshwaters. *Environmental Toxicology and Chemistry*, **20**, 2841–2847.

Van Aerle, R., Pounds, N., Hutchinson, T.H., et al. (2002). Window of sensitivity for the estrogenic effects of ethinylestradiol in early life-stages of fathead minnow, *Pimephales promelas. Ecotoxicology*, **11**, 423–434.

Vanbeneden RJ (1994) Molecular analysis of bivalve tumors – Models for environmental genetic interactions. *Environmental Health Perspectives*, **102**, 81–83.

Vethaak, A.D., Jol, J.G., & Pieters, J.P. (2009). Long-term trends in the prevalence of cancer and other major diseases among flatfish in the southeastern North Sea as indicators of changing ecosystem health. *Environmental Science and Technology*, **43**, 2151–2158.

Vlahogianni, T., Dassenakis, M., Scoullos, M.J., & Valavanidis, A. (2007). Integrated use of biomarkers (superoxide dismutase, catalase and lipid peroxidation) in mussels *Mytilus galloprovincialis* for assessing heavy metals pollution in coastal areas from the Saronikos Gulf of Greece. *Marine Pollution Bulletin*, **54**, 1361–1371.

Watanabe, K.H., Li, Z., Kroll, K.J., et al. (2009). A computational model of the hypothalamic-pituitary-gonadal axis in male fathead minnows exposed to 17alpha-ethinylestradiol and 17beta-estradiol. *Toxicological Sciences*, **109**, 180–192.

Wells, R.S., Tornero, V., Borrell, A., et al. (2005). Integrating life-history and reproductive success data to examine potential relationships with organochlorine compounds for bottlenose dolphins (*Tursiops truncatus*) in Sarasota Bay, Florida. *Science of the Total Environment*, **349**, 106–119.

Wright, S.L., Rowe, D., Thompson, R.C., & Galloway, T.S. (2013a). Microplastic ingestion decreases energy reserves in marine worms. *Current Biology*, **23**, R1031–R1033.

Wright, S.L., Thompson, R.C., & Galloway, T.S. (2013b). The physical impacts of microplastics on marine organisms: a review. *Environmental Pollution*, **178**, 483–492.

Wu, R.S., Zhou, B.S., Randall, D.J., et al. (2003). Aquatic hypoxia is an [endocrine] disrupter and impairs fish reproduction. *Environmental Science and Technology*, **37**, 1137–1141.

Yang, G., Kille, P., & Ford, A.T. (2008). Infertility in a marine crustacean: have we been ignoring pollution impacts on male invertebrates? *Aquatic Toxicology*, **88**, 81–87.

CHAPTER 6

Nitrogen stress in the marine environment: from scarcity to surfeit

Erica B. Young and John A. Berges

6.1 Introduction

Although nitrogen is not a particularly rare element on our planet, its diverse redox states and the fact that its common gaseous form, N_2, is difficult for most organisms to acquire predisposes it to being a limiting factor. The large biological requirements for nitrogen in critical structural to catalytic roles are balanced by the paradox that most metazoans must excrete large amounts of nitrogen as metabolic waste (Sterner and Elser, 2002). In marine systems nitrogen ranges from being critically limiting for oceanic primary production to reaching levels verging on toxicity in eutrophic coastal systems. This chapter will focus on the effects of nitrogen on macrofauna, macrophytes, protists, and the cyanobacterial photoautotrophs which form the phytoplankton, but it will not include specific analysis of nitrogen stress in prokaryotic heterotrophs. Clearly the archaeal, bacterial and viral components of ecosystems are essential to nitrogen cycling, not only in terms of atmospheric N_2 fixation but also for transformation of nitrogen compounds, which ultimately affect the forms and availability of nitrogen for eukaryotic organisms; these processes are beyond the scope of the present review and we refer the reader to relevant chapters in the recently published *Nitrogen in the Marine Environment* (Capone et al., 2008). However, there is increasing recognition of the influence of bacterial microbiomes and bacterial biofilms in the physiology of macroscopic organisms, as they can mediate and modulate access to nutrients and responses to stresses including eutrophication (Wahl et al., 2012; Kelly et al., 2014), in the groups of organisms considered below. It is also of importance that heterotrophic prokaryotes are typically richer in nitrogen relative to carbon than many primary producers. Therefore, when microbes assimilate organic matter, there is expected to be an excess of carbon lost as CO_2 which likely contributes a significant term in global C budgets (see Bauer et al., 2013).

Stresses related to nitrogen have often been discussed in broader ecological context e.g. selection for competitive growth of particular organisms, or inhibition of others (See Capone et al., 2008), or the anthropogenic effects on biogeochemical nitrogen cycling e.g. significant industrial output of fixed nitrogen from the Haber process (Gu et al., 2013). However, this chapter will focus on the *physiological* processes by which such effects are mediated, that is, how and why nitrogen affects physiological functions (see Table 6.1 for a summary of these). Other recent reviews have focused on finding mathematical expressions for processes with a view to modelling and prediction (e.g. Glibert et al., 2013), but we will consider primarily the physiological processes, and not the model equations or parameters evaluated.

Inherent in our exploration is the belief that physiological acclimation plays a critical role in organism responses to environmental changes. This is in contrast to a view that would consider that environmental conditions simply select for the best-adapted organisms within a community with diverse physiological capacities. Clearly, neither perspective is exclusive, though those working with larger, longer-lived organisms might tend to favour the former, while those familiar with the smaller planktonic species that turn over rapidly would be more comfortable with the latter. Reynolds (1984) and others have suggested the concept of a diverse phytoplankton assemblage whose structure is shaped by periodic environmental fluctuations. Indeed, the strategy of investing energy and resources in acclimating to an environmental change when there are other species present better adapted to respond to the new conditions might seem futile. Yet

Stressors in the Marine Environment. Edited by Martin Solan and Nia M. Whiteley
© Oxford University Press 2016. Published in 2016 by Oxford University Press.

Table 6.1 Summary of physiological stresses due to nitrogen in different groups of marine organisms.

Stress	Organisms (chapter section)	Environment	Direct physiological effects	Indirect/secondary effects
Low nitrogen	microalgae (6.2.1.1) Case Study 1	oligotrophic oceans	declines in photosynthetic capacity, electron transport, carbon fixation, growth	encystment
	macroalgae (6.2.2.1)	seasonal	declines in protein synthesis, catabolism of storage products, limitation of growth	
	angiosperms (6.2.3.1)	salt marshes, sea grass beds, mangroves; water column and sediment	changes to shoot/root biomass, limitation of growth	anoxia
	animals (6.3.1)	intertidal, zooplankton, fishes (aquaculture)	effects of food quality on growth rate, reproductive output related to C:N ratio; importance of critical compounds (e.g. essential amino acids)	species selection
High nitrogen	microalgae (6.2.1.2)	estuaries, coastal embayments	ionic effects on membrane proton gradients including photophosphorylation	species selection
	macroalgae (6.2.2.2) Case Study 2	estuaries	ammonium toxicity (spores > adults)	eutrophication, species selection
	angiosperms (6.2.3.2) Case Study 3	salt marshes, sea grass beds, mangroves	changes in composition (% N), effects on proton gradients, imbalance between N acquisition and photosynthetic C fixation, die-offs	light limitation from increased epiphytes, competition with macroalgae, decrease in below-ground biomass, sediment anoxia
	Animals (6.3.2) Case Study 4	coastal	toxic ammonium, nitrite, urea; acidosis, disruption of oxidative phosphorylation, osmoregulation	species selection

even so, physiological responses demonstrate a phenotypic plasticity that is ecologically relevant as well as informative, sharing enough similarities among groups that useful generalizations emerge; moreover, even information from laboratory cultures of simple species can be informative (Reynolds, 1984; Fogg and Thake, 1987).

6.1.1 Nitrogen forms that need to be considered

Biogeochemical cycling of nitrogen in the ocean is complex and beyond the scope of this review (see Gruber, 2008), but it is important to appreciate the major nitrogen forms available to organisms, both primary producers and those further up the food chain. The atmosphere is a major reservoir for gaseous N_2, but only some prokaryotes can directly access it. Dissolved

inorganic nitrogen (DIN) forms (nitrate, nitrite, and ammonium) constitute much of the inventory and are principal resources for autotrophs (Gruber, 2008). Dissolved organic forms of nitrogen (DON) can also constitute major proportions of available nitrogen (Gruber, 2008), and there is increasing awareness of the diversity and availability of compounds (Aluwihare and Meador, 2008), particularly urea (Glibert et al., 2006). Heterotrophs obtain nitrogen mainly from dietary protein, though mixotrophic organisms, which can make use of both autotrophy and heterotrophy, can clearly supplement DIN uptake with nitrogen acquisition from organic sources.

6.1.2 The spectrum of nitrogen stress

From a physiological standpoint, deficiency of nitrogen is an obvious stress; virtually all metabolic processes and growth and reproduction depend on nitrogen. The direct physiological effects of high levels of nitrogen (as distinguished from physiological stress imposed by conditions caused by elevated nitrogen, such as anoxia) are less clear, but certainly toxicity of some forms (such as ammonia) are well understood. The potential for nitrogen stress likely follows nitrogen availability, from highly oligotrophic open oceans where fixed nitrogen is scarce and in limited supply (Howarth, 1988) (**low-N stress**, Fig. 6.1) to potentially hypereutrophic coastal regions and estuaries where inputs from land (predominantly from anthropogenic sources) results in nitrogen oversupply (Howarth and Marino, 2006) (**high-N stress,** Fig. 6.2). There are also gradients from planktonic waters (where processes such as stratification and sedimentation lead to nitrogen depletion) to benthic (where regeneration in sediments favour higher concentrations).

The vast majority of our understanding of nitrogen stress in photoautotrophs is based on studies of nitrogen limitation, and we know comparatively little about the physiology of high-N stress (see e.g. Collos et al.,

1997); high nutrient concentrations in ecosystems have been regarded as 'non-ecological'. However, with escalating anthropogenic nitrogen loadings into marine ecosystems from terrestrial and atmospheric sources (Howarth and Marino, 2006; Seitzinger and Harrison, 2008; Voss et al., 2013), high-N stress is becoming more common and relevant to understanding of human impacts on marine ecosystems.

6.1.3 Some important concepts and definitions

Stress can most generally be defined in terms of effects on an organism's function to the point where chances of survival are reduced (cf. Barton, 2002); note that some authors prefer to distinguish the effect on the organism ('stress') from the cause ('stressor'), but in this chapter we will not use this convention.

When considering a stress caused by low N, a wide number of terms may be applied, many of which have been poorly defined and even used in contradictory ways. For example, determining whether 'nutrient limitation' exists or the identification of which nutrient

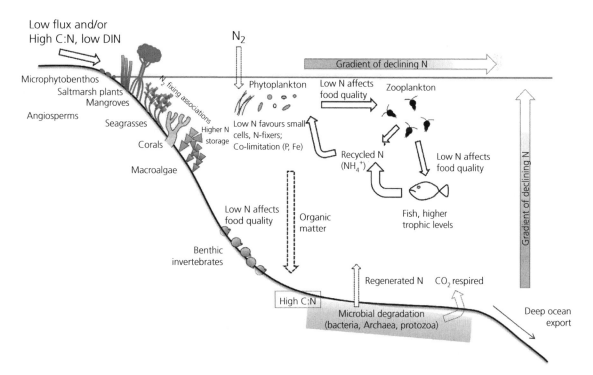

Figure 6.1 Conceptual representation of major components and fluxes in marine ecosystems under stress due to low nitrogen conditions.

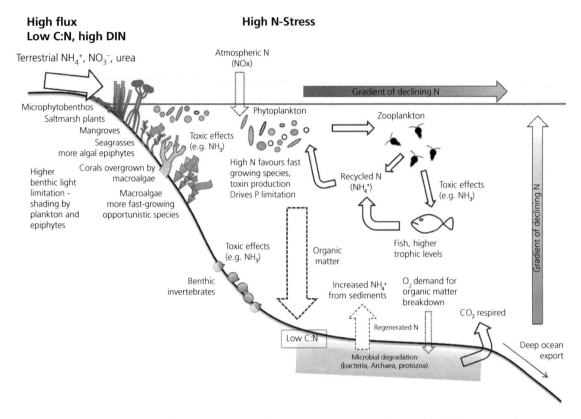

Figure 6.2 Conceptual representation of major components and fluxes in marine ecosystems under stress due to high nitrogen conditions.

is 'most limiting' in a given ecosystem at a given point in time depend on how these terms are defined and the specific (often experimental) criteria used (Howarth and Marino, 2006). Even the results of the most conceptually simple enrichment experiment (i.e. the addition of a putative limiting nutrient) can challenge interpretation depending on what process is considered (e.g. productivity, growth rate, or reproduction rate) and what end point is measured (biomass yield, carbon fixation, or photosynthetic capacity) (see reviews by Beardall et al., 2001; Moore et al., 2013).

Paradoxically, systems in which nitrogenous nutrients are below detection limits, and so might be defined as low-N stressed, may actually have very low biomass growing rapidly (e.g. Laws et al., 1987; Behrenfeld and Boss 2014) and supplied by nutrients from high grazing and/or cell lysis rates (Brussaard et al., 1995; Fuhrman, 1999; Cáceres et al., 2013). Additional organism-specific measures of stress (e.g. C:N or N:P ratios, or biochemical markers of specific deficiencies like P or Fe; Beardall et al., 2001) and/or more

sophisticated molecular approaches, such as those involving measuring expression of specific genes (e.g. Lindell et al., 2005) or protein abundances (Saito et al., 2014), are needed to better understand physiological limitation.

Much of our understanding of marine autotrophs, especially phytoplankton, is governed by models developed from batch cultures in which cells first deplete an external nutrient to the point where they reach a limit to physiological acquisition (Falkowski, 1975), and then use up internal nutrient stores (e.g. Hockin et al., 2012). This sequence represents a complex and continuously changing state of increasing stress (Geider et al., 1998; Young and Beardall, 2003); it is unclear at what point terms like 'starvation' might be appropriate. In contrast, cells might also be cultured in continuous chemostat culture, in which they are exposed to a near-zero nutrient concentration and *rate* of nutrient supply limits growth. Under such conditions, a degree of acclimation is possible and stress may be less (see Harrison et al., 1977; Parkhill et al., 2001). The relevance of these distinct conditions to natural ecosystems is not always clear (Harrison et al., 1977), but it is important to consider what is meant by nutrient limitation in a particular context (Moore et al., 2013).

It is also critical to understand that limitation by a single element may be marginal or for only a limited period of time. Some oceanic ecosystems show evidence of oscillations between nitrogen and phosphorus limitation on seasonal or shorter timescales (Eppley et al., 1973; Perry, 1976; Moore et al., 2013). There is good evidence that trace metals, and even organic compounds like vitamins, can co-limit acquisition of other nutrients, and that subtle differences in bioavailability are critical, challenging traditional definitions (see Saito et al., 2008; Moore et al., 2013; Saito et al., 2014). For example, a N_2-fixing organism might appear to be N-deficient, but if it were provided with sufficient iron, it could increase nitrogen fixation and overcome any nitrogen limitation; such metals limitations may be common in the oceans (Hutchins and Fu, 2008).

In considering high-N stress, the term 'eutrophication' arises, though it is critical to distinguish between increases in supply of organic matter, and nutrient pollution (Nixon, 2009). The former is what we should truly define as eutrophication and represents a change in the energy sources of an ecosystem. In this chapter we will deal more explicitly with the nutrients and not with organic loading aspects of eutrophication.

6.1.4 Our approach and major paradigms

In this review, we will consider low-N and high-N stress first in major groups of marine algae, plants, and animals, and also in symbioses and associations which may include both diverse partners. Autotrophs and heterotrophs deal with nitrogen availability in fundamentally different ways; for autotrophs, in which energy comes from independent pathways (e.g. photosynthetic electron transport), restriction of nitrogen either represents lack of a substrate for building new cells, or impairs the capacity for harvesting energy or fixing carbon. In heterotrophic animals, nitrogen is a substrate, but it is also directly coupled with energy supply from food sources through catabolism. Moreover, catabolism of food containing nitrogen creates nitrogenous wastes which need to be excreted. Consumers also appear to have markedly lower tolerance to variations in elemental stoichiometry, demonstrating tighter homeostasis of composition than autotrophs. A common feature of the physiological responses in autotrophs and heterotrophs is the

differential (typically greater) sensitivity of larval, juvenile, or recruiting stages compared to adult stages to nitrogen stress, especially high-N stress. Low-N stress may arise from greater nitrogen demand to build cells in developing organisms. High-N stress in larvae and juveniles may be due to greater sensitivity to toxins of developing compared with mature organ systems, some of which arises from the higher surface area/volume ratios, which favour diffusion for gas exchange but also promote diffusion of toxic nitrogen forms into smaller larvae; in crustaceans, less calcified exoskeletons allow faster diffusion of toxins to target organs (Hutchinson et al., 1998).

6.2 Nitrogen stress in photoautotrophs

Our understanding of physiological responses of photoautotrophs to nitrogen is derived predominantly from studies on microalgae (reviewed in Carpenter and Capone, 1983; Capone et al., 2008). But there are common features of DIN uptake and assimilation pathways with relatively minor variations between photoautotrophs, recently reviewed by Mullholland and Lomas (2008).

6.2.1 Microalgae

Much of our understanding of nitrogen nutritional physiology of microalgae is based on laboratory culture experiments with a restricted number of (typically easier to grow) taxa, and predominantly in response to low-N stress (reviewed by Mullholland and Lomas, 2008). The physiological stress of high N has only more recently been appreciated as ecologically relevant (e.g. Collos et al., 1997; Glibert et al., 2013; Collos and Harrison, 2014) and the relative lack of physiological data on phytoplankton responses to high-N stress is a clear research gap which needs to be addressed.

6.2.1.1 Low-N stress in microalgae

Deprivation of nitrogen induces profound changes in photosynthetic capacity in phytoplankton, including declines in chlorophyll pigments as well as specific proteins within the photosynthetic light harvesting and electron transport apparatus, impairing cell capacity to harvest energy and suppressing photosynthetic quantum yield (Geider et al., 1998; Young and Beardall, 2003). Another effect is reduced abundance of the enzyme Rubisco which catalyses photosynthetic carbon fixation, leading to suppression of P_{max} and growth (reviewed by Turpin, 1991). Low-N stress can also up-regulate active photosynthetic carbon acquisition via a carbon concentrating mechanism (Young and Beardall, 2005; Raven et al., 2011). The imperative for nitrogen acquisition to restore photosynthetic capacity and cell growth drives physiological acclimation, to increase nitrogen uptake capacity but also to divert ATP and reductant use from C fixation to acquisition of any nitrogen encountered by cells (Turpin, 1991; Young and Beardall, 2003). Cell growth becomes increasingly uncoupled from DIN uptake with increasing nitrogen deprivation (Goldman and Glibert, 1983; Moore et al., 2013).

Success of many taxa, including diatoms, in low-N stress conditions of oligotrophic oceans relate to capacity to rapidly take up DIN supplied episodically to the photic zone in upwelling zones, or in nutrient micropatches following zooplankton or fish excretion, through capacity for luxury and/or surge nutrient uptake (McCarthy and Goldman, 1979; Goldman and Glibert, 1983; Cochlan and Harrison, 1991; **Case Study 1**). Excess nutrients taken up during nutrient availability can be stored in vacuoles (Dortch, 1982) to support growth and metabolism during nitrogen depletion periods. However, the extents to which phytoplankton are capable of storing excess nitrogen is unclear. Rubisco can be a nitrogen storage protein, although the Rubisco localized to pyrenoids in green microalgae or carboxysomes in cyanobacteria is probably active in CO_2 fixation rather than inactive protein storage (Meyer and Griffiths, 2013). The evolution of flexibility in nutrient uptake in phytoplankton has clear adaptive benefits in oligotrophic areas (Bonachela et al., 2011). Reduced DIN availability can stimulate increased affinity and uptake capacity of cells, physiologically assessed through Michaelis–Menten uptake kinetics (Goldman and Glibert, 1983, but see Bonachela et al., 2011), which change with expression of plasma membrane nitrate and/or ammonium transporters (Hildebrand and Dahlin, 2000; Mullholland and Lomas, 2008). Elevated ammonium availability can suppress nitrate uptake and the transcription of genes encoding the enzyme nitrate reductase (NR) (Goldman and Glibert, 1983; Berges, 1997). Uptake of nitrate by phytoplankton has traditionally been viewed as inducible, whereas ammonium uptake is typically held to be constitutive (Jenkins and Zehr, 2008; Mullholland and Lomas, 2008). Therefore, under extended deprivation of external nitrate, cells can lose capacity for nitrate uptake (Dortch et al., 1984), which can be induced after re-exposure to nitrate (Hildebrand and Dahlin, 2000; Young and Beardall, 2003).

As a special category of microalgae, benthic microalgae show similar physiological responses to low-N and high-N stress as phytoplankton. In N-limited water, microphytobenthos (MPB) have greater access to nutrients released from sediments after microbial remineralization (Joye and Anderson, 2008), although fluxes from some marine sediments may be relatively depleted of nitrogen (Howarth, 1988). MPB communities can be limited for nitrogen supply (Sundbäck et al., 2000), although benthic production is often limited by light (McGlathery, 2008).

Another response to nitrogen deficiency is resting stage formation. Several phytoplankton groups produce resting stages, but these are best characterized in diatoms and dinoflagellates. There is considerable complexity in such stages because they may or may not be associated with reproductive cycles and there are subtle distinctions. For example, McQuoid and Hobson (1996) distinguish distinct non-sexual diatom resting stages: spores (which involve distinct morphological changes, energy storage, and requirement for a distinct

---

Case Study 1 Oligotrophic open ocean— North Pacific Subtropical Gyre

Setting:
Open ocean regions with chronically low N supply (typically NO_3^- < 50 nM, though dissolved organic N may be 5–6 µM), where physical processes determine N availability (isolated from coastal waters with minimal horizontal advection, N supply is via vertical eddy diffusion and episodic/stochastic events).

Observations and physiology:

- N limitation stimulates phytoplankton N-uptake capacity, favours picoplankton and diazotrophy (pico-cyanobacteria).
- Picocyanobacterial N_2 fixation can be limited by supply of micronutrients (Fe, Co, and Zn).

Significance, gaps in knowledge, and emerging questions:

- Such regions constitute vast areas of the marine environment.
- Standing biomass may be low but how rapid is growth rate and turnover?
- How significant is microbial access to DON for carbon fluxes in the food web?
- Role of zooplankton in mineralization of N is poorly understood.

References: Coale et al., 1996; Karl, 1999; Selph et al., 2005; Moore et al., 2013; Behrenfeld and Boss, 2014.

germination event), resting stages (morphologically similar, but physiologically distinct cells), and winter stages (involving cell morphology changes, but lacking energy storage of spores and not requiring germination for renewed growth). There is perhaps even greater variation in dinoflagellate resting stages (see Dale, 1983) which is beyond the scope of this review. Resting stage formation can be triggered by a number of factors, but rapid nitrogen depletion is frequently implicated (McQuoid and Hobson, 1996). However, making generalizations is difficult because complex patterns are species-specific; in three Baltic dinoflagellate species, nitrogen depletion alone triggered cyst formation in only one species, but while elevated temperature induced encystment for one other species, nitrogen depletion altered its cell size (Kremp et al., 2009). The physiology of excystment has also been studied largely independent of nutrients, more typically exploring effects of cold periods ('vernalization') and light to induce cyst germination (McQuiod and Hobson, 1996).

6.2.1.2 High-N stress in microalgae

Open oceans rarely experience particularly high nitrogen concentrations; in terms of DIN forms, nitrate is typically < 30 µM, nitrite < 0.1 µM, and ammonium < 0.3 µM, while DON is in the order of 5 µM (Gruber, 2008). In contrast, recent surveys of UK estuaries and harbours have demonstrated much higher ammonium concentrations, in the range 2–11 µM (Cole et al., 1999), and there are reports of measurements in eutrophied seagrass sediments of >100 µM ammonium (Hauxwell et al., 2001).

Negative effects of high concentrations of DIN on phytoplankton in culture have been reported, though physiological mechanisms have rarely been explored. Ammonium is more toxic than nitrate and so has received more attention. Chlorophytes appear to be more tolerant to high ammonium than other taxa and dinoflagellates most sensitive (reviewed by Collos and Harrison, 2014). Some coastal phytoplankton species are affected by 100–250 µM, but most tolerate > 1000 µM ammonium (McLachlan, 1973; Harrison and Berges, 2005). In contrast, there are reports of growth of oceanic species compromised at just 25 µM (Keller et al., 1987). Ammonium uptake by cells at low concentrations follows saturating uptake kinetics, but with increased linear uptake at elevated concentrations (Collos et al., 1997), suggesting cells may not be able to exclude ammonium and thus avoid inhibitory effects of elevated intracellular ammonium.

It is unclear whether inhibitory effects of ammonium are specific or whether some are generic effects of high ionic concentration (e.g. destabilizing membrane proton gradients). External ammonium concentrations required for intracellular ammonia sufficient to uncouple photosynthetic electron transport are in the mM range (Lee et al., 1990; Falkowski and Raven, 1997). There is one report of urea activating specific pathways of cell death in a cyanobacterium, but this involved concentrations of 50 mM urea (Sakamoto et al., 1998). Because culture media often contain mM concentrations of nitrogen (see Anderson et al., 2005), there is physiological relevance to such levels, but their meaning in natural environments is questionable.

6.2.2 Marine macroalgae

6.2.2.1 Low-N stress in marine macroalgae

Macroalgae are often limited by nitrogen supply, but, compared to unicells, macroalgae can have a greater capacity to store internal nitrogen, and benthic habit may provide greater access to nutrients released from sediments than for water column algae. In temperate oceans, where DIN is higher in winter and depleted during summer, winter nitrogen uptake by seaweeds is stored internally as unassimilated ammonium or nitrate, presumably in vacuoles (Philips and Hurd, 2003; Young et al., 2007a), and seaweeds also store nitrogen as amino acids and proteins (Gomez and Wiencke, 1998), and more specifically as Rubisco in pyrenoids (Ekman et al., 1989, but see Meyer and Griffiths, 2013). The Arctic kelp, *Laminaria solidungula*, stores nitrogen in specialized cytoplasmic protein bodies which disappear with N-starvation (Puschel and Korb, 2001). A degree of cellular specialization in some seaweed species may influence nutrient uptake by macroalgae, and translocation of nitrogen between cells and cell types also allows reallocation of scarce nitrogen to meristems (Lobban and Harrison, 1997). Consequently, it can be harder to experimentally induce low-N stress in macroalgae than in microalgae. Growth-limiting nitrogen had no effect on photosynthetic quantum yield in *Caulerpa prolifera* (Malta et al., 2005) or in *Fucus* species (Young et al., 2009). In red algae, nitrogen deprivation initially decreased total proteins and phycobiliproteins in thalli, while Rubisco levels were preserved until more severe nitrogen starvation (García-Sánchez et al., 1993). Nitrogen limitation may also differentially impair synthesis of chloroplastic versus cytosolic proteins (Vergara et al., 1995). Unlike the typical case for microalgae, there is little evidence to support the idea that macroalgae prefer ammonium to nitrate or that ammonium inhibits nitrate uptake, although uptake rates of nitrogen sources do acclimate in response to nitrogen availability (Lobban and Harrison, 1997; Young et al., 2009). The absence of ammonium suppression of NR activity also suggests diverse regulation of nitrogen assimilation in macroalgae (Berges, 1997; Young et al., 2009). Some species have been shown to also take up urea (Philips and Hurd, 2003). Lower maximum nitrogen uptake rates in macroalgae than microalgae, and variations in uptake rates between macroalgal species, can be related to surface area–volume relationships (Lobban and Harrison, 1997; Valiela et al., 1997).

The intertidal habitat for many marine macroalgae brings the additional stress of tidal emersion–immersion cycles which cause desiccation and restrict time of exposure to dissolved nutrients. We speculated that differences in the regulation of the nitrate assimilating enzyme NR in response to light availability could relate to position in the intertidal (Young et al., 2007b). Both mild desiccation and higher intertidal zonation seem to stimulate higher nutrient uptake rates and nitrogen storage (Lobban and Harrison, 1997; Philips and Hurd, 2003). In contrast, tidal flushing and high turbulence in coastal margins improve nutrient uptake as water movement breaks down diffusion boundary layers which impair nutrient uptake (Stevens and Hurd, 1997).

6.2.2.2 High-N stress in marine macroalgae

Recent high-profile examples of macroalgal blooms ('green-' and 'golden tides') highlight problems of coastal eutrophication (**Case Study 2**). However, few generalizations about the response of macroalgal-dominated ecosystems to nutrient enrichment can be made, other than the observation that macroalgae do better with high-N stress than seagrasses or corals (Fong, 2008). Increases in nitrogen supply stimulate macroalgal growth, nitrogen uptake rates, nitrogen content of thalli, and photosynthetic parameters, which promote higher productivity, including P_{max} (Lobban and Harrison, 1997; Valiela et al., 1997). Changes in macroalgal species composition observed with nitrogen enrichment may relate to differential tolerance to or capacity to rapidly take up nitrogen and grow fast (Pedersen and Borum, 1996; Valiela et al., 1997). Increased water column light attenuation associated with eutrophication can induce light limitation to benthic organisms, but macroalgae are relatively tolerant to low light conditions (Valiela et al., 1997). Ammonia toxicity may be higher to macroalgal spore survival, germination, and growth than to adult thalli (Sousa et al., 2007). Growth bioassays of ammonia toxicity using germinating *Hormosira banksii* spores suggested EC_{50} values of 2–200 µM (Myers et al., 2007).

Case Study 2 Macroalgal 'green tides', Yellow Sea, Qingdao, China

Setting:
Quiet coastal environments receiving very high inorganic N loading.
Observations and physiology:
- Conspicuous macroalgal biomass accumulations.
- Dominated by *Ulva*—high nutrient uptake capacity, efficient photosynthesis, fast growth rate, and diverse reproductive strategies.
- Growth rates closely correlated with N inputs from terrestrial sources.
- Biomass decomposition causes oxygen depletion, toxicity to animals, and stimulation of red tides.

Significance, gaps in knowledge, and emerging questions:
- Blooms compromise human use of coastal environments and may cause health issues.
- Are 'green tides' seeded from *Porphyra* farms?
- Do other aspects of eutrophication (organic C and N) contribute to blooms?
- While growth depends on light and nutrients, how do locations of biomass accumulations depend on hydrodynamics and coastal topography?

References: Teichberg et al., 2009; Shen et al., 2012; Wang et al., 2012; Lui et al., 2013; Smetacek and Zingone, 2013.

6.2.3 Marine angiosperms—saltmarsh grasses, seagrasses, and mangroves

Among the most productive ecosystems on the planet, coastal salt marshes, mangroves, and seagrass meadows are structured and dominated by marine angiosperms. As well as being highly productive, these ecosystems provide habitat for diverse organisms, and in oligotrophic tropical and subtropical oceans mangroves and seagrasses can concentrate nitrogen into organic sources to support food webs (Holguin et al., 2001). These vascular plants generally have high C:N ratios relative to macroalgae due to high structural carbon and fibre content associated with vascular structural support (Mann, 1988). Nitrogen is typically limiting in these ecosystems but these plants have access to sediment nitrogen via roots. Compared with algae and terrestrial plants, little is known about the nutritional physiology of most seagrass and saltmarsh photoautotrophs. The biogeochemistry and ecology of nitrogen in seagrass meadows, mangroves, and salt marsh ecosystems has been more comprehensively examined (Burkholder et al., 2007; Hopkinson and Giblin, 2008; McGlathery, 2008; Reef et al., 2010) than the physiological effects of nitrogen on these vascular plants.

6.2.3.1 Low-N stress in marine angiosperms

The most important nitrogen source in seagrass ecosystems is recycled nitrogen from sediment remineralization, releasing bioavailable nitrogen to below-ground roots/rhizomes, but nitrogen from the water column is also taken up by leaves. Because of nitrogen uptake from sediments (e.g. Lee and Dunton, 1999), seagrasses may be less vulnerable to low-N stress than macroalgae, and more limited by light, although in calcareous sediments phosphorus can also become limiting (McGlathery, 2008). Experimental supply of additional nitrogen sources to sediments or water column has been shown in some species to stimulate seagrass growth, root or leaf nitrogen uptake, internal storage, and induction or stimulation of N assimilation enzyme activity, particularly NR (Burkholder et al., 1994, 2007; Pedersen et al., 1997; Udy and Dennison, 1997; van Katwijk et al., 1997; Touchette and Burkholder, 2001). Elevated water column and sediment nutrients also decrease the proportion of below-ground biomass (Perez et al., 1994). The ability of seagrasses to take up (scarce) nitrogen within the water column is affected by shoot morphology and meadow density, with sparser shoot density allowing increased flow rates within the canopy, improving nutrient uptake (Morris et al., 2008). Possibly as a buffer against low external nitrogen availability, nitrate can be stored in leaf vacuoles (Touchette and Burkholder, 2000).

Saltmarsh plants are exposed to extreme physiological stresses of salinity gradients, physical stress from tidal water movement, and tidal water level variability, including waterlogging and low oxygen sediments. Anoxia can also affect redox conditions for microbial nitrogen transformations, especially organic nitrogen breakdown to ammonium. In saltmarsh sediments ammonium dominates the DIN fraction, but most of the nitrogen is as organic and particulate forms. Studies of saltmarsh ecosystems suggested nitrogen limitation of plant productivity and biomass production in *Spartina*-dominated marshes in the NE United States (Hopkinson and Giblin, 2008; Morris et al., 2013; Sullivan, 2014), in US west coast marshes dominated by *Sarcocornia pacifica* (Ryan and Boyer, 2012), and in Mediterranean salt marshes (Quintana et al., 1998). Experimental nitrogen fertilization of potted *Spartina* plants

showed increased above-ground production, and a decrease in the root-to-shoot biomass allocation with increasing nitrogen availability, suggesting the plants were limited by nitrogen supply (Valiela et al., 1976). Experimentally increasing nitrogen availability to *Spartina* plants resulted in higher maximum photosynthesis rates (Sullivan, Friedman, Johnson, and Robertson, unpubl.), and increased allocation of nitrogen to Rubisco protein (Qing et al., 2012), which should promote higher C fixation and growth, or could indicate nitrogen storage. Other experiments on nitrogen uptake kinetics suggest that ammonium concentrations around saltmarsh plant roots should be sufficient to saturate uptake (Morris, 1982). The additional physiological stresses of anaerobic respiration and high sodium ion concentrations may impair DIN uptake (Jefferies, 1981; Morris, 1984; Bradley and Morris, 1990). Anoxia and hydrogen sulphide exposure in sediments may also limit energy metabolism (Koch et al., 1990), so active uptake of DIN ions may be constrained by respiratory ATP synthesis. An additional nitrogen demand in halophytic saltmarsh plants is in production of N-containing osmoregulatory compounds; in response to high salinity, *Spartina alterniflora* allocates up to 30% of nitrogen to cytoplasmic accumulation of glycine betaine and the amino acids asparagine and proline (Cavalieri and Huang, 1981; Colmer et al., 1996). Macroalgae in coastal saltmarsh ecosystems can contribute significant primary productivity, and recent studies show that macroalgae in saltmarsh ecosystems may be important in transferring limiting nitrogen from water to saltmarsh plants (Gulbransen and McGlathery, 2013; Newton and Thornber, 2013). Even in relatively high-nutrient coastal regions, saltmarsh plants may be limited by nitrogen supply (Morris et al., 2013; Newton and Thornber, 2013).

Mangrove species show a range of N-conserving adaptations including efficient N resorption prior to leaf abscission, high root/shoot ratios, phenotypic plasticity in relation to nutrient acquisition, some of the highest nitrogen use efficiencies of angiosperms (Reef et al., 2010), and maximal photosynthesis rates despite nitrogen limiting conditions of tropical and subtropical oceans (Lovelock et al., 2006). Mangroves predominantly take up ammonium which is the dominant DIN form in often anoxic sediments (Alongi et al., 1992). Limitation of mangrove growth by nitrogen versus phosphorus varies by ecosystem and depends upon species, sediment type, tidal range, and flooding regime, as well as biotic factors (Lovelock et al., 2006; Reef et al., 2010). High salinity (ionic concentration) in intertidal mangrove ecosystems can interfere with uptake of ions including potassium and nitrogen. The organic loading to mangrove sediments promotes microbial activity, including nitrogen fixation, which can contribute significantly to mangrove ecosystem nitrogen budgets (Alongi et al., 1992; Holguin et al., 2001).

6.2.3.2 High-N stress in marine angiosperms

Seagrass ecosystems face many anthropogenic threats, including physical alteration of coastal habitats, environmental change, sedimentation, and nutrient-driven eutrophication (Waycott et al., 2009). Seagrasses are more likely to experience stress from high-N, for example, *Zostera marina* meadow shoot density declined as the nitrogen percentage in leaves increased (Lee et al., 2004). Responses of seagrasses to experimental nutrient enrichment or eutrophication vary from die-off, no effect, or increased growth, depending on the species, ecosystem, and nutrient enrichment forms, and water column or sediment (Table 1 in Burkholder et al., 2007). Mechanisms for the responses include nutrient limitation, co-limitation with phosphorus, onset of phosphorus limitation, light limitation due to increased attenuation (from increased algal growth), competition with algae, or apparent specific inhibition by nitrate or ammonium.

Chronic exposure to water column ammonium (25 µM) was toxic to *Zostera marina* and *Ruppia maritima* leaves (van Katwijk et al., 1997; Burkholder et al., 2007; van der Heide et al., 2008). Plants apparently lack the ability to restrict ammonium uptake, with linear uptake kinetics at high ammonium concentrations (Short and McRoy, 1984). Degradation of macroalgal blooms near the sediment surface can result in new seagrass shoots experiencing toxic ammonium concentrations of >100 µM (Hauxwell et al., 2001). van Katwijk et al. (1997) speculated that ammonium toxicity relates to buffering of protons released in ammonium assimilation, or to limitation of assimilation imposed by enzyme activities and/or availability of C skeletons from photosynthesis, both of which may be limiting at low light. Nitrogen assimilation clearly depends on photosynthesis and C metabolism (Touchette and Burkholder, 2001) but NR activity may be preserved in the dark under elevated nitrate (Burkholder et al., 2007). In contrast, there was little short-term effect of ammonium on photosynthetic quantum yield in two tropical species, possibly due to faster nitrogen assimilation into amino acids (Christianen et al., 2011). Toxicity of ammonium to seagrasses is exacerbated under high temperature, low light, pH, and low shoot density (van Katwijk et al., 1997; Brun et al., 2008; van der Heide et al., 2008; Villazan et al., 2013). Seagrass species also show differences in responses to elevated nitrate; *Halodule wrightii* and *Ruppia maritima* shoot growth was stimulated by mild but chronic nitrate enrichment in mesocosms over several weeks, but growth was suppressed in *Z. marina* (Burkholder et al., 1994). Growth suppression in N-enrichment mesocosms could relate to onset of phosphorus limitation in *Z. marina*, highlighting the importance of N:P availability ratios and difficulties in interpreting N-limitation bioassays (Burkholder et al., 1994; Beardall et al., 2001). Mismatches between nitrogen uptake and incorporation rates suggest that some seagrasses can take up DIN forms and excrete DIN and organic nitrogen (Lee and Dunton, 1999), possibly providing a mechanism for dealing with high nitrogen uptake influx without excess intracellular accumulation of inorganic nitrogen.

With the higher nutrient and particulate loads associated with eutrophication, seagrasses experience higher macroalgal epiphyte loads and increased water column light attenuation, driving greater light limitation of seagrasses (Burkholder et al., 2007; Waycott et al., 2009). Seagrasses do relatively better in low-N, high light conditions but increasing nutrient loads favour a seagrass-to-macroalgae and/or phytoplankton phase-shift (reviewed by Burkholder et al., 2007; Fong, 2008). High-N stress can also result in macroalgal and microalgal blooms and this additional biomass load to the sediment surface is more likely to result in anoxia stress for seagrasses. Therefore, as macroalgal biomass increases with increased nutrient loads, the interactions between seagrasses and macroalgae do not seem to relate to nitrogen physiology, but result from these other effects related to light and sediment biomass loads (Burkholder et al., 2007).

The location of salt marshes at the margins of terrestrial habitats makes them vulnerable to escalating nitrogen loads from terrestrial run-off and coastal eutrophication, but they also potentially act as nitrogen retention buffers between terrestrial and coastal marine habitats, including seagrass meadows (Hopkinson and Giblin, 2008; Kinney and Valeila, 2013). The ecological implications of escalating nitrogen loads on coastal salt marshes are being examined (see Deegan et al., 2012; **Case Study 3**). However, less is known about the physiological effects, though some of the competitive interactions and zonation of saltmarsh species in New England marshes may relate to species differences in physiological responses to limiting versus excess nitrogen supply. Levine et al. (1998) suggested that, under low-N stress, species with greater below-ground biomass allocation (*Spartina patens*, *Juncus gerardii*, and *Distichlis spicata*) compete better, presumably through

Case Study 3 Coastal eutrophication and New England salt marsh ecosystems

Setting:
The Plum Island Sound LTER–TIDE project, a long-term *in situ* N fertilization experiment to examine the responses of salt marsh ecosystems to coastal eutrophication (http://www.mbl.edu/tide/).

Observations and physiology:

- N fertilization leads to stimulation of grass production, light-saturated photosynthetic rates, reduced N translocation and retention, and altered perennial grass species competition.
- Organic loading increases sediment microbial N mineralization and anoxia.
- Long-term enrichment leads to loss of below-ground biomass leading to collapse of marsh vegetation.

Significance, gaps in knowledge, and emerging questions:

- Salt marshes can act as buffers of terrestrial N run-off; however, high N and organic loading to sediments contributes to loss of below-ground biomass and breakdown of marsh stability.
- What are the prospects for long-term stability and recovery of damaged ecosystems?

References: Levine et al., 1998; Drake et al., 2008, 2009; Deegan et al., 2012; Morris et al., 2013; Sullivan, 2014.

with biotic stress factors; *S. densiflora* marsh plants exposed to increased nutrient (combined NPK fertilizer) loading showed increased susceptibility to fungal infection (Daleo et al., 2013).

Mangroves, like other coastal ecosystems, are threatened by a range of anthropogenic factors, including high nutrient loading associated with eutrophication (Reef et al., 2010). Like salt marshes, they have been viewed as a buffer between sources of land-derived nutrients and the oceans, but the adaptations which allow mangroves to tolerate low nutrients also limit their capacity to retain nutrient pollutants. High-N stress can result in stunted growth and higher mortality, and increasing shoot/root ratios can increase vulnerability to other stresses (Lovelock et al., 2009). Higher nitrogen loadings can also increase herbivore pressure (Reef et al., 2010). Increased nitrogen loading to mangrove sediments can result in microbial conversion to and release of the potent greenhouse gas N_2O (Reef et al., 2010).

6.3 Nitrogen stress in animals

Heterotrophic consumers derive nitrogen from their diets, predominantly from proteins. Thus, conceptually, animals could experience limitations in either food quantity (i.e. energy) or food quality (e.g. content of elements such as nitrogen). At the other end of the spectrum, high levels of some forms of dissolved nitrogen can be toxic. Because the literature is smaller and generalizations emerge across different taxonomic groups of animals, we will consider them together.

6.3.1 Low-N stress in animals

A logical question is whether, in systems where primary producers are limited for nitrogen (see Section 6.2), the higher trophic levels suffer nutritional limitation, versus simply a scarcity of biomass. For benthic animals, this question has been approached (e.g. Tenore, 1988), but answers remain surprisingly elusive. Evidence that nutrient effects do transfer up trophic steps is available from enrichment studies. For example, Hemmi and Jormalainen (2002) found that when the brown algal *Fucus vesiculosus* was nutrient enriched, herbivorous isopod grazers grew faster, consumed more food, and laid more and larger eggs. Nutrient enrichment apparently affected algal primary and secondary metabolism as enriched plants showed increased insoluble sugars, lower total carbon content, and lower physical toughness. In work on benthic invertebrate grazers in Western Australia, Hatcher (1994)

higher root nitrogen uptake capacity, whereas under nitrogen fertilization species with lower below-ground biomass allocation do better, with nutrient enrichment favouring *Spartina alterniflora*. Nitrogen fertilization of coastal marshland can stimulate *Spartina* grass production even after >10 years elevated nitrogen exposure (Morris et al., 2013). Specific physiological responses to nitrate fertilization include lower nitrate uptake efficiency and lower nitrogen retranslocation out of leaves before shedding than in control plants (Drake et al., 2008), implying the fertilized plants were N-saturated (Drake et al., 2009). This suggests a down-regulation in nitrogen acquisition under N-replete conditions. Under high nitrogen loading to *S. alterniflora*, the combined physiological stresses of high salinity and ammonium concentrations can stimulate physiological responses to oxidative stress (Hessini et al., 2013). High nutrient stress with eutrophication may also interact demonstrated strong links between phosphorus content of food and nutritional quality. However, there are limits to these types of studies. In a review focused on marine detritus feeders (where such limitations would presumably be most severe, given the poorer nutritional quality of their food), Phillips (1984) pointed out that most research has dealt only with caloric and protein requirements, but not those of specific nutrients (e.g. sterols or essential amino acids), though this is of interest for aquaculture. Nutritional studies in aquaculture have often had surprising results; for many fish species, supplementation of supposedly 'non-essential' amino acids (e.g. taurine, glutamine, and glycine) can improve processes ranging from immunity to larval metamorphosis and reproduction (Li et al., 2009). Yet work on nutrition of herbivorous fishes has tended to focus on ecological impacts on reefs or how algal species deter grazing, rather than physiological effects of food composition or nutrient acquisition and assimilation (Clements et al., 2009). Without such knowledge it remains difficult to assess the degree of nitrogen stress animals are experiencing in nature.

Somewhat clearer information is available from studies in the plankton. The concepts of ecological stoichiometry (*sensu* Sterner and Elser, 2002) have provided unique perspectives on elemental limitation. Some of the clearest examples are in fresh waters, where zooplankton with high phosphorus requirements (e.g. *Daphnia* species) frequently appear to be limited by food P content and not simply food quantity (Sterner and Hessen, 1994). Long-term data sets examining eutrophication of San Francisco Bay illustrate that changes in nutrient loadings led to repeated shifts in both phytoplankton and zooplankton groups, and altered the stoichiometry of nutrient uptake, organism composition, and the ratios of regenerated nutrients (Glibert et al., 2011; **Case Study 4**). Such analyses need to be applied more broadly in marine systems to improve understanding of how nutrient element limitations affect food webs, but there is still relatively poor understanding of the underlying physiology.

Steinberg and Saba (2008) have provided a comprehensive review specific to nitrogen consumption and metabolism in marine zooplankton and show that: both quantity and quality of food affect feeding, growth, egg production, and hatching success; food quality is more complex than elemental composition; an organism may be protein limited or even amino-acid limited without being N-limited; and zooplankton have several compensatory mechanisms to acclimate to poorer food quality (increasing feeding rates, changing migration patterns, or feeding selectively). They

---

Case Study 4 Eutrophication of San Francisco Bay estuary long-term data set

Setting:
A heavily exploited estuary prone to eutrophication and invasive species where over 30 years of data demonstrate how changing N:P nutrient stoichiometry affects multiple components of the food web.

Observations and physiology:

- Increasing ammonium (NH_4^+) associated with decline in phytoplankton, especially diatoms and green algae, and increasing non-N_2 fixing cyanobacteria (*Microcystis*).
- Increased availability of reduced nitrogen also correlated with increases in invasive macrophytes (*Eichornia*, *Egeria*) and an invasive mollusc (*Corbula*).
- Changes in N:P ratio of phytoplankton result in changes in N:P ratio of zooplankton and shifts in species composition—decrease in calanoid copepods, increase in cyclopoid copepods and cladocerans.
- Corresponding decrease in planktivorous fish relative to piscivores.

Significance, gaps in knowledge, and emerging questions:

- Ecosystem-level effects of eutrophication are mediated not just by quantities of nutrients but also by changes in nutrient stoichiometry.
- What roles do nutritional qualities of food (beyond nutrient ratios) play in changes?
- Do such changes facilitate establishment of invasive species?

References: Glibert et al., 2011.

---

concluded that critical gaps exist in the areas of gelatinous zooplankton, and in specific biochemical nutritional requirements of zooplankton. Complicating matters of nutritional quality are specific allelopathic compounds in prey, such as aldehydes that may have unpredictable longer-term effects (Paffenhofer, 2002; Ianora et al., 2004), and also issues of properly scaling measurements across a wide range of organism size (e.g. Berges et al., 1993).

There is evidence that animals have mechanisms controlling their net composition. Miller and Roman (2008) fed the calanoid copepod *Acartia tonsa* a range of food concentration and C:N ratio from 5 to 25. Nitrogen excreted varied much less than that ingested, suggesting a degree of homeostasis, with implications for trophodynamics. Homeostasis is also evident in freshwater research. Ventura et al. (2008) examined the effects of nutrient enrichment on stoichiometry in mesocosms designed to mimic pond ecosystems. Enrichment with nitrogen decreased C:N in macrophytes, epiphytes, and seston, but there were only weak or no effects on C:N of consumers (insects, leeches, molluscs, or crustaceans). What physiological trade-offs are used to achieve homeostasis are largely unknown and results can seem counter-intuitive: Augustin and Boersma (2006) found that feeding copepods of the genus *Acartia* algal diets with lower N (C:N > 10) did not change ingestion rates relative to N-balanced diets, but actually significantly *increased* egg production, and yet had no effects on hatching rates of the eggs. Relatively little is known about the capacity of marine invertebrates to tolerate elemental imbalances, but for insects quite radical nutritional imbalances can be survived, albeit with slower development and skewed body composition (Raubenheimer and Jones, 2006). Although the concept of a 'point of no return' in planktonic larval metabolism is much discussed, many seemingly contradictory examples exist. Larvae of the bivalve *Meretrix meretrix* can develop and even reach metamorphosis without feeding, though growth is reduced (Tang et al., 2006), and Moran and Manahan (2004) showed that even 33 d of food deprivation in oyster larvae did not decrease their swimming, or their ability to feed when algal cells were provided, again demonstrating the resilience of invertebrate species. Other examples include Southern Ocean pteropods. When deprived of food for up to two weeks, some species of these planktonic molluscs reduced their respiration rates by 20% yet did not alter ammonium excretion rates (Maas et al., 2011). The physiological basis of these changes is poorly understood. Fish species, as well, can show considerable tolerance of food deprivation (e.g. Gronquist and Berges, 2013), and they often display energy-conserving behaviours such as seeking cooler water strata to lower metabolic rates (see Sogard and Olla, 1996).

Under unfavourable environmental conditions, many planktonic algae encyst (see Section 6.2.1.1). The comparable response for invertebrates involves some form of dormancy or metabolic arrest in response to poor food quantity or quality. Radzikowski (2013) has recently reviewed resistance of dormant stages of planktonic invertebrates and distinguished between that driven directly by external conditions ('quiescence') and that involving internal mechanisms ('diapause'), with diapause the most interesting physiologically. The majority of such mechanisms involve desiccation or temperature resistance, for which

polar copepods provide a good example: Antarctic calanoid copepods such as *Calanoides acutus* survive long periods of food deprivation during winter, when there is no primary production, through deep-water diapause, subsisting on stored lipid (Schnackschiel and Hagen, 1995). However, the mechanism of diapause may also be based on nitrogen status. In order to maintain buoyancy at depth, the copepods use ion exchange, using ammonium in haemolymph in place of heavier cations (e.g. Mg^{2+}). It is hypothesized that in order to prevent the formation of large concentrations of ammonia, the pH of the haemolymph is reduced. The acidic pH results in significant metabolic depression and may be responsible for diapause (Schrunder et al., 2013).

6.3.2 High-N stress in animals

Of nitrogen forms that induce stress at high concentrations, ammonia is generally the most toxic for invertebrates and fishes, and therefore where greatest attention has been focused. Toxic effects are especially strong in organisms with extensive gill surfaces like fishes and bivalves. Toxic actions include proliferation of gill tissue, progressive acidosis, uncoupling of oxidative phosphorylation, and disruption of osmoregulation (Russo, 1985; Camargo and Alonso, 2006). There are considerably more data available for freshwater species (a reflection of the number of river and stream systems prone to nutrient pollution), though generalizations hold for marine species. The proportion of ions existing as ammonium versus ammonia varies with pH and temperature, so that, typically, concentrations are expressed such as 'total ammonia nitrogen' and normalized for pH and temperature. Thus, in fresh waters, the US Environmental Protection Agency has suggested limits for ammonium concentrations of about 1200 μM (acute exposure) and 136 μM (chronic exposure) (USEPA, 2013); at seawater pH ~ 8.2, these values fall within ammonia limits suggested for a range of marine species: 3.5–25 μM (acute) and 0.7–1.4 μM (chronic) (Camargo and Alonso, 2006), and close to the Environmental Quality Standard of 0.021 mg ammonia N/l (1.5 μM) proposed for the protection of saltwater fish and shellfish (Cole et al., 1999). Environmental quality guidelines developed for estuarine and marine waters by Batley and Simpson (2009), based on a range of species, support somewhat higher limits for ammonia, of the order of 11 μM for pristine waters. Nixon et al. (1995) reviewed ammonium effects on estuarine and marine benthic organisms and reported slightly higher 'No Observed Effects Concentrations' (of the order of 7 μM ammonia or 240 μM ammonium), but effects varied considerably and concentrations were higher at lower salinity. Importantly, few data exist for early life stages where toxicity appears to be greatest (Nixon et al., 1995). In contrast, for fishes, ammonia is actually less toxic at lower salinity, and there is also an interaction with oxygen such that ammonium becomes more toxic as dissolved oxygen decreases (Nixon et al., 1995).

For oxidized forms of DIN, harmful effects have not been reported below 5.7–25 μM nitrite and 140 μM nitrate (Camargo and Alonso, 2006). One effect of nitrate in invertebrates is conversion of oxygen-carrying pigments to forms that are not capable of carrying oxygen (Camargo et al., 2005); many marine invertebrates show little effect at 1.4 mM nitrate (about an order of magnitude greater than most freshwater species), but early developmental stages have not been tested extensively. In fishes, nitrite is a well-characterized toxicant that is taken up across the gills in competition with chloride (Kroupova et al., 2005). Effects include disruption of ion regulation, excretion, and respiration, due to oxidation of haemoglobin. Toxicity is affected by pH, and oxygen, but because of the competition with chloride ions, salinity moderates toxicity considerably, suggesting that marine species are likely to be less prone to negative effects; nitrite LD$_{50}$ concentrations are in the mM range (Kroupova et al., 2005).

The toxicity of organic forms of nitrogen is much less certain. While often a major constituent of total N, individual compounds in the DON pool occur at very low concentrations, and are highly variable in space and time (Pehlivanoglu-Mantasa and Sedlak, 2006). Urea is one of the few commonly measured DON forms, but few reports of toxicity are available. Urea is unlikely to constitute a major toxic hazard at typical environmental concentrations; based on physiological studies for marine fishes, it is of the order of 25–30-fold less toxic than ammonia (Handy and Poxton, 1993).

6.4 Nitrogen stress in symbioses, associations, and mixotrophy

6.4.1 Coral–zooxanthellae symbioses

Much of the physiological research on coral symbioses relates to light and temperature stress, with much less focus on nutrient physiology. However, coral symbioses are potentially interesting systems that might be prone to effects of nitrogen limitation. Generally, physiological experimental data have suggested that the host cnidarian provides ammonium to the dinoflagellate symbiont. This led to early speculation that the host cnidarian might control the relationship through nitrogen supply, i.e. by limiting the flow of nitrogen to the dinoflagellate (see arguments summarized by Rees et al., 1991). In undisturbed coral reefs, DIN supply is usually extremely low and can limit coral production and reef macroalgal biomass, although evidence is variable (Szmant, 2002). Furthermore, recent work suggests that, if anything, the symbionts may be phosphorus limited (Miller and Yellowlees, 1989). As evidence for nitrogen storage, in experiments using pulse-chase ^{15}N labelling, Kopp et al. (2013) showed that corals provided with DIN or organic nitrogen formed uric acid crystals in dinoflagellate endosymbionts within < 1 h, with subsequent remobilization of that nitrogen after 6 h.

As with seagrass ecosystems (Section 6.2.3.2), increasing nutrient and particulate loads with eutrophication tend to result in a phase-shift from corals to macroalgae (McCook, 1999), especially as nitrogen availability exceeds the > 1 μM DIN needed for maximum macroalgal growth (Lapointe, 1997). Physiological bases for this shift are not well characterized, but, as with seagrasses, coral reef ecosystems are well adapted to oligotrophic conditions, whereas macroalgae are less stressed by high N and many fast-growing species can capitalize on high nutrient loads (see Case Study 2). Coral reef ecosystems exposed to high-N stress from terrestrial run-off or groundwater influxes are likely to also be experiencing other physiologically stressful factors, including reduced salinity and increased water column light attenuation due to organic matter and sediment loadings, or physical disturbance (Lapointe, 1997; Szmant, 2002). Elevated nutrients may increase susceptibility to disease and to bleaching in corals (O'Niell and Capone, 2008). The ecological impressions of a three-year *in situ* experimental ammonium and/or phosphate enrichment experiment on coral reefs have been summarized by O'Niell and Capone (2008). Elevated-N stress had the greatest impact on reproductive processes, but the physiological bases of these responses have not been characterized. Two to six times ambient ammonium in the water column was traced using δ^{15}N and was incorporated not only within zooxanthellae, but also into coral structures (Hoegh-Guldberg et al., 2004), suggesting excess nitrogen can be used by both symbionts. Elevated nitrogen supply to algae may reduce carbon transfer to the host and disrupt physiological control of the host over algal symbionts (O'Niell and Capone, 2008). As well as tightly controlled symbioses, more loosely associated bacteria and fungi in coral microbiomes can supply fixed nitrogen and transform nitrogen forms, as well as respond to changes in nitrogen availability (Wegley et al., 2007; Kelly et al., 2014) (see Section 6.1).

6.4.2 Nitrogen-fixing associations

While symbioses with N_2-fixing organisms are an ecological rather than a physiological adaptation, their prevalence among a broad range of organisms in the marine environment and the possibility of facultative gain/loss of symbionts makes them worth considering. Most marine eukaryotes have developed symbioses with cyanobacteria versus other diazotrophs, and, interestingly, the degree of integration of the metabolisms of host and symbiont does not seem to correlate well with either how early the symbiosis evolved, or with how dependent the host is on the symbiont for survival (see Usher et al., 2007). N_2-fixing cyanobacteria are associated with ascidians, sponges, seagrasses, and macroalgae (Gerard et al., 1990; Foster and O'Mullan, 2008; McGlathery, 2008), which can inhabit nutrient-poor environments. Examples of non-cyanobacterial associations are anaerobic N_2-fixers in guts of zooplankton and Teredinidae shipworms (Foster and O'Mullan, 2008), where the symbiosis is presumably based on the high cellulose, nitrogen-poor shipworm host diet. In coastal sediments under low-N stress, including mangroves, bacterial nitrogen-fixing activity associated with roots or just within sediments can be an important source of nitrogen (Holguin et al., 2001) (see Section 6.2.3.1).

6.4.3 Mixotrophy

Mixotrophy is a strategy to use photoautotrophy and/or heterotrophy depending on the light and food availability (Matantseva and Skarlato, 2013). Mixotrophy is an important and possibly underestimated contributor to phytoplankton production in phytoplankton communities (Flynn et al., 2013). Many organisms associated with harmful algal blooms (HABs), especially dinoflagellates and diatoms, show adaptations of flexible nutrient acquisition strategies, including mixotrophy and osmotrophy. Mixotrophy under nitrogen deprivation has been examined in relatively few species (Kudela et al., 2010). However, it is clear that under low-N stress conditions in oligotrophic waters, osmotrophy of organic compounds and/or phagocytosis of prey cells can provide photoautotrophs with alternatives to acquire significant nitrogen (Glibert et al., 2009; Carvalho and Granéli, 2010; Matantseva and Skarlato, 2013). Symbiotic *Symbiodinium*, which

typically receive nitrogen from their coral host, can survive as free-living mixotrophs, acquiring nitrogen from bacterial prey (Jeong et al., 2012). Expression of cell surface proteolytic enzymes in mixotrophic dinoflagellate species also indicates capacity for acquisition of alternative nitrogen sources (Stoecker and Gustafson, 2003; Salerno and Stoecker, 2009). Significance of mixotrophy as a response strategy to low-N stress depends not only on species, but on carbon and light supply (Skovgaard et al., 2003). In mixed cultures of different dinoflagellates, competition for limiting nitrogen can influence mixotrophy, but allelopathy can also be a physiological mechanism determining success (Li et al., 2012). Mixotrophy is significant for HAB species studied in eutrophic conditions, where the high particulate organic nutrient loadings (Nixon, 2009) make grazing an efficient nutrient acquisition strategy (Burkholder et al., 2008). Several of the toxins associated with dinoflagellate, cyanobacterial, and diatom HABs (saxitoxin, anatoxin, and domoic acid) contain nitrogen; saxitoxin production can respond to nitrate, ammonium, and urea supply, and to internal nitrogen status of cells (Kudela et al., 2010; Murata et al., 2012). More detailed understanding of the physiological responses of HAB species to low-N and high-N stress is important for managing and predicting HABs in the context of increasing eutrophication (Kudela et al., 2008).

6.5 Research challenges

6.5.1 Physiology versus ecology

This review has demonstrated many gaps in knowledge at the physiological level; for example, observations that elevated nitrogen concentrations lead to large ecosystem-level changes, but lack of data to understand the specific physiological mechanisms driving the changes, or the observation that dietary changes in C:N or protein affect zooplankton egg production and viability, while the specific nutritional physiology underlying those changes remains unknown. In some cases, the inability to cultivate organisms under the conditions necessary to conduct physiological experiments contributes to the gaps. However, the almost 'cultural' divide between ecological approaches (focus on relevance of environmental conditions, and concerns about applicability of single-species experiments) and physiological ones (focus on tight control/manipulation of variables, and concerns about near-impossibility of interpreting complex natural systems) (cf. Pankhurst and Herbert, 2013) also contribute to the lack of progress. Broader collaboration between ecologists and physiologists, and more integration of ideas from progress in marine and freshwater studies, are desperately needed (Allen and Polimene, 2011; Németh et al., 2013).

6.5.2 Interactions among nitrogen stress and other factors

For marine organisms, nitrogen stress rarely if ever occurs in isolation and nitrogen stress effects are modified by other conditions, as outlined for some cases in the sections of this chapter. It is beyond the scope of this chapter to review such interactions in detail, but a few that are highly relevant to the current picture of global change are temperature and ocean acidification.

In terms of temperature, predicted decreases in phytoplankton cell size and increases in C:N ratios in response to higher temperatures (Peter and Sommer, 2013) would modify nitrogen requirements and thus change N stresses. There are specific metabolic effects of temperature on some nitrogen acquisition processes. Breitbarth et al. (2007) suggested that *Trichodesmium* fixes N_2 maximally between 24 and 30 °C and so it is likely that net N_2 fixation may decrease in some global warming scenarios. As an interesting example of direct temperature–nitrate interactions, Lomas and Glibert (1999) suggested that under low temperature and high light stress, marine diatoms may take up nitrate in excess of nutritional requirements and, by reducing nitrate to nitrite/ammonium/amino acids and then excreting those sources, provide an additional energy and reductant sink, reducing potential for damaging effects of high photosynthetic electron flow. For both saltmarsh and coral ecosystems there is evidence that high-N stress can be exacerbated by high temperature (see also Chapter 4 this volume).

Potential effects of ocean acidification have been identified (Doney et al., 2009; Chapter 3 in this volume). In terms of potential ammonia toxicity, lower ocean pH may represent reduced risk as the equilibrium will shift towards ammonium. However, Saba et al. (2012) demonstrated that artificially elevated CO_2 in water of the Eastern Antarctic Peninsula resulted in acidification that significantly elevated key metabolic enzymes (e.g. lactate and malate dehydrogenases) and decreased urea excretion in the keystone zooplankton species, *Euphausia superba*. A more acidic environment apparently incurred increased physiological costs associated with regulating internal acid–base equilibria, and such changes in excretion could have real implications for nutrient cycling and availability (Saba et al., 2012).

So far, most of the research on the effects of these and other stressors has tended to work with, at the most, two distinct factors, but progressive marine environmental change will certainly involve multiple stressors, challenging the validity of experimental assessments based on manipulations of single factors. It has been very clear for decades in fields like toxicology that the effects of multiple stressors are more than additive (e.g. Heugens et al., 2001), yet some of the basic vocabulary and methods involved in the risk assessment techniques designed to deal with such problems (Landis et al., 2013) remain foreign to physiologists and ecologists.

Nitrogen compounds may also mediate responses to stress caused by other factors. One example is a class of compounds collectively 'reactive nitrogen species' (RNS), which include compounds such as nitrite and nitric oxide (NO). RNS species are detectible in diverse oceanic regions and associated with the presence of other free radicals (Naqvi et al., 1998). Within cells, it has long been recognized that RNS can serve as signals and are associated with oxidative stress, DNA damage, disruption of calcium homeostasis and mitochondrial function, and cell death (Brune et al., 1999; Durner and Klessig, 1999). However, more recently, evidence has accumulated that RNS forms are also present in diverse phytoplankton species (Okamoto and Hastings, 2003; Kim et al., 2006), and it has even been hypothesized that forms like NO may function as secondary messengers in signal transduction of stress in some marine diatoms (Vardi et al., 2006, 2008). The degree to which such responses occur in other marine taxa is simply unknown.

6.5.3 Complex feedbacks and interactions among individuals and populations

It is becoming increasingly well recognized that the effects of stress are modulated not simply by the physical environment but also by their biotic and even social environment. In bacteria, such concepts have been examined for some time; for example, the process of quorum sensing, whereby cell–cell communication mediated by small molecules such as N-acyl homoserine lactone (AHL) allows populations to coordinate processes including biofilm formation and toxin production (Dobretsov et al., 2009). Quorum sensing also has roles in stabilizing population size under nutrient-limiting conditions, and in establishing stress resistance (Atkinson and Williams, 2009). Only relatively recently has the ability of marine eukaryotic organisms to sense, exploit, and even disrupt such prokaryotic signalling for their own benefit become clear (e.g. Joint et al., 2002; Tait et al., 2009). The degree to which marine eukaryotes communicate within populations, or across the eukaryote–prokaryote boundary, is not clear, but there are significant physiological and ecological implications (Tait et al., 2009; Goercke et al., 2010). We understand little about whether such forms of communication may play roles in stabilizing populations under growth-limiting conditions by limiting growth rate or by improving stress tolerance.

6.6 Recent and emerging approaches

Recent technological developments in molecular biology have opened up new possibilities for exploring some of these challenging questions. For example, there have been impressive advances in characterizing multi-gene expression responses to nitrogen deficiency in cyanobacteria and coupling key genes with sensitive reporter systems to allow *in situ* diagnosis of low-N stress (e.g. Lindell et al., 2005; Scanlan and Post, 2008; Scanlan et al., 2009). Challenges remain in expanding these approaches to eukaryotes and to finding more generic genetic nitrogen stress responses across multiple taxa, as well as finding ways to use such reporter systems in ecologically meaningful ways. Now that we have many full genomes of marine organisms sequenced and annotated, and large-scale sequencing and bioinformatics are becoming widely available and cost-effective, approaches to understanding physiological stress along the gradient from nitrogen scarcity to surfeit involving use of transcriptomics (e.g. Dyhrman et al., 2012; Hockin et al., 2012), and community meta-transcriptomics, are becoming feasible. Physiological insights can also be gained from proteomic (e.g. Sandh et al., 2011; Wurch et al., 2011) and metabolomic approaches (McKew et al., 2013; Zhang et al., 2014). To date, much of this work has been limited to laboratory cultures of clonal organisms, but at least some thought is being devoted to translating the molecular-level understanding of individual taxa to biogeochemical cycles in the field (Melack et al., 2011; Saito et al., 2014).

While there is a divide between ecologists and physiologists, which impairs the understanding of physiological stress and its relevance to ecology, an equally urgent problem is that application of molecular biological approaches are happening without much consideration of basic physiological processes. Application of cutting-edge genetic and genomics methods can offer little insight if they are applied to experiments in which physiological states are ambiguously defined because of poor experimental design. We observe a sharp divide between expertise in molecular and bioinformatics methodologies, and physiological competence, and ecological expertise (though of course there are brilliant exceptions that prove the rule). These exciting new approaches will provide much more meaningful insights if there is more interdisciplinary collaboration and training for students and post-doctoral researchers; physiological understanding can potentially provide a central integrating focus for exciting interdisciplinary research.

References

Allen, J.I. & Polimene, L. (2011). Linking physiology to ecology: towards a new generation of plankton models. *Journal of Plankton Research*, **33**, 989–997.

Alongi, D.M., Boto, K.G., & Robertson, A.I. (1992). Nitrogen and phosphorus cycles. In: Robertson, A.I. & Alongi, D.M. (eds), *Tropical Mangrove Ecosystems*, pp. 251–292. American Geophysical Union, Washington, DC.

Aluwihare, L.I. & Meador, T. (2008). Chemical composition of marine dissolved organic nitrogen. In: Capone, D.G., Bronk, D.A., Mulholland, M.R., & Carpenter, E.J. (eds), *Nitrogen in the Marine Environment*, 2nd edition, pp. 95–140. Academic Press, Boston.

Andersen, R.A., Berges, J.A., Harrison, P.J. & Watanabe, M.M. (2005). Appendix A. Recipes for freshwater and seawater media. In: Andersen, R.A. (ed.), *Algal Culturing Techniques*, pp. 429–538. Academic Press, Boston.

Atkinson, S. & Williams, P. (2009). Quorum sensing and social networking in the microbial world. *Journal of the Royal Society Interface*, **6**, 959–978.

Augustin, C.B. & Boersma, M. (2006). Effects of nitrogen stressed algae on different *Acartia* species. *Journal of Plankton Research*, **28**, 429–436.

Barton, B.A. (2002). Stress in fishes: a diversity of responses with particular reference to changes in circulating corticosteroids. *Integrative and Comparative Biology*, **42**, 517–525.

Batley, G.E. & Simpson, S. L. (2009). Development of guidelines for ammonia in estuarine and marine water systems. *Marine Pollution Bulletin*, **58**, 1472–1476.

Bauer, J.E., Cai, W.J., Raymond, P.A., et al. (2013). The changing carbon cycle of the coastal ocean. *Nature*, **504**, 61–70.

Beardall, J., Young, E., & Roberts, S. (2001). Interactions between nutrient uptake and cellular metabolism: approaches for determining algal nutrient status. *Aquatic Sciences*, **63**, 44–69.

Behrenfeld, M. & Boss, E. (2014). Resurrecting the ecological underpinnings of ocean plankton blooms. *Annual Review of Marine Science*, **6**, 167–194.

Berges, J.A. (1997). Algal nitrate reductases. *European Journal of Phycology*, **32**, 3–8.

Berges, J.A., Roff, J.C., & Ballantyne, J.S. (1993). Enzymatic indices of respiration and ammonia excretion: relationships to body size and food levels. *Journal of Plankton Research*, **15**, 239–254.

Bonachela, J.A., Raghib, M., & Levin, S.A. (2011). Dynamic model of flexible phytoplankton nutrient uptake. *Proceedings of the National Academy of Sciences of the USA*, **108**, 20633–20638.

Bradley, P.M. & Morris, J.T. (1990). Influence of oxygen and sulfide concentration on nitrogen uptake kinetics in *Spartina alterniflora*. *Ecology*, **71**, 282–287.

Breitbarth, E., Oschlies, A., & LaRoche, J. (2007). Physiological constraints on the global distribution of *Trichodesmium*: effect of temperature on diazotrophy. *Biogeosciences*, **4**, 53–61.

Brun, F.G., Olive, I., Malta, E.J., et al. (2008). Increased vulnerability of *Zostera noltii* to stress caused by low light and elevated ammonium levels under phosphate deficiency. *Marine Ecology Progress Series*, **365**, 67–75.

Brune, B., von Knethen, A., & Sandau, K. B. (1999). Nitric oxide (NO): an effector of apoptosis. *Cell Death and Differentiation*, **6**, 969–975.

Brussaard, C.P.D., Riegman, R.M. Noordeloos, A.A.M., et al. (1995). Effects of grazing, sedimentation and phytoplankton cell lysis on the structure of a coastal pelagic food-web. *Marine Ecology Progress Series*, **123**, 259–271.

Burkholder, J.M., Glasgow, H.B., & Cooke, J.E. (1994). Comparative effects of water-column nitrate enrichment on eelgrass *Zostera marina*, shoalgrass *Halodule wrightii*, and widgeongrass *Ruppia maritima*. *Marine Ecology Progress Series*, **105**, 121–138.

Burkholder, J.M., Glibert, P.M., & Skelton, H.M. (2008). Mixotrophy, a major mode of nutrition for harmful algal species in eutrophic waters. *Harmful Algae*, **8**, 77–93.

Burkholder, J.M., Tomasko, D.A., & Touchette, B.W. (2007). Seagrasses and eutrophication. *Journal of Experimental Marine Biology and Ecology*, **350**, 46–72.

Cáceres, C., Taboada, F.G., Hofer, J., & Anadon, R. (2013). Phytoplankton growth and microzooplankton grazing in the subtropical Northeast Atlantic. *PLoS ONE*, **8**, e69159.

Camargo, J.A. & Alonso, A. (2006). Ecological and toxicological effects of inorganic nitrogen pollution in aquatic ecosystems: a global assessment. *Environment International*, **32**, 831–849

Camargo, J.A., Alonso, A., & Salamanca, A. (2005). Nitrate toxicity to aquatic animals: a review with new data for freshwater invertebrates. *Chemosphere*, **58**, 1255–1267.

Capone, D.G., Bronk, D.A., Mulholland, M.R., & Carpenter, E.J. (eds) (2008). *Nitrogen in the Marine Environment*, 2nd edition. Academic Press, Boston.

Carpenter, E.J. and Capone, D.G. (1983). *Nitrogen in the Marine Environment*. Academic Press, San Diego, CA.

Carvalho, W.F. and Granéli, E. (2010). Contribution of phagotrophy versus autotrophy to *Prymnesium parvum* growth under nitrogen and phosphorus sufficiency and deficiency. *Harmful Algae*, **9**, 105–115.

Cavalieri, A.J. & Huang, H.C. (1981). Accumulation of proline and glycine betaine in *Spartina alterniflora* Loisel. in response to NaCl and nitrogen in the marsh. *Oecologia*, **49**, 224–228.

Christianen, M.J.A., Van Der Heide, T., Bouma, T.J., et al. (2011). Limited toxicity of NH_x pulses on an early and late successional tropical seagrass species: interactions with pH and light level. *Aquatic Toxicology*, **104**, 73–79.

Clements, K.D., Raubenheimer, D., & Choat, J. H. (2009). Nutritional ecology of marine herbivorous fishes: ten years on. *Functional Ecology*, **23**, 79–92.

Coale, K.H., Johnson, K.S., Fitzwater, S.E., et al. (1996). A massive phytoplankton bloom induced by an ecosystem-scale iron fertilization experiment in the equatorial Pacific Ocean. *Nature*, **383**, 495–501.

Cochlan, W.P. & Harrison, P. J. (1991). Uptake of nitrate, ammonium, and urea by nitrogen-starved cultures of *Micromonas pusilla* (Prasinophyceae): transient responses. *Journal of Phycology*, **27**, 673–679.

Cole, S., Codling, I.D., Parr, W., & Zabel, T. (1999). *Guidelines for managing water quality impacts within UK European marine sites*. Prepared for the UK Marine SAC Project. http://www.ukmarinesac.org.uk/pdfs/water_quality.pdf

Collos, Y. & Harrison, P.J. (2014). Acclimation and toxicity of high ammonium concentrations to unicellular algae. *Marine Pollution Bulletin*, **80**, 8–23.

Collos, Y., Vaquer, A., Bibent, B., et al. (1997). Variability in nitrate uptake kinetics of phytoplankton communities in a Mediterranean coastal lagoon. *Estuarine, Coastal and Shelf Science*, **44**, 369–375.

Colmer, T.D., Fan, T.W.M., Lauchli, A., & Higashi, R.M. (1996). Interactive effects of salinity, nitrogen and sulphur on the organic solutes in *Spartina alterniflora* leaf blades. *Journal of Experimental Botany*, **47**, 369–375.

Dale, B. (1983). Dinoflagellate resting cysts: 'benthic plankton'. In: Fryxel, G.A. (ed.), *Survival Strategies of the Algae*, pp. 69–137. Cambridge University Press, Cambridge, UK.

Daleo, P., Alberti, J., Pascual, J., & Iribarne, O. (2013). Nutrients and abiotic stress interact to control ergot plant disease in a SW Atlantic salt marsh. *Estuaries and Coasts*, **36**, 1093–1097.

Deegan, L.A., Johnson, D.S., Warren, R.S., et al. (2012). Coastal eutrophication as a driver of salt marsh loss. *Nature*, **490**, 388–392.

Dobretsov, S., Teplitski, M., & Paul, V. (2009). Mini-review: quorum sensing in the marine environment and its relationship to biofouling. *Biofouling*, **25**, 413–427.

Doney, S.C., Fabry, V.J., Feely, R.J., & Kleypas, J.A. (2009). Ocean acidification: the other CO_2 problem. *Annual Review of Marine Science*, **1**, 169–192.

Dortch, Q. (1982). Effect of growth conditions on accumulation of internal nitrate, ammonium, amino acids and protein in three marine diatoms. *Journal of Experimental Marine Biology and Ecology*, **61**, 243–264.

Dortch, Q., Clayton Jr, J.R., Thoressen, S.S., & Ahmed, S.I. (1984). Species differences in accumulation of nitrogen pools in phytoplankton. *Marine Biology*, **81**, 237–250.

Drake, D.C., Peterson, B.J., Deegan, L.A., et al. (2008). Plant nitrogen dynamics in fertilized and natural New England salt marshes: a paired ^{15}N tracer study. *Marine Ecology Progress Series*, **354**, 35–46.

Drake, D.C., Peterson, B.J., Galván, K.A., et al. (2009). Salt marsh ecosystem biogeochemical responses to nutrient enrichment: a paired ^{15}N tracer study. *Ecology*, **90**, 2535–2546.

Durner, J. & Klessig, D.F. (1999). Nitric oxide as a signal in plants. *Current Opinions in Plant Biology*, **2**, 369–374.

Dyhrman, S.T., Jenkins, B.D., Rynearson, T.A., et al. (2012). The transcriptome and proteome of the diatom *Thalassiosira pseudonana* reveal a diverse phosphorus stress response. *PloS ONE*, **7**, e33768.

Ekman, P., Lignell, Å., & Pedersén, M. (1989). Localization of ribulose-1,5-bisphosphate carboxylase/oxygenase in *Gracilaria secundata* (Rhodophyta) and its role as a nitrogen storage pool. *Botanica Marina*, **32**, 527–534.

Eppley, R.W., Renger, E.H., Venrick, E.L., & Mullin, M.M. (1973). A study of plankton dynamics and nutrient cycling in the Central Gyre of the North Pacific Ocean. *Limnology and Oceanography*, **18**, 534–551.

Falkowski, P.G. (1975). Nitrate uptake in marine phytoplankton: comparison of half-saturation constants from seven species. *Limnology and Oceanography*, **20**, 412–417.

Falkowski, P.G. & Raven, J.A. (1997). *Aquatic Photosynthesis*. Blackwell Science, Oxford.

Flynn, K.J., Stoecker, D.K., Mitra, A., et al. (2013). Misuse of the phytoplankton zooplankton dichotomy: the need to assign organisms as mixotrophs within plankton functional types. *Journal of Plankton Research*, **35**, 3–11.

Fogg, G.E. & Thake, B. (1987). *Algal Cultures and Phytoplankton Ecology*, 3rd edition. University of Wisconsin Press, Madison.

Fong, P. (2008). Macroalgal-dominated ecosystems. In: Capone, D.G., Bronk, D.A., Mulholland, M.R., & Carpenter, E.J. (eds), *Nitrogen in the Marine Environment*, 2nd edition, pp. 917–948. Academic Press, Boston.

Foster, R.A. & O'Mullan, G.D. (2008). Nitrogen-fixing and nitrifying symbioses in the marine environment. In: Capone, D.G., Bronk, D.A., Mulholland, M.R., & Carpenter, E.J. (eds), *Nitrogen in the Marine Environment*, 2nd edition, pp. 1197–1218. Academic Press, Boston.

Fuhrman, J.A. (1999). Marine viruses and their biogeochemical and ecological effects. *Nature*, **399**, 541–548.

García-Sánchez, M.J., Fernández, J.A., & Niell, F. X. (1993). Biochemical and physiological responses of *Gracilaria tenuistipitata* under two different nitrogen treatments. *Physiologia Plantarum*, **88**, 631–637.

Geider, R.J., Macintyre, H.L., Graziano, L.M., & Mckay, R.M.L. (1998). Responses of the photosynthetic apparatus of *Dunaliella tertiolecta* (Chlorophyceae) to nitrogen and phosphorus limitation. *European Journal of Phycology*, **33**, 315–332.

Gerard, V.A., Dunham, S.E., & Rosenberg, G. (1990). Nitrogen-fixation by cyanobacteria associated with *Codium fragile* (Chlorophyta): environmental effects and transfer of fixed nitrogen. *Marine Biology*, **105**, 1–8.

Glibert, P.M., Burkholder, J.M., Kana, T.M., et al. (2009). Grazing by *Karenia brevis* on *Synechococcus* enhances its growth rate and may help to sustain blooms. *Aquatic Microbial Ecology*, **55**, 17–30.

Glibert, P.M., Fullerton, D., Burkholder, J.M., et al. (2011). Ecological stoichiometry, biogeochemical cycling, invasive species, and aquatic food webs: San Francisco estuary and comparative systems. *Reviews in Fisheries Science*, **19**, 358–417.

Glibert, P.M., Harrison, J., Heil, C., & Seitzinger, S. (2006). Escalating worldwide use of urea—a global change contributing to coastal eutrophication. *Biogeochemistry*, **77**, 441–463.

Glibert, P.M., Kana, T.M., & Brown, K. (2013). From limitation to excess: the consequences of substrate excess and stoichiometry for phytoplankton physiology, trophodynamics and biogeochemistry, and the implications for modeling. *Journal of Marine Systems*, **125**, 14–28.

Goecke, F., Labes, A., Wiese, J., & Imhoff, J.F. (2010). Chemical interactions between marine macroalgae and bacteria. *Marine Ecology Progress Series*, **409**, 267–299.

Goldman, J.C. & Glibert, P.M. (1983). Kinetics of inorganic nitrogen uptake by phytoplankton. In: Carpenter, E.J. & Capone, D.G. (eds), *Nitrogen in the Marine Environment*, pp. 233–274. Academic Press, San Diego, CA.

Gomez, I. & Wiencke, C. (1998). Seasonal changes in C, N and major organic compounds and their significance to morpho-functional processes in the endemic Antarctic brown alga *Ascoseira mirabilis*. *Polar Biology*, **19**, 115–124.

Gronquist, D. & Berges, J.A. (2013). Effects of aquarium-related stressors on the zebrafish: a comparison of behavioral, physiological, and biochemical indicators. *Journal of Aquatic Animal Health*, **25**, 53–65.

Gruber, N. (2008). The marine nitrogen cycle: overview and challenges. In: Capone, D.G., Bronk, D.A., Mulholland, M.R., & Carpenter, E.J. (eds), *Nitrogen in the Marine Environment*, 2nd edition, pp. 1–50. Academic Press, Boston.

Gu, B.J., Chang, J., Min, Y., et al. (2013). The role of industrial nitrogen in the global nitrogen biogeochemical cycle. *Scientific Reports*, **3**, 2579.

Gulbransen, D. & McGlathery, K. (2013). Nitrogen transfers mediated by a perennial, non-native macroalga: a N-15 tracer study. *Marine Ecology Progress Series*, **482**, 299–304.

Handy, R.D. & Poxton, M.G. (1993). Nitrogen pollution in mariculture: toxicity and excretion of nitrogenous compounds by marine fish. *Reviews in Fish Biology and Fisheries*, **3**, 205–241.

Harrison, P.J. & Berges, J.A. (2005). Marine culture media. In: Andersen, R.A. (ed.), *Algal Culturing Techniques*, pp. 21–33. Academic Press, Boston.

Harrison, P.J., Conway H.L., Holmes, R.W., & Davis, C.O. (1977). Marine diatoms grown in chemostats under silicate or ammonium limitation. III. Cellular chemical composition and morphology of *Chaetoceros debilis*, *Skeletonema costatum*, and *Thalassiosira gravida*. *Marine Biology*, **43**, 19–31

Hatcher, A. (1994). Nitrogen and phosphorus turnover in some benthic marine-invertebrates—implications for the use of C:N ratios to assess food quality. *Marine Biology*, **121**, 161–166.

Hauxwell, J., Cebrián, J., Furlong, C., & Valiela, I. (2001). Macroalgal canopies contribute to eelgrass (*Zostera marina*) decline in temperate estuarine ecosystems. *Ecology*, **82**, 1007–1022.

Hemmi, A. & Jormalainen, V. (2002). Nutrient enhancement increases performance of a marine herbivore via quality of its food alga. *Ecology*, **83**, 1052–1064.

Hessini, K., Ben Hamed, K., Gandour, M., et al. (2013). Ammonium nutrition in the halophyte *Spartina alterniflora* under salt stress: evidence for a priming effect of ammonium? *Plant and Soil*, **370**, 163–173.

Heugens, E.H.W., Hendriks, A.J., Dekker, T., et al. (2001). A review of the effects of multiple stressors on aquatic organisms and analysis of uncertainty factors for use in risk assessment. *Critical Reviews in Toxicology*, **31**, 247–284.

Hildebrand, M. & Dahlin, K. (2000). Nitrate transporter genes from the diatom *Cylindrotheca fusiformis* (Bacillariophyceae): mRNA levels controlled by nitrogen source and by the cell cycle. *Journal of Phycology*, **36**, 702–713.

Hockin, N.L., Mock, T., Mulholland, F., et al. (2012). The response of diatom central carbon metabolism to nitrogen starvation is different from that of green algae and higher plants. *Plant Physiology*, **158**, 299–312.

Hoegh-Guldberg, O., Muscatine, L., Goiran, C., Siggaard, D., and Marion, G. (2004). Nutrient induced perturbations to $\delta^{13}C$ and $\delta^{15}N$ in symbiotic dinoflagellates and their coral hosts. *Marine Ecology Progress Series*, **280**, 105–114.

Holguin, G., Vasquez, P., and Bashan, Y. (2001). The role of sediment microorganisms in the productivity, conservation, and rehabilitation of mangrove ecosystems: an overview. *Biology and Fertility of Soils*, **33**, 265–278.

Hopkinson, C.S. and Giblin, A.E. (2008). Nitrogen dynamics of coastal salt marshes. In: Capone, D.G., Bronk, D.A., Mulholland, M.R., & Carpenter, E.J. (eds), *Nitrogen in the Marine Environment*, 2nd edition, pp. 991–1036. Academic Press, Boston.

Howarth, R.W. (1988). Nutrient limitation of net primary production in marine ecosystems. *Annual Reviews of Ecology and Systematics*, **19**, 89–110.

Howarth, R.W. & Marino, R. (2006). Nitrogen as the limiting nutrient for eutrophication in coastal marine ecosystems: evolving views over three decades. *Limnology and Oceanography*, **51**, 364–376.

Hutchins, D.A. & Fu, F.-X. (2008). Linking the oceanic biogeochemistry of iron and phosphorus with the marine nitrogen cycle. In: Capone, D.G., Bronk, D.A., Mulholland, M.R., & Carpenter, E.J. (eds), *Nitrogen in the Marine Environment*, 2nd edition, pp. 1627–1666. Academic Press, Boston.

Hutchinson, T.H, Solbe, J., & Kloepper-Sams, P. (1998). Analysis of the ecetoc aquatic toxicity (eat) database iii—comparative toxicity of chemical substances to different life stages of aquatic organisms. *Chemosphere*, **36**, 129–142.

Ianora, A., Miralto, A., Poulet, S. A., et al. (2004). Aldehyde suppression of copepod recruitment in blooms of a ubiquitous planktonic diatom. *Nature*, **429**, 403–407.

Jefferies, R.L. (1981). Osmotic adjustment and the response of halophytic plants to salinity. *BioScience*, **31**, 42–46.

Jenkins, B.D. & Zehr, J.P. (2008). Molecular approaches to the nitrogen cycle. In: Capone, D.G., Bronk, D.A., Mulholland, M.R., & Carpenter, E.J. (eds), *Nitrogen in the Marine Environment*, 2nd edition, pp. 1303–1344. Academic Press, Boston.

Jeong, H.J., Yoo, Y.D., Kang, N.S., et al. (2012). Heterotrophic feeding as a newly identified survival strategy of the dinoflagellate *Symbiodinium*. *Proceedings of the National Academy of Sciences of the USA*, **109**, 12604–12609.

Joint, I., Tait, K., Callow, M.E., et al. (2002). Cell-to-cell communication across the prokaryote-eukaryote boundary. *Science*, **298**, 1207–1207.

Joye, S.B. & Anderson, I.C. (2008). Nitrogen cycling in coastal sediments. In: Capone, D.G., Bronk, D.A., Mulholland, M.R., & Carpenter, E.J. (eds), *Nitrogen in the Marine Environment*, 2nd edition, pp. 867–916. Academic Press, Boston.

Karl, D.M. (1999). A sea of change: biogeochemical variability in the North Pacific Subtropical Gyre. *Ecosystems*, **2**, 181.

Keller, M.D., Selvin, R.C., Claus, W., & Guillard, R.R.L. (1987). Media for the culture of oceanic ultraphytoplankton. *Journal of Phycology*, **23**, 633–638.

Kelly L.W., Williams, G.J., Barott, K.L. et al. (2014). Local genomic adaptation of coral reef-associated microbiomes to gradient of natural variability and anthropogenic stressors. *Proceedings of the National Academy of Sciences of the USA*, **111**, 10227–10232.

Kim, D., Yamaguchi, K., & Oda, T. (2006). Nitric oxide synthase-like enzyme mediated nitric oxide generation by harmful red tide phytoplankton, *Chattonella marina*. *Journal of Plankton Research*, **28**, 613–620.

Kinney, E.L. & Valiela, I. (2013). Changes in delta N-15 in salt marsh sediments in a long-term fertilization study. *Marine Ecology Progress Series*, **477**, 41–66.

Koch, M.S., Mendelssohn, I.A., & McKee, K.L. (1990). Mechanism for the hydrogen sulfide-induced growth limitation in wetland macrophytes. *Limnology and Oceanography*, **35**, 399–408.

Kopp, C., Pernice, M., Domart-Coulon, I., et al. (2013). Highly dynamic cellular-level response of symbiotic coral to a sudden increase in environmental nitrogen. *MBio*, **4**(3), e00052-13.

Kremp, A., Rengefors, K., & Montresor, M. (2009). Species-specific encystment patterns in three Baltic cold-water dinoflagellates: the role of multiple cues in resting cyst formation. *Limnology and Oceanography*, **54**, 1125–1138.

Kroupova, H., Machova, A., & Svobodova, Z. (2005). Nitrite influence on fish: a review. *Veterinarni Medicina*, **50**, 461–471.

Kudela, R.M., Ryan, J.P., Blakely, M.D., et al. (2008). Linking the physiology and ecology of *Cochlodinium* to better understand harmful algal bloom events: a comparative approach. *Harmful Algae*, **7**, 278–292.

Kudela, R.M., Seeyave, S., & Cochlan, W.P. (2010). The role of nutrients in regulation and promotion of harmful algal blooms in upwelling systems. *Progress in Oceanography*, **85**, 122–135.

Landis, W.G., Durda, J.L., Brooks, M.L., et al. (2013). Ecological risk assessment in the context of global climate change. *Environmental Toxicology and Chemistry*, **32**, 79–92.

Lapointe, B.E. (1997). Nutrient thresholds for bottom-up control of macroalgal blooms on coral reefs in Jamaica and southeast Florida. *Limnology and Oceanography*, **42**, 1119–1131.

Laws, E.A., Ditullio, G.R., & Redalje, D.G. (1987). High phytoplankton growth and production rates in the north Pacific subtropical gyre. *Limnology and Oceanography*, **32**, 905–918.

Lee, K.S. & Dunton, K.H. (1999). Inorganic nitrogen acquisition in the seagrass *Thalassia testudinum*: development of a whole-plant nitrogen budget. *Limnology and Oceanography*, **44**, 1204–1215.

Lee, C.B., Rees, D., & Horton, P. (1990). Non-photochemical quenching of chlorophyll fluorescence in the green alga *Dunaliella*. *Photosynthesis Research*, **24**, 167–173.

Lee, K.S., Short, F.T., & Burdick, D.M. (2004). Development of a nutrient pollution indicator using the seagrass, *Zostera marina*, along nutrient gradients in three New England estuaries. *Aquatic Botany*, **78**, 197–216.

Levine J., Brewer, S.J., & Bertness, M.D. (1998). Nutrient availability and the zonation of marsh plant communities. *Journal of Ecology*, **86**, 285–292.

Li, J., Glibert, P.M., Alexander, J.A., & Molina, M.E. (2012). Growth and competition of several harmful dinoflagellates under different nutrient and light conditions. *Harmful Algae*, **13**, 112–125.

Li, P., Mai, K., Trushenski, J., & Wu, G. (2009). New developments in fish amino acid nutrition: towards functional and environmentally oriented aquafeeds. *Amino Acids*, **37**, 43–53.

Lindell, D., Penno, S., Al-Qutob, M., et al. (2005). Expression of the nitrogen stress response gene *ntcA* reveals nitrogen-sufficient *Synechococcus* populations in the oligotrophic northern Red Sea. *Limnology and Oceanography*, **50**, 1932–1944.

Lobban, C.S. & Harrison, P.J. (1997). *Seaweed Ecology and Physiology*. Cambridge University Press, Cambridge, UK.

Lomas, M.W. & Glibert, P.M. (1999). Temperature regulation of nitrate uptake: a novel hypothesis about nitrate uptake and reduction in cool-water diatoms. *Limnology and Oceanography*, **44**, 556–572.

Lovelock, C.E., Ball, M.C., Martin, K.C., & Feller, I.C. (2009). Nutrient enrichment increases mortality of mangroves. *PLoS ONE*, **4**, e5600.

Lovelock, C.E., Feller, I.C., Ball, M.C., et al. (2006). Differences in plant function in phosphorus- and nitrogen-limited mangrove ecosystems. *New Phytologist*, **172**, 514–522.

Lui, D., Keesing, J.K., He, P., et al. (2013). The world's largest macroalgal bloom in the Yellow Sea China: formation and implications. *Estuarine, Coastal and Shelf Science*, **129**, 2–10.

Maas, A.E., Elder, L.E., Dierssen, H.M., & Seibel, B.A. (2011). Metabolic response of Antarctic pteropods (Mollusca: Gastropoda) to food deprivation and regional productivity. *Marine Ecology Progress Series*, **441**, 129–139.

Malta, E.J., Ferreira, D.G., Vergara, J.J., & Perez-Llorens, J.L. (2005). Nitrogen load and irradiance affect morphology, photosynthesis and growth of *Caulerpa prolifera* (Bryopsidales: Chlorophyta). *Marine Ecology Progress Series*, **298**, 101–114.

Mann, K.H. (1988). Production and use of detritus in various freshwater, estuarine and coastal marine ecosystems. *Limnology and Oceanography*, **33**, 910–930.

Matantseva, O.V. & Skarlato, S.O. (2013). Mixotrophy in microorganisms: ecological and cytophysiological aspects. *Journal of Evolutionary Biochemistry and Physiology*, **49**, 377–388.

McCarthy, J.J. & Goldman, J.C. (1979). Nitrogenous nutrition of marine phytoplankton in nutrient-depleted waters. *Science*, **203**, 670–672.

McCook, L.J. (1999). Macroalgae, nutrients and phase shifts on coral reefs: scientific issues and management consequences for the Great Barrier Reef. *Coral Reefs*, **18**, 357–367.

McGlathery, K.J. (2008). Seagrass habitats. In: Capone, D.G., Bronk, D.A., Mulholland, M.R., & Carpenter, E.J. (eds), *Nitrogen in the Marine Environment*, 2nd edition, pp. 1037–1071. Academic Press, Boston.

McKew, B.A., Lefebvre, S.C., Achterberg, E.P., et al. (2013). Plasticity in the proteome of *Emiliania huxleyi* CCMP 1516 to extremes of light is highly targeted. *New Phytologist*, **200**, 61–73.

McLachlan, J. (1973). Growth media—marine. In: Stein, J. (ed.), *Handbook of Phycological Methods: Culture Methods and Growth Measurements*, pp. 25–51. Cambridge University Press, Cambridge, UK.

McQuoid, M.R. & Hobson, L.A. (1996). Diatom resting stages. *Journal of Phycology*, **32**, 889–902.

Melack, J.M., Finzi, A.C., Siegel, D., et al. (2011). Improving biogeochemical knowledge through technological innovation. *Frontiers in Ecology and the Environment*, **9**, 37–43.

Meyer, M. & Griffiths, H. (2013). Origins and diversity of eukaryotic CO_2-concentrating mechanisms: lessons for the future. *Journal of Experimental Botany*, **64**, 769–786.

Miller, C.A. & Roman, M.R. (2008). Effects of food nitrogen content and concentration on the forms of nitrogen excreted by the calanoid copepod, *Acartia tonsa*. *Journal of Experimental Marine Biology and Ecology*, **359**, 11–17.

Miller, D.J. & Yellowlees, D. (1989). Inorganic nitrogen uptake by symbiotic marine cnidarians—a critical review. *Proceedings of the Royal Society of London Series B*, **237**, 109–125.

Moore, C.M., Mills, M.M., Arrigo, K.R. et al. (2013). Processes and patterns of oceanic nutrient limitation. *Nature Geoscience*, **6**, 701–710.

Moran, A.L. & Manahan, D.T. (2004). Physiological recovery from prolonged 'starvation' in larvae of the Pacific oyster *Crassostrea gigas*. *Journal of Experimental Marine Biology and Ecology*, **306**, 17–36.

Morris, J.T. (1982). A model of growth responses by *Spartina alterniflora* to nitrogen limitation. *Journal of Ecology*, **70**, 25–42.

Morris, J.T. (1984). Effects of oxygen and salinity on ammonium uptake by *Spartina alterniflora* Loisel and *Spartina patens* (Aiton) Muhl. *Journal of Experimental Marine Biology and Ecology*, **78**, 87–98.

Morris, E.P., Peralta, G., Brun, F.G., et al. (2008). Interaction between hydrodynamics and seagrass canopy structure: spatially explicit effects on ammonium uptake rates. *Limnology and Oceanography*, **53**, 1531–1539.

Morris, J.T., Sundberg, K., & Hopkinson, C.S. (2013). Salt marsh primary production and its responses to relative sea level and nutrients in estuaries at Plum Island, Massachusetts, and North Inlet, South Carolina, USA. *Oceanography*, **26**, 78–84.

Mulholland, M.R. and Lomas, M.W. (2008). Nitrogen uptake and assimilation. In: Capone, D.G., Bronk, D.A., Mulholland, M.R., & Carpenter, E.J. (eds), *Nitrogen in the Marine Environment*, 2nd edition, pp. 303–384. Academic Press, Boston.

Murata, A., Nagashima, Y., & Taguchi, S. (2012). N:P ratios controlling the growth of the marine dinoflagellate *Alexandrium tamarense*: content and composition of paralytic shellfish poison. *Harmful Algae*, **20**, 11–18.

Myers, J.H., Duda, S., Gunthorpe, L., & Allinson, G. (2007). Evaluation of the *Hormosira banksii* (Turner) desicaine germination and growth inhibition bioassay for use as a regulatory assay. *Chemosphere*, **69**, 955–960.

Naqvi, S.W.A., Yoshinari, T., Jayakumar, D.A., et al. (1998). Budgetary and biogeochemical implications of N_2O isotope signatures in the Arabian Sea. *Nature*, **394**, 462–464.

Németh, Z., Bonier, F., & Macdougall-Shackleton, S.A. (2013). Coping with uncertainty: integrating physiology, behavior, and evolutionary ecology in a changing world. *Integrative and Comparative Biology*, **53**, 960–964.

Newton, C. & Thornber, C. (2013). Ecological impacts of macroalgal blooms on salt marsh communities. *Estuaries and Coasts*, **36**, 365–376.

Nixon, S.W. (2009). Eutrophication and the macroscope. *Hydrobiologia*, **629**, 5–19.

Nixon, S.C., Gunby, A., Ashley, S.J., et al. (1995). *Development and testing of General Quality Assessment schemes: dissolved oxygen and ammonia in estuaries*. Environment Agency R&D Project Record PR 469/15/HO.

Okamoto, O.K. & Hastings, J.W. (2003). Genome-wide analysis of redox-regulated genes in a dinoflagellate. *Gene*, **321**, 73–81.

O'Neil, J.M. & Capone, D.G. (2008). Nitrogen cycling in coral reef environments. In: Capone, D.G., Bronk, D.A., Mulholland, M.R., & Carpenter, E.J. (eds), *Nitrogen in the Marine Environment*, 2nd edition, pp. 949–990. Academic Press, Boston.

Paffenhofer, G.A. (2002). An assessment of the effects of diatoms on planktonic copepods. *Marine Ecology Progress Series*, **227**, 305–310.

Pankhurst, N.W. & Herbert, N.A. (2013). Fish physiology and ecology: the contribution of the Leigh Laboratory to the collision of paradigms. *New Zealand Journal of Marine and Freshwater Research*, **47**, 392–408.

Parkhill, J.P., Maillet, G., & Cullen, J.J. (2001). Fluorescence-based maximal quantum yield for PSII as a diagnostic of nutrient stress. *Journal of Phycology*, **37**, 517–529.

Pedersen, M.F. & Borum, J. (1996). Nutrient control of algal growth in estuarine waters. Nutrient limitation and the importance of nitrogen requirements and nitrogen storage among phytoplankton and species of macroalgae. *Marine Ecology Progress Series*, **142**, 261–272.

Pedersen, M.F., Paling, E.I., & Walker, D.I. (1997). Nitrogen uptake and allocation in the seagrass *Amphibolis antarctica*. *Aquatic Botany*, **56**, 105–117.

Pehlivanoglu-Mantas, E. & Sedlak, D.L. (2006). Wastewater-derived dissolved organic nitrogen: analytical methods, characterization, and effects—a review. *Critical Reviews in Environmental Science and Technology*, **36**, 261–285.

Perez, M., Duarte, C.M., Romero, J., et al. (1994). Growth plasticity in *Cymodocea nodosa* stands—the importance of nutrient supply. *Aquatic Botany*, **47**, 249–264.

Perry, M.J. (1976). Phosphate utilization by an oceanic diatom in phosphorus-limited chemostat culture and in the oligotrophic waters of the central North Pacific. *Limnology and Oceanography*, **21**, 88–107.

Peter, K.H. & Sommer, U. (2013). Phytoplankton cell size reduction in response to warming mediated by nutrient limitation. *PLoS ONE*, **8**, e71528.

Phillips, N.W. (1984). Role of different microbes and substrates as potential suppliers of specific, essential nutrients to marine detritivores. *Bulletin of Marine Science*, **35**, 283–298.

Phillips, J.C. & Hurd, C.L. (2003). The nitrogen ecophysiology of intertidal seaweeds from New Zealand: patterns of N uptake, storage and utilisation in relation to shore position and season. *Marine Ecology Progress Series*, **264**, 31–48.

Pueschel, C.M. & Korb, R.E. (2001). Storage of nitrogen in the form of protein bodies in the kelp *Laminaria solidungula*. *Marine Ecology Progress Series*, **218**, 107–114.

Qing, H., Cai, Y., Xiao, Y., et al. (2012). Leaf nitrogen partition between photosynthesis and structural defense in invasive and native tall form *Spartina alterniflora* populations: effects of nitrogen treatments. *Biological Invasions*, **14**, 2039–2048.

Quintana, X.D., Moreno-Amich, R., & Comin, F.A. (1998). Nutrient and plankton dynamics in a Mediterranean salt marsh dominated by incidents of flooding. Part 1: differential confinement of nutrients. *Journal of Plankton Research*, **20**, 2089–2107.

Radzikowski, J. (2013). Resistance of dormant stages of planktonic invertebrates to adverse environmental conditions. *Journal of Plankton Research*, **35**, 707–723.

Raubenheimer, D. & Jones, S.A. (2006). Nutritional imbalance in an extreme generalist omnivore: tolerance and recovery through complementary food selection. *Animal Behaviour*, **71**, 1253–1262.

Raven, J.A., Giordano, M., Beardall, J., & Maberly, S.C. (2011). Algal and aquatic plant carbon concentrating mechanisms in relation to environmental change. *Photosynthesis Research*, **109**, 281–296.

Reef, R., Feller, I.C., & Lovelock, C.E. (2010). Nutrition of mangroves. *Tree Physiology*, **30**, 1148–1160.

Rees, T.A.V., Pick, U., Avron, M., & Degani, H. (1991). Are symbiotic algae nutrient deficient? *Proceedings of the Royal Society of London Series B*, **243**, 227–233.

Reynolds, C.S. (1984). Phytoplankton periodicity: the interactions of form, function and environmental variability. *Freshwater Biology*, **14**, 111–142.

Russo, R.C. (1985). Ammonia, nitrite, and nitrate. In: Rand, G.M. & Petrocelli, S.R. (eds), *Fundamentals of Aquatic Toxicology and Chemistry*, pp. 455–471. Hemisphere Publishing Corp., Washington, DC.

Ryan, A.B. & Boyer, K.E. (2012). Nitrogen further promotes a dominant salt marsh plant in an increasingly saline environment. *Journal of Plant Ecology*, **5**, 429–441.

Saba, G.K., Schofield, O., Torres, J.J., et al. (2012). Increased feeding and nutrient excretion of adult Antarctic krill, *Euphausia superba*, exposed to enhanced carbon dioxide (CO_2). *PLoS ONE*, **7**, e52224.

Saito, M.A., Goepfert, T.J., & Ritt, J.T. (2008). Some thoughts on the concept of colimitation: three definitions and the importance of bioavailability. *Limnology and Oceanography*, **53**, 276–290.

Saito, M.A., McIlvin, M.R., Moran, D.M., et al. (2014). Multiple nutrient stresses at intersecting Pacific Ocean biomes detected by protein markers. *Science*, **345**, 1173–1177.

Sakamoto, T., Delgaizo, V., & Bryant, D. (1998). Growth on urea can trigger death and peroxidation of the cyanobacterium *Synechococcus* sp. strain PCC 7002. *Applied and Environmental Microbiology*, **64**, 2361–2366.

Salerno, M. & Stoecker, D.K. (2009). Ectocellular glucosidase and peptidase activity of the mixotrophic dinoflagellate *Prorocentrum minimum* (Dinophyceae). *Journal of Phycology*, **45**, 34–45.

Sandh, G., Ran, L.A., Xu, L.H., et al. (2011). Comparative proteomic profiles of the marine cyanobacterium *Trichodesmium erythraeum* IMS101 under different nitrogen regimes. *Proteomics*, **11**, 406–419.

Scanlan, D.J., Ostrowski, M., Mazard, S., et al. (2009). Ecological genomics of marine picocyanobacteria. *Microbiology and Molecular Biology Reviews*, **73**, 249–299.

Scanlan, D.J. & Post, A.F. (2008). Aspects of marine cyanobacterial nitrogen physiology and connection to the nitrogen cycle. In: Capone, D.G., Bronk, D.A., Mulholland, M.R., & Carpenter, E.J. (eds), *Nitrogen in the Marine Environment*, 2nd edition, pp. 1072–1096. Academic Press, Boston.

Schnackschiel, S.B. & Hagen, W. (1995). Life-cycle strategies of *Calanoides acutus*, *Calanus propinquus*, and *Metridia gerlachei* (Copepoda, Calanoida) in the eastern Weddell Sea, Antarctica. *ICES Journal of Marine Science*, **52**, 541–548.

Schründer, S., Schnack-Schiel, S.B., Auel, H., & Sartoris, F.J. (2013). Control of diapause by acidic pH and ammonium accumulation in the hemolymph of Antarctic copepods. *PLoS ONE*, **8**, e77498.

Seitzinger, S.P. & Harrison, J.A. (2008). Land-based nitrogen sources and their delivery to coastal systems. In: Capone, D.G., Bronk, D.A., Mulholland, M.R., & Carpenter, E.J. (eds), *Nitrogen in the Marine Environment*, 2nd edition, pp. 469–510. Academic Press, Boston.

Selph, K.E., Shacat, J., & Landry, M.R. (2005). Microbial community composition and growth rates in the NW Pacific during spring 2002. *Geochemistry Geophysics Geosystems*, **6**, Q12M05.

Shen, Q., Li, H.Y., Li, Y., et al. (2012). Molecular identification of green algae from the rafts based infrastructure of *Porphyra yezoensis*. *Marine Pollution Bulletin*, **64**, 2077–2082.

Short, F.T. & McRoy, C.P. (1984). Nitrogen uptake by leaves and roots of the seagrass *Zostera marina* L. *Botanica Marina*, **27**, 547–555.

Skovgaard, A., Legrand, C., Hansen, P.J., & Granéli, E. (2003). Effects of nutrient limitation on food uptake in the toxic haptophyte *Prymnesium parvum*. *Aquatic Microbial Ecology*, **31**, 259–265.

Smetacek, V. & Zingone, A. (2013). Green and golden seaweed tides on the rise. *Nature*, **504**, 84–88.

Sogard, S.M. & Olla, B.L. (1996). Food deprivation affects vertical distribution and activity of a marine fish in a thermal gradient: potential energy-conserving mechanisms. *Marine Ecology Progress Series*, **133**, 43–55.

Sousa, A.I., Martins, I., Lillebø, A.I., et al. (2007). Influence of salinity, nutrients and light on the germination and growth of *Enteromorpha* sp. spores. *Journal of Experimental Marine Biology and Ecology*, **341**, 142–150.

Steinberg, D.K. & Saba, G.K. (2008). Nitrogen consumption and metabolism in marine zooplankton. In: Capone, D.G., Bronk, D.A., Mulholland, M.R., & Carpenter, E.J. (eds), *Nitrogen in the Marine Environment*, 2nd edition, pp. 1135–1196. Academic Press, Boston.

Sterner, R.W. & Elser, J.J. (2002). *Ecological Stoichiometry: The Biology of Elements from Molecules to the Biosphere*. Princeton University Press, Princeton.

Sterner, R.W. & Hessen, D.O. (1994). Algal nutrient limitation and the nutrition of aquatic herbivores. *Annual Review of Ecology and Systematics*, **25**, 1–29.

Stevens C.L. & Hurd, C.L. (1997). Boundary-layers around bladed aquatic macrophytes. *Hydrobiologia*, **346**, 119–128.

Stoecker, D.K. & Gustafson, D.E. (2003). Cell-surface proteolytic activity of photosynthetic dinoflagellates. *Aquatic Microbial Ecology*, **30**, 175–183.

Sullivan, H. (2014). *The effects of nitrate fertilization on the photosynthetic performance of the salt marsh cordgrass, Spartina alterniflora*. MS thesis, Clark University.

Sundbäck, K., Miles, A., & Göransson, E. (2000). Nitrogen fluxes, denitrification and the role of microphytobenthos in microtidal shallow-water sediments: an annual study. *Marine Ecology Progress Series*, **200**, 59–76.

Szmant, A.M. (2002). Nutrient enrichment on coral reefs: is it a major cause of coral reef decline? *Estuaries*, **25**, 743–766.

Tait, K., Williamson, H., Atkinson, S., et al. (2009). Turnover of quorum sensing signal molecules modulates cross-kingdom signalling. *Environmental Microbiology*, **11**, 1792–1802.

Tang, B.J., Liu, B.Z., Wang, G.D., et al. (2006). Effects of various algal diets and starvation on larval growth and survival of *Meretrix meretrix*. *Aquaculture*, **254**, 526–533.

Teichberg, M., Fox, S.E., Olsen, Y.S., et al. (2009). Eutrophication and macroalgal blooms in temperate and tropical coastal waters: nutrient enrichment experiments with Ulva spp. *Global Change Biology*, **16**, 2624–2637.

Tenore, K.R. (1988). Nitrogen in benthic food chains. In: Blackburn, T.H. & Sorensen, J. (eds), *Nitrogen Cycling in Coastal Marine Environments*, pp. 191–206. John Wiley and Sons Ltd, London.

Touchette, B.W. & Burkholder, J. (2001). Nitrate reductase activity in a submersed marine angiosperm: controlling influences of environmental and physiological factors. *Plant Physiology and Biochemistry*, **39**, 583–593.

Touchette, B.W. & Burkholder, J.M. (2000). Review of nitrogen and phosphorus metabolism in seagrasses. *Journal of Experimental Marine Biology and Ecology*, **250**, 133–167.

Turpin, D.H. (1991). Effects of inorganic N availability on algal photosynthesis and carbon metabolism. *Journal of Phycology*, **27**, 14–20.

Udy, J.W. & Dennison, W.C. (1997). Growth and physiological responses of three seagrass species to elevated sediment nutrients in Moreton Bay, Australia. *Journal of Experimental Marine Biology and Ecology*, **217**, 253–277.

U.S. Environmental Protection Agency (2013). *Aquatic Life Ambient Water Quality Criteria for Ammonia Freshwater 2013*. EPA-822-R-13-001. Office of Water 4304T.

Usher, K.M., Bergman, B., & Raven, J.A. (2007). Exploring cyanobacterial mutualisms. *Annual Review of Ecology Evolution and Systematics*, **38**, 255–273.

Valiela, I., McClelland, J., Hauxwell, J., et al. (1997). Macroalgal blooms in shallow estuaries: controls and ecophysiological and ecosystem consequences. *Limnology and Oceanography*, **42**, 1105–1118.

Valiela, I., Teal, J.M., & Persson, N.Y. (1976). Production and dynamics of experimentally enriched salt marsh vegetation: belowground biomass. *Limnology and Oceanography*, **21**, 245–252.

Van der Heide, T., Smolders, A., Rijkens, B., et al. (2008). Toxicity of reduced nitrogen in eelgrass (*Zostera marina*) is highly dependent on shoot density and pH. *Oecologia*, **158**, 411–419.

Van Katwijk, M.M., Vergeer, L.H.T., Schmitz, G.H.W., & Roelofs, J.G.M. (1997). Ammonium toxicity in eelgrass *Zostera marina*. *Marine Ecology Progress Series*, **157**, 159–173.

Vardi, A., Bidle, K.D., Kwityn, C., et al. (2008). A diatom gene regulating nitric-oxide signaling and susceptibility to diatom-derived aldehydes. *Current Biology*, **18**, 895–899.

Vardi, A., Formiggini, F., Casotti, R., et al. (2006). A stress surveillance system based on calcium and nitric oxide in marine diatoms. *PLoS Biology*, **4**, 411–419.

Ventura, M., Liboriussen, L., Lauridsen, T., et al. (2008). Effects of increased temperature and nutrient enrichment on the stoichiometry of primary producers and consumers in temperate shallow lakes. *Freshwater Biology*, **53**, 1434–1452.

Vergara, J.J., Bird, K.T., & Niell, F.X. (1995). Nitrogen assimilation following NH_4^+ pulses in the red alga *Gracilariopsis lemaneiformis*: effect on C metabolism. *Marine Ecology Progress Series*, **122**, 253–263.

Villazan, B., Pedersen, M.F., Brun, F.G., & Vergara, J.J. (2013). Elevated ammonium concentrations and low light form a dangerous synergy for eelgrass *Zostera marina*. *Marine Ecology Progress Series*, **493**, 141–154.

Voss, M., Bange, H.W., Dippner, J.W., et al. (2013). The marine nitrogen cycle: recent discoveries, uncertainties and the potential relevance of climate change. *Philosophical Transactions of the Royal Society Series B*, **368**, 20130121.

Wahl, M., Goecke, F., Labes, A., et al. (2012). The second skin: ecological role of epibiontic biofilms on marine organisms. *Frontiers in Microbiology*, **3**, 292.

Wang, C., Yu, R.C., & Zhou, M.J. (2012). Effects of the decomposing green macroalga *Ulva* (*Enteromorpha*) *prolifera* on the growth of four red-tide species. *Harmful Algae*, **16**, 12–19.

Waycott, M., Duarte, C.M., Carruthers, T.J.B., et al. (2009). Accelerating loss of seagrasses across the globe threatens coastal ecosystems. *Proceedings of the National Academy of Sciences of the USA*, **106**, 12377–12381.

Wegley, L., Edwards, R., Rodriguez-Brito, B., et al. (2007). Metagenome analysis of the microbial community associated with the coral *Porites astreoides*. *Environmental Microbiology*, **9**, 2707–2719.

Wurch, L.L., Bertrand, E.M., Saito, M.A., et al. (2011). Proteome changes driven by phosphorus deficiency and recovery in the brown tide-forming alga *Aureococcus anophageffferens*. *PLoS ONE*, **6**, e28949.

Young, E. & Beardall, J. (2003). Photosynthetic function in *Dunaliella tertiolecta* during a nitrogen starvation and recovery cycle. *Journal of Phycology*, **39**, 897–905.

Young, E.B. & Beardall, J. (2005). Modulation of photosynthesis and inorganic carbon acquisition in a marine microalga by nitrogen, iron and light availability. *Canadian Journal of Botany*, **83**, 917–928.

Young, E.B., Berges, J.A., & Dring, M.J. (2009). Physiological responses of intertidal marine algae to nitrogen deprivation and resupply of nitrate and ammonium. *Physiologia Plantarum*, **135**, 400–411.

Young, E.B., Dring, M.J., Birkett, D.A., et al. (2007a). Seasonal variation in nitrate reductase activity and internal N pools in intertidal brown algae correlate with ambient nitrate availability. *Plant, Cell and Environment*, **30**, 764–774.

Young, E.B., Dring, M.J., & Berges, J.A. (2007b). Distinct patterns of nitrate reductase activity in brown algae: light and ammonium sensitivity in Laminaria digitata is absent in Fucus species. *Journal of Phycology*, **43**, 1200–1208.

Zhang, W., Tan, N.G.J., & Li, S.F.Y. (2014). NMR-based metabolomics and LC-MS/MS quantification reveal metal-specific tolerance and redox homeostasis in *Chlorella vulgaris*. *Molecular Biosystems*, **10**, 149–160.

CHAPTER 7

The cellular responses of marine algae and invertebrates to ultraviolet radiation, alone and in combination with other common abiotic stressors

David J. Burritt and Miles D. Lamare

7.1 Introduction

The ultraviolet radiation (UV-R) component of the solar spectrum reaching the Earth's surface consists of UV-B (290–320 nm) and UV-A (320–400 nm), and makes up ~1% of the total incident irradiance levels. Incident UV-B and UV-A irradiances average ~ 1 and 40 W m^{-2} respectively, but vary with latitude, season, and interannually. UV-irradiances are greatest and seasonally least variable in tropical regions (~ 50–70 W m^{-2}), lowest in polar regions (0 to ~ 35 W m^{-2}), and intermediate and most variable at temperate latitudes (~ 20–55 W m^{-2}). UV-R is rapidly attenuated by the water column at a rate depending on the optics of various water bodies. In the clearest oceanic waters, 10% of incident UV-B and UV-A may penetrate to depths as great as 16 and 46 m respectively, while turbid coastal or ice covered seas have penetrations < 5 m (Tedetti and Sempere, 2006).

Recent changes in atmospheric ozone, aerosol, and cloud cover have increased UV-B irradiances reaching the Earth's surface (Herman, 2010). The increases are most pronounced at high latitudes where UV-B levels in springtime can be elevated by between 22 and 130%, but are also evident in mid- and lower-latitudes where UV-B levels have increased 4–7% depending on season and latitude (Madronich et al., 1998). While the Montreal Protocol has succeeded in lowering atmospheric levels of chlorofluorocarbons and other ozone-depleting substances and largely halted the decline in stratospheric ozone levels, it has been projected that, during the twenty-first century, while upper stratospheric ozone levels will increase, lower stratosphere ozone levels will decrease (Williamson et al., 2014). In addition, increasing concentrations of carbon dioxide (CO_2) and other greenhouse gases have the potential to influence the levels of stratospheric ozone and cloud cover, and hence influence global UV exposure patterns. Current models predict that by 2100, UV levels will increase in the tropics and decrease at polar latitudes (Williamson et al., 2014). However, because of the large number of factors that can influence global UV exposure patterns there is considerable uncertainty regarding the global UV patterns that will emerge in the future.

Increases in UV-R entering the ocean may also result from other climate change outcomes (Häder et al., 2007), including loss of sea ice, reduced primary production and concentrations of dissolved organic matter (CDOM), which attenuates UV-B more effectively that UV-A (Tedetti and Sempere, 2006), while greater UV-doses may occur in pelagic organisms as a result of increased ocean stratification (Gao et al., 2012a, 2012b).

In this chapter the key physiological responses of marine algae and invertebrates to UV-R exposure are outlined, and the interaction of UV-R with other environmental stressors including pollutants, temperature, pH, and salinity on these responses are examined. Special emphasis is placed on oxidative stress and antioxidant metabolism as a common stress response to UV-R in marine species. Lastly, using oxidative stress responses to UV-R as an example, the concept of cross protection versus greater harm when exposed to multiple environmental stressors is explored.

Stressors in the Marine Environment. Edited by Martin Solan and Nia M. Whiteley
© Oxford University Press 2016. Published in 2016 by Oxford University Press.

7.2 Key physiological responses to UV-B

There are a number of key photobiological outcomes that result from exposure to solar radiation, with UV-R thought to be responsible for ≈50% of all photobiological reactions (Fig. 7.1). The importance of UV-R in these processes is due to the inverse relationship between radiation wavelength and energy, and the tendency for key biomolecules (i.e. DNA and proteins) to absorb more strongly in the shorter wavelengths. As a consequence of this relationship, the physiological responses to UV-R are wavelength dependent, and hence, dependent on the ambient spectral irradiance. This includes key photobiological stresses observed in animals and plants: DNA damage, oxidative damage in both, and photoinhibition in plants (Fig. 7.1).

7.2.1 DNA damage

UV-R damage to DNA is both direct and indirect, with the resulting lesions interrupting DNA transcription and replication, and promoting mutation and apoptosis (Fig. 7.1). DNA pyrimidine dimers, in the form of cyclobutane pyrimidine dimers (CPDs) and 6–4-photoproducts, are the most common form of DNA damage and can result in interruption of transcription and mutation (Sinha and Häder, 2002). 6–4-photoproducts form 10% of DNA dimers, but are more cytotoxic as they are not restored by photoreactivation and have a much stronger inhibition of DNA polymerization. DNA dimer frequencies in marine animals exposed to UV-R *in situ* have been reported in the range of 1–35 CPD Mb^{-1} DNA (reviewed by Dahms and Lee, 2010a), while CPD frequencies are often negligible or low in marine algae (Van de Poll et al., 2001; Bischof et al., 2002).

Distortion of DNA in the form of strand breaks, linkages, DNA crossovers, and changes in hydration also result from direct UV-B exposure. Indirect damage of DNA, in the form of oxidative modification of nucleotide bases (e.g. 8-hydroxy-2'-deoxyguanosine (8-OHdG)), results from the production of ROS by UV-A exposure, and in some cases can be the most important form of damage in marine organisms (Lesser, 2010).

7.2.2 Direct damage to proteins

For proteins, the absorption of UV-R by chromophoric amino acid residues (namely tryptophan, tyrosine, and phenylalanine) has the potential to degrade the protein through peptide breakages, the cross-linking of amino acids, and the disruption of salt bridges. Key proteins directly damaged by UV-R in this direct manner include those involved in photosynthesis (Lidon et al., 2012).

7.2.3 Oxidative stress

UV-R can indirectly damage marine organisms through the production of reactive oxygen species (ROS). ROS are a group of molecules that includes singlet oxygen ($^1O_2$), superoxide ($^\bullet O_2^-$), hydrogen peroxide (H_2O_2) and the hydroxyl radical ($HO^\bullet$). This group of highly reactive oxygen intermediates is produced by the partial, single-electron reduction of oxygen or by the alteration of oxygen electron spin states by photoactivation, which results in the production of singlet oxygen. In living cells ROS production can be initiated through the absorption of UV-R by photosensitive molecules such as photosynthetic pigments, or as a by-product of electron transport, associated with normal mitochondrial or chloroplast function (Halliwell and Gutteridge, 2007). When not under stress, marine organisms produce relatively low and steady levels of ROS, which can be readily neutralized by the levels of cellular antioxidant defences found in healthy cells. Lesser (2006) provides an excellent overview of antioxidant defences, which include non-enzymatic antioxidants such as ascorbate, glutathione, and α-tocopherol, and enzymatic antioxidants such as catalase (CAT), glutathione peroxidase (GPx), glutathione reductase (GR), and superoxide dismutase (SOD) in algae and animals, and ascorbate peroxidase (APX) in algae. Other molecules with antioxidant capacities, such as phenolics and polyamines, are also found in marine organisms (Schweikert et al., 2011). These antioxidants are able to regulate ROS levels, minimize oxidative damage, and maintain a stable cell redox potential. When exposed to UV-R marine organisms can overproduce ROS that can result in oxidative damage and stress (Lesser, 2006), with DNA, proteins, and lipids critical cellular macromolecules that can be damaged by ROS over-production induced by UV-R. Increases in antioxidant activity in response to UV-R exposure have been shown for a range of marine species including macroalgae (Bischof et al., 2003), hermatypic corals (Lesser and Lewis, 1996) and sea urchin embryos (Lister et al., 2010a, b). In a clear example of environmental influences on oxidative stress, Lister et al. (2010a) observed that oxidative damage and antioxidant enzyme activities in embryos of the Antarctic sea urchin *Sterechinus neumayeri* were the outcome of exposure to elevated UV-B irradiances that occur during ozone depletion and when no sea ice was present.

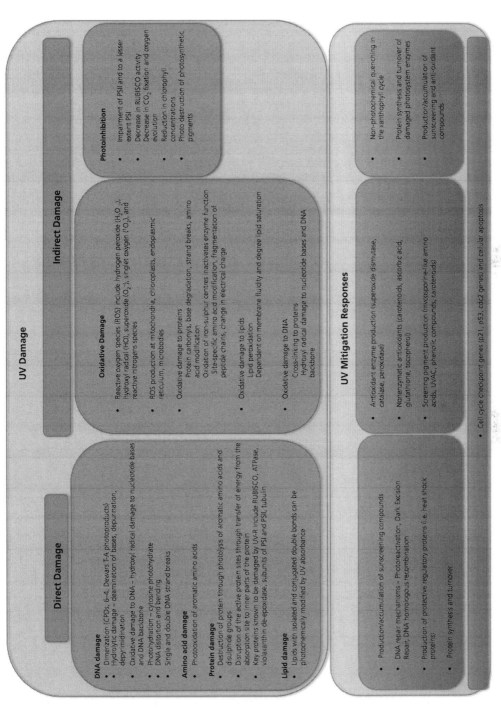

Figure 7.1 Major responses to the ultraviolet radiation (UV-R) exposure observed at the molecular and cellular level. Absorption of UV-B can cause direct damage to DNA and proteins, while indirect damage results from the production of reactive oxygen species, driving the oxidation of proteins, lipids, and DNA. Photoinhibition can be the result of both direct damage to chromophores and proteins within the photocentres and indirect damage associated with excess production of reactive oxygen species. Slower photosynthetic rates also result from the photo-degradation of RubisCO. UV-mitigation strategies at the cellular level include the production of protective compounds (sunscreens and antioxidants), repair (DNA restoration and protein turnover), and minimizing damage (cell-cycle genes and apoptosis).

7.2.4 Decreased photosynthesis

Unlike non-photosynthetic organisms, photosynthetic organisms require solar radiation to survive and hence both daily and seasonal fluctuations in UV-R levels provide the constant potential for UV-R induced stress. Numerous studies have been conducted on the photosynthetic performance of a wide range of marine algae, with or without directly considering the influences of UV-R on photosynthetic processes. These allow some generalizations regarding the impact of UV-R on photosynthesis to be made. Photosynthesis in many macroalgae is depressed under full sunlight and, under clear-sky conditions, the photosynthetic efficiency of O_2 evolution is higher in the morning than in the afternoon (Xu and Gao, 2012a). It is important to note that under natural solar radiation UV-A and UV-B can be less damaging than excessive PAR, but at midday photoinhibition in macroalgae is significantly enhanced under a full spectrum of solar radiation that includes UV-R. This clearly indicates the negative influence of total solar UV-R, in the presence of high PAR levels, on photosynthesis (Xu and Gao, 2012a). The reduction in photosynthetic rate due to UV-R exposure has generally been attributed to two mechanisms. The first is via inhibition of photosystem II (PSII) activity that impedes the electron chain pathway. This is possibly through damage to either the PSII electron donor or acceptor, or to the PSII core protein, D1 (Vincent and Neale, 2000). Damage to D1 may be less significant, however, as any UV-damaged protein is rapidly replaced in order to maintain PSII activity (Vincent and Neale, 2000). The second mechanism of photoinhibition may be through UV-induced degradation of the carbon-fixing enzyme Ribulose-1, 5-bisphosphate carboxylase oxygenase (RubisCO) and hence reducing carbon fixation. Bischof et al. (2000) noted a reduction in the maximum quantum yield in five species of algae exposed to UV-B that was associated with a loss of RubisCO activity. Similarly, a loss of photosynthetic activity in *Ulva* exposed to UV-R was attributed in part a reduction in RubisCO concentrations in algae thalli (Bischof et al., 2002). However, it is important to note that inhibition of PSII repair, caused by exposure to UV-B, could also result in decreased photosynthetic performance (Takahashi and Murata, 2008).

While the negative influence of total solar UV-R on photosynthesis is well proven, the relative impacts of UV-A and UV-B are less clearly defined. For example, Häder et al. (2001) demonstrated that high levels of UV-A radiation at midday caused photosynthetic inhibition in some macroalgae, but other studies have shown that low levels of UV-A radiation enhance the growth of *Fucus gardneri* embryos (Henry and Van Alstyne, 2004) and photosynthetic CO_2 fixation by phytoplankters (Gao et al., 2007). In addition, in a recent study of *Gracilaria lemaneiformis*, Gao and Xu (2008) demonstrated that while UV-B had a negative impact on photosynthetic efficiency under clear-sky conditions at midday, UV-A enhanced photosynthetic efficiency during sunrise. Transfer of absorbed UV-A energy to chlorophyll *a* has been found to occur in a diatom (Orellana et al., 2004), and a similar process could occur in other algal species; hence under the appropriate environmental conditions UV-A exposure may be beneficial.

7.2.5 Apoptosis and cell-cycle genes

Apoptosis (as apposed to necrosis) is an important response to UV-B exposure, where irreparably damaged cells undergo a process of programmed cell death. Associated with the process is a suite of cell-cycle and stress response genes, including *p53*, *p21*, and cdc2 genes, and the 14–3–3 protein family. Of these *p53* is the best studied and is activated by DNA damage. The gene is involved in a range of cellular functions associated with cell cycle regulation, transcription, and DNA repair (Kulms and Schwarz, 2000). Following exposure to UV-R, expression of *p53* increases to arrest cell division at the G1/S phase to allow cells to either repair DNA damage or to undergo the process of apoptosis (Kulms and Schwarz, 2000).

Apoptosis and increases in cell cycle genes after exposure to ambient UV-R exposure have been reported in a range of marine organisms including invertebrate and vertebrate larvae (Lesser et al., 2001, 2003) and hermatypic corals (Lesser and Farrell, 2004). Lesser et al. (2001) exposed developmental stages of the sea urchin *Strongylocentrotus droebachiensis* to a range of UV-B and UV-A irradiances, noting that with increasing UV-dose relative expression of *p53* and *p21* increased, while cdc-2 expression decreased. Relative increases in *p53* expression were also apparent in the Caribbean coral *Montastraea faveolata* when exposed to high levels of solar radiation (Lesser and Farrell, 2004).

7.3 Spectral irradiances and biological weighting functions

The key biological responses to UV-R outlined have a wavelength-specific action spectrum. This is especially important when considering the affects of UV-R on marine organisms because of the spectral changes that occur with increasing depth in the water column, and also among water masses of different optical properties (Neale, 2000). Therefore, quantifying the physiological responses of an organism to UV-R in its natural environment requires weighting processes against the incident spectral irradiances using biological weighting functions.

Biological weighting functions (BWFs) exist for a number of UV-R responses in marine species, principally for photoinhibition (compared in Neale, 2000),

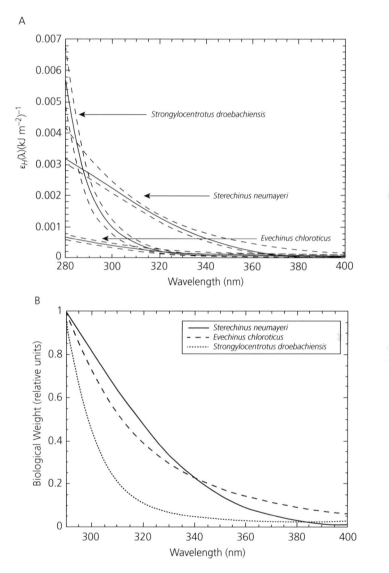

Figure 7.2 (A) Biological weighting functions (± 95% CI) for DNA damage in embryos from three species of sea urchin (*Evechinus chloroticus, Sterechinus neumayeri,* and *Strongylocentrotus droebachiensis*) exposed to UVR. The BWFs represent the net accumulation of DNA lesions (i.e. damage versus repair), so the variation in functions between species may reflect differences in sunscreen and repair capacities. For example, *Sterechinus neumayeri* is sensitive to DNA damage across the entire UV-B and UV-A spectrum, and may reflect the low sunscreen concentrations observed in this species and a slower rate of DNA repair (Lamare et al., 2006). (B) Biological weighting functions for DNA damage in the three species normalized to 1 at 290 nm and combined with measured ambient spectral irradiances can be used to estimate species-specific biologically effective doses, which was found to be greatest in *Evechinus chloroticus* (0.043 kJ m^{-2}) and similar in *S. neumayeri* (0.0126 kJ m^{-2}) and *S. droebachiensis* (0.013 kJ m^{-2}). Used with permission from Lesser et al., 2006, *J. Exp. Mar. Biol. Ecol.* 328: 10–21.

but also for DNA lesions (Lesser et al., 2006) and developmental abnormalities (Hunter et al., 1979). As well as being process-specific, weighting functions can be species-specific, and can vary depending on the level of protection and repair among species (i.e. sunscreen concentrations and repair capacities). For example, Lesser et al. (2006) constructed BWFs for CPD accumulation in the embryos of three sea urchin species (Fig. 7.2). In the study, weighting functions differed among the three species, with the most sensitive embryos belonging to the Antarctic species *Sterechinus neumayeri*, which likely reflects the comparatively low concentrations of sunscreens and slow rate of CPD photorepair observed in the species (Lamare et al., 2006). Species-specific differences in biological weighting functions for UV-photoinhibition in phytoplankton are also apparent, especially in the UV-A region, and will reflect differences in mitigation capacities such as repair and sunscreening (Neale, 2000).

Biological weighting functions, in conjunction with measurements of ambient irradiance, allow biologically effective doses to be calculated for the physiological response of interest, important for understanding the detrimental affects of UV-R *in situ*. Biological weighting functions are equally important when quantifying the affects of ozone depletion because of the enhancement of shorter wavelengths (<315 nm) that occurs under lower column ozone concentrations. Neale et al. (1998) used BWFs for photosynthetic responses in six Antarctic phytoplankton species to UV-R to determine which species were more susceptible to changes in spectral irradiances associated with ozone depletion. Differences among species and phytoplankton assemblages were apparent and were attributed to acclimation (such as changes in chlorophyll intermediates, i.e. chlorophyllide) and different capacities to repair UV-damage.

7.4 Interactive effects of UV-R with other common environmental stressors

The cellular responses to UV-R exposure will vary among organisms depending on their phylogeny, physiological status, and life-history stage (Dahms and Lee, 2010b). It will also be influenced by the interaction with other environmental stressors, both natural and anthropogenic. Indeed, spatial and temporal variations in incident UV-B irradiance are associated with variation in a range of other environmental stressors, including several related to climate change. Häder et al. (2007) reviewed the affects of UV-radiation on

aquatic systems and interactions with climate change and noted at the organism level UV-R often acts as an additional physiological stressor. Despite these observations, most studies on the affects of UV-radiation have been undertaken in isolation despite the potential for important synergistic interactions among the variables. Examples in marine invertebrates and algae where UV-B interactions with temperature, pollution, ocean pH, desiccation, and salinity have been explored are outlined here (Fig. 7.3, Table 7.1).

7.4.1 Interactions between UV-R and temperature

As temperature alters physiological rates and causes stress at the thermal limits of a species, temperature can alter or add to the response to UV-R in processes such as oxidative stress, DNA damage, and DNA repair. Several studies have examined the interaction between temperature and UV-R in tropical corals, with results indicating synergistic negative affects on the survival of planulae larvae and polyps (Zeevi-Ben-Yosef and Benayahu, 2008), coral bleaching (Gleason and Wellington, 1993), and reductions in zooxanthallae productivity (photosynthetic efficiency, Ferrier-Pages et al., 2007), responses largely attributed to oxidative stress. For four species of hermatypic corals, Ferrier-Pages et al. (2007) noted that exposure to UV-R (20.5 and 1.2 W m^{-2} UV-B) decreased photosynthetic efficiency (*Fv/Fm*), as did increases in temperature (up to 34 °C) for two of the species. However, when stressors were applied together *Fv/Fm* decreases were greater in all species, ranging between 25 and 40%. Interestingly, in the same study, prolonged exposure to UV-R at ambient temperatures (27 °C) prior to exposing corals to elevated temperatures appeared to benefit corals in terms of acute thermal stress by inducing the production of the sunscreen and antioxidant amino acid, mycosporine-glycine.

For intertidal organisms, where a number of stressors exist, UV-R can significantly increase the effects of temperature. Przeslawski et al. (2005) examined survival and development of embryos in intertidal egg masses deposited by *Bembicium nanum* and *Siphonaria denticulata*, and noted that across three temperature treatments (18, 21, and 26 °C) full spectrum light significantly lowered survival and slowed development rates compared with treatments lacking UV-R, with the UV-treatment affect greatest at the highest temperature. Similarly, in the trematode *Maritrema novaezealandensis* that parasitizes intertidal snails, the

UV Radiation

Elevated CO₂ and Ocean Acidification

- Responses to OA (reduced calcification, changes in metabolism, intracellular pH change) adds to or alters UV-R responses

- Additional CO₂ may enhance photosynthesis, and reduce costs of CO₂ uptake, countering UV-photoinhibition in algae

- Indirectly, ocean acidification may result in lower concentrations of UV-B absorbing compounds (CDOM), resulting in greater penetration of UV-B into aquatic ecosystems

- Moderate levels of UV-A stimulate the activity of the extracellular CA and CO₂ acquisition process as well as the synthesis of other periplasmic proteins, having a positive affect on photosynthesis

- Enriched CO₂ and lower pH increases photoinhibition of the electron chain pathway

- Increased CO₂, down-regulates CCMs and may act synergistically with UV-R to trigger additional light stress and photodamage, stimulating cellular defenses and enhancing mitochondrial respiration and photorespiration, which is known to play crucial protective roles against photodamage caused by reactive oxygen species

- UV-photoinhibition may reduce energy and optimal conditions required for CaCo₃ production, 'exacerbating acidification-induced reduction in calcification in tropical corals

- Opaque calcified deposits may have a photoprotective role, so reduction in calcification may increase tissue exposure to UV-B (i.e. corals and coccolithophorids)

- UV-photoinhibition and increases in cellular respiration may further increase pCO₂ and lower intracellular and extracellular pH, enhancing ocean acidification affects on physiology (i.e. Acidosis)

Salinity

- Osmotic-stress adds to alters UV-R affects or responses

- Photosynthetic pigments and sunscreen concentrations altering by osmotic stress

- Osmotic-stress may enhance the affects of UV-R exposure by reducing the capacity of organisms to repair or mitigate the UV - damage (i.e. through denaturing of proteins and enzymes associated with UV - repair and oxidative stress)

- Osmotic stress provides an additional source of damage (i.e. oxidative stress, DNA strand breaks)

- Hyposalinity reductions in photosynthesis may result in excess ROS production, additional to UV-induced photoinhibiton and oxidative stress

Pollution

- UV-R alters chemical structure of pollutants

- Cytotoxic affects of pollutants alters UV-responses

- Nitrogen-based pollutants may alter the physiological response of algae to UV-R

- Pollution enhances UV-B affects directly through the chemical disruption of cellular processes associated with UV mitigation (i.e. cadmium induced inhibition of the DNA excision repair enzyme, endonucleases)

- Pollutants may add to UV-R affects, such as copper-induced photoinhibition adding to UV photoinhibition

- UV-induced increase in toxicity of pollutants to organisms through the photo-degradation of compounds to more reactive and soluble breakdown products

- Nitrogen enrichment alleviates UV-induced inhibition of photosynthesis and growth

- Increased ammonia and nitrate concentrations promoted the production of the UV-absorbing phycoerythrin and phenolics, reducing affects of UV exposure

Temperature

- DNA damage temperature independent, repair temperature dependent

- Thermal stress add/alter UV-R affects or responses

- Thermal adaptations in physiology (i.e. cell wall fluidity) alter UV-R affects or responses

- Damage to DNA temperature independent but repair of DNA slower under colder temperatures. In this respect, temperature affects on catalytic rate influence cellular responses to UV-B repair and damage

- Photoinhibition greater in cold temperature due to: (1) slower repair of photosynthesis enzymes such as D1 protein in PSII; (2) slow down in the formation of zeaxanthin by de-epoxidation of xanthophyll cycle components and the dissipation of excess energy in photosystem II turnover; and; (3) activity of RUBISCO slowed

- Temperature-induced changes in Calvin Cycle enzyme activities results in the generation of ROS, promoting the degradation of the D1 protein in the PSII. D1 of PSII is also degraded under UV. Replacement of D1 into PSII limited by lateral diffusion through the thylakoid membrane, in turn influenced by the temperature-altered fluidity of the membrane

- The rapid induction of soluble phlorotannins triggered by UV-R minimize the effects of oxidative stress to maintain photosynthesis during short-term thermal stress

- Under warmer temperatures, electrons from the electron transport chain are used more efficiently, reducing the likelihood for UV-induced photoinhibition

- UV-B irradiance enhanced high temperature bleaching of corals through photoinhibition of zooxanthellae by light and heat through accumulation of ROS

Figure 7.3 The observed, or potential, interaction of ultraviolet radiation with four environmental variables. Key affects of each variable on UV-R are given, as well as observed or potential outcomes of the interactions for molecular and cellular responses in marine algae and invertebrates. More complex interactions (such as three-way interactions between temperature, ocean acidification, and UV-R) are not given in the figure, but will be important to consider under climate change scenarios. These are discussed in Section 7.4.

Table 7.1 Selected studies directly investigating the interaction of UV-R and other stressors.

Reference	Year	Organism	Stressor
Gleason and Wellington	1993	Coral	Temperature
Ferrier-Pages et al.	2007	Coral	Temperature
Zeevi-Ben-Yosef and Benayahu	2008	Coral larvae and polyps	Temperature
Przeslawski et al.	2005	Mollusc egg masses	Temperature
Studer and Poulin	2013	Trematode	Temperature
Hoffman et al.	2003	Brown algal spores	Temperature
Rothausler et al.	2011	Brown algae	Temperature
Cruces et al.	2012	Brown algae	Temperature
Rautenberger and Bischof	2006	Green algae	Temperature
Hernandez and Helbling	2010	Crab larvae	Temperature
Chen and Gao	2011	Red algae	pCO_2/ocean acidification
Swanson and Fox	2007	Brown algae	pCO_2/ocean acidification
Gao et al.	2009	Calcifying phytoplankton	pCO_2/ocean acidification
Gao and Zheng	2010	Calcified algae	pCO_2/ocean acidification
Duquesne and Liess	2003	Amphipods	Pollution (heavy metals)
Preston et al.	1999	Rotifers	Pollution (heavy metals)
Steevens et al.	1999	Echinoderm larvae	Pollution (PAHS)
Peachy et al.	2005	Crab larvae	Pollution (PAHS)
Pelletier et al.	1997	Juvenile mussels	Pollution (PAHS)
Pelletier et al.	1997	Mysid shrimps	Pollution (PAHS)
Xu and Gao	2012b	Red algae	Pollution (nitrates/ammonia)
Cabello-Pasini et al.	2011	Green algae	Pollution (nitrates/ammonia)
Huovinen et al.	2010	Brown algae	Pollution (heavy metals and nitrates)
Przeslawski et al.	2005	Mollusc egg masses	Salinity/desiccation
Russell and Phillips	2009a, b	Mollusc egg masses	Salinity/desiccation
Studer and Poulin	2013	Trematode	Salinity/desiccation
Nygard and Ekelund	2006	Brown algae	Salinity/desiccation
Karsten	2003	Red algae	Salinity/desiccation

activity of free-living infective stages (cercariae) decreases under elevated temperatures (30 °C) with the response greater in the presence of UV-R (Studer and Poulin, 2013). Interestingly, the negative affect of UV-R on survival was greater at low temperature (20 °C) and may reflect a slowing of enzyme activities involved in defence and repair at lower temperatures (i.e. antioxidant enzymes).

Studies on several algal species have also illustrated the importance of considering UV-R temperature interactions when predicting the potential impacts of UV-R on marine organisms. Cruces et al. (2012) found that the decrease in photochemical processes of three intertidal seaweed species, due to temperature, was slightly exacerbated by the presence of UV-R, while Hoffman et al. (2003) found that spores of *Alaria marginata* could survive and grow at 10 °C in the absence of UV-R, but could not survive at that temperature in the presence of high levels of UV-R. However, at 15 °C spores could survive and grow at all of the UV-R levels tested. They stated that their results support the hypothesis that the net biological effect of UV-R can be mediated by temperature, and vice versa. Rothausler et al. (2011), in a study of large *Macrocystis pyrifera* rafts floating along temperate coasts of Chile and exposed to large variations in environmental conditions such as UV-R and temperature, found that the species could physiologically acclimate to UV-R. They suggested that this was due to decreasing pigment contents and by dynamic photoinhibition, but differences in PSII repair could also be important (Takahashi and Murata, 2008). However, at an elevated temperature (20 °C) acclimation to UV-R occurred at the expense of growth, lower biomass gains and blade elongation rates, and decreased reproductive output. In a study of the interactive effects of UV-B and increased temperature on two *Ulva* species, *U. bulbosa* from Antarctica and *U. clathrata* from southern Chile, Rautenberger and Bischof (2006) reported that at 0 °C, UV-induced inhibition of photosynthesis was greater in *U. clathrata* than in *U. bulbosa*, but at 10 °C the negative effects of UV-B exposure were lessened in both species. They noted that under all the conditions tested, the activity of the antioxidant enzyme SOD was higher in *U. bulbosa* than in *U. clathrata*, and suggested that the higher degree of cold adaptation of *U. bulbosa* resulted in a higher UV-tolerance at 0 °C than in the less cold-adapted *U. clathrata*.

As a final comment with respect to temperature, it has been noted that UV damage is largely temperature independent whereas repair may increase with warming (Hoffman et al., 2003). In this respect, while damage to DNA in the form of CPDs is temperature independent, the repair of the dimers by the enzyme photolyase is temperature dependent (Pakker et al., 2000; Lamare et al., 2006). For example, Hernandez and Helbling (2010) noted that hatched Zoea I larvae of three crab species off the Patagonia coast exposed to UV-B for 8–10 h survived longer at 20 °C than at 15 °C. This difference in survival was attributed to a slower accumulation of UV damage in warmer temperatures, where repair processes may be stimulated.

7.4.2 Interactions between UV-R, pCO_2, and ocean acidification

Although increases in ocean pCO_2 (ocean acidification) is recognized as one of the most important challenges facing marine species, little is known of the interaction of UV-R with increased pCO_2 and the wider effects of ocean acidifications. Responses to UV-R and pCO_2 can work synergistically to alter the photosynthetic performance of marine algae. For the red tide algae *Phaeocystis globosa*, enriched pCO_2 and moderate levels of UV-A (4 J d^{-1}) enhanced photosynthetic output and growth by $\approx$8 and 29.7%, respectively (Chen and Gao, 2011). Under higher irradiances (10 J d^{-1}), however, elevated pCO_2 had a detrimental affect on growth (4.8% reduction) and photosynthetic efficiency (24% reduction). Chen and Gao (2011) noted that while UV-B (1.5–3 J d^{-1}) consistently slowed growth and promoted photoinhibition in *P. globosa*, ocean acidification conditions increased UV-B photoinhibition of growth and nonphotochemical quenching, as well as elevating metabolic costs associated with increasing concentrations of carotenoids and chlorophylls. Chen and Gao (2011) and (Gao et al., 2012a) discussed several mechanisms that would explain the positive affects of moderate UV-A exposure, including the UV-A repair of damage, UV-A stimulation of carbonic anhydrase activity, and carbon fixation. In contrast, phytoplankton down-regulate their CO_2-concentrating mechanisms (CCMs) under increased CO_2 levels, reducing costs of carbon acquisition, but this may result in additional light stress when cells are exposed to high UV-irradiances (Fig. 7.4).

Using outdoor tanks, Swanson and Fox (2007) investigated the direct influences over a 55-day period of CO_2 and UV-B, at past and future levels, on *Saccharina latissima* (*Laminaria saccharina*) and *Nereocystis luetkeana* development and phlorotannin concentrations. In addition they investigated the indirect influences of these two environmental factors on interactions of the kelp with the marine herbivore *Tegula funebralis* and on intertidal detritivores. This was relevant as changes in metabolite levels in the kelp, brought about in response to UV-R, could influence other trophic levels. They found both CO_2 and UV-B influenced growth, although the response was species-specific. For *N. luetkeana* high CO_2 ameliorated the negative growth influence of UV-B, while in *S. latissima* UV-B ameliorated the negative influence of high CO_2 concentrations. Phlorotannins, like other phenolic molecules, are thought to be important antioxidant molecules, particularly as they can undergo redox interactions with metal ions. Phlorotannins also act as sunscreens and these compounds increased in response to increasing UV-B levels, with the increase enhanced by elevated CO_2. In brown algae the polyphenolic phlorotannins can also act as anti-herbivory defensive compounds,

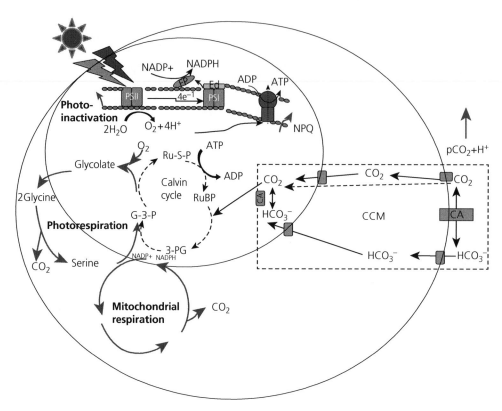

Figure 7.4 The interaction of light and ocean acidification with phytoplankton physiology. Metabolic pathways up-regulated (solid) and down-regulated (dotted) in phytoplankton under ocean acidification and solar radiation. Phytoplankton down-regulate their CO_2-concentrating mechanisms (CCMs) under increased CO_2 levels reducing costs on carbon acquisition, but may result in additional light stress when cells are exposed to high irradiances. At the same time cellular defences are enhanced, including increased non-photochemical quenching (NPQ), enhanced mitochondrial respiration, and photorespiration. However, these protective activities may not fully compensate for photodamage, thus carbon fixation (the Calvin cycle) and growth rate are ultimately decreased under sunlight. PSI: Photosystem I; PSII: Photosystem II; CA: carbonic anhydrase; Fd: ferredoxin. Used with permission from (Gao et al., 2012a), *MEPS* 470: 167–189, with modifications.

and changes in concentrations have the potential to alter algae/herbivore interactions. However, while increased CO_2 levels had a significant direct influence on kelp physiology, no clear interactions between CO_2 and UV-B for detritivore consumers or herbivores were found.

For calcifying phytoplankton such as coccolitho-phorids, the interaction of ocean acidification and UV-R can be significant. Gao et al. (2009) showed that for *Emiliania huxleyi*, in the absence of UV-R, lower seawater pH reduces calcification, but enhances pho-tosynthetic carbon fixation. Exposure to UV-R in the presence of lower seawater pH, however, inhibited calcification by up to 99%, and photosynthesis by 15%. The authors suggest that UV-R may damage the mo-lecular mechanisms for calcification, further slowing

calcification rates associated with lower seawater pH. Reduced calcification increased UV-transmission through the cell by 26%, a response that would en-hance the synergistic relationship between UV-R and acidification.

Although few studies have quantified the interac-tive affects of UV-R on responses to ocean acidifica-tion in marine invertebrates, ocean acidification is largely seen as having a negative affect on these ani-mals across a range of physiological and development processes (Kroeker et al., 2013), including a number that are also negatively affected by UV-R exposure (i.e. fertilization, embryonic and larval development, and growth). For hermatypic coral species, the direct role of light in mediating responses to ocean acidifica-tion in terms of calcification and growth is equivocal

(Chan and Connolly, 2013; Sugget et al., 2013), and these studies do not specifically examine the role of UV-R in the process. Ocean acidification does reduce calcification in corals and calcifying algae (Chan and Connolly, 2013), and it is possible that the loss of UV-opaque calcified structures will enhance UV-exposure of coral tissues. This mechanism has been proposed for the calcified algae *Corallina sessilis*, where the observed increases in UV-absorbing compounds (UVAC) and MAAs under high pCO_2 may, in part, be due to thinning of the photo-protective calcified layer (Gao and Zheng, 2010). In addition, the indirect affects of UV-B photoinhibition on coral zooxanthallae may have synergistic effects on ocean acidification responses, given the importance of photosynthesis for coral calcification (Gattuso et al., 1999).

An additional consideration is the indirect outcome of increasing pCO_2 on UV-B penetration in the ocean. Ocean acidification may result in lower concentrations of UV-B absorbing compounds (CDOM) in the future, with the result being exposure of aquatic organisms to greater UV-doses (Williamson et al, 2014). (Gao et al., 2012a) also point out that in addition to ocean acidification, increased stratification in warmer oceans will reduce the mixing layer, exposing phytoplankton communities to higher light levels (Fig. 7.5). This observation, along with (Gao et al., 2012b) finding that at elevated pCO_2, photoinhibition in phytoplankton communities in the South China Sea occurs at lower light thresholds, led the authors to speculate that future declines in ocean productivity may result from the interaction of these climate change processes. It is therefore likely that the deleterious affects of elevated pCO_2 will be accentuated by the exposure to UV-B in their natural environment.

7.4.3 Interactions between UV-R and pollution

For organisms living in coastal ecosystems, nutrient loading and chemical contamination are increasingly problematic. An important question is how will these pollutants interact with other environmental stressors such as UV-R. The interaction between UV-R and any given pollutant is likely to depend on whether the pollutant has the potential to promote growth, e.g. nitrogen containing effluents or agricultural run-off, or if

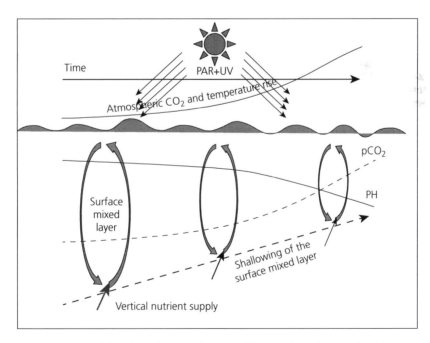

Figure 7.5 Interaction of UV-R, increased thermal stratification, and ocean acidification in driving future conditions for marine plankton. Atmospheric CO_2 and sea surface temperatures increase (solid lines) elevating pCO_2 (dashed line) in surface waters, decreasing seawater pH (solid line). Increases in stratification shoaling the surface mixed layer reduces vertical transport of nutrients (decreasing size of upward arrows) and increases plankton exposure to greater solar photosynthetically active radiation (PAR) and UV-radiation doses. Used with permission from (Gao et al., 2012a), *MEPS* 470: 167–189.

it has toxic properties that will act to inhibit growth and/or cause genetic damage, e.g. heavy metals and organic pollutants. The interactive effect of UV-B with potentially toxic pollutants (pesticides, heavy metals, and hydrocarbons) may be intrinsic (i.e. cadmium induced inhibition of the DNA excision repair enzymes, endonucleases (Schroder et al., 2005)), or extrinsic (i.e. changes in the pollutant chemistry and its toxicity by UV-R induced photochemical processes).

In many cases, the sensitivity of a species to pollutants is increased in the presence of UV-B (reviewed by Pelletier et al., 2006). This includes increased sensitivity to heavy metals in amphipods (Duquesne and Liess, 2003) and rotifers (Preston et al., 1999), to polycyclic aromatic hydrocarbons (PAHs) in echinoderm and crab larvae (Steevens et al., 1999; Peachy et al., 2005), juvenile mussels, and mysid shrimps (Pelletier et al., 1997). In this later example, toxicity increased up to 50 000times in the presence of UV-radiation. For PAHs, UV-R can increase toxicity directly through the photodegradation of hydrocarbons to more toxic compounds, or indirectly by photosensitizing PAHs, which generates reactive oxygen species and promotes oxidative stress.

The potential influence of pollutants on marine animals may also occur via their influences on the productivity of some marine primary producers. Xu and Gao (2012a) used *Gracilaria lemaneiformis* to test the hypothesis that nitrogen availability may alter the sensitivity of the algae to UV-R and have a significant influence on productivity. They found that while UV-B significantly inhibited net photosynthesis and growth, this inhibition could be alleviated by enrichment with ammonia. Furthermore, while both UV-A and UV-B stimulated the accumulation of UV screening compounds, levels of these compounds were not influenced by enrichment with ammonia. In contrast, levels of phycoerythrin (PE), a photosynthetic pigment that strongly absorbs UV-R, only increased in response to ammonia enrichment in the absence of UV-R. With respect to nitrogen uptake and metabolism, ammonia uptake and nitrate reductase activity were repressed by UV-R, but UV-R did not significantly increase nitrate uptake. Xu and Gao (2012a) concluded that the increased PE content associated with ammonia enrichment played a protective role against UV-R in *G. lemaneiformis*.

The resynthesis of proteins and the repair of DNA is an energy intensive process requiring enhanced respiration (Xu and Gao, 2009). Xu and Gao (2012a) demonstrated that respiration was enhanced under low levels of NH_4^+ but not with seawater enriched with NH_4^+, which caused a decrease in respiration irrespective of UV-R treatment. Such a decrease is most likely related to the assimilation of ammonia, as it has been suggested that assimilation of nitrogen as ammonia has a lower energetic cost compared to nitrogen assimilated as nitrate (Flores et al., 2005). UV-R and nitrate concentrations have also been shown to be important for the green intertidal alga *Ulva rigida*. Cabello-Pasini et al. (2011) demonstrated that high nitrate levels could partly mitigate the negative influence of UV-R on photosynthesis in this species and help to maintain the activities of enzymes involved in key metabolic processes. Increased levels of phenolic compounds were also found in the thallus of *U. rigida* under high nitrate conditions, and the authors stated that an increase in phenolics could provide increased photoprotection when algae are exposed to high UV-R levels at low tide. As mentioned in the previous section, polyphenolic phlorotannins can be induced by UV-R exposure, which has led to the suggestion that they may also have a photoprotective and antioxidant role, just as phenolics do in higher plants. As many pollutants, including heavy metals, induce oxidative stress in marine algae, elevated phlorotannin levels as a result of UV-exposure may influence the responses of algae to oxidative stress-inducing pollutants.

Huovinen et al. (2010) studied the interactive effects of copper, nitrates, and UV-R on *Macrocystis pyrifera*, *Lessonia nigrescens*, and *Durvillaea antarctica* from coastal Chile. It was found that *M. pyrifera* accumulated more copper in its tissues than the other macroalgae and that copper levels decreased under nitrate-enriched conditions. Inhibition of photosynthesis by copper occurred in all three species, but nitrate enrichment mitigated the inhibitory effect of copper on photosynthesis. They also found that soluble phlorotannin content decreased under copper and/or nitrate-enriched conditions with the addition of short-term UV-R exposure. This response was especially strong in *D. antarctica*. Huovinen et al. (2010) stated that: 'the antioxidant responses of soluble phlorotannins to copper stress may limit the antioxidant action in response to UV-radiation or their chemical transformation allowing polymerization in the cell wall'.

7.4.4 Interactions between UV-B and salinity

Marine environments that experience the greatest UV irradiances, such as intertidal and estuarine habitats, are also zones where fluctuations in salinity, temperature, and desiccation are the greatest. While much is known of the role of these latter stressors, there are few studies examining their interaction with UV-radiation. Przeslawski et al. (2005) noted that the effects of UV-R on embryonic development and survival rates in the egg capsules of four intertidal molluscs were significantly greater in salinities above and below 35 ppt. Synergistic affects of salinity, temperature, and UV-B on the cercaria stages of intertidal trematode parasites are also apparent (Studer and Poulin, 2013). Similarly, desiccation can act synergistically with UV-R on intertidal egg masses. Both and Przeslawski (2005) Russell and Phillips (2009a, b) and noted that the level of UV-induced mortality in intertidal mollusc egg masses was greater when desiccated, although Russel and Phillips (2009a) noted the affect of UV-R and desiccation on mortality varied among species with one species tolerant of both UV-R and desiccation.

Relatively few investigations of the interaction of UV-R and desiccation/salinity in algae have been undertaken, and those that have been conducted suggest that when interactions exist they are complex. Few species of algae live in the Baltic Sea due to its low salinity. One species that is found is *Fucus vesiculosus*, although individuals found in the Baltic are much smaller than those found in the Atlantic Ocean. Nygard and Ekelund (2006) studied photosynthesis and UV-B tolerance of *F. vesiculosus* specimens from the Baltic Sea and the Atlantic Ocean at different seawater salinities and UV-B doses. Their results showed a significantly higher initial photosynthetic capacity for Atlantic compared to Baltic specimens. Baltic specimens were also found to be more sensitive to UV-B with a 40–50% decrease of P_{max} and a slowed rate of recovery from exposure compared to the Atlantic specimens. Increasing salinity (35) had a positive influence on Baltic specimens, with both an increase in P_{max} and an increase in UV-B tolerance. Lower salinity (5) caused a decrease in P_{max} and a lower tolerance to UV-B for Atlantic specimens. The concentrations of photosynthetic pigments violaxanthin, diadinoxanthin, and zeaxanthin were very sensitive to changes in salinity and changes in the concentrations of these molecules could at least in part be responsible for the differences in UV-B tolerances observed.

In addition to altering UV-R sensitivity by influencing the concentrations of pigments associated with photosynthetic apparatus, extremes of salinity have also been shown to influence the tissue levels of UV-R absorbing molecules. Karsten (2003) investigated the influence of hyposaline and hypersaline media on the ability of two species of red algae, *Devaleraea ramentacea* and *Palmaria palmata*, from the Arctic Kongsfjord (Spitsbergen) to cope with exposure to artificial UV-R. The quantum yield of photosynthesis and the capacity to synthesize and accumulate UV-R absorbing mycosporine-like amino acids (MAAs) were determined. *D. ramentacea* acclimated well to the UV-R and exhibited euryhaline characteristics, while *P. palmata* showed a limited ability to acclimate to changing UV-R and was considered a stenohaline alga, due to its high mortality under hyposalinity (15). While both species synthesized and accumulated MAAs in response to UV-R, only *D. ramentacea* showed a correlation between elevated MAA levels and lower UV-R sensitivity.

It is also worth noting that while changes in salinity and desiccation may not directly interact with UV-R to increase initial UV-induced damage, it is likely that the associated osmotic stress not only provides an additional source of damage (i.e. DNA strand breaks, Burg et al., 2007) but also reduces the capacity of organisms to repair or mitigate the UV damage (i.e. through denaturing of proteins and enzymes associated with repair and oxidative stress).

7.5 UV-B and climate change (multiple-stressor interactions)

Both natural variation in the marine environment and longer-term climate change processes are characterized by simultaneous changes in a range of abiotic factors (Häder et al., 2007). For example, climate changes in the intertidal may involve increases in sea and air temperatures and salinities, and reductions in seawater pH and dissolved gas concentrations, along with increases in UV-B irradiances. The role of UV-R during these changes may be an important consideration, given that UV-B can have important synergistic effects as outlined in Section 7.4. To fully understand the role of UV-B during climate change processes, however, multifactorial studies are required. Examples include the study by Przeslawski et al. (2005) who noted an important role of UV-R on the affects of temperature and salinity on mortality of intertidal molluscs, and Gao et al. (2012a, 2012b) who examined the influence of UV-R and PAR on responses to ocean acidification and warming in phytoplankton growth and photoinhibition. In the Gao et al. (2012a) study, results indicated that the net effects of ocean acidification on phytoplankton were largely dependent on photobiological conditions, including UV stress. They further noted that the response to ocean acidification would include interactions with warming and other variables such as nutrient concentrations. Studer and Poulin (2013) also explored interactions of temperature, salinity, and UV-B on the intertidal trematode parasite *Maritrema novaezealandensis*, noting that all three variables had a significant affect, and that UV-B and temperature significantly interacted. Importantly, all three studies highlight that the potential influence of UV-B during climate change could not have been predicted by examining the environmental variables in isolation.

Climate change processes may also interact indirectly with UV-B, with, for example, the loss of sea ice exposing polar species to higher incident UV-B (Lister et al., 2010b), increased thermal stratification resulting in phytoplankton and zooplankton spending more time in shallower UV-illuminated waters (Häder et al., 2007; Gao et al., 2012a, Fig. 7.5), and UV-B driven reductions in primary production altering trophic processes within marine food webs (Flores et al., 2012).

7.6 Possible sites for interaction in marine organisms exposed to UV-R and other stressors

In Sections 7.4 and 7.5 we have shown that UV-R and four of what could be the most serious additional stressors in marine environment—elevated CO_2, salinity, pollutants, and temperature—clearly interact with each other and influence the biology of marine algae and invertebrates. Significant changes in both primary and secondary metabolism occur, but how these changes are regulated is still unclear. While interactions between UV-R and other abiotic stressors could have the greatest impacts on marine organisms living in coastal areas the potential impacts on those living in the open ocean should not be underestimated. For example pollution has become a global issue with increasing pollution of the open ocean (Jackson, 2010). Nikinmaa and Tjeerdema (2013) recently suggested that changes in UV-R could affect the responses of sea surface organisms to pollutants and modify the structure of potential toxicants, an area worthy of further investigation.

One common response of marine organisms to UV-R, and to the other stressors detailed earlier in this chapter, appears to be changes in the levels of ROS, which can if not scavenged result in oxidative stress. Increased cellular levels of ROS are known to activate stress-responsive genes in many organisms, especially genes that encode enzymatic antioxidants such as SOD and those encoding the enzymes responsible for the biosynthesis of the non-enzymatic antioxidants. For example, any abiotic stressor that induces the accumulation of H_2O_2 production could enhance expression of

ROS scavenging enzymes and hence provide a degree of cross-tolerance. In addition ROS and cellular antioxidants interact not only to maintain cellular integrity, but also cellular redox balance. The redox status of a cell is essential for the correct functioning of many enzymes and while numerous possible molecules exist that could serve as common signals associated with different abiotic stressors, it has been suggested that alteration of the redox status within a cell could be a key common signalling mechanism associated with stress responses in many organisms (Halliwell and Gutteridge, 2007). For example one of the best-known redox-sensitive molecules is glutathione that exists in the reduced form as GSH and oxidized as the dimeric glutathione disulphide (GSSG). All of the stressors detailed above could potentially influence the balance between cellular GSH and GSSG and influence the cellular redox state. In higher plants and animals, the cellular redox state is known to influence key enzymes such as ribonucleotide reductase and thioredoxin reductase, and to regulate multiple transcription factors, resulting in significant changes in the expression of genes involved in primary and secondary metabolism.

More in-depth studies profiling the changes in gene expression, metabolism, and physiology are required for us to understand the complexities associated with the cellular interactions found in marine organisms. For example, proteomic studies in the sea urchin *Strongylocentrotus purpuratus* have quantified the molecular and physiological responses to UV-R (Campanale et al., 2011) as well and other climate change processes such as ocean acidification (Tomanek, 2012) in isolation, but are lacking for multiple stressor responses (i.e. UV-R and ocean acidification) despite the insight they would provide. What is clear is that if we are to fully understand how marine organisms will respond to UV-R under future climate change scenarios, such studies must involve experimental designs that include the potential interactions of multiple stressors.

7.7 Conclusions

Ultraviolet radiation accounts for approximately 50% of all photobiological processes, the majority of which are detrimental. In isolation, UV-R can directly damage key biomolecules, and indirectly damage cell machinery through the generation of reactive oxygen species. Key responses include DNA damage, oxidative damage, and photoinhibition.

Interacting with other environmental variables, UV-R can also act as an important moderator of biological responses. This interaction can be additive, with the detrimental affects of UV-R adding to those caused by another environmental stressor (i.e. oxidative stress associated with exposure to elevated temperatures and UV-R irradiances) or synergistic, with the response to a variable enhanced under UV-R. The synergistic effects are well demonstrated by the interaction between UV-R and pollution, where toxicity is often greater under UV-R. In some cases exposure to low levels of UV-R can be beneficial and reduce the affects of co-variables, a clear example being the UV-induced increases in sunscreens (e.g. mycosporine-glycine) that have antioxidant properties.

An appreciation of these interactions is even more important when quantifying the future responses of marine species to climate change. In this respect, climate change involves the interaction of a number of important changes (e.g. sea temperatures, ocean pH, and coastal pollution) and research into climate change responses will often examine their synergistic effects, most commonly between lower pH and increasing sea temperatures. There is less known on the synergistic role of UV-R during environmental changes, although recent research, such as by Gao et al. (2012a) and others, provides clear evidence that UV-R has the potential to modify responses during climate change.

Although not dealt with extensively here, UV-R responses at the molecular and physiological level will have important consequences at the organismal level. This may include reductions in the key life-history traits of growth, survival, and reproductive output, as well as behavioural or tropism responses. Similarly, at the population level ultraviolet radiation, in isolation or interacting with other variables, can play an important role in terms of trophic processes and species interactions.

References

Bischof, K., Hanelt, D., & Wiencke, C. (2000). Effects of ultraviolet radiation on photosynthesis and related enzyme reactions of marine macroalgae. *Planta*, **211**, 555–562.

Bischof, K., Janknegt, P.J., Buma, A.G.J., et al. (2003). Oxidative stress and enzymatic scavenging of superoxide radicals induced by solar UV-B radiation in Ulva canopies from southern Spain. *Scientia Marina*, **67**, 353–359.

Bischof, K., Krabs, G., Wiencke, C., & Hanelt, D. (2002). Solar ultraviolet radiation affects the activity of ribulose-1, 5-bisphosphate carboxylase-oxygenase and the composition of photosynthetic and xanthophyll cycle pigments in the intertidal green alga *Ulva lactuca* L. *Planta*, **215**, 502–509.

Burg, M.B., Ferraris, J.D., & Dmitrieva, N.I. (2007). Cellular response to hyperosmotic stresses. *Physiological Reviews*, **87**, 1441–1474.

Cabello-Pasini, A., Macias-Carranza, V., Abdala, R., et al. (2011). Effect of nitrate concentration and UVR on photosynthesis, respiration, nitrate reductase activity, and phenolic compounds in *Ulva rigida* (Chlorophyta). *Journal of Applied Phycology*, **23**, 363–369.

Campanale, J.P., Tomanek, L., & Adams, N.L. (2011). Exposure to ultraviolet radiation causes proteomic changes in embryos of the purple sea urchin, *Strongylocentrotus purpuratus*. *Journal of Experimental Marine Biology and Ecology*, **397**, 106–120.

Chan, N.C.S. & Connolly, S.R. (2013). Sensitivity of coral calcification to ocean acidification: a meta-analysis. *Global Change Biology*, **19**, 282–290.

Chen, S.W. & Gao, K.S. (2011). Solar ultraviolet radiation and CO_2-induced ocean acidification interacts to influence the photosynthetic performance of the red tide alga *Phaeocystis globosa* (Prymnesiophyceae). *Hydrobiologia*, **675**, 105–117.

Cruces, E., Huovinen, P., & Gomez, I. (2012). Phlorotannin and antioxidant responses upon short-term exposure to UV radiation and elevated temperature in three south Pacific kelps. *Photochemistry and Photobiology*, **88**, 58–66.

Dahms, H.U. & Lee, J.S. (2010a). UV radiation in marine ectotherms: molecular effects and responses. *Aquatic Toxicology*, **97**, 3–14.

Dahms, H.U. & Lee, J.S. (2010b). UVR effects on biota with emphasis on polar regions. *Molecular and Cellular Toxicology*, **6**, 28–28.

Duquesne, S. & Liess, M. (2003). Increased sensitivity of the macroinvertebrate *Paramorea walkeri* to heavy-metal contamination in the presence of solar UV radiation in Antarctic shoreline waters. *Marine Ecology Progress Series*, **255**, 183–191.

Ferrier-Pages, C., Richard C., Forcioli, D., et al. (2007). Effects of temperature and UV radiation increases on the photosynthetic efficiency in four Scleractinian coral species. *Biological Bulletin*, **213**, 76–87.

Flores, H., Atkinson, A., Kawaguchi, S., et al. (2012). Impact of climate change on Antarctic krill. *Marine Ecology Progress Series*, **458**, 1–19.

Flores, E., Frias, J.E., Rubio, L.M., & Herrero, A. (2005). Photosynthetic nitrate assimilation in cyanobacteria. *Photosynthesis Research*, **83**, 117–133.

Gao, K., Helbling, W.E., Donat, P., et al. (2012a). Responses of marine primary producers to interactions between ocean acidification, solar radiation, and warming. *Marine Ecology Progress Series*, **470**, 167–189.

Gao, K.S., Ruan, Z.X., Villafane, V.E., et al. (2009). Ocean acidification exacerbates the effect of UV radiation on the calcifying phytoplankter *Emiliania huxleyi*. *Limnology and Oceanography*, **54**, 1855–1862.

Gao, K.S., Wu, Y.P., Li, G., et al. (2007). Solar UV radiation drives CO_2 fixation in marine phytoplankton: a double-edged sword. *Plant Physiology*, **144**, 54–59.

Gao, K. & Xu, J. (2008). Effects of solar UV radiation on diurnal photosynthetic performance and growth of *Gracilaria lemaneiformis* (Rhodophyta). *European Journal of Phycology*, **43**, 297–307.

Gao, K.S., Xu, J.T., Gao, G., et al. (2012b). Rising CO_2 and increased light exposure synergistically reduce marine primary productivity. *Nature Climate Change*, **2**, 519–523.

Gao, K.S. & Zheng, Y.Q. (2010). Combined effects of ocean acidification and solar UV radiation on photosynthesis, growth, pigmentation and calcification of the coralline alga *Corallina sessilis* (Rhodophyta). *Global Change Biology*, **16**, 2388–2398.

Gattuso, J.P., Allemand, D., & Frankignoulle, M. (1999). Photosynthesis and calcification at cellular, organismal and community levels in coral reefs: a review on interactions and control by carbonate chemistry. *American Zoologist*, **39**, 160–183.

Gleason, D.F. & Wellington, G.M. (1993). Ultraviolet radiation and coral bleaching. *Nature*, **365**, 836–838.

Häder, D.P., Kumar, H.D., Smith, R.C., & Worrest, R.C. (2007). Effects of solar UV radiation on aquatic ecosystems and interactions with climate change. *Photochemical and Photobiological Sciences*, **6**, 267–285.

Häder, D.P., Porst, M., & Lebert, M. (2001). Photosynthetic performance of the Atlantic brown macroalgae, *Cystoseira abies-marina*, *Dictyota dichotoma* and *Sargassum vulgare*, measured in Gran Canaria on site. *Environmental and Experimental Botany*, **45**, 21–32.

Halliwell, B. & Gutteridge, J.M.C. (2007). *Free Radicals in Biology and Medicine*. Oxford University Press, Oxford.

Henry, B.E. & Van Alstyne, K.L. (2004). Effects of UV radiation on growth and phlorotannins in *Fucus gardneri* (Phaeophyceae) juveniles and embryos. *Journal of Phycology*, **40**, 527–533.

Herman, J.R. (2010). Global increase in UV irradiance during the past 30 years (1979–2008) estimated from satellite data. *Journal of Geophysical Research*, **115**, D04203.

Hernández, M.R.D. & Helbling, E.W. (2010). Combined effects of UVR and temperature on the survival of crab larvae (Zoea I) from Patagonia: the role of UV-absorbing compounds. *Marine Drugs*, **8**, 1681–1698.

Hoffman, J.R., Hansen, L.J., & Klinger, T. (2003). Interactions between UV radiation and temperature limit inferences from single-factor experiments. *Journal of Phycology*, **39**, 268–272.

Hunter, J.R., Taylor, J.H., & Moser, H.G. (1979). Effects of ultraviolet irradiation on eggs and larvae of the northern anchovy, *Engaulis mordax*, during the embryonic stage. *Photochemistry and Photobiology*, **29**, 325–338

Huovinen, P., Leal, P., & Gomez, I. (2010). Interacting effects of copper, nitrogen and ultraviolet radiation on the physiology of three south Pacific kelps. *Marine and Freshwater Research*, **61**, 330–341.

Jackson, B.C. (2010). The future of the oceans past. *Philosophical Transactions of the Royal Society Series B*, **365**, 3765–3778.

Karsten, U., Dummermuth, A., Hoyer, K., & Wiencke, C. (2003). Interactive effects of ultraviolet radiation and salinity on the ecophysiology of two Arctic red algae from shallow waters. *Polar Biology*, **26**, 249–258.

Kroeker, K.J., Kordas, R.L., Crim, R.N., et al. (2013). Impacts of ocean acidification on marine organisms: quantifying sensitivities and interaction with warming. *Global Change Biology*, **19**, 1884–1896.

Kulms, D. & Schwarz, T. (2000). Molecular mechanisms of UV-induced apoptosis. *Photodermatology, Photoimmunology and Photomedicine*, **16**, 195–201.

Lamare, M.D., Barker, M.F., Lesser, M.P., & Marshall, C. (2006). DNA photorepair in echinoid embryos: effects of temperature on repair rate in Antarctic and non-Antarctic species. *Journal of Experimental Biology*, **209**, 5017–5028.

Lesser, M.P. (2006). Oxidative stress in marine environments: biochemistry and physiological ecology. *Annual Review of Physiology*, **68**, 253–278.

Lesser, M.P. (2010). Depth-dependent effects of ultraviolet radiation on survivorship, oxidative stress and DNA damage in sea urchin (*Strongylocentrotus droebachiensis*) embryos from the Gulf of Maine. *Photochemistry and Photobiology*, **86**, 382–388.

Lesser, M.P., Barry, T.M., Lamare, M.D., & Barker, M.F. (2006). Biological weighting functions for DNA damage in sea urchin embryos exposed to ultraviolet radiation. *Journal of Experimental Marine Biology and Ecology*, **328**, 10–21.

Lesser, M.P. & Farrell, J.H. (2004). Exposure to solar radiation increases damage to both host tissues and algal symbionts of corals during thermal stress. *Coral Reefs*, **23**, 367–377.

Lesser, M.P., Farrell, J.H., & Walker, C.W. (2001). Oxidative stress, DNA damage and p53 expression in the larvae of Atlantic cod (*Gadus morhua*) exposed to ultraviolet (290–400 nm) radiation. *Journal of Experimental Biology*, **204**, 157–164.

Lesser, M.P. & Lewis, S. (1996). Action spectrum for the effects of UV radiation on photosynthesis in the hermatypic coral *Pocillopora damicornis*. *Marine Ecology Progress Series*, **134**, 171–177.

Lesser, M.P., Kruse, V.A., & Barry, T.M. (2003). Exposure to ultraviolet radiation causes apoptosis in developing sea urchin embryos. *Journal of Experimental Biology*, **206**, 4097–4103.

Lidon, F.J.C., Teixeira, M., & Ramalho, J.C. (2012). Decay of the chloroplast pool of ascorbate switches on the oxidative burst in UV-B-irradiated rice. *Journal of Agronomy and Crop Science*, **198**, 130–144.

Lister, K.N., Lamare, M.D., & Burritt, D.J. (2010a). Oxidative damage in response to natural levels of UV-B radiation in larvae of the tropical sea urchin *Tripneustes gratilla*. *Photochemistry and Photobiology*, **86**, 1091–1098.

Lister, K.N., Lamare, M.D., & Burritt, D.J. (2010b). Sea ice protects the embryos of the Antarctic sea urchin *Sterechinus neumayeri* from oxidative damage due to naturally enhanced levels of UV-B radiation. *Journal of Experimental Biology*, **213**, 1967–1975.

Madronich, S., McKenzie, R.L., Björn, L.O., & Caldwell, M.M. (1998). Changes in biologically active ultraviolet radiation reaching the Earth's surface. *J. Photochem. Photobiol. B: Biology*, **46**, 5–19.

Neale, P. (2000). Spectral weighting functions for quantifying effects of UV radiation in marine ecosystem. In: de

Mora, S., Demers, S., & Vernet, M. (eds), *The Effects of UV Radiation in the Marine Environment*, pp. 72–100. Cambridge University Press, Cambridge.

Neale, P.J., Davis, R.F., & Cullen, J.J. (1998). Interactive effects of ozone depletion and vertical mixing on photosynthesis of Antarctic phytoplankton. *Nature*, **392**, 585–589.

Nikinmaa, M. & Tjeerdema, R. (2013). Environmental variations and toxicological responses. *Aquatic Toxicology*, **127**, 1.

Nygard, C.A. & Ekelund, N.G.A. (2006). Photosynthesis and UV-B tolerance of the marine alga *Fucus vesiculosus* at different sea water salinities. *Journal of Applied Phycology*, **18**, 1.

Orellana, M.V., Petersen, T.W., & Van den Engh, G. (2004). UV-excited blue autofluorescence of Pseudo-nitzschia multiseries (Bacillariophyceae). *Journal of Phycology*, **40**, 705–710.

Pakker, H., Beekman, C.A.C., & Breeman, A.M. (2000). Efficient photoreactivation of UVBR-induced DNA damage in the sublittoral macroalga *Rhodymenia pseudopalmata* (Rhodophyta). *European Journal of Phycology*, **35**, 109–114.

Peachey, R.B.J. (2005). The synergism between hydrocarbon pollutants and UV radiation: a potential link between coastal pollution and larval mortality. *Journal of Experimental Marine Biology and Ecology*, **315**, 103–114.

Pelletier, M.C., Burgess, R.M., Ho, K.T., et al. (1997). Phototoxicity of individual polycyclic aromatic hydrocarbons and petroleum to marine invertebrate larvae and juveniles. *Environmental Toxicology and Chemistry*, **16**, 2190–2199.

Pelletier, E., Sargian, P., Payet, J., & Demers, S. (2006). Ecotoxicological effects of combined UVB and organic contaminants in coastal waters: a review. *Photochemistry and Photobiology*, **82**, 981–993.

Preston, B.L., Snell, T.W., & Kneisel, R. (1999). UV-B exposure increases acute toxicity of pentachlorophenol and mercury to the rotifer *Brachionus calyciflorus*. *Environmental Pollution*, **106**, 23–31.

Przeslawski, R., Davis, A.R., & Benkendorff, K. (2005). Synergistic effects associated with climate change and the development of rocky shore molluscs. *Global Change Biology*, **11**, 515–522.

Rautenberger, R. & Bischof, K. (2006). Impact of temperature on UV-susceptibility of two Ulva (Chlorophyta) species from Antarctic and Subantarctic regions. *Polar Biology*, **29**, 988–996.

Rothausler, E., Gomez, I., Karsten, U., et al. (2011). UV-radiation versus grazing pressure: long-term floating of kelp rafts (*Macrocystis pyrifera*) is facilitated by efficient photoacclimation but undermined by grazing losses. *Marine Biology*, **158**, 127–141.

Russell, J. & Phillips, N. (2009a). Species-specific vulnerability of benthic marine embryos of congeneric snails (*Haminoea* spp.) to ultraviolet radiation and other intertidal stressors. *Biological Bulletin*, **217**, 65–72.

Russell, J. & Phillips, N.E. (2009b). Synergistic effects of ultraviolet radiation and conditions at low tide on egg masses of limpets (*Benhamina obliquata* and *Siphonaria australis*) in New Zealand. *Marine Biology*, **156**, 579–587.

Schröder, H.C., Janipour, N., Müller, W.E.G., et al. (2005). DNA damage and developmental defects after exposure to UV and heavy metals in sea urchin cells and embryos compared to other invertebrates. In: Müller, W.E.G., Jeanteur, P., Kuchino, Y., et al. (eds), *Echinodermata*, pp. 111–137. Springer, Berlin/Heidelberg.

Schweikert, K., Sutherland, J.E.S., Hurd, C.L., & Burritt D.J. (2011). UV-B radiation induces changes in polyamine metabolism in the red seaweed *Porphyra cinnamomea*. *Plant Growth Regulation*, **65**, 389–399.

Sinha, R.P. & Häder, D.-P. (2002). UV-induced DNA damage and repair: a review. *Photochemistry and Photobiology Sciences*, **1**, 225–236.

Steevens, J.A., Slattery, M., Schlenk, D., et al. (1999). Effects of ultraviolet-B light and polyaromatic hydrocarbon exposure on sea urchin development and bacterial bioluminescence. *Marine Environmental Research*, **48**, 439–457.

Studer, A. & Poulin, R. (2013). Cercarial survival in an intertidal trematode: a multifactorial experiment with temperature, salinity and ultraviolet radiation. *Parasitology Research*, **112**, 243–249

Suggett, D. J., Dong, L.F., Lawson, T., et al. (2013). Light availability determines susceptibility of reef building corals to ocean acidification. *Coral Reefs*, **32**(2), 327–337.

Swanson, A.K. & Fox, C.H. (2007). Altered kelp (Laminariales) phlorotannins and growth under elevated carbon dioxide and ultraviolet-B treatments can influence associated intertidal food webs. *Global Change Biology*, **13**, 1696–1709.

Takahashi, S. & Murata, N. (2008). How do environmental stresses accelerate photoinhibition? *Trends in Plant Science*, **13**, 178–182.

Tedetti, M. & Sempere, R. (2006). Penetration of ultraviolet radiation in the marine environment. A review. *Photochemistry and Photobiology*, **82**, 389–397.

Tomanek, L. (2012). Environmental proteomics of the mussel *Mytilus*: implications for tolerance to stress and change in limits of biogeographic ranges in response to climate change. *Integrative and Comparative Biology*, **52**, 648–664.

Van de Poll, W.H., Eggert, A., Buma, A.G.J., & Breeman, A.M. (2001). Effects of UV-B-induced DNA damage and photoinhibition on growth of temperate marine red macrophytes: habitat-related differences in UV-B tolerance. *Journal of Phycology*, **37**, 30–37.

Vincent, W.F. & Neale, P. J. (2000). Mechanisms of UV damage to aquatic organisms. In: de Mora, S., Demers, S., & Vernet, M. (eds), *The Effects of UV Radiation in the Marine Environment*, pp. 149–176. Cambridge University Press, Cambridge.

Williamson, C.E., Zepp, R.G., Lucas, R.M., et al. (2014). Solar ultraviolet radiation in a changing climate. *Nature Climate Change*, **4**, 434–441.

Xu, Z.G. & Gao, K.S. (2009). Impacts of UV radiation on growth and photosynthetic carbon acquisition in *Gracilaria lemaneiformis* (Rhodophyta) under phosphorus-limited and replete conditions. *Functional Plant Biology*, **36**, 1057–1064.

Xu, Z.G. & Gao, K.S. (2012a). NH_4^+ enrichment and UV radiation interact to affect the photosynthesis and nitrogen uptake of *Gracilaria lemaneiformis* (Rhodophyta). *Marine Pollution Bulletin*, **64**, 99–105.

Xu, J.T. & Gao, K.S. (2012b). Future CO_2-induced ocean acidification mediates the physiological performance of a green tide alga. *Plant Physiology*, **160**, 1762–1769.

Zeevi-Ben-Yosef, D. & Benayahu, Y. (2008). Synergistic effects of UVR and temperature on the survival of azooxanthellate and zooxanthellate early developmental stages of soft corals. *Bulletin of Marine Science*, **83**, 401–414.

CHAPTER 8

Physiological effects of noise on aquatic animals

Natacha Aguilar de Soto and Caitlin Kight

8.1 Introduction

The oceans are full of natural sounds creating sound-scapes that provide essential information for marine fauna. Sounds with both biological and non-biological origins range from the 20-Hz song of the fin whale that was recorded by hydrophones deployed during the Second World War, to the characteristic acoustic signa-ture of snapping shrimps and sea-urchins in reefs, to the sound of the wind and waves. Signals within these soundscapes indicate, for example, where larvae should settle during dispersal, whether a dangerous preda-tor is approaching, and if a conspecific is a desirable mate. However, any sound can be considered as 'noise' when it masks, or introduces ambiguity or equivoca-tion in the reception and interpretation of, signals of interest by animals, or when it induces harmful behav-ioural or physiological responses. Human activities such as blasting, pile-driving, seismic surveys, sonar, and shipping have increasingly introduced noise into the marine environment, to the extent that noise is now considered a contaminant of emerging concern (CEC) in the oceans (Weis, 2014). Unfortunately, while re-searchers are aware that anthropogenic noise pollution can have physiological, behavioural, and population-level effects in terrestrial wildlife, our understanding of the impacts of marine acoustic pollution is relatively rudimentary. This gap must be addressed so that ap-propriate mitigation, conservation, and management measures can be designed to protect marine fauna.

As the most obvious impact of noise pollution is audi-tory fatigue, current efforts to regulate noise pollution often focus on identifying the sound levels that prod-uce measurable noise-induced hearing loss (NIHL) in species of concern (e.g. Southall et al., 2007). NIHL may present as either temporary or permanent threshold shifts (TTS and PTS, respectively), otherwise known as deafness. TTS is considered a low-level impact because baseline thresholds are recoverable. Thus, guidelines for thresholds of safe exposure based on the onset of TTS have traditionally been viewed as adequate for the protection of marine fauna. It has also been pro-posed that levels causing the onset of PTS can be used for defining safe exposure limits (Southall et al., 2007). This approach, implemented by some government agencies, has received criticism from the scientific community (e.g. Tougaard et al., 2015). Using TTS or PTS falls short because it does not consider that noise can induce a range of effects at levels tens of decibels lower than those causing TTS, including physiological stress responses which may result in observable aver-sive behaviours (Lucke et al., 2009; Miller et al., 2015; Tougaard et al., 2015). In some cases, these responses may lead to indirect mortality, described as death due to secondary effects of noise impacts, in contrast to direct mortality caused by noise-induced injuries (re-viewed in Popper et al., 2014). For example, the most accepted scientific explanation for some mass mortal-ities of charismatic megafauna such as beaked whales and giant squid is anthropogenic noise pollution from naval sonar and seismic surveys, respectively, causing escape responses resulting in abnormal physiological processes with lethal consequences (Jepson et al., 2003; Cox et al., 2006; Guerra et al., 2011) (Figs 8.1 and 8.4).

Little is known about long-term physiological effects of marine noise pollution in marine fauna. However, the potential for long-term effects of noise acquire a greater conservation relevance when they affect re-productive success and, potentially, population stabil-ity (Kight and Swaddle, 2011). In their recent review of how anthropogenic noise affects fish and turtles, Popper et al. (2014) used an integrative approach in their discussion of a wide range of sublethal, non-auditory impacts of noise pollution. Although this is an important step forward in terms of the management

Stressors in the Marine Environment. Edited by Martin Solan and Nia M. Whiteley
© Oxford University Press 2016. Published in 2016 by Oxford University Press.

Figure 8.1 Beaked whales are especially vulnerable to noise. Several mass mortalities have occurred in association to naval exercises using sonar or underwater blasts (Frantzis, 1998; Jepson et al., 2003), such as these Cuvier's beaked whales (*Ziphius cavirostris*) stranded in Greece. Photo © L. Aggelopoulos/Pelagos Research Institute. [PLATE 4]

implications associated with marine acoustic pollution, the field still suffers from a general scarcity of data (Hawkins et al., 2014). This is particularly true for long-term and/or cumulative effects of noise on genes, cells, tissues, or physiological processes associated with stress responses, which can occur in the absence, or after the recovery, of hearing loss. An example of this is provided by experimental work on the effects of low-frequency underwater noise on humans. Divers experienced short-term negative effects such as tingling and loss of tactile sensation caused by overexposure of the pressure-sensitive corpuscles of Paccini in the dermis, pain in the digestive system due to gas vibrations, and low or moderate TTS. This was in spite of some of the affected divers describing subjectively the received sound exposures as low level (Clark et al., 1996; Steevens et al., 1999). Most importantly, noise-exposed divers suffered short- and long-term neural damage expressed as headaches, dizziness, blurred vision, and other symptoms apparently linked to effects in the vestibular system. These symptoms would probably pass unnoticed in experimental studies with laboratory or wild animals, underlying the potential for these studies to underestimate noise effects.

A primer on the physics of acoustics and noise pollution

(i) Frequency is the number of cycles per second (called Hertz, abbreviation Hz) in a sound wave. Sounds with lower frequency have a longer wavelength (the wavelength of a sound is given by the sound-speed divided by the frequency) and travel much farther than higher frequency signals in the water. This is because at very low frequencies the transmission loss by absorption is negligible (Urick, 1983).

The psychoacoustic property of sound related to frequency is pitch. The maximum range of frequencies over which humans can hear and discriminate frequencies is 20 Hz to 20 kHz; sounds at frequencies below and above that range are named infrasonic and ultrasonic, respectively. Animals tend to be most susceptible to hearing damage or behavioural responses when exposed to sounds within their most sensitive frequency range (Amoser and Ladich, 2003; Tougaard et al., 2015). However, animals can be affected by sounds outside of this range also.

(ii) Rise time is the time from the start of a sound until it reaches its peak pressure, in msec, or the rate of increase to this peak pressure (decibels (dB)/s). Impulsive signals with a fast rise time are related to stronger non-habituating startle responses and have a higher potential to create barotraumas if the signal is of high enough intensity (McCauley et al., 2003; Götz and Janik, 2010).

(iii) Duty cycle and duration of exposure. Duty cycle is the proportion of time that the sound is 'on', for example, a sound comprising one 200 ms long pulse every second has a 20% duty cycle. Duration of exposure is the amount of time that animals experience the acoustic noise. There are a number of reasons why these two measures are not necessarily identical. For example, wildlife may move within hearing range midway through a noise event. Animals may be impacted by instantaneous high-pressure waves, but damage increases in a non-linear fashion with total acoustic energy received for a given type of sound stimulus (Halvorsen et al., 2012; Casper et al., 2013; Kastelein et al., 2014).

(iv) Sound pressure level (SPL) and sound exposure level (SEL). $SPL = 20 \log 10(\text{pressure/reference pressure})$, where pressure and the reference pressure can be quantified in several ways:

- Maximum: this can refer to the peak-to-peak (p–p) amplitude, i.e. the difference between the highest and lowest instantaneous pressure of the sound wave, or to the peak (0–peak) amplitude, i.e. the difference between the ambient pressure and the highest positive or negative peak of the wave.
- RMS (root mean square) is an averaging measure (the standard deviation for continuous sounds) of the sound over a defined time window.

In a continuous sinusoid sound wave, the SPL difference between 0–p and p–p is 6 dB, corresponding to a doubling of the pressure, and the RMS is 9 dB lower than the p–p. For transient sounds, the difference between RMS and p–p levels is usually higher, but this depends on the averaging duration used for the calculation of RMS. To standardize this measure, the RMS averaging window should be chosen to include either 90% or 97% of the signal's energy, depending on the signal-to-noise ratio (Madsen, 2005). Shorter windows result in measures closer to the maximum, while longer windows result in low RMS levels as they may include samples with little or no signal. This has resulted in some authors calling for fixed averaging windows (Tougaard et al., 2015).

SEL is a measure of the acoustic energy of the signal. It can be calculated as $20.\log_{10}(\text{RMS pressure}) + 10.\log_{10}(\text{duration})$. For transients it is common to provide the SEL of a single pulse (SELss) and of the entire duration of the exposure [cumulative SEL (SELcum)].

Underwater SPL is reported as dB re 1 µPa where the nature of the measure (pp or rms) should be noted, while SEL is reported as dB re 1 µPa²s. Madsen (2005) describes how a sperm whale click with the same SPL p–p as a longer duration sonar ping can have 1000 times (~30 dB) lower SEL. This exemplifies the importance of appropriately describing sound signals.

(v) Noise damage may be produced by the pressure component and/or by the particle motion component of the sound wave. The amplitude of the pressure component is given by the height of the wave, while particle velocity is dictated by the movement of each particle due to the pressure wave. In a free medium and while in the far field of an acoustic source, i.e. at distances greater than several wavelengths, where the sound can be considered a point source (Urick, 1983), the pressure (p) and particle velocity (v) components of the sound wave are related as $p = v^*z$, where z is the characteristic impedance of the medium (1.5×10^6 Rayls in seawater), so the magnitude of particle velocity can in some ideal cases be inferred from a pressure measurement. In the near field of the acoustic transducer, this relationship no longer holds and animals will likely experience higher particle motions than would be expected at the same pressure level in the far field. This is relevant if levels of particle motion in the near field are sufficient to rupture tissues even when pressure levels may not be as high, and should be considered in laboratory experiments.

(vi) Differences in the acoustic impedance of air and water and in the reference levels chosen to report measures in these two media (20 µPa in air and 1 µPa in water), mean that a signal with the same intensity (i.e. the product of pressure and particle velocity) in air and water will have a far-field SPL that is 62 dB higher in water than in air. This is why it is incorrect to compare the sensation level of underwater sounds with terrestrial noises, such as jets taking off. While correcting for the 62 dB difference may suffice to compare sound levels in air and water with respect to the potential mechanical damage of the sounds, there are many reasons to be cautious. The acoustic sensitivity of marine animals may be adapted to the high acoustic impedance of water and thus these animals may be more sensitive than expected. Also, the range of effects of some intense sound sources, such as underwater explosions, is magnified in water. For these reasons, it is incorrect to simply add 62 dB to a measurement in air and perform direct comparisons of sound levels in aquatic and terrestrial environments when studying the effects of noise on fauna.

8.2 Effects of noise on the hearing system

Noise exposure may cause noise-induced hearing loss associated with either temporary or permanent threshold shifts (TTS or PTS) in the sensitivity of an animal to all or part of its sonic frequencies. PTS results from non-recoverable damage to hair cell bodies or to their mechano-sensory cilia (Liberman and Dodds, 1984). There is no loss of hair cells in TTS, although there may be swelling and damage to the sensory cilia. PTS may result from repetitive events causing TTS or be produced by a single exposure to intense sound. TTS can be produced by: (i) metabolic exhaustion caused by long noise exposures; (ii) mechanical damage produced by exposure to intense sound waves; or (iii) a combination of both of the above (Slepecky et al., 1982). Hair cell damage may be at the level of the sensory cilia (kinocilia or stereocilia), or at the level of the cell-body, or both.

Noise exposure may lead to loss of cilia or changes in their structure and rigidity, which can be reversed in some cases. For example, Solé et al. (2013) found indications of replacement of kinocilia in cephalopods, and the replacement of stereocilia has been observed in fish (Popper and Hoxter, 1984; McCauley et al., 2003), in the organ of Corti of mammals, and in the inner ear and organ of Corti of birds (Corwin and Cotanche, 1988). Birds, reptiles, amphibians, and fishes spontaneously undergo hair cell regeneration following hair cell loss due to either acoustic or ototoxic trauma. In contrast, dead hair cells do not recover in the auditory or in the vestibular systems of terrestrial mammals. Instead, the dead hair cells are replaced by supporting cells which prevent leaking endolymph from contacting the spiral neurons and causing axonal degeneration (Raphael, 2002). In some cases hair cell replacement in mammals may be induced by exogenous growth hormone treatments (Sun et al., 2011). Noise may cause cell necrosis (including swelling and rupture of the membrane), and cell apoptosis (programmed cell death including chromatin condensation and rupture of the cell body). In terrestrial mammals, Hu et al. (2000) proposed that apoptosis is related to more intense sound exposures. Removal of apoptotic cells by macrophages reduces the release of cytotoxic substances to the tissue and thus mitigates swelling, which is involved in the observed increasing degeneration of the sensory epithelium with time after noise exposure. This delayed damage has been observed in terrestrial mammals as well as in cephalopods (Solé, 2012) and fish (Hastings et al., 1996; McCauly et al., 2003).

An additional mechanism causing delayed damage is noise-induced sensorineural hearing loss due to delayed death of neurons. In terrestrial mammals noise-induced loss of spiral ganglion cells, i.e. the afferent neurons innervating the hair cells of the cochlea, can appear months after exposure and progress for years (Kujawa and Liberman, 2009). Noise exposure can result in degeneration of the dendrites of the spiral ganglion cells innervating the inner hair cells (Fig. 8.2). In mice, the loss of synaptic connections between the cochlear nerve and the hair cells happens within 24 h, but the subsequent death of the affected neurons may not occur for months or sometimes even for a year or more. The process is selective and the first affected dendrites are those involved in hearing in noisy conditions; this means that neural damage raises supra-thresholds of sensitivity and, thus, the ability to detect signals in complex acoustic environments. For this reason, the damage is not detected by usual measures of threshold shifts–behavioural assays or auditory brain response (ABR) techniques, because TTS and PTS are measured in quiet environments, i.e. they evaluate the minimum sound pressure level that an animal is able to detect for a given frequency or frequency band.

Delayed noise-induced sensorineural hearing loss, as described previously, was only recently identified experimentally in mice (Kujawa and Liberman, 2009); damage was produced by exposure to one octave band (8–16 kHz) at 100 dB re 20 µPa SPL for 2 h. This did not produce PTS, but resulted in strong TTS (20–40 dB). Surprisingly, stereocilia bundles appeared normal when this TTS was recorded, suggesting that TTS was caused by swelling or metabolic exhaustion and not by mechanical damage to the cilia. Moreover, while neural damage worsened with time, the mice returned to their pre-trial hearing thresholds. These results are important because they show that sensorineural hearing loss may be present in the absence of the indicators usually employed to identify hearing effects, such as damage to sensory cilia or TTS. Furthermore, these findings challenge previous assumptions about the recovery of noise-damaged neural terminals. The latter were based on initial observation of vacuolization or swelling of cochlear nerve terminals after noise exposure, and the absence of damaged terminals in the long term. Unfortunately, as is often the case for research on marine fauna, this work lacked detail and failed to show the full extent of noise effects; for example, there were no neural counts to detect terminal removal.

Recently, swollen and vacuolized terminals of afferent neurons were found in cephalopods exposed for 2 h to low-frequency sound (50–400 Hz) at approximately

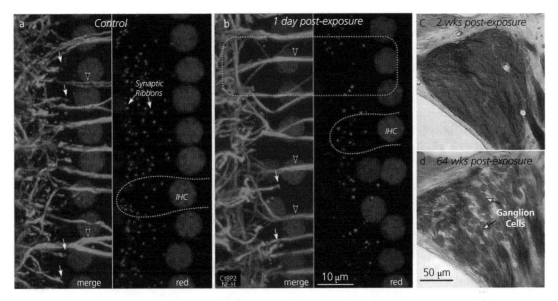

Figure 8.2 Noise-induced loss of synapses and degeneration of the spiral ganglion cells innervating hair cells of mice. Courtesy of Kujawa and Liberman (2009). These effects were first evidenced in terrestrial fauna, but damage to afferent dendrites of the hair cells has also been observed in cephalopods (Solé, 2012). [PLATE 5]

160 dB re 1 μPa RMS (175 dB peak; the particle velocity of the sound exposure received by the animals is unknown) (Solé, 2012). While the study did not investigate delayed dendrite degeneration, the observation of neural swelling suggests that that sensorineural hearing loss may have occurred, albeit unnoticed, in the noise-exposed cephalopods—or in fish from previous experiments where noise exposure resulted in extensive damage to hair cell bundles (e.g. McCauley et al., 2003).

8.2.1 Invertebrates

Very little is known about hearing in the more than 170 000 described species of marine invertebrates (not counting protistas) (Groombridge and Jenkins, 2000; Bouchet, 2006), and this lack of knowledge extends to the potential impacts of noise on their acoustic sensitivity. Invertebrates have mechanoreceptors facilitating the detection of particle motion in the surrounding medium. In most cases it is unknown if the animals translate the vibration sensation to what we understand as auditory signals, but they certainly use the information encoded in the hydrodynamic sound wave.

Most studies to date have focused on cephalopods where statocyst organs are responsible for hearing and balance. Statocysts are contained in a cavity within the

cranial cartilage. A statolith is suspended in the cavity and its motion stimulates the sensory epithelium with hair cells covering the inner wall of the cavity. There is a linear gravity-acceleration sensor, the *macula-statolith* system, and a *crista-cupula* system sensitive to angular accelerations. In the apex of the hair cells there are long kinocilia, which are highly motile special cilia, surrounded by shorter microvilli. Extensive statocyst damage has been observed in giant squid (*Architeuthis dux*) found in a mass-stranding that coincided with seismic explorations (Guerra et al., 2004, 2011). This inspired experimental work on four species of coastal cephalopods (*Sepia officinalis*, *Loligo vulgaris*, *Illex coindetii*, and *Octopus vulgaris*). Damage to the sensory epithelium was produced by exposure to 2 h of low-frequency sweeps at 100% duty cycle (André et al., 2011; Solé, 2012; Solé et al., 2013) (Fig. 8.3). Specifically, this noise regime resulted in morphological damage to the kinocilia and the hair cell bodies. Kinocilia were either lost, bent, or fused, while cell swelling ruptured the cell membrane and resulted in partial or complete extrusion of the cell contents. Apoptotic cells were found in the damaged sensory epithelium. Other effects of noise exposure included cell vacuolization, mitochondrial damage, and alteration of the surface of the endoplasmic reticulum. Hypertrophy and vacuolization of the afferent dendrites were also observed.

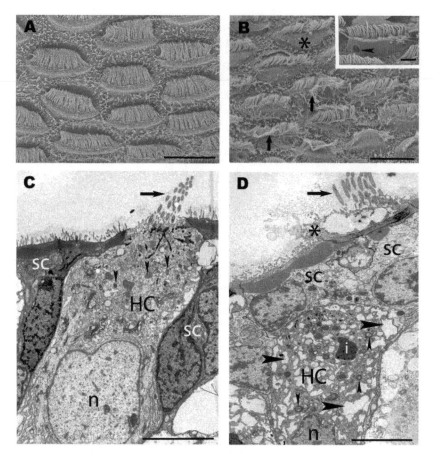

Figure 8.3 Electron microscope images of the *macula crista* of the statocyst organ of common cuttlefish (*Sepia officinalis*). Left: non-exposed sensory epithelia (a) and hair cell with organized cilia (c). Right: damaged cilia and hair cell bodies showing holes marking sites of extrusion of the cytoplasm (b); disorganized cilia and vacuolized cell body (d). Images courtesy of Solé (2012) and André et al. (2011).

8.2.2 Fish

Noise may affect fish hearing by damaging the vestibular system, the lateral line, or the swim bladder—the last of which may tear or rupture in response to intense sound exposure (Popper and Hastings, 2009; Casper et al., 2013). Effects on hair cells of the inner ear have been observed in several fish species exposed to tones over 180 dB re 1 μPa (Hastings et al., 1996), or to seismic pulses (e.g. pink snapper, *Pagrus auratus*) (McCauley et al., 2003). Sensory cilia are constantly replaced in fishes (Popper and Hoxter, 1984), and hair cells can also recover (Smith et al., 2006). However, McCauley et al. (2003) observed that the damage in the epithelium of the inner ear in pink snapper 58 days after exposure to seismic pulses was more extensive than the damage observed 18 h after exposure. This pattern,

where the extent of acoustic injury to the sensory epithelium is positively related to time after exposure, is similar to the response shown by cephalopods (Solé, 2012) and terrestrial mammals (Kujawa and Liberman, 2009), and seems to be caused by delayed effects of swelling and excitotoxicity in the cells.

Immediate TTS after noise exposure has also been recorded, for example, in both goldfish (*Carassius auratus*) and catfish (*Pimelodus pictus*) examined in studies using auditory brain response (ABR) techniques (Amoser and Ladich, 2003). Recovery of 6–20 dB after TTS has been observed in freshwater species (rainbow trout and salmon) within 24 h (Popper et al., 2005). In contrast with this fast rate of recovery, other species, such as fathead minnow (*Pimephales promelas*), did not recover even two weeks after exposure (Scholik and

Yan, 2002). This variability in both the level of impact and the recovery time is caused by differences in sound exposure and in the sensitivity of the species at the frequencies of the exposure. This is influenced by the ecophysiological adaptations of the species, in terms of how they use sound and the extent to which they experience ambient noise in their habitat. This explains why differences in vulnerability to noise are observed even between species with similar anatomical adaptations for sound detection. For example, Amoser and Ladich (2003) found variations in hearing loss and recovery time between two otophysine species exposed to the same acoustic stimuli. Otophysine fishes have specialized hearing, as they are able to detect the pressure component of sound because of a chain of Weberian ossicles connecting their swim bladder to the inner ear. These differences in hearing loss may reflect differences in the behaviour and ecophysiology of the species: the less impacted species lives in a noisier habitat and does not use acoustic signals for communication. To account for the variability in the vulnerability of the species, it has been proposed that thresholds of hearing damage are corrected by the sensitivity of the species (dBht) or of groups of species clustered by their anatomic characters influencing noise effects (presence/absence of swim bladder and connections between this and the ear) (reviewed in Popper et al., 2014).

8.2.3 Turtles

To our knowledge, there are no studies on noise-induced hearing loss for sea turtles. The avian and reptilian middle ear is a columellar system, a single bone instead of the ossicular chain of mammals. This limits the sensitivity of birds and reptiles to signals with frequencies up to not much more than 10 kHz (Saunders et al., 2000). Aquatic turtles are sensitive to low frequencies, with a hearing range from some 50 to 1200 Hz (Lavender et al., 2012) and higher sensitivity from 100 to 700 Hz (Ridgway et al., 1969; Bartol et al., 1999; Martín et al., 2012).

8.2.4 Marine mammals

The structure of the inner ear and neural innervations of the auditory system are shared between terrestrial and aquatic mammals. Thus, it is possible to extrapolate from the many previous studies on terrestrial mammals in order to predict the mechanisms of impact of noise exposure on aquatic mammals. In terrestrial mammals, high amplitude and fast rise blasts result in increased pressure of the cerebrospinal fluid that can, in turn, cause ruptures of the tympanic, oval, and round window membranes, and lead to the breakage or detachment of hearing ossicles (Henderson et al., 2008). Similar injuries have been observed in humpback whales exposed to underwater blasts (Ketten et al., 1993).

Most studies on the effects of noise on marine mammals have aimed to uncover the acoustic characteristics (e.g. level, frequency, or duty cycle) responsible for the onset and recovery of hearing after TTS. Usually, behavioural or ABR assays are used on a limited number of captive animals which may have performed psychoacoustic tests for many years. Unfortunately, these experiments are not designed to evaluate whether the animals experience sensorineural damage from the current experiment or suffer from cumulative impacts of the many experiments performed over each animal's lifetime (e.g. Kujawa and Liberman, 2009). Furthermore, it is not always clear whether the responses are influenced by behavioural and/or physiological habituation/sensitization to previous acoustic tests; however, strong aversive reactions of captive cetaceans subjected to several noise exposure experiments (Finneran et al., 2002) suggest that animals are, at least, able to remember previous exposures.

It is not possible to study PTS in marine mammals because of ethical constraints. The onset of PTS is assumed to be some 15–20 dB above the onset of TTS, and TTS has been studied in a range of cetaceans and pinnipeds exposed to different types of sound (reviewed in Southall et al., 2007). However, Kastelein et al. (2013a) found that higher levels above onset of TTS were required to cause PTS to harbour seals (*Phoca vitulina*). The largest TTS reported for marine mammals (23 dB) was caused by exposing captive bottlenose dolphins to 3 kHz tones. The exposures were 4–128 s long, with 100–200 dB re 1 µPa SPL and 106–218 dB SEL (Finneran et al., 2010). These authors found that the level of TTS was better predicted by SPL and duration than by SEL, because at the same total energy received (same SEL value), longer exposures produced higher TTS. The influence of the duration of noise exposure on the extent of hearing damage had been observed before (e.g. 40 versus 20 min in pinnipeds, Schusterman et al., 2000). TTS may not manifest until an animal has been exposed to a series of sound pulses (Kastelein et al., 2013b) increasing both the duration and the total energy received by the animal.

Other factors add complexity to models predicting the onset and extent of TTS; for instance, TTS is also affected by variations in inter-pulse interval (Kastelein et al., 2014). This may be due in part to the ability of animals to protect their hearing system from expected or repetitive intense sound exposure. Pre-exposure conditioning exerts a protective effect over the mammalian inner ear. In terrestrial mammals there is growing evidence that cholinergic and dopaminergic control by the efferent feedback from the olivocochlear system (OC) modulates the excitability of the auditory nerve. OC fibres protect nerve dendrites from excitotoxicity and also reduce TTS and PTS (Rajan, 1988; Reiter and Liberman, 1995; Darrow et al., 2007). This type of hearing protection may influence the result of studies on masked-hearing TTS (MTTS) in marine mammals. MTTS experiments measure TTS in response to a test signal played either while the focal animal is exposed to masking noise that removes fluctuations in background noise during the experiment (e.g. Schlundt et al., 2000; Finneran et al., 2002), or while the animal is experiencing TTS from a previous exposure (Lucke et al., 2009). Masking noise itself may induce a rise in the hearing threshold, which is interpreted as a defence mechanism in the ear. But this has the potential to effectively 'mask' the TTS induced by the test signal. For example, Finneran et al. (2000) exposed bottlenose dolphins (*Tursiops truncatus*) and beluga whales (*Delphinapterus leucas*) to masking noise and found hearing thresholds 20 dB above the published values for these species. The continuation of the experiment included exposures to pulses simulating explosions with max (p–p) levels of SEL 188 dB re 1 µPa^2 s. These exposures did not exceed a MTTS of 6 dB in the dolphins or 7 dB in the belugas, although the authors argue that these values may have been influenced by previous TTS from the masking noise.

Both the occurrence of, and recovery from, TTS vary according to the characteristics of the sound exposure, the extent of TTS, and the auditory sensitivity of each species (Finneran et al., 2002; Kastelein et al., 2013a). Lucke et al. (2009) found that harbour porpoises experienced high TTS levels of 15 dB after exposure to single seismic pulses with a SEL of 166 dB re 1 µPa^2 s. This was experienced at 4 kHz in spite of the low frequency characteristic of seismic pulses. Karstelein et al. (2015) observed TTS at 4–8 kHz in the same species, but not at frequencies below or above these. The results of these two studies are explained by the ototopic configuration of the mammal cochlea, which results in TTS at frequencies of higher sensitivity within the spectrum of the noise exposure. In addition, higher level exposures elicit TTS at higher frequencies. This has been observed in terrestrial mammals and also in bottlenose dolphins experiencing TTS at test frequencies above

those of the fatiguing signal (Nachtigall et al., 2004; see also Finneran et al., 2010). Many studies on noise effects describe the sound exposure using a frequency weighting to account for the different sensitivity of the species in the spectral range; this is called mammal weighting in the case of marine mammals (Southall et al., 2007).

8.2.5 Seabirds

Diving birds may be exposed to underwater noise when feeding and to airborne noise when nesting in their colonies. To our knowledge, there are no studies on the effects of underwater noise on the hearing of seabirds. However, there are many records of noise-induced damage to the hair cells in the basilar papilla of terrestrial birds (e.g. Nakagawa et al., 1997; Hu et al., 2000). Birds are able to replace hair cells of their cochlea and vestibular system (Corvin and Cotanche, 1988) and thus may be more resilient to auditory damage than mammals.

8.3 Mechanical stress and barotrauma

Impulsive sounds such as industrial or military underwater explosions, seismic pulses for geophysical exploration, or pile driving, may induce barotrauma and mechanical stress in the tissues due to high pressure differences created by fast rise sound waves (Casper et al., 2013). Popper et al. (2014) summarize two main mechanisms inducing barotrauma: (i) formation of bubbles due to changes in solubility of the gases in the blood, which may result in gas emboli with rupture of blood vessels and organic damage; (ii) cavitation and resonances in naturally occurring air bubbles and cavities, such as the swim bladder, producing mechanical stress in the surrounding organs and tissues with eventual rupture of the swim bladder. Intense sound waves may dislocate or rupture ossicular bones and tear tissues (e.g. Klima et al., 1988; Casper et al., 2013) due to mechanical forces such as cavitation and to other effects such as increased pressure of the cerebrospinal liquid (e.g. Ketten et al., 1993). Underwater detonation magnifies the blasting effect of explosives. For example, while accidental in-air detonations of fireworks produce burnings, underwater detonations have caused brain trauma and life-threatening lung haemorrhages (Nguyen et al., 2014). Barotrauma will occur at high received sound levels usually found in the vicinity of powerful sound sources, at ranges dependent of the source level and the transmission properties of the medium. It is important to further understand the mechanical effects of noise and consider these effects in environmental law. Current legislation on the environmental impact assessment of activities producing intense impulsive noises with the potential of causing barotrauma is variable among nations, and does not provide free access of information in most countries in the case of naval exercises and some industrial activities.

8.3.1 Invertebrates

Two atypical mass-strandings involving nine giant squids were associated with seismic surveys co-occurring in nearby underwater canyons where this species concentrates (Guerra et al., 2004, 2011). Two specimens suffered extensive multiorganic damage to internal muscle fibres, gills, ovaries, stomach, and digestive tract (Fig. 8.4). Other squids were probably disoriented due to extensive damage in their statocysts. The authors propose that squids without signs of barotrauma may have died by asphyxiation if they floated towards the surface while disorientated: moving from deep, cold waters to warmer, shallower waters will cause oxygen desaturation because haemocyanin, the oxygen carrier protein in cephalopods, has lower affinity for oxygen at higher temperatures (requiring temperatures below 10 °C to reach P_{50} or 50% oxygen saturation of the haemocyanin) (Brix et al., 1994).

There are two enigmatic aspects of the reported damage to the circular and radial muscle fibres of the mantle in this giant squid: the damage was clearly localized in a discrete area of the mantle, and the collagen tunics that surround the musculature were intact (Fig. 8.4). This can be due to complex patterns of sound radiation and particle velocity in the interfaces of tissues with different densities and elastic properties. Damage to the mantle was found in the individual suffering the most severe barotrauma, suggesting that this animal might have been swimming within a zone of convergence (Urick, 1983) of the seismic sound waves reflected by the sea surface/sea floor, and possibly by the walls of the steep underwater canyons in the area where the seismic survey took place. The variability in the extent of barotrauma experienced by different individuals stranding at the same time, in coincidence with the same seismic survey, underlines the difficulties inherent in predicting noise-induced damage to animals in the wild. Here, some giant squid suffered direct mortality from barotrauma, while the death of others seemed to be caused by indirect effects of physiological and behavioural responses to noise exposure.

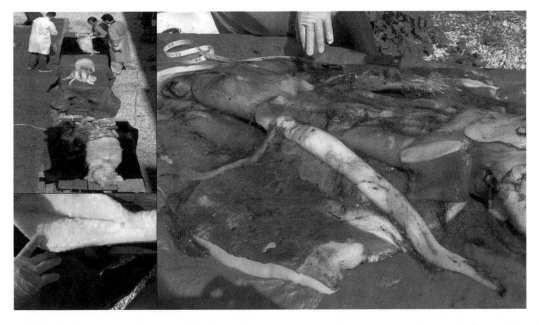

Figure 8.4 Atypical mass stranding of giant squid (*Architeutix dux*) after a seismic survey. The necropsy showed multiorganic damage. Rupture of the internal muscular fibres of the mantle can be observed, surprisingly concentrated in a discrete area 43 cm long and not affecting the external collagenous tunic of the mantle (Guerra et al., 2004, 2011). Images from A. Guerra and A. González (CSIC). [PLATE 6]

8.3.2 Fish

Explosive removal of oil platforms has been related to mortality of thousands of individuals of several fish species (Gitschlag and Herczeg, 1994). Subsequent necropsies of impacted animals found minor to severe damage in the abdominal organs and gas bladders (Klima et al., 1988).

Physostomous fishes have a pneumatic channel connecting the swim bladder and the oesophagus. These fishes can rapidly control the volume of air in the swim bladder by gulping or exhaling air. A rapid reduction of air volume in response to noise exposure is the likely explanation for the lower level of acoustic trauma found in a physostomous species (Chinook salmon *Oncorhynchus tshawytscha*) as compared to a physoclistous fish lacking a pneumatic channel (hybrid striped bass: white bass *Morone chrysops* × striped bass *Morone saxatilis*) (Casper et al., 2013). The majority of modern fishes, teleosts, are physoclistous.

The severity of acoustic trauma varies with size for a given species and sound exposure. Yelverton et al. (1975) found lower mortality in larger fishes exposed to underwater blasts. In contrast, Casper et al. (2013) found that in hybrid striped bass exposed to simulations of pile-driving pulses at close ranges, larger animals suffered more intense acoustic baro-trauma. The authors hypothesize that a larger swim bladder may produce increased vibrations and, therefore, greater mechanical stress in the surrounding tissues. The exposed bass suffered damage classified as 'mortal injury', such as a ruptured swim bladder and renal haemorrhaging, and also damage classified as 'moderate injury', such as extensive tissue haemorrhages and haematomas. These fishes were exposed to very high sound levels up to 213 dB re 1 µPa²s SEL-cum. Fishes were held in captivity during recovery, and throughout this period, the number of injuries per large fish decreased (Fig. 8.5). However, for small fish the number of injuries decreased and then increased throughout the recovery period—a pattern attributed to delayed development of liver haematomas. The authors argue that recovery in a laboratory setting, free of predators and with a secure food supply, may not be indicative of the chances of recovery in the wild, as discussed by Popper et al. (2014) and Aguilar de Soto (2015). Behavioural monitoring of impacted fish did not reveal a tendency towards abnormal swimming or abnormal control of buoyancy in fish that had previously experienced acoustic trauma (Casper et al.,

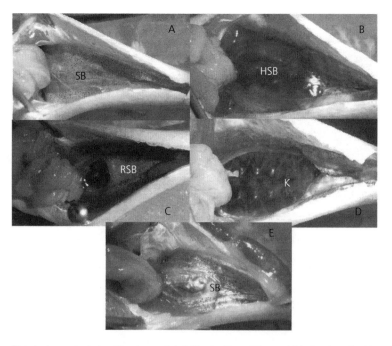

Figure 8.5 Examples of injuries from noise-induced barotrauma in hybrid striped bass: (A) control fish showing a healthy swim bladder; (B) HSB: herniated swim bladder; (C) RSB: ruptured swim bladder; (D) kidney haemorrhages; (E) healed swim bladder. Images courtesy of Casper et al. (2013). [PLATE 7]

Plate 1 Ecological structures predicted to form in place of the coral reefs for three different scenarios of global climate change, the Coral Reef Scenario (CRS)-A, CRS-B, and CRS-C. The typical anticipated ecological structures are illustrated using extant examples of reefs from the Great Barrier Reef. The atmospheric CO_2 concentration and temperature increases are shown for each Coral Reef Scenario (note that these conditions do not refer to the values measured at the photographed locations). CRS-A scenario assumes that the atmospheric CO_2 concentrations have stabilized at ~380 ppmv (note that as of September 2013, the atmospheric CO_2 levels have already passed that point, reaching ~395 ppmv). CRS-B scenario assumes an increase in CO_2 levels to approximately 500 ppmv, which is slightly below the predictions of a conservative IPCC B1 scenario for the year 2100, at ~550 ppmv. CRS-C scenario assumes an increase of CO_2 to levels above 500 ppmv. For comparison, a moderate IPCC A2 emission scenario predicts atmospheric CO_2 levels of ~800 ppmv by the year 2100, and the current trajectory of CO_2 increase indicates that it is a conservative estimate likely to be exceeded. (A) Reef slope communities at Heron Island. (B) Mixed algal and coral communities associated with inshore reefs around St. Bees Island near Mackay. (C) Reefs not dominated by corals illustrated by an inshore reef slope around the Low Isles near Port Douglas. Reprinted from Science, Vol. 318, by Hoegh-Guldberg et al. 'Coral reefs under rapid climate change and ocean acidification', pp. 1737–1742, Copyright 2007, with permission of the American Association for the Advancement of Science (See also Figure 3.1 on page 37).

Plate 2 Effects of elevated CO_2 levels on growth of decapod crustaceans. A, B—the American lobster, *Homarus americanus*, raised under normocapnia (400 ppmv CO_2; A) and elevated CO_2 levels (2850 ppmv; B). C, D—the blue crab, *Callinectes sapidus*, raised under normocapnia (400 ppmv CO_2; C) and elevated CO_2 levels (2850 ppmv; D). Higher biomineralization rates of the decapod crustaceans observed at elevated CO_2 levels were associated with faster growth as shown by larger sizes of the representative crustaceans shown on the photo. Photo credit: Justin B. Ries (Northeastern University, USA). Reproduced with permission from J. B. Ries (See also Figure 3.3 on page 42).

Plate 3 Sampling for marine microplastics; (a) Neuston nets towed on the surface; (b) sampling the strandline by hand; (c) a Ponar sediment grab used for sampling marine sediments. Photos by Ceri Lewis (See also Figure 5.3 on page 84).

Plate 4 Beaked whales are especially vulnerable to noise. Several mass mortalities have occurred in association to naval exercises using sonar or underwater blasts (Frantzis, 1998; Jepson et al., 2003), such as these Cuvier's beaked whales (*Ziphius cavirostris*) stranded in Greece. Photo © L. Aggelopoulos/Pelagos Research Institute (See also Figure 8.1 on page 136).

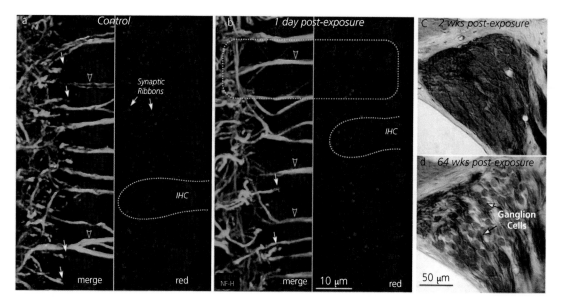

Plate 5 Noise-induced loss of synapses and degeneration of the spiral ganglion cells innervating hair cells of mice. Courtesy of Kujawa and Liberman (2009). These effects were first evidenced in terrestrial fauna, but damage to afferent dendrites of the hair cells has also been observed in cephalopods (Solé, 2012) (See also Figure 8.2 on page 139).

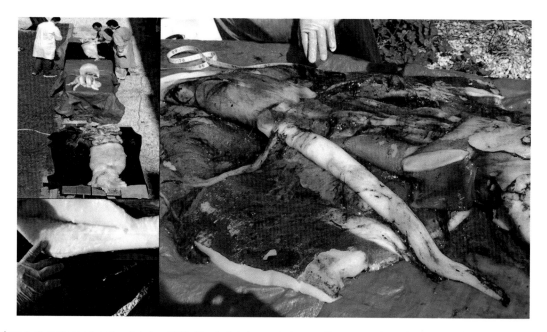

Plate 6 Atypical mass stranding of giant squid (*Architeutix dux*) after a seismic survey. The necropsy showed multiorganic damage. Rupture of the internal muscular fibres of the mantle can be observed, surprisingly concentrated in a discrete area 43 cm long and not affecting the external collagenous tunic of the mantle (Guerra et al., 2004, 2011). Images from A. Guerra and A. González (CSIC) (See also Figure 8.4 on page 143).

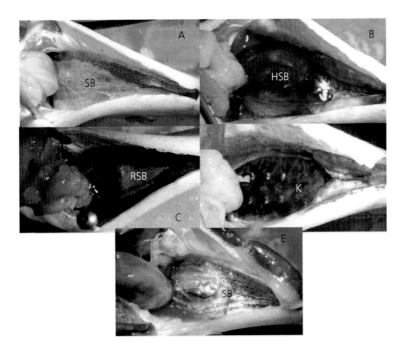

Plate 7 Examples of injuries from noise-induced barotrauma in hybrid striped bass: (A) control fish showing a healthy swim bladder; (B) HSB: herniated swim bladder; (C) RSB: ruptured swim bladder; (D) kidney haemorrhages; (E) healed swim bladder. Images courtesy of Casper et al. (2013) (See also Figure 8.5 on page 144).

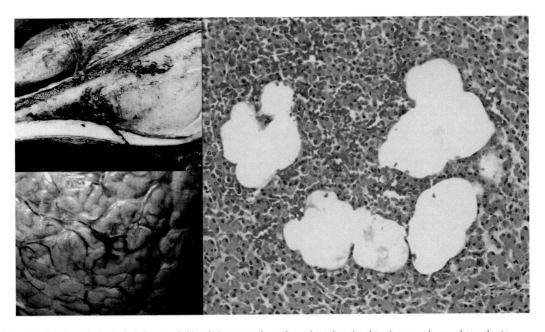

Plate 8 Lesions found in beaked whales stranded in relation to naval exercises using submarine-detection sonar: haemorrhages due to intravascular bubbles (emboli) in the lower jaw and cerebral cortex, and gas-bubble like dilatations in the liver. Images courtesy of Fernández et al. (2005) (See also Figure 8.6 on page 149).

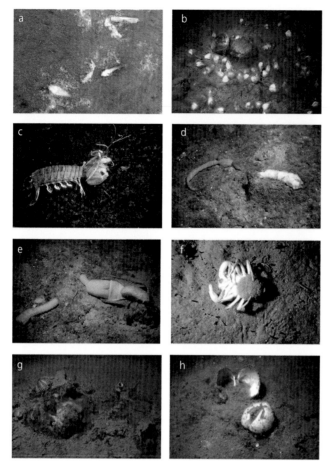

Plate 9 Behavioural responses and mortality of benthic macrofauna after oxygen depletion in the Northern Adriatic Sea, Mediterranean: (a) emerged and dead organisms on beach including from lower left to upper right flat fish, burrowing shrimp, gobiid, and bivalve; (b) emerged bivalves (*Corbula gibba*) on the sediment, in the centre dead crab (carapace) and large bivalve; (c) nocturnal mantis shrimp *Squilla mantis* emerges during day and also swims into water column—note dark colour of the sediment; (d) emerged, moribund sipunculid and dead and decomposing gobiid fish; (e) emerged bivalve with cast-off siphon; (f) dead female swimming crab with eggs—multi-generational impact of hypoxia; (g) 'Black spot' indicates remains of former multi-species clump—note empty sea urchin test; (h) post-anoxia condition—bivalve and sea urchin test as potential substrate for future epigrowth. Photos: M. Stachowitsch, except for f (department photo archive, author unknown). Time-lapse films showing the effects of oxygen depletion on benthic macrofauna during and after experimentally induced hypoxia available at: http://phaidra.univie.ac.at/o:87923 and http://phaidra.univie.ac.at/o:262380 (See also Figure 10.5 on page 185).

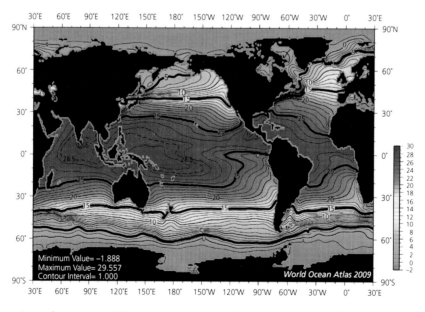

Plate 10 Mean annual sea surface temperature (°C) climatology on a one-degree latitude-longitude grid. Bold lines indicate limits to major thermal biogeographic zones: tropical (> 25 °C), subtropical (25–15 °C), temperate (15–5 °C in the northern hemisphere, or 15–2 °C in the southern hemisphere), and polar (< 5 °C in the northern hemisphere or < 2 °C in the southern hemisphere) (Lalli and Parsons, 1997). Adapted from Locarini et al. (2010) (See also Figure 12.1 on page 214).

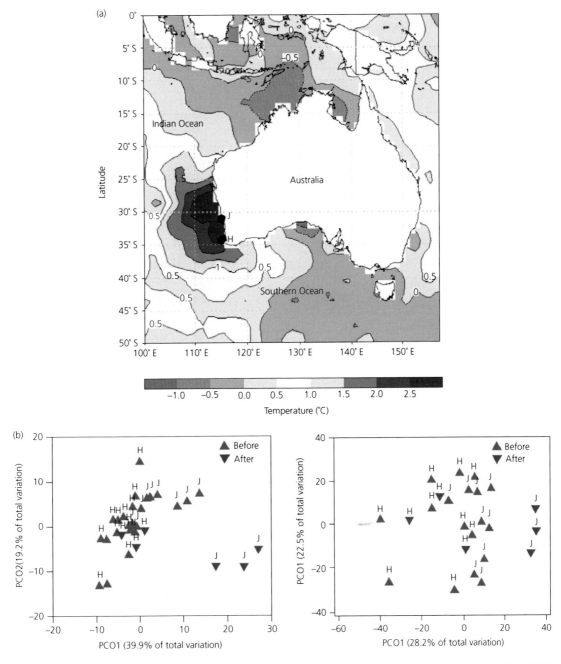

Plate 11 (a) The 2011 heat wave in the southeast Indian Ocean (relative to 1971–2000 baseline). Increased warming was observed (> 2.5 °C) along the west coast of Australia; position of Jurien Bay (J) and Hamelin Bay (H) indicated. (b, c) The ecological structure of marine communities before and after the heat wave of 2011. Principal coordinates analysis of (b) benthic (invertebrates and macroalgae) and (c) fish community structure on the rocky reefs at each study location, before and after the 2011 warming event. PCO1 and PCO2 are the first and second principal coordinate axes, indicating percentage of variation explained by each axis. Reproduced from Wernberg et al. (2013) (See also Box 12.2 Figure 1 on page 220).

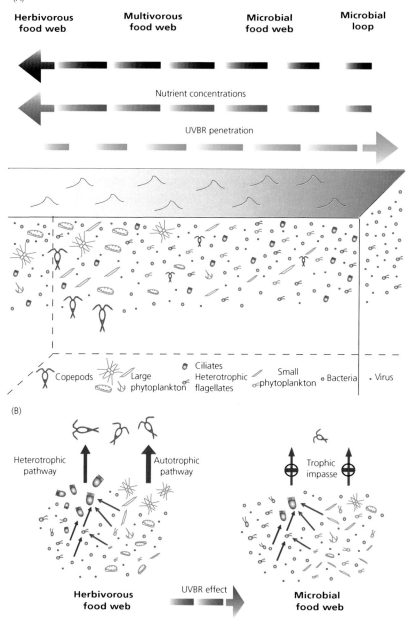

Plate 12 (A) Schematic representation of the continuum between the herbivorous, multivorous, and microbial food webs and the microbial loop. The UVBR penetration is usually higher when the microbial food web or the microbial loop are dominant because of the lower dissolved and particulate matter present within the water column due to oligotrophic conditions. On the other hand, the herbivorous food web is usually more present when nutrients concentrations are higher. In addition, the production of the system, the sedimentation of organic matter and the export of organic carbon to higher trophic levels are higher when the system is dominated by the herbivorous food web. In the herbivorous food web, large copepods graze on large phytoplankton cells, so that predators are mostly metazoans. In the multivorous food web, both copepods (metazoans) and ciliates (protozoans) graze on both large and small prey. In the microbial food web, all components are microorganisms and small phytoplankton and heterotrophic bacteria are predated by protozoans (ciliates and flagellates). Finally, in the microbial loop all components are heterotrophic microorganisms (bacteria, flagellates, and ciliates). It should be noted that viruses are present in all types of food webs. (B) Based on the results of Mostajir et al. (1999) and Ferreyra et al. (2006), UVBR seem to drive planktonic communities from being dominated by a herbivorous food web to being dominated by a microbial food web and, therefore, towards less productive systems with less food transfer to higher trophic levels. The green and blue arrows represent 'predation' (See also Figure 15.3 on page 266).

2013). This indicates that absence of behavioural responses does not necessarily mean absence of physical damage.

Bolle et al. (2012) exposed larvae of common sole (*Solea solea*) to a similar regime of simulated pile-driving pulses. Exposed stage 2 larvae had a 50% reduction in survivorship than larvae in the control group. The difference in mortality was not significant, though this may have been influenced by the experiment's relatively small sample size. The larvae were exposed to very high pressure levels of 206 dB re 1 µPa²s SELcum (100 strikes of a 186 dB re 1 µPa²s SEL). In contrast, a single discharge of a relatively small airgun (with peak RL up to 232 dB re 1 Pa) resulted in a significant increase in the mortality of early stage trout (*Salvelinus namaycush*) embryos located at very short distances (0.1 m) from the noise source (Cox et al., 2012). Increased mortality was not observed in more advanced embryos of other salmonid species subjected to the same treatment. These conflicting patterns underline the complexity of noise-effect studies; both species and ontogenetic stage appear to impact vulnerability to the same stimulus.

8.3.3 Marine turtles

The only information on noise-induced barotrauma in turtles is related to blasting activities for oil platform removal. In the Gulf of Mexico, turtle strandings appear to increase during periods of higher blasting activity in the Gulf of Mexico (Klima et al., 1988). These strandings were recorded at a radius of up to several km from the removal locations. Post-mortem examination showed that some turtles had no tissue damage characteristic of shock blasts, while other individuals showed signs consistent with impacts from underwater explosions. This included bloated carcasses, lung haemorrhage, and ruptures of the heart atria. The analysed individuals, Kemp's ridley (*Lepidochelys kempii*) and loggerhead (*Caretta caretta*) turtles, were members of a group of 51 turtles stranded after blasting (n = 22 blasts in 17 days).

Captive turtles (from a recovery centre) were experimentally exposed to a platform removal involving up to four simultaneous blasts of 23-kg buried charges. The turtles were located at distances of 230–900 m from the platform, resulting in estimated received levels from 221 to 209 dB re 1 µPa. Immediate responses varied among animals and included the following: temporal unconsciousness of most of the turtles exposed at the largest distances; evertion of the cloacal lining through the anal opening at the closest distance;

and dilation of blood vessels at the base of the throat and flippers, leading to a pink coloration that lasted for three weeks. The turtles were kept in captivity until apparent full recovery, but it is unknown if the animals would have recovered without assistance.

8.3.4 Marine mammals

Barotrauma has been found in whales and dolphins exposed to underwater blasts during naval exercises or industrial activities. In 2011, three long-beaked common dolphins (*Delphinus capensis*) were killed by underwater explosions used for military training (Danil and Leger, 2011). Acute fracture of the right tympanic plate and disruption of the ossicles was observed in one individual. The dolphins had intravascular gas bubbles in several tissues and acute haemorrhages were observed in the lungs, trachea, oesophagus, and in the acoustic fat along the mandible. In the Gulf of Mexico, dolphin mortalities have been related in some cases to underwater blasts for removal of oil platforms, including the stranding of 40 bottlenose dolphins (*Tursiops truncatus*) who were impacted by the same short period of disturbance that also affected the turtles mentioned in Section 8.3.3 (Klima et al., 1988). Nearly 40% of the stranded dolphins were smaller than the usual length at birth for this species and were thus considered foetuses or neonates. This suggests that their deaths may have not been caused directly by barotrauma, but by a stress response of the mothers resulting in premature birth.

Surprisingly, at least some marine mammals in some contexts may not leave an area in spite of receiving noise exposures causing serious barotrauma. Humpback whales (*Megaptera novaeangliae*) exposed to industrial underwater explosions (typical levels of 150 dB re 1 µPa measured at 1 mile from the source) were observed to continue their use of the affected area without significantly altering their surface behaviours (n = 71 whales sighted in 19 days) (Lien et al., 1993). However, local fishermen reported unusually high net collision rates by humpback whales and two animals abnormally recollided with the fishing gear and died. When the ears of these whales were analysed, there was evidence of mechanical trauma such as bilateral periotic fractures, rupture of the round window, disruption of the ossicular chain in the middle ear and disruption and sero-sanguinous effusion of peribullar spaces (Ketten et al., 1993). These observations are consistent with blast injury symptoms in humans, where victims sustain damage from massive, precipitous increases in cerebrospinal fluid pressure.

Many studies assume that lack of behavioural responses can be indicative of a lack of internal damage. Disturbed animals are expected to protect themselves from physiological or mechanical damage via avoidance responses. The studies above show that this is not true in all cases, which might be explained here by site fidelity driven by the evolutionary adaptation of some marine mammals to specific breeding and foraging areas.

8.3.5 Seabirds

Diving seabirds will be at higher risk of noise impact from underwater sources if they are attracted to feed on dead or disoriented fish in the vicinity of intense impulsive noise sources such as seismic arrays, pile driving, or explosives. For example, Gitschlag and Herczeg (1994) estimated 63 500 dead fish due to explosive removal of oil platforms in the Gulf of Mexico in 1993. The authors only quantified fish found floating at the surface. Seabirds may be attracted to this easy source of food, but underwater blasts damage birds. This was shown in a series of experiments performed by Yelverton et al (1973). He reported lethal effects in ducks closer than 12 m to experimental blasts of one pound of TNT. These birds had extensive pulmonary haemorrhage, ruptured livers and kidneys, as well as coronary air embolism. Half of the ducks had ruptured air sacs and eardrums. Sublethal effects included coma, respiratory alterations, and an inability to control their movement. Surviving ducks appeared to recover after 14 days. Birds at longer distances were not strongly affected by the blasts. More recently (2006), an underwater detonation from military training activities resulted in 70 western grebes (*Aechmophorus occidentalis*) being accidentally killed by subsequent sequential detonations in the same training exercise (Danil and Leger, 2011). Ten of the 70 western grebes impacted were necropsied, verifying primary blast injury as cause of death.

8.4 Stress responses

There are studies showing primary (e.g. plasma cortisol), secondary (e.g. blood glucose and haematological measures), and tertiary (e.g. growth, behaviour, and mortality) stress responses related to noise exposure in a range of marine fauna. Lethal and sublethal effects may occur as direct consequences of noise exposure, or indirectly due to behavioural responses induced by noise stimuli. These may result in animals stranding, demonstrating disorientation behaviour, reducing their foraging efficiency, or disrupting other biological functions important for homeostasis. Stress may have effects on the immune and reproductive systems via hormonal pathways involving the release by the adrenal medulla of catecholamines (adrenaline, noradrenaline, and dopamine), and the release of gluco/ mineral corticoids from the pituitary–hypothalamic axes (Romano et al., 2004). Stress responses may be short term, such that individuals can, in many cases, make a complete recovery; however, they may also cause long-term effects and thus have the potential to influence population dynamics if exposure is chronic or affects a significant portion of the population.

8.4.1 Invertebrates

Noise can alter the reproduction, growth, and development of marine invertebrates and these effects are probably explained by stress responses. This was observed initially in aquarium-dwelling brown shrimp (*Crangon crangon*) exposed to ambient noise of some 30 dB higher than normal at 25–400 Hz (Lagardere, 1982; Regnault and Lagardere, 1983). Noise-exposed shrimps consumed less food but increased their metabolic rate (higher rates of oxygen consumption and ammonia excretion) within a few hours of the start of the exposure. Ammonia is generated via oxidation of the amino group which is removed when proteins are converted to carbohydrates to provide energy; thus, these two results indicate that noise-stressed shrimp were utilizing higher levels of energy. The imbalance between energy gain and energy expenditure was associated with reduced body growth in the noise-exposed animals as compared to control individuals. The results were significant for both sexes, but particularly obvious among females. Noise-exposed groups had also lower reproductive rates (50% versus 80%) and fewer egg-bearing females (70% versus 92%). This indicates that the noise-stressed individuals may not have had the resources necessary for reproduction (Lagardere, 1982; Regnault and Lagardere, 1983). Shrimps did not seem to habituate throughout the experiment.

In larval Dungeness crabs (*Metacarcinus magister*, previously *Cancer magister*), however, there were no significant differences in survival nor in time-to-moult between a control group and a group exposed to a single pulse from a seven-airgun array, even at the higher received level of 231 dB re 1 µPa (Pearson et al., 1994). Other crustacean larvae seem more sensitive to noise, as significant differences in time-to-moult were recorded in crab larvae exposed to playbacks of noise from tidal turbines (Pine et al., 2012). *Austrohelice crassa* and *Hemigrapsus crenulatus* megalopae experienced significant (≥18 h) increases in the median time to metamorphosis (TTM) when exposed to either simulated tidal turbine or sea-based wind turbine sound. In contrast, median TTM decreased by approximately 21–31% when individuals were subjected to natural habitat sound.

The apparent contradiction in the larval responses from different species of crabs may be due, among other things, to the experimental set-up (wild versus laboratory, one pulse versus a continuous exposure), the biology of the species, or the characteristics of the sound treatment. Pine et al. (2012) argue that the spectral content of the exposures was more influential than the sound pressure levels.

Cellular and humoral immune responses of marine invertebrates to noise have also been examined. In the European spiny lobster (*Palinurus elephas*), exposure to sounds resembling shipping noise in the laboratory affected various haematological and immunological parameters considered to be potential health or disease markers in crustaceans (Celi et al., 2014). Noise-exposed lobsters showed a decreased total haemocyte count (THC) and decreased phenoloxidase (PO) activity in the haemolymph, while haemolymphatic protein concentration and heat shock protein 27 (Hsp27) expression in haemocyte lysate increased significantly. These responses had been related with stress in previous studies and indicate that noise acts as a stressor for lobsters.

In molluscs, recent studies have shown delayed and abnormal development as well as an increase in mortality rates in eggs and larvae of two species. New Zealand scallop larvae (*Pecten novaezelandiae*) exposed to playbacks of seismic pulses in the laboratory showed significant developmental delays, and 46% developed body abnormalities (Aguilar de Soto et al., 2013). The number of eggs of sea hares (*Stylocheilus striatus*) that failed to develop at the cleavage stage, as well as the number that died shortly after hatching, were significantly higher in a group exposed to boat noise playback compared to playback of ambient noise (Nedelec et al., 2014). The sound pressure levels received by the animals in these experiments may be encountered in the wild, although the particle velocity was higher than expected for the same SPL levels in the far field. Malformations appeared in the most advanced developmental stage (D-veliger) examined in the scallops, perhaps due to the cumulative exposure attained by this stage or to a greater vulnerability of D-veliger to sound-mediated physiological or mechanical stress. The authors proposed two mechanisms leading to the development of bulges and other abnormalities in the scallop D-larvae: morphogenetic changes mediated by homeobox genes, which are implicated in cell stress responses (e.g. oxidative, thermal, or hydric stress) including mediating tumour control or progression; or a process of system-wide calcium deregulation, which has been linked to noise-induced physiological stress in pregnant rats exposed to noise, giving birth to offspring with greater fluctuating asymmetry or lower dental calcification (reviewed in Kight and Swaddle, 2011). But many other explanations are possible. Nedelec et al. (2014) discuss other potential mechanisms explaining the increased mortality of sea-hare larvae, such as disrupted tissue formation, tissue damage, or altered gene expression, due to barotrauma or stress.

Squid (*Sepioteuthis australis*) exposed to seismic pulses from a single airgun showed signs of stress such as significant increases in the number of startle and alarm responses, with ink ejection in many cases, increased activity, and changing position in the water column (Fewtrell and McCauley, 2012); successive shots elicited fewer alarm responses. These patterns may have been caused by hearing loss from the first exposure, habituation, or exhaustion. These explanations may also explain why captive squid became immobile after being exposed to low-frequency noise in tanks (Solé, 2012).

8.4.2 Fish

An early study by Banner and Hyatt (1973) observed increased mortality in fish eggs and embryos of sheepshead minnow *Cyprinodon variegatus* and longnose killifish *Fundulus similis* exposed to ambient noise levels raised by 15 dB re 1 μPa. The surviving fry had slower growth rates. A significant decrease in the survival of anchovy (*Engraulis mordax*) eggs was observed after airgun exposure (Holliday et al., 1987). It is difficult to resolve if these effects are due to barotrauma, physiological stress, or a combination of both. The same applies to early experiments of Rucker (1973) showing an increased mortality of trout (*Salmo clarkii*) and salmon (*Oncorhynchus tshawytscha*) eggs exposed to shock waves. The lethal effects of sound were apparently restricted to the embryonic stages of the species. Other experiments have not shown effects in the hatching success of eggs, fry survival, or fry length/weight after exposure in the laboratory to intermittent playbacks of boat recordings in a lake fish (*Neolamprologus pulcher*) (Bruintjes and Radford, 2014).

Sublethal effects of noise-induced stress are varied and can influence growth and condition. Seahorses (*Hippocampus erectus*) exposed for one month to louder aquarium noise (123–137 dB re 1 µPa RMS versus 111–120 dB in the control group) had a lower weight and condition than the control group (Anderson et al., 2011). This was associated with a higher heterophil-to-lymphocyte ratio (H:L) and greater cortisol levels, consistent with a stress response. Noise-exposed seahorses were also significantly more affected by kidney parasites. Higher plasma cortisol and glucose levels were also found in the goldfish (*Carrasius auratus*), a hearing specialist, when kept in conditions of high noise exposure (white noise, 160–170 dB re 1 µPa) in comparison with a control group kept at 110–125 dB re 1 µPa (Smith 2004). Airgun noise was also reported to induce stress response in captive fish (Santulli et al., 1999).

Fewtrell and McCauley (2012) observed signs of anxiety in pink snapper and striped jack/trevally (*Pseudocaranx dentex*) exposed to single shots of a seismic airgun. These responses increased linearly with level of exposure from received levels of 147 dB re 1 µPa^{2}s. Fish seemed to respond to noise with a standard anti-predator behaviour in these species, swimming towards the bottom, even though noise levels were approximately 12 dB higher at the lower end of their large holding tank. Airgun signals may also cause vestibular damage to the ears, resulting in displayed aberrant and disoriented swimming behaviour (McCauley et al., 2003).

A more subtle effect of noise is loss of attention and ability to concentrate. This has been shown primarily in terrestrial mammals (Theakston, 2011), but also in marine fauna where noise reduced performance solving foraging tasks among three-spined sticklebacks (*Gasterosteus aculeatus*) (Purser and Radford, 2011). More information is necessary to learn about the development of this response in case of chronic exposure and how this might affect foraging efficiency and thus fitness.

8.4.3 Turtles

Stress responses to noise are not well studied in reptiles and amphibians. Experimental exposure to traffic noise in White's treefrogs (*Litoria caerulea*) resulted in significantly higher levels of corticosterone and lower sperm counts and viability than in frogs exposed to control playbacks of conspecific chorus sound at the same level (Kaiser et al., 2015). Although we are not aware of studies measuring physiological indicators of noise-induced stress in marine reptiles, there are records of behavioural indications. Captive turtles exposed to seismic pulses of 174–189 dB re 1 µPa^{2}s showed behavioural signs of stress such as a tendency

towards faster and more erratic swimming during airgun noise than during control periods, as well as startle responses (O'Hara and Wilcox, 1990; McCauley et al., 2003). In the wild, startle responses and diving (suggestive of avoidance) in response to airguns shots from a seismic survey were observed in loggerhead turtles (*Caretta caretta*) (DeRouter and Larbi Doukara, 2012) but not in olive ridley sea turtles (*Lepidochelys olivacea*) (Weir, 2007).

8.4.4 Marine mammals

Stress responses may have lethal effects in some extreme cases, for example when they elicit behaviours that alter the normal functioning of physiological mechanisms required to survive. This seems to explain the mass mortalities of beaked whales stranded with gas/fat emboli when exposed to submarine-detection naval sonar (Jepson et al., 2003; Fernández et al., 2005). Whales had severe and diffuse congestion and haemorrhages, especially around the acoustic jaw fat, ears, brain, and kidneys (Fig. 8.6). Most researchers agree that a 'fight or flight'-type stress response is responsible for the deaths of whales following noise disturbances (Cox et al., 2006). Interruption of foraging and avoidance at high speed has been found in different species of beaked whales subject to playbacks of naval sonar at 1/3rd octave RMS received levels as low as 89–127 dB re 1 µPa (Tyack et al., 2011; DeRuiter et al., 2013; Miller et al., 2015). A Cuvier's beaked whale (*Ziphius cavirostris*)was observed interrupting a deep foraging dive during the passage of a large ship that caused a 15-dB increase in noise at the same ultrasonic click frequencies as those utilized by hunting whales (Aguilar de Soto et al., 2006). This resulted in the noisy dive containing half of the number of prey capture attempts recorded in other dives.

The discovery of gas/fat emboli in beaked whales stranded in relation to naval sonar led to the hypothesis that these species may suffer decompression sickness in certain circumstances (Fernández et al., 2005). There is still a good deal of uncertainty over how air-breathing deep-diving mammals cope with the challenging problem of gas management (Hooker et al., 2011). The long periods that beaked whales spend at depth (Tyack et al., 2006) increase the potential risk of accumulating nitrogen in slow absorbing tissues such as fat. However, lung compression during dives limits gas interchange with the blood while cetaceans are at depth (Ridgway and Howard, 1979). Aguilar de Soto (2006) argues that beaked whales' usual ascent rate from deep dives is slow at depth, but not near the

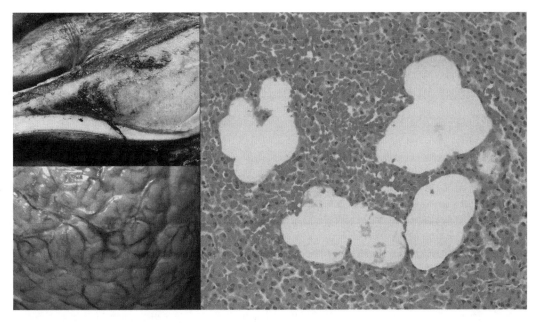

Figure 8.6 Lesions found in beaked whales stranded in relation to naval exercises using submarine-detection sonar: haemorrhages due to intravascular bubbles (emboli) in the lower jaw and cerebral cortex, and gas-bubble like dilatations in the liver. Images courtesy of Fernández et al. (2005). [PLATE 8]

surface, when cetaceans re-expand their lungs. It is unknown how nitrogen may return from the tissues to the blood flow before the expansion of the lungs enables passive diffusion of gases from the blood to the lungs and thus reduces nitrogen partial pressure in the blood. This creates an enigma: how the slow deep ascent behaviour of beaked whales might function for progressive gas interchange as the slow ascent of human divers does to prevent decompression sickness. Solving this enigma requires a consideration of the changes in tissue–gas affinity with temperature and research into the potential role of intramuscular and intra-bone lipids in nitrogen regulation. However, other functions of the slow deep ascent are possible, such as predator avoidance (Aguilar de Soto et al., 2012).

There are inherent challenges in the monitoring of short- and long-term responses of free-swimming cetaceans to sound sources. Essential anatomo-pathological information is gathered when animals strand and die, but even in these cases it is often difficult to identify without ambiguity the original cause of the stranding and the mechanisms of death. Gas/fat emboli found in beaked whales provided exceptional diagnostic evidence linking mortalities of these species to noise exposure (Jepson et al., 2003; Fernández et al., 2005). Beaked whales (Ziphiidae) perform surprisingly deep

and long dives for their size, and they show a stereotyped diving pattern which is very different from that recorded for other cetaceans (Tyack et al., 2006). This may explain why other cetacean species do not seem to develop gas/fat emboli syndrome in response to sonar exposure. However, many other species, such as harbour porpoises, long- and short-finned pilot whales (*Globicephala melas, G. macrorhynchus*), melon-headed whales (*Peponocephala electra*), dwarf sperm whales (*Kogia sima*) and long- and short-beaked common dolphins (*Delphinus capensis, D. delphis*) have stranded in large numbers or risk stranding by entering shallow areas in temporal and spatial association with naval activities in the past years (reviewed in Jepson et al., 2013). A systematic review of potential causes eliciting these strandings points to naval exercises as the most likely cause of the behavioural responses leading to stranding, but there is not enough baseline information to be certain. In these cases cetaceans may be rescued and returned to sea or die on the beach due to drowning or pathological consequences of the stranding-stress process. If the initial cause of the stranding is a panic response from acoustic sources used in the naval exercises, this suggests that non-stranding animals exposed offshore may suffer stress responses with unknown consequences. Some whales seem to survive

(Claridge, 2013) while others apparently die at sea. This was evidenced in a beaked whale mass stranding coinciding with naval exercises using sonar more than 100 km offshore, where animals arrived on the beach well after death but showed the same pathologies found in sonar-related atypical strandings (Fernández et al., 2005, 2012).

Mortalities or aberrant behaviours of marine mammals have been recorded in temporal coincidence with other intense sound sources, such as seismic pulses, but the lack of a comprehensive analysis of the circumstances of these events, and the pathologies of the animals, prevents firm conclusions about a cause–effect relationship. As in the case of sonar, beaked whales are the main taxa involved in the mortalities, with several mass strandings of ziphiids recorded in coincidence with seismic surveys (Castellote and Llorents, 2013). Also, a pantropical striped dolphin with aberrant behaviour was observed to sink motionless near a ship performing a seismic survey (Gray and Van Waerebeek, 2011). The dolphin entered into this apparently akinetic state after a period of intense swimming keeping its head out of the water, interpreted as an intent to avoid acoustic exposure.

Other stress responses of marine mammals to noise are more benign. Tagged North Atlantic right whales (*Eubalaena glacialis*) interrupted foraging when exposed to novel alarm sounds but did not react to playbacks of ship noise (Nowacek et al., 2004). This lack of an observable short-term behavioural response is surprising given that ship noise is known to increase chronic stress in this species. This was evidenced in a long-term study of right whales in the Bay of Fundy (Canada) showing significantly lower baseline levels of stress-related faecal hormone metabolites (glucocorticoids) after a reduction in shipping resulted in a 6 dB decrease of background noise in the study area at frequencies below 150 Hz (Rolland et al., 2012).

Other indications of noise-induced stress in marine mammals are changes in heart rate (HR). This was first observed in a beluga exposed to playbacks of noise from 140 to 160 dB for 1, 3, or 10 min, at frequencies from 19 to 108 kHz. The onset of noise coincided with a dramatic HR increase to 100 beats/min. Tachycardia was sometimes replaced by bradycardia during 10-min-long exposures (Lyamin et al., 2011). Further work by the same authors (Lyamin et al., 2015) observed interindividual variability in responses and suggested that young or naïve animals may be more susceptible than older individuals. While a young naïve animal showed a startle response to noise exposure, with a heart rate increasing up to 210% above baseline level, in older

individuals the onset of noise caused both a decrease in heart rate (to as low as 5 beats min[-1]) and apneas, resembling the diving response.

Exposure of a captive harbour porpoise to playbacks of pile-driving pulses (SEL 146 dB re 1 µPa^2 s) during several 1-h sessions was related to an increase in level of exertion and anxiousness of the animal, evidenced by higher swimming speed, respiration rate (increased by some 7%), and number of jumps (Kastelein et al., 2015). These effects might be lessened in the wild if the animals leave the area, for example as observed by Tougaard et al. (2009). However, avoidance is also considered a stress response—one with potential ecological consequences—and the physiological responses of the escaping animals were not measured in these studies.

Stress may have consequences in the functioning of the immune system. A beluga and a bottlenose dolphin kept in captivity had an increase in stress indicators such as the release of catecholamines (beluga) and a reduction in monocytes and increase in aldosterone (bottlenose dolphin) after exposure to a single pulse of a seismic airgun (198–226 dB re 1 µPa) or a single 3 kHz pure tone of 1 s with levels ranging from 130 to 201 dB re 1 µPa (Romano et al., 2004). Monocyte decrease has been identified as a consequence of stress in humans (Weisse et al., 1990) and aldosterone is a better indicator of stress in cetaceans than cortisol (Thomson and Geraci, 1986).

8.4.5 Seabirds

There are no studies about reactions of diving birds to underwater noise, but there are records of strong behavioural reactions of seabirds to airborne noise. The most dramatic example is when approximately 7000 king penguins (*Aptenodytes patagonicus*) died from asphyxiation because of a stampede. Brown (1990) concluded that the most likely explanation for this event was a strong alarm response to the approach of a large aircraft flying at low altitude (250 m) near the colony. Most of the dead penguins were chicks, and in this case immediate effects and population-level effects are clearly correlated even when the noise elicited a stress (escape) response that was not directly lethal.

Other responses in penguin colonies were reported by Wilson et al. (1991). Aircraft caused Adélie penguins (*Pygoscelis adeliae*) to panic at distances greater than 1000 m, and 3 days' exposure to a helicopter inhibited birds that had been foraging from returning to their nests, caused bird numbers in the colonies to decrease by 15%, and resulted in an active nest mortality of 8%. A reduction in the response of king penguins to flights above their colony as a controlled exposure experiment progressed suggests that some animals may habituate with time (Hughes et al., 2008).

Escape or startle responses were observed also in crested terns (*Sterna bergii*) exposed to helicopter and fixed-wing aircraft noise, although only in part of the colony. For the same received level, birds were more sensitive to the noise produced by the helicopter (Brown, 1990). Noise may also provoke a response in birds in flight. Gollop et al. (1974) showed that, when presented with gas compressor station noise, individual snow geese altered their flight direction (61% of which made adjustments of more than 90 degrees). In addition, snow goose flocks avoided landing in response to decoys while in the presence of such noise. Brant and Canada geese were also observed taking flight after aircraft flew over their colonies (Ward et al., 1999).

It is unknown if diving birds are as sensitive to underwater noise as to airborne sound. Penguins seem to be especially vulnerable to noise and they spend long periods foraging at sea and diving to forage. If underwater noise elicits avoidance responses from preferred feeding areas, this would have the potential of reducing foraging efficiency, thereby potentially reducing survival and fitness.

Can animals survive physiological effects of noise? Differences in laboratory and field conditions.

Animals exposed to noise and allowed to recover in the laboratory are fed and free of predators in an environment with clean filtered water reducing the probability of infections. In the wild the picture is very different: (i) disoriented or weak animals are easy prey; (ii) reduced foraging abilities limit the energy stores available for recovery, and (iii) animals may be more prone to sickness if the acoustic exposure elicited a stress response depressing the immune system.

8.5 Conclusions

Although we do not have a full tally of the marine animals impacted by anthropogenic noise pollution annually, the number is likely to be significant; the World Health Organization recently estimated that among humans alone, more than a million years of our collective health are lost every year as a result of the negative impacts of noise (Smith, 1991; Theakston, 2011). Documenting and measuring the effects of our activities are important first steps towards applying mitigation measures that reduce the impacts of anthropogenic noise not only on humans, but also the fauna with which we share both terrestrial and marine habitats.

The studies described in this review are just the tip of the iceberg in terms of what we need to know; further, while they show some important trends that are useful for making management decisions, there are also some conflicting patterns (e.g. strong behavioural responses indicating stress at very low received sound levels versus no behavioural signs of large physiological damage; high inter- and intra-species variability in the vulnerability to physiological effects versus inter-taxa commonality in mechanisms of noise induced damage). This complexity in quantifying dose-effect mechanisms of noise damage challenges our understanding and thus the management of the impact of acoustic pollution.

Some inter-taxa similarities in the basic mechanisms by which noise impacts a variety of faunal features and processes include neuroendocrine, immune, and reproductive systems; development; metabolism; cardiovascular health; stress pathways; and gene integrity (as reviewed in Wright et al., 2007; Kight and Swaddle, 2011). Although these effects have been studied mainly in terrestrial fauna, a growing body of research suggests that at least some of these findings may also be applicable to marine fauna. For example, both terrestrial mammals and cephalopods experience noise-induced damage to neural terminals innervating sensory hair cells (Kujawa and Liberman, 2009; Solé, 2012). In terrestrial mammals, this results from glutamate excitotoxicity; the evolutionarily basal origins of glutamate suggest that it may drive noise-induced excitotoxicity in other animals also. This emphasizes how comparative analyses on mechanisms of noise effects can yield results that are broadly applicable across the animal kingdom (e.g. Wright et al., 2007). That said, it is also true that the same acoustic stimuli are either known, or suspected, to have vastly different impacts on animals depending on species, gender, age, life stage, and individual identity (Regnault and Lagardere, 1983; Cox et al., 2006; Anderson et al., 2011; Aguilar de Soto et al., 2013). Thus, while it may be possible to generalize some results, we must be cautious about doing so too extensively.

Similar physiological responses to noise are most likely when it produces barotrauma (a simple mechanical effect of high amplitude/fast rise pressure waves) or acts as a stressor (activating hormonal pathways that are often shared among many species) (Wright et al., 2007). On the other hand, heterogeneity in responses to noise is expected for many reasons. Species-specific adaptations to ambient noise levels in the habitat may result in different tolerances of, or sensitivities to, introduced noise, while ontogenetic processes may make particular species, age classes, or sexes more susceptible to noise pollution, either for behavioural reasons (e.g. noise interferes with a young animal's ability to use, detect, or learn acoustic cues important for its development; Schroeder et al., 2012) or for molecular/genetic reasons (e.g. noise induces variations in gene expression, as recently observed in mice cochlea; Gratton et al., 2011).

In summary, we are confronted with two main avenues by which noise can impact animals, both with potential population-level effects: bottom-up responses, where genetic, cellular, and physiological level responses affect the individual; and top-down effects, where these base mechanisms of response are modulated by the behavioural ecology and Grinnellian niche (i.e. the sum of the adaptations of an individual and species for a given 'lifestyle', with specific habitat requirements, Grinnell, 1917) of an animal. It is important to note that these two modes of influence

are not mutually exclusive; rather, they interact and may modulate each other (Fig. 8.7). However, while we should expect bottom-up responses/mechanisms to mainly be a function of the received sound and, therefore, to be shared across many taxa, top-down responses will vary according to the behavioural traits of the animal, and thus should be expected to show more variation both within and among species. It is important to recognize these two avenues modulating noise impact because bottom-up mechanisms enable us to make broad predictions of mechanisms of noise effects applicable to many taxa; for instance, a stressing noise is very likely to depress the immune system. In contrast, learning about top-down mechanisms contributes to our understanding of the complexity of responses to noise and prevents biases in our interpretation of diverse observations. For example, it is intriguing that noise-sensitive beaked whales are resident in areas holding naval exercise ranges (Claridge, 2013) when they have strong aversive reactions to naval sonar (Tyack et al., 2011; Miller et al., 2015), resulting in mass strandings in some cases (Jepson et al., 2003). The answer may lay in the high site-fidelity of the affected species. However, differences in the population structure of beaked whales resident in a naval

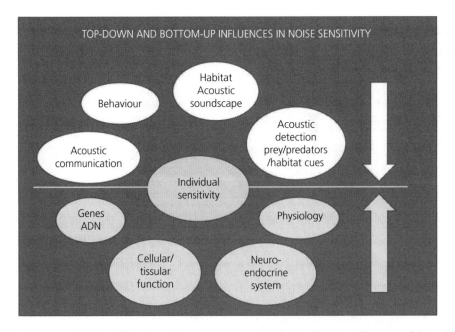

Figure 8.7 Interactions among 'top-down' (driven by environment, behaviour, and ecology) and 'bottom-up' (genetic, cellular, and physiological) mechanisms can explain why certain species or individuals are especially vulnerable to noise despite the fact that there are large inter-taxa similarities in the basic biochemical and physiological pathways of noise damage.

exercise range and in a nearby area indicate that individuals staying in a noisy area may suffer long-term population-level consequences (Claridge, 2013).

Behavioural flexibility may enable some species or individuals to mitigate the negative impacts of noise—though avoidance of noisy areas may be associated with potential fitness costs, depending on the scale of the displacement and the availability of alternative habitats with similar resources. Avoidance may, however, be impeded by factors such as site fidelity, territoriality, or physical ability (or lack thereof) to move. When animals stay in areas of chronic noise pollution they may change their behaviours (e.g. alter their vocalizations; Kight and Swaddle, 2015) and physiological parameters (e.g. experience shifts in peak cortisol levels) to cope with stress and masking, and so thrive in noisy areas that offer abundant resources, as observed in urban house sparrows (Crino et al., 2013) but not in tits (Schroeder et al., 2012). This species-specific selective adaptation to noise and other types of pollution favours opportunistic species and may reduce richness and biodiversity in an area. Species may not leave a noisy area but carry chronic stress and decreases to fitness (Barber et al., 2010; Kight et al., 2012; Schroeder et al., 2012; Claridge, 2013). Even temporary exposures to stressors in early life stages may have later fitness consequences (Blas et al., 2007).

It is important to note that the characteristics of each animal's response to anthropogenic noise regimes will also be influenced by spatial and temporal variations in exposure to acoustic pollution. Some sources of noise—such as ship engines—are mobile, whereas others—such as pile-driving equipment—are static; some may be problematic year-round (e.g. well-used cargo and recreational boat routes), while others are associated with only periodic activity (e.g. seasonal fisheries). These environmental factors, as well as details of the physical nature of the active acoustic space in which each focal animal lives, should be taken into consideration when evaluating physiological responses to noise pollution regimes.

Though further research is required to more fully understand the physiological impacts of noise on marine fauna, the studies performed to date clearly justify the application of mitigation measures such as ship quieting techniques, bubble curtains for pile driving, seismic vibrators, and thermal and acoustic detection techniques for marine mammals enabling shut-down of noise sources when animals are detected. Future researchers should aim to design projects that will yield information that can be used by engineers to streamline current technology or develop new methods that are more

economical, protect a wider array of species, have benefits over a larger portion of the habitat, and/or work more efficiently. It would also be valuable to understand ways by which we might alter our temporal or spatial use of marine habitats so as to protect particularly vulnerable individuals or species—for example, those that are in the midst of a breeding season, or are at an especially sensitive life stage. The oceans offer humans an invaluable reservoir of biodiversity and food resources; thus, measures to reduce the impact of underwater acoustic pollution will benefit wildlife and humans alike.

References

Aguilar de Soto, N. (2006). *Acoustic and foraging behavior of short-finned pilot whales* (Globicephala macrorhynchus) *and Blainville's beaked whales* (Mesoplodon densirostris) *in the Canary Islands. Implications for impacts of man-made noise and ship collisions*. PhD thesis, La Laguna University, Canary Islands, Spain. 253 pp.

Aguilar de Soto, N. (2015). Peer-reviewed papers on the effects of noise on marine invertebrates. In: Popper, A.N. & Hawkins, A. (eds), *Effects of Noise on Aquatic Life II*. Springer, New York.

Aguilar de Soto, N., Delorme, N., Atkins, J., et al. (2013). Anthropogenic noise causes body malformations and delays development in marine larvae. *Scientific Reports*, **3**, 2831.

Aguilar de Soto, N., Johnson, M., Madsen, P.T.M., et al. (2006). Does intense ship noise disrupt foraging in deep-diving Cuvier's beaked whales (*Ziphius cavirostris*)? *Marine Mammal Science*, **22**(3), 690–699.

Aguilar de Soto, N., Johnson, M., Tyack, P., et al. (2012). No shallow talk: deep social communication of Blainville's beaked whales. *Marine Mammal Science*, **28**(2), E75–E92.

Amoser, S. & Ladich, F. (2003). Diversity in noise-induced temporary hearing loss in otophysine fishes. *Journal of the Acoustical Society of America*, **113**, 2170–2179.

Anderson, P.A., Berzins, I.K., Fogarty, F., et al. (2011). Sound, stress and seahorses: the consequences of a noisy environment to animal health. *Aquaculture*, **311**(1–4), 129–138.

André, M., Solé, M., Lenoir, M., et al. (2011). Low-frequency sounds induce acoustic trauma in cephalopods. *Frontiers in Ecology and the Environment*, **9**(9), 489–493.

Banner, A. & Hyatt, M. (1973). Effects of noise on eggs and larvae of two estuarine fishes. *Transactions of the American Fisheries Society*, **102**(1), 134–136.

Barber, J.R., Crooks, K.R., & Fristrup, K.M. (2010). The costs of chronic noise exposure for terrestrial organisms. *Trends in Ecology and Evolution*, **25**, 180–189.

Bartol, S.M., Musick, J.A., & Lenhardt, M.L. (1999). Auditory evoked potentials of the loggerhead sea turtle. *Copeia*, **3**, 836–840.

Blas, J., Bortolotti, G.R., Tella, J.L., et al. (2007). Stress response during development predicts fitness in a wild, long lived vertebrate. *Proceedings of the National Academy of Sciences of the USA*, **104**, 8880–8884.

Bolle, L.J., de Jong, C.A.F., Bierman, S.M., et al. (2012). Common sole larvae survive high levels of pile-driving sound in controlled exposure experiments. *PLoS ONE*, **7**(3), e33052.

Bouchet, P. (2006). The magnitude of marine biodiversity. In: Duarte, C.M. (ed.), *The Exploration of Marine Biodiversity: Scientific and Technological Challenges*, pp. 32–64. Fundacion BBVA, Madrid.

Brix, O., Colosimo, A., & Giardina, B. (1994). Temperature dependence of oxygen binding to cephalopod haemocyanins: ecological implications. In: Pörtner, H.O., O'Dor, R.K., & Macmillan, D.L. (eds), *Physiology of Cephalopod Molluscs: Lifestyle and Performance Adaptations*, pp. 149–162. Gordon and Breach Science Publishers, Basel, Switzerland.

Brown, A. (1990). Measuring the effect of aircraft noise on lated jet aircraft noise on heart rate and behaviour of sea birds. *Environment International*, **16**, 587–592.

Bruintjes, R. & Radford, A.N. (2014). Chronic playback of boat noise does not impact hatching success or post-hatching larval growth and survival in a cichlid fish. *PeerJ*, **2**. doi:10.7717/peerj.594.

Casper, B.M., Halvorsen, M.B., Matthews, F., et al. (2013). Recovery of barotrauma injuries resulting from exposure to pile driving sound in two sizes of hybrid striped bass. *PLoS ONE*, **8**(9), e73844.

Castellote, M. & Llorents, C. (2013). *Review of the effects of offshore seismic surveys in cetaceans: are mass strandings a possibility?* Proceedings of the 3rd Conference on the Effects of Noise on Aquatic Life, Budapest.

Celi, M., Filiciotto, F., Vazzana, M., et al. (2014). Shipping noise affecting immune responses of European spiny lobster (*Palinurus elephas*). *Canadian Journal of Zoology*, **93**, 113–121.

Claridge, D. (2013). *Population ecology of Blainville's beaked whales* (Mesoplodon densirostris). PhD thesis, University of St Andrews. 312 pp.

Clark, J.B., Russel, K.L., Knafele, M.E., & Stevens, C.C. (1996). Assessment of vestibular function of divers exposed to high intensity low frequency underwater sound. *Undersea Hyperbaric Medicine*, **23**, 33.

Corwin, J.T. & Cotanche, D.A. (1988). Regeneration of sensory hair cells after acoustic trauma. *Science*, **240**, 1772–1774.

Cox, B., Dux, A., Quist, M., & Guy, C. (2012). Use of a seismic airgun to reduce survival of nonnative lake trout embryos: a tool for conservation? *North American Journal of Fishing Management*, **32**(2), 292–298.

Cox, T.M., Ragen, T.J., Read, A.J., et al. (2006). Understanding the impacts of anthropogenic sound on beaked whales. *Journal of Cetacean Research and Management*, **7**(3), 177–187.

Crino, O.L., Johnson, E.E., Blickley, J.L., et al. (2013). Effects of experimentally elevated traffic noise on nestling white-crowned sparrow stress physiology, immune function and life history. *Journal of Experimental Biology*, **216**, 2055–2062.

Danil, K. & Leger, J.A. (2011). Seabird and dolphin mortality associated with underwater detonation exercises. *Marine Technology Society Journal*, **45**(6), 89–95.

Darrow, K.N., Maison, S.F., & Liberman, M.C. (2007). Selective removal of lateral olivocochlear efferents increases vulnerability to acute acoustic injury. *Journal of Neurophysiology*, **97**(2), 1775–1785.

DeRuiter, S. & Larbi Doukara, K. (2012). Loggerhead turtles dive in response to airgun sound exposure. *Endangered Species Research*, **16**, 55–63.

DeRuiter, S.L., Southall, B.L., Calambokidis, J., et al. (2013). First direct measurements of behavioural responses by Cuvier's beaked whales to mid-frequency active sonar. *Biology Letters*, **9**, 20130223.

Fernández, A., Edwards, J.F., Rodríguea, F., et al. (2005). Gas and fat embolic syndrome involving a mass stranding of beaked whales (family Ziphiidae) exposed to anthropogenic sonar signals. *Veterinary Pathology*, **42**, 446–457.

Fernández, A., Sierra, E., Martín, V., et al. (2012). Last 'atypical' beaked whales mass stranding in the Canary Islands (July, 2004). *Journal of Marine Science*, **2**, 2.

Fewtrell, J.L. & McCauley, R.D. (2012). Impact of air gun noise on the behaviour of marine fish and squid. *Marine Pollution Bulletin*, **64**(5), 984–993.

Finneran, J., Carder, D., Schlundt, C., & Dear, R. (2010). Growth and recovery of temporary threshold shift at 3 kHz in bottlenose dolphins: experimental data and mathematical models. *Journal of the Acoustical Society of America*, **127**(5), 3256.

Finneran, J.J., Schlundt, C.E., Carder, D.A., et al. (2000). Auditory and behavioral responses of bottlenose dolphins (*Tursiops truncatus*) and a beluga whale (*Delphinapterus leucas*) to impulsive sounds resembling distant signatures of underwater explosions. *Journal of the Acoustical Society of America*, **108**(1), 417–431.

Finneran, J.J., Schlundt, C.E., Dear, R., et al. (2002). Temporary shift in masked hearing thresholds in odontocetes after exposure to single underwater impulses from a seismic watergun. *Journal of the Acoustical Society of America*, **111**(6), 2929–2940.

Frantzis, A. (1998). Does acoustic testing strand whales? *Nature*, **392**, 29.

Gitschlag, G. & Herczeg, B.A. (1994). Sea turtle observations at explosive removals of energy structures. *Marine Fisheries Review*, **56**(2), 1–8.

Gollop, M.A., Black, J., Felsje, B., & Davis, R. (1974). Disturbance studies of breeding black brant, common eiders, glaucous gulls, and Arctic terns at Nunaluk Spit and Philips Bay, Yukon. In: Gunn, W. & Livingston, J. (eds), *Disturbance to Birds by Gas Compressor Noise Simulators, Aircraft and Human Activity in the Mackenzie Valley and the North Slope*. Arctic Gas Biol. Rept. Ser. No. 14, pp. 153–200. Canadian Arctic Gas Study Ltd, Calgary, Alberta.

Götz, T. & Janik, V.M. (2010). Aversiveness of sounds in phocid seals: psycho-physiological factors, learning processes and motivation. *Journal of Experimental Biology*, **213**, 1536–1548.

Gratton, M.A., Eleftheriadou, A., Garcia, J., et al. (2011). Noise-induced changes in gene expression in the cochleae of mice differing in their susceptibility to noise damage. *Hearing Research*, **277**, 211–226.

Gray, H. & Van Waerebeek, K. (2011). Postural instability and akinesia in a pantropical spotted dolphin, *Stenella attenuata*, in proximity to operating airguns of a geophysical seismic vessel. *Journal for Nature Conservation*, **19**, 363–367.

Grinnell, J. (1917). The niche relationship of the California Thrasher. *Auk*, **34**, 427–233.

Groombridge, B. & Jenkins, M.D. (2000). *Global Biodiversity: Earth's Living Resources in the 21st century*. World Conservation Press, Cambridge, UK.

Guerra, Á., González, Á.F., Pascual, S., & Dawe, E.G. (2011). The giant squid *Architeuthis*: an emblematic invertebrate that can represent concern for the conservation of marine biodiversity. *Biological Conservation*, **144**(7), 1989–1997.

Guerra, Á., González, Á.F., & Rocha, F. (2004). A review of records of giant squid in the northeastern Atlantic and severe injuries in *Architeuthis dux* stranded after acoustic exploration. *ICES*, **29**, 1–17.

Halvorsen, M.B., Casper, B.M., Matthews, F., et al. (2012). Effects of exposure to pile-driving sounds on the lake sturgeon, Nile tilapia and hogchoker. *Proceedings of the Royal Society of London Series B*, **279**(1748), 4705–4714.

Hastings, M.C., Popper, A.N., Finneran, J.J., & Lanford, P.J. (1996). Effects of low frequency underwater sound on hair cells of the inner ear and lateral line of the teleost fish *Astronotus ocellatus*. *Journal of the Acoustical Society of America*, **99**(3), 1759–1766.

Hawkins, A.D., Pembroke, A.E., & Popper, A.N. (2014). Information gaps in understanding the effects of noise on fishes and invertebrates. *Reviews in Fish Biology and Fisheries*, **25**, 39–64.

Henderson, D., Hu, B., & Bielfeld, E. (2008). Patterns and mechanisms of noise-induced cochlear pathology. In: Schacht, J., Popper A., & Fay R. (eds), *Auditory Trauma, Protections and Repair*, pp 195–217. Springer, New York.

Holliday, D.V., Pieper, R.E., Clarke, M.E., & Greenlaw, C.F. (1987). *The Effects of Airgun Energy Releases on the Eggs, Larvae and Adults of the Northern Anchovy (Engraulis Mordax)*. API Publication No. 4453. American Petroleum Institute, Washington, DC.

Hooker, S.K., Fahlman, A., Moore, M.J., et al. (2011). Deadly diving? Physiological and behavioural management of decompression stress in diving mammals. *Proceedings of the Royal Society of London Series B*, **279**(1731), 1041–1050.

Hu, B.H., Guo, W., Wang, P.Y., et al. (2000). Intense noise-induced apoptosis in hair cells of guinea pig cochleae. *Acta Otolaryngol*, **120**, 19–24.

Hughes, K.A., Waluda, C.M., Stone, R.E., et al. (2008). Short-term responses of king penguins *Aptenodytes patagonicus*

to helicopter disturbance at South Georgia. *Polar Biology*, **31**, 1521–1530.

Jepson, P.D., Arbelo, M., Deaville, R., et al. (2003). Gas-bubble lesions in stranded cetaceans. Was sonar responsible for a spate of whale deaths after an Atlantic military exercise? *Nature*, **425**(6958), 575–576.

Jepson, P.D., Deaville, R., Acevedo-Whitehouse, K., et al. (2013). What caused the UK's largest common dolphin (*Delphinus delphis*) mass stranding event? *PLoS ONE*, **8**(4), e60953.

Kaiser, K., Devito, J., Jones, C.G., et al. (2015). Effects of anthropogenic noise on endocrine and reproductive function in White's treefrog, Litoria caerulea. *Conservation Physiology*, **3**, cou061. doi:10.1093/conphys/cou061

Kastelein, R.A., Gransier, R., & Hoek, L. (2013a). Comparative temporary threshold shifts in a harbor porpoise and harbor seal, and severe shift in a seal. *Journal of the Acoustical Society of America*, **134**, 13–16.

Kastelein, R.A., Gransier, R., Marijt, M.A.T., & Hoek. L. (2015). Hearing frequency thresholds of harbor porpoises (*Phocoena phocoena*) temporarily affected by played back offshore pile driving sounds. *Journal of the Acoustical Society of America*, **137**(2), 556–564.

Kastelein, R.A., Hoek, L., Gransier, R., et al. (2014). Effect of level, duration, and inter-pulse interval of 1–2 kHz sonar signal exposures on harbor porpoise hearing. *Journal of the Acoustical Society of America*, **136**(1), 412–422.

Kastelein, R.A., van Heerden, D., Gransier, R., & Hoek, L. (2013b). Behavioral responses of a harbor porpoise (*Phocoena phocoena*) to playbacks of broadband pile driving sounds. *Marine Environmental Research*, **92**, 206–214.

Ketten, D.R., Lien, J., & Todd, S. (1993). Blast injury in humpback whale ears: evidence and implications. *Journal of the Acoustical Society of America*, **94**(3), 1849–1850.

Kight, C.R., Saha, M.S., & Swaddle, J.P. (2012). Anthropogenic noise is associated with reductions in the productivity of breeding Eastern Bluebirds (*Sialia sialis*). *Ecological Applications*, **22**, 1989–1996.

Kight, C.R. & Swaddle, J.P. (2011). How and why environmental noise impacts animals: an integrative, mechanistic review. *Ecology Letters*, **14**(10), 1052–1061.

Kight, C.R. & Swaddle, J.P. (2015). Male bluebirds alter their song in response to anthropogenic changes in the acoustic environment. *Integrative and Comparative Biology*, **55**(3), 418–431 doi: 10.1093/icb/icv070.

Klima, E.F., Gitschlag, G.R., & Renaud, M.L. (1988). Impacts of the explosive removal of offshore petroleum platforms on sea turtles and dolphins. *Marine Fisheries Review*, **50**(3), 33–42.

Kujawa, S.G. & Liberman, M.C. (2009). Adding insult to injury: cochlear nerve degeneration after 'temporary' noise-induced hearing loss. *Journal of Neuroscience*, **29**(45), 14077–14085.

Lagardere, J.P. (1982). Effects of noise on growth and reproduction of *Crangon crangon* in rearing tanks. *Marine Biology*, **71**, 177–185.

Lavender, A.L., Bartol, S.M., & Bartol, I.K. (2012). Hearing capabilities of loggerhead sea turtles (*Caretta caretta*) throughout ontogeny. In: Popper, A.N. & Hawkins, A. (eds), *The Effects of Noise on Aquatic Life*, pp. 89–92. Springer, New York.

Liberman, M.C. & Dodds, L.W. (1984). Single-neuron labeling and chronic cochlear pathology. III. Stereocilia damage and alterations of threshold tuning curves. *Hearing Research*, **16**, 55–74.

Lien, J., Todd, S., Stevick, P., et al. (1993). The reaction of humpback whales to underwater explosions: orientation, movements, and behavior. *Journal of the Acoustical Society of America*, **94**(3), 1849–1849.

Lucke, K., Siebert, U., Lepper, P.A., & Blanchet, M. (2009). Temporary shift in masked hearing thresholds in a harbor porpoise (*Phocoena phocoena*) after exposure to seismic airgun stimuli. *Journal of the Acoustical Society of America*, **125**(6), 4060–4070.

Lyamin, O.I., Korneva, S.M., Rozhnov, V.V., & Mukhametov, L.M. (2011). Cardiorespiratory changes in beluga in response to acoustic noise. *Doklady Biological Sciences*, **440**(1), 275–278.

Lyamin, O., Rozhnov, V., & Mukhametov, L. (2015). Cardiorespiratory responses to acoustic noise in belugas. In: Popper, A.N. & Hawkins, A. (eds), *Effects of Noise on Aquatic Life II*. Springer, New York.

Madsen, P.T. (2005). Marine mammals and noise: problems with root mean square sound pressure levels for transients. *Journal of the Acoustical Society of America*, **117** (6), 3952–3957.

Martin, K.J., Alessi, S.C., Gaspard, J.C., et al. (2012). Underwater hearing in the loggerhead turtle (*Caretta caretta*): a comparison of behavioral and auditory evoked potential audiograms. *Journal of Experimental Biology*, **215**, 3001–3009.

McCauley, R.D., Fewtrell, J., & Popper, A.N. (2003). High intensity anthropogenic sound damages fish ears. *Journal of the Acoustical Society of America*, **113**(1), 631–642.

Miller, P.J.O., Kvadsheim, P.H., Lam, F.P.A., et al. (2015). First indications that northern bottlenose whales are sensitive to behavioural disturbance from anthropogenic noise. *Royal Society Open Science*, **2**, 140484.

Nachtigall, P.E., Supin, A.Y., Pawloski, J., & Au, W.W.L. (2004). Temporary threshold shifts after noise exposure in the bottlenose dolphin (*Tursiops truncatus*) measured using evoked auditory potentials. *Marine Mammal Science*, **20**(4), 673–687.

Nakagawa, T., Yamane, H., Shibata, S., et al. (1997). Two modes of auditory hair cell loss following acoustic overstimulation in the avian inner ear. *ORL Journal for Otorhinolaryngology and Its Related Specialties*, **59**, 303–310.

Nedelec, S.L., Radford, A.N., Simpson, S.D., et al. (2014). Anthropogenic noise playback impairs embryonic development and increases mortality in a marine invertebrate. *Scientific Reports*, **4**, 5891.

Nguyen, N., Hunt, J.P., Lindfors, D., & Greiffenstein, P. (2014). Aerial fireworks can turn deadly underwater: magnified blast causes severe pulmonary contusion. *Injury Extra*, **45**, 32–34.

Nowacek, D., Johnson, M., & Tyack, P.L. (2004). North Atlantic right whales (*Eubalaena glacialis*) ignore ships but respond to alarm stimuli. *Proceedings of the Royal Society of London Series B*, **271**, 227–231.

O'Hara, J. & Wilcox, J.R. (1990). Avoidance responses of loggerhead turtles, *Caretta caretta*, to low frequency sound. *Copeia*, **1990**, 564–567.

Pearson, W.H., Skalski, J.R., Sulkin, S.D., & Malme, C.I. (1994). Effects of seismic energy releases on the survival and development of zoeal larvae of Dungeness crab (*Cancer magister*). *Marine Environmental Research*, **38**(2), 93–113.

Pine, M.K., Jeffs, A.G., & Radford, C.A. (2012). Turbine sound may influence the metamorphosis behaviour of estuarine crab megalopae. *PLoS ONE*, e51790. doi:10.1371/journal.pone.0051790.

Popper, A. & Hastings, M. (2009). The effects of anthropogenic sources of sound on fishes. *Journal of Fish Biology*, **75**(3), 455–489.

Popper, A.N., Hawkins, A.D., Fay, R.R., et al. (2014). *ASA S3/SC1.4 TR-2014 Sound Exposure Guidelines for Fishes and Sea Turtles: A Technical Report Prepared by ANSI-Accredited Standards Committee S3/SC1*, pp. 33–51. Springer International Publishing. http://link.springer.com/10.1007/978–973–319–06659–2_7.

Popper, A.N. & Hoxter, B. (1984). Growth of a fish ear: quantitative analysis of sensory hair cell and ganglion cell proliferation. *Hearing Research*, **15**, 133–142.

Popper, A.N., Smith, M.E., Cott, P.A., et al. (2005). Effects of exposure to seismic airgun use on hearing of three fish species. *Journal of the Acoustical Society of America*, **117**(6), 3958–3971.

Purser, J. & Radford, A.N. (2011). Acoustic noise induces attention shifts and reduces foraging performance in threespined sticklebacks (*Gasterosteus aculeatus*). *PLoS ONE*, **6**(2), e17478.

Rajan, R. (1988). Effect of electrical stimulation of the crossed olivocochlear bundle on temporary threshold shifts in auditory sensitivity. I. Dependence on electrical stimulation parameters. *Journal of Neurophysiology*, **60**, 549–568.

Raphael Y. (2002). Cochlear pathology, sensory cell death and regeneration. *British Medical Bulletin*, **63**, 25–38.

Regnault, M. & Lagardere, J.P. (1983). Effects of ambient noise on the metabolic level of *Crangon crangon* (Decapoda, Natantia). *Marine Ecology Progress Series*, **11**(1), 71–78.

Reiter, E.R. & Liberman, M.C. (1995). Efferent-mediated protection from acoustic over-exposure: relation to slow effects of olivocochlear stimulation. *Journal of Neurophysiology*, **73**, 506–514.

Ridgway, S.H. & Howard, R. (1979). Dolphin lung collapse and intramuscular circulation during free diving: evidence from nitrogen washout. *Science*, **206**, 1182–1183.

Ridgway, S.H., Wever, E.G., McCormick, J.G., et al. (1969). Hearing in the giant sea turtle, *Chelonia mydas*. *Proceedings of the National Academy of Sciences of the USA*, **64**, 884–890.

Romano, T.A., Keogh, C.K., Feng, P., et al. (2004). Anthropogenic sound and marine mammal health: measures of the nervous and immune systems before and after intense sound exposure. *Canadian Journal of Fishery and Aquatic Science*, **61**, 1124–1134.

Rolland, R.M., Parks, S.E., Hunt, K.E., et al. (2012). Evidence that ship noise increases stress in right whales. *Proceedings of the Royal Society of London Series B*, **279**(1737), 2363–2368.

Rucker, R.S. (1973). *Effect of Sonic Boom on Fish*. DTIC report AD0758239 online. http://oai.dtic.mil/oai/oai?verb= getRecord&metadataPrefix=html&identifier= AD0758239.

Santulli, A., Modica, A., Messina, C., et al. (1999). Biochemical responses of European sea bass (*Dicentrarchus labrax*) to the stress induced by off shore experimental seismic prospecting. *Marine Pollution Bulletin*, **38**(12), 1105–1114.

Saunders, J.C., Duncan, R.K., Doan, D.E., & Werner, Y.L. (2000). The middle ear of reptiles and birds. In: Dooling, R., Popper, A., & Fay, R. (eds). *Comparative Hearing: Birds and Reptiles*, pp 13–69. Springer, New York.

Schlundt, C.E., Finneran, J.J., Carder, D.A., & Ridgway, S.H. (2000). Temporary shift in masked hearing thresholds of bottlenose dolphins, *Tursiops truncatus*, and white whales, *Delphinapterus leucas*, after exposure to intense tones. *Journal of the Acoustical Society of America*, **107**(6), 3496–3508.

Scholik, A.R. & Yan, H.Y. (2002). The effects of noise on the auditory sensitivity of the bluegill sunfish, *Lepomis macrochirus*. *Comparative Biochemistry and Physiology Part A*, **133**(1), 43–52.

Schroeder, J., Nakagawa, S., Cleasby, I.R., & Burke, T. (2012). Passerine birds breeding under chronic noise experience reduced fitness. *PLoS ONE*, **7**, e39200.

Schusterman, R.J., Kastak, D., Levenson, D.H., et al. (2000). Why pinnipeds don't echolocate. *Journal of the Acoustical Society of America*, **107**(4), 2256–2264.

Slepecky, N., Hamernik, R., Henderson, D., & Coling, D. (1982). Correlation of audiometric data with changes in cochlear hair cell sterocilia resulting from impulse noise trauma. *Acta Otolaryngol*, **93**, 329–340.

Smith, A. (1991). A review of the non-auditory effects of noise on health. *Work and Stress*, **5**, 49–62.

Smith, M.E. (2004). Noise-induced stress response and hearing loss in goldfish (*Carassius auratus*). *Journal of Experimental Biology*, **207**(3), 427–435.

Smith, M.E., Coffin, A.B., Miller, D.L., & Popper, A.N. (2006). Anatomical and functional recovery of the goldfish (*Carassius auratus*) ear following noise exposure. *Journal of Experimental Biology*, **209**, 4193–4202.

Solé Carbonell, M. (2012). *Statocyst sensory epithelia ultrastructural analysis of Cephalopods exposed to noise*. PhD thesis, University of Cataluña. 183 pp.

Solé, M., Lenoir, M., Durfort, M., et al. (2013). Does exposure to noise from human activities compromise sensory information from cephalopod statocysts? *Deep Sea Research Part II*, **95**, 160–181.

Southall, B.L., Bowles, A.E., Ellison, W.T., et al. (2007). Exposure criteria to noise for marine mammals. *Aquatic Mammals*, **33**(4), 411–414.

Steevens, C.C., Russel, K.L., Knafele, M.E., et al. (1999). Noise-induced neurological disturbances in divers exposed to intense water-borne sound: two case reports. *Undersea Hyperbaric Medicine*, **26**, 261–265.

Sun, H., Lin, C.-H., & Smith, M.E. (2011). Growth hormone promotes hair cell regeneration in the zebrafish (*Danio rerio*) inner ear following acoustic trauma. *PLoS ONE*, **6**, e28372.

Theakston, F. (ed.) (2011). *Burden of Disease from Environmental Noise: Quantification of Healthy Life Years Lost in Europe*. World Health Organization, Regional Office for Europe, Copenhagen.

Thomson, C.A. & Geraci, J.R. (1986). Cortisol, aldosterone, and leucocytes in the stress response of bottlenose dolphins, *Tursiops truncatus*. *Canadian Journal of Fisheries and Aquatic Science*, **43**, 1010–1016.

Tougaard, J., Carstensen, J., Teilmann, J., et al. (2009). Pile driving zone of responsiveness extends beyond 20 km for harbour porpoises (*Phocoena phocoena*). *Journal of the Acoustical Society of America*, **126**, 11–14.

Tougaard, J., Wright, A.J., & Madsen, P.T. (2015). Cetacean noise criteria revisited in the light of proposed exposure limits for harbour porpoises. *Marine Pollution Bulletin*, **90**, 196–208.

Tyack, P.L., Johnson, M., Aguilar de Soto, N., et al. (2006). Extreme diving of beaked whales. *Journal of Experimental Biology*, **209**, 4238–4253.

Tyack, P.L., Zimmer, W.M.X., Moretti, D., et al. (2011). Beaked whales respond to simulated and actual navy sonar. *PLoS ONE*, **6**, e17009.

Urick, R.J. (1983). *Principles of Underwater Sound*. McGraw-Hill, New York.

Ward, D.H., Stehn, R.A., Erickson, W.P., & Derksen, D.V. (1999). Response of fall-staging Brant and Canada geese to aircraft overflights in Southwestern Alaska. *Journal of Wildlife Management*, **63**, 373–381.

Weir, C.R. (2007). Observations of marine turtles in relation to seismic airgun sound off Angola. *Marine Turtle Newsletter*, **116**, 17–20.

Weis, J.S. (2014). *Physiological, Developmental and Behavioral Effects of Marine Pollution*. Springer Netherlands.

Weisse, C.S., Pato, C.N., McAllister, C.G., et al. (1990). Differential effects of controlable and uncontrollable acute stress on lymphocyte proliferation and leukocyte percentages in humans. *Brain Behaviour and Immunology*, **4**, 339–351.

Wilson, R.P., Culik, B., Danfeld, R., & Adelung, D. (1991). People in Antarctica, how much do adelie penguins, *Pygoscelis adeliae,* care? *Polar Biology,* **11**, 363–370.

Wright, A.J., Aguilar de Soto, N., Baldwin, A.L., et al. (2007). Are marine mammals stressed by anthropogenic noise? *International Journal of Comparative Psychology,* **20**, 274–316.

Yelverton, J.T., Richmond, D.R., Hicks, W., et al. (1975). *The Relationship between Fish Size and Their Response to Underwater Blast*. Report DNA 3677T, US Ministry of Commerce, Washington, DC.

Yelverton, J.T., Richmond, D.R., Royce, E, & Jones, R. (1973). *Safe Distances from Underwater Explosions for Mammals and Birds*. NTIS. Report DNA 3114T, US Ministry of Commerce, Washington, DC.

PART II

Ecological Responses

CHAPTER 9

Effects of changing salinity on the ecology of the marine environment

Katie Smyth and Mike Elliott

9.1 Salinity as a driver affecting marine ecology

9.1.1 A brief introduction to salinity

The salinity of any water body is the result of an equilibrium between the environmental ionic inputs and removals. The open oceans are characterized by a remarkable stability and similarity of their composition, not only in their total salinity but for the relative proportions of the constituents within and between them; this is referred to as *constancy of composition* (Sverdrup et al., 1942; Wright and Colling, 1995). In contrast, the inshore and nearshore marine environment and transitional water bodies such as fjords, estuaries, and lagoons are subjected to the influences of tides, freshwater inputs, evaporation, and, increasingly, anthropogenic contaminants (Elliott and Whitfield, 2011; Wolanski and Elliott, 2015). This increases the salinity variability, on daily, weekly, spring–neap, lunar, and seasonal cycles. The salinity as the amount of dissolved salts (Sverdrup et al., 1942) may behave conservatively, i.e. changing just by dilution or, in the case of semi-tropical areas, by evaporation (Potter et al., 2010). Prior to 1978 when the practical salinity scale (PSS) was introduced as the dimensionless way of measuring salinity (Lewis, 1980; Lewis and Perkin, 1980; UNESCO, 1985), salinity was measured in parts per thousand (ppt, ‰) of dissolved salt per volume of water, based on a standard measurement for seawater, the International Association for Physical Sciences of the Ocean (IAPSO) Copenhagen Water (Lewis, 1980). The PSS generally equates to ‰; however, it is not *exactly* the same. To avoid confusion, salinities mentioned in this chapter will use the unit (or no units if using the PSS) as chosen by the original author of any work cited.

9.1.2 The importance of salinity to aquatic organisms

Salinity is regarded as an environmental master factor for aquatic organisms such that they can be described according to their salinity tolerances; hence a stenohaline organism has a restricted salinity tolerance in contrast to a euryhaline organism (see Section 9.1.3). Thus any change to ambient salinity has the potential to affect the ability of animals to carry out vital biological processes and thus their ability to survive and thrive. Environmental salinity is a major factor in reproduction, larval dispersal and recruitment, geographical distribution, and behaviour of marine species (Anger, 1991, 1996; Spivak and Cuesta, 2009; Smyth et al., 2014). Because of this, salinity changes will influence the structure of communities and the boundaries of species distribution. There is also an energetic cost required in adapting to changing ambient salinity which may have resultant impacts on organisms (see Section 9.1.3).

9.1.3 Osmo-/ionoregulation effects

Osmoregulation is the ability of an aquatic animal to maintain its internal fluid concentrations at an acceptable level for biological function, in response to changes in the concentration of the external media. Osmoregulation is covered in more detail elsewhere; however, it is one of the key factors in determining how salinity change affects ecology (see Chapter 1 in this volume). The ability to osmoregulate is especially important in organisms that are faced with regular salinity challenges, such as those in estuarine and brackish areas, and directly affects which species can inhabit such environments and hence the ecology. Aquatic animals are either osmoconformers or osmoregulators with regard to

Stressors in the Marine Environment. Edited by Martin Solan and Nia M. Whiteley
© Oxford University Press 2016. Published in 2016 by Oxford University Press.

salinity change (Kinne, 1971; Schubart and Deisel, 1999; Karleskint et al., 2009):

1. **Osmoconformer**: an animal that allows its internal fluid concentrations to change in line with those of the environment. They can only control the concentration of their body fluids by behavioural means, such as avoidance. Long-term exposure to unfavourable salinity will often result in the death of an osmoconformer.
2. **Osmoregulator**: an animal that maintains its internal fluid concentrations at a level acceptable to the animal regardless of environmental changes such as decreasing or increasing salinity. In extremes of salinity they may need to supplement their physiological control methods by behavioural means, for example a bivalve closing its valves until conditions improve. Osmoregulators can regulate hyper- or hypoosmotically depending on species, or achieve both. Although the animal can regulate its body fluids, long-term exposure to unfavourable salinity can still result in the death of the animal and may have impacts on its ability to grow or reproduce.

Aquatic organisms can also be classified according to their degree of salinity tolerance:

1. **Euryhaline**: an animal which can tolerate a range of environmental salinities and salinity fluctuations within a specified range; estuarine species are usually euryhaline and as such differ from diadromous organisms migrating through estuaries and which are freshwater or saline-adapted, i.e. they are not necessarily euryhaline but just have an ability to switch from being fresh- or saline-water adapted to the other type (Elliott and Hemingway, 2002).
2. **Stenohaline**: an animal which can tolerate only a small range of environmental salinities, usually taken to mean marine levels and so often relating to marine species; this is not the same as a freshwater organism that cannot tolerate any salinity.

Often osmoconformers are euryhaline, but this is not always the case. Osmoregulation is very important in estuarine intertidal species, but not so in fully marine species. Primary marine inhabitants (those that evolved in the sea) are mainly stenohaline—they live in the open sea and therefore encounter little osmotic stress resulting in a poor osmotic regulation ability. This is due to the inability of their cells to cope with or adapt to any change in body fluid composition, especially when coupled with their high cell permeability to ions and water (Péqueux, 1995). Their euryhaline

relatives are found in the coastal and estuarine zones where salinities can change regularly and hence rely heavily on osmotic control to regulate the concentrations of internal body fluids (Davenport, 1985). Hence the osmoregulatory ability of aquatic species is key to their geographical distribution. Each species is likely to have an optimal salinity range and thus any deviation from that will impair the organism; that optimum creates the fundamental niche for the organism. This impairment may eventually affect the population and hence potentially alter the community and ecosystem within which it resides.

Tolerance of aquatic organisms to changing salinity is not dependent on salinity alone; it also decreases at temperatures different from the optimum, thus unfavourable salinities are best tolerated at the temperature optimum (Kinne, 1964, 1971). Temperature changes may significantly affect the osmoregulatory capability of aquatic species; hence what may be an acceptable salinity in winter months may be intolerable in the summer and vice versa. Changes in the ionic concentration of the ambient medium have also been shown to affect the tolerance of organisms to cold and heat. Methods for surviving adverse salinity conditions include movement away from the unfavourable environment, osmoregulation or osmoconforming, and in the case of sessile and sedentary organisms, burrowing or closing shell valves. This behavioural control of internal osmotic concentrations is used by many euryhaline decapod crustaceans as a way of reducing their exposure to stressful conditions (Davenport, 1985; Laverack, 1985; Smyth et al., 2014).

The effort needed by animals not moving away from unfavourable salinity change to maintain their bodies at a suitable osmolarity may have adverse implications for their energy budget, and thus growth, which may be affected by many factors, including the chemical nature of its environment. Changes or cessation to the growth rate can affect maturity, ageing, and reproductive potential, as well as having implications for population structure through changes to birth and death rates (Moriarty, 1993). The amount of energy left for growth after all other energy requirements are met is known as 'scope for growth' and is defined as the difference between the energy content of the food consumed and all energy losses apart from growth (Moriarty, 1993; Mazik et al., 2013). Environmental stressors such as salinity change may result in aquatic species needing to put more of their energy resources into regulating their internal osmolality so that their metabolic systems can function properly. Any lower growth rate (Shirley and Stickle, 1982a, b; Garton, 1984) could, in the case of commercially important species, give a lower product yield.

9.1.4 The effects of salinity coupled with temperature

There may be synergistic or antagonistic interactions between different environmental stressors. Water has a higher oxygen saturation level at colder temperatures and lower salinities (Table 9.1); hence warm hypersaline areas, such as those seen in mangrove swamps, maintain some of the lowest oxygen saturation values.

As temperature increases, metabolic rate increases. In aquatic species, stress caused by sub- or supra-ambient salinities, when coupled with an increase in water temperature, may mean that larger than normal amounts of metabolic energy are used to maintain homeostasis, with the concomitant impacts on growth, reproductive potential/success, and species range/distribution. For example, the interaction between salinity and temperature on metabolic rates has been suggested as a factor controlling the distribution of species such as the freshwater prawn *Macrobrachium rosenbergii* (Nelson et al., 1977); the marine invasive clam *Potamocorbula amurensis* varies between aerobic and anaerobic metabolism at different salinity and temperature combinations as found in its estuarine habitat (Miller et al., 2014). Growth can also be affected by the interaction between salinity and temperature and hence the combination of factors is likely to influence species distributions. For example, the desert pupfish (*Cyprinodon macularius*) shows growth rates that are dependent on salinity and temperature, with growth decreasing as temperature decreases from 30 to 15 °C and when salinity increases or decreases from 35‰. At 15 and 20 °C, pre-adult growth is fastest in fresh water, but at 25, 30, and 35 °C growth is fastest at 35 and 55 ‰ (Kinne, 1960). Juvenile turbot also show lower growth

rates and food conversion when salinities increase from 15‰ to 33.5‰ in the temperature range 10–22 °C and the optimal temperature for growth varies with salinity such that conversion efficiency is best in intermediate salinities in the upper temperature range (Imsland et al., 2001). As well as animal metabolic processes being affected by salinity and temperature, the seagrass *Zostera marina* reduces photosynthesis as salinities deviate from ambient, reducing to almost 0 in distilled fresh water or 200% seawater. Photosynthesis increases with temperature up to 35 °C in intertidal *Z. marina* and up to 30 °C in subtidal *Z. marina* (Biebl and McRoy, 1971).

9.1.5 pH effects

The main anion in seawater, Cl^-, alone does not affect the pH of the water; however, seawater naturally includes other constituents, including bicarbonate ions which make the water slightly more alkaline with a pH to which marine organisms are adapted. The pH is reduced by increased dissolved CO_2 and anthropogenic influences such as ocean acidification (Elliott et al., 2015). The main metabolic waste product of aquatic organisms is ammonia, which exists as the unionized fraction (NH_3), which is more toxic to aquatic animals, and the ionized fraction ammonium (NH_4^+), which is less toxic; the ratio of NH_3 to NH_4^+ is affected by pH and temperature. In general, ammonia is more in the toxic form at higher alkaline pH values and as the temperature of the water increases, although a shift towards acidic pH does not mean the survival of more animals. As many marine and freshwater organisms, e.g. molluscs and crustaceans, use calcium carbonate ($CaCO_3$) for shells and carapaces, suboptimal pH has implications for shell thickness and hardness, in turn affecting resistance to environmental stressors and thus survival and distribution. For example, when the mussel *Mytilus galloprovincialis* was exposed to different pH (8.07 and 7.25) and CO_2 gradients, in acidified seawater the calcite layer of the shells became thinner and began to lack structure in the orientation of the calcite prisms. Furthermore the nacreous layer of the shells became thinner and partly dissolved, due to a combination of the lower pH and higher CO_2 and resultant food availability (Hahn et al., 2012). In estuarine waters, the oyster *Crassostrea virginica* showed a significant decrease in biocalcification when natural environmental salinity was reduced by 0.5 pH units; however, the complex relationship between salinity, pH, and temperature is demonstrated by the fact that decreases in calcification were mitigated by a higher temperature and a higher

Table 9.1 The effect of temperature (°C) and salinity on oxygen saturation (mg.l^{-1}) of sea water. Created from data in Salinity Nomogram in Richards and Corwin (1956). Original table expanded from Smyth (2011).

Salinity → Temp. (°C) ↓	5	15	25	35	45	55
5	12.0	11.3	10.6	9.9	9.1	8.4
10	10.6	10.0	9.4	8.8	8.1	7.5
15	9.5	8.9	8.4	7.9	7.3	6.8
20	8.6	8.1	7.7	7.2	6.7	6.2

salinity which increased calcification (Waldbusser et al., 2011). At a population level, a higher density of individuals may increase carapace thickening in the shore crab *Carcinus maenas* as a means of defence (Souza et al., 2011) and any punctures in the carapace are quickly repaired by a cuticle of 20% $CaCO_3$ (Dillaman and Roer, 1980). Therefore factors that affect the ability to calcify the exoskeleton in crustaceans, or the shell in molluscs, such as pH changes and salinity changes, are likely to have impacts on the survival of populations in affected areas.

9.1.6 Effects on ecosystem function

Salinity is a driver of ecosystem function. In the Baltic Sea, for example, the presence of strong estuarine-like salinity gradients has contributed (together with temperature gradients and anthropogenic influences) to a variation in the number of functional benthic species groups. In the northernmost areas of the Baltic (where salinities are around 1–3 ppt) only 1–3 groups are found which are also poor in functional complexity, compared to 8–20 complex groups in the south (where salinities are around 25–30 ppt) (Bonsdorff and Pearson, 1999). Salinity is also the dominating factor on the microbial scale, affecting archaeal community structure and ecological function from the North River tributary of the Pearl River (China), its estuary, and the coastal South China Sea, which is consistent with salinity control on microbial diversity in other regions of the world (Xie et al., 2014).

9.2 Salinity change as an environmental problem

In this section we provide examples of salinity differences worldwide, ways this affects the local ecology and the adaptations of species to survive in these areas.

9.2.1 Ice melt/formation in polar regions

Water reaches its maximum density at around 4 °C and decreases in density towards the freezing point, causing the water near freezing to rise to the surface. Water expands further upon freezing which makes it less dense than the liquid water so it floats and water bodies freeze from the surface down. Hence many aquatic species can survive periods of freezing by moving to the unfrozen waters below. The freezing of sea ice can increase environmental salinity and hence aquatic polar species require to be adapted. In the polar regions, high salinity brines commonly form in pockets

within the sea ice. The phenomenon occurs as the freezing process removes the pure water from solution to form ice, leaving behind concentrated salts which are generally unable to be incorporated into the ice lattice due to their size or electrical conductivity. Salinity in these transition areas can vary from almost fresh water where ice melts to pockets of brine four times the ambient salinity in areas of ice formation (Horner et al., 1992; Petrich and Eicken, 2010). Formation and melt of sea ice helps to control ocean $\leftrightarrows$ atmosphere exchanges of CO_2 (Eicken, 1992; Rysgaard et al., 2011) and uptake of CO_2 by sea ice is driven by a number of factors: (1) rejection during sea ice development and sinking of CO_2 loaded brine into oceanic water masses; (2) blocking of air–sea CO_2 exchange during winter; (3) release of CO_2-depleted meltwater with excess total alkalinity during sea ice melt; and (4) biological CO_2 drawdown during primary production in sea ice and surface oceanic waters (Rysgaard et al., 2011).

This complex environment provides a habitat for many of the organisms that form the basis of polar food webs. Bacteria and algae here are commonly termed 'extremophiles', as they are adapted to live in extreme conditions. Within the sea ice, the only liquid environment available is pockets of highly saline brine formed from the freezing of the water (Eicken, 1992; Thomas and Dieckmann, 2002). Brines with low concentrations of CO_2 and elevated O_2, high concentrations of dissolved organic matter, low levels of nutrients, high ammonia, and elevated pH are a result of biological activity within these pockets. The basis of the polar food chain (excluding tundra areas) are heterotrophic bacteria and autotrophic algae; the latter contribute significantly to the total primary production in the Arctic (25%) and in the Antarctic (20%) (Legendre et al., 1992), and further support secondary production and hence the higher trophic levels such as marine mammals. Physiological and biochemical adaptations of the sea-ice biota accommodate swift transitions between water and ice and the subsequent rapid physical and chemical changes which can occur within a few hours or days as they become trapped within, or freed from, ice as it freezes and thaws (Thomas and Dieckmann, 2002).

9.2.2 Dilution in estuarine and coastal environments

Arguably the ecological features most affected by salinity change are those in estuaries and other transitional water bodies where the community is defined according to the salinity tolerances of the component organisms. Potter et al. (2010) define an estuary and included hypersaline estuarine systems:

An estuary is a partially enclosed coastal body of water that is either permanently or periodically open to the sea and which receives at least periodic riverine discharge, and thus, while its salinity is typically less than that of natural seawater and varies temporally and along its length, it can become hypersaline in regions when evaporative water loss is high and freshwater and tidal inputs are negligible.

Whitfield and Elliott (2011) and Elliott and Whitfield (2011) further summarized the estuarine features including its characteristic biota and emphasized the structuring feature of salinity.

The distribution of organisms along an estuary from seawater to freshwater has long been described according to a classical Remane Curve/Diagram (Remane, 1934). Derived from studies in the Baltic Sea in the early 1900s, the Remane Diagram (Fig. 9.1) shows

the minimum species complement in the freshwater–seawater interface with the 5–7 salinity area being dominated by a small number of true brackish/estuarine species, perhaps the most stressful part of the estuary (Whitfield et al., 2012). In the classical Remane curve, the dominant block of marine organisms will be stenohaline in character whereas those denoted 'brackish animals' will be euryhaline. However, given the relatively small number of organisms whose salinity preferences have been determined experimentally, such a characterization is based on circumstantial evidence of the salinities in the area where an organism is found.

It is this variable salinity which makes estuarine and coastal environments so challenging for those aquatic species not adapted to the conditions, in that for non-adapted species this is a stress whereas for adapted species the conditions are a subsidy and allow tolerant species to thrive (Elliott and Quintino, 2007).

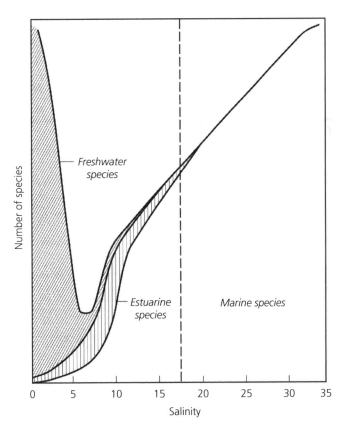

Figure 9.1 A redrawn version of the original Remane diagram for aquatic species distribution (Remane, 1934). The diagonally slashed area indicates freshwater species, vertically slashed area indicates brackish species, and white area below the curve represents marine species. The vertical dashed line represents a salinity of approximately 50% seawater.

The salinity at any one place in an estuary will vary greatly, possibly even between fresh and seawater, with tidal, daily, fortnightly, monthly (lunar), seasonal, and equinoctial cycles (Wolanksi and Elliott, 2015). Similarly run-off and rainfall in intertidal coastal areas can dilute estuarine waters and tidal pools randomly. Species colonizing such environments will be adapted to withstand and accommodate such changes and communities within these areas reflect this adaptation, with the dominant salinity for the area being a key factor in determining the colonizing assemblages.

Elliott and Whitfield (2011) gave the characteristics of estuaries as a set of paradigms, the third of which is:

Hydromorphology is the key to understanding estuarine functioning but these systems are always influenced by salinity (and the resulting density/buoyancy currents) as a primary environmental driver.

Hydromorphology is the principal determinant for the residence time of water within an estuary and has a major impact on the water characteristics and ecological functioning of systems worldwide (Chícharo and Chícharo, 2006; Haines et al., 2006; Wolanski and Elliott, 2015). Salinity is the primary factor (with its dominance suggesting everything else is secondary) and most estuary-associated species are highly euryhaline (see Section 9.1.3, and also Bulger et al., 1993). Also of high importance is the input from river flow—especially in the context of estuarine morphology—which is a key component in the understanding of hydromorphology (Gómez-Gesteira et al., 2003) and the emerging field of estuarine ecohydrology (Wolanski and Elliott, 2015). Based on the hydromorphology and the estuarine characteristics, pressures on the biota are thus shaped by the relative combination of tidal and river inputs, the creation of a freshwater–seawater interface and, where sediment resuspension allows, a turbidity maximum zone which may limit the photic zone and thus primary production (Burford et al., 2011; Wolanski and Elliott, 2015).

Although of differing strength of influence, density/buoyancy driven currents, produced by the interaction of waterbodies of different salinity (and/or temperature), are central to the hydrodynamic functioning of all estuaries (Wolanski et al., 2004; Uncles, 2010; Wolanski and Elliott, 2015). Furthermore, the net result of the tidal and freshwater influences determines the estuarine flushing rate and residence time (Monsen et al., 2002), both of which, in combination with flow rates, influence estuarine ecological forcing factors such as the salinity, the ability to retain nutrients, the dispersal of certain stages of benthic organisms and plankton, and the wider connectivity between systems (de Brauwere et al., 2011). Some migrating fish species follow salinity strata to achieve an energetically and physiologically optimal transport between the freshwater and sea (Elliott and Hemingway, 2002).

In addition, freshwater delivery into the estuary and the resulting changes to the salinity there will occur both from anthropogenic activities, such as water abstraction, barriers, and water interception, and from climatic and global factors such as the North Atlantic Oscillation or El Nino Southern Oscillations (Elliott et al., 2015). Hence, hydromorphology is a major driver of estuarine ecosystem functioning in that it changes salinity conditions and/or the physical removal of organisms (Wolanksi and Elliott, 2015), such as marine fishes removed from linear estuaries (Whitfield and Harrison, 2003), even though they can survive in freshwater or oligohaline waters for prolonged periods (Ter Morshuizen et al., 1996), hence showing the greater physical rather than osmoregulatory influence. Such evidence may indicate a hierarchy of the influence of physical factors and that the hydromorphological influence may be more of a global paradigm than that of salinity for certain systems (Marais, 1982). However, given the interlinked nature of all of these factors, it is difficult to separate them.

As with all estuarine biotic components (McLusky and Elliott, 2004), benthic macroinvertebrate species behaviour, morphology, and physiology and community distributions reflect estuarine saline penetration (Ysebaert et al., 2003; Attrill and Thomas, 2006). For example, male shore crabs, *Carcinus maenas*, exhibit different carapace colours according to their salinity tolerance, with green males more tolerant of salinity fluctuations and aerial exposure than red males. Conversely, red males compete more successfully for mates and food. It is hypothesized that some crabs remain in inter-moult longer than others, developing stronger and thicker carapaces and chelae, and are thus more likely to win mating conflicts; however, this advantage is gained at the cost of reduced tolerance to changes in the surrounding environment (Reid et al., 1997; Styrishave et al., 2004). Two similar prawn species, *Palaemon adspersus* and *P. elegans*, exhibit different estuarine distributions in the Baltic Sea, with *P. adspersus* being more estuarine and *P. elegans* being more marine. Although adults of both can survive in experimental conditions at 10, 7.5, and 5‰, *P. elegans* has a significantly reduced reproductive success in terms of numbers of eggs hatching and larvae survival when compared with *P. adspersus* (Berglund, 1985).

As with the fauna, salinity tolerance is reflected in coastal and estuarine macrophyte and macroalgae distributions, e.g. spatial salinity segregation of plant species is driven by freshwater marsh plants, which are competitively superior, displacing salt-tolerant plants to the more physically harsh brackish habitats, but they are limited from living in salt marshes by physical factors such as high salinity (Crain et al., 2004). Similar factors appear to be present in mangrove systems with varying salinities, where the low soil salinities of 8–12 mg.l⁻¹ salt at 2 km from an estuary mouth in Sri Lanka support the largest species richness and diversity of the estuary (Perera et al., 2013).

Bulger et al. (1993) showed, as did Remane, that biological changes do occur along a salinity gradient, that estuarine species are not evenly distributed across the salinity gradient, and that a salinity of 24 rather than 30 is the lower limit of salinity tolerance for fully marine species. They also suggested that a salinity of 4 was the upper tolerance limit for stenohaline (*sic*) freshwater fishes. Attrill (2002) further strongly argued for salinity variations to be the primary factor in determining both species diversity and distribution within estuaries, rather than overall salinity tolerance. His linear model gave the mean salinity range at a given point of the Thames Estuary, UK, versus its mean subtidal faunal diversity. This model showed the species diversity minimum occurred in approx. 10 mean salinity, a region subject to large fluctuations in salinity (Attrill and Rundle, 2002). This minimum is relatively close to the minimum in the Remane diagram (Fig. 9.1); however, in the Thames Estuary this minimum represents the lower ends of two ecoclines, one from the freshwater and the other from the marine side. The study also proposed that true estuarine species do not in fact exist (at least in the case of the Thames Estuary); hence there is the fundamental difference from the original work by Remane.

Despite the above, recent papers show that estuaries are a complex of ecotones, some of which are salinity driven, and that species are specifically adapted to those conditions across boundaries (Basset et al., 2013). Similarly, the communities are the net result of the stress–subsidy continuum and those inhabiting estuaries are tolerant of and therefore benefit from the stressful conditions; furthermore, they can absorb the effects of environmental stressors, what has been termed environmental homeostasis (Elliott and Quintino, 2007; Elliott and Whitfield, 2011). Whitfield et al. (2012) more recently used studies based on the Remane Curve (Fig. 9.1) to redevelop a conceptual but wider species diversity change model for estuaries, as opposed to the Remane Curve developed for brackish seas and the macrobenthos (Fig. 9.2, Table 9.2).

9.2.3 Naturally hypersaline areas

Evaporation will increase salinity, especially in tropical climates, mangrove swamps or estuaries with a restricted mouth and low freshwater input, intertidal rockpools, and terrestrial saline waters such as brine

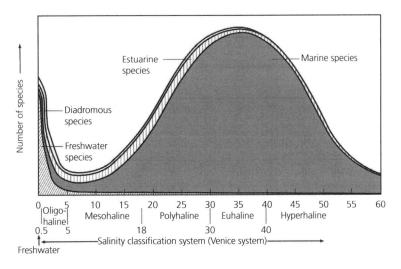

Figure 9.2 Proposed conceptual model for estuarine biodiversity (species) changes covering the salinity continuum from freshwater to hyperhaline condition (after Whitfield et al., 2012).

Table 9.2 The characteristics of salinity distributions along an estuary (after Whitfield et al., 2012).

Features of salinity-dependent biodiversity change and species distribution along an estuarine gradient, including fresh and hypersaline conditions.
1   The freshwater assemblage in a particular region is generally not as species rich as marine taxa in the same region.
2   Most freshwater species do not penetrate saline waters, i.e. they are mainly confined to the freshwater area above the estuary, with only a few taxa extending into mesohaline, polyhaline, and euhaline areas.
3   Marine species extend into oligohaline estuarine and sometimes even fresh waters but in relatively small numbers when compared to their diversity in more saline waters.
4   Marine species overwhelmingly dominate (in terms of taxa) the mesohaline, polyhaline, euhaline, and hyperhaline estuarine waters.
5   Estuarine/brackish species are most diverse in mesohaline and polyhaline waters, but can also be present in limnetic, oligohaline, euhaline, and even hyperhaline waters.
6   Biotic diversity in estuaries starts to decline above a salinity of about 40, with most species unable to survive in salinities above 50.
7   There are very few freshwater, estuarine, and marine species that can be termed holohaline, i.e. those species that are capable of occupying waters with a salinity range of 0–100 +.

pools (e.g. Potter et al., 2010). In all ecosystems, but especially the small and restricted ones, salinity is influenced by climate, water inputs, and tidal inundation (Chan, 2000). Temperatures can increase > 10 °C from ambient due to sunlight, and dissolved oxygen can fluctuate between very high (> 30 ppm) in daylight when algae are photosynthesizing to extremely low (< 4 ppm) during the night when respiration dominates (Morris and Taylor, 1983; Hugget and Griffiths, 1986). CO_2 tends to follow the converse of the oxygen concentration and higher CO_2 concentration gives lower pH. Tidal pools also can be salinity stratified, thus giving an extremely challenging environment which requires adaptation. This adaptation can be physiological, as in the case of the harpacticoid copepod *Tigriopus brevicornis*, which is the dominant fauna in tidal pools above mean high water spring tides in NE Europe (McAllen and Taylor, 2001); the species reduces its activity and oxygen consumption when exposed to higher salinities, and vice versa; hence it shows salinity-mediated metabolic rate changes (McAllen and Taylor, 2001). Physiological mechanisms to cope with salinity increases are common and many marine crabs in the families Ocypodidae, Grapsidae, and Varunidae regulate haemolymph NaCl at levels below the medium, apparently by excreting salts across the gills (Mantel and Farmer, 1983). However some organisms change their behaviour to accommodate short periods of hypersalinity; for instance, intertidal pulmonate limpets *Siphonaria capensis* reoccupy scars on rocks to avoid osmotic stress due to evaporation and dilution of tide pool water, as well as minimizing desiccation when tidally exposed (Branch and Cherry, 1985).

Natural brine lakes (and salt pans, see Section 9.2.4), include the Great Salt Lake in Utah, USA, and the Dead Sea in the Levant, and highly saline brines include deep ground waters, deep stable oceanic pools, and near salt springs (Anati, 1999). Salinity here correlates negatively with species richness (Williams et al., 1990), e.g. while the Dead Sea contains no macrofauna due to the high salinity, the Great Salt Lake does support populations of the brine shrimps *Artemia gracilis* (Jensen, 1918) and *Artemia franciscana*, as well as brine fly species (Ephydridae) (Vorhies, 1917; Stephens and Gardener, 2007), rotifers, algae, and, in the less saline regions where there is a freshwater input, some species of fish (Stephens and Gardener, 2007). These then support large water bird populations similar to anthropogenic salt pans (see Section 9.2.4). *Artemia* spp. are particularly well adapted to thrive in the challenging environments provided by saline lakes. They are powerful hypoosmoregulators in salinities above 30% seawater, tolerating hypersaline conditions by excreting NaCl through specialized salt glands in nauplii and through gills in adults (Croghan, 1958) and can tolerate up to 10× the concentration of ordinary seawater (Gajardo and Beardmore, 2012). When conditions become too extreme, for instance in drought conditions, they produce an embryonic cyst which can survive severe dehydration until rehydrated (Gajardo and Beardmore, 2012).

Deep oceanic pools of brine often have high levels of methane or hydrogen sulphide as well as an extremely high salinity, and hence are inhabited by chemosynthetic extremophiles that gain their energy from the methane (Xu, 2010; Marlow et al., 2014). Such microorganisms keep hazardous levels of greenhouse gases from escaping into the atmosphere (Marlow et al., 2014) and form the basis of many deep oceanic food chains, such as those at hydrothermal vents (Van Dover and Fry, 1994).

9.2.4 Anthropogenic causes of environmental salinity change

Anthropogenic salinity changes are also of concern, including dilutions by wastewater discharges and increases (hypersalinity) from desalination plants, salinas (salt pans), and solute mining (Wolanski and Elliott, 2015). Desalination supplies fresh water for domestic, agricultural, and industrial uses and results in a hypersaline brine discharge (Meerganz von Medeazza, 2005; Raventos et al., 2006), not only in arid areas but in all water shortage areas worldwide (Romeril, 1977; Raventos et al. 2006). This has occurred in the UK since 2010 (Li et al., 2011). Similarly, dissolution of underlying salt deposits through the injection of heated seawater (solute mining), to create gas and CO_2 storage cavities in salt strata adjacent to coasts, also discharges hypersaline brine (Bérest et al., 2001; Dusseault et al., 2001; Shi and Durucan, 2005; Quintino et al., 2008). The brine generated by desalination and solute mining is subsequently discharged into surrounding coastal or estuarine environments, potentially increasing the ambient salinity and hence affecting marine fauna (Smyth, 2011; Smyth et al., 2014).

The salinity of brine produced by desalination plants is in the range of 44–90 (Fernández-Torquemada et al., 2005), and solute mining discharges can be up to eight times that of normal seawater (Smyth, 2011). Quintino et al. (2008) showed an adverse effect of this on stenohaline organisms in terms of adult survival, larval abnormal development, and sperm fertilization. Echinoderms also disappeared from a *Posidonia oceanica* seagrass meadow affected by a brine discharge that raised bed salinity to 38.5 near Urbanova, Spain (Fernández-Torquemada et al., 2005), indicating that even relatively small changes in salinity can have major impacts on the local ecology. The effects can be exacerbated in naturally or anthropogenically poorly mixed (stratified) waters with the dense brine on the bed.

Not only do desalination plants increase the salinity of the receiving environment, but those that operate by thermal desalination also increase the temperature of the receiving environment. This decreases the dissolved oxygen and increases the metabolic rate and thus changes behaviour and physiology (Miri and Chouikhi, 2005). Furthermore, high salinity can lead to an increase in water turbidity and thus reduction in the photic depth (Miri and Chouikhi, 2005).

Anthropogenic brine discharges usually discharge into shallow waters, often the sites of sea grass beds which can be particularly affected by changing environmental salinity. Given that these angiosperms (Posidoniaceae, Zosteraceae, Hydrocharitaceae, or Cymodoceaceae) are ecosystem engineers with rhizomes and leaves which stabilize the sediment and reduce boundary layer friction (Wolanksi and Elliott, 2015), then any impact of lowered salinity has wide-reaching ecological effects.

Desalination plants discharge salinities above 37 and reduce both the germination and growth of young Mediterranean *Posidonia oceanica* (Fernández-Torquemada and Sánchez-Lizaso, 2013), and there is a predicted impact threshold above salinity 39.3 (Gacia et al., 2007). Hypersalinity can remove seed coats of the sea grass *Cymodocea nodosa* found in shallow Mediterranean and adjoining Atlantic coasts (Caye et al., 1992). Germination naturally occurs in late Spring to early Summer (Pirc et al., 1986) and whereas at 38 ‰ (ambient salinity for the Mediterranean) the seed coats are not removed, stopping germination, at 20 ‰ shedding of the seed coat is stimulated. Artificial removal of seed coats at 38 ‰ did not prompt germination (Caye et al., 1992). Reduced salinity will therefore facilitate germination in *Cymodocea nodosa* and the seagrass's ecology depends ultimately on seasonally reduced salinity. The sea grass *Zostera marina* also depends on reduced salinity for successful germination with the optimum (laboratory) salinity being 5–10 ‰ (Coolidge Churchill, 1983). Anthropogenic (unnatural) brine discharges in areas containing these species therefore can cause severe ecological impacts.

Anthropogenic salinity change is not always detrimental to local ecology. Anthropogenic salt pans/salterns have long produced salt for both industrial and food commercial purposes. These shallow, often warm-water pools provide unique and important habitat and feeding areas for many species of waterbird, often migratory, threatened, and/or endangered species (Takekawa et al., 2001; Geslin et al., 2002; Pedro and Ramos, 2009). In South Africa, large waders and water birds are attracted to ponds of lower salinity which contain benthic macrofauna similar to that of adjoining estuaries, and smaller shorebirds are attracted to the higher salinity ponds where amphipods and chironomid larvae proliferate (Velasquez, 1992). The Guérande salt pans in France represent a major breeding site for the bluethroat (*Luscinia svecica*), and management of the pans has enhanced the breeding success of the species, which is considered endangered in Europe (Geslin et al., 2002). In the Ria de Averio, Portugal, the wading bird dunlin (*Calidris alpina*) over-winters on both natural mudflats and artificial salt pans. Those that winter on the natural intertidal mudflats feed only at low tide, but those that winter on the salt pans feed during both low and high tide, suggesting either a lower food resource or possibly a greater feeding period (Luís et al., 2002).

Conversely, anthropogenic activities can also result in reduced salinity. As well as the well-understood point sources of dilution, such as storm drain outfalls and similar, there are also larger-scale potential effects. Recent climate predictions have suggested decreases in the salinity of marine ecosystems (Houghton et al., 2001), likely through increased precipitation and polar and glacial melt. For example, according to Houghton et al. (2001), precipitation has increased by 0.5 to 1% per decade in the twentieth century over most mid- and high latitudes of the Northern Hemisphere continents, and it is likely that rainfall has increased by 0.2 to 0.3% per decade over the tropical (10 °N to 10 °S) land areas. Furthermore, there has been widespread glacial retreat in the twentieth century in non-polar regions, and northern hemisphere sea-ice extent has decreased by up to 15% since 1950, with a 40% decrease in thickness during summer in recent decades. All of these mean that freshwater inputs to the environment have increased.

Marine effects of sea level rise and salinity changes due to climate change are difficult to predict as both are known to affect oceanic circulation. However, the direction of the North Atlantic Oscillation has been hypothesized to affect outbreaks of marine disease in the Caribbean (Hayes et al., 2001) and climate change in general is expected to have an impact on disease risk for marine biota (Harvell et al., 2002), although in four species of harmful North Sea algae, reducing salinity from 31 to 29 showed no significant changes to their growth rates (Peperzak, 2003). There are also potential effects on species distributions from reductions in salinity such as the large-scale loss of taxa seen in places like the Baltic Sea where the more saline reaches of the sea support more taxa than the inner, more freshwater reaches. This is thought to have occurred through climatic change (Bonsdorff, 2006); hence there could be a large-scale loss of taxa in regions that suffer increased precipitation through climate changes. Elliott et al. (2015) illustrate the linked effects of climate change on the physico-chemical processes in estuarine and coastal areas including the changes in salinity due to climate change effects; they emphasize that while there

is a good conceptual knowledge of such changes and the repercussions on the biota, there is the need for more long-term evidence to quantify such effects.

9.3 Summary

Changing salinity is an environmental master factor in the distribution of marine and estuarine species and thus it creates the assemblage structure and functioning. Some species are adapted to tolerate fluctuating environmental salinity such as those in estuarine environments, and hence they capitalize on such conditions, producing large populations. In contrast, many stenohaline marine species are limited by their physiological tolerances.

The removal of species, changes in their behaviour, or limitations to reproduction and germination can have far-reaching effects on the wider ecology of certain habitats, by altering the structure and functioning of the ecosystem. The effects of changing salinity on the ecology of different habitats is driven ultimately by the underlying physiology of organisms and how well their physiological systems can cope with salinity fluctuations in both long- and short-term scales.

There are natural and anthropogenically created habitats which experience wide variations in salinity or consistent hyper- or hyposaline conditions. Each of these will create its own characteristic community even though some of the conditions are only present for a short time. However, anthropogenic and climate-driven changes to water balances and the resultant effects on salinity will influence the ecological structure and community functioning.

Salinity acts synergistically and antagonistically with other environmental stressors in affecting ecology. Factors such as nutrient availability, temperature, pH, and dissolved oxygen have all been shown potentially to interact to produce a dynamic and changing environment where one factor can exacerbate or negate to some extent the detrimental effects of another. Human activities and anthropogenic effluents can modify local salinity regimes and their effects depend on the assimilative capacity of both the area and the species. Future studies into the effects of salinity on ecological processes need to take into account some of these factors to ensure a holistic representation of natural environmental functioning.

References

Anati, D. (1999). The salinity of hypersaline brines: concepts and misconceptions. *International Journal of Salt Lake Research*, **8**, 55–70.

Anger, K. (1991). Effects of temperature and salinity on the larval development of the Chinese mitten crab *Eriocheir sinensis* (Decapoda: Grapsidae). *Marine Ecology Progress Series*, **72**, 103–110.

Anger, K. (1996). Salinity tolerance of the larvae and first juveniles of a semiterrestrial grapsid crab, *Armases miersii* (Rathbun). *Journal of Experimental Marine Biology and Ecology*, **202**, 205–223.

Attrill, M.J. (2002). A testable linear model for diversity trends in estuaries. *Journal of Animal Ecology*, **71**, 262–269.

Attrill, M.J. & Rundle, S.D. (2002). Ecotone or ecocline: ecological boundaries in estuaries. *Estuarine, Coastal and Shelf Science*, **55**, 929–936.

Attrill, M. & Thomas, R. (2006). Long-term distribution patterns of mobile estuarine invertebrates (Ctenophora, Cnidaria, Crustacea: Decopoda) in relation to hydrological parameters. *Marine Ecology Progress Series*, **143**, 25–36.

Basset, A., Barbone, E., Elliott, M., et al. (2013). A unifying approach to understanding transitional waters: fundamental properties emerging from ecotone ecosystems. *Estuarine, Coastal and Shelf Science*, **132**, 5–16.

Bérest, P., Bergues, J., Brouard, B., et al. (2001). A salt cavern abandonment test. *International Journal of Rock Mechanics and Mining Sciences*, **38**, 357–368.

Berglund, A. (1985). Different reproductive success at low salinity determines the estuarine distribution of two *Palaemon* prawn species. *Holarctic Ecology*, **8**, 49–52.

Biebl, R. & McRoy, C.P. (1971). Plasmatic resistance and rate of respiration and photosynthesis of *Zostera marina* at different salinities and temperatures. *Marine Biology*, **8**, 48–56.

Bonsdorff, E. (2006). Zoobenthic diversity-gradients in the Baltic Sea: continuous post-glacial succession in a stressed ecosystem. *Journal of Experimental Marine Biology and Ecology*, **330**, 383–391.

Bonsdorff, E. & Pearson, T.H. (1999). Variation in the sublittoral macrozoobenthos of the Baltic Sea along environmental gradients: a functional-group approach. *Australian Journal of Ecology*, **24**, 312–326.

Branch, G.M. & Cherry, M.I. (1985). Activity rhythms of the pulmonate limpet *Siphonaria capensis* Q&G as an adaptation to osmotic stress, predation and wave action. *Journal of Experimental Marine Biology and Ecology*, **87**, 153–168.

Bulger, A.J., Hayden, B.P., Monaco, M.E., et al. (1993). Biologically-based estuarine salinity zones derived from a multivariate analysis. *Estuaries*, **16**, 311–322.

Burford, M.A., Revill, A.T., Palmer, D.W., et al. (2011). River regulation alters drivers of primary productivity along a tropical river-estuary system. *Marine and Freshwater Research*, **62**, 141–151.

Caye, G., Bulard, C., Meinesz, A., & Loquès, F. (1992). Dominant role of seawater osmotic pressure on germination in *Cymodocea nodosa*. *Aquatic Botany*, **42**, 187–193.

Chan, B.K.K. (2000). Diurnal physico-chemical variations in Hong Kong rock pools. *Asian Marine Biology*, **17**, 43–54.

Chícharo, L. & Chícharo, M.A. (2006). Applying the ecohydrology approach to the Guadiana estuary and coastal areas: lessons learned from dam impacted ecosystems. *Estuarine, Coastal and Shelf Science*, **70**, 1–2.

Coolidge Churchill, A. (1983). Field studies on seed germination and seedling development in *Zostera marina* L. *Aquatic Botany*, **16**, 21–29.

Crain, C., Stillman, B., Bertness, S., & Berntness, M. (2004). Physical and biotic drivers of plant distribution across estuarine salinity gradients. *Ecology*, **85**, 2539–2549.

Croghan, P.C. (1958). The mechanism of osmotic regulation in *Artemia salina* (L.): the physiology of the branchiae. *Journal of Experimental Biology*, **35**, 219–233.

Davenport, J.D. (1985). Osmotic control in marine animals. In: Laverack, M.S. (ed.), *Physiological Adaptations of Marine Animals*. Society of Experimental Biology Symposium 39, pp. 207–244. The Company of Biologists Ltd, Cambridge, UK.

de Brauwere, A., de Brye, B., Blaise, S., & Deleersnijder, E. (2011). Residence time, exposure time and connectivity in the Scheldt Estuary. *Journal of Marine Systems*, **84**, 85–95.

Dillaman, R.M. & Roer, R.D. (1980). Carapace repair in the green crab, *Carcinus maenas* (L.). *Journal of Morphology*, **163**, 135–155.

Dusseault, M.B., Bachu, S., Davidson, B.C. (2001). *Carbon dioxide sequestration potential in salt solution caverns in Alberta, Canada*. Technical paper, Solution Mining Research Institute Fall 2001, Technical Meeting, Albuquerque, New Mexico, USA, October 8–9–10, pp. 1–10.

Eicken, H. (1992). The role of sea ice in structuring Antarctic ecosystems. *Polar Biology*, **12**, 3–13.

Elliott, M., Borja, Á., McQuatters-Gollop, A., et al. (2015). *Force majeure*: will climate change affect our ability to attain Good Environmental Status for marine biodiversity? *Marine Pollution Bulletin*, **95**, 7–27.

Elliott, M. & Hemingway, K. (2002). *Fishes in Estuaries*. Blackwell Science Ltd, Oxford.

Elliott, M. & Quintino, V. (2007). The estuarine quality paradox, environmental homeostasis and the difficulty of detecting anthropogenic stress in naturally stressed areas. *Marine Pollution Bulletin*, **54**, 640–645.

Elliott, M. & Whitfield, A. (2011). Challenging paradigms in estuarine ecology and management. *Estuarine, Coastal and Shelf Science*, **94**, 306–314.

Fernández-Torquemada, Y. & Sánchez-Lizaso, J.L. (2013). Effects of salinity on seed germination and early seedling growth of the Mediterranean seagrass *Posidonia oceanica* (L.) Delile. *Estuarine, Coastal and Shelf Science*, **119**, 64–70.

Fernández-Torquemada, Y., Sánchez-Lizaso, J.L., & González-Correa, J.M. (2005). Preliminary results of the monitoring of the brine discharge produced by the SWRO desalination plant of Alicante (SE Spain). *Desalination*, **182**, 395–402.

Gacia, E., Invers, O., Manzanera, M., et al. (2007). Impact of the brine from a desalination plant on a shallow seagrass (*Posidonia oceanica*) meadow. *Estuarine, Coastal and Shelf Science*, **72**, 579–590.

Gajardo, G.M. & Beardmore, J.A. (2012). The brine shrimp *Artemia*: adapted to critical life conditions. *Frontiers in Physiology*, **3**, 185.

Garton, D.W. (1984). Relationship between multiple locus heterozygosity and physiological energetics of growth in the estuarine gastropod *Thais haemastoma*. *Physiological Zoology*, **57**, 530–543.

Geslin, T., Lefeuvre, J.C., Le Pajolec, Y., et al. (2002). Salt exploitation and landscape structure in a breeding population of the threatened bluethroat (*Luscinia svecica*) in salt-pans in western France. *Biological Conservation*, **107**, 283–289.

Gómez-Gesteira, M., deCastro, M., & Prego, R. (2003). Dependence of the water residence time in Ria of Pontevedra (NW Spain) on the seawater inflow and the river discharge. *Estuarine, Coastal and Shelf Science*, **58**, 567–573.

Hahn, S., Rodolfo-Metalpa, R., Griesshaber, E., et al. (2012). Marine bivalve shell geochemistry and ultrastructure from modern low pH environments: environmental effect versus experimental bias. *Biogeosciences*, 9, 1897–1914.

Haines, P.E., Tomlinson, R.B., & Thom, B.G. (2006). Morphometric assessment of intermittently open/closed coastal lagoons in New South Wales, Australia. *Estuarine, Coastal and Shelf Science*, **67**, 321–332.

Harvell, C.D., Mitchell, C.E., Ward, J.R., et al. (2002). Climate warming and disease risks for terrestrial and marine biota. *Science*, **296**, 2158–2162.

Hayes, M., Bonaventura, J., Mitchell, T., et al. (2001). How are climate and marine biological outbreaks functionally linked? *Hydrobiologia*, **460**, 213–220.

Horner, R., Ackley, S.F., Dieckmann, G.S., et al. (1992). Ecology of sea ice biota. *Polar Biology*, **12**, 417–427.

Houghton, J.T., Ding, Y., Griggs, D.J., et al. (2001). *Climate Change 2001: The Scientific Basis*. Contribution of Working Group I to the Third Assessment Report of the Intergovernmental Panel on Climate Change. Cambridge University Press, Cambridge.

Hugget, J. & Griffiths, C. (1986). Some relationships between elevation, physico-chemical variables and biota of intertidal rock pools. *Marine Ecology Progress Series*, **29**, 189–197.

Imsland, A.K., Foss, A., Gunnarsson, S., et al. (2001). The interaction of temperature and salinity on growth and food conversion in juvenile turbot (*Scophthalmus maximus*). *Aquaculture*, **198**, 353–367.

Jensen, A.C. (1918). Some observations on *Artemia gracilis*, the brine shrimp of Great Salt Lake. *Biological Bulletin*, **34**, 18–32.

Karleskint, P., Turner, R., & Small, J. (2009). *Introduction to Marine Biology*, 3rd edition. Brooks/Cole, California, USA.

Kinne, O. (1960). Growth, food intake, and food conversion in a euryplastic fish exposed to different temperatures and salinities. *Physiological Zoology*, **33**, 288–317.

Kinne, O. (1964). The effects of temperature and salinity on marine and brackish water animals. II. Salinity and temperature combinations. Oceanography and Marine Biology - An Annual Review., 2, 281–339.

Kinne, O. (1971). *Marine Ecology. A Comprehensive, Integrated Treatise on Life in Oceans and Coastal Waters. Volume 1: Environmental Factors, Part 2*. Wiley Interscience, John Wiley and Sons Ltd, London, New York.

Laverack, M. (1985). *Physiological Adaptations of Marine animals*. Society of Experimental Biology Symposium 39. The Company of Biologists Ltd, Cambridge, UK.

Legendre, L., Ackley, S.F., Dieckmann, G.S., et al. (1992). Ecology of sea ice biota. *Polar Biology*, **12**, 429–444.

Lewis, E. (1980). The practical salinity scale 1978 and its antecedents. *IEEE Journal of Oceanic Engineering*, **5**, 3–8.

Lewis, E. & Perkin, R. (1980). The Practical Salinity Scale 1978: fitting the data. *IEEE Journal of Oceanic Engineering*, **5**, 9–16.

Li, H., Russell, N., Sharifi, V., & Swithenbank, J. (2011). Techno-economic feasibility of absorption heat pumps using wastewater as the heating source for desalination. *Desalination*, **281**, 118–127.

Luís, A., Goss-Custard, J.D., & Moreira, M.H. (2002). The feeding strategy of the dunlin (*Calidris alpina* L.) in artificial and non-artificial habitats at Ria de Aveiro, Portugal. *Hydrobiologia*, **475–476**, 335–343.

Mantel, L.H. & Farmer, L.L. (1983). Osmotic and ionic regulation. In: Mantel, L.H. (ed.), *The Biology of Crustacea. Volume 5: Internal Anatomy and Physiological Regulation*, pp. 53–161. Academic Press, New York.

Marais, J.F.K. (1982). The effects of river flooding on the fish populations of two eastern Cape estuaries. *South African Journal of Zoology*, **17**, 96–104.

Marlow, J.J., Steele, J.A., Ziebis, W., et al. (2014). Carbonate-hosted methanotrophy represents an unrecognized methane sink in the deep sea. *Nature Communications*, **5**. doi 10.1038/ncomms6094

Mazik, K., Hitchman, N., Quintino, V., et al. (2013). Sublethal effects of a chlorinated and heated effluent on the physiology of the mussel, *Mytilus edulis* (L): a reduction in fitness for survival? *Marine Pollution Bulletin*, **77**, 123–131.

McAllen, R. & Taylor, A. (2001). The effect of salinity change on the oxygen consumption and swimming activity of the high-shore rockpool copepod *Tigriopus brevicornis*. *Journal of Experimental Marine Biology and Ecology*, **263**, 227–240.

McLusky, D. & Elliott, M. (2004). *The Estuarine Ecosystem. Ecology, Threats and Management*, 3rd edition. Oxford University Press, Oxford.

Meerganz von Medeazza, G.L. (2005). Direct and socially-induced environmental impacts of desalination. *Desalination*, **185**, 57–70.

Miller, N., Chern, X., & Stillman, J. (2014). Metabolic physiology of the invasive clam, *Potamocorbula amurensis*: the interactive role of temperature, salinity, and food availability *PLoS ONE*, **9**, e91604.

Miri, R. & Chouikhi, A. (2005). Ecotoxicological marine impacts from seawater desalination plants. *Desalination*, **182**, 403–410.

Monsen, N.E., Cloern, J.E., Lucas, L.V., & Monismith, S.G. (2002). A comment on the use of flushing time, residence time, and age as transport time scales. *Limnology and Oceanography*, **47**, 1545–1553.

Moriarty, F. (1993). *Ecotoxicology: The Study of Pollutants in Ecosystems*, 2nd edition. Academic Press Ltd, London.

Morris, S. & Taylor, A.C. (1983). Diurnal and seasonal variation in physico-chemical conditions within intertidal rock pools. *Estuarine, Coastal and Shelf Science*, **17**, 339–355.

Nelson, S.G., Armstrong, D.A., Knight, A.W., & Li, H.W. (1977). The effects of temperature and salinity on the metabolic rate of juvenile *Macrobrachium rosenbergii* (Crustacea: Palaemonidae). *Comparative Biochemistry and Physiology Part A*, **56**, 533–537.

Pedro, P. & Ramos, J.A. (2009). Diet and prey selection of shorebirds on salt pans in the Mondego estuary, western Portugal. *Ardeola*, **56**, 1–11.

Peperzak, L. (2003). Climate change and harmful algal blooms in the North Sea. *Acta Oecologica*, **24**(Suppl. 1), S139–S144.

Péqueux, A. (1995). Osmotic regulation in crustaceans. *Journal of Crustacean Biology*, **15**, 1–60.

Perera, K.A.R.S., Amarasinghe, M.D., & Somaratna, S. (2013). Vegetation structure and species distribution of mangroves along a soil salinity gradient in a micro tidal estuary on the north-western coast of Sri Lanka. *American Journal of Marine Science*, **1**, 7–15.

Petrich, C. & Eicken, H. (2010). Growth, structure and properties of sea ice. *Sea Ice*, **2**, 23–77.

Pirc, H., Buia, M.C., & Mazzella, L. (1986). Germination and seedling development of *Cymodocea nodosa* (Ucria) Ascherson under laboratory conditions and 'in situ'. *Aquatic Botany*, **26**, 181–188.

Potter, I.C., Chuwen, B.M., Hoeksema, S.D., & Elliott, M. (2010). The concept of an estuary: a definition that incorporates systems which can become closed to the ocean and hypersaline. *Estuarine, Coastal and Shelf Science*, **87**, 497–500.

Quintino, V., Rodrigues, A.M., Freitas, R., & Ré, A. (2008). Experimental biological effects assessment associated with on-shore brine discharge from the creation of gas storage caverns. *Estuarine, Coastal and Shelf Science*, **79**, 525–532.

Raventos, N., Machpherson, E., & García-Rubíes, A. (2006). Effect of brine discharge from a desalination plant on macrobenthic communities in the NW Mediterranean. *Marine Environmental Research*, **62**, 1–14.

Reid, D., Abelló, P., Kaiser, M., & Warman, C. (1997). Carapace colour, inter-moult duration and the behavioural and physiological ecology of the shore crab *Carcinus maenas*. *Estuarine, Coastal and Shelf Science*, **44**, 203–211.

Remane, A. (1934). Die Brackwasserfauna. *Verhandlungen Der Deutschen Zoologischen Gesellschaft*, **36**, 34–37.

Richards, F.A. & Corwin, N. (1956). Some oceanographic applications of recent determinations of the solubility of oxygen in sea water. *Limnology and Oceanography*, **1**, 263–267.

Romeril, M.G. (1977). Heavy metal accumulation in the vicinity of a desalination plant. *Marine Pollution Bulletin*, **8**, 84–87.

Rysgaard, S., Bendtsen, J., Delille, B., et al. (2011). Sea ice contribution to the air–sea CO_2 exchange in the Arctic and Southern Oceans. *Tellus B*, **63**, 823–830.

Schubart, C.D. & Deisel, R. (1999). Osmoregulation and the transition from marine to freshwater and terrestrial life. *Archives of Hydrobiology*, **145**, 331–347.

Shi, J.Q. & Durucan, S. (2005). Stockage du CO_2 dans les cavernes et les mines. *Oil & Gas Science and Technology*, **60**, 569–571.

Shirley, T.C. & Stickle, W.B. (1982a). Responses of *Leptasterias hexactis* (Echinodermata: Asteroidea) to low salinity. *Marine Biology*, **69**, 155–163.

Shirley, T.C. & Stickle, W.B. (1982b). Responses of *Leptasterias hexactis* (Echinodermata: Asteroidea) to low salinity. *Marine Biology*, **69**, 147–154.

Smyth, K. (2011). Effects of hypersalinity on the behaviour, physiology and survival of commercially important North Sea crustaceans. PhD thesis, Institute of Estuarine and Coastal Studies, University of Hull.

Smyth, K., Mazik, K., & Elliott, M. (2014). Behavioural effects of hypersaline exposure on the lobster *Homarus gammarus* (L) and the crab *Cancer pagurus* (L). *Journal of Experimental Marine Biology and Ecology*, **457**, 208–214.

Souza, A.T., Ilarri, M.I., Campos, J., et al. (2011). Differences in the neighborhood: structural variations in the carapace of shore crabs *Carcinus maenas* (Decapoda: Portunidae). *Estuarine, Coastal and Shelf Science*, **95**, 424–430.

Spivak, E.D. & Cuesta, J.A. (2009). The effect of salinity on larval development of *Uca tangeri* (Eydoux, 1835) (Brachyura: Ocypodidae) and new findings of the zoeal morphology. *Scientia Marina*, **73**, 297–305.

Stephens, D. & Gardener, J. (2007). *Great Salt Lake Utah*. Report by the United States Geological Survey. http://pubs.usgs.gov/wri/wri994189/PDF/WRI99-4189.pdf.

Styrishave, B., Rewitz, K., & Andersen, O. (2004). Frequency of moulting by shore crabs *Carcinus maenas* (L.) changes their colour and their success in mating and physiological performance. *Journal of Experimental Marine Biology and Ecology*, **313**, 317–336.

Sverdrup, H., Johnson, M., & Fleming, R. (1942). *The Oceans: Their Physics, Chemistry, and General Biology*. Prentice-Hall, New York.

Takekawa, J., Lu, C., & Pratt, R. (2001). Avian communities in baylands and artificial salt evaporation ponds of the San Francisco Bay estuary. In: Melack, J., Jellison, R., & Herbst, D. (eds), *Saline Lakes*, pp 317–328. Springer Netherlands, Dordrecht.

Ter Morshuizen, L.D., Whitfield, A.K., & Paterson, A.W. (1996). Influence of freshwater flow regime on fish assemblages in the great fish river and estuary. *Southern African Journal of Aquatic Sciences*, **22**, 52–61.

Thomas, D. & Dieckmann, G. (2002). Antarctic sea ice—a habitat for extremophiles. *Science*, **295**, 641–644.

Uncles, R.J. (2010). Physical properties and processes in the Bristol Channel and Severn Estuary. *Marine Pollution Bulletin*, **61**, 5–20.

UNESCO (1985). *The International System of Units (SI) in Oceanography*. UNESCO Technical Papers No. 45, IAPSO Pub. Sci. No. 32. Paris, France.

Van Dover, C.L. & Fry, B. (1994). Microorganisms as food resources at deep-sea hydrothermal vents. *Limnology and Oceanography*, **39**, 51–57.

Velasquez, C.R. (1992). Managing artificial saltpans as a waterbird habitat: species' responses to water level manipulation. *Colonial Waterbirds*, **15**, 43–55.

Vorhies, C.T. (1917). Notes on the fauna of Great Salt Lake. *American Naturalist*, **51**, 494–499.

Waldbusser, G., Voigt, E., Bergschneider, H., et al. (2011). Biocalcification in the eastern oyster (*Crassostrea virginica*) in relation to long-term trends in Chesapeake Bay pH. *Estuaries and Coasts*, **34**, 221–231.

Whitfield, A. & Elliott, M. (2011). 1.07: ecosystem and biotic classifications of estuaries and coasts. In: Wolanski, E. & McLusky, D. (eds), *Treatise on Estuarine and Coastal Science*, pp 99–124. Academic Press, Waltham.

Whitfield, A.K., Elliott, M., Basset, A., et al. (2012). Paradigms in estuarine ecology—a review of the Remane diagram with a suggested revised model for estuaries. *Estuarine, Coastal and Shelf Science*, **97**, 78–90.

Whitfield, A.K. & Harrison, T.D. (2003). River flow and fish abundance in a South African estuary. *Journal of Fish Biology*, **62**, 1467–1472.

Williams, W., Boulton, A., & Taaffe, R. (1990). Salinity as a determinant of salt lake fauna: a question of scale. In: Melack, J., Jellison, R., & Herbst, D. (eds), *Saline Lakes*, pp. 257–266. Springer Netherlands, Dordrecht.

Wolanski, E., Boorman, L.A., Chícharo, L., et al. (2004). Ecohydrology as a new tool for sustainable management of estuaries and coastal waters. *Wetlands Ecology and Management*, **12**, 235–276.

Wolanksi, E. & Elliott, M. (2015). *Estuarine Ecohydrology: An Introduction*, 2nd edition. Elsevier Science, Amsterdam.

Wright, J. & Colling, A. (1995). *Seawater: Its Composition, Properties and Behaviour*, 2nd edition. Pergamon Press, Oxford.

Xie, W., Zhang, C., Zhou, X., & Wang, P. (2014). Salinity-dominated change in community structure and ecological function of Archaea from the lower Pearl River to coastal South China Sea. *Applied Microbiology and Biotechnology*, **98**, 7971–7982.

Xu, H. (2010). Synergistic roles of microorganisms in mineral precipitates associated with deep sea methane seeps. In: Barton, L.L., Mandl, M., & Loy, A. (eds), *Geomicrobiology: Molecular and Environmental Perspective*, pp. 325–346. Springer Netherlands, Dordrecht.

Ysebaert, T., Herman, P., Meire, P., et al. (2003). Large-scale spatial patterns in estuaries: estuarine macrobenthic communities in the Schelde estuary, NW Europe. *Estuarine, Coastal and Shelf Science*, **57**, 335–355.

CHAPTER 10

The ecological consequences of marine hypoxia: from behavioural to ecosystem responses

Bettina Riedel, Robert Diaz, Rutger Rosenberg, and Michael Stachowitsch

10.1 Welcome to the Anthropocene!

The Greek philosopher Heraclitus of Ephesus is reported to have said 'nothing endures but change', and, indeed, the biological world is changing rapidly. The rapidity and magnitude of these changes are mirrored in a suggested new, currently informal, designation for this period in Earth history - the Anthropocene (Crutzen and Stoemer, 2000). This is because the pace and sheer spectrum of drivers behind changes in the biosphere can be traced to a single hub, humankind. Our fingerprint is everywhere, or to use a new catchword, our footprint is enormous and has tended to flatten everything in its path, both figuratively and literally. Our influence has gone beyond the habitats we actually inhabit to encompass the entire globe and even beyond (man-made space debris orbiting our planet).

Humans have altered marine environments and affected marine organisms from the shoreline to the deep sea. When looking at Halpern et al.'s (2008) global map of human impacts, it is difficult to find any part of the world's oceans that has not been affected. It is not only our activities in the marine environment that affect life in the sea—it's also our actions and behaviour on land. Combined effects of habitat destruction, overfishing, introduced and invasive species, warming, acidification, pollution, and massive run-off of anthropogenic nutrients are transforming complex ecosystems, clear and productive coastal seas, and complex food webs into monotonous seabeds, dead zones (areas with insufficient oxygen to sustain most metazoan life, Diaz and Rosenberg, 2008), and into simplified, microbially dominated ecosystems (Jackson, 2008).

Instead of describing how ecosystems work, marine ecologists are increasingly describing why things now work differently: nearly half of all the scientific publications in a range of subfields are devoted to describing the deterioration and suggesting solutions (Rose et al.,

2011). Our appreciation of how ecosystems are changing is complicated by the 'shifting baseline syndrome'—the problem of recognizing natural, relatively unspoilt reference habitats against which all current and future ecosystem states can be measured—addressed by Sheppard (1995). Meanwhile, however, due to our vast footprint, both spatially and temporally, experts question the existence and/or use of natural reference points today. A particular example is the long history and tremendous impact of modern industrial fisheries (since about 1880) on marine ecosystems (e.g. Callum, 2007); this also mirrors new knowledge in the field of historical ecology that humans significantly altered many coastal marine environments centuries and even millennia ago (Rick and Erlandson, 2008). Oxygen deficiencies have become a major global stress factor to estuarine and marine environments, based on their potential to act as a primary driver of ecosystem collapse and on the rapidity of their effect. The present contribution summarizes key findings to provide an overview of the far-reaching consequences of anthropogenically induced deoxygenation in our seas, with special focus on coastal systems.

10.2 Coastal hypoxia and anoxia: turning 'normal' into 'extreme' habitats

'When you can't breathe, nothing else matters.'

This old motto by the American Lung Association expresses best the seriousness of oxygen depletion as an environmental issue (Diaz, 2001). Oxygen is fundamental for sustaining aerobic life, and despite the discoveries of numerous subsystems that bypass aerobic pathways, this life physically and biologically structures most marine habitats. Thus, the worldwide increase in hypoxic (low dissolved oxygen DO; in coastal bottom waters traditionally ≤ 2 ml l^{-1}, but see Section 10.2.2 for discussion on different hypoxia thresholds)

Stressors in the Marine Environment. Edited by Martin Solan and Nia M. Whiteley
© Oxford University Press 2016. Published in 2016 by Oxford University Press.

and anoxic (0 ml DO l⁻¹) areas in coastal marine ecosystems over the past decades (Diaz and Rosenberg, 2008) is an alarming signal. It underlines the loss of marine biodiversity and complex and profound changes in ecosystem properties/functions (e.g. productivity, decomposition rates, nutrient cycling, and resistance or resilience to perturbations) and services (e.g. climate regulation, recreation, and food supply) (see Sala and Knowlton, 2006; Worm et al., 2006; Palumbi et al., 2008 and references therein). The signs are clear in nearshore waters and, alarmingly, now also in the open ocean (mid-water oxygen minimum zones, OMZs; Stramma et al., 2010; see Section 10.2.1). Ongoing eutrophication and/or climate change is predicted to further exacerbate the situation in the future (Rabalais et al., 2009).

Although some areas are already experiencing the worst-case scenario many other of the world's coastal ecosystems may increasingly be approaching tipping points (Fig. 10.1). The result will be a new set of extreme habitats (Stachowitsch et al., 2012). Traditional 'extreme' habitats such as deep-sea hydrothermal vents or hot springs are typically characterized by a single inhibitive parameter at the far end of the natural range (e.g. temperature, salinity, pH, or hydrogen sulphide). Such parameters, however, are typically remain stable here over long periods of time, i.e. decades to millennia enabling adaptation by the often complex communities inhabiting them (e.g. Wharton, 2002).

In contrast, the 'off-on' or 'all-or-nothing' property of typical, recurring oxygen depletion with a comparatively shorter duration (i.e. days to months) makes affected areas more unstable than 'extreme' habitats: little adaptation is possible and a minimalistic set of opportunistic and tolerant species tend to dominate.

10.2.1 Hypoxia: occurrence, development, and crux of the matter

Life on Earth emerged under anaerobic conditions, and low-oxygen environments are ancient phenomena: the modern oxic world apparently represents an exceptional state of the atmosphere–ocean system on our planet (Strauss, 2006). Nonetheless, the most recent geological eon, the Phanerozoic (542 mya to the present; rise of the metazoan world), is also marked by a long series of widespread oxygen deficiencies in the global ocean or parts of it. These were often associated with mass extinction events, for example the Late Permian mass extinction (Wignall and Twitchett, 1996), series of short-lived abiotic recoveries related to shifts back to ocean anoxia in the Triassic (Grasby et al., 2012), or multiple black shale depositions during the Cretaceous. The duration of such oxygen deficiency episodes ranged from a few thousand to millions of years (see references in Gooday et al., 2009, on both geological and more recent historical records).

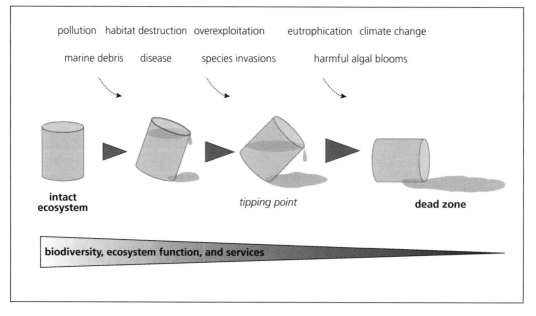

Figure 10.1 The interactive and cumulative effects of multiple stressors increasingly force marine coastal systems to a critical tipping point and qualitatively altered state.

Low oxygen waters also occur naturally in the world's oceans today. Oxygen minimum zones (OMZs; DO concentration < 0.5 ml l^{-1}), on the one hand, are widespread and stable (periods greater than decadal scales) low-oxygen features at intermediate depths (typically 100–1200 m) in the open ocean, intercepting the continental margins primarily along the eastern Pacific, off southwest Africa, in the Arabian Sea and in the Bay of Bengal (Levin, 2003; Rabalais et al., 2010). OMZs cover about 30 million km^2 of open ocean, but their depth and thickness of occurrence, i.e. the upper and lower depth boundaries, are variable and depend on natural processes and cycles, e.g. circulation and oxygen content of the respective ocean region (Helly and Levin, 2004). The principal formation drivers are high surface and photic (light) zone productivity, a limited circulation, and old water masses (Levin et al., 2009; Rabalais et al., 2010): the bacterial degradation of decaying organic matter that sinks from the productive shallow oceans to the deep increases the oxygen consumption and thus decreases oxygen concentration in the water column, while the poor water exchange restricts oxygen supply from surface waters. Often OMZs are associated with western boundary upwelling regions where oxygen-depleted but nutrient-rich water fuels primary production—the primary contributor to persistent hypoxia under upwelling regions. Where oxygen minima impinge on the seafloor, organisms have adapted their behaviour, morphology, and/or their metabolism/enzymatic processes over geological time to either permanently or temporarily (diel vertical migration) survive at (for more information see Childress and Seibel, 1998; Levin, 2003; Levin et al., 2009). The temporary encounter with hypoxic waters from OMZs or upwelling systems in coastal areas (e.g. off the coast of Chile, west Africa, or India), however, causes similar effects as human-induced coastal hypoxia (see Section 10.3) with severe impacts on local organisms (e.g. Levin et al., 2009 and references therein).

Most coastal oxygen depletions (critical DO concentration ≤ 2 ml l^{-1} versus < 0.5 ml l^{-1} in OMZs), on the other hand, are smaller, variable in size, seasonal (for other duration classes and coastal hypoxic threshold concentrations see Section 10.2.2) and typically associated with a particular bottom topography and/or hydrogeomorphology. Combined with a density or thermal stratification of the water column because of freshwater discharge from rivers (pycnocline: transition layer that separates less saline surface water from more saline/denser bottom water) or heating of the ocean surface layer during calm and warm periods (thermocline), this restricts water exchange and prevents oxygen-rich surface waters from mixing to the bottom. The duration of the stagnation is decisive: when oxygen consumption rates of bottom organisms exceed those of resupply, hypoxia and subsequent anoxia develop. Examples of coastal areas susceptible to the formation of oxygen depletion include shallow and semi-enclosed seas, estuaries, deep fjords, and silled basins (e.g. fjords in the Skagerrak area, northern Europe, Rosenberg, 1980; central parts of the Baltic Sea, Zillén et al., 2008).

In contrast to what occurs in OMZs, much of the coastal oxygen depletion has developed in the last 50 years and is closely associated with anthropogenic activities. Coastal hypoxia was a rather localized problem until the 1950s, generally associated with sewage discharges or industrial outflows in areas with a semi-enclosed hydrogeomorphology. Since the 1960s, however, reported cases of hypoxic sites have doubled in each decade (Fig. 10.2) and, increasingly, shallow continental seas were affected (e.g. Gulf of Mexico and Kattegat) (Diaz, 2001; Diaz and Rosenberg, 2008). Increasing population growth, intensified development, and expansion of agricultural activities (e.g. increasing use of industrially produced fertilizers), especially in the temperate zone, led to a twofold increase in the global flux of nitrogen (N; Galloway et al., 2004) and a two- to threefold greater phosphorus flux (P; Howarth et al., 1995) to coastal waters. This excess nutrient input to aquatic ecosystems (eutrophication) stimulates primary production and enhances the flow of organic water to the seafloor, fuelling microbial respiration and amplifying oxygen depletion (Rabalais et al., 2002). Importantly, shallow water communities are, as opposed to those in OMZs, less adapted to low DO and have, except for a few, highly resistant species, much higher oxygen tolerance thresholds (see Section 10.2.2). Anthropogenic eutrophication has multipronged impacts. On the one hand it fuels positive feedback loops such as hypoxia-induced enhanced ammonium (NH$_4^+$) and phosphate fluxes to the overlying water, which in turn sustain high rates of primary production (changes in biogeochemical processes upon hypoxia are summarized in Kemp et al., 2009, or Middelburg and Levin, 2009). On the other hand, the cycle of decomposition of dead organic material fuels microbial respiration, which causes further mortalities. Thus, eutrophication of coastal seas has the potential to exacerbate conditions in areas already predisposed to oxygen depletion or to tip previously unaffected systems into hypoxia (e.g. Conley et al., 2009).

Several studies have demonstrated a correlation over time between anthropogenic eutrophication and the occurrence of hypoxia/anoxia in coastal areas. The shallow (< 50 m), semi-enclosed Northern Adriatic, Mediterranean Sea, provides a good example for this connection. Oxygen depletions, often associated with massive marine snow events (decaying organic material that sinks from higher in the water column), have been noted here periodically for centuries (Danovaro et al., 2009 and references therein), but their frequency and severity have markedly increased in the second half of the twentieth century due to high anthropogenic input of nutrients via the Po River, Italy. In the Northern Adriatic, echoing the global phosphorus flux noted above, the inputs of N and P increased by five and three times, respectively, between 1945 and 1985. This was reflected in an average long-term decrease in water body transparency (by 4–6 m), an increase in surface-water oxygen concentrations produced by phytoplankton blooms, and decreasing bottom oxygen concentrations (Justić et al., 1987; Justić, 1988). The result has been severe bottom oxygen deficiencies, with impacted areas ranging from several km² (Stachowitsch, 1992) to approximately 4000 km² (Stefanon and Boldrin, 1982).

Early this century, dead zones appeared on top of the list of emerging environmental challenges (UNEP, 2004), and, in 2008, more than 400 marine systems worldwide (total area > 245 000 km², i.e. larger than the United Kingdom) had been identified as being eutrophication-related hypoxic (Diaz and Rosenberg, 2008; Fig. 10.2). The latest global assessment has brought the total to over 600 coastal areas affected (Diaz et al., 2013). Hotspots of oxygen deficiencies include Chesapeake Bay and the Gulf of Mexico, USA, Scandinavian and Baltic waters, and the Black Sea, as well as Chinese/Korean/Japanese waters (e.g. UNEP, 2006; STAP, 2011; Diaz et al., 2013).

Single areas have reached enormous dimensions. The Baltic Sea and the Gulf of Mexico, for example, harbour the two largest anthropogenic hypoxic zones worldwide, covering a mean area of 49 000 and 17 000 km² (maximum 67 000 km²/22 000 km²), respectively (Turner et al., 2008; Savchuk, 2010). At the other end of the range are small-scale hypoxias in unexpected locations. In McMurdo Sound (Southern Antarctic), for example, the untreated sewage from the US Antarctic research centre led to organic enrichment in the 1990s with subsequent benthos reduction (Conlan et al., 2004); recently, hypoxia most likely triggered by climate change was reported from the Eastern Antarctic (Powell et al., 2012). These polar examples point to

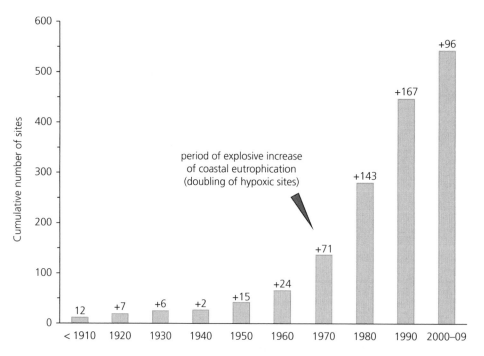

Figure 10.2 Cumulative increase of hypoxia in coastal areas over time reported in the scientific literature
Adapted from Diaz and Rosenberg (2008).

a close match between the distribution of dead zones and the widening global human footprint. It is now in east and south Asia where the United Nations Environment Programme (UNEP) expects the steepest increase in the number of coastal hypoxic sites in the upcoming years due to population growth and increases in the magnitude and changes in nutrient ratios in river export (STAP, 2011).

Climate change can make marine systems more susceptible to the development of hypoxia due to direct effects on the water column stratification, precipitation patterns, and temperature. Gilbert et al. (2010), for example, determined a faster decline in oxygen rates over the last three decades within a 30 km band near the coast than in the open ocean (> 100 km from shoreline) between 0 and 300 m water depth. However, there is also evidence that oxygen concentrations in the global open ocean are declining as well (Helm et al., 2011). Stramma et al. (2010), for example, reported horizontally and vertically expanding OMZs in the tropical Pacific, Atlantic, and Indian Oceans during the past few decades. This has been attributed most likely to lower oxygen solubility due to surface-layer warming and the weakening of ocean overturning and convection (Keeling et al., 2010). Recently, oxygen decrease and shoaling of the oxic–hypoxic boundary has been observed on the southern Californian shelf (Bograd et al., 2008). Off the Oregon coast, previously unreported hypoxia has been documented on the inner shelf since 2000 (Chan et al., 2008). Importantly, other stressors such as rising temperatures and ocean acidification will team up and create synergistic effects with deoxygenation, exacerbate the frequency, duration, and intensity of both naturally and eutrophication-related hypoxia (Gruber, 2011), and amplify the stress on resident organisms in both the coastal and open ocean (e.g. Harley et al., 2006; Stramma et al., 2010; Bopp et al., 2013; for examples see Section 10.4). Thus, understanding hypoxia and its effects on ecosystems requires several perspectives, starting at the local level, moving to regional scale, and finally to a global perspective (Diaz et al., 2013).

10.2.2 Definitions and thresholds: we can agree to disagree

The terminology and units used to define hypoxia and anoxia, as well as the conventional definition of when a critical oxygen concentration is reached, vary between and within earth and life sciences and their subdisciplines (Rabalais et al., 2010). This has led to the selected use of a wide range of terms (e.g. oxic, dysoxic, suboxic, anoxic; aerobic, dysaerobic, quasi-aerobic, anaerobic; hypoxic, normoxic), depending on whether these expressions are applied to environments, biofacies, or the physiological responses of living organisms, and whether additional stress factors are involved, such as the development of sulphidic (euxinic) conditions (Tyson and Pearson, 1991).

Aquatic ecologists traditionally determine the onset of hypoxic conditions as the point when most organisms become physiologically stressed and/or show sublethal avoidance reactions (see Section 10.3). Breitburg et al. (2009), for example, define hypoxia mechanistically as 'oxygen concentrations that are sufficiently reduced that they affect the growth, reproduction, or survival of exposed animals, or result in avoidance behaviors'. The traditional critical oxygen concentration defining hypoxia in shallow coastal seas and estuaries is thereby often set at at ≤ 2 ml l^{-1}, equivalent to 2.8 mg DO l^{-1} or 91.4 µM (see for example Diaz and Rosenberg, 1995; Wu, 2002). Others determine hypoxia at DO ≤ 2 mg l^{-1} (1.4 ml DO l^{-1} or 63 µM; e.g. Rabalais et al., 2001; Gray et al., 2002; Vaquer-Sunyer and Duarte, 2008). For the most recent compilation of thresholds, ranges, and technical terms, see Table 1 in Altenbach et al. (2012). The sensitivity of organisms, however, varies with physical factors such as temperature or salinity, and with biological/taxon-specific factors such as life habits, mobility, life cycle stage, or physiological adaptations to hypoxia or tolerance to hydrogen sulphide (H$_2$S) (e.g. Hagerman, 1998; Vaquer-Sunyer and Duarte, 2010, 2011). Moreover, the duration, intensity, extent, and frequency of hypoxia/anoxia can also modulate the critical oxygen range limits of organisms.

According to the temporal scale, hypoxia types are classified into (from Diaz and Rosenberg, 1995, 2008):

- Seasonal: yearly events lasting from weeks to months, typically associated with summer or autumnal water column stratification (e.g. Chesapeake Bay Mainstem, USA, Officer et al., 1984; Gulf of Mexico shelf, USA, Rabalais et al., 2007)
- Periodic: hours to days, i.e. diel cycles in shallow estuaries and tidal creeks caused by 24-h cycles of water column photosynthesis and respiration, or lunar neap–spring tidal cycles modulating the stratification intensity (e.g. York River estuary, USA, Nestlerode and Diaz, 1998; Delaware coastal bays, USA, Tyler et al., 2009)
- Episodic: infrequent events at intervals > 1 year, prevailing for weeks, often the first indication that an

ecosystem is severely stressed (e.g. Gullmarsfjord, Sweden, Nilsson and Rosenberg, 2000), or

• Permanent: years to decades to centuries (e.g. parts of the Baltic Sea, Conley et al., 2011).

A meta-analysis by Vaquer-Sunyer and Duarte (2008) shows the enormous variability in oxygen thresholds for lethal responses to oxygen depletion among benthic organisms: median lethal concentration (LC_{50}) thresholds ranged from 8.6 mg DO l^{-1} for the first larval zoea stage of the crustacean *Cancer irroratus*, to persistent resistance to anoxia for the oyster *Crassostrea virginica* at a temperature of 20 °C. Accordingly, no 'conventional' threshold is universally applicable or can fully cover the range of biological/environmental stages or situations. In fact, 'biological stress' occurs well above the traditional oxygen limits noted above, with effects of hypoxia on growth and behaviour observed at oxygen concentrations near 6.0 mg l^{-1} and impacts on other metabolic components near 4 mg DO l^{-1} (Gray et al., 2002). Vaquer-Sunyer and Duarte (2008) determined, in 90% of the 872 experiments analysed, a median lethal oxygen concentration at < 4.59 mg DO l^{-1}.

10.3 Impact of hypoxia in coastal environments: the ecological implications

Responses at the molecular level, leading to physiological and metabolic adaptions, are the initial organismic responses to low oxygen concentrations. Then, cascading effects, direct and indirect, can trigger changes at levels ranging from behaviour, growth, recruitment, species diversity, biological interactions, trophic dynamics, and community structure to sediment geochemistry and habitat complexity (e.g. summarized in Diaz and Rosenberg, 1995; Wu, 2002; Gray et al., 2002; Middelburg and Levin, 2009; Levin et al., 2009) (Fig. 10.3). Every component of the marine environment is affected—fauna and flora, pelagos and benthos, from bacteria to macrofauna organisms. Hypoxia therefore threatens biodiversity and alters ecosystem structure and function (Solan et al., 2004; Worm et al., 2006). The response to oxygen depletion can go beyond replacing a given community with another one (i.e. species replaced by others with similar ecological roles). In the worst-case scenario, mass mortalities change formerly vital and rich coastal ecosystems into taxonomically and functionally depauperated, microbial ones (Sala and Knowlton, 2006). Finally, apart from the ecological consequences, hypoxia also negatively influences the social and economic activities related to ecosystem services provided by marine ecosystems, e.g. tourism and fisheries (e.g. STAP, 2011; Diaz et al., 2013).

Here we present the ecological consequences of hypoxia/anoxia at the various levels. Importantly, these biological responses act in concert and therefore always require a holistic approach in both understanding and responding to this phenomenon.

10.3.1 Individual- and population-level effects

Environmental change or stress, even if it influences the performance and fitness of organisms, is not inherently

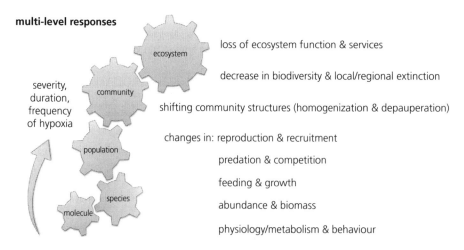

Figure 10.3 Cascading effects of oxygen depletion, triggering individual responses which in turn affect population attributes, community dynamics, and ultimately biodiversity and ecosystem integrity.

bad. It is part of life and in fact a motor of evolution. Disturbances in environmental conditions at certain levels and frequencies (e.g. intermediate disturbance hypothesis; Connell, 1978) can set positive impulses, but in most other cases the outcome for the resident species and communities will be negative (e.g. Pearson and Rosenberg, 1978; Harley et al., 2006; Halpern et al., 2008).

In terms of hypoxia, marine organisms have evolved two general and non-exclusive coping mechanisms. Firstly, they can respond with physiological adaptions. These can include lowered metabolism, effective use of respiratory pigments, increased ventilation rate and/or blood flow, and switch to anaerobic metabolism (Hagerman, 1998). These strategies are treated separately in Chapter 2 of this volume. Secondly, and generally the first visible sign of defence, animals can change their behaviour to either increase oxygen supply or decrease oxygen demand. The sensitivity, mobility, and tolerance of marine organisms determine their exposure and reaction to intermittent or chronic hypoxia. Depending on the animal's lifestyle, such behaviour primarily involves avoidance strategies: 'run for your lives' (migration of more mobile fauna to a normoxic refuge). If this is not possible, it's 'only the strong survive' (with specific behavioural responses of sessile and less mobile fauna to reduce hypoxic stress, Fig. 10.4).

Migration to more oxygenated areas can occur both vertically and horizontally. The former mirrors the oxygen gradient in the water column, the latter the

areal extent affected. Accordingly, the distances may range from centimetres to metres to kilometres. Meiofauna, for example, typically responds with a decrease in burial depth (e.g. Alve and Bernhard, 1995) and sometimes even floats up into the water column until normoxic conditions are re-established (Wetzel et al., 2001). Macroinfauna typically emerges onto the sediment surface. Well-known examples of this behaviour include sea urchins (i.e. *Echinocardium cordatum*: Nilsson and Rosenberg, 1994; *Schizaster canaliferus*: Riedel et al., 2014), polychaetes (Pihl et al., 1992), bivalves (i.e. *Solecurtus strigilatus, Laevicardium* sp., *Ensis ensis*: Hrs-Brenko et al., 1994), and burrowing crustaceans (i.e. the mantis shrimp *Squilla mantis*: Stachowitsch, 1992). The latter were even observed swimming in the water column (Stachowitsch, 1984). Note that the mantis shrimp and other commercially important burrowing crustaceans such as *Nephrops norvegicus* normally emerge onto the sediment surface only during the night (and are then fished), but under hypoxic conditions are also present on the sediment during the daytime (Baden et al., 1990).

For less mobile epifauna, unable to escape or avoid hypoxic waters, hypoxia typically induces a range of species-specific behavioural changes that involve changing the body shape or position. This includes raising the respiratory structures higher in the water column to avoid low oxygen concentrations near the sediment surface (i.e. arm-tipping brittle stars,

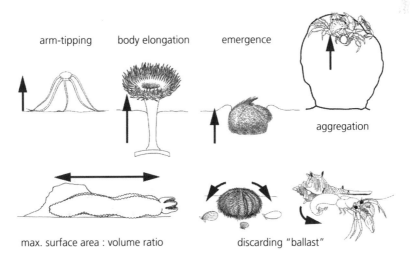

arm-tipping   body elongation   emergence

aggregation

max. surface area : volume ratio   discarding "ballast"

Figure 10.4 Typical avoidance responses of sessile and less mobile benthic macrofauna to hypoxia. Upper left to lower right: arm-tipping brittle star, body elongation in sea anemones, emergence from sediment by infaunal sea urchins, aggregation of cryptic species (e.g. small crabs) on elevated substrate, increase in surface area:volume ratio in holothurians, epifaunal sea urchins discard protecting/camouflaging material (i.e. shells and debris), and hermit crabs leave their protective shells.

siphon-stretching bivalves, or tiptoeing crustaceans; e.g. review by Diaz and Rosenberg, 1995), or maximizing the surface area:volume ratio to minimize the diffusion distances within tissues and improve oxygen diffusion (i.e. sea anemones: Shick, 1991). Organisms may try to reach any elevated substrate above the sediment surface (i.e. bioherm-associated crabs such as *Pilumnus spinifer* or *Pisidia longicornis* emerge from hiding and aggregate on top of sponges and ascidians; Stachowitsch, 1984; Haselmair et al., 2010). This is because the oxygen gradients on the seafloor are often very steep, and moving only a few centimetres can mean the difference between tolerable and lethal conditions. Other behaviours may serve to increase mobility and/or reduce oxygen demand by discarding 'additional ballast' when exposed to environmental stress. The decorator crab *Ethusa mascarone* and the sea urchin *Psammechinus microtuberculatus*, for example, discarded their protecting camouflage when confronted with low DO concentrations (Haselmair et al., 2010; Riedel et al., 2014). Similarly, hermit crabs emerge from their shells and move about fully exposed (Stachowitsch, 1984; Pretterebner et al., 2012).

Numerous laboratory and field observations worldwide underline the similarity of such key behavioural responses to hypoxia. Thus, the responses of representatives with certain life habits in one system provide a window into comparable interpretations of benthic health/status elsewhere. This would be one further new and intriguing aspect to the much-discussed early concept of 'parallel level-bottom communities' (Thorson, 1958). Arm-tipping in ophiuroids is a case in point. The same posture has been observed for *Amphiura chiajei*, *A. filiformis*, and *Ophiura albida* in the Kattegat (Rosenberg et al., 1991; Vistisen and Vismann, 1997), *Ophiura texturata* in the North Sea (Dethlefsen and von Westernhagen, 1983), *Ophiothrix quinquemaculata* in the Northern Adriatic (Stachowitsch, 1984; Riedel et al., 2014), and brittle stars in the Gulf of Mexico (Rabalais et al., 2001).

The more mobile taxa (mostly fish and crustaceans) can escape to surface waters overlying the bottom hypoxic layer or to the margins of the hypoxic zone. In Tolo Harbour, Hong Kong, for example, hypoxia occurs seasonally (strong stratification during the wet season, monsoon) and most benthic epifauna escapes to more oxygenated open waters in summer and returns again in winter (Fleddum et al., 2011). Sometimes a migration is so extensive that it leads to a (not mortality-related) defaunation of a population in the hypoxic site, as reported for the blue crab *Callinectes sapidus* and flounders *Paralichthys dentatus* and *P. lethostigma* escaping from chronic and episodic (upwelling events) hypoxia in the Neuse River Estuary, North Carolina, USA (Bell and Eggleston, 2005). In the Gulf of Mexico, in the absence of physical barriers to movement, evading organisms (i.e. brown shrimp *Farfantepenaeus aztecus* and several finfishes) aggregated just beyond (1–3 km) the margins of the hypoxic zone (2.0 mg l^{-1} contour; Craig, 2012). This example points to another, indirect lethal effect of hypoxia, namely the enhanced susceptibility of brown shrimp and non-target finfishes to the commercial shrimp trawl fishery. This effect is probably most intense within a relatively narrow region along the hypoxic edge (Craig, 2012), mirroring a more general observation that fishers often detect hypoxia/anoxia events based on unusual catches of species normally not caught together (Stefanon and Boldrin, 1982). Elsewhere, fish come to the surface and into shallowest waters. In Mobile Bay, Alabama, in the Gulf of Mexico, for example, such hypoxia-induced aggregation events have historically been termed 'jubilees' because local residents could collect the still-living fish in knee-deep water along the beach (Loesch, 1960).

Importantly, while avoidance reactions may help organisms to escape or survive hypoxia longer, they rarely promote individual performance or population structure. Unusual and unexpected migrations (or altered, typically extended, migration paths), for example, can quickly reduce an organism's energy reserves, especially if accompanied by reduced food acquisition. Potentially altered abiotic factors (i.e. temperature or salinity) or food availability in the new refuge area may further lower body condition and fitness. In the Columbia River Estuary on the Oregon–Washington border, USA, the Dungeness crab *Cancer magister* avoided, i.e. migrated away from, water with < 50% oxygen saturation and reduced its food intake in such hypoxic waters; this effect, coupled with reduced salinity (the animals are weak osmoregulators), caused the crabs to reduce feeding, ingestion, and pumping water over their gills, impairing performance and growth (Roegner et al., 2011).

Any change or shift in the spatial distribution of benthic organisms—whether it be migration to a new region, shallower burial depth, or exposure atop an elevated substrate—may adversely affect survival and render stressed fauna more vulnerable to predators (and human exploitation). Apart from fishers, who may change their fishing patterns to take advantage of faunal aggregations at the margins of hypoxic zones (e.g. Tolo Harbour, Hong Kong; Fleddum et al., 2011), natural predators with a higher tolerance to low DO conditions than prey organisms (and/or that can swim faster) can also exploit the moribund fauna in hypoxic areas (Pihl et al., 1992). Sometimes, unusual behaviours that do not involve migration or full exposure to predators can also increase predation pressure. Well-known examples include abnormal extension of siphons or of the palps of bivalves and polychaetes above the sediment surface, which are then bitten off by foraging predators (e.g. Sandberg et al., 1996). It sometimes pays to look upwards as well: vertical escape to near-surface waters can also increase the predation risk from diving birds (see references in Domenici et al., 2007). However, it is not always the predator that can benefit from hypoxia. Oxygen deficiencies may also create predation refuges for more tolerant prey species, e.g. enhanced prey survivorship of more hypoxia-tolerant quahog clam *Mercenaria mercenaria*, softshell clam *Mya arenaria*, and blue mussel *Mytilus edulis* from less tolerant predators such as crabs, the sea star *Asterias forbesi*, or the oyster drill *Urosalpinx cinerea* (Altieri, 2008). An intermediate scenario also exists when both predator and prey are affected. Here, relative tolerance will govern predation efficiency (Breitburg et al., 1994). In the Northern Adriatic, for example, more tolerant anemones were able to utilize a 'window of opportunity' during sinking oxygen concentrations to feed on moribund brittle stars (Riedel et al., 2008, 2014). A weakened intraspecific interaction strength can also increase predation-induced mortality. For example, altered fish school structure (e.g. increasing school volume, i.e. space between individual fish) and dynamics (e.g. changing the shuffling behaviour within the school) may affect school integrity in terms of synchronization and execution of antipredator manoeuvres (Domenici et al., 2002). Thus, different tolerance levels and changes in both predator and prey behaviour at low DO may modify their relationships and potentially alter the population structure of communities.

Hypoxic effects can also manifest themselves as reduced feeding and growth (fish: Brandt et al., 2009; diatom *Skeletonema costatum*: Wu et al., 2012) and can lower reproductive capacity (crustacean: Brown-Peterson et al., 2008). In most aquatic species, early life history stages are the most sensitive in a life cycle—one reason why the use of early life stages (of fish, for example) has been suggested in establishing water-quality criteria or screening chemicals. Reduced survival in larvae and juveniles under hypoxia (e.g. the oyster *Crassostrea virginica*: Widdows et al., 1989; Baker and Mann, 1992) can shift the overall population age in favour of adult life stages, affecting the sustainability of

a population. In Bornholm Basin, Southern Baltic, for example, only adult individuals of the bivalve *Astarte borealis* were found after a decade-long stagnation and largely hypoxic conditions (Leppäkoski, 1969); in Kiel Bay a similar age-class shift was observed for *Astarte elliptica*, *Mya arenaria*, and *Cyprina islandica* (Rosenberg, 1980).

Tolerance to hypoxia may be also based upon allometric issues, with smaller individuals in and between species reported to be more tolerant of hypoxia (e.g. in yellow perch *Perca flavescens*, Robb and Abrahams, 2003). Nonetheless, the organism size (or weight) effect on hypoxia tolerance is mixed. Shimps et al. (2005), for example, investigated how vulnerability to hypoxia changes across fish size in spot (*Leiostomus xanthurus*) and Atlantic menhaden (*Brevoortia tyrannus*). While for the Atlantic menhaden, smaller fish were consistently more tolerant to hypoxia than larger fish, the converse effect was observed for spots (i.e. larger individuals were more vulnerable than their smaller conspecifics). The authors attribute this to a mix of mechanisms and therefore call for an approach integrating lethal and sublethal effects, direct and indirect effects, and fish behaviour. Other studies reported no correlation between hypoxia tolerance and size (e.g. Wagner et al., 2001). Thus, size can affect hypoxia tolerance, but the effect may be species-specific. Note also that the presence of smaller individual sizes, both of single species and of entire communities, can also reflect early stages of growth during the recovery process after mass mortalities.

10.3.2 Community-level effects

Changes at the community level emerge from processes operating at the individual level. Here, the impact of hypoxia and anoxia on a particular species triggers cascading series of direct and indirect effects on the overall biota, which by definition are linked in one way or the other (O'Gorman et al., 2011). Thus, beyond reducing growth, disrupting life cycles, altering behaviour, and changing biological interactions, hypoxia may significantly reduce abundance, biomass, and diversity in marine ecosystems. As noted above, there is a high intra- and interspecific variability in tolerance to hypoxia, with some individuals/taxa already experiencing mortality at an oxygen concentration where others do not even show visible signs of stress (Vaquer-Sunyer and Duarte, 2008; Riedel et al., 2012; see Section 10.3.1). In a generalised pattern of community response, mild (approximate

DO concentration > 1 ml l^{-1}) and short-term hypoxia may therefore primarily involve quantitative losses—in the sense of punctuated, species-specific mortalities—rather than fundamentally altered overall community structure and composition. Longer-lasting or more intense hypoxia, however, threatens to depauperize and functionally homogenize benthic communities (Sala and Knowlton, 2006); in a worst-case scenario, the macrofauna is eliminated entirely (Stachowitsch, 1984).

Going from species to next higher taxonomic levels, distinct patterns of sensitivity/tolerance emerge. The literature reveals, as a broad generalization, that fish are the most sensitive group, followed by crustaceans, polychaetes, echinoderms, sea anemones, molluscs, hydro-/scyphozoans, and ascidians (Vaquer-Sunyer and Duarte, 2008; Riedel et al., 2012). The broad consensus is that:

- larval stages are more vulnerable to low oxygen concentrations than their corresponding adults (Gray et al., 2002);
- macrofauna is more sensitive than meiofauna (Josefson and Widbom, 1988);
- epifauna is more vulnerable than infauna because tolerance to low oxygen and sulphide in marine species is closely correlated to the *in situ* levels they are generally exposed to (Vistisen and Vismann, 1997); and
- mobile forms are more vulnerable than sessile forms, in line with their apparently decreased tolerance for both physiological and physical stress (Sagasti et al., 2001).

The effects on the individual and species level—ranging from benthic invertebrates to fishes—change the complex interactions between species and can derail community dynamics. Even if the benthic community does not suffer extensive mortality, functionally important biological interactions such as trophic dynamics may be fundamentally altered (see above). Sturdivant et al. (2014), for example, assessed the effects of seasonal hypoxia on macrobenthic production in Chesapeake Bay and its three main tributaries from 1996 to 2004. The results showed 90% lower macrobenthic production during hypoxia (≤ 2 mg DO l^{-1}) relative to normoxia. The biomass loss of ca. 7000–13 000 metric tons of organic carbon over an area of 7700 km^2 was estimated to equate to a 20–35% displacement of the Bay's macrobenthic productivity during the summer. While the higher consumers may benefit from easy access to stressed prey in some areas, the large spatial and temporal extent of hypoxia markedly limits macrobenthic production and thus transfer to higher trophic levels. The decrease in macrobenthic production may be particularly detrimental during critical periods when epibenthic and demersal predators have high-energy demands. A more detailed picture of how oxygen depletion can alter the relative importance of different trophic pathways within a community is provided by Munari and Mistri (2011). The authors studied the effect of short-term hypoxia (24 h, < 2 mg DO l^{-1}) on predation by the muricid gastropod *Rapana venosa* (hypoxia-tolerant) on three bivalve species: *Scapharca inaequivalvis* (adapted to hypoxic conditions with anaerobic metabolism), *Tapes philippinarum* (responding with a decrease in burial depth and siphon extension), and *Cerastoderma glaucum* (hypoxia-sensitive). Under normoxia, the gastropod had a marked preference for *S. inaequivalvis*. Under hypoxia, however, the predator switched to the more easy-to-catch *T. philippinarum*. Thus, hypoxia facilitates the coexistence of two more-hypoxia-tolerant bivalves because the predator switches prey. The hypoxia-sensitive *C. glaucum* is 'the net loser of the game': it is predated—if not preferentially—under normoxia, and becomes moribund/dies at hypoxia.

To minimize competition with superior individuals or to avoid being targeted by predators, (prey) species generally use avoidance strategies or maintain a safe distance. With ongoing duration of oxygen depletion, however, these intra- and interspecific interaction strengths can weaken. The unusual, dense aggregation of organisms on top of elevated substrates as a hypoxia-avoidance mechanism is a good example. The crab *Pilumnus spinifer*, for example, is territorial and preferentially preys upon small adult crabs and brittle stars in the Adriatic. Target organisms usually avoid the sponges and ascidians inhabited by the crab or flee. The almost uniform avoidance response to hypoxia, i.e. climbing on higher substrates to reach more oxygenated areas, however, diminishes this safe distance within and between species. It apparently outweighs normal agonistic behaviour and predator–prey relationships (Haselmair et al., 2010).

Many natural communities are characterized by, and depend on, a single or a functional group of habitat-forming foundation species (ecosystem engineers) that provide the framework for the entire community (Crain and Bertness, 2006). Stachowitsch (1984) provides an example of how the loss of one such species can trigger a cascade of secondary mortalities with major impacts on community functioning and stability: in the northeastern Adriatic, the benthic macroepifauna largely consists of interspecific aggregations termed multi-species clumps (Fedra et al., 1976). Their presence and distribution depends on the initial presence of shelly material for the settlement of larval invertebrates. Here, the settled sessile filter- and suspension feeders (mostly sponges, ascidians, or bivalves) serve as an elevated substrate for additional mobile and hemi-sessile organisms, i.e. brittle stars, holothurians, and decapods. These 'biodiversity

hotspots' form a discrete community from the surrounding soft sediment habitat and provide substrate for larval settlement and epigrowth, shelter, and food. The mortality of key component species, such as a large sponge, accelerates mortality of the associated fauna in a positive feedback loop during hypoxia/anoxia (Fig. 10.5). Such rapid mortality stands in contrast to the recovery, which can be delayed for decades even after significant increases in bottom water oxygen concentrations (e.g. Northern Adriatic: Stachowitsch,

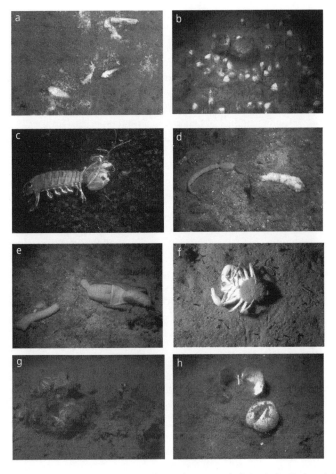

Figure 10.5 a–h Behavioural responses and mortality of benthic macrofauna after oxygen depletion in the Northern Adriatic Sea, Mediterranean: (a) emerged and dead organisms on beach including from lower left to upper right flatfish, burrowing shrimp, gobiid, and bivalve; (b) emerged bivalves (*Corbula gibba*) on the sediment, in the centre dead crab (carapace) and large bivalve; (c) nocturnal mantis shrimp *Squilla mantis* emerges during day and also swims into water column—note dark colour of the sediment; (d) emerged, moribund sipunculid and dead and decomposing gobiid fish; (e) emerged bivalve with cast-off siphon; (f) dead female swimming crab with eggs—multi-generational impact of hypoxia; (g) 'Black spot' indicates remains of former multi-species clump—note empty sea urchin test; (h) post-anoxia condition—bivalve and sea urchin test as potential substrate for future epigrowth. Photos: M. Stachowitsch, except for f (department photo archive, author unknown). Time-lapse films showing the effects of oxygen depletion on benthic macrofauna during and after experimentally induced hypoxia available at: http://phaidra.univie.ac.at/o:87923 and http://phaidra.univie.ac.at/o:262380 [PLATE 9]

1991; case studies from the Adriatic Sea, the Black Sea, Danish estuaries, and Delaware Bay, USA, summarized in Steckbauer et al., 2011). Community development is further impaired by harmful fishing activities, which remove or destroy this benthic 3D complexity, and by renewed oxygen depletions. The net result is marked longer-term community degradation (i.e. biodiversity loss) and decreased habitat complexity. Such interactions between habitat/community degradation and hypoxia have also been observed in other biogenic habitats. Oyster beds, for example, create physically complex habitats, important to estuarine biodiversity and fishery production. Deeper parts of the beds can experience hypoxia events, whereby mass mortality of the bivalves impacts the abundance, distribution, and diversity of associated fish and invertebrates (Lenihan and Peterson, 1998). Hypoxia-induced habitat degradation can in turn impact remote, undisturbed surrounding habitats (i.e. through the movement and abnormal concentration of refugee organisms that have subsequent strong trophic impacts, Lenihan et al., 2001), triggering a domino effect of widespread ecosystem degradation.

Finally, lower-diversity communities (including reduced competition for resources and habitat in defaunated areas) provide a niche for more rapidly growing, opportunistic colonizers (e.g. Pearson and Rosenberg, 1978). They are also less resistant to invasive species; in the Chesapeake Bay, for example, the cover by the tunicates *Botryllus schlosseri* and *Molgula manhattensis*, the polychaete *Ficopomatus enigmaticus*, and the anemone *Diadumene lineata*—all four now found worldwide—was highest on settling plates exposed to moderately low oxygen (DO 2-4 mg l^{-1}) (Jewett et al., 2005).

10.3.3 Ecosystem-level effects

The increasing intensity of human disturbance in productive coastal marine ecosystems is triggering unprecedented habitat and biodiversity loss. This raises concerns about whether ecosystem function and ecosystem services provided to humanity can be maintained under such severe disturbance (UNEP, 2006).

The complexity of marine ecosystem structure and dynamics is based on the interactions of the component species, the food web connections across trophic levels, and the landscape modifications induced by biotic–abiotic interactions. Hypoxia affects all three levels. Intuitively, high diversity maintains high complexity of interactions and feedbacks among species, promoting stability and resistance to invasion or other forms of disturbance. Often, however, ecosystem performance

may depend more on the presence of key functional types than on species richness itself (Thrush et al., 2006). This calls for examining the potential effect of hypoxia on key functional groups, e.g. the ecosystem engineers mentioned above.

Ecosystem engineers alter their physical surroundings or change the flow of resources through their presence or activity (e.g. feeding or burrowing), thereby creating or modifying habitats and influencing all associated species. They therefore significantly contribute to and influence ecosystem biodiversity, functions, and stability. The role of such ecosystem engineers may persist on time scales longer than their individual lifetimes. On structurally less complex soft-bottom surfaces, for example, dominant sessile organisms can create new habitat enabling the establishment of entire epifaunal communities that would otherwise be unable to persist. Examples include the above-mentioned bivalve/oyster beds, coral reefs, the Northern Adriatic sponge-, bivalve-, and ascidian-dominated multi-species clumps, or tube-building polychaetes. Among the infauna, abundant bioturbators—such as the irregular sea urchin *Schizaster canaliferus* in the Northern Adriatic—may determine the 3D complexity in the sediment (Schinner et al., 1997).

Benthic suspension feeders play a key role in the energy transfer, nutrient (re)cycling, and sedimentation or resuspension of particulate organic matter (POM) in coastal ecosystems. In shallow systems with moderate water exchange, the feedback processes between the benthic and the pelagic subsystems are more immediate. Such a quicker and more direct interaction, specifically that involving nutrient exchanges, has been termed benthic–pelagic coupling (reviewed by Graf, 1992). When the benthic and pelagic subsystems closely adjoin, the macrobenthos—consisting largely of filter- and suspension-feeders such as sponges, bivalves, ascidians, tube worms, and brittle stars—directly controls the pelagic biomass and production through grazing. Depending on water depth, they can filter the entire volume of a basin within days to weeks (e.g. 3 d, Laholm Bay, Sweden, Loo and Rosenberg, 1989; 20 d, Gulf of Trieste, Northern Adriatic, Ott and Fedra, 1977). In the context of hypoxia/anoxia, such benthic communities play a regulatory role that has been termed a 'natural eutrophication control' (Officer et al., 1982). In removing most of the suspended material in the overlying water column, they convert pelagic biomass into benthic biomass with a lower respiration/biomass ratio, thus stabilizing the entire system and controlling water quality (Hily, 1991). Moreover, biodeposits (i.e. faeces, pseudofaeces, and sedimented POM) are a source of food for bacteria and meio- and macrofauna organisms, promoting secondary production in the benthos and increasing the nutrient turnover (Graf, 1992). Most suspension feeders are relatively long-lived, with a low mobility, and characterized by spatial variability but temporal stability. The loss of such stabilizing compartments due to oxygen crises unbalances ecosystem dynamics and makes the whole system more vulnerable to additional perturbations.

Hypoxia and the corresponding loss of biodiversity also significantly affect bioturbation, another major process conducted by benthic ecosystem-engineering macrofauna (Sturdivant et al., 2012). Bioturbation includes all transport processes and their physical effects on the substrate, along with particle reworking and burrow ventilation (reviewed by Kristensen et al., 2012). Bioturbators affect the sediment permeability and water content, break up chemical gradients in pore water, and subduct organic matter. This, in turn, influences organismic biomass, remineralization rates, and the inorganic nutrient efflux—all vital for primary production. The burrowing and ventilation activities lead to a complex mosaic of reduced and oxidized zones in the sediment, substantially affecting biogeochemical properties and processes (for a description of the complex interactions of biogeochemical cycles that accompany eutrophication and hypoxia, see Middelburg and Levin, 2009).

Key functional processes translate into positive and negative effects cascading from bacteria/microalgae, to meiofauna, and macrofauna/-algae, potentially extending in the food chain to fish and birds (Kristensen et al., 2012). Eutrophication-induced die-offs of benthic suspension feeders and bioturbators disrupt important benthic–pelagic fluxes and significantly impair ecosystem function and services (Solan et al., 2004). Ultimately, broad-scale losses may trigger a vicious cycle via accumulation of organic material on the sea floor, increased microbial decomposition, decreased bottom water DO, nutrient release from the sediment to water column, accelerated primary production, and—back to step one—increased flow of organic material to the sea floor.

Diversity loss (i.e. reduced number of genes, species, or functional groups) due to oxygen depletion is determined by multiple factors ranging from sensitivity to low DO to the presence of additional stressors. The multiple layers of interactions and feedbacks, hidden drivers, and emergent properties, however, make the consequences of species loss for ecosystem function difficult to predict. High-diversity communities, comprising a wide range of functional traits, can efficiently capture biologically relevant resources (e.g. nutrients, light, or prey), produce biomass, decompose, and recycle essential nutrients. The loss of biodiversity may initially only minimally impact ecosystem function: when some individuals of one species die, then individuals of another species but same functional trait can fill the gap (quantitative loss but compensatory response). Ultimately, however, the rates of change will accelerate. When dynamics are shaped by engineering species, a few strong interactions dominate (Ellison et al., 2005). Here, the loss of even a single key species can immediately and profoundly affect other species and ecosystem processes, whereby the order in which species are lost is also important (Solan et al., 2004). Systems with little functional redundancy are therefore relatively fragile and susceptible to even small perturbations: the loss of key engineers is likely to lead to rapid, possibly irreversible shifts in biodiversity and to system-wide changes in structure and function (Ebenman and Jonsson, 2005). Importantly, substituting one species with another from the same functional guild does not imply uninterrupted, full efficiency: ecosystem structure and functional processes may weaken or change long before the engineering compartment itself disappears completely. The suspension-feeding capacity is a case in point. A particular species typically filters a certain range of particle sizes. When a species is lost, that particle size range may no longer be filtered out, which—from the view of the overall filter-feeding capacity—represents a qualitative change in function (i.e. altered spectrum of particle sizes filtered out by remaining species) (Riedel et al., 2012).

Hypoxia can therefore alter community composition, influence overall ecosystem properties and ultimately trigger a regime shift (an often irreversible shift between two alternate stable environmental states) (e.g. Conley et al., 2009). Suspension feeders might be replaced by deposit feeders, macrobenthos by meiobenthos, and bioturbators may be lost, causing functional homogenization at the community level (Thrush et al., 2006). In a study on eight million years of fluctuating hypoxia during the Late Jurassic, for example, Caswell and Frid (2013) report that less intense hypoxia was associated with significant changes in species composition but not in biological traits, implying the retention of ecological function. In periods of extremely different oxygen conditions, however, traits suggestive of opportunists (e.g. more surface-living and shallow-burrowing species, and thinner skeletons) and altered function occurred. More generally, under hypoxic stress Wu (2002) suggests a general shift from K-selected (i.e. large body size, long life expectancy, with few offspring) to r-selected, often opportunistic species (i.e. high fecundity, small body size, early maturity onset, and fast generation times), and from complex to simple food chains. Such scenarios represent undisputable worst-case situations for biodiversity and ecosystem function. The result is local extinction (Solan et al., 2004) and large-scale (functional) homogenization at increasingly lower levels ('microbialization'; Sala and Knowlton, 2006).

10.4 Interplay and synergy of multiple stressors

Despite the gravity of hypoxia and anoxia as stressors of marine systems, they remain only one of the many listed at the onset of this contribution. Most marine ecosystems are clearly affected by the interactive and cumulative effects—additive, synergistic, and antagonistic—of multiple human stressors (Crain et al., 2008). The multi-level effects (i.e. physiology, morphology, ecology, individual- to ecosystem-level, etc.) of even a single stressor on marine ecosystems are often difficult enough to understand. Determining the cumulative effects of multiple stressors, however, is a major challenge (but note, for example, interactions between effects of oxygen depletion and chemicals such as heavy metals or pesticides reviewed in Holmstrup et al., 2010). We restrict ourselves here to stressors that are increasingly being studied in combination with the effects of exposure to oxygen deficiency such as hydrogen sulphide (H_2S), higher temperatures, and/or a lowered pH.

Sulphate, for example, is always present in the marine anoxic sediment layers, and certain bacteria (e.g. *Beggiatoa* spp. or *Desulfovibrio* spp.) typically use organic compounds such as lactate, pyruvate, and short-chain fatty acids as energy and carbon sources to reduce sulphate to sulphide (Jørgensen and Fenchel, 1974). In general, the level and distribution of H_2S largely depends on the sediment type (i.e. muddy sediment is easily stratified and is anoxic a few millimetres below the surface, compared to well-oxygenated coarser sediments) and on the amount of organic material present. Infaunal species therefore typically encounter not only transient low dissolved oxygen conditions but also elevated H_2S levels. Some organisms have evolved physiological and metabolic adaptions to survive days or even weeks in sulphidic habitats (reviewed in Grieshaber and Völkel, 1998). For most aerobic organisms,

however, even micromolar (µMol) concentrations of hydrogen sulphide are highly toxic; they inhibit cytochrome c oxidase and impair pulmonary function and oxygen transport, negatively affecting the aerobic energy supply (Nicholls and Kim, 1982). During oxygen depletion events, the reduced layer of the sediment migrates towards the sediment surface and sulphate reduction is fuelled by the abundant dead organic material on the sea floor. This can result in a relatively rapid build-up of H_2S in the sediment and the overlying water column, reducing the survival times in marine benthic communities by an average of 30% (meta-analysis by Vaquer-Sunyer and Duarte, 2010, involving 30 macrofauna species from 10 groups). The authors also reported a higher effect of sulphide on the survival of eggs than for juvenile or adult stages. More examples of the combined effects of hypoxia, hydrogen sulphide, and/or ammonia (also acutely toxic to marine organisms) are reviewed by Gray et al. (2002).

Equally, exposure to higher temperatures may also make marine benthic organisms more vulnerable to hypoxia because of the lower oxygen solubility (with increasing temperature and salinity) but increasing metabolic rates and thus oxygen requirement (Brown et al., 2004). Juvenile Atlantic sturgeon (*Acipenser oxyrinchus*), for example, displayed a more severe mortality response to hypoxia (< 4 mg DO l^{-1}) at 26 °C (mortality 92%) compared to 19 °C (22%) (Secor and Gunderson, 1998). A correlation between hypoxia tolerance and temperature was also found in spot (*Leiostomus xanthurus*), Atlantic menhaden (*Brevoortia tyrannus*) (Shimps et al., 2005), and Atlantic cod *Gadus morhua* (Schurmann and Steffensen, 1992); in juvenile summer flounder *Paralichthys dentatus*, the combination of low DO levels (50-70% air saturation) and high temperature (30 °C) severely reduced growth rates (Stierhoff et al., 2006). A meta-analysis assessing the effects of an expected maximum temperature increase of 4 °C during the twenty-first century on hypoxia thresholds for coastal benthic macrofauna (Vaquer-Sunyer and Duarte, 2011) highlights this threat: the prediction is a decrease in survival time under hypoxia by 74% and an increase in the threshold oxygen concentrations for mortality by 25.5%. Although these predictions represent worst-case scenarios, they generate concerns for the future of coastal communities in a globally changing world where considerable warming of both air and sea is projected (IPCC, 2007).

Organisms exposed to hypoxia are commonly confronted simultaneously with acidification (hypercapnia) stress: respiration reduces oxygen, leads to elevated carbon dioxide (CO_2) levels in the water, and lowers the pH, thereby causing a significant acidosis in tissues (e.g. Burnett and Stickle, 2001). Melzner et al. (2013) recently illustrated that hypoxic coastal areas are already characterized by CO_2 partial pressure (pCO_2) values that will probably not be reached by ocean acidification in the surface ocean in the next few hundred years (i.e. depending on salinity, >1700–3200 µatm), with the potential for a 50–100% increase within this century. Consequently, during future climate change, the simultaneously acting stressors ocean warming, acidification, and deoxygenation (e.g. Gruber, 2011) will markedly increase the sensitivity to environmental extremes relative to a change in just one of these parameters (Pörtner et al., 2005). For more information on climate change impacts on marine ecosystems see Doney et al. (2012) and Poloczanska et al. (2013) and references therein; for a model approach on how multiple factors including temperature, pH, dissolved oxygen, or primary productivity may evolve in different marine regions over the course of the twenty-first century see Bopp et al. (2013).

10.5 Conclusions

Coastal hypoxia and anoxia have become a global key stressor to marine ecosystems. Prediction attempts and societal adaptation strategies critically require precisely determining and understanding key processes of oxygen depletion in marine systems (cf Zhang et al., 2010). Correlations, and in some cases clear cause-and-effects, between anthropogenic nutrient enrichment and oxygen-depleted zones have been offered for several seas. The gravity of eutrophication warrants priority status and immediate action. The problem is not primarily one of marine science, but requires interdisciplinary and international efforts along with political ability and will. Action must be taken soon to reduce nutrient inputs despite any 'impracticalities' of withdrawing specific environmentally damaging compounds from the market. We know the ultimate effects of most stressors—whether they be as different as radioactivity, eutrophication, or marine debris—and know how to mitigate them. Wetlands and other habitats that can retain nutrients coming from terrestrial sources should be restored. More holistic approaches involving Marine Protected Areas or ocean zoning are needed to make marine management more effective and improve environmental governance. Our common sense should prompt us to more definitively tackle the problem. 'Good oxygen, good life. Poor oxygen, poor life. No oxygen, no life.' Things can be be so simple sometimes.

Acknowledgements

We are grateful to anonymous reviewers for insightful comments that helped improve the quality of the chapter.

References

Altenbach, A.V., Bernhard, J.M., & Seckbach, J. (2012). Stepping into the book of anoxia and eukaryotes. xi-xvii. In: Altenbach, A.V., Bernhard, J.M., Seckbach, J. (eds). *Anoxia. Evidence for Eukaryote Survival and Paleontological Strategies*, pp. 648. Springer, Dordrecht.

Altieri, A.H. (2008). Dead zones enhance key fisheries species by providing predation refuge. *Ecology*, **89**, 2808–2818.

Alve, E. & Bernhard, J.M. (1995). Vertical migratory response of benthic foraminifera to controlled oxygen concentrations in an experimental mesocosm. *Marine Ecology Progress Series*, **116**, 137–151.

Baden, S.P., Pihl, L., & Rosenberg, R. (1990). Effects of oxygen depletion on the ecology, blood physiology and fishery of the Norway lobster *Nephrops norvegicus*. *Marine Ecology Progress Series*, **67**, 141–155.

Baker, S.M. & Mann, R. (1992). Effects of hypoxia and anoxia on larval settlement, juvenile growth, and juvenile survival of the oyster *Crassostrea virginica*. *Biological Bulletin*, **182**, 265–269.

Bell, G.W. & Eggleston, D.B. (2005). Species-specific avoidance responses by blue crabs and fish to chronic and episodic hypoxia. *Marine Biology*, **146**, 761–770.

Bograd, S.J., Castro, C.G., Di Lorenzo, E., et al. (2008). Oxygen declines and the shoaling of the hypoxic boundary in the California Current. *Geophysical Research Letters*, **35**, L12607.

Bopp, L., Resplandy, L., Orr, J.C., et al. (2013). Multiple stressors of ocean ecosystems in the 21st century: projections with CMIP5 models. *Biogeosciences*, **10**, 6225–6245.

Brandt, S.B., Gerken, M., Hartman, K.J., & Demers, E. (2009). Effects of hypoxia on food consumption and growth of juvenile striped bass (*Morone saxatilis*). *Journal of Experimental Marine Biology and Ecology*, **381**, S143–S149.

Breitburg, D.L, Hondorp, L., Davis, W., & Diaz, R.J. (2009). Hypoxia, nitrogen and fisheries: integrating effects across local and global landscapes. *Annual Review of Marine Science*, **1**, 329–350.

Breitburg, D.L., Steinberg, N., DuBeau, S., et al. (1994). Effects of low dissolved oxygen on predation on estuarine fish larvae. *Marine Ecology Progress Series*, **104**, 235–246.

Brown, J.H., Gillooly, J.F., Allen, A.P., et al. (2004). Toward a metabolic theory of ecology. *Ecology*, **85**, 1771–1789.

Brown-Peterson, N.J., Manning, C.S., Patel, V., et al. (2008). Effects of cyclic hypoxia on gene expression in a grass shrimp *Palaemonetes pugio*. *Biological Bulletin*, **214**, 6–16.

Burnett, L.E. & Stickle, W.B. (2001). Physiological responses to hypoxia. In: Rabalais, N.N. & Turner, R.E. (eds). *Coastal Hypoxia: Consequences for Living Resources and Ecosystems.*

Coastal and Estuarine Studies 58, pp. 101–114. American Geophysical Union, Washington, DC.

Callum, R. (2007). *The Unnatural History of the Sea*. Island Press, Washington, DC.

Caswell, B.A. & Frid, C.L.J. (2013). Learning from the past: functional ecology of marine benthos during eight million years of aperiodic hypoxia, lessons from the Late Jurassic. *Oikos*, 122, 1687–1699.

Chan, F., Barth, J.A., Lubchenco, J., et al. (2008). Emergence of anoxia in the California large marine ecosystem. *Science*, 319, 920.

Childress, J.J. & Seibel, B.A. (1998). Life at stable low oxygen levels: adaptations of animals to oceanic oxygen minimum layers. *Journal of Experimental Biology*, 210, 1223–1232.

Conlan, K.E., Kim, S.L., Lenihan, H.S., & Oliver, J.S. (2004). Benthic changes over ten years of organic enrichment by McMurdo Station, Antarctica. *Marine Pollution Bulletin*, 49, 43–60.

Conley, D.J., Carstensen, J., Aigars, J., et al. (2011). Hypoxia is increasing in the coastal zone of the Baltic Sea. *Environmental Science and Technology*, 45, 6777–6783.

Conley, D.J., Carstensen, J., Vaquer-Sunyer, R., & Duarte, C.M. (2009). Ecosystem thresholds with hypoxia. *Hydrobiologia*, 629, 21–29.

Connell, J.H. (1978). Diversity in tropical rain forests and coral reefs. *Science*, 199, 1302–1310.

Craig, J.K. (2012). Aggregation on the edge: effects of hypoxia avoidance on the spatial distribution of brown shrimp and demersal fishes in the Northern Gulf of Mexico. *Marine Ecology Progress Series*, 445, 75–95.

Crain, C.M. & Bertness, M.D. (2006). Ecosystem engineering across environmental gradients: implications for conservation and management. *BioScience*, 56, 211–218.

Crain, C.M., Kroeker, K., & Halpern, B.S. (2008). Interactive and cumulative effects of multiple human stressors in marine systems. *Ecology Letters*, 11, 1304–1315.

Crutzen, P.J., & Stoemer, E.F. (2000). The Anthropocene. *IGBP Newsletter*, 41, 12.

Danovaro, R., Fonda Umani, S., & Pusceddu, A. (2009). Climate change and the potential spreading of marine mucilage and microbial pathogens in the Mediterranean Sea. *PLoS ONE*, 4, e7006.

Dethlefsen, V. & von Westernhagen, H. (1983). Oxygen deficiency and effects on bottom fauna in the eastern German Bight. *Meeresforschung*, 30, 42–53.

Diaz, R.J. (2001). Overview of hypoxia around the world. *Journal of Environmental Quality*, 30, 275–281.

Diaz, R.J., Eriksson-Hägg, H., & Rosenberg, R. (2013). Hypoxia. In: Noone, K.J., Sumaila, U.R., & Diaz, R.J. (eds), *Managing Ocean Environments in a Changing Climate: Sustainability and Economic Perspectives*, pp. 67–96. Elsevier, Amsterdam.

Diaz, R.J. & Rosenberg, R. (1995). Marine benthic hypoxia: a review of its ecological effects and the behavioural responses of benthic macrofauna. *Oceanography and Marine Biology*, 33, 245–303.

Diaz, R. & Rosenberg, R. (2008). Spreading dead zones and consequences for marine ecosystems. *Science*, 321, 926–929.

Domenici, P., Ferrari, R.S., Steffensen, J.F., & Batty, R.S. (2002). The effects of progressive hypoxia on school structure and dynamics in Atlantic herring *Clupea harengus*. *Proceedings of the Royal Society of London Series B*, 269, 2103–2111.

Domenici, P., Lefrancois, C., and Shingles, A. (2007). Hypoxia and the antipredator behaviours of fishes. *Philosophical Transactions of the Royal Society Series B*, 362, 2105–2121.

Doney, S.C., Ruckelshaus, M., Duffy, J.E., et al. (2012). Climate change impacts on marine ecosystems. *Annual Review of Marine Science*, 4, 11–37.

Ebenman, B. & Jonsson, T. (2005). Using community viability analysis to identify fragile systems and keystone species. *Trends in Ecology and Evolution*, 20, 568–575.

Ellison, A.M., Bank, M.S., Clinton, B.D., et al. (2005). Loss of foundation species: consequences for the structure and dynamics of forested ecosystems. *Frontiers in Ecology and the Environment*, 3, 479–486.

Fedra, K., Ölscher, E.M., Scherübel, C., et al. (1976). On the ecology of a North Adriatic benthic community: distribution, standing crop and composition of the macrobenthos. *Marine Biology*, 38, 129–145.

Fleddum, A., Cheung, S.G., Hodgson, P., & Shin, P.K.S. (2011). Impact of hypoxia on the structure and function of benthic epifauna in Tolo Harbour, Hong Kong. *Marine Pollution Bulletin*, 63, 221–229.

Galloway, J.N., Dentener, F.J., Capone, D.G., et al. (2004). Nitrogen cycles: past, present, and future. *Biogeochemistry*, 70, 153–226.

Gilbert, D., Rabalais, N.N., Diaz, R.J., & Zhang, J. (2010). Evidence for greater oxygen decline rates in the coastal ocean than in the open ocean. *Biogeosciences*, 7, 2283–2296.

Gooday, A.J., Jorissen, F., Levin, L.A., et al. (2009). Historical records of coastal eutrophication-induced hypoxia. *Biogeosciences*, 6, 1707–1745.

Graf, G. (1992). Benthic-pelagic coupling: a benthic view. *Oceanography and Marine Biology*, 30, 149–190.

Grasby, S.E., Beauchamp, B., Embry, A., & Sanei, H. (2012). Recurrent Early Triassic ocean anoxia. *Geology*, 41, 175–178.

Gray, J.S., Wu, R.S.S., & Or, Y.Y. (2002). Effects of hypoxia and organic enrichment on the coastal marine environment. *Marine Ecology Progress Series*, 238, 249–279.

Grieshaber, M.K. & Völkel, S. (1998). Animal adaptations for tolerance and exploitation of poisonous sulphide. *Annual Review of Physiology*, 60, 33–53.

Gruber, N. (2011). Warming up, turning sour, losing breath: ocean biogeochemistry under global change. *Philosophical Transactions of the Royal Society Series A*, 369, 1980–1996.

Hagerman, L. (1998). Physiological flexibility; a necessity for life in anoxic and sulphidic habitats. *Hydrobiologia*, 375/376, 241–254.

Halpern, B.S., Walbridge, S., Selkoe, K.A., et al. (2008). A global map of human impact on marine ecosystems. *Science*, **319**, 948–952.

Harley, C.D.G., Hughes, A.R., Hultgren, K.M., et al. (2006). The impacts of climate change in coastal marine systems. *Ecology Letters*, **9**, 228–241.

Haselmair, A., Stachowitsch, M., Zuschin, M., & Riedel, B. (2010). Behaviour and mortality of benthic crustaceans in response to experimentally induced hypoxia and anoxia *in situ*. *Marine Ecology Progress Series*, **414**, 195–208.

Helly, J.J. & Levin, L.A. (2004). Global distribution of naturally occurring marine hypoxia on continental margins. *Deep Sea Res Part I*, **51**, 1159–1168.

Helm, K.P., Bindoff, N.L., & Church, J.A. (2011). Observed decreases in oxygen content of the global ocean. *Geophysical Research Letters*, **38**, L23602.

Hily, C. (1991). Is the activity of benthic suspension feeders a factor controlling water quality in the Bay of Brest? *Marine Ecology Progress Series*, **69**, 179–188.

Holmstrup, M., Bindesbøl, A.-M., Oostingh, G.J., et al. (2010). Interactions between effects of environmental chemicals and natural stressors: a review. *Science of the Total Environment*, **408**, 3746–3762.

Howarth, R.W., Jensen, H., Marino, R., & Postma, H. (1995). Transport to and processing of phosphorus in near-shore and oceanic waters. In: Tiessen, H. (ed.), *Phosphorus in the Global Environment: Transfers, Cycles, and Management*, Vol. 54, pp. 323–345. Scope, Wiley, New York.

Hrs-Brenko, M., Medakovic, D., Labura, Z., & Zahtila, E. (1994). Bivalve recovery after a mass mortality in the autumn of 1989 in the northern Adriatic Sea. *Periodicum Biologorum*, **96**, 455–458.

IPCC (2007). *Climate Change 2007: Synthesis Report*. Contribution of Working Groups I, II and III to the Fourth Assessment Report of the Intergovernmental Panel on Climate Change, Pachauri, R.K. & Reisinger, A. (eds). IPCC, Geneva, Switzerland.

Jackson, J.B. (2008). Ecological extinction and evolution in the brave new ocean. *Proceedings of the National Academy of Sciences of the USA*, **105**, 11458–11465.

Jewett, E.B., Hines, A.H., & Ruiz, G.M. (2005). Epifaunal disturbance by periodic low levels of dissolved oxygen: native vs invasive species response. *Marine Ecology Progress Series*, **304**, 31–44.

Jørgensen, B.B. & Fenchel, T. (1974). The sulfur cycle of a marine sediment model system. *Marine Biology*, **24**, 189–201.

Josefson, A.B. & Widbom, B. (1988). Differential response of benthic macrofauna and meiofauna to hypoxia in the Gullmar Fjord basin. *Marine Biology*, **100**, 31–40.

Justić, D. (1988). Trend in the transparency of the Northern Adriatic Sea 1911–1982. *Marine Pollution Bulletin*, **19**, 32–35.

Justić, D., Legović, T., & Rottini-Sandrini, L. (1987). Trends in oxygen content 1911–1984 and occurrence of benthic mortality in the Northern Adriatic Sea. *Estuarine, Coastal and Shelf Science*, **25**, 435–445.

Keeling, R.F., Körtzinge, A., & Gruber, N. (2010). Ocean deoxygenation in a warming world. *Annual Review of Marine Science*, **2**, 199–229.

Kemp, W.M., Testa, J., Conley, D.J., et al. (2009). Temporal responses of coastal hypoxia to nutrient loading and physical controls. *Biogeosciences*, **6**, 2985–3008.

Kristensen, E., Penha-Lopes, G., Delefosse, M., et al. (2012). What is bioturbation? The need for a precise definition for fauna in aquatic sciences. *Marine Ecology Progress Series*, **446**, 285–302.

Lenihan, H.S. & Peterson, C.H. (1998). How habitat degradation through fishery disturbance enhances impacts of hypoxia on oyster reefs. *Ecological Applications*, **8**, 128–140.

Lenihan, H.S., Peterson, C.H., Byers, J.E., et al. (2001). Cascading of habitat degradation: oyster reefs invaded by refuge fishes escaping stress. *Ecological Applications*, **11**, 764–782.

Leppäkoski, E. (1969). Benthic recolonization of the Bornholm Basin after extermination by oxygen deficiency. *Cahiers der Biologie Marine*, **10**, 163–172.

Levin, L.A. (2003). Oxygen minimum zone benthos: adaptation and community response to hypoxia. *Oceanography and Marine Biology*, **41**, 1–45.

Levin, L.A., Ekau, W., Gooday, A.J., et al. (2009). Effects of natural and human-induced hypoxia on coastal benthos. *Biogeosciences*, **6**, 2063–2098.

Loesch, H. (1960). Sporadic mass shoreward migrations of demersal fish and crustaceans in Mobile Bay, Alabama. *Ecology*, **41**, 292–298.

Loo, L.-O. & Rosenberg, R. (1989). Bivalve suspension feeding dynamics and benthic-pelagic coupling in an eutrophicated marine bay. *Journal of Experimental Marine Biology and Ecology*, **130**, 253–276.

Melzner, F., Thomsen, J., Koeve, W., et al. (2013). Future ocean acidification will be amplified by hypoxia in coastal habitats. *Marine Biology*, **160**, 1875–1888.

Middelburg, J.J. & Levin, L.A. (2009). Coastal hypoxia and sediment biogeochemistry. *Biogeosciences*, **6**, 1273–1293.

Munari, C. & Mistri, M. (2011). Short-term hypoxia modulates *Rapana venosa* (Muricidae) prey preference in Adriatic lagoons. *Journal of Experimental Marine Biology and Ecology*, **407**, 166–170.

Nestlerode, J.A. & Diaz, R.J. (1998). Effects of periodic environmental hypoxia on predation of a tethered polychaete, *Glycera americana*: implications for trophic dynamics. *Marine Ecology Progress Series*, **172**, 185–195.

Nicholls, P. & Kim, J.K. (1982). Sulfide as an inhibitor and electrondonor for the cytochrome-c oxidase system. *Canadian Journal of Biochemistry*, **60**, 613–623.

Nilsson, H.C. & Rosenberg, R. (1994). Hypoxic response of two marine benthic communities. *Marine Ecology Progress Series*, **115**, 209–217.

Nilsson, H.C. & Rosenberg, R. (2000). Succession in marine benthic habitats and fauna in response to oxygen deficiency: analysed by sediment profile-imaging and by grab samples. *Marine Ecology Progress Series*, **197**, 139–149.

Officer, C.B., Biggs, R.B., Taft, J.L., et al. (1984). Chesapeake Bay anoxia: origin, development, and significance. *Science*, **223**, 22–27.

Officer, C.B., Smayda, T.J., & Mann, R. (1982). Benthic filter feeding: a natural eutrophication control. *Marine Ecology Progress Series*, **9**, 203–210.

O'Gorman, E.J., Yearsley, J.M., Crowe, T.P., et al. (2011). Loss of functionally unique species may gradually undermine ecosystems. *Proceedings of the Royal Society of London Series B*, **278**, 1886–1893.

Ott, J. & Fedra, K. (1977). Stabilizing properties of a high biomass benthic community in a fluctuating ecosystem. *Helgolander Wissenschaftliche Meeresuntersuchungen*, **30**, 485–494.

Palumbi, S.R., Sandifer, P.A., Allan, J.D., et al. (2008). Managing for ocean biodiversity to sustain marine ecosystem services. *Frontiers in Ecology and the Environment*, **7**, 204–211.

Pearson, T.H. & Rosenberg, R. (1978). Macrobenthic succession in relation to organic enrichment and pollution of the marine environment. *Oceanography and Marine Biology*, **16**, 229–311.

Pihl, L., Baden, S.P., Diaz, R.J., & Schaffner, L.C. (1992). Hypoxia-induced structural changes in the diet of bottom-feeding fish and crustacea. *Marine Biology*, **112**, 349–361.

Poloczanska, E.S., Brown, C.J., Sydeman, W.J., et al. (2013). Global imprint of climate change on marine life. *Nature Climate Change*, **3**, 919–925.

Pörtner, H.-O., Langenbuch, M., & Michaelidis, B. (2005). Synergistic effects of temperature extremes, hypoxia, and increases in CO_2 on marine animals: from earth history to global change. *Journal of Geophysical Research*, **110**, C09S10.

Powell, S.M., Palmer, A.S., Johnstone, G.J., et al. (2012). Benthic mats in Antarctica; biophysical coupling of sea-bed hypoxia and sediment communities. *Polar Biology*, **35**, 107–116.

Pretterebner, K., Riedel, B., Zuschin, M., & Stachowitsch, M. (2012). Hermit crabs and their symbionts: reactions to artificially induced anoxia on a sublittoral sediment bottom. *Journal of Experimental Marine Biology and Ecology*, **411**, 23–33.

Rabalais, N.N., Diaz, R.J., Levin, L.A., et al. (2010). Dynamics and distribution of natural and human-caused hypoxia. *Biogeosciences*, **7**, 585–619.

Rabalais, N.N., Harper, D.E., & Turner, R.E. (2001). Responses of nekton and demersal and benthic fauna to decreasing oxygen concentrations. In: Rabalais, N.N. & Turner, R.E. (eds), *Coastal Hypoxia: Consequences for Living Resources and Ecosystems*. Coastal and Estuarine Studies 58, pp. 115–128. American Geophysical Union, Washington, DC.

Rabalais, N.N., Turner, R.E., Diaz, R.J., & Justić, D. (2009). Global change and eutrophication of coastal waters. *ICES Journal of Marine Science*, **66**, 1–10.

Rabalais, N.N., Turner R.E., Gupta, B.S., et al. (2007). Hypoxia in the northern Gulf of Mexico: does the science support the plan to reduce, mitigate, and control hypoxia? *Estuaries and Coasts*, **30**, 753–772.

Rabalais, N.N., Turner, R.E., & Wiseman, W.J. (2002). Gulf of Mexico hypoxia, AKA "The dead zone". *Annual Review of Ecology and Systematics*, **33**, 235–263.

Rick, T.C. & Erlandson, J.M. (2008). *Human Impacts on Ancient Marine Ecosystems: A Global Perspective*. University of California Press, Berkeley.

Riedel, B., Pados, T., Pretterebner, K., et al. (2014). Effect of hypoxia and anoxia on invertebrate behaviour: ecological perspectives from species to community level. *Biogeosciences*, **11**, 1491–1518.

Riedel, B., Stachowitsch, M., & Zuschin, M. (2008). Sea anemones and brittle stars: unexpected predatory interactions during induced *in situ* oxygen crises. *Marine Biology*, **153**, 1075–1085.

Riedel, B., Zuschin, M., & Stachowitsch, M. (2012). Tolerance of benthic macrofauna to hypoxia and anoxia in shallow coastal seas: a realistic scenario. *Marine Ecology Progress Series*, **458**, 39–52.

Robb, T. & Abrahams, M.V. (2003). Variation in tolerance to hypoxia in a predator and prey species: an ecological advantage of being small. *Journal of Fish Biology*, **62**, 1067–1081.

Roegner, G.C., Needoba, J.A., & Baptista, A. (2011). Coastal upwelling supplies oxygen-depleted water to the Columbia River Estuary. *PLoS ONE*, **6**, e18672.

Rose, N.A., Janiger, D., Parsons, E.C.M., & Stachowitsch, M. (2011). Shifting baselines in scientific publications: a case study using cetacean research. *Marine Policy*, **35**, 477–482.

Rosenberg, R. (1980). Effects of oxygen deficiency on benthic macrofauna in fjords. In: Freeland, H.J., Farmer, D.M., & Levings, C.D. (eds), *Fjord Oceanography*, pp. 499–514. Plenum Publishing Corp., New York.

Rosenberg, R., Hellman, B., & Johansson, B. (1991). Hypoxic tolerance of marine benthic fauna. *Marine Ecology Progress Series*, **79**, 127–131.

Sagasti, A., Schaffner, L.C., & Duffy, J.E. (2001). Effects of periodic hypoxia on mortality, feeding and predation in an estuarine epifaunal community. *Journal of Experimental Marine Biology and Ecology*, **258**, 257–283.

Sala, E. & Knowlton, M. (2006). Global marine biodiversity trends. *Annual Review of Environment and Resources*, **31**, 93–122.

Sandberg, E., Tallqvist, M., & Bonsdorff, E. (1996). The effects of reduced oxygen content on predation and siphon cropping by the brown shrimp, *Crangon crangon*. *Marine Ecology*, **17**, 411–423.

Savchuk, O.P. (2010). Large-scale dynamics of hypoxia in the Baltic Sea. In: Yakushev, E. (ed.), *Chemical Structure of Pelagic Redox Interfaces: Observation and Modelling*. The Handbook of Environmental Chemistry, p. 24. Springer, Berlin, Heidelberg.

Schinner, F., Stachowitsch, M., & Hilgers, H. (1997). Loss of benthic communities: warning signal for coastal ecosystem management. *Aquatic Conservation*, **6**, 343–352.

Schurmann, H. & Steffensen, J.F. (1992). Lethal oxygen levels at different temperatures and the preferred temperature during hypoxia of the Atlantic cod, *Gadus morhua* L. *Journal of Fish Biology*, **41**, 927–934.

Secor, D.H. & Gunderson, T.E. (1998). Effects of hypoxia and temperature on survival, growth, and respiration of juvenile Atlantic sturgeon, *Acipenser oxyrinchus. Fishery Bulletin*, **96**, 603–613.

Sheppard, C. (1995). The shifting baseline syndrome. *Marine Pollution Bulletin*, **30**, 766–767.

Shick, J.M. (ed.) (1991). *A Functional Biology of Sea Anemones*. Chapman & Hall, New York.

Shimps, E.L., Rice, J.A., & Osborne, J.A. (2005). Hypoxia tolerance in two juvenile estuary-dependent fishes. *Journal of Experimental Marine Biology and Ecology*, **325**, 146–162.

Solan, M., Cardinale, B.J., Downing, A.L., et al. (2004). Extinction and ecosystem function in the marine benthos. *Science*, **306**, 1177–1180.

Stachowitsch, M. (1984). Mass mortality in the Gulf of Trieste: the course of community destruction. *Marine Ecology*, **5**, 243–264.

Stachowitsch, M. (1991). Anoxia in the Northern Adriatic Sea Rapid death, slow recovery. In: Tyson, R.V. & Pearson, T.H. (eds), *Modern and Ancient Continental Shelf Anoxia*. Geol Soc Spec Publ 58, pp. 119–129. The Geological Society, London.

Stachowitsch, M. (1992). Benthic communities: Eutrophication's 'memory mode'. In: Vollenweider, R.A., Marchetti, R., & Viviani, R. (eds), *Marine Coastal Eutrophication*, pp. 1017–1028. Sci Total Environ Suppl. Elsevier, Amsterdam.

Stachowitsch, M., Riedel, B., & Zuschin, M. (2012). The return of shallow shelf seas as extreme environments: anoxia and macrofauna reactions in the Northern Adriatic Sea. In: Altenbach, A., Bernhard, J., & Seckbach, J. (eds), *Anoxia: Evidence for Eukaryote Survival and Paleontological Strategies; Cellular Origins, Life in Extreme Habitats and Astrobiology*, Vol. 21, pp. 353–368. Springer, Netherlands.

STAP (2011). *Hypoxia and Nutrient Reduction in the Coastal Zone. Advice for Prevention, Remediation and Research*. A STAP Advisory Document. Global Environment Facility, Washington, DC.

Steckbauer, A., Duarte, C.M., Carstensen, J., et al. (2011). Ecosystem impact of hypoxia: thresholds of hypoxia and pathways to recovery. *Environmental Research Letters*, **6**, 025003.

Stefanon, A. & Boldrin, A. (1982). The oxygen crisis of the northern Adriatic Sea waters in late fall 1977 and its effects on benthic communities. In: Blanchard, J., Mair, J., & Morrison, I. (eds), *Proceedings of the 6th Symposium of the Confederation Mondiale des Activites Subaquatique*, pp. 167–175. Natural Environmental Research Council, Swindon.

Stierhoff, K.L., Targett, T.E., & Miller, K.L. (2006). Ecophysiological responses of juvenile summer flounder and winter flounder to hypoxia: experimental and modeling analyses of effects on estuarine nursery quality. *Marine Ecology Progress Series*, **325**, 255–266.

Stramma, L., Schmidtko, S., Levin, L.A., & Johnson, G.C. (2010). Ocean oxygen minima expansion and their biological impacts. *Deep Sea Res Part I*, **57**, 587–595.

Strauss, H. (2006). Anoxia through time. In: Neretin, L.N. (ed.), *Past and Present Water Column Anoxia*, pp. 3–19. Springer, Dordrecht.

Sturdivant, S.K., Diaz, R.J., & Cutter, G.R. (2012). Bioturbation in a declining oxygen environment, *in situ* observations from Wormcam. *PLoS ONE*, **7**, e34539.

Sturdivant, S.K., Diaz, R.J., Llansó, R., & Dauer, D.M. (2014). Relationship between hypoxia and macrobenthic production in Chesapeake Bay. *Estuaries and Coasts*, **37**, 1219–1232.

Thorson, G. (1958). Parallel level-bottom communities, their temperature adaptation, and their 'balance' between predators and food animals. In: Buzzata-Traverso, A.A. (ed.), *Perspectives in Marine Biology*, pp. 67–86. University of California Press, Berkeley, California.

Thrush, S.F., Hewitt, J., Gibbs, M., et al. (2006). Functional role of large organisms in intertidal communities: community effects and ecosystem function. *Ecosystems*, **9**, 1029–1040.

Turner, R.E., Rabalais, N.N., & Justić, D. (2008). Gulf of Mexico hypoxia: alternate states and a legacy. *Environmental Science and Technology*, **42**, 2323–2327.

Tyler, R.M., Brady, D.C., & Targett, T.E. (2009). Temporal and spatial dynamics of diel-cycling hypoxia in estuarine tributaries. *Estuaries and Coasts*, **32**, 123–145.

Tyson, R.V. & Pearson, T.H. (eds) (1991). *Modern and Ancient Continental Shelf Anoxia*. Geol Soc Spec Publ 58. Geological Society, London.

UNEP (United Nations Environment Programme) (2004). *Geo Year Book 2003*. GEO Section/UNEP, Nairobi.

UNEP (2006). *Marine and Coastal Ecosystems and Human Wellbeing: A Synthesis Report Based on the Findings of the Millennium Ecosystem Assessment*. UNEP, Nairobi.

Vaquer-Sunyer, R. & Duarte, C.M. (2008). Thresholds of hypoxia for marine biodiversity. *Proceedings of the National Academy of Sciences of the USA*, **105**, 15452–15457.

Vaquer-Sunyer, R. & Duarte, C.M. (2010). Sulfide exposure accelerates hypoxia-driven mortality. *Limnology and Oceanography*, **55**, 1075–1082.

Vaquer-Sunyer, R. & Duarte, C.M. (2011). Temperature effects on oxygen thresholds for hypoxia in marine benthic organisms. *Global Change Biology*, **17**, 1788–1797.

Vistisen, B. & Vismann, B. (1997). Tolerance to low oxygen and sulfide in *Amphiura filiformis* and *Ophiura albida* (Echinodermata: Ophiuroidea). *Marine Biology*, **128**, 241–246.

Wagner, E.J., Arndt, R.E., & Brough, M. (2001). Comparative tolerance of four stocks of cutthroat trout to extremes in temperature, salinity, and hypoxia. *Western North American Naturalist*, **61**, 434–444.

Wetzel, M.A., Fleeger, J.W., & Powers, S. (2001). Effects of hypoxia and anoxia on meiofauna: a review with new data from the Gulf of Mexico. In: Rabalais, N.N. & Turner, R.E. (eds), *Coastal Hypoxia: Consequences for Living Resources and Ecosystems*. Coastal and Estuarine Studies

58, pp. 165–184. American Geophysical Union, Washington, DC.

Wharton, D.A. (2002). *Life at the Limits: Organisms in Extreme Environments*. Cambridge University Press, Cambridge.

Widdows, J., Newell, R.I.E., & Mann, R. (1989). Effects of hypoxia and anoxia on survival, energy metabolism, and feeding of oyster larvae (*Crassostrea virginica*, Gmelin). *Biological Bulletin*, **177**, 154–166.

Wignall, P.B. & Twitchett, R.J. (1996). Oceanic anoxia and the End Permian mass extinction, *Science*, **272**, 1155–1158.

Worm, B., Barbier, E.B., Beaumont, N., et al. (2006). Impacts of biodiversity loss on ocean ecosystem services. *Science*, **314**, 787–790.

Wu, R.S.S. (2002). Hypoxia: from molecular responses to ecosystem responses. *Marine Pollution Bulletin*, **45**, 35–45.

Wu, R.S.S., Wo, K.T., & Chiu, J.M.Y. (2012). Effects of hypoxia on growth of the diatom *Skeletonema costatum*. *Journal of Experimental Marine Biology and Ecology*, **420–421**, 65–68.

Zhang, J., Gilbert, D., Gooday, A.J., et al. (2010). Natural and human-induced hypoxia and consequences for coastal areas: synthesis and future development. *Biogeosciences*, **7**, 1443–1467.

Zillén, L., Conley, D.J., Andrén, T., et al. (2008). Past occurrences of hypoxia in the Baltic Sea and the role of climate variability, environmental change, and human impact. *Earth Science Reviews*, **91**, 77–92.

CHAPTER 11

Ecological effects of ocean acidification

M. Débora Iglesias-Rodriguez, Katharina E. Fabricius, and Paul McElhany

11.1 Introduction

The accelerated increase in carbon dioxide (CO_2) levels since the Industrial Revolution, caused by human burning of fossil fuels, is altering marine inorganic carbon chemistry through a process termed 'ocean acidification' (Caldeira and Wickett, 2003). The rising atmospheric CO_2 increases the partial pressure of CO_2 (pCO_2) in ocean surface waters, leading to a number of chemical changes including increased hydrogen ion concentrations (lower pH, more acidic) and reduced carbonate ion concentrations. These changes have important implications for marine organisms, affecting physiological processes such as calcification, photosynthesis, and neurophysiological pathways. In turn, changes in the physiology of individuals can alter the dynamics of their populations and ultimately affect entire ecosystems.

Ocean acidification alters sources and sinks of carbon within important ecophysiological processes. For example, as CO_2 increases in the surface waters of the oceans, a proportion of it combines with H_2O to form H_2CO_3, which dissociates into protons and bicarbonate ions ($H_2CO_3 \Rightarrow H^+ + HCO_3^-$) (Fig. 11.1). The excess protons in seawater combine with carbonate ions (CO_3^{2-}) to increase HCO_3^- ions even further ($H^+ + CO_3^{2-} \Rightarrow HCO_3^-$). These chemical shifts in the inorganic carbon system, specifically a decline in pH and CO_3^{2-} ions and an increase in CO_2 and HCO_3^- ions, can have profound effects on physiological performances and, hence, on ecosystem functions (Fig. 11.1).

A study of the physiological mechanisms that drive organismal responses to ocean acidification has been used to develop 'first hypotheses' about which species in an ecosystem are likely to increase or decrease in abundance as a direct consequence of ocean acidification. However, additionally to the direct physiological and neurological effects, the indirect effects of ocean acidification mediated through altered habitats and food web dynamics are also important. Species directly affected by ocean acidification serve as prey, predators, symbionts, competitors, or shelter for other species. When the abundance, distribution, or function of organisms that are sensitive to ocean acidification changes, the impact will likely ripple throughout the food webs. For example, larvae of clownfish appear to exhibit altered responses to olfactory cues under low pH (Munday et al., 2009), and response to visual cues of risk were impaired by elevated CO_2 in juvenile damselfish *Pomacentrus amboinensis* (Ferrari et al., 2012). Also, predator–prey relationships have been found to be affected by ocean acidification in dottyback fish (*Pseudochromis fuscus*), specifically, by elevated activity levels of predators under high CO_2 and reduced olfactory ability, and enhanced visual detection of food (Cripps et al., 2011).

Predictions of major ecosystem upheaval from ocean acidification are derived from five primary sources of information: (1) laboratory experiments on individual species that are extrapolated to changes in populations, which are then incorporated into food web models; (2) the paleontological record of ecosystem responses to major changes in pCO_2, using these past events as analogues of ocean acidification; (3) observations of community differences along CO_2 gradients *in situ*, either steep gradients like those at volcanic CO_2 vents or less steep gradients that occur as a result of natural variation in marine pCO_2; (4) *in situ* CO_2 enrichment experiments in benthic chambers or large floating mesocosms; and (5) time-series analyses of community change in response to acidification that has already occurred. This review touches on all of these approaches except the paleontological record (which is addressed in Hönisch et al., 2012). However, our

Stressors in the Marine Environment. Edited by Martin Solan and Nia M. Whiteley
© Oxford University Press 2016. Published in 2016 by Oxford University Press.

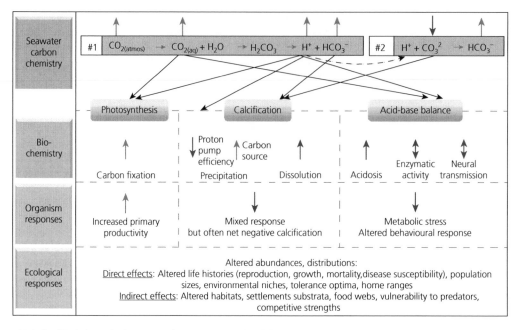

Figure 11.1 Simplified relationship between anthropogenic CO_2-induced changes in seawater carbon chemistry and direct population effects on marine organisms caused by effects on photosynthesis, calcification, and acid–base balance. Both photosynthesis and calcification could be considered acid–base balance processes, but they are separated here because of their distinct relationships to the external seawater carbon chemistry. In seawater carbon chemistry equation #1, increases in atmospheric CO_2 result in increased hydrogen and bicarbonate ion concentrations. In equation #2, many of the hydrogen ions generated via equation #1 combine with carbonate ions already in the ocean to reduce carbonate concentrations. The grey arrows pointing upwards indicate chemical species that increase in concentration with ocean acidification and are processes that tend to increase fitness in response to OA. The grey arrow pointing downwards indicates a decrease and is a process that tends to decrease fitness in response to OA. The solid black arrows indicate which carbon chemistry species affect each of the biochemical processes. The biochemical changes from OA can lead to changes at the organism level, which translate into population level responses.

primary focus is on the first approach, translating data from experiments on individual species combined with data on regional carbon chemistry to predict ecosystem responses.

In moving from laboratory experiments to ecosystem models, uncertainty increases at each step in organizational complexity such that ecosystem-level predictions tend to be rather speculative. However, confidence in ecosystem-level predictions can be improved through an improved understanding of the physiological mechanisms that drive ocean acidification responses. This chapter reviews this topic at several levels of organization, from molecular and biogeochemical to ecological and evolutionary, to explore the effects of ocean acidification on marine ecosystems. We start with a discussion of the biochemical and organismal processes directly affected by altered concentrations of the various dissolved inorganic carbon (DIC) species; we then discuss the tools and strategies to predict the ecological impacts of ocean acidification;

next, we discuss the ecological impacts of ocean acidification in specific regions; and finally, we summarize some of the types and findings of ecosystem models of ocean acidification effects.

11.2 Biochemical and organism responses to increased CO_2

11.2.1 Photosynthesis

Increasing CO_2 availability at the ocean surface has the potential to enhance photosynthetic carbon fixation (Paasche, 1964; Nimer and Merrett, 1993). For phytoplankton, the rate of photosynthesis is limited by the rate of diffusion, mass transport rate, and partial pressure/concentration of CO_2 in seawater, the dehydration kinetics of bicarbonate to CO_2, and the low affinity of ribulose-1,5-bisphosphate carboxylase/oxygenase (rubisco) for its substrate, CO_2 (Reinfelder, 2011). The increase in CO_2 from ocean acidification generally enhances photosynthetic carbon fixation when sufficient light, trace elements, and macro- and micronutrients are available. This has been shown in coccolithophores (Riebesell et al., 2000; Iglesias-Rodriguez et al., 2008), diatoms (Tortell et al., 2002), cyanobacteria (Fu et al., 2007), and some corals (Marubini and Thake, 1999; Schneider and Erez, 2006; Herfort et al., 2008). Photosynthesis may also be boosted by an increase in HCO_3^- ions, since many taxa have developed a carbon concentrating mechanism that enables them to utilize bicarbonate to overcome CO_2 limitation (Badger et al., 1998; Thoms et al., 2001; Rost et al., 2008). However, photosynthetic responses to elevated CO_2 appear to vary widely across taxa of marine macrophytes (seagrasses and macroalgae), with seagrasses often showing strong gains, whereas some but not all macrophytes appear to be limited by CO_2 availability.

11.2.2 Calcification

A decline in the concentration of carbonate ions (CO_3^{2-}) is generally found to decrease the rate of biosynthesis of $CaCO_3$ in organisms forming calcareous plates, skeletons, and shells, and/or increase the rate of $CaCO_3$ dissolution (Doney et al., 2009; Iglesias-Rodriguez et al., 2010; Kroeker et al., 2013). Although the equation representing the calcification process can be expressed as $Ca^{2+} + CO_3^{2-} \rightarrow CaCO_3$, the biological mechanisms of calcium carbonate formation are more complex. Biological membranes appear to be impermeable to carbonate ions (Raven, 2011) such that organisms are forced to acquire an alternative form of inorganic carbon (HCO_3^- or CO_2) and subsequently convert it into CO_3^{2-}. Specifically, most calcifiers appear to use HCO_3^- as the external DIC source for calcification, and HCO_3^- enrichment appears to promote calcification in many calcifiers including some coccolithophore strains (Iglesias-Rodriguez et al., 2008), some crustaceans (Ries et al., 2009), and some coral species (e.g. Marubini and Thake, 1999; Herfort et al., 2008; Jury et al., 2010), although there are examples of corals that appear to increase calcification in response to both HCO_3^- and CO_3^{2-} addition (Comeau et al., 2013).

Given that HCO_3^- constitutes 85–90% of seawater DIC, HCO_3^- utilization represents an evolutionary advantage in adapting[1] to ocean acidification conditions.

When HCO_3^- ions enter the calcifying fluid, the conversion of HCO_3^- into CO_3^{2-} is required to precipitate $CaCO_3$. This process generates protons ($HCO_3^- \Rightarrow CO_3^{2-} + H^+$) and proton pumps are required to excrete them from the calcification reservoir to prevent acidification. The energetic cost needed to extrude protons from the calcification fluid, in some cases reported to be supplied by photosynthesis (Jokiel et al., 2014), has been argued to be a key factor controlling calcification (Ries, 2011). In the coccolithophore species *Coccolithus pelagicus*, a plasma membrane voltage-gated H^+ channel provides pH homeostasis (Taylor et al., 2011). Similar mechanisms have been found in other calcifiers like corals, where a Ca^{2+}/H^+ exchanger removes protons from the calcification reservoir and imports Ca^{2+} into the calcification fluid (Cohen and Holcomb, 2009). Some argue that this generation of protons increases the intracellular concentration of CO_2, which promotes photosynthesis in coccolithophores (Buitenhuis et al., 1999), calcareous algae (Ries et al., 2010), and corals (Gattuso et al., 1999), while others have found that photosynthesis and calcification are decoupled (Gattuso et al., 2000; Trimborn et al., 2007; Leonardos et al., 2009).

The precipitation of $CaCO_3$ represents a source of CO_2 to the environment because the removal of HCO_3^- from seawater generates CO_2 ($2HCO_3^- + Ca^{2+} \rightarrow CaCO_3 + CO_2 + H_2O$). Therefore, the relative contribution of photosynthesis (CO_2 sink) and calcification (CO_2 source) is important in representing the magnitude and direction of carbon fluxes and energy budgets in ecosystems. In corals, the algal endosymbionts can provide healthy corals with over 100% of their daily metabolic energy requirements via photosynthetic carbon fixation at high irradiance (Muscatine et al., 1981). In those coral species where algal carbon fixation increases with acidification, this might help maintain calcification rates under acidification despite increasing energy demands for calcification (Brading et al., 2011), at least in shallow waters. However, a meta-analysis of ocean acidification studies across many species of corals reported a mean 15–22% decline in net calcification under CO_2 conditions predicted for the end of this century under business-as-usual scenarios (Chan and Connolly, 2013). The observed variability in responses between studies was attributed to species-specific differences in tolerances and differences in other environmental conditions (Chan and Connolly, 2013), indicating that there will be disruptions in species abundances and physiological performances.

Observed declines in net calcification with increased pCO_2 are a consequence of decreased precipitation or increased dissolution or both. On the precipitation side

---

[1] By adaptation we refer to the process of differential survival or reproduction of individuals displaying specific phenotypes. As a result, a population evolves towards those phenotypes that are best suited to the present environmental conditions.

of the equation, calcification rates are likely to be largely dependent on the efficiency of a carbon-concentrating mechanism and the energy availability to control the flux of H^+ during $CaCO_3$ precipitation. The dissolution of $CaCO_3$ is affected by the saturation state of calcite (Ω_{Cal}) or aragonite (Ω_{Ara}), two main polymorphs of calcium carbonate. The saturation state, Ω, is a function of the calcium ion concentration $[Ca^{2+}]$, the carbonate ion concentration $[CO_3^{2-}]$, and the temperature, salinity, and pressure-dependent solubility constant K^*_{sp} ($\Omega = [Ca^{2+}] \times [CO_3^{2-}]/K^*_{sp}$). When $\Omega < 1$, dissolution of $CaCO_3$ is favoured. Given that the concentration of Ca^{2+} ions in open ocean seawater is semiconservative, a decline in CO_3^{2-} ions will cause a decrease in Ω, which will have important consequences for calcifiers. Changes in the calcium carbonate (calcite or aragonite) 'saturation horizon', defined as the depth at which calcite/aragonite dissolution equals precipitation (i.e. $\Omega = 1$), could be detrimental to biotic carbonate accretion of deep sea organisms or of vertically migrating taxa. It has been shown that the carbonate saturation horizon is shoaling due to ocean acidification (Fabry et al., 2008; Feely et al., 2008) and, therefore, organisms that live in depths that become undersaturated ($\Omega < 1$) will be particularly vulnerable.

Another factor affecting the susceptibility of carbonate to dissolution is the type of $CaCO_3$ mineral and its composition (e.g. percentage of Mg in carbonate). Aragonite is more susceptible to dissolution than low-magnesium calcite, and the solubility of calcite increases with increased concentration of Mg (Mucci and Morse, 1983). The relative rank order of solubility of calcium carbonate is therefore: amorphous > high-Mg calcite > aragonite > low-Mg calcite. For example, echinoderms precipitate high-Mg calcite, which is more susceptible to dissolution than calcite or aragonite skeletons (Bertram et al., 1991). Moreover, unstable, amorphous $CaCO_3$ is formed during larval stages; e.g. in echinoderms (Beniash et al., 1997; Raz et al., 2003) and molluscs (Weiss et al., 2002), which makes early developmental stages highly susceptible to acidification (Lebrato et al., 2010). Of course, $CaCO_3$ mineralogy on its own is not sufficient to predict vulnerability to ocean acidification, as the responses to acidification vary between organisms producing the same polymorph of $CaCO_3$; for example, among aragonite-producing corals (e.g. Reynaud et al., 2003; Cohen et al., 2009; Ries et al., 2009; Holcomb et al., 2010; Rodolfo-Metalpa et al., 2010) and calcite-producing coccolithophores (e.g. Riebesell et al., 2000; Iglesias-Rodriguez et al., 2008; Langer et al., 2009). A mechanism that may provide some protection of the biominerals against dissolution is the synthesis of organic membranes covering the $CaCO_3$ skeletal structures such as those in echinoids, molluscs (Marin et al., 2008), corals (Allemand et al., 2004), and calcifying algae (Young and Henriksen, 2003; Ries et al., 2009).

11.2.3 Other physiological and neurological effects of increased pCO$_2$

Changes in pH and pCO_2 can also have important physiological and pathological effects on acid–base systems in organisms, with consequences for their energy budgets (Melzner et al., 2013). For example, many enzymes in aquatic environments have maximum hydrolysing capacity within relatively narrow pH ranges (El-Shahed et al., 2006; Yamada and Sukumura, 2010). This suggests that acidification could alter some catalytic performances controlling physiological processes in marine organisms, potentially leading to changes in the cycling of organic matter, particularly in lipids and proteins (Yamada and Suzumura, 2010).

In higher animals, abnormally elevated CO_2 levels (hypercapnia) can cause blood acidosis (Roos and Boron, 1981; Somero, 1985; Cameron, 1986; Truchot, 1987; Walsh and Milligan, 1989). CO_2-driven changes in extracellular acid–base equilibria can alter the relative proportions of carbonate species in blood, which can cause pathologies such as hypercalcification. This has been observed in cephalopods, crustaceans, and teleost fish (Gutowska et al., 2010). Elevated CO_2 can also affect the neurophysiology of organisms by altering their internal acid–base balance and affecting the pH-sensitive neutrotransmitter GABA (Munday et al., 2009; Briffa et al., 2012; Nilsson et al., 2012); this can for example lead to pathological behaviour in teleost fish (Munday et al., 2009, 2014; Dixson et al., 2010; Nilsson et al., 2012) and invertebrates (Watson et al., 2014). Indeed, the range of intracellular pH to ensure the metabolic health of an organism and the range of pH tolerance are relatively narrow in marine organisms. Therefore, when an organism undergoes hypercapnia, changes in different cellular pools require alterations in ion exchange between different cellular compartments and between the organism and the external medium, which can involve energy reallocation (e.g. Dupont et al., 2010; Stumpp et al., 2011).

11.2.4 Variability in the vulnerability within and across taxa

Sensitivity to ocean acidification appears to be highly variable within and among taxa (Kroeker et al., 2010,

2013), and responses can be species- and even strain-specific (Langer et al., 2006, 2009). Corals, echinoderms, and molluscs were found to be more sensitive to ocean acidification than crustaceans and fish larvae if all life stages and types of responses are jointly considered, although sensitivities are often not uniform within taxonomic groups (Wittmann and Pörtner, 2013). A meta-analysis of ocean acidification data from experimental studies found average negative effects of increased CO_2 in at least one response attribute (survival, calcification, growth, or abundance) for all calcifying groups (calcareous algae, corals, coccolithophores, molluscs, and echinoderms) except crustaceans, for which the trend was negative, but responses were highly variable (Kroeker et al., 2013).

Metadata analysis of primary producers showed that calcareous algae had on average reduced photosynthesis, whereas diatoms and some fleshy algae and seagrasses increased photosynthetic carbon fixation (Kroeker et al., 2010) and/or growth under higher CO_2. It is important to emphasize that the purpose of such meta-analyses is to determine general trends; however, there is some variability within and between taxa. For example, in coccolithophores, the effect of ocean acidification is known to vary between and even within the species concept (e.g. Iglesias-Rodriguez et al., 2008; Langer et al., 2009). This variability, combined with the relatively small number of species that have been studied, make it challenging to predict the extent to which populations will respond to increasing acidification. Also, this diversity of observations might suggest that some representatives of calcifiers may have mechanisms to counteract $CaCO_3$ dissolution; for example, via organic coating of the biomineral (Marin et al., 2008; Godoi et al., 2009; Ries et al., 2009), and in some cases it has been established that maintenance or increase in calcification can take place, often at the expense of other important metabolic processes (e.g. Wood et al., 2008).

11.2.5 Susceptibility of different life stages to ocean acidification

Metadata analysis of global data has revealed substantial variability of responses to ocean acidification (Kroeker et al., 2013), but differing sensitivities during the life cycle of an organism may be masked in some of these studies (Dupont et al., 2010, but see also Wittmann and Pörtner, 2013). Larval development appears to be often more susceptible to ocean acidification than in adult stages based on performance measures (Kurihara and Shirayama, 2004; Ross et al., 2011). For example, echinoderms have a vulnerable pelagic larval stage that can last for hours to months before settling into benthic adult forms (Dupont et al., 2010). An alteration of the length of larval life and changes in calcification are likely detrimental to many populations because these impacts directly affect larval dispersal, survival (e.g. exposure duration to predators), and settlement at the end of their larval stage (Ross et al., 2011).

Importantly, the mineralogy of structures may differ across life stages and body parts. For example, skeletal rods, adult tests, teeth, and spicules of larvae and adult echinoderms are precipitated from an amorphous calcite precursor which is ~30 times more susceptible to dissolution than low-magnesium calcite (Politi et al., 2004). Such variability within the individual (e.g. at different life stages) was initially not accounted for in studies of the physiological effects of ocean acidification on organism responses. Warming may further complicate the responses; for example one study observed increasing growth in adult sea stars both with temperature and CO_2 (Gooding et al., 2009), while another study showed that temperature dramatically reduced their gastrulation success and impaired early development of a sea urchin, while elevated CO_2 had no major effect on the early developmental stages (Byrne et al., 2009). A number of studies have also shown a detrimental effect of ocean acidification on animal reproduction (e.g. Kurihara, 2008; Dupont et al., 2010), transition between life-cycle stages (Dupont et al., 2013), and trans-generational plasticity (e.g. Parker et al., 2012), although some results may indicate a degree of adaptation to fluctuations in pH and temperature (Byrne, 2011). Other amplification effects, such as the sensitivity of obligatory symbionts to heat stress, could further contribute to organism responses (Hoegh-Guldberg et al., 2007).

In organisms with high generation turnover, such as marine photoautotrophic microbes, experiments have provided insight into adaptation to ocean acidification (Elena and Lenski, 2003; Lohbeck et al., 2012; Jin et al., 2013; Reusch and Boyd, 2013; Sunday et al., 2014). For example, there is considerable intraspecific variation in physiological traits (Brand et al., 1981; Gallagher, 1982; Wood and Leathem, 1992) and in genetic identity (Gallagher, 1982; Medlin et al., 2000; Rynearson and Armbrust, 2000; Iglesias-Rodriguez et al., 2006; Härnström et al., 2011) of many phytoplankton species. As a result, the rapid frequency shifts among genotypes within a time scale of hundreds of generations can result in changes in the mean population response (Reusch and Boyd, 2013).

Although Darwinian fitness in asexually reproducing exponentially growing organisms is used as a proxy for adaptation in selection experiments (Elena and Lenski, 2003), from an ecological point of view, functional responses are equally important (Reusch and Boyd, 2013). For example, alteration in cell size, changes in biomineralization (e.g. silicate or calcium carbonate cell surface structures), and in the chemical composition of the cell and hence its food quality, can impact on higher trophic levels (Reusch and Boyd, 2013). Indeed, it is likely that observed shifts in patterns result from interactions between community change (ecology) and adaptive changes (evolution) (Collins and Gardner, 2009; Reusch and Boyd, 2013).

In addition, physiological or morphological acclimation[2] responses may be important (Reusch and Boyd, 2013), for example, via alterations in grazing preference and preservation of carbon through the water column. However, these experiments select for individuals with high fitness and, therefore, it can be a challenge to extrapolate results to the field and identify the mechanisms of adaptation to increasing CO_2 levels (Sunday et al., 2014).

11.3 Tools and strategies to predict the ecological effects of ocean acidification

Laboratory experiments are essential to assess the direct effects of CO_2 on the physiology of specific taxa. However, most laboratory experiments are short relative to the longevity of the study organisms, the time scales of the manipulations are abrupt compared to what populations would experience in the field, and it is challenging to simulate natural levels of food, light, and species interactions. Long-term field, mesocosm, and laboratory studies have therefore been used to predict the ecological effects of ocean acidification, such as species interactions, recruitment processes, and altered habitat quality, and to assess acclimation and adaptation.

Although ocean acidification is a global-scale phenomenon, the effects of ocean acidification on carbonate chemistry can vary in response to latitude, temperature, depth/pressure, oceanographic circulation, and biological processes (e.g. photosynthesis, respiration, and calcification). Some areas are characterized by low pH conditions, such as cold environments; for

example, high latitude and deep sea waters. High latitudes harbour seawater with higher CO_2 concentrations because CO_2 is more soluble in cold waters, and therefore these environments are theoretically more susceptible to the effect of increasing atmospheric CO_2. Indeed, modelling studies have suggested that Arctic and Southern Ocean waters will experience prolonged aragonite undersaturation ($\Omega < 1$) by the middle of this century (Orr et al., 2005; Gehlen et al., 2007; Cao and Caldeira, 2008). Some areas already experiencing seasonal undersaturation of calcium carbonate ($\Omega < 1$) have been used as study sites to predict responses to ocean acidification; for example, planktonic foraminifera in the Southern Ocean display reduced calcification (Moy et al., 2009) and Antarctic pteropods have thinner shells as ocean acidification increases (Bednaršek et al., 2012; Teniswood et al., 2013). There are, however, some conflicting results; for example, long-term time series (~50 years) of densities of calcareous plankton in the central North Sea showed increasing densities of foraminiferans, coccolithophores, and echinoderm larvae, but declining densities of pteropods and bivalves over time, without a statistically significant relationship between organism abundances and the pH trends (Beare et al., 2013).

In situ observations in naturally acidified waters, like in volcanic vents that release CO_2 into the surrounding seawater, are being used as analogues of how ecosystems may respond to future 'high CO_2 oceans'. Their geological age implies that these ecosystems have been exposed for decades or centuries, such that most sessile or territorial organisms found at the sites have been exposed to high CO_2 (at natural conditions of predation, grazing, light, plankton, etc.) throughout their post-larval life times. For example, vents in shallow rocky shores of the Mediterranean and in tropical coral reefs of Papua New Guinea both revealed profound ecosystem alterations under elevated CO_2 (Hall-Spencer et al., 2008; Fabricius et al., 2011). In both studies, a significant decrease in biodiversity was observed and non-calcifying photosynthetic organisms (e.g. macroalgae and seagrasses) thrived under elevated CO_2 while calcareous plants and animals, such as coralline algae, benthic foraminifera, and some sea urchins, had reduced abundances. Importantly, seep studies also demonstrated that the indirect effects of ocean acidification can severely alter the community composition of benthic ecosystems, and that these ecological effects compound the direct physiological effects of CO_2 on organisms. For example, structurally complex corals that are essential habitat for reef organisms are sparse in coral reefs around seeps, resulting

---

[2] We refer to acclimation as any alteration of physiological or morphological characteristics through phenotypic plasticity in response to environmental selection pressure, following Munday et al. (2014).

in significant losses of crustaceans that are known for their highly CO_2-tolerant physiology (Fabricius et al., 2014). Often, 'acidified' waters have characteristics that make them imperfect analogues of ocean acidification; for example, unlike in a future high CO_2 ocean, recruitment at the seeps is possible from nearby waters with present-day CO_2 concentrations. Similarly, areas exposed to upwelling of high CO_2 water in western boundary regions have been used as analogues for a high CO_2 ocean, but the upwelled water is also cold and typically rich in nutrients. Nevertheless, these field observations, where organisms and ecosystems have been exposed to high CO_2 for decades or centuries, are extremely valuable in understanding whether and how biota acclimate and adapt to ocean acidification.

Ocean acidification conditions can also be created artificially in 'Free Ocean CO_2 Enrichment' (FOCE) experiments (Gattuso et al., 2014) within partial enclosures that are erected on the seafloor, in which parameters such as light and plankton as food source vary naturally. Multi-species laboratory mesocosm experiments have also been used to test for the direct and indirect effects of ocean acidification, and the interaction between ocean acidification and ocean warming. For example, Alsterberg et al. (2013) demonstrated the importance of testing the effects of ocean acidification and warming on mesograzers and their mediating effects on microalgal biomass in a system that also contained sediment infauna, macroalgae, and seagrass. A mesocosm experiment also revealed how the active dissolution of calcium carbonate substratum by photosynthetic bioeroding microalgae in coral reefs can accelerate under combined exposure to ocean acidification and warming (Reyes-Nivia et al., 2013).

Other areas have naturally higher pH variability due to biological metabolism, with high pH during daylight photosynthetic carbon draw-down (e.g. seagrass beds and phytoplankton blooms), and reduced pH at night due to respiration (especially in eutrophic areas with high rates of heterotrophy). For example, along eastern boundary currents, in areas undergoing seasonal upwelling or in intertidal zones, organisms are exposed to dramatic short-term fluctuations in pH and temperature. Their responses can be useful in providing clues for the mechanisms of adaptation to ocean acidification in the context of environmental change. The upper ocean pH can vary by 0.02 to 1.43 pH units in areas of upwelling (Hoffman et al., 2011), indicating that organisms inhabiting these waters must bear the appropriate acclimatory and evolutionary makeup to withstand shifts in water chemistry. It will be critically important to understand the evolutionary speed of adaptation of organisms with both fast and slow generation times, because the rates of ocean acidification changes are now faster than observed at any time within the last 65 million years, and possibly 300 million years (Hoenisch et al., 2012).

11.4 State of knowledge about ecological responses to ocean acidification

11.4.1 Polar regions

Polar regions are characterized by naturally low levels of CO_3^{2-} ions as a result of the higher solubility of CO_2 at low temperatures compared with temperate and tropical regions, which makes polar ecosystems global sinks of CO_2 (Sabine et al., 2004). For example, the Arctic Ocean is an important CO_2 sink, contributing globally to 5–14% of the oceans' CO_2 sequestration (Bates and Mathis, 2009). This, in addition to other climate-driven factors, including the contraction of marginal ice zone biomes and the expansion of subpolar biomes (Sarmiento et al., 2004), makes polar ecosystems particularly susceptible to climate change. For example, parts of the Arctic Ocean have already shown seasonal $CaCO_3$ undersaturation (Tyrrell et al., 2008; Bates et al., 2009), especially in coastal and shelf zones (e.g. Orr et al., 2005; Steinacher et al., 2009). However, there is great seasonal variability in pH and carbon chemistry conditions. For example, extremely low CO_2 partial pressure (e.g. 130–150 µatm; Bates, 2006; Fransson et al., 2009), far below atmospheric in parts of the Arctic, is likely a result of seasonally high inorganic carbon uptake due either to enhanced productivity or to freshwater inputs such as ice melt. Thus, a seasonal increase in primary production removes CO_2 from sea water and increases the concentration of CO_3^{2-} (Merico et al., 2006), such that primary production can play a central role in controlling the extent to which polar regions act as a sink of CO_2 (Takahashi et al., 2002).

In the Arctic, ocean acidification and other climate-relevant environmental stressors including warming, rapid loss of sea-ice, consequent changes in salinity, and increased coastal erosion will likely act additively or synergistically (McGuire et al., 2006; Serreze and Francis, 2006; Maslanik et al., 2007; Arrigo et al., 2008; Pabi et al., 2008; McGuire et al., 2009; Wang and Overland, 2009; Sunda and Cai, 2012), suggesting that the impact of acidification may be amplified in these fast-changing environments. Specifically, factors such as increased stratification due to ocean warming and freshening (rain, ice melting), and a subsequent increase in irradiance and ultraviolet exposure, affect photosynthetic organisms (Gao et al., 2012). The recent decreases in ice cover in the Arctic could theoretically have a beneficial impact on primary producers by increasing light penetration in the water column and promoting algal growth at the beginning of the spring. As a result of freshwater input, there are important chemical changes including a dilution of dissolved inorganic carbon and total alkalinity in surface waters, which could affect photosynthetic carbon fixation, and a decline in the $[CO_3^{2-}]$ and in the buffering capacity of these waters (Chierici and Fransson, 2009), which will likely have an impact on calcifying organisms. In the near future, the Arctic Ocean is expected to undergo further sea-ice loss and increases in phytoplankton primary production, which are expected to increase the uptake of CO_2 by surface waters (although mitigated to some extent by warming) (Bates and Mathis, 2009).

Like in the Arctic Ocean, large parts of Antarctic ecosystems are poorly known. These include the deep-sea fauna of the Southern Ocean (Brandt et al., 2007), where the shelf break is typically closer to the shore and the shelf is deeper than the global average (Snelgrove, 2001; Clarke and Johnston, 2003). A study on Antarctic benthic calcifiers has revealed that the skeletal $CaCO_3$ content decreased with decreasing seawater temperature and Ω in all representatives of bivalve and gastropod molluscs, brachiopods, and echinoids although a latitudinal trend in shell mass and thickness could not be established for all the taxa (Watson et al., 2012). The Southern Ocean is an interesting study area because it has oceanographic boundaries that restrict dispersal of individuals. Therefore, the ecological resilience in these relatively isolated environments is critically dependent upon the natural variation within populations and their capacity of adaptation. It has been hypothesized that, unlike in the Arctic, prey taxa in Antarctica have evolved thin shells in the absence of 'crushing predators' for tens of millions of years, and predation on calcifiers may return as a result of fast climate disturbances (Aronson et al., 2007; Fabry et al., 2009). This may have contributed to the evolution of the thin, weakly calcified shells in Antarctic marine benthic invertebrates (Aronson et al., 2007; McClintock et al., 2008). For example, large populations of king crabs native to the Southern Hemisphere were recently found in deep-water, continental-slope environments off the Antarctic Peninsula (Thatje and Arntz, 2004; Thatje and Lörz, 2005; Aronson et al., 2007). The implications are that the combined thinning of shells and the change in the predator community could be detrimental for high latitude calcifiers including molluscs, brachiopods, and echinoderms (Aronson et al., 2007).

Ocean acidification has also been found to cause alterations in the phytoplankton community structures of polar ecosystems (Boyd and Doney, 2002). Indeed, mesocosm experiments revealed a shift towards the smallest phytoplankton groups as a result of ocean acidification (Brussaard et al., 2013). These results have been interpreted as having the potential to decrease the carbon accumulation in cells and increasing the pool of dissolved organic carbon, with consequences for the biological carbon pump (Brussaard et al., 2013). On the other hand, the community composition of Arctic bacteria appear to be unaffected by ocean acidification, and only nutrient addition appeared to be the primary driver of bacterial community structure (Roy et al., 2013). In contrast, the effect of ocean acidification on pelagic consumers does not appear to be as strong as on primary producers. For example, in a mesocosm study with varying CO_2 and pH conditions, ocean acidification did not affect the diversity, abundance, and phenology of a post-bloom high Arctic (Svalbard) coastal microzooplankton community (Aberle et al., 2013), nor the mesozooplankton abundance and taxonomic composition, but altered the phytoplankton community composition; specifically, increasing CO_2 promoted the growth of autotrophic dinoflagellates (Niehoff et al., 2013).

11.4.2 Temperate and tropical regions

Carbon dioxide solubility in the warm waters of tropical and subtropical regions is low, leading to typically very high concentrations of CO_3^{2-} and seawater carbonate oversaturation. This facilitates the widespread 'hypercalcification' found in many tropical marine organisms and ecosystems. Calcification observed in corals, giant clams, large benthic foraminifera, and many other tropical taxa also depends upon their symbiosis with endosymbiotic dinoflagellates, which transfer organic sugars derived from photosynthetically fixed carbon to their hosts. These endosymbionts utilize a type of enzyme with particularly low affinity for inorganic carbon for carbon fixation (ribulose biphosphate carboxylase/oxygenase type II; Whitney et al., 1995; Lilley et al., 2010). Some, but not all, studies demonstrate that their gross photosynthesis may therefore benefit from additional CO_2 (in the absence of other co-limiting factors such as other nutrients, light, or temperature ranges that promote carbon fixation; e.g. Uthicke and Fabricius, 2012). However, an improved supply of inorganic carbon for photosynthesis does not seem to compensate for the increased energetic demand for calcification at elevated CO_2 in these organisms: controlled laboratory and mesocosm experiments have shown that net production typically remains unaltered and reef calcification declines with rising CO_2 (Langdon et al., 2003). A recent meta-analysis of studies integrating light and dark calcification suggests that under business-as-usual future atmospheric CO_2 scenarios, coral growth rates will decline by ~15% on average by the end of this century due to ocean acidification alone (Chan and Connolly, 2013). Additionally, coral growth rates also decline due to thermal stress (De'ath et al., 2009; Tanzil et al., 2009; Manzello, 2010; Cooper et al., 2012), and at present insufficient information is available to assess the interactive effects of multiple stressors on the corals' physiology, and their ability to acclimatize to such high future CO_2 levels.

Calcification of corals (temperate as well as tropical and high latitude) appears to be controlled by the saturation state of aragonite, and a species-specific threshold in seawater $[CO_3^{2-}]$ exists, below which calcification becomes impaired (Ries et al., 2010). However, some temperate species of scleractinian corals exhibit either no response and/or a nonlinear response to CO_2-induced decreases in seawater $[CO_3^{2-}]$ (Holcomb et al., 2010; Rodolfo- Metalpa et al., 2010). Results on both temperate and tropical corals (Reynaud et al., 2003; Cohen et al., 2009; Ries et al., 2009; Holcomb et al., 2010; Jury et al., 2010; Rodolfo-Metalpa et al., 2010; Edmunds et al., 2013) reveal variable responses, suggesting that the coral calcification response to ocean acidification may be rather complex. Acidification is likely to affect a range of calcifiers, and this will have a variety of indirect ecological effects that will add to the direct physiological impact. For example, the decline of temperate vermetid reefs from acidification may have significant ecological consequences for their many associated organisms, and may also have socioeconomic ramifications as vermetid reefs provide coastal protection from erosion (Milazzo et al., 2014). Other anomalies have been found in cold water corals from temperate latitudes. For example, *Lophelia pertusa* and *Madrepora oculata* appeared to be insensitive to ocean acidification in both short- and long-term experiments, indicating that long-term acclimation may occur (Maier et al., 2013). It is also becoming apparent that heterogeneity due to local environmental conditions and species composition is likely to be significant in assessing ocean acidification impacts (Chan and Connolly, 2013).

In the temperate zones, pelagic ecosystems are generally variable when it comes to calcification. For example, a recent remote sensing study has revealed variations in the areal extent and net $CaCO_3$ production of surface coccolithophore blooms in the North Atlantic over a 10-year period (1998–2007; Shutler et al., 2013). The inter-annual fluctuation in surface area and net production of these blooms are correlated with the El Niño Southern Oscillation (ENSO), and this variability can result in differences in the extent of their contribution to the annual North Atlantic net sink of atmospheric CO_2 (Shutler et al., 2013). A North Atlantic study between 1997 and 2004 has revealed that > 89% of variation in surface coverage of the coccolithophore species *Emiliana huxleyi* can be explained by variations in physical conditions of solar radiation, mixed layer depth, and water temperature (Raitsos et al., 2006; Shutler et al., 2013). A long-term study (1960–2009) by Beaugrand et al. (2013), conducted in the North Atlantic, reported that calcifying plankton primarily responded to climate-driven warming, overriding the signal from the effects of ocean acidification. Specifically, foraminifera, coccolithophores, both pteropod and non-pteropod molluscs, and echinoderms exhibited drastic shifts circa 1996, coinciding with significant warming (Beaugrand et al., 2013).

The future of coral reefs as ocean acidification and warming continue to increase depends strongly on the ability of corals to acclimate and adapt to future conditions (Palumbi et al., 2014). Existing data on the global distribution of coral reefs suggest that coral reefs can only exist in areas where aragonite saturation exceeds ~3.2 (Kleypas et al., 1999a, b; Hoegh-Guldberg et al., 2007). Over the next century, the surface seawater saturation state of aragonite in tropical oceans is predicted to decline from a pre-industrial mean of ~4 to ~2, and the pH from ~8.05 to ~7.8 (Orr et al., 2005). Reef calcification is not only reduced due to the decline in coral cover and declining rates of coral calcification by the remaining coral colonies, as other groups that contribute to reef accretion (e.g. crustose coralline algae and other calcareous benthic macroalgae) appear even more susceptible to ocean acidification than corals (Price et al., 2011; Diaz-Pulido et al., 2012), and reef night-time dissolution may accelerate with global warming (Langdon et al., 2000; Yates et al., 2007). These factors in combination make it likely that it will be increasingly difficult for reef calcification to compensate for losses from increasing frequencies and intensities of climate-related disturbances, especially in places where coral cover has declined below 20%. Photosynthetic microbioeroders accelerate their rates of dissolution of carbonate substrata under high CO_2 and warming temperatures (Reyes-Nivia et al., 2013). Dissolution is predicted to exceed reef accretion at atmospheric carbon dioxide concentrations of 450 to 560 ppm (Hoegh-Guldberg et al., 2007; Silverman et al., 2009). The losses will therefore lead to a progressive contraction in the geographic extent of coral reefs towards the oceanic regions with highest carbonate saturation state (Kleypas et al., 1999a; Hoegh-Guldberg et al., 2007; Silverman et al., 2009; Hoegh-Guldberg, 2014).

'Hot spots' for ocean acidification studies have given insight into the selection of organisms under high CO_2. For example, studies around volcanic CO_2 seeps, where the benthos has been exposed to elevated CO_2 for decades to millennia, suggest that long-term exposure to increased pCO_2 can lead to reduced structural complexity, increased bioerosion, reduced coral recruitment, or altered balances between primary producers and grazers, and the complete loss of entire groups such as benthic foraminifera (Fabricius et al., 2011; Uthicke et al., 2013). The seep studies also indicate that tropical seagrasses and macroalgae are among the likely winners in a high CO_2 world, with both tropical and temperate sites predicting substantially increased growth rates and biomass as their carbon limitation is alleviated (Hall-Spencer et al., 2008; Fabricius et al., 2011). Some coral species such as massive *Porites* are also highly tolerant to ocean acidification (Fabricius et al., 2011). However, many questions of how ocean acidification will change the future state and performance of tropical benthic and pelagic ecosystems remain unanswered to date.

11.4.3 Ocean acidification and multiple stressors

Climate change together with terrestrial run-off of nutrients and destructive fishing practices have led to the degradation of ecosystems; for example, loss of coral cover on many reefs around the world (Bruno and Selig, 2007). On the Great Barrier Reef, severe storms and heat periods leading to coral bleaching have caused 42% and 10% of these losses, respectively (De'ath et al., 2012). The combined effects of ocean acidification and other environmental stressors, such as warming and deoxygenation, can act in an additive or synergistic fashion (Pörtner et al., 2005; Hoegh-Guldberg et al., 2007). Synergistic effects of ocean acidification and warming can also lead to a narrowing of thermal tolerance windows and reduced performance (reproduction, behaviour, and growth), with potential implications

for geographic ranges and for species interactions (Pörtner, 2008; Gaylord et al., 2015). In phytoplankton, synergistic impacts of climate-relevant variables and nutrient availability appear to affect short-term physiological responses. For example, in coccolithophores, results of multifactorial experiments using varying nutrient and CO_2 concentrations suggest that the competitive ability for nutrient uptake might be altered in future high-CO_2 oceans (Rouco et al., 2013). Some environments such as coastal ecosystems and estuaries are particularly interesting because they are already undergoing significant acidification, on top of eutrophication and pollution (Feely et al., 2008; Cai et al., 2011). Ocean acidification combined with altered nutrient availability or temperature can also dramatically increase the toxicity of some harmful diatom and dinoflagellate algal groups (Fu et al., 2012). These results suggest that, although taxonomy cannot fully explain the diversity of future responses, global changes that include ocean acidification, warming, and coastal eutrophication will undoubtedly alter the structure and functional properties of many marine ecosystems.

11.5 Ecosystem models of ocean acidification effects

Models are being used to estimate how ocean acidification will affect entire ecosystems. For example, Busch et al. (2013) used a food web model (Ecosim) to explore how species abundances in Puget Sound (USA) would change as a result of acidification. They simulated scenarios in which the productivity of calcifying organisms declined over a 50-year period and observed both positive and negative indirect effects on predators, prey, and competitors, although in general indirect effects were weak. In a similar study, Kaplan et al. (2010) used the Atlantis ecosystem model to evaluate how reduction in benthic calcifiers might indirectly affect harvested fish species. Also using the Atlantis model, Griffith et al. (2011, 2012) predicted synergistic effects of ocean acidification and fishing in southeast Australia, especially in the demersal food web. Ainsworth et al. (2011) used an Ecosim model of the California current to simulate multiple ocean acidification and climate change effects (i.e. primary productivity, species range shifts, zooplankton community size structure, ocean acidification, and ocean deoxygenation). The model projected changes in range shifts primarily driven by temperature, but potential ocean acidification effects may not have been well characterized in the model.

These and other ecosystem modelling studies of ocean acidification have some severe limitations. Model results are highly sensitive to the choice of input values, and models typically operate on relatively coarsely lumped 'functional groups' rather than individual species. As revealed by CO_2 exposure experiments, there is a great deal of variability in CO_2 response among closely related species or even within species. Thus, scenarios that assume a direct effect of ocean acidification on an entire functional group will miss a great deal of ecological complexity. There is also the fundamental challenge of assuming that short-term experiments conducted under laboratory conditions on a few individuals can be translated into projected changes in entire populations. For example, the paucity of experiments on key species has been identified as a major shortcoming in a qualitative ecosystem assessment of ocean acidification impacts to the Baltic Sea (Havenhand, 2012). Finally it should be noted that that baseline ecosystem food web models (before applying ocean acidification scenarios) are themselves highly uncertain—it is not clear how well they can predict the response to perturbations outside observed conditions (Busch et al., 2013). Despite these limitations, ecosystem models remain a key tool for representing and predicting how ocean acidification will affect entire ecosystems. Although current models generally do not provide precise predictions of direct effects of OA on species abundances, the models demonstrate the importance of indirect effects in driving ecosystem change.

Most modelling studies on primary production have so far focused on warming rather than ocean acidification, and changes in chlorophyll concentrations as a result of warming appear to differ between ocean basins (Sarmiento et al., 2004). One study based on remote sensing observations combined with ocean chlorophyll records suggest that global ocean annual primary production has declined more than 6% since the early 1980s (Gregg et al., 2003) and almost 70% occurred in high latitudes. This was largely a result of warming and decreases in atmospheric iron deposition (northern hemisphere) and wind stress (Antarctic waters) (Gregg et al., 2003). Recent studies based on models and observations have suggested a general decline of ~1% of the global median net primary production per year (Boyce et al., 2010) largely due to warming. This is important because phytoplankton contributes ~ 50% of the biosphere's net primary production and is central in the cycling of carbon between the upper ocean, biota, and inorganic stocks.

Ecosystem impacts can also be modelled using bioenvelope models, where knowledge of physiological tolerance is used to predict changes in the distribution range of a species in response to a changing environment (Cheung et al., 2009). Shifts in distribution can be complicated and species must simultaneously respond to multiple physiological demands. For example, the temperature-facilitated poleward range expansion of species in response to climate change may be offset by a contraction of areas with suitable seawater carbonate chemistry towards the equator where the CO_3^{2-} saturation state is higher. Couce et al. (2013) developed a bioenvelope model to predict changes in coral distribution under climate change and ocean acidification, and concluded that temperature changes will have a greater effect on reefs than ocean acidification, with the net result of a poleward range shifts in corals, as is already observed (e.g. Japan: up to 14 km yr^{-1}; Yamano et al., 2011). An important limitation of species envelope models that are based solely on physiological tolerance is the potential mismatch in range shifts among trophically linked components of the ecosystem, that could lead to geographic separation of predators and prey (Hoegh-Guldberg, 2014). The biovelope models also do not consider the potential for adaptation and evolution. Even with these limitations, models based on temperature changes alone predict massive rates of local extinctions and species invasions in many regions, and species turnovers greater than 60% of the present biodiversity (Cheung et al., 2009).

In addition to the higher trophic level models (e.g. Ecosim and Atlantis), lower and intermediate trophic level models of ocean acidification effects have been developed, often as part of biogeochemical modelling efforts. Lower trophic level models indicate that plankton changes from ocean acidification can have important effects on biogeochemical cycles and climate. For example, the Medusa model explicitly explores the role of plankton in the carbon 'biological pump' and indicates complex feedbacks in the biological production and sinking of $CaCO_3$ that affects the rate of acidification (Yool et al., 2013). Modelling by Six et al. (2013) indicates that ocean acidification-induced changes in phytoplankton will lead to reduced production of dimethylsulphide (DMS), which will result in a positive feedback on climate change. Like in all models, small changes in assumptions about species productivity or the predicted environment can lead to very different predictions at global scales. Furthermore, species at lower trophic levels tend to be genetically diverse and have short generation times, which facilitates evolution—a factor that is difficult to confidently incorporate into models because of a lack of data on adaptive potential.

11.6 Conclusions

Investigations of the ecological effects of ocean acidification, especially in the context of other global stressors, indicate that responses will require both physiological acclimation (e.g. through energetic trade-offs), and genetic and ecological adaptation (e.g. selection of resilient genotypes with altered pH tolerance ranges, and reorganization of ecosystems). Continued efforts to conduct single-species multi-generational laboratory and mesocosm manipulations are important to understand acclimation and adaptation response mechanisms. However, these experiments must be combined with multi-species observations and experiments *in situ* in ecosystems already experiencing acidification, to better predict responses in population structures, species interactions, and ecosystem functions. Nevertheless, even today much of the existing body of research indicates that ocean acidification already is, and will increasingly become, a driver for significant shifts in species composition and functions in marine ecosystems (Wittmann and Pörtner, 2013). Current models face the challenge of incorporating biological complexity in an ecological and functionally meaningful way.

References

Aberle, N., Schulz, K.G., Stuhr, A., et al. (2013). High tolerance of microzooplankton to ocean acidification in an Arctic coastal plankton community. *Biogeosciences*, **10**, 1471–1481.

Ainsworth, C.H., Samhouri, J.F., Busch, D.S., et al. (2011). Potential impacts of climate change on Northeast Pacific marine foodwebs and fisheries. *ICES Journal of Marine Science*, **68**, 1217–1229.

Allemand, D., Ferrier-Pagès, C., Furla, P., et al. (2004). Biomineralisation in reef-building corals: from molecular mechanisms to environmental control. *Comptes Rendus Palevol*, **3**, 453–467.

Alsterberg, C., Eklöf, J.S., Gamfeldt, L., et al. (2013). Consumers mediate the effects of experimental ocean acidification and warming on primary producers. *Proceedings of the National Academy of Sciences of the USA*, **110**, 8603–8608.

Aronson, R.B., Thatje, S., Clarke, A., et al. (2007). Climate change and invasibility of the Antarctic benthos. *Annual Review of Ecology, Evolution, and Systematics*, **38**, 129–154.

Arrigo, K.R., van Dijken, G., & Pabi, S. (2008). Impact of a shrinking Arctic ice cover on marine primary production. *Geophysical Research Letters*, **35**, 19.

Badger, M.R., Andrews, T.J., Whitney, S.M., et al. (1998). The diversity and coevolution of Rubisco, plastids, pyrenoids, and chloroplast-based CO_2-concentrating mechanisms in algae. *Canadian Journal of Botany*, **76**, 1052–1071.

Bates, N.R. (2006). Air-sea CO_2 fluxes and the continental shelf pump of carbon in the Chukchi Sea adjacent to the Arctic Ocean. *Journal of Geophysical Research*, **111**, C10013.

Bates, N.R. & Mathis, J.T. (2009). The Arctic Ocean marine carbon cycle: evaluation of air-sea CO_2 exchanges, ocean acidification impacts and potential feedbacks. *Biogeosciences*, **6**, 2433–2459.

Bates, N.R., Mathis, J.T., & Cooper, L. (2009). The effect of ocean acidification on biologically induced seasonality of carbonate mineral saturation states in the Western Arctic Ocean. *Journal of Geophysical Research*, **114**(C11). doi:10.1029/2008JC004862.

Beare, D., McQuatters-Gollop, A., van der Hammen, T., et al. (2013). Long-term trends in calcifying plankton and pH in the North Sea. *PLoS ONE*, **8**, e61175.

Beaugrand, G., McQuatters-Gollop, A., Edwards, M., & Goberville, E. (2013). Long-term responses of North Atlantic calcifying plankton to climate change. *Nature Climate Change*, **3**, 263–267.

Bednaršek, N., Tarling, G.A., Bakker, D.C.E., et al. (2012). Extensive dissolution of live pteropods in the Southern Ocean. *Nature Geoscience*, **5**, 881–885.

Beniash, E., Aizenberg, J., Addadi, L., & Weiner, S. (1997). Amorphous calcium carbonate transforms into calcite during sea urchin larval spicule growth. *Proceedings of the Royal Society of London Series B*, **264**, 461–465.

Bertram, M.A., Mackenzie, F.T., Bishop, F.C., & Bischoff, W.D. (1991). Influence of temperature on the stability of magnesian calcite. *American Mineralogist*, **76**, 1889–1896.

Boyce, D.G., Lewis, M.R., & Worm, B. (2010). Global phytoplankton decline over the past century. *Nature*, **466**, 591–596.

Boyd, P.W. & Doney, S.C. (2002). Modelling regional responses by marine pelagic ecosystems to global climate change. *Geophysical Research Letters*, **29**. doi:10.1029/2001GL014130.

Brading, P., Warner, M.E., Davey, P., et al. (2011). Differential effects of ocean acidification on growth and photosynthesis among phylotypes of Symbiodinium (Dinophyceae). *Limnology and Oceanography*, **56**, 927–938.

Brand, L.E., Guillard, R.R.L., & Murphy, L.S. (1981). A method for the rapid and precise determination of acclimated phytoplankton reproduction rates. *Journal of Plankton Research*, **3**, 193–201.

Brandt, A., Gooday, A.J., Brandão, S.N., et al. (2007). First insights into the biodiversity and biogeography of the Southern Ocean deep sea. *Nature*, **447**, 307–311.

Briffa, M., de la Haye, K., & Munday, P.L. (2012). High CO_2 and marine animal behaviour: potential mechanisms and ecological consequences. *Marine Pollution Bulletin*, **64**, 1519–1528.

Bruno, J. & Selig, E. (2007). Regional decline of coral cover in the Indo-Pacific: timing, extent, and subregional comparisons. *PLoS ONE*, **2**, e711.

Brussaard, C.P.D., Noordeloos, A.A.M., Witte, H., et al (2013). Arctic microbial community dynamics influenced by elevated CO_2 levels. *Biogeosciences*, **10**, 719–731.

Buitenhuis, E.T., de Baar, H.J.W., & Veldhuis, M.J.W. (1999). Photosynthesis and calcification by *Emiliania huxleyi* (Prymnesiophyceae) as a function of inorganic carbon species. *Journal of Phycology*, **35**, 949–959.

Busch, D.S., Harvey, C.J., & McElhany, P. (2013). Potential impacts of ocean acidification on the Puget Sound food web. *ICES Journal of Marine Science*, **70**, 823–833.

Byrne, M. (2011). Impact of ocean warming and ocean acidification on marine invertebrate life history stages: vulnerability and potential for persistence in a changing ocean. In: Gibson, R.N., Atkinson, R.J.A., Gordon, J.D.M., Smith, I.P., & Hughes, D.J. (eds), *Oceanography and Marine Biology: An Annual Review*, Vol. 49, pp. 1–42. CRC Press, Boca Raton.

Byrne, M., Ho, M., Selvakumaraswamy, P., et al. (2009). Temperature, but not pH, compromises sea urchin fertilization and early development under near-future climate change scenarios. *Proceedings of the Royal Society of London Series B*, **276**, 1883–1888.

Cai, W.J., Hu, X., Huang, W.J., et al. (2011). Acidification of subsurface coastal waters enhanced by eutrophication. *Nature Geosciences*, **4**, 766e770.

Caldeira, K. & Wickett, M.E. (2003). Anthropogenic carbon and ocean pH. *Nature*, **425**, 365.

Cameron, J.N. (1986). Acid–base equilibria. In: Heisler, N. (ed.), *Acid–Base Regulation in Animals*, pp. 357–394. Elsevier, Amsterdam.

Cao, L. & Caldeira, K. (2008). Atmospheric CO_2 stabilization and ocean acidification. *Geophysical Research Letters*, **35**, L19609.

Chan, N.C.S. & Connolly, S.R. (2013). Sensitivity of coral calcification to ocean acidification: a meta-analysis. *Global Change Biology*, **19**, 282–290.

Cheung, W.W.L., Lam, V.W.Y., Sarmiento, J.L., et al. (2009). Projecting global marine biodiversity impacts under climate change scenarios. *Fish and Fisheries*, **10**, 235–251.

Chierici, M. & Fransson, A. (2009). Calcium carbonate saturation in the surface water of the Arctic Ocean: undersaturation in freshwater influenced shelves. *Biogeosciences*, **6**, 2421–2431.

Clarke, A. & Johnston, N.M. (2003). Antarctic marine benthic diversity. *Oceanography and Marine Biology*, **41**, 47–114.

Cohen, A.L. & Holcomb, M. (2009). Why corals care about ocean acidification: uncovering the mechanism. *Oceanography*, **22**, 118–127.

Cohen, A.L., McCorkle, D.C., de Putron, S., et al. (2009). Morphological and compositional changes in the skeletons of new coral recruits reared in acidified seawater: insights into the biomineralization response to ocean acidification. *Geochemistry, Geophysics, Geosystems*, **10**, Q07005.

Collins, S. & Gardner, A. (2009). Integrating physiological, ecological and evolutionary change: a Price equation approach. *Ecology Letters*, **12**, 744–757.

Comeau, S., Edmunds, P.J., Spindel, N.B., & Carpenter, R.C. (2013). The responses of eight coral reef calcifiers to increasing partial pressure of CO_2 do not exhibit a tipping point. *Limnology and Oceanography*, **58**, 388–398.

Cooper, T.F., O'Leary, R.A., & Lough, J.M. (2012). Growth of Western Australian corals in the Anthropocene. *Science*, **335**, 593–596.

Couce, E., Irvine, P.J., Gregorie, L.J., et al. (2013). Tropical coral reef habitat in a geoengineered, high-CO_2 world. *Geophysical Research Letters*, **40**, 1799–1805.

Cripps, I.L., Munday, P.L., & McCormick, M.I. (2011). Ocean acidification affects prey detection by a predatory reef fish. *PLoS ONE*, **6**, e22736.

De'ath, G., Fabricius, K.E., Sweatman, H., & Puotinen, M. (2012). The 27-year decline of coral cover on the Great Barrier Reef and its causes. *Proceedings of the National Academy of Sciences of the USA*, **109**, 17995–17999.

De'ath, G., Lough, J.M., & Fabricius, K.E. (2009). Declining coral calcification on the Great Barrier Reef. *Science*, **323**, 116–119.

Diaz-Pulido, G., Anthony, K.R.N., Kline, D.I., et al. (2012). Interactions between ocean acidification and warming on the mortality and dissolution of coralline algae. *Journal of Phycology*, **48**, 32–39.

Dixson, D.L., Munday, P.L., & Jones, G.P. (2010). Ocean acidification disrupts the innate ability of fish to detect predator olfactory cues. *Ecology Letters*, **13**, 68–75.

Doney, S.C., Fabry, V.J., Feely, R.A., & Kleypas, J.A. (2009). Ocean acidification: the other CO_2 problem. *Annual Review of Marine Science*, **1**, 169–192.

Dupont, S., Dorey, N., Stumpp, M., et al. (2013). Long-term and trans-life-cycle effects of exposure to ocean acidification in the green sea urchin *Strongylocentrotus droebachiensis*. *Marine Biology*, **160**, 1835–1843.

Dupont, S., Dorey, N., & Thorndyke, M. (2010). What meta-analysis can tell us about vulnerability of marine biodiversity to ocean acidification? *Estuarine, Coastal and Shelf Science*, **89**(2), 182–185.

Edmunds, P.J., Carpenter, R.C., & Comeau, S. (2013). Understanding the threats of ocean acidification to coral reefs. *Oceanography*, **26**, 149–152.

Elena, S.F. & Lenski, R.E. (2003). Evolution experiments with microorganisms: the dynamics and genetic bases of adaptation. *Nature Reviews Genetics*, **4**, 457–469.

El-Shahed, A.M., Ibrahim, H., & Abd-Elnaeim, E. (2006). Isolation and characterization of phosphatase enzyme from the freshwater macroalga *Cladophora glomerata* Kützing (Chlorophyta). *Pakistan Journal of Biological Sciences*, **9**, 2456–2461.

Fabricius, K.E., De'ath, G., Noonan, S., & Uthicke, S. (2014). Ecological effects of ocean acidification and habitat complexity on reef-associated macroinvertebrate communities. *Proceedings of the Royal Society of London Series B*, **281**, 20132479.

Fabricius, K.E., Langdon, C., Uthicke, S., et al. (2011). Losers and winners in coral reefs acclimatized to elevated carbon dioxide concentrations. *Nature Climate Change*, **1**, 165–169.

Fabry, V.J., McClintock, J.S.B., Mathis, J.T., & Grebmeier, J.M. (2009). Ocean acidification at high latitudes: the Bellwether. *Oceanography*, **22**, 160–171.

Fabry, V.J., Seibel, B.A., Feely, R.A., & Orr, J.C. (2008). Impacts of ocean acidification on marine fauna and ecosystem processes. *ICES Journal of Marine Science*, 65, 414–432.

Feely, R.A., Sabine, C.L, Hernandez-Ayon, J.M., et al. (2008). Evidence for upwelling of corrosive 'acidified' water onto the continental shelf. *Science*, 320, 1490–1492.

Ferrari, M.C.O., McCormick, M.I., Munday, P.L., et al. (2012). Effects of ocean acidification on visual risk assessment in coral reef fishes. *Functional Ecology*, 26, 553–558.

Fransson, A., Chierici, M., & Nojiri, Y. (2009). New insights into the spatial variability of the surface water carbon dioxide in varying sea ice conditions in the Arctic Ocean. *Continental Shelf Research*, 29, 1317–1328.

Fu, F.X., Tatters, A.O., & Hutchins, D.A. (2012). Global change and the future of harmful algal blooms in the ocean. *Marine Ecology Progress Series*, 470, 207–233.

Fu, F.-X., Warner, M.E., Zhang, Y., et al. (2007). Effects of increased temperature and CO_2 on photosynthesis, growth and elemental ratios of marine *Synechococcus* and *Prochlorococcus* (cyanobacteria). *Journal of Phycology*, 43, 485–496.

Gallagher, J.C. (1982). Physiological variation and electrophoretic banding patterns of genetically different seasonal populations of *Skeletonema cosfatum* (Bacillariophyceae). *Journal of Phycology*, 18, 148–162.

Gao, K., Helbling, E.W., Häder, D.-P., & Hutchins, D.A. (2012). Responses of marine primary producers to interactions between ocean acidification, solar radiation, and warming. *Marine Ecology Progress Series*, 470, 167–189.

Gattuso, J.-P., Allemand, D., & Frankignoulle, M. (1999). Photosynthesis and calcification at cellular, organismal and community levels in coral reefs: a review on interactions and control by carbonate chemistry. *American Zoologist*, 39, 160–183.

Gattuso, J.-P. & Kirkwood, W. (2014). xFOCE: long-term impacts of ocean acidification *in situ*. *Limnology and Oceanography Bulletin*, 23, 15–16.

Gattuso, J.-P., Reynaud-Vaganay, S., Bourge, I., et al. (2000). Calcification does not stimulate photosynthesis in the zooxanthellate scleractinian coral *Stylophora pistillata*. *Limnology and Oceanography*, 45, 246–250.

Gaylord, B., Kroeker, K.J., Sunday, J.M., et al (2015). Ocean acidification through the lens of ecological theory. *Ecology*, 96, 3–15.

Gehlen, M., Gangsto, R., Schneider, B., et al. (2007). The fate of pelagic $CaCO_3$ production in a high CO_2 ocean: a model study. *Biogeosciences*, 4, 505–519.

Godoi, R.H.M., Aerts, K., Harlay, J., et al. (2009). Organic surface coating on coccolithophores—*Emiliana huxleyi*: its determination and implication in the marine carbon cycle. *Microchemical Journal*, 91, 266–271.

Gooding, R.A., Harley, C.D.G., & Tang, E. (2009). Elevated water temperature and carbon dioxide concentration increase the growth of a keystone echinoderm. *Proceedings of the National Academy of Sciences of the USA*, 106, 9316–9321.

Gregg, W.W., Conkright, M.E., Ginoux, P., et al. (2003). Ocean primary production and climate: global decadal changes. *Geophysical Research Letters*, 30. doi:10.1029/2003GL016889.

Griffith, G.P., Fulton, E.A., Gorton, R., & Richardson, A.J. (2012). Predicting interactions among fishing, ocean warming, and ocean acidification in a marine system with whole-ecosystem models. *Conservation Biology*, 26, 1145–1152.

Griffith, G.P., Fulton, E.A., & Richardson, A.J. (2011). Effects of fishing and acidification-related benthic mortality on the southeast Australian marine ecosystem. *Global Change Biology*, 17, 3058–3307.

Gutowska, M.A., Melzner, F., Portner, H.O., & Meier, S. (2010). Cuttlebone calcification increases during exposure to elevated seawater pCO_2 in the cephalopod *Sepia officinalis*. *Marine Biology*, 157, 1653–1663.

Hall-Spencer, J.M., Rodolfo-Metalpa, R., Martin, S., et al. (2008). Volcanic carbon dioxide vents show ecosystem effects of ocean acidification. *Nature*, 454, 96–99.

Härnström, K., Ellegaard, M., Andersen, T. J., & Godhe, A. (2011). Hundred years of genetic structure in a sediment revived diatom population. *Proceedings of the National Academy of Sciences of the USA*, 108, 4252–4257.

Havenhand, J.N. (2012). How will ocean acidification affect Baltic sea ecosystems? An assessment of plausible impacts on key functional groups. *Ambio*, 41, 637–644.

Herfort, L., Thake, B., & Taubner, I. (2008). Bicarbonate stimulation of calcification and photosynthesis in two hermatypic corals. *Journal of Phycology*, 44, 91–98.

Hoegh-Guldberg, O. (2014). Coral reef sustainability through adaptation: glimmer of hope or persistent mirage? *Current Opinion in Environmental Sustainability*, 7, 127–133.

Hoegh-Guldberg, O., Mumby, P.J., Hooten, A.J., et al. (2007). Coral reefs under rapid climate change and ocean acidification. *Science*, 318, 1737–1742.

Hofmann, G.E., Smith, J.E., Johnson, K.S., et al. (2011). High-frequency dynamics of ocean pH: a multi-ecosystem comparison. *PLoS ONE*, 6, e28983.

Holcomb, M., McCorkle, D.C., & Cohen, A.L. (2010). Long-term effects of nutrient and CO_2 enrichment on the temperate coral Astrangia. *Journal of Experimental Marine Biology and Ecology*, 386, 27–33.

Hönisch, B., Ridgwell, A., Schmidt, D.N., et al. (2012). The geological record of ocean acidification. *Science*, 335, 1058–1063.

Iglesias-Rodriguez, M.D., Anthony, K.R.N., Bijma, J., et al. (2010). Towards an integrated global ocean acidification observation network. In: Hall, J., Harrison, D.E., & Stammer, D. (eds), *Proceedings of the 'OceanObs'09: Sustained Ocean Observations and Information for Society' Conference (Vol. 2), Venice, Italy, 21–25 September 2009*, p. 24. ESA Publication WPP-306. doi: 10.5270/OceanObs09.

Iglesias-Rodriguez, M.D., Halloran, P.R., Rickaby, R.E.M., et al. (2008). Phytoplankton calcification in a high-CO_2 world. *Science*, 320, 336–340.

Iglesias-Rodriguez, M.D., Schofield, O.M., Batley, J., et al. (2006). Intraspecific genetic diversity in the marine coccolithohore *E. huxleyi* (Prymnesiophyceae): the use of microsatellite analysis in marine phytoplankton population studies. *Journal of Phycology*, **42**, 526–536.

Jin, P., Gao, K., & Beardall, J. (2013). Evolutionary responses of a coccolithophorid *Gephyrocapsa oceanica* to ocean acidification. *Evolution*, **67**, 1869–1878.

Jokiel, P.L., Jury, C.P., & Rodgers, K.S. (2014). Coral-algae metabolism and diurnal changes in the CO_2-carbonate system of bulk sea water. *PeerJ*, **2**, e378.

Jury, C.P., Whitehead, R.F., & Szmant, A.M. (2010). Effects of variations in carbonate chemistry on the calcification rates of *Madracis auretenra* (= *Madracis mirabilis* sensu Wells, 1973): bicarbonate concentrations best predict calcification rates. *Global Change Biology*, **16**, 1632–1644.

Kaplan, I.C., Burden, M., Levin, P.S., & Fulton, E.A. (2010). Fishing catch shares in the face of global change: a framework for integrating cumulative impacts and single species management. *Canadian Journal of Fisheries and Aquatic Sciences*, **67**, 1968–1982.

Kleypas, J.A., Buddemeier, R., Archer, D., et al. (1999a). Geochemical consequences of increased atmospheric carbon dioxide on coral reefs. *Science*, **284**, 118–120.

Kleypas, J.A., McManus, J.W., & Menez, L.A.B. (1999b). Environmental limits to coral reef development: where do we draw the line? *American Zoologist*, **39**, 146–159.

Kroeker, K.J., Kordas, R.L., Crim, R., et al. (2013). Impacts of ocean acidification on marine organisms: quantifying sensitivities and interaction with warming. *Global Change Biology*, **19**, 1884–1896.

Kroeker, K.J., Kordas, R.L., Crim, R.N., & Singh, G.G. (2010). Meta-analysis reveals negative yet variable effects of ocean acidification on marine organisms. *Ecology Letters*, **13**, 1419–1434.

Kurihara, H. (2008). Effects of CO_2-driven ocean acidification on the early developmental stages of invertebrates. *Marine Ecology Progress Series*, **373**, 275–284.

Kurihara, H. & Shirayama, Y. (2004). Effects of increased atmospheric CO_2 on sea urchin early development. *Marine Ecology Progress Series*, **274**, 161e169.

Langdon, C., Broecker, W.S., Hammond, D.E., et al. (2003). Effect of elevated CO_2 on the community metabolism of an experimental coral reef. *Global Biogeochemical Cycles*, **17**, 1011.

Langdon, C., Takahashi, T., Sweeney, C., et al. (2000). Effect of calcium carbonate saturation state on the calcification rate of an experimental coral reef. *Global Biogeochemical Cycles*, **14**, 639–654.

Langer, G., Geisen, M., Baumann, K.-H., et al. (2006). Species-specific responses of calcifying algae to changing seawater carbonate chemistry. *Geochemistry, Geophysics, Geosystems*, **7**. doi: 10.1029/2005GC001227.

Langer, G., Nehrke, G., Probert, I., et al. (2009). Strain-specific responses of *Emiliania huxleyi* to changing seawater carbonate chemistry. *Biogeosciences*, **6**, 2637–2646.

Lebrato, M., Iglesias-Rodríguez, M.D., Feely, R.A., et al. (2010). Global contribution of echinoderms to the marine carbon cycle. *Ecological Monographs*, **80**, 441–467.

Leonardos, N., Read, B., Thake, B., & Young, J.R. (2009). No mechanistic dependence of photosynthesis on calcification in the coccolithophorid *Emiliania huxleyi* (Haptophyta). *Journal of Phycology*, **45**, 1046–1051.

Lilley, R.M., Ralph, P.J., & Larkum, A.W.D. (2010). The determination of activity of the enzyme Rubisco in cell extracts of the dinoflagellate alga *Symbiodinium* sp by manganese chemiluminescence and its response to short-term thermal stress of the alga. *Plant Cell and Environment*, **33**, 995–1004.

Lohbeck, K.T., Riebesell, U., & Reusch, T.B.H. (2012). Adaptive evolution of a key phytoplankton species to ocean acidification. *Nature Geoscience*, **5**, 346–351.

Maier, C., Schubert, A., Berzunza Sànchez, M.M., et al. (2013). End of the century pCO$_2$ levels do not impact calcification in Mediterranean cold-water corals. *PLoS ONE*, **8**, e62655.

Manzello, D.P. (2010). Coral growth with thermal stress and ocean acidification: lessons from the eastern tropical Pacific. *Coral Reefs*, **29**, 749–758.

Marin, F., Luquet, G., Marie, B., & Medakovic, D. (2008). Molluscan shell proteins: primary structure, origin and evolution. *Current Topics in Developmental Biology*, **80**, 209–276.

Marubini, F. & Thake, B. (1999). Bicarbonate addition promotes coral growth. *Limnology and Oceanography*, **44**, 716–720.

Maslanik, J.A., Fowler, C., Stroeve, J., et al. (2007). A younger, thinner Arctic ice cover: increased potential for rapid, extensive sea-ice loss. *Geophysical Research Letters*, **34**. doi:10.1029/2007GL032043.

McClintock, J., Ducklow, H., & Fraser, B. (2008). Ecological impacts of climate change on the Antarctic Peninsula. *American Scientist*, **96**, 302–310.

McGuire, A., Chapin III, D.F.S., Walsh, J.E., & Wirth, C. (2006). Integrated regional changes in arctic climate feedbacks: implications for the global climate system. *Annual Review of Environmental Resources*, **31**, 61–91.

McGuire, A.D., Anderson, L.G., Christensen, T.R., et al. (2009). Sensitivity of the carbon cycle in the Arctic to climate change. *Ecological Monographs*, **79**, 523–555.

Medlin, L.K., Lange, M., & Nothig, E.V. (2000). Genetic diversity in the marine phytoplankton: a review and a consideration of Antarctic phytoplankton. *Antarctic Science*, **72**, 325–333.

Melzner, F., Thomsen, J., Koeve, W., et al. (2013). Future ocean acidification will be amplified by hypoxia in coastal habitats. *Marine Biology*, **160**, 1875–1888.

Merico, A., Tyrrell, T., & Cokacra, T. (2006). Is there any relationship between phytoplankton seasonal dynamics and the carbonate system? *Journal of Marine Systems*, **59**, 120–142.

Milazzo, M., Rodolfo-Metalpa, R., Chan, V.B.S., et al. (2014). Ocean acidification impairs vermetid reef recruitment. *Scientific Reports*, **4**, 4189.

Moy, A.D., Howard, W.R., Trull, T.W., & Bray, S. (2009). Reduced calcification in modern Southern Ocean planktonic foraminifera. *Nature Geoscience*, **2**, 276–280.

Mucci, A. & Morse, J.W. (1983). The incorporation of Mg^{2+} and Sr^{2+} into calcite overgrowths: influences of growth rate and solution composition. *Geochimica et Cosmochimica Acta*, **47**, 217–233.

Munday, P.L. (2014). Transgenerational acclimation of fishes to climate change and ocean acidification. *F1000Prime Reports*, **6**, 99.

Munday, P.L., Cheal, A.J., Dixson, D.L., et al. (2014). Behavioural impairment in reef fishes caused by ocean acidification at CO_2 seeps. *Nature Climate Change*, **4**, 487–492.

Munday, P.L., Dixson, D.L., Donelson, J.M., et al. (2009). Ocean acidification impairs olfactory discrimination and homing ability of a marine fish. *Proceedings of the National Academy of Sciences of the USA*, **106**, 1848–1852.

Muscatine, L., McCloskey, L.R., & Marian, R.E. (1981). Estimating the daily contribution of carbon from zooxanthellae to coral animal respiration. *Limnology and Oceanography*, **26**, 601–611.

Niehoff, B., Schmithüsen, T., Knüppel, N., et al. (2013). Mesozooplankton community development at elevated CO_2 concentrations: results from a mesocosm experiment in an Arctic fjord. *Biogeosciences*, **10**, 1391–1406.

Nilsson, G.E., Dixson, D.L., Domenici, P., et al. (2012). Near-future carbon dioxide levels alter fish behaviour by interfering with neurotransmitter function. *Nature Climate Change*, **2**, 201–204.

Nimer, N.A. & Merrett, M.J. (1993). Calcification rate in *Emiliania huxleyi* Lohmann in response to light, nitrate and inorganic carbon availability. *New Phytologist*, **123**, 673–677.

Orr, J.C., Fabry, V.J., Aumont, O., et al. (2005). Anthropogenic ocean acidification over the twenty-first century and its impact on calcifying organisms. *Nature*, **437**, 681–686.

Paasche, E. (1964). A tracer study of the inorganic carbon uptake during coccolith formation and photosynthesis in the coccolithophorid *Coccolithus huxleyi*, *Physiologia Plantarum Supplementum*, **3**, 1–82.

Pabi, S., van Dijken, G.L., & Arrigo, K.R. (2008). Primary production in the Arctic Ocean, 1998–2006. *Journal of Geophysical Research*, **113**, C8.

Palumbi, S.R., Barshis, D.J., Traylor-Knowles, N., & Bay, R.A. (2014). Mechanisms of reef coral resistance to future climate change. *Science*, **344**, 895–898.

Parker, L.M., Ross, P.M., O'Connor, W.A., et al. (2012). Adult exposure influences offspring response to ocean acidification in oysters. *Global Change Biology*, **18**, 82–92.

Politi, Y., Arad, T., Klein, E., et al. (2004). Sea urchin spine calcite forms via a transient amorphous calcium carbonate phase. *Science*, **306**, 1161–1164.

Pörtner, H.-O. (2008). Ecosystem effects of ocean acidification in times of ocean warming: a physiologist's view. *Marine Ecology Progress Series*, **373**, 203–217.

Pörtner, H.-O., Langenbuch, M., & Michaelidis, B. (2005). Synergistic effects of temperature extremes, hypoxia, and increases in CO_2 on marine animals: from Earth history to global change. *Journal of Geophysical Research*, **110**, C09S10.

Price, N.N., Hamilton, S.L., Tootell, J.S., & Smith, J.E. (2011). Species-specific consequences of ocean acidification for the calcareous tropical green algae *Halimeda*. *Marine Ecology Progress Series*, **440**, 67–78.

Raitsos, D.E., Lavender, S. J., Pradhan, Y., et al. (2006). Coccolithophore bloom size variation in response to the regional environment of the subarctic North Atlantic. *Limnology and Oceanography*, **51**, 2122–2130.

Raven, J.A. (2011). Effects on marine algae of changed seawater chemistry with increasing atmospheric CO_2. *Royal Irish Academy*, **111**. doi:10.3318/BIOE.2011.01.

Raz, S., Hamilton, P.C., Wilt, F.H., et al. (2003). The transient phase of amorphous calcium carbonate in sea urchin larval spicules: the involvement of proteins and magnesium ion in its formation and stabilization. *Advanced Functional Materials*, **13**, 480–486.

Reinfelder, J.R. (2011). Carbon concentrating mechanisms in eukaryotic marine phytoplankton. *Annual Review of Marine Science*, **3**, 291–315.

Reusch, T.B.H. & Boyd, P.W. (2013). Experimental evolution meets marine phytoplankton. *Evolution*, **67**, 1849–1859.

Reyes-Nivia, C., Diaz-Pulido, G., Kline, D.I., et al. (2013). Ocean acidification and warming scenarios increase microbioerosion of coral skeletons. *Global Change Biology*, **19**, 1919–1929.

Reynaud, S., Leclercq, N., Romaine-Lioud, S., et al. (2003). Interacting effects of CO_2 partial pressure and temperature on photosynthesis and calcification in a scleractinian coral. *Global Change Biology*, **9**, 1660–1668.

Riebesell, U., Zondervan, I., Rost, B., et al. (2000). Reduced calcification of marine plankton in response to increased atmospheric CO_2. *Nature*, **407**, 364–367.

Ries, J.B. (2011). A physicochemical framework for interpreting the biological calcification response to CO_2-induced ocean acidification. *Geochimica et Cosmochimica Acta*, **75**, 4053–4064.

Ries, J.B., Cohen, A.L., & McCorkle, D.C. (2009). Marine calcifiers exhibit mixed responses to CO_2-induced ocean acidification. *Geology*, **37**, 1131–1134.

Ries, J.B., Cohen, A.L., & McCorkle, D.C. (2010). A nonlinear calcification response to CO_2-induced ocean acidification by the coral *Oculina arbuscula*. *Coral Reefs*, **29**, 661–674.

Rodolfo-Metalpa, R., Martin, S., Ferrier-Pagès, C., & Gattuso, J.-P. (2010). Response of the temperate coral *Cladocora caespitosa* to mid- and long-term exposure to $pCO2$ and temperature levels projected for the year 2100. *Biogeosciences*, **7**, 289–300.

Roos, A. & Boron, W.F. (1981). Intracellular pH. *Physiological Reviews*, **61**, 296–434.

Ross, P.M., Parker, L., O'Connor, W.A., & Bailey, E.A. (2011). The impact of ocean acidification on reproduction, early development and settlement of marine organisms. *Water*, **3**, 1005–1030.

Rost, B., Zondervan, I., & Wolf-Gladrow, D. (2008). Sensitivity of phytoplankton to future changes in ocean carbonate chemistry: current knowledge, contradictions and research directions. *Marine Ecology Progress Series*, **373**, 227–237.

Rouco, M., Branson, O., Lebrato, M., & Iglesias-Rodríguez, M.D. (2013). The effect of nitrate and phosphate availability on *Emiliania huxleyi* (NZEH) physiology under different CO_2 scenarios. *Frontiers in Microbiology*, **4**. doi: 10.3389/fmicb.2013.00155.

Roy, A.-S., Gibbons, S.M., Schunck, H., et al. (2013). Ocean acidification shows negligible impacts on high-latitudebacterial community structure in coastal pelagic mesocosms. *Biogeosciences*, **10**, 555–566.

Rynearson, T.A. & Armbrust, E.V. (2000). DNA fingerprinting reveals extensive genetic diversity in a field population of the centric diatom *Ditylum brightwellii*. *Limnology and Oceanography*, **45**, 1329–1340.

Sabine, C.L., Feely, R.A., Gruber, N., et al. (2004). The oceanic sink for anthropogenic CO_2. *Science*, **305**, 367–371.

Sarmiento, J.L., Slater, R., Barber, R., et al. (2004). Response of ocean ecosystems to climate warming. *Global Biogeochemical Cycles*, **18**, GB3003. doi:10.1029/2003GB002134.

Schneider, K. & Erez, J. (2006). The effect of carbonate chemistry on calcification and photosynthesis in the hermatypic coral *Acropora eurystoma*. *Limnology and Oceanography*, **51**, 1284–1293.

Serreze, M.C. & Francis, J.A. (2006). The Arctic amplification debate. *Climatic Change*, **76**, 241–264.

Shutler, J.D., Land, P.E., Brown, C.W., et al. (2013). Coccolithophore surface distributions in the North Atlantic and their modulation of the air-sea flux of CO_2 from 10 years of satellite Earth observation data. *Biogeosciences*, **10**, 2699–2709.

Silverman, J., Lazar, B., Cao, L., et al. (2009). Coral reefs may start dissolving when atmospheric CO_2 doubles. *Geophysical Research Letters*, **36**, L05606.

Six, K.D., Kloster, S., Ilyina, T., et al. (2013). Amplified global warming by altered marine sulfur emissions induced by ocean acidification. *Nature Climate Change*, **3**, 975–978.

Snelgrove, P.V.R. (2001). Marine sediments. In: Levin, S.A. (ed.), *Encyclopedia of Biodiversity*, pp. 71–84. Academic Press, San Diego.

Somero, G.N. (1985). Intracellular pH, buffering substances and proteins: imidazole protonation and the conservation of protein structure and function. In: Gilles, R. & Gilles-Baillien M. (eds), *Transport Processes, Iono- and Osmoregulation*, pp. 454–468. Springer, Berlin.

Steinacher, M., Joos, F., Frolicher, T.L., et al. (2009). Imminent ocean acidification in the Arctic projected with the NCAR global coupled carbon cycle-climate model. *Biogeosciences*, **6**, 515–533.

Stumpp, M., Dupont, S., Thorndyke, M.C., & Melzner, F. (2011). CO_2 induced seawater acidification impacts sea urchin larval development II: gene expression patterns in pluteus larvae. *Comparative Biochemistry and Physiology Part A*, **160**(3), 320–330.

Sunda, W.G. & Cai, W.-J. (2012). Eutrophication induced CO_2-acidification of subsurface coastal waters: interactive effects of temperature, salinity, and atmospheric pCO_2. *Environmental Science and Technology*, **46**, 10651–10659.

Sunday, J.M., Calosi, P., Dupont, S., et al. (2014). Evolution in an acidifying ocean. *Trends in Ecology and Evolution*, **29**, 117–125.

Takahashi, T., Sutherland, S.C., Sweeney, C., et al. (2002). Global sea-air CO_2 flux based on climatological surface ocean pCO_2, and seasonal biological and temperature effects. *Deep Sea Research Part II*, **49**, 1601–1622.

Tanzil, J.T.I., Brown, B.E., Tudhope, A.W., & Dunne, R.P. (2009). Decline in skeletal growth of the coral *Porites lutea* from the Andaman Sea, South Thailand between 1984 and 2005. *Coral Reefs*, **28**, 519–528.

Taylor, A.R., Chrachri, A., Wheeler, G., et al. (2011). A voltage-gated H^+ channel underlying pH homeostasis in calcifying coccolithophores. *PLoS Biology*, **9**, e1001085.

Teniswood, C.M.H., Roberts, D., Howard, W.R., & Bradby, J.E. (2013). A quantitative assessment of the mechanical strength of the polar pteropod *Limacina helicina antarctica* shell. *ICES Journal of Marine Science*, **70**(7), 1499–1505.

Thatje, S. & Arntz, W.E. (2004). Antarctic reptant decapods: more than a myth? *Polar Biology*, **27**, 195–201.

Thatje, S. & Lörz, A.N. (2005). First record of lithodid crabs from Antarctic waters off the Balleny Islands. *Polar Biology*, **28**, 334–337.

Thoms, S., Pahlow, M., & Wolf-Gladrow, D.A. (2001). Model of the carbon concentrating mechanism in chloroplasts of eukaryotic algae. *Journal of Theoretical Biology*, **208**, 295–313.

Tortell, P.D., DiTullio, G.R., Sigman, D.M., & Morel, F.M.M. (2002). CO_2 effects on taxonomic composition and nutrient utilization in an equatorial Pacific phytoplankton assemblage. *Marine Ecology Progress Series*, **236**, 37–43.

Trimborn, S., Langer, G., & Rost, B. (2007). Effect of varying calcium concentrations and light intensities on calcification and photosynthesis in *Emiliania huxleyi*. *Limnology and Oceanography*, **52**, 2285–2293.

Truchot, J.-P. (1987). *Comparative Aspects of Extracellular Acid–Base Balance*. Zoophysiology, Vol. 20. Springer, Berlin, Heidelberg.

Tyrrell, T., Schneider, B., Charalampopoulou, A., & Riebesell, U. (2008). Coccolithophores and calcite saturation state in the Baltic and Black Seas. *Biogeosciences*, **5**, 485–494.

Uthicke, S. & Fabricius, K. (2012). Productivity gains do not compensate for reduced calcification under near-future ocean acidification in the photosynthetic benthic foraminifera *Marginopora vertebralis*. *Global Change Biology*, **18**, 2781–2791.

Uthicke, S., Momigliano, P., & Fabricius, K.E. (2013). High risk of extinction of benthic foraminifera in this century due to ocean acidification. *Scientific Reports*, **3**, 1769.

Walsh, P.J. & Milligan, C.L. (1989). Coordination of metabolism and intracellular acid-base status: ionic regulation and metabolic consequences. *Canadian Journal of Zoology*, **67**, 2994–3004.

Wang, M. & Overland, J.E. (2009). A sea ice free summer Arctic within 30 years? *Geophysical Research Letters*, **36**. doi:10.1029/2009GL037820.

Watson, S.-A., Lefevre, S., McCormick, M.I., et al. (2014). Marine mollusc predator-escape behaviour altered by near-future carbon dioxide levels. *Proceedings of the Royal Society of London Series B*, **281**, 20132377.

Watson, S.-A., Peck, L.S., Tyler, P.A., et al. (2012). Marine invertebrate skeleton size varies with latitude, temperature and carbonate saturation: implications for global change and ocean acidification. *Global Change Biology*, **18**, 3026–3038.

Weiss, I.M., Tuross, N., Addadi, L., & Weiner, S. (2002). Mollusc larval shell formation: amorphous calcium carbonate is a precursor phase for aragonite. *Journal of Experimental Zoology*, **293**, 478–491.

Whitney, S., Shaw, D., & Yellowlees, D. (1995). Evidence that some dinoflagellates contain a ribulose -1,5-bisphosphate carboxylase/oxygenase related to that of the alpha-proteobacteria. *Proceedings of the Royal Society of London Series B*, **259**, 271–275.

Wittmann, A.C. & Pörtner, H.-O. (2013). Sensitivities of extant animal taxa to ocean acidification. *Nature Climate Change*, **3**, 995–1001.

Wood, A.M. & Leatham, T. (1992). The species concept in phytoplankton ecology. *Journal of Phycology*, **28**, 723–729.

Wood, H.L., Spicer, J.I., & Widdicombe, S. (2008). Ocean acidification may increase calcification rates, but at a cost. *Proceedings of the Royal Society of London Series B*, **275**, 1767–1773.

Yamada, N. & Suzumura, M. (2010). Effects of seawater acidification on hydrolytic enzyme activities. *Journal of Oceanography*, **66**, 233–241.

Yamano, H., Sugihara, K., & Nomura, K. (2011). Rapid poleward range expansion of tropical reef corals in response to rising sea surface temperatures. *Geophysical Research Letters*, **38**, L04601.

Yates, K.K., Dufore, C., Smiley, N., et al. (2007). Diurnal variation of oxygen and carbonate system parameters in Tampa Bay and Florida Bay. *Marine Chemistry*, **104**, 110–124.

Yool, A., Popova, E. E., & Anderson, T.R.E. (2013). MEDUSA-2.0: an intermediate complexity biogeochemical model of the marine carbon cycle for climate change and ocean acidification studies. *Geoscientific Model Development*, **6**, 1767–1811.

Young, J.R. & Henriksen, K. (2003). Biomineralization within vesicles: the calcite of coccoliths. In: Dove, P.M., De Yoreo, J.J., & Weiner, S. (eds), *Biomineralisation. Volume 54: Reviews in Mineralogy and Geochemistry*, pp. 189–215. Mineralogical Society of America, Chantilly, VA.

CHAPTER 12

Effects of temperature stress on ecological processes

Elizabeth A. Morgan, Alastair Brown, Benjamin J. Ciotti, and Anouska Panton

12.1 Summary

Temperature is a fundamental determinant of chemical and biological rates, with profound influences on living systems at all levels of organization. Investigating how ocean temperature affects marine ecosystems is critical to understanding consequences of global climate change. This chapter describes influences of temperature on ecological processes from organism to community levels. Specifically we (i) outline global spatial and temporal temperature patterns and projected changes, (ii) examine direct and indirect influences of these temperature variations on marine organisms and consider resulting implications for ecological processes and species distributions, (iii) use examples from several habitats to demonstrate emergent consequences of temperature-driven changes in ecological processes for the structure and function of marine communities, and (iv) highlight the importance of interactions between temperature and other abiotic variables.

12.2 Global spatial trends in mean temperature and temperature variability

Heat input to the global oceans is delivered predominantly by absorption of infrared solar radiation in the surface metre of water, and consequently temperature varies across latitude and depth (Lalli and Parsons, 1997). The intensity of the radiation (insolation) reaching the sea surface depends on the distance the solar radiation has to travel through the atmosphere and the surface area over which energy will be spread, which are determined by the angle of the sun with respect to the Earth's surface. Insolation also varies because the Earth's rotational axis is not perpendicular to its

orbital plane, driving the seasonal cycle. Ocean circulation determines how heat input at the surface is mixed throughout the ocean (Gage and Tyler, 1991). Wind-driven currents distribute heat across latitudes, with currents from low latitudes transporting warm waters polewards, and currents from high latitudes transporting cool water equatorwards (Brown et al., 2001). Deep ocean circulation is at present driven by downward penetration of heat at low latitudes due to turbulent mixing, and density-driven sinking of bottom waters at high latitudes (thermohaline circulation); seawater density increases during cooling and as salinity increases, which occurs as salt is rejected during sea ice formation and growth (Brown et al., 2001). Halo-thermal circulation (density-driven sinking at low latitudes caused by increased salinity due to evaporation) has dominated during other times in Earth history (McClain and Hardy, 2010). Deep water temperature at the Equator is within a few degrees of deep water in polar regions, where water columns are approximately isothermal (Gage and Tyler, 1991).

Latitudinal and bathymetric gradients in ocean temperature occur in both hemispheres and have contributed to latitudinal and bathymetric biodiversity patterns (see Brown and Thatje, 2014). At low latitudes, open ocean sea surface temperatures can exceed 30 °C (Lalli and Parsons, 1997). Sea surface temperatures at high latitudes reach the freezing point of seawater: −1.9 °C (Gage and Tyler, 1991). The global distribution of sea surface temperature characterizes four major biogeographic zones: tropical (>25 °C), sub-tropical (25–15 °C), temperate (15–5 °C in the northern hemisphere, or 15–2 °C in the southern hemisphere), and polar (<5 °C in the northern hemisphere or <2 °C in the southern hemisphere) (Fig. 12.1; Lalli and Parsons, 1997). Wind and tides force vertical mixing,

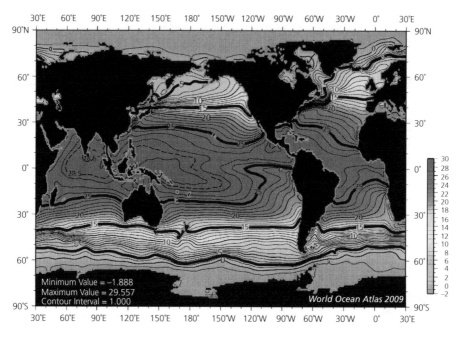

Figure 12.1 Mean annual sea surface temperature (°C) climatology on a one-degree latitude-longitude grid. Bold lines indicate limits to major thermal biogeographic zones: tropical (>25 °C), subtropical (25–15 °C), temperate (15–5 °C in the northern hemisphere, or 15–2 °C in the southern hemisphere), and polar (<5 °C in the northern hemisphere or <2 °C in the southern hemisphere) (Lalli and Parsons, 1997). Adapted from Locarini et al. (2010). [PLATE 10]

transferring heat downwards from the surface, establishing a surface mixed layer of between a few metres and several hundred metres deep at low and mid latitudes (Lalli and Parsons, 1997). Temperature decreases with depth below this layer in the permanent thermocline, reaching ~5 °C at ~1000 m and only decreasing slowly beyond this depth to reach ~2 °C (Gage and Tyler, 1991). Seasonal thermoclines form in the temperate surface layer during summer, as increased solar radiation elevates surface temperature and winds lessen, reducing turbulent mixing (Gage and Tyler, 1991).

Temperature in the ocean can vary at a wide range of temporal scales (Thompson et al., 2013). The high specific heat capacity of water buffers variation in water temperature; a large change in heat is required to deliver a small change in temperature. Consequently, mean daily variation in open ocean surface temperatures is typically small (< 0.3 °C), and decreases rapidly with depth: even in coastal waters daily surface temperature change is typically less than 2 °C (Lalli and Parsons, 1997). Annual surface temperature variations are larger, ranging from less than 2 °C in tropical and polar seas, to nearly 10 °C at mid latitudes; annual water temperature fluctuations in coastal areas reflect

air temperature variation and may exceed 10 °C (Lalli and Parsons, 1997). Intertidal habitat can experience considerably greater temperature changes over tidal cycles, with additional differences between day and night that vary seasonally. Deep sea temperatures are more stable, with little variation on daily and annual scales (Gage and Tyler, 1991), but have varied over geological periods as deep ocean circulation shifts between thermohaline and halothermal modes (McClain and Hardy, 2010), possibly by as much as 17 °C (Cramer et al., 2011).

Temperature experienced by marine biota can differ remarkably from regional conditions due to local-scale processes. For example, maximum and minimum ocean surface temperatures in the Antarctic intertidal zone are determined by several factors including water movement by tidal forces, wave action, air temperature, sunlight intensity, shore lithology, and the presence of ice and snow in the area (Kuklinski and Balazy, 2014). In addition, temperature experienced by intertidal biota will be highly variable due to local oceanographic and meteorological processes such as tidal elevation, air temperature, solar radiation, wind speed, wave height, and the timing of low tide. Temperature changes at local scales can therefore be highly complex (Helmuth et al., 2011).

12.2.1 Recent temperature observations

Anthropogenic impacts on radiative forcing agents (e.g. carbon dioxide, methane, halocarbons, nitrous oxide, ozone, stratospheric water vapour from methane, surface albedo, contrails, aerosol–radiation interactions, and aerosol–cloud interactions) have increased radiative forcing of climate, leading to an uptake of energy by the climate system (Rhein et al., 2013). Ocean warming dominates the total planetary energy change, accounting for ~93% of the total on average from 1971 to 2010 (Rhein et al., 2013). Over this period the strongest warming has occurred in the surface ocean (0.11 °C decade^{-1} in the upper 75 m of the ocean), and warming decreases with depth (~0.015 °C decade^{-1} at 700 m) (Rhein et al., 2013). As a result, thermal stratification between 0 and 200 m water depth has increased by about 4% (Rhein et al., 2013). Sea surface temperatures have increased more rapidly in coastal locations than in oceanic regions off the continental shelf (Lima and Wethey, 2012). Widespread long-term warming has shifted isotherms polewards in almost all regions (Sen Gupta et al., 2015), although the pattern is reversed in some places (e.g. Pinksy et al., 2013). However, speeds vary among regions, seasons, and over time, because the speed of isotherm movement depends on local sea surface temperature gradients and there is heterogeneity in ocean warming (Sen Gupta et al., 2015).

Analysis by Lima and Wethey (2012) indicates that the incidence of extreme temperature events has also increased, particularly in coastal areas. Between 1982 and 2010, almost three quarters of coastal areas experienced a significant increase in sea surface temperature (mean rate of 0.25 °C decade^{-1}), whereas the annual number of extremely hot days (sea surface temperature exceeding the 95th percentile of daily sea surface temperatures from 1982 to 2010) over this period increased in nearly 40% of the world's coastal areas (mean rate of 13.8 days decade^{-1}). These changes correlate with the intensity of warming. Concurrently, the number of yearly extreme cold days (sea surface temperature below the 5th percentile of daily sea surface temperatures from 1982 to 2010) significantly decreased in almost half of coastal areas (mean rate of 13.1 days decade^{-1}). As a consequence, approximately three quarters of coastlines experienced a simultaneous increase in extreme hot days and decrease in extreme cold days. However, there is considerable variability in trends; ~10% of coastlines experienced a simultaneous increase in the frequency of both hot days and cold days. Further, significant cooling occurred in ~7% of coastal areas between 1982 and 2010, and ~1% of the world's coastlines experienced a decrease in the number of yearly extremely hot days (mean rate of 12.9 days decade^{-1}) (Lima and Wethey, 2012).

12.2.2 Temperature projections

The ocean mixed layer is projected to respond rapidly to external forcing compared to the deep ocean (Collins et al., 2013). Globally averaged surface and near-surface ocean temperatures are predicted to warm during the twenty-first century (Collins et al., 2013). Projected surface warming varies between emission scenarios, with regional variations in amplitude; the strongest surface and near-surface warming is projected at low and mid latitudes (Fig. 12.2; Collins et al., 2013). Increases in sea surface temperature are projected to be greater in summer than in winter in most ocean regions, and therefore net poleward isotherm speeds are expected to be faster in summer than in winter (Sen Gupta et al., 2015). Median isotherm speeds during the twenty-first century may be seven times faster than during the last century (Sen Gupta et al., 2015). Although temperature extremes are challenging to forecast, it is projected that the frequency and magnitude of warm days (on which the maximum temperature exceeds the 90th percentile for that day relative to the period 1961–1990), and the duration, frequency, and intensity of warm spells (≥ 6 consecutive warm days) or heat waves (≥ 5 days on which the maximum temperature exceeds 5 °C more than temperature recorded for that day in the period 1961–1990) will increase (Seneviratne et al., 2012).

The predicted timescale of temperature changes in deep ocean environments is much slower than in surface waters. Mixing and circulation are projected to transfer additional heat to deeper levels of about 2000 m by the end of the twenty-first century; global ocean warming between 0.5 and 1.5 °C will reach a depth of about 1 km by the end of the century. Warming at depths between 200 and 1000 m depth is strongest in northern hemisphere high latitudes (Fig. 12.2; Collins et al., 2013); in deeper water, warming is most pronounced in the Southern Ocean where deep water formation is projected to decline (Collins et al., 2013). In contrast, slight cooling is projected for parts of the northern mid- and high-latitudes below 1000 m, linked to the projected decrease in the strength of the northern hemisphere deep circulation (Collins et al., 2013). Deep ocean temperatures are projected to continue changing beyond the twenty-first century; coupled

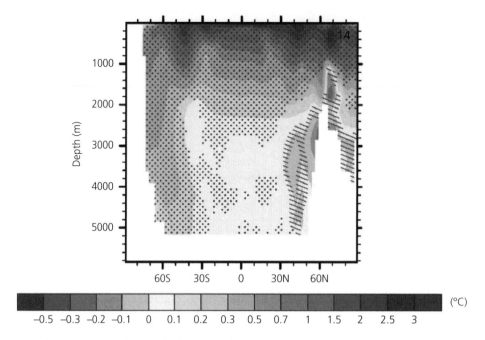

Figure 12.2 Annual mean temperature change in the global oceans for 2081–2100 relative to 1986–2005. Coupled Model Intercomparison Project Phase 5 multi-model changes in annual mean temperature under the most severe representative concentration pathway (>8.5 W/m² by 2100 and rising). Hatching indicates mean change less than one standard deviation of internal variability; stippling indicates mean change is greater than two standard deviations of internal variability and at least 90% of models agree on the sign of change. Reproduced from Collins et al. (2013).

ocean–atmosphere general circulation model simula-tions suggest that the deep ocean will not reach equi-librium with external forcing for several millennia (Stouffer and Manabe, 2003).

12.3 Influences of thermal stress on ecological processes

12.3.1 Organism-level influences

Temperature influences processes at all levels of bio-logical organization, ranging across enzyme activi-ties, metabolic rates, individual performance traits (e.g. body size, growth rate, reproductive output, and immune competence) and demographic pro-cesses controlling population dynamics (Atkinson et al., 2003; Gardner et al., 2011; Kordas et al., 2011). Therefore, temperature is a fundamental determinant of the function, occurrence, and abundance of organ-isms in nature, and thus their ecological role. Warming increases performance at the lower end of an organ-ism's thermal range but has the reverse effect above

a thermal optimum (Walther et al., 2002; Pörtner and Farrell, 2008; Kordas et al., 2011; Rogers et al., 2011). Temperature change therefore has different outcomes depending on the spatial or seasonal context, such as the location or time of year. For example, warming in-creased growth rates of juvenile cod (*Gadhus morhua*) on the Norwegian coast in spring but had the oppo-site effect in summer when thermal optima were ex-ceeded (Rogers et al., 2011). The speed and severity of temperature changes are also critical: sudden, extreme events are likely to have greater impacts than gradual changes in mean temperature alone (Smale and Wern-berg, 2013; Thompson et al., 2013).

Thermal sensitivities, tolerance ranges, and op-tima can all vary among species, life stages, and even populations (Pörtner and Farrell, 2008; Byrne, 2011; Kordas et al., 2011). Species from warmer (e.g. upper intertidal) or more stable (e.g. tropical) environments live close to upper acute thermal tolerance limits, and are therefore particularly susceptible to warming (Walther et al., 2002; Stillman, 2003). For example, due to their extreme thermal sensitivity, tropical corals

undergo catastrophic bleaching and mortality following relatively moderate, short-term temperature increases (Hughes et al., 2003). Temperature responses
may vary within species: smaller individuals of some
Antarctic invertebrate species survive to higher temperatures under acute warming (Peck et al., 2009).
Variation in sensitivity to thermal stress among life
stages means that the timing of life history events,
such as spawning, fertilization, and development,
relative to seasonal temperature dynamics are also
critical (Pörtner and Farrell, 2008; Byrne, 2011; Sommer et al., 2012). Thermal stress at specific times of
year associated with thermally sensitive life-history
stages or population bottlenecks may have greater
impacts than changes in mean or maximum temperatures (Sommer et al., 2012), and warming of surface
waters in winter and spring has disrupted sensitive
processes of reproduction and growth (Poloczanska
et al., 2013).

12.3.2 Influences on biotic interactions

The spatial distribution, abundance, and function
of marine organisms depends not only on the direct effects of temperature but also on indirect effects
mediated by changes in competitors, mutualists, predators, prey, or pathogens (Kordas et al., 2011). Since
thermal responses vary among species, the outcome
of biotic interactions will be highly sensitive to temperature. This can occur if the strength or sign of an
individual's interaction with the community ('per
capita effects'), or the total number of individuals in
a population ('density effects') changes due to temperature (Kordas et al., 2011). For example, feeding
rate (per capita effect) and intertidal densities (density effect) of the ochre sea star *Pisaster ochraceus* were
higher under warm conditions, thus increasing predation on intertidal mussels (Sanford, 1999). Barnacles
from the rocky intertidal illustrate the temperature
dependence of competitive interactions: on both sides
of the north Atlantic, species with lower temperature
ranges competitively exclude species with higher temperature ranges from the mid and high intertidal, but
not when the species with lower temperature ranges
are excluded from these tidal levels by thermal stress
(summarized by Kordas et al., 2011). Similarly, higher
temperatures reduced ecological and behavioural performance in the Mediterranean rainbow wrasse *Coris
julis*, relative to the peacock wrasse *Thalassoma pavo*,
a more thermally tolerant competitor (Milazzo et al.,
2013). Facilitation among marine organisms will also
be affected: warming-related declines in fucoid algal

canopies represent losses of cool, moist microhabitats
for associated invertebrate taxa (Harley et al., 2012).
There is clear evidence that temperature changes can
alter disease risk in marine organisms through influences on the host, the pathogen, or even on disease
vectors (Altizer et al., 2013). The wider Caribbean region is emerging as a disease hotspot, with intense and
widespread deterioration of corals due to temperature-
induced stress as well as the emergence of new diseases (Burge et al., 2014).

The outcome of thermally driven changes in the
distribution, abundance, and interactions of species
are likely to have complex, but potentially profound,
community-level implications. Changes in interactions involving important groups, such as keystone
species or ecosystem engineers, can dramatically
alter the structure and function of communities and
ecosystems. Temperature-dependent predation by *P.
ochraceus* on rocky shores in the north east Atlantic, for
example, shifts monocultures of competitively dominant mussels to diverse algal and invertebrate assemblages (Sanford, 1999). Similarly, thermal bleaching
and mortality in tropical hard corals has ramifications
for entire reef communities, since these corals play
numerous ecological roles, not least as sources of biogenic habitats for a huge diversity of fishes and invertebrates (Jones et al., 2004). Finally, warming-related
declines in large fucoid algae are likely to have far-
reaching consequences on wave-protected and semi-
exposed British shores (Harley et al., 2006). These
algae are important competitors for space and are
ecosystem engineers, providing a cool, moist microhabitat. Increased physiological stress and abundance
of warm-water grazers resulting from rising air temperatures are expected to reduce algal cover. Reduced
algal cover, coupled with distribution shifts in barnacle species that form the primary substrate, will inhibit
establishment of new algal propagules. The resulting
decline in subcanopy habitat and benthic primary
productivity will affect algae-associated invertebrate
communities as well as the algal detritus food web
of the strand line and the birds that forage on these
invertebrate resources. Insights into the broader ecological consequences of temperature change gained
from whole community manipulations are discussed
in Section 12.4.

12.3.3 Influences on species distributions

By influencing physiological performance and biotic interactions, temperature ultimately determines
how marine organisms are distributed in time and

space (Kordas et al., 2011). Distributions shift at small scales when organisms move into thermal refugia: this occurs in environments with steep temperature gradients, such as the intertidal or thermally stratified waters (Helmuth et al., 2002). For example, upper limits of vertical intertidal zones decrease due to long-term warming (Harley, 2011) and extreme temperature events (Hawkins and Hartnoll, 1985). Similarly, some North Sea fish species respond to climatic variation by local movement offshore or into pockets of deeper water (Perry et al., 2005). Biotic interactions may mediate influences of temperature on fine-scale

---

Box 12.1 Thermal restriction of intertidal organisms

Zonation patterns of intertidal organisms are useful model systems for examining the physiological limits and biotic interactions underlying ecological responses to temperature change. In general, upper intertidal limits are determined by thermal tolerance, but distributions are also influenced by temperature-sensitive biotic interactions. Harley (2011) examined changes in the vertical extent of mussel beds in the Salish Sea on the west coast of North America: an area characterized by a west to east gradient of increasing temperature. While upper limits of intertidal mussels and barnacles were correlated with site temperature, the upper foraging limit of predatory sea stars was not. This caused predator-free space to be restricted and zones of sessile invertebrates were compressed and even-

tually excluded along the spatial temperature gradient. A 52-year period of warming during the late twentieth century and early twenty-first century produced similar zone shifts, resulting in 50% reductions in the extent of vertical zones and local extinctions. Spatial and temporal changes in sessile invertebrate distributions also had broader ecological implications, apparently influencing species richness of associated mesofaunal communities. This demonstrates that temperature change can dramatically alter species distributions and community structure at both local (i.e. along the intertidal gradient) and regional (i.e. along west to east temperature gradient) scales, due to both the physiological tolerances and biotic interactions of keystone species.

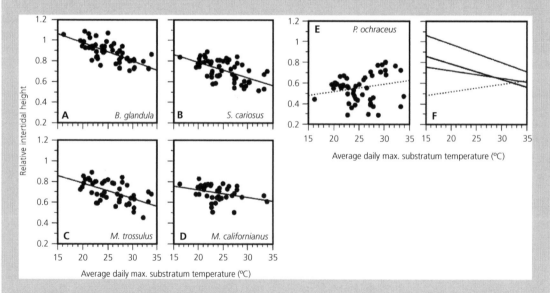

Box 12.1, Fig. 1. Relation between intertidal zonation patterns and temperature in intertidal fauna in the Salish Sea, across a spatial thermal gradient. Upper distributional limits of the barnacles *Balanus glandula* (A) and *Semibalanus cariosus* (B) and the mussels *Mytilus trossulus* (C) and *M. californianus* (D) are negatively related to mid-intertidal substratum temperature, whereas the upper foraging limit of the predatory sea star *Pisaster ochraceus* (E) is not. The *y* axes represent scaled intertidal heights, where 0 = extreme low water and 1 = extreme high water for the year in which the data were collected (2006). Regression lines for A–E (solid lines = significant, dotted lines = not significant; α = 0.05) are superimposed in (F); note that the regression lines for *S. cariosus* and *M. trossulus* overlap. Reproduced from Harley (2011).

distributions. For example, warming led to the competitive exclusion of *C. julis* from suboptimal habitats (Milazzo et al., 2013), a shift that did not occur in the absence of competition. Studies of intertidal invertebrates on the west coast of North America have illustrated the complex interactions between physiological tolerance limits and predator–prey interactions responsible for temperature-related shifts in species distributions and community structure, and how local-scale distributional shifts dictate broader-scale range boundaries (Box 12.1).

At geographic scales, considerable evidence now exists that global warming is causing poleward shifts in species distributions (Pearson and Dawson, 2003; Harley et al., 2006; Burrows et al., 2011; Kordas et al., 2011; Box 12.2). Despite a slower rate of warming, these shifts appear to be faster in the ocean than on land (Burrows et al., 2011). Spatial distributional responses to climate change are complex because species dispersal, habitat availability, biotic interactions, and evolutionary change complicate expectations based on physiological models or existing distributions (Davis et al., 1998; Pearson and Dawson, 2003). Furthermore, changes in thermal stress may not vary consistently with latitude, and subregional or even habitat-specific changes in thermal stress can produce unexpected or spatially complex distribution shifts (Helmuth et al., 2002; Burrows et al., 2011).

In addition to spatial shifts, thermally sensitive biological processes can be moved along seasonal temperature clines. Climate warming is causing earlier springs and longer summers with consequences for the timing of reproduction, larval release, migration, and population blooms of marine organisms (Poloczanska et al., 2013). As with biogeographic shifts, phenological changes in the ocean appear to be rapid, relative to the rate of warming (Edwards and Richardson,

2004; Burrows et al., 2011; Poloczanska et al., 2013). However, the effect of temperature on the timing of biological events varies between species and in some cases these events do not respond to thermal cues at all (Edwards and Richardson, 2004; Kordas et al., 2011).

Changes in spatial or temporal distribution along temperature clines expose organisms to new abiotic conditions or habitat features and mediate overlap with biotic factors including prey, competitors, predators, and pathogens. Phenological changes in temperate marine ecosystems, for example, control the extent to which life cycles of planktonic organisms are matched to seasonal maxima of food availability (Sommer et al., 2012). Warming-related seasonal progression of physiological development, larval release, and reproduction of many planktonic organisms in the North Sea has become mismatched to light-cued blooms in prey populations, disrupting energy flow up the food chain (Edwards and Richardson, 2004). Similarly, mild winters are thought to produce poor recruitment of bivalves in European estuaries by increasing the seasonal overlap between bivalve early life-history stages and shrimp predators (Philippart et al., 2003). Temperature also mediates variations in spatial overlaps, increasing the spatial overlap between intertidal mussels and the predator, *P. ochraceus*, for example (Harley, 2011) (Box 12.1). Spatial or temporal distribution shifts will produce novel combinations of species or cause species drop-outs with consequences for community structure and functioning (Kordas et al., 2011; Sommer et al., 2012; Brown and Thatje, 2015). Temperature-related range expansions in pathogens, for example, have created new host–pathogen pairings, inflicting widespread damage to coral reef ecosystems and abalone populations (Altizer et al., 2013; Burge et al., 2014).

Box 12.2 Evidence from West Australia of tropical species range expansion

The influence of extreme climatic events on a species' range can result in compositional changes to fish and invertebrate communities. Extreme ocean warming on the west coast of Australia during the 2011 La Niña event led to the expansion of tropical species' distributional ranges, changing community composition in typically temperate Jurien Bay (Wernberg et al., 2013). Evidence from West Australia suggests that the ability of tropical fish to overwinter in temper-

ate waters and thus undergo range expansion is increasing in line with seawater warming. The local environmental conditions deteriorated, shifting the community structure from a diverse kelp bed to a depauperate ecological state of simple algal turfs. Extreme warming events such as these are predicted to increase in frequency and magnitude as a direct result of global warming, contributing to species distribution shifts and ecosystem restructuring.

Box 12.2 *Continued*

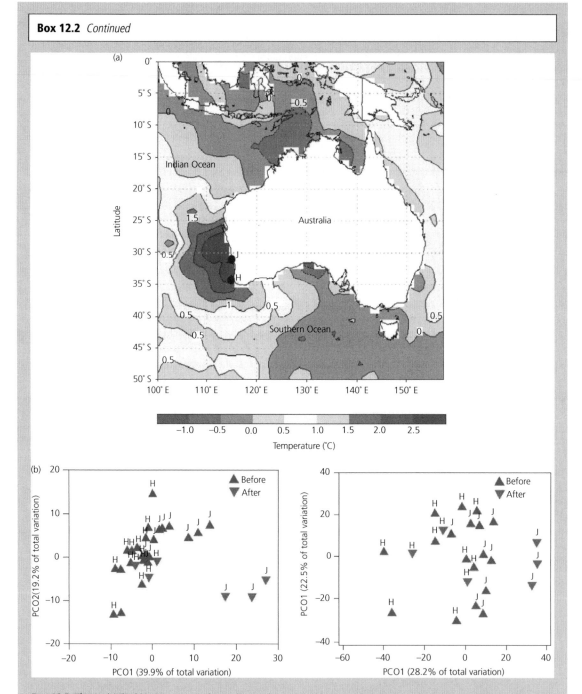

Box 12.2, Fig. 1. (a) The 2011 heat wave in the southeast Indian Ocean (relative to 1971–2000 baseline). Increased warming was observed (>2.5 °C) along the west coast of Australia; position of Jurien Bay (J) and Hamelin Bay (H) indicated. (b, c) The ecological structure of marine communities before and after the heat wave of 2011. Principal coordinates analysis of (b) benthic (invertebrates and macroalgae) and (c) fish community structure on the rocky reefs at each study location, before and after the 2011 warming event. PCO1 and PCO2 are the first and second principal coordinate axes, indicating percentage of variation explained by each axis. Reproduced from Wernberg et al. (2013). [PLATE 11]

12.4 Community responses to temperature stress

While evidence is accumulating for effects of temperature change/stress on organismal performance, species interactions, and species' distributions, predicting how this will ultimately influence the structure and function of whole communities is challenging due to the complexity and interspecific variability of direct and indirect impacts (Fig. 12.3). To meet this challenge, the influences of thermal stress in a variety of habitats have been studied at multiple trophic levels, both in the field and in laboratory-based manipulations (examples are discussed below and summarized in Table 12.1). The ultimate aim of these studies is to identify the impact of temperature change on ecosystem-level processes such as energy transfer, habitat complexity, and community structure.

12.4.1 Phytoplankton and the microbial loop

Laboratory and mesocosm-based studies on tropical, temperate, and high-latitude planktonic communities (both pelagic and benthic) suggest that both gross photosynthesis rates and phytoplankton biomass decrease as water temperature increases beyond physiologically

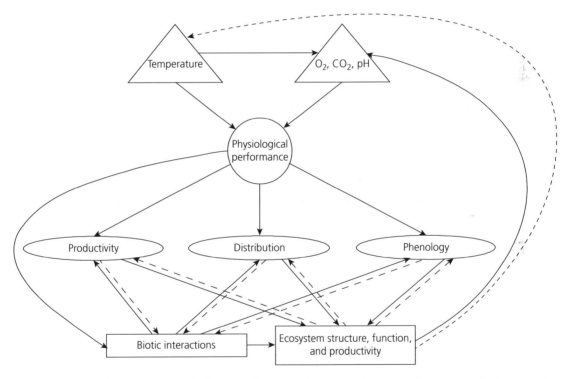

Figure 12.3 Schematic diagram representing the direct and indirect effects of temperature change on marine ecosystems. Triangles represent abiotic factors; other shapes represent hierarchical biological organizational levels: circle represents individuals, ellipses represent species, and rectangles represent communities and ecosystems. Solid arrows represent direct consequences of temperature change; dashed arrows represent feedbacks. Temperature change may impact the physiological performance of individuals both directly and indirectly through temperature-dependence of oxygen and carbon dioxide supply, and pH. The effect of fluctuations in abiotic factors varies between species and therefore affects the outcome of biotic interactions. Adjustments to physiological performance may cause independent or concomitant shifts in species parameters (productivity, distribution, and phenology), and/or in biotic interactions, which all affect ecosystem structure, function, and productivity. Effects on biotic interactions may directly influence species parameters, but adjustments to species parameters may also feedback to biotic interactions. Biotic interactions determine ecosystem structure, function, and productivity, which may feedback to species parameters, and may also influence environmental oxygen and carbon dioxide supply, and pH. Modifications to ecosystem structure, function, and productivity may affect environmental temperature too, e.g. through loss of ecosystem engineers that ameliorate thermal conditions. Shifts in ecosystem structure, function, and productivity may result in changes in the vulnerability and resilience of ecosystems and biodiversity. Adapted from Hollowed et al. (2013).

Table 12.1 Examples of the impact of temperature stress at a community level.

Stressor	Habitat type	Species affected	Impact	Impact on community	Ecological consequence	Reference
Increased temperature	Rocky intertidal	Mussels	Decreased upper distributional limit	Decreased predator-free space	Exclusion of mussels and associated species, shift in community structure	Harley (2011)
Marine heat wave	Rocky reef	Seaweed	Decreased range of distribution	Loss of habitat-forming canopy	Degraded ecosystem, decreased productivity, and energy transfer	Smale and Wernberg (2013)
Increased temperature events	Rocky benthos	Multiple	Mass mortality	Loss of multiple invertebrate species	Decreased biodiversity	Crisci et al. (2011)
Thermal discharge	Rocky benthos	Reef fish	Decreased habitat complexity	Decreased species richness	Altered community composition	Teixeira et al. (2009)
Increased temperature	Coral reef	Coral	Mortality	Decreased food and shelter for juvenile fish, increased competition, mortality of specialist species	Altered interspecific interactions, shift in community structure	McCormick (2012)
Thermal discharge	Subtidal	Kelp	Switch in dominant species	Changes in abundance of invertebrate and fish species	Altered community composition	Schiel et al. (2004)

optimal temperature, but that community respiration rates are stimulated by a similar increase in temperature. The result is a decrease in the carbon fixation to respiration ratio (Hancke and Glud, 2004; Müren et al., 2005; Kordas et al., 2011). Bacterial production and respiration rates increase with temperature in the absence of substrate limitation and, as grazing rates increase at a lower rate than bacterial production, bacterial abundance also increases (Sarmento et al., 2010). In addition, the size distribution of phytoplankton and zooplankton communities shift to domination by smaller organisms, and zooplankton growth and grazing rates typically increase (Keller et al., 1999; Müren et al., 2005; Aberle et al., 2007). Subsequently, the quality and quantity of organic matter sinking in particulate form to the benthos will decrease. The effects at an ecosystem level are a shortened duration of the spring bloom and an earlier switch from autotrophy to heterotrophy. As the spring bloom can constitute the bulk food supply for many benthic organisms, changes in the pelagic zone may also impact benthic ecosystem functioning even if the water temperature does not increase in these regions (e.g. in the deep-sea; Jones et al., 2014). Ultimately, the ability of the system to act as a biological sink for carbon dioxide is impaired (Kritzberg

et al., 2010). However, in the context of global climate change, warming surface waters are predicted to be accompanied by changes in water column stability and inorganic nutrient concentrations. Primary production in mid latitudes is expected to decrease as a result (e.g. Hollowed et al., 2013), but primary production in high latitudes is expected to increase (e.g. Doney, 2006). The timing of warming with respect to the seasonal cycle is important, and evidence suggests that this is particularly true in high-latitude ecosystems where factors such as low substrate availability in winter months can limit respiration rates despite increasing water temperature (Vaquer-Sunyer et al., 2010).

12.4.2 Habitat complexity

Some marine habitats are particularly sensitive to fluctuations in water temperature as the habitat itself is created by a small number of species with relatively narrow thermal tolerances, such as coral, macroalgae (kelp or other seaweeds), or seagrasses. These foundation species not only provide a physical structure but also control important variables such as light availability, turbidity, and dissolved oxygen concentration. Such reefs, 'meadows', or 'forests' are some of the most diverse marine communities globally. While some species do exhibit plasticity and resilience to cope with thermal change (Kemp et al., 2011), evidence from a number of studies suggests that temperature stress negatively impacts upon these habitat-forming species. The result is typically a complete loss of local habitat structure and a switch to a degraded ecosystem with low diversity and decreased energy transfer. As larval forms of marine organisms may also be reliant upon these ecosystems, the impacts may be more far-reaching than just the local environment.

12.4.3 Community structure

The response of marine communities to changes in temperature varies according to the direct and indirect impacts arising from thermal stress. A simulated heatwave (9.5 °C above mean temperature for three days) resulted in a decrease in total adult species richness and a shift in adult community composition from native to non-native species in a subtidal epibenthic fouling community (California, USA), but no detectable impact on juvenile species richness (Sorte et al., 2010). Recovery time, recruitment success, and intraspecific competition were all important factors in determining species survival. Observation of the epilithic community in a thermal flume of a nuclear power plant in Sweden over a period of a year demonstrated an increase in the temporal turnover of three different functional groups of benthic organisms (diatoms, macrophytes, and macroinvertebrates) with warming (Hillebrand et al., 2010). The taxa that benefited from increased temperature in this brackish environment tended to display opportunistic reproduction strategies, such as asexual reproduction or short generation times, or displayed seasonally independent reproduction patterns.

12.5 Interactions of temperature with other stressors

Influences of temperature on ecological processes are modified by complex interactions with other abiotic factors. Temperature modulates levels of other important abiotic variables, such as the partial pressures of gases, pH, and salinity, and biological responses to temperature may be compounded by additional variables such as anthropogenic pollution. As a result, it is difficult to tease apart the impacts of temperature from other abiotic stressors because changes in environmental conditions associated with temperature result

Box 12.3 Ocean circulation and ecosystem engineers-the complex example of the East Australia Current

Sea surface temperature and salinity off the east coast of Tasmania have increased steadily over the last 50 years as a result of intensification of the East Australia Current (EAC), leading to oceanic warming at a rate 3–4 times the global average (Ridgway, 2007). This warm, nutrient-poor boundary current originating from the South Pacific Gyre now makes stronger and more frequent extensions into the region, replacing the cooler, nutrient-rich sub-Antarctic water masses as the dominant oceanographic signature (Johnson et al., 2011). Evidence from multi-decadal data sets of physical and biological parameters suggest that this shift in ocean circulation, in conjunction with other stressors, has had far-reaching direct and indirect impacts on multiple trophic levels. This has resulted in observed changes in phytoplankton abundance, as lower dissolved silicate concentrations in the EAC are believed to be restricting diatom growth and ultimately primary productivity (Thompson et al., 2009). A decline in phytoplankton spring bloom biomass of approximately 50% was observed between 1997 and 2007 (Thompson et al., 2009). Furthermore, distinct shifts in zooplankton community composition with increased abundance of warm water 'signature species' have been observed when comparing data from the 1970s and 2000s (Johnson et al., 2011). In addition to these changes in the planktonic community, a decline in the extent of giant kelp forests (*Macrocystis pyrifera*) observed since 1980 has been most pronounced in the northeast of Tasmania. Some sites have lost >90% of average canopy extent since the 1940s (Johnson et al., 2011). Increased water temperature and low nutrient levels are attributed as causes of these declines although overgrazing by sea urchins is also a contributing factor. Increased water temperature provides more favourable conditions for larval development and recruitment of the barrens-forming sea urchin *Centrostephanus rodgersii* (Ling and Johnson, 2009). Overfishing of large individuals of the southern rock lobster (*Jasus edwardsii*), the primary predator of *C. rodgersii*, has allowed urchin population size to expand, leading to destructive overgrazing of macroalgae. There has been a poleward range expansion of this ecosystem engineer over the past 40 years, from New South Wales along the entire eastern coast of Tasmania (Ling et al., 2009). More research is required to improve understanding of how these stressors affect responding species, but there is overwhelming evidence that this large-scale oceanographic shift has had profound impacts on the marine ecosystems of eastern Tasmania.

in simultaneous shifts in other environmental parameters and complicated cause and effect relationships. For example, increasing anthropogenic pollution in the Baltic Sea combined with rising water temperatures have increased the frequency, severity, and duration of hypoxic and anoxic episodes in sea basins (Kotilainen et al., 2014). These episodes exacerbate environmental problems by mobilizing toxic heavy metals and nutrients from seafloor sediments, which amplify eutrophication and toxicity problems (Kotilainen et al., 2014). In addition, atmospheric warming over the Baltic Sea is predicted to increase precipitation, increase river water run-off, and shorten the sea-ice season. These temperature mediated changes in oxygen depletion and biogeochemical processes are expected to lead to large changes in the Baltic Sea ecosystem. However, local adaptation can moderate stress responses, even in organisms that are broadly tolerant and highly plastic (Schulte, 2014). The saltmarsh Atlantic killifish, *Fundulus heteroclitus*, encounters highly variable environmental conditions. Photosynthesis increases water oxygen levels during the day, and respiration decreases them at night. Rates of both photosynthesis and respiration increase with temperature and, as a result, both extreme hyperoxia and hypoxia are more common in the summer (Layman et al., 2000) and less common in the winter (Schulte, 2014). *F. heteroclitus* has demonstrated tolerance to changing temperature, salinity, and oxygenation. However, killifish interpopulation variation in sensitivity to environmental stressors and acclimation ability (Shulte, 2014) indicates that populations that are currently intolerant of environmental extremes may have the capacity to adapt.

12.6 Conclusion

Temperature is a key driver of ecosystem change, affecting the function of organisms and ecosystems globally (Clarke, 2003; Kordas et al., 2011; Alcaraz et al., 2014; Brown and Thatje, 2014). Predicting how changes in temperature will impact marine ecosystems over short- and long-term timescales requires understanding of how communities will respond to thermal stress. Research needs to focus on key areas such as identifying critical species within an ecosystem and the life stages most vulnerable to change, and the interactions between these organisms and the wider ecosystem (Russell et al., 2012). The capacity to acclimate and adapt are fundamental for resolving organism-level responses to long-term temperature change, but remain little understood (Hughes et al., 2003; Harley et al., 2006; Blois et al., 2013; De Block et al., 2013) and need

further exploration. Critically, potential environmental stressors often occur in synchrony, making knowledge of interactive effects essential in determining consequences for marine organisms (Helmuth et al., 2010).

References

Aberle, N., Lengfellner, K., & Sommer, U. (2007). Spring bloom succession, grazing impact and herbivore selectivity of ciliate communities in response to winter warming. *Oecologia*, **150**, 668–681.

Alcaraz, M., Felipe, J., Grote, U., et al. (2014). Life in a warming ocean: thermal thresholds and metabolic balance of arctic zooplankton. *Journal of Plankton Research*, **36**, 3–10.

Altizer, S., Ostfeld, R.S., Johnson, P.T.J., et al. (2013). Climate change and infectious diseases: from evidence to a predictive framework. *Science*, **341**, 514–519.

Atkinson, D., Ciotti, B.J., & Montagnes, D.J.S. (2003). Protists decrease in size linearly with temperature: ca. 2.5% degrees C^{-1}. *Proceedings of the Royal Society of London Series B*, **270**, 2605–2611.

Blois, J.L., Zarnetske, P.L., Fitzpatrick, M.C., & Finnegan, S. (2013). Climate change and the past, present, and future of biotic interactions. *Science*, **341**, 499–504.

Brown, A. & Thatje, S. (2014). Explaining bathymetric diversity patterns in marine benthic invertebrates and demersal fishes: physiological contributions to adaptation of life at depth. *Biological Reviews*, **89**, 406–426.

Brown, A. & Thatje, S. (2015). The effects of changing climate on faunal depth distributions determine winners and losers. *Global Change Biology*, **21**(1), 173–80.

Brown, E., Colling, A., Park, D., et al. (2001). *Ocean Circulation*, 2nd edition. Elsevier Butterworth-Heinemann, Oxford.

Burge, C.A., Eakin, C.M., Friedman, C.S., et al. (2014). Climate change influences on marine infectious diseases: implications for management and society. *Annual Review of Marine Science*, **6**, 249–277.

Burrows, M.T., Schoeman, D.S., Buckley, L.B., et al. (2011). The pace of shifting climate in marine and terrestrial ecosystems. *Science*, **334**, 652–655.

Byrne, M. (2011). Impact of ocean warming and ocean acidification on marine invertebrate life history stages: vulnerabilities and potential for persistence in a changing ocean. *Oceanography and Marine Biology*, **49**, 1–42.

Clarke, A. (2003). Costs and consequences of evolutionary temperature adaptation. *Trends in Ecology and Evolution*, **18**, 573–581.

Collins, M., Knutti, R., Arblaster, J., et al. (2013). Long-term climate change: projections, commitments and irreversibility. In: Stocker, T.F., Qin, D., Plattner, G.-K., et al. (eds), *Climate Change 2013: The Physical Science Basis. Contribution of Working Group I to the Fifth Assessment Report of the International Panel on Climate Change*. Cambridge University Press, Cambridge and New York.

Cramer, B.S., Miller, K.G., Barrett, P.J., & Wright, J.D. (2011). Late Cretaceous-Neogene trends in deep ocean temperature and continental ice volume: reconciling records of benthic foraminiferal geochemistry (δ^{18}O and Mg/Ca) with sea level history. *Journal of Geophysical Research*, **116**, C12023.

Crisci, C., Bensoussan, N., Romano, J.-C., & Garrabou, J. (2011). Temperature anomalies and mortality events in marine communities: insights on factors behind differential mortality impacts in the NW Mediterranean. *PLoS One*, **6**(9), e23814.

Davis, A.J., Jenkinson, L.S., Lawton, J.H., et al. (1998). Making mistakes when predicting shifts in species range in response to global warming. *Nature*, **391**, 783–786.

De Block, M., Pauwels, K., Van den Broeck, M., et al. (2013). Local genetic adaptation generates latitude-specific effects of warming on predator-prey interactions. *Global Change Biology*, **19**, 689–696.

Doney, S. (2006). Plankton in a warmer world. *Nature*, **444**, 695–696.

Edwards, M. & Richardson, A.J. (2004). Impact of climate change on marine pelagic phenology and trophic mismatch. *Nature*, **430**, 881–884.

Gage, J.D. & Tyler, P.A. (1991). *Deep-Sea Biology: A Natural History of Organisms at the Deep-Sea Floor*. Cambridge University Press, Cambridge.

Gardner, J.L., Peters, A., Kearney, M.R., et al. (2011). Declining body size: a third universal response to warming? *Trends in Ecology and Evolution*, **26**, 285–291.

Hancke, K. & Glud, R.N. (2004). Temperature effects on respiration and photosynthesis in three diatom-dominated benthic communities. *Aquatic Microbial Ecology*, **37**, 265–281.

Harley, C.D.G. (2011). Climate change, keystone predation, and biodiversity loss. *Science*, **334**, 1124–1127.

Harley, C.D.G., Anderson, K.M., Demes, K.W., et al. (2012). Effects of climate change on global seaweed communities. *Journal of Phycology*, **48**, 1064–1078.

Harley, C.D.G., Hughes, A.R., Hultgren, K.M., et al. (2006). The impacts of climate change in coastal marine systems. *Ecology Letters*, **9**, 228–241.

Hawkins, S.J. & Hartnoll, R.G. (1985). Factors determining the upper limits of intertidal canopy-forming algae. *Marine Ecology Progress Series*, **20**, 265–271.

Helmuth, B., Broitman, B.R., Yamane, L., et al. (2010). Organismal climatology: analyzing environmental variability at scales relevant to physiological stress. *Journal of Experimental Biology*, **213**, 995–1003.

Helmuth, B., Harley, C.D.G., Halpin, P.M., et al. (2002). Climate change and latitudinal patterns of intertidal thermal stress. *Science*, **298**, 1015–1017.

Helmuth, B., Yamane, L., Lalwani, S., et al. (2011). Hidden signals of climate change in intertidal ecosystems: what (not) to expect when you are expecting. *Journal of Experimental Marine Biology and Ecology*, **400**, 191–199.

Hillebrand, H., Soininen, J., & Snoeijs, P. (2010). Warming leads to higher species turnover in a coastal ecosystem. *Global Change Biology*, **16**, 1181–1193.

Hollowed, A.B., Barange, M., Beamish, R.J., et al. (2013). Projected impacts of climate change on marine fish and fisheries. *ICES Journal of Marine Science*, **70**, 1023–1037.

Hughes, T.P., Baird, A.H., Bellwood, D.R., et al. (2003). Climate change, human impacts, and the resilience of coral reefs. *Science*, **301**, 929–933.

Johnson, C.R., Banks, S.C., Barrett, N.S., et al. (2011). Climate change cascades: shifts in oceanography, species' ranges and subtidal marine community dynamics in eastern Tasmania. *Journal of Experimental Marine Biology and Ecology*, **400**, 17–32.

Jones, D.O.B., Yool, A., Wei, C.-L., et al. (2014). Global reductions in seafloor biomass in response to climate change. *Global Change Biology*, **20**, 1861–1872.

Jones, G.P., McCormick, M.I., Srinivasan, M., & Eagle, J.V. (2004). Coral decline threatens fish biodiversity in marine reserves. *Proceedings of the National Academy of Sciences of the USA*, **101**, 8251–8253.

Keller, A.A., Oviatt, C.A., Walker, H.A., & Hawk, J.D. (1999). Predicted impacts of elevated temperature on the magnitude of the winter-spring phytoplankton bloom in temperate coastal waters: a mesocosm study. *Limnology and Oceanography*, **44**(2), 344–356.

Kemp, D.W., Oakley, C.A., Thornhill, D.J., et al. (2011). Catastrophic mortality on inshore coral reefs of the Florida Keys due to severe low-temperature stress. *Global Change Biology*, **17**, 3468–3477.

Kordas, R.L., Harley, C.D.G., & O'Connor, M.I. (2011). Community ecology in a warming world: the influence of temperature on interspecific interactions in marine systems. *Journal of Experimental Marine Biology and Ecology*, **400**, 218–226.

Kotilainen, A.T., Arppe, L., Dobosz, S., et al. (2014). Echoes from the past: a healthy Baltic Sea requires more effort. *Ambio*, **43**, 60–68.

Kritzberg, E.S., Duarte, C.M., & Wassman, P. (2010). Changes in Arctic marine bacterial carbon metabolism in response to increasing temperature. *Polar Biology*, **33**, 1673–1682.

Kuklinski, P. & Balazy, P. (2014). Scale of temperature variability in the maritime Antarctic intertidal zone. *Journal of Sea Research*, **85**, 542–546.

Lalli, C.M. & Parsons, T.R. (1997). *Biological Oceanography: An Introduction*, 2nd edition. Elsevier Butterworth-Heinemann, Oxford.

Layman, C.A., Smith, D.E and Herod, J.D. (2000) Seasonally varying importance of abiotic and biotic factors in marsh-pond fish communities. *Marine Ecology Progress Series*, **207**, 155–169.

Lima, F.P. & Wethey, D.S. (2012). Three decades of high-resolution coastal sea surface temperatures reveal more than warming. *Nature Communications*, **3**, 704.

Ling, S.D. & Johnson, C.R. (2009). Population dynamics of an ecologically important range-extender: kelp beds versus sea urchin barrens. *Marine Ecology Progress Series*, **374**, 113–125.

Ling, S.D., Johnson, C.R., Ridgway, K., et al. (2009). Climate-driven range extension of a sea urchin: inferring future trends by analysis of recent population dynamics. *Global Change Biology*, **15**, 719–731.

Locarini, R.A., Mishonov, A.V., Antonov, J.I., et al. (2010). *World Ocean Atlas 2009. Volume 1: Temperature*. S. Levitus, Ed. NOAA Atlas NESDIS 68, US Government Printing Office, Washington, DC.

McClain, C.R. & Hardy, S.M. (2010). The dynamics of biogeographic ranges in the deep sea. *Proceedings of the Royal Society of London Series B*, **277**, 3533–3546.

McCormick, M.I. (2012). Lethal effects of habitat degradation on fishes through changing competitive advantage. *Proceedings of the Royal Society of London Series B*, **279**, 3899–3904.

Milazzo, M., Mirto, S., Domenici, P., & Gristina, M. (2013). Climate change exacerbates interspecific interactions in sympatric coastal fishes. *Journal of Animal Ecology*, **82**, 468–477.

Müren, U., Berglund, J., Samuelsson, K., & Andersson, A. (2005). Potential effects of elevated sea-water temperature on pelagic food webs. *Hydrobiologia*, **545**, 153–166.

Pearson, R.G. & Dawson, T.P. (2003). Predicting the impacts of climate change on the distribution of species: are bioclimate envelope models useful? *Global Ecology and Biogeography*, **12**, 361–371.

Peck, L.S., Clark, M.S., Morley, S.A., et al. (2009). Animal temperature limits and ecological relevance: effects of size, activity and rates of change. *Functional Ecology*, **23**, 248–256.

Perry, A.L., Low, P.J., Ellis, J.R., & Reynolds, J.D. (2005). Climate change and distribution shifts in marine fishes. *Science*, **308**, 1912–1915.

Philippart, C.J.M., van Aken, H.M., Beukema, J.J., et al. (2003). Climate-related changes in recruitment of the bivalve *Macoma balthica*. *Limnology and Oceanography*, **48**, 2171–2185.

Pinksy, M.L., Worm, B., Fogarty, M.J., et al. (2013). Marine taxa track local climate velocities. *Science*, **341**, 1239–1242.

Poloczanska, E.S., Brown, C.J., Sydeman, W.J., et al. (2013). Global imprint of climate change on marine life. *Nature Climate Change*, **3**, 919–925.

Pörtner, H.-O. & Farrell, A.P. (2008). Physiology and climate change. *Science*, **322**, 690–692.

Rhein, M., Rintoul, S.R., Aoki S, et al. (2013). In: Stocker, T.F., Qin, D., Plattner, G.-K., et al. (eds), *Climate Change 2013: The Physical Science Bases. Contribution of Working Group I to the Fifth Assessment Report of the Intergovernmental Panel on Climate Change*. Cambridge University Press, Cambridge and New York.

Ridgway, K.R. (2007). Long-term trend and decadal variability of the southward penetration of the East Australian Current. *Geophysical Research Letters*, **34**, L13613.

Rogers, L.A., Stige, L.C., Olsen, E.M., et al. (2011). Climate and population density drive changes in cod body size throughout a century on the Norwegian coast. *Proceed-*

ings of the National Academy of Sciences of the USA, **108**, 1961–1966.

Russell, B.D., Harley, C.D.G., Wernberg, T., et al. (2012). Predicting ecosystem shifts requires new approaches that integrate the effects of climate change across entire systems. *Biology Letters*, **8**, 164–166.

Sanford, E. (1999). Regulation of keystone predation by small changes in ocean temperature. *Science*, **283**, 2095–2097.

Sarmento, H., Montoya, J.M., Vázquez-Domínguez, E., et al. (2010). Warming effects on marine microbial food web processes: how far can we go when it comes to predictions? *Philosophical Transactions of the Royal Society Series B*, **365**, 2137–2149.

Schiel, D.R., Steinbeck, J.R., & Foster, M.S. (2004). Ten years of induced ocean warming causes comprehensive changes in marine benthic communities. *Ecology*, **85**(7), 1833–1839.

Schulte, P.M. (2014). What is environmental stress? Insights from fish living in a variable environment. *Journal of Experimental Biology*, **217**, 23–34.

Seneviratne, S.I., Nicholls, N., Easterling, D., et al. (2012). Changes in climate extremes and their impacts on the natural physical environment. In: Field, C.B., Barros, V., Stocker, T.F., et al. (eds), *Managing the Risks of Extreme Events and Disasters to Advance Climate Change Adaptation*. A Special Report of Working Groups I and II of the Intergovernmental Panel on Climate Change (IPCC). Cambridge University Press, Cambridge and New York.

Sen Gupta, A., Brown, J.N., Jourdain, N.C., et al. (2015). Episodic and non-uniform shifts of thermal habitats in a warming ocean. *Deep Sea Research Part II*, **113**, 59–72.

Smale, D.A. & Wernberg, T. (2013). Extreme climatic event drives range contraction of a habitat-forming species. *Proceedings of the Royal Society of London Series B*, **280**, 20122829.

Sommer, U., Adrian, R., Bauer, B., & Winder, M. (2012). The response of temperate aquatic ecosystems to global warming: novel insights from a multidisciplinary project. *Marine Biology*, **159**, 2367–2377.

Sorte, C.J.B., Fuller, A., & Bracken, M.E.S. (2010). Impacts of a simulated heat wave on composition of a marine community. *Oikos*, **119**, 1909–1918.

Stillman, J.H. (2003). Acclimation capacity underlies susceptibility to climate change. *Science*, **301**, 65–65.

Stouffer, R.J. & Manabe, S. (2003). Equilibrium response of thermohaline circulation to large changes in atmospheric CO_2 concentration. *Climate Dynamics*, **20**, 759–773.

Teixeira, T.P., Neves, L.M., & Araújo, F.G. (2009). Effects of a nuclear power plant thermal discharge on habitat complexity and fish community structure in Ilha Grande Bay, Brazil. *Marine Environmental Research*, **68**, 188–195.

Thompson, P.A., Baird, M.E., Ingleton, T., & Doblin, M.A. (2009). Long-term changes in temperate Australian coastal waters: implications for phytoplankton. *Marine Ecology Progress Series*, **394**, 1–19.

Thompson, R.M., Beardall, J., Beringer, J., et al. (2013). Means and extremes: building variability into community-level climate change experiments. *Ecology Letters*, **16**, 799–806.

Vaquer-Sunyer, R., Duarte, C.M., Santiago, R., et al. (2010). Experimental evaluation of planktonic respiration response to warming in the European Arctic Sector. *Polar Biology*, **33**, 1661–1671.

Walther, G.R., Post, E., Convey, P., et al. (2002). Ecological responses to recent climate change. *Nature*, **416**, 389–395.

Wernberg, T., Smale, D.A., Tuya, F., et al. (2013). An extreme climatic event alters marine ecosystem structure in a global biodiversity hotspot. *Nature Climate Change*, **3**, 78–82.

CHAPTER 13

Chemical pollutants in the marine environment: causes, effects, and challenges

Catriona K. Macleod, Ruth S. Eriksen, Zanna Chase, and Sabine E. Apitz

13.1 Introduction

Any toxic or contaminating substance that is likely to have an adverse effect on the natural environment or life can be considered a pollutant. Chemical pollutants, present in every marine environment from the tropics to the poles, can enter marine ecosystems in many ways, but the most significant sources are terrestrial in origin. Consequently, understanding and controlling human and land-based inputs are critical for effective management.

Human settlements have historically been based around rivers and estuaries; these same water bodies provide a means of waste removal, with most drains and sewage treatment plants still discharging waste into inland and coastal waters. With industrialization, deposits into estuaries and rivers have become more complex, with pollutants such as pharmaceutical residues, fertilizers, radioactive chemicals, and plastics presenting new challenges. There are frequently hotspots of pollution adjacent to ports, harbours, and urban centres, and in areas subject to intensive agricultural activity (Haynes and Johnson, 2000).

The impact of such inputs on the marine environment will be determined by the inherent properties of the adjacent ecosystems. Until relatively recently the overarching view was that the oceans possessed a vast capacity to assimilate wastes and the well-known engineer's adage that 'the solution to pollution is dilution' was frequently applied. Unfortunately this approach has left some significant legacy issues and challenges for management.

13.2 Types of pollutant

13.2.1 Nutrients

Nutrients are a major issue in the marine environment, with inputs coming from a broad range of sources, some obvious (i.e. human and animal wastes, agricultural fertilizers, and industrial effluents) and some more obscure (e.g. glacial melt, Hood et al., 2009; large animal mortalities, Smith and Baco, 2003; and volcanic activity, Langmann et al., 2010). However, the consequences are generally similar, albeit potentially on different scales. The impacts and challenges of chemical contamination are dealt with explicitly in Chapter 5.

13.2.2 Pharmaceuticals and personal care products

Pharmaceuticals have had a hugely beneficial impact on society, ensuring the health and well-being of both people and farmed stock. Pharmaceuticals and personal care products (PPCPs) comprise a diverse group of chemicals including human and veterinary drugs (both prescription and over-the-counter), neutraceuticals, diagnostic agents, and other consumer chemicals such as fragrances, make-up, and sun-screen products (Daughton, 2002). The effects of many PPCPs may be subtle or poorly understood, but there are increasing examples of ecological effects of common PPCPs at low-levels (Halling-Sørensen et al., 1998; Daughton and Ternes, 1999; Thomas and Hilton, 2004). PPCPs are widespread and pseudo-persistent (due to continual inputs), with the potential for cumulative effects.

There are many modes of action and toxicity associated with PPCPs, the wastes produced during their manufacture, and the excreted by-products.

Pharma- and neutraceuticals enter sewage systems as a result of both human waste and the domestic disposal of unused medicines (Fig. 13.1); treatment processes are relied upon to remove them, their metabolites, and transformation products. Unfortunately no effluent treatment technology works well for all compounds (Kümmerer, 2009a), and pharmaceutical discharges are expected to increase as the global population ages and improved standards of living and affordability make medicines more accessible (Kümmerer, 2009b).

Effects of pharmaceuticals in the marine environment can be species-specific or may affect broad biological, ecological, or functional groups. Impacts may be direct and physiological, or result from ancillary changes to communities and ecosystem ecology. There is potential for inadvertent and unforeseen impacts on non-target species or communities; several striking examples in terrestrial and aquatic domains exist. For instance, diclofenac, a non-steroidal anti-inflammatory commonly used in humans and livestock, was found to have negative effects on Asian vultures when used in cattle in India and Pakistan (Oaks et al., 2004). The oriental white-backed vulture was driven close to extinction as a result of its unforeseen and acute sensitivity to diclofenac accumulated by scavenging dead livestock (Kümmerer, 2010). More recently, low levels of a common anti-anxiety drug in wastewater were found to have a serious effect on European perch; they became bolder, more anti-social and ate more (Brodin et al., 2013), with these behavioural changes potentially restructuring the local food web. On occasion, breakdown (metabolic) products may have a greater impact than the original contaminant. For instance, mutagenic properties have been associated with photolysis and photo-oxidation processes (Lee et al., 2007).

The potential for negative interactions with ecosystem processes is a key concern (Kümmerer, 2010). While many pharmaceuticals can be detected in the marine environment, they are generally at levels well below therapeutic doses (Zuccato et al., 2006), and as

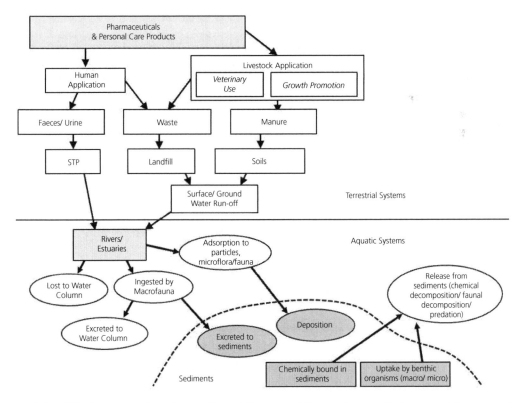

Figure 13.1 Potential environmental pathways for assimilation of pharmaceuticals in the environment. NB. Factors only define the general pathways and may be affected by the various forms in which chemicals may appear (STP—sewage treatment plant).

such present little risk of direct adverse effect on humans, but PCPPs, or secondary metabolites generated in the body or the environment, may influence marine ecology and ecosystem processes. Although knowledge is increasing, the chronic/long-term effects, effects on more sensitive species or lifecycle stages, and impacts of therapeutant/contaminant cocktails remains unclear. Research thus far has generally focused on specific drugs, entry/effluent mechanisms of interest, standard ecotoxicological test species, and/or species of particular social or commercial importance.

A few common PPCPs have been studied more extensively—for example: triclosan, an antibacterial and antifungal agent found in soaps and detergents; the non-steroidal anti-inflammatories (NSAIDs) naproxen and ibuprofen; antidepressants; and caffeine (Zarelli et al., 2014). It has been suggested that triclosan concentrations are both prevalent and increasing in the marine environment (Maruya et al., 2014), with triclosan resistance and adverse impacts on marine bacteria being significant concerns (Drury et al., 2013). Naproxen and ibuprofen are regularly detected in marine systems. While it has been suggested that their specific ecotoxicity might be relatively low (Cleuvers, 2004), ibuprofen was found to stimulate marine cyanobacteria (Zuccato et al., 2006) and degradation products of naproxen can be more toxic than parent compounds (Isidori et al., 2005). Antidepressants have been shown to affect both physiological (e.g. affecting ovarian and testicular growth) and neurological function (e.g. foot detachment in snails), with effects often observed at environmentally relevant concentrations (De Lange et al., 2006; Guler and Ford, 2010; Schultz et al., 2011; Fong and Hoy, 2012; Di Poi et al., 2013; Fong and Molnar, 2013; Franzellitti et al., 2013; Fong and Ford, 2014). Although there is currently no evidence that caffeine presents any large-scale threat to marine ecosystems, it is ubiquitous in the marine environment, and has been shown to be a useful tracer for sewage outputs, with detectable levels in samples collected far from human settlement (Weigel et al., 2004).

Most regulatory programmes include a list of substances of priority concern, and these are often linked to risk-based environmental performance or compliance action levels. However, this approach does not necessarily ensure protection against ecotoxicological stress or address the issue of chemical interactions. Brack et al. (2007) reviewed the priority toxicants in European River Basins and, while a number of these did have some role in observed effects, they often only explained a small proportion of measured effects.

The key mechanisms by which pharmaceuticals are removed from the environment are sorption and biodegradation (Fig. 13.1), although photodegradation and hydrolysis can also be important (Mendez-Arriaga et al., 2008; Kümmerer, 2010). In the aquatic environment suspended material and colloids represent important removal vectors (Maskaoui and Zhou, 2010). In sewage treatment and in seawater, bacteria are responsible for most biodegradation processes (Kümmerer, 2010).

Antibiotics are used extensively, in both human and livestock therapies. Livestock usage is mostly terrestrially based with marine and coastal effects associated with land-use run-off; however, the relatively recent use of antibiotics (antimicrobials) in aquaculture is important as inputs are directly into the aquatic system. The primary environmental concerns regarding antibiotic usage are similar irrespective of source or use: are there toxicity effects on non-target organisms, do they persist in the environment, and what is the potential for the development of antibiotic resistance?

While there is a substantial literature on the ecotoxicological effects of antibiotics with respect to human health, information on effects in aquatic systems is more limited, and is largely focused on environmental persistence and antibiotic resistance. Chemical residues in fauna and sediments can result in bioaccumulation or biomagnification, and may contribute to the development of bacterial resistance that both can threaten therapeutic regimes and may potentially be transferred back to the human food chain (Smith et al., 1994; Schmidt et al., 2001). The rising prevalence of antibiotic-resistant species ('super bugs') may cause shifts in community structure, as more resistant species become dominant, which may affect important microbial processes such as decomposition and mineralization (Kumar et al., 2005).

13.2.3 Agricultural chemicals (pesticides and fertilizers)

Agricultural effluents are a major source of pollutants to marine and coastal environments. They contribute significant levels of nutrients, with estimates suggesting around 50% of total ammonium (NH_4^+) and nitrate (NO_3^-) in marine surface waters is agriculturally derived (Islam and Tanaka, 2004). This has significant flow-on effects for eutrophication.

A pesticide is described as 'any substance or mixture of substances intended to prevent, destroy, repel, or mitigate an undesirable organism' (FAO, 2002). Pesticide use has been a feature of agricultural practice since sulphur was first used in Ancient Greece to control crop pests (Rajinder et al., 2009). The list of pesticides (herbicides, insecticides, fungicides, acaricides, nematocides, soil fumigants, and rodenticides) used in terrestrial agriculture worldwide is significant, although the largest use appears to be associated with a relatively small number of products (http://pesticide.umd.edu) In the developed world, herbicides are the most significant products in terms of volume of production, and these products tend to have lower acute toxicity than do insecticides (Rajinder et al., 2009). In developing countries the use of insecticides tends to be more widespread, and consequently there is a greater risk of acute toxicity. Pesticides and pesticide residues (particularly DDT, aldrin, and lindane) have been identified in environmental samples from the equator to the Arctic.

Non-persistent pesticides, such as pyrethroids and organophosphates, are designed to break down relatively quickly in the environment, and as a result the potential for environmental impacts is reduced. However, many pesticides are designed to be long lasting (persistent), and consequently both the active ingredients and their breakdown products may eventually make their way into waterways. Persistent pesticides take a very long time to degrade; these chemicals, for example DDT (possibly the most renowned persistent organic pollutant), have been shown to leave a legacy of environmental impacts (Zhang et al., 2011; Hellou et al., 2013) and are now banned in many countries. Persistent pesticides can remain longer, and in higher levels, in food and water sources, and as a result may bioaccumulate or biomagnify, potentially reaching toxic levels in individual organisms and the food chain. Most pesticides are highly toxic to humans and will have some environmental impact that may persist beyond the period of application (Zhang et al., 2011). There are many documented incidents of serious environmental pollution associated with pesticides.

Pesticides have many different modes of action: some work by affecting the nervous system (e.g. organophosphate, organochlorine, avermectins, and pyrethroids), some inhibit growth and development (e.g. chitin synthesis inhibitors, insect growth regulators, and non-specific growth regulators), others affect energy production (e.g. electron transport and oxidative phosphorylation) and metabolism. It is suspected that many of the products developed as terrestrial insecticides will have detrimental effects on marine crustacean fauna. Although data from marine systems are limited, there are several freshwater studies showing marked depletion in aquatic fauna or reduction

in ecosystem function in systems with elevated insecticide loads (Peters et al., 2013; Pisa et al., 2014). The overall impact of any particular pesticide in the marine environment will be a function of its specific mode of action, its persistence, and any indirect effects that might result from associated changes in community composition. Many pesticides are designed to affect growth or development; consequently the environmental impacts may be chronic (e.g. physiological and cellular impacts such as endocrine disruption and impaired reproduction). Such sublethal impacts on ecosystem balance, structure, and function may only become apparent when exacerbated by other toxicants or environmental conditions. As with antibiotics, a key concern regarding pesticide residues is the potential for the development of resistance in target species.

13.2.4 Antifoulants

Antifouling products are a specific class of pesticides, and are typically applied as paints to prevent settlement and growth of fouling communities (e.g. slimes, macro- and microalgae, encrusting invertebrates, etc.) in industrial and marine situations (Finnie, 2006). Antifoulants are designed to slowly release toxic agents and are generally non-selective. Treatments are designed to be long lasting (months to years), but they often require periodic removal and re-application to maintain effectiveness, which can result in the discharge of particles, contaminated biological material, and liquid waste into the coastal environment, with hotspots around industrialized and maritime areas (ports, harbours, shipyards, slipways, etc.; Batley and Maher, 2001). As a consequence antifoulants may persist in the environment and have the potential to accumulate in sediment and biota.

Copper- and zinc-based products—Copper is one of the earliest known antifoulants, with records showing sheets attached to boat hulls to discourage marine growth many centuries ago. However, both copper and zinc are acutely toxic to a range of aquatic organisms, and as a consequence they are commonly incorporated into antifouling paint formulations. A major issue with both copper and zinc is that they cannot be degraded by biotic or abiotic processes and thus are persistent; the ecological ramifications of this are increasingly being recognized and acknowledged as a significant environmental concern for our coastal ecosystems (Dafforn et al., 2011).

TBT—Organotin compounds such as tributyltin (TBT) are highly effective biocides. TBT was first introduced as an antifoulant in the late 1960s. However, the adverse effects of TBT on the broader environment became clear when Alzieu et al. (1981, 1986) demonstrated a link between shell deformities and TBT exposure in the pacific oyster (*Crassostrea gigas*). Since then numerous studies have identified the acute and chronic toxicity of TBT. Exposure to TBT has been linked to imposex (irreversible development of male physical characteristics) in dogwhelks, resulting in significant population declines due to sterilization (Gibbs et al., 1991) and reproductive failure has been associated with hormonal imbalances and neurotoxic effects (endocrine disruption).

Many nations had partial bans in place for TBT by the late 1990s, with the International Maritime Organization (IMO) introducing protocols to completely prohibit the use of organotin-based antifouling paints by 2008 (IMO, 2002). Many countries now have codes of practice for antifouling and in-water hull cleaning (e.g. ANZECC, 1997; EU Biocidal Products Directive (98/8/EC); US Environmental Protection Agency, USEPA, 2003).

While the impact of TBT has been significantly reduced due to legislative change, it is persistent in the environment and long-term recovery will depend upon the rates at which TBT can be broken down into less toxic di- and monobutyl tin forms (De Mora et al., 1995). Batley (1995) estimated the half-life of TBT in sediments to be between two and five years; the half-life in water is significantly shorter, of the order of weeks or months (USEPA, 2003). However, research in Portugal has shown a significant decrease in imposex in the dog whelk *Nucella lapillus* since the ban on TBT was implemented (Galante-Oliviera et al., 2009).

Alternative biocides—As the use of these traditional antifoulants has become more restricted, new antifouling products have been developed. Many of these have been purposely designed to have broad spectrum toxicity to a wide range of aquatic organisms, and this presents significant environmental concerns. Herbicides such as Diuron and Irgarol-1051 have been employed as active ingredients to suppress photosynthesis. These products are highly water soluble (van Dam et al., 2012), and have been shown to have adverse impacts on seagrass, corals, and mangroves (Moss et al., 2005). Diuron is carcinogenic, and bioaccumulation and human health issues are a significant concern (Evans and Nicholson, 2000). Elevated levels of Diuron have been recorded around boat washing areas (Stewart et al., 2009). Approval for Diuron in the UK was revoked in 2000, and several other countries have restricted use to vessels > 25 m in length (Thomas et al., 2002). Thiram is another very effective broad spectrum

biocide that has been used in commercial antifoulants; it is toxic to algae, fish, crustacea, and amphibians at concentrations < 1 µg/l (ANZECC, 2000). However, it has been shown to have teratogenic properties, producing developmental and neurological malformations in fish (Van Boxtel et al., 2010). Little is known about the impact or levels of Thiram contamination in the broader marine environment.

Recent research efforts have focused on the potential for more targeted (species-specific) modes of action (Yan Ting et al., 2014); for instance, medetomidine is highly effective against barnacle larvae but does not affect microfouling communities such as bacteria and periphyton (Ohlauson et al., 2012). Unfortunately, for this approach to be effective in the real world it is likely to require a combination of active ingredients, and that would present a complex mix of interactions and potential issues.

13.2.5 Disinfectants

Chemical disinfectants (chlorine, chloramines, ozone, and chlorine dioxide) are frequently used in medical facilities, water treatment plants, and industry as biocides to control pathogenic microorganisms, inhibit biofilm formation, and oxidize reduced inorganic solutes, such as sulphide and ferrous iron (Agus et al., 2009). Chlorine is the most common disinfectant, with chloramines and chlorine dioxide becoming increasingly common. Some industries, such as the paper industry, produce chlorine-containing by-products, but by far the greatest proportion of chlorine discharged into marine waters has been used to prevent biofouling of cooling water systems. Electricity generating stations in particular can release large amounts of chlorinated effluents into coastal waters (Abarnou and Miossec, 1992).

The key issue with disinfectants is that biocidal function of chlorinated waters continues for as long as they retain their oxidizing capacity. Fortunately, impacts tend to be restricted to a discernible mixing/dilution zone. The severity of impact will largely depend upon the ecological sensitivity of receiving waters. Disinfectants, particularly chlorine-based chemicals, are commonly included in sediment and water quality guidelines, and as a result discharge is generally regulated.

13.2.6 Hydrocarbons/oil spills

There are many potential sources of hydrocarbons in the marine environment: there are natural sources (e.g. forest fires, volcanic activity, and plant decay) but by far the greatest sources are petroleum products. Land-based sources (i.e. urban run-off and coastal refineries), oil transport and offshore extraction, atmospheric deposition, and natural seeps (GESAMP, 1993; NRC, 2003) are key contributors. Naturally occurring oil seeps can deliver significant amounts of crude oil to the marine environment, but these oil releases tend to be slow, and as a consequence the environment and local ecology in such areas are usually well adapted. Chronic inputs from land-based human activity (e.g. motor oil use) delivered through run-off, sewerage, and atmospheric transport represent approximately one third of all oil inputs to the ocean, although this is expected to decrease as oil recycling programmes are adopted (Farrington, 2013). Although discharges and accidental spills from ships (especially tankers), offshore platforms, and pipelines are among the most obvious causes of hydrocarbon/oil pollution, these account for less than 10% of marine contamination.

There are different types of oils and their impacts differ; oils can be sticky (crude oil and bunker fuels) or non-sticky (refined petroleum products). Refined petroleum products do not persist in the marine environment for as long as crude oil or bunker fuel, and consequently they are less likely to adhere to birds or other animals; however, they tend to be more acutely toxic (US Department of Health and Human Services, 1998).

The Amoco Cadiz (Brittany, France, 1978), Exxon Valdez (Alaska, 1989), Prestige (Spain, 2002), and more recent Deepwater Horizon/BP (Gulf of Mexico, 2010) oil spills have focused attention sharply on environmental impacts and methods of prevention, response, and remediation (Penela Arenaz et al., 2009; NRC, 2013; Bodkin et al., 2014). Even small spills can have severe effects on marine wildlife. The extent of impact will depend on the type of oil, the timing and location of the spill, the natural ecology of the area, and the prevailing weather conditions. Location can have a major influence, for instance an oil spill in Antarctica could have catastrophic consequences, due to remoteness, lack of personnel, and weather constraints (Ruoppolo et al., 2012), while warm seas and high winds can encourage lighter oils to vaporize, potentially reducing impacts. Impacts of a spill may also change over time; oil may initially be more toxic, but become stickier with weathering (AMSA, 2012) affecting outcome, risk, and dispersion potential.

Marine mammals and birds are often the most obvious casualties of oil spills, with many potential impacts well documented. Oil sticking to fur and feathers causes many problems: hypothermia from reduced insulation (Paine et al., 1996; Peterson, 2001; Peterson et al., 2003; Votier et al., 2005); and reduced waterproofing adversely affects buoyancy and swimming and flying ability, and can make birds easy prey (Piatt et al., 1990). Ingestion of oil can damage airways and lungs and inhalation of oil droplets or fumes/gas can also cause mortalities (Paine et al., 1996; Peterson, 2001; Peterson et al., 2003). Toxic hydrocarbons are introduced to higher trophic levels when animals attempt to clean their fur and feathers, or are eaten (Piatt et al., 1990; Peterson, 2001). Indirect effects may be equally significant. For instance, oils spills during the breeding season may result in seal pups and other young mammals being unable to recognize their mother's scent, or may cause some species to change their breeding behaviours (Peterson, 2001). Physical damage to important marine habitats and breeding areas can interfere with reproductive cycles. Animals subject to oil spills will be stressed, and there may be other sublethal effects such as irritation and ulceration of skin, mouth, or nasal cavities, and inflammation, infection, and ulceration where oil has been ingested (Peterson, 2001). Damage to eyes (ulcers, conjunctivitis, and blindness) will affect foraging ability. Ingestion can lead to damage to red blood cells, organ damage, thinning of birds' egg shells, damage to fish eggs, larvae, and young fish, tainting of fish, crustaceans, molluscs, and algae, while immune system impairment can lead to secondary bacterial or fungal infections (Peterson, 2001).

Polycyclic aromatic hydrocarbons (PAHs) are of particular concern for the marine environment as they adsorb to particulate matter and are readily taken up by biota (CCME, 1999). Many PAHs are acutely toxic, with some forming carcinogenic metabolites (MPMMG, 1998). Several studies have linked environmental PAH levels to abnormalities in fish (Malins et al., 1988; Vethaak and Rheinallt, 1992). While higher organisms can metabolize PAHs, there is also the potential for bioaccumulation, and therefore a trophic interaction risk to higher order species and humans via the food chain (WHO, 1998). Environmental conditions also need to be considered when determining effects, as some PAHs are more toxic upon exposure to UV light (Arfsten et al., 1996). Most countries have some form of water/sediment quality guidelines that regulate the loads of total PAHs.

13.2.7 Plastics (macro and micro)

Mass production of plastics began in the 1940s and has increased exponentially; by 2009 more than 230 million tonnes of plastic had been produced globally and plastic production accounted for approximately 8% of global oil consumption (Thompson et al., 2009). Although a relatively new group of pollutants, they have had major social, economic, and environmental impacts (Gregory, 2009; Cole et al., 2011). The 'throwaway' approach to packaging has resulted in plastics comprising about 10% of all municipal waste (Barnes et al., 2009). It is estimated that approximately 10% ends up in the oceans (Thompson, 2006). Macroplastics clutter coastlines, can be a hazard to navigation and marine industries (e.g. fishing, aquaculture, and energy production), pose a threat to wildlife via entanglement and ingestion (Gregory, 2009), and can be a mechanism for translocation of non-native species (Barnes, 2002). On the seafloor they can create artificial substrates and interfere with normal sediment water exchange processes (Gregory, 2009).

In recent years, research has revealed the significant environmental impacts associated with microplastics. These tiny plastic granules, found in cosmetics and air blasting powders, but also generated as macroplastics breakdown, are now common in the marine environment (Cole et al., 2011) and in our oceans (Kukulka et al., 2012). Microplastics are small enough to be bioavailable through the food chain, and readily absorb dissolved organic pollutants and leach toxic plasticizers (Cole et al., 2011); many of these plasticizers act as endocrine disruptors (Vethaak and Legler, 2013). Bisphenol-A (BPA), bis(2-ethylhexyl)phthalate (DEHP), and dibutyl phthalate (DBP) are frequently found in hard plastics (Yang et al., 2011); BPA is widely used in the production of disposable food and drink containers and has been shown to be a potent environmental estrogen (Rubin, 2011).

13.2.8 Radioactive substances

Radionuclides can occur naturally as a result of weathering or the action of cosmic rays (Eisenbud and Gesell, 1997), but those radioactive substances that might be considered pollutants in the marine environment are generally man-made. Anthropogenic sources of radioactivity include nuclear power plants and fuel reprocessing plants, medical wastes, and by-products of oil and gas production and the phosphate fertilizer industry (OSPAR Commission, 2010), as well as nuclear weapons testing and fallout from nuclear accidents (such as those at Three Mile Island, USA, in 1979, Chernobyl, Ukraine, in 1986, and Fukushima, Japan, in 2011). The main medical radiation source, iodine-131, primarily enters the marine environment as medical waste via sewage treatment plants, and effects from such sources are likely to be relatively minor as iodine-131 has a very short half-life (eight days). Oil and gas inputs generally come from the water used to descale the insides of pipes and from oil reservoirs (OSPAR Commission, 2010), and are the largest non-nuclear contributors of radionuclides to the marine environment.

The potential impacts of radioactive particles depend on the source. Particles released at low temperature and pressure differ from those released during high-temperature and high-pressure events (IAEA, 2011). The main impacts of ionizing radiation in the marine environment are genetic, reproductive, and cancer-causing effects in marine organisms (Salbu et al., 1997). Impacts can occur at ecosystem, community, and population levels, and have the potential to affect human health through the food chain. An excellent example of such a food chain impact occurred in the early 1960s, where people living close to the Windscale nuclear power plant in the UK were found to have elevated radiation levels—the critical radiation exposure pathway turned out to be consumption of Laver bread made using a locally gathered seaweed (Preston and Jeffries, 1967, 1969).

As a result of the long half-lives of many radionuclides, discharged radiation can persist long after the initial contamination has ceased, with the potential for long-term environmental and ecological impacts (Salbu, 2007, 2011). In most countries there are guidelines with respect to low-level radioactive contamination of sediments and water; however, as with oil pollution, these guidelines are not designed to deal with 'spill' style events where major and long-term effects are likely. Environmental quality criteria are one way to manage for adverse effects of radioactive substances, but unfortunately there is currently insufficient information to develop reliable risk-based standards for most marine species (Pentreath, 2002; Brown et al., 2006).

13.2.9 Metals

Metals such as copper, zinc, lead, cadmium, and mercury can enter coastal waters through industrial and domestic waste, atmospheric deposition, stormwater run-off, and the use of antifouling paints (see Section 13.2.4). The toxicity of metals in marine sediments and waters depends upon their bioavailability, and this is principally controlled by the geochemistry of the associated sediments (Peng et al., 2004; Gillan et al., 2012; Hoffman et al., 2012) or waters (Money et al., 2011; Hoffman et al., 2012; Sinoir et al., 2012). The presence of sulphide is one of the major factors controlling the interaction between metals in sediments and the biota that live in and above sediments. Metals such as copper and zinc can form insoluble complexes with sulphide, which binds the metals and makes them unavailable under stable redox conditions (Naylor et al., 2004). Consequently, organic carbon content, pH, particle size, and redox status of sediments can influence metal bioavailability (Peng et al., 2004).

There are a range of biological effects that can result from metal pollution, from acute toxicity to chronic sublethal effects, such as deformations and reduced reproductive capacity. Ecosystem effects such as shifts in community structure and associated changes in functional groups have also been observed (Morrisey et al., 1996). It is not possible to list all of the potential impacts, interactions, or risks associated with metal contamination, so we have selected a few to highlight some key mechanisms.

While antifoulants (see Section 13.2.4) are an important source of copper (Cu), there are also terrestrial sources: Cu in dust has been shown to negatively impact phytoplankton communities (Paytan et al., 2009). Cyanobacteria have mixed responses to Cu; some are highly sensitive (Brand et al., 1986; Moffet et al., 1997—*Synechococcus*), while others thrive in Cu-contaminated waters (Mann et al., 2002—*Skeletonema costatum, Prochlorococcus*). These differential sensitivities can result in community shifts.

In addition, the mechanisms employed to deal with metal contamination can have negative outcomes for other species/ecosystems. For example, it has been reported that the diatom *Pseudo-nitzschia* produces the toxin domoic acid (the agent responsible for amnesic shellfish poisoning) as a means to complex copper (Maldonado et al., 2002)—providing an interesting potential link between water quality and harmful algal blooms.

Bioaccumulation and the potential for adverse consequences for human and ecosystem health are major concerns with any metal contamination situation. However, this is a particular concern for mercury (Hg), where the human health risks are significant (Hg is associated with damage to the brain, kidneys, and lungs, and especially prenatal neurological development). It is primarily the highly toxic methylated form (MeHg) which is of concern. Most MeHg is produced in sediments, where less toxic inorganic Hg is methylated by bacteria (Ekstrom et al., 2003). Human activity is estimated to have increased both atmospheric mercury emissions and mercury levels in ocean surface waters by a factor of three when compared to pre-anthropogenic levels (Lamborg et al., 2014). Fitzgerald et al. (2007) provide an excellent review of mercury cycling in the marine environment.

Advances in sampling and analysis over the last 30 years have clarified trace metal dynamics. Trace metal concentrations in most open ocean surface waters are extremely low, but many have important functions in limiting biological processes. For example iron (Fe) plays a critical role in phytoplankton dynamics and nitrogen fixation. Because of the importance of phytoplankton as the base of the marine food chain and in the global carbon cycle, Fe can be critical in the regulation of ocean biogeochemistry. Several studies have identified the presence of trace metal pollutants in the oceans, and even the potential for remediation. For instance it has been shown that while the atmospheric deposition of lead (Pb) from leaded fuel elevated dissolved concentrations in oceans well above 'natural' levels (Wu and Boyle, 1997), since leaded fuel was phased out (between 1986 and 2002), oceanic lead levels have been steadily declining (Boyle et al., 2014).

13.2.10 Freshwater (environmental flows)

It may seem odd to consider freshwater as a potential pollutant, but for many marine and estuarine ecosystems changes in freshwater inputs and the attendant changes in salinity profiles, sedimentation, and nutrient dynamics can be significant. In particular, estuaries, the interfaces between freshwater and marine systems, tend to have naturally high productivity and species diversity and are characterized by ecological processes that are highly dependent on seasonal flow dynamics. Consequently, anthropogenic interference such as the introduction of dams for electricity generation or freshwater management can result in significant changes in nutrient inputs, salinity regimes, and toxic chemical concentrations.

The Three Gorges Dam in China has interrupted natural fish migration pathways and resulted in alterations to the chemical balance, temperature, and velocity of the river and downstream estuary (Hvistendahl, 2008). The 'Iron Gates' dam has resulted in a reduction in silica discharge from the Danube, which is believed to have led to changes in primary production in the Black Sea (Humborg et al., 1997). The Aswan Dam has reduced the quantity and quality of discharge from the Nile to the extent that very little water and virtually no sediment escape to the Mediterranean, with reduced fish catches and shoreline erosion cited as likely outcomes (Milliman, 1997). While flow is often measured as part of environmental monitoring programmes, there are very few flow guidelines. Climate change is also affecting rainfall patterns globally, and the resultant environmental flow effects cannot yet be predicted.

13.2.11 Dredging waste/sediment (as a source and sink)

Deposited sediments provide a site of potential contaminant exposure and, if inputs decrease over time, a mechanism for exposure reduction via burial (Apitz, 2011). Natural (e.g. ocean chemistry, species range, and weather patterns) and anthropogenic processes (e.g. construction, remedial and maintenance dredging, boating and fisheries practices) can influence depositional conditions, in turn affecting the risk associated with contaminants in sediments and pore waters. While the disposal of contaminated dredged material is strictly regulated via international agreements such as OSPAR and the London Convention (IMO, 2009), there is disparity internationally as to how dredged materials are assessed and managed (PIANC, 2009). Furthermore, the dredging process has in itself potential to impact, as it can expose previously buried contaminants (Bridges et al., 2008). Management processes for dredging (and other drivers of sediment disturbance) generally involve assessment of 'downstream' exposure pathways and pose a significant source of uncertainty in long-term sediment risk assessment (Apitz and Black, 2010).

13.2.12 Emerging contaminants

New contaminants and/or new risks will continue to emerge. A currently emerging contaminant risk is nanomaterials, substances with nanoscale structures (i.e. between 1 and 1000 nm). While there are naturally occurring nanomaterials (for example clays and natural colloids, or even viruses), it is generally the man-made substances that are considered pollutants. The use of engineered nanoparticles is growing rapidly—these materials are designed to impart distinctive mechanical, electrical, optical, magnetic, catalytic, and biological properties, and their function or benefit is associated with having very high surface area. The fate and risk of these materials in environmental and biological systems is poorly understood (Klaine et al., 2012). Metal oxide nanoparticles (nano-TiO_2, nano-ZnO, and Nano-Ag) are used in cosmetics, coatings, and pigments, with the potential for water discharge (Gottschalk et al., 2009). While there are some data on potential toxicological impacts of nanoparticles (Klaine

et al., 2012), there is little information on real-world effects, environmental prevalence, and transformations (Nowack and Bucheli, 2007). Environmental factors such as pH, temperature, UV light, natural organic matter, and other contaminants have been shown to strongly affect these materials (Wong et al., 2013) and models suggest they may be present in surface waters in high enough concentrations to impact aquatic ecosystems (Gottschalk et al., 2009; Wong et al., 2013). While the ecological risks of nanomaterials in marine systems are difficult to predict, their growing use and ubiquity has led to calls for action on risk assessment and management (Anon, 2011, 2012).

13.3 Pollutant effects

The three most important considerations in determining the potential impact of any contaminant are persistence, potential to bioaccumulate, and toxicity. Rachel Carson's book 'Silent Spring' highlighted the persistence and potential adverse effects of chemical contaminants in the environment (Carson, 1994). The longer a chemical remains in the environment the greater the potential for negative impacts. The contaminants of greatest concern are highly persistent, for example plastics, organo-pollutants (e.g. PCBs, PAHs, and their derivatives), and metals; being resistant to environmental degradation through chemical, biological, and photolytic processes, they can remain in the environment for a very long time. Persistent contaminants can be transported over greater distances and can both bioaccumulate and bioconcentrate through the food chain; as a result they have the potential to significantly impact human health and the environment.

Bioaccumulation and bioconcentration of contaminants in seafood represent major environmental exposure pathways. Consequently, monitoring and management of contaminants is often focused on fitness for human consumption rather than species protection (e.g. Food Standards). Bioaccumulation and bioconcentration assessments can serve as useful management tools, with many instances of species being used as 'biomonitors' of pollutant loads. For example, mussels have been used to trace spatial and temporal variability in contaminant loads around the US coast since 1986 (e.g. Smith et al., 1986; Konar and Stephenson, 1995; Luengen et al., 2004; Flegal et al., 2007).

While the ecotoxicological and broader ecological effects of many common pollutants have been extensively studied, effects of less common and emerging chemicals, and synergistic or antagonistic effects of contaminant cocktails, are less well known (Beiras et al.,

2012). Many of the contaminants (particularly PPCPs) found in the marine environment come through waste treatment plants and as a result will generally be encountered in conjunction with organic enrichment. Similarly, agricultural chemicals such as fertilizers and pesticides are often washed into catchments after rainfall and may co-occur with increased freshwater flows and organic materials. Sometimes combined contaminants exacerbate effects; for example, in antifouling paints the combination of zinc pyrithione and copper causes greater toxicity than would be the case with the individual compounds (Boa et al., 2008). Alternatively, some contaminants mitigate the effect of others. For example mercury toxicity in mammals may be ameliorated by selenium (Ullrich et al., 2001), and copper and manganese interact positively in phytoplankton growth assays (Stauber and Florence, 1987). Such interactions should be taken into account in risk assessments.

Typically, aquatic pollution does not receive much attention unless threshold levels are exceeded, there is potential for direct effects on human health and wellbeing, or an immediate adverse ecological effect is observed. For major events or continuous point source pollution, this may happen relatively quickly, but for diffuse or intermittent discharges it may be some time before physiological or ecological responses are observed (Islam and Tanaka, 2004).

Endocrine disrupters (EDs) are worth highlighting specifically, as they cause chronic effects by interfering with sexual development and hormone systems. Many studies have identified pollutant-induced endocrine disruption in aquatic biota (e.g. Pulliainen et al., 1992; Barker et al., 1994; Jobling et al., 1996; Leblanc et al., 1997; Collier et al., 1998; Allen et al., 1999; Christiansen et al., 2000; Monteiro et al., 2000; Fossi et al., 2001, 2002; Porte et al., 2006). The effects of endocrine disruption in marine organisms include inappropriate manifestation of both male and female gonadal characteristics (Jobling et al., 2003), increased female hormones in male fish (Folmar et al., 1996; Janssen et al., 1997), and degeneration of gonadal tissue (Janssen et al., 1997; Lye et al., 1999). Effects are frequently estrogenic, affecting reproductive systems, although several physiological functions can be affected.

Only a very few EDs have been purposefully designed to that end (e.g. the human contraceptive pill). In most cases the ED effect is an unintended side-effect either of the initial chemical or its breakdown products. Organic agricultural products (e.g. PCBs and organochlorine pesticides) are key sources of EDs in the marine environment, but EDs are increasingly being identified in plastics (Yang et al., 2011), chlorinated hydrocarbons (Oberdörster and Cheek, 2001), pesticides, degradation products of high-use industrial surfactants (Arcand-Hoy and Benson, 1998), and pharmaceutical products (Bortone and Cody, 1999; Nakari and Erkomaa, 2003; Porte et al., 2006). The potential for ED bioaccumulation in marine biota is largely unknown, but there could be significant direct and transmitted impacts. ED impacts are not necessarily restricted to individuals, as impaired reproductive capacity will operate at population level, and hence community interactions and broader ecosystem functionality may also be compromised.

13.4 Challenges for management and research

Management of chemical contaminants in the marine environment is complex. Our scientific understanding of marine pollution is incomplete, making it difficult to ensure that management strategies are effective (Islam and Tanaka, 2004). As the majority of pollutants in the marine environment are derived from terrestrial sources (Widdicome and Somerfield, 2012), management must extend beyond coastal zones and into catchments. Given the connectivity of coastal and marine systems, this requires integration of legislation in local, regional, and international jurisdictions (Apitz et al., 2006). The interconnectedness of systems, complicated environmental and chemical interactions, uncertainty of impacts or sources, conflicts between users and polluters, and financial and logistical constraints on the implementation of management actions all serve to complicate the situation.

There are several significant international agreements which provide underpinning frameworks for environmental management. The 'Earth Summit' series (Dodds, 2014) highlights the level of collaboration required to regulate and manage contaminants in the marine environment at an international level. The signatories committed to protect ecosystems, to prevent environmental dumping, and to determine the ongoing viability of activities with potential for significant environmental harm. However, to be successful individual partners must both develop and implement the management strategies and controls to achieve these goals. While the development of an internationally compatible management framework may seem ambitious, in most developed countries there are similar initiatives and objectives—often it is just the details of the guidelines (e.g. water and sediment quality guidelines and discharge guidelines) that differ.

The initial focus of marine and coastal management was predominantly on development of targets and objectives for regulation of particular impacts, such as nutrient discharges. However, emphasis has shifted over time towards ecosystem-based indicators and the protection of biodiversity. However, recent management strategies have started to focus on actually tackling impacts by seeking to reduce and/or eliminate chemical pollutant inputs into coastal and offshore waters (Rogers and Greenaway, 2005).

Understanding the effect pathways for contaminants is critical to limiting impacts on marine communities. To really understand the impacts of chemical pollutants in the environment requires consideration of both the stressors and effects. This necessitates the integrated monitoring of chemical and biological components, considering the form and fate of contaminants and how they enter the marine environment, as well as their various breakdown products and pathways. Understanding sediment and water conditions, trophic interactions, and effects of contaminants on community structure, and hence biodiversity, are crucial for determining the magnitude of short- and long-term effects. In Europe the OSPAR and HELCOM initiatives are good examples of integrated management approaches which seek to bring stakeholders together to collectively reduce pollution and improve ecological condition, with the EU Marine Strategy Framework Directive (Marine Strategy Framework Directive, 2008) providing the necessary policy guidance (OSPAR Commission, 2010). These processes need continual review and update to remain best practice; for example, recent work has provided guidance on integrated marine environmental monitoring of chemicals and their effects within OSPAR (Davies and Vethaak, 2012).

Sustainable management requires a clear understanding of issues, their likely effects, indicators or trigger levels, as well as management options and expected responses. A healthy, functioning environment can potentially process, detoxify, or reduce the impact of many contaminants, as long as threshold levels are not exceeded. Therefore many monitoring and assessment programmes include some measure of ecological performance or risk. This can be guided by assessment of indicators for the driving forces, pressures, states, impacts, and responses (DPSIR) (Elliott, 2002; EEA, 2005; Atkins et al., 2011). These indicators are most useful where they define a clear protective status and allow for both the possibility of synergistic/antagonistic effects and additional inputs.

Water quality is often a primary criterion for management, and there are examples of international (e.g. Water Framework Directive), national (e.g. USEPA National Recommended Water Quality Criteria), and ecosystem-specific (e.g. USEPA National Guidance Water Quality Standards for Wetlands) water quality guidelines, where if levels are exceeded managers are prompted to take action. In all of these, concentrations and trigger values for pollutants (i.e. sediment loads, nutrients, pesticides, and pharmaceuticals) have been established with a view to protecting and maintaining marine species and ecosystem health. Ecological quality standards (EQS)—the concentration below which a substance is not believed to be detrimental to aquatic life—or ecological assessment criteria (EACs)—the concentrations above which there may be impacts on biota—are regularly employed as biological standards in environmental assessments (Rogers and Greenaway, 2005). In North America many water and sediment quality guidelines are similarly based upon no effects, threshold effects, and probable effects-based criteria (Wenning et al., 2005). However, there is general consensus that measurement of the pollutant alone (chemical criteria) should only be one line of evidence in any management decision framework (Chapman et al., 2002).

Ecological exposure and/or effects in marine ecosystems can be assessed using biomarkers/bioindicators (Porte et al., 2006). While there is growing use of biomarkers in marine programmes (ICES, 2000; PIANC, 2006a, b; Apitz, 2011), these methods tend to lack the standardization and quality control that is associated with more established chemical assessments of contaminants. Diagnostic tools for specific groups of chemicals or types of response are needed (Cajaraville et al., 2000). Emerging disciplines such as genomics, proteomics, and biodiversity informatics may provide solutions (Porte et al., 2006; Aricò, 2008; Steinberg et al., 2008; Mearns et al., 2008).

Ecosystem-based management approaches, whereby the combined effects of all activities (natural and human-induced) within an ecosystem are collectively monitored and managed, are increasingly being recommended as a means to better understand and assess the interactions and the impact of human activities on marine organisms. The concept of monitoring 'ecosystem services' (ES)—the benefits provided by a functioning ecosystem such as flood regulation, coastal protection, and water purification—is increasingly being proposed as the gold standard for management. But it may take years or even decades to develop sufficient understanding of ecosystem function to be able to identify appropriate indicators and standards for ecosystem services.

13.4.1 Monitoring

Regulation and monitoring provide the necessary structure and controls for management. Monitoring underpins sustainable management, by providing information on contaminant levels and determining the effectiveness of management actions. Monitoring may also be undertaken to increase knowledge about the environmental conditions and effects, to improve design, management, and monitoring in the future.

Most regulatory monitoring programmes can roughly be categorized into three types (PIANC, 2006b):

* surveillance monitoring to assess spatial and temporal changes to selected parameters over space and time;
* feedback/adaptive monitoring in which predictable variables are forecast and then monitored; and
* compliance monitoring to ensure compliance with contractual or regulatory restrictions.

Unfortunately, it is not possible to monitor for the presence and/or impacts of all possible contaminants and pollutant combinations; therefore, it is important in designing monitoring programmes to be very clear about the objectives and choose the monitoring approach carefully—in particular, ensuring that the target systems, chemical and biological sensitivities, and spatial and temporal scales are appropriate. Where possible, monitoring programmes should be adaptable, such that they can be improved as our understanding of the complexity of chemical pollutants evolves.

The cost of monitoring in the marine environment is significant, thus monitoring programs often target indicators (biological or chemical) which are of particular economic or ecological significance, for example fishery decline, shellfish deformations, or toxin levels in recreationally and commercially important species. It is important that such programmes be reviewed regularly to ensure best practice and that the monitoring remains 'fit for purpose'.

13.5 Recommendations for the future

Minimizing the impacts of chemical pollutants in the marine environment (especially those that bioaccumulate, have toxic breakdown products, or do not biodegrade) is an important goal. One way to achieve this is to reduce or eliminate their use; something that requires the participation of the various polluters.

While it is not possible to completely eliminate chemicals from the marine environment, providing the knowledge and understanding to identify the risks and support companies in improving waste management practices is an important step towards reducing the overall impacts. In addition, it is worth regularly reviewing needs and usage with a view to identifying substitution and harm minimization strategies (Fig. 13.2). Such reviews should encompass three key objectives:

1. Define the **need** for each chemical and where possible propose more benign alternatives.
2. Where a chemical cannot be eliminated, then review the **quantities** and consider recovery, reuse, or recycling to reduce waste.
3. Finally consider **treatment** and review disposal strategies.

An important factor of this approach is that it puts the responsibility for management onto the polluter, rather than the regulator or the environment.

Comprehensive elimination of toxic chemicals will require a systematic multi-jurisdictional approach, encompassing international policy and law as well as national

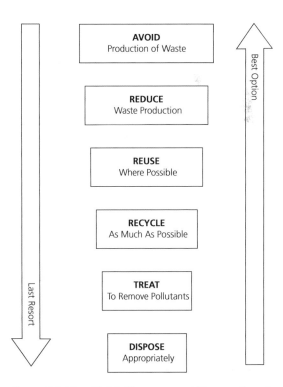

Figure 13.2 Hierarchical decision support model for waste disposal options.

and state regulation. While rare, such consensus is achievable where there is collective will, a good example being TBT which is now banned worldwide as a result of the demonstrated adverse environmental impacts.

It is impossible to identify all potential contaminants in the marine environment, and it is unclear whether those currently regulated and monitored are those causing the most risk (Wagner, 1995; Apitz and Agius, 2013). To address the problem of emerging contaminants it might be useful to consider the effects rather than the pollutants per se, with a view to building management responses around those effect types most hazardous to ecological status (e.g. cytotoxicity, neurotoxicity, mutagenicity, endocrine disruption, or photosynthesis inhibiting potencies). All of these ecological impacts are difficult to detect using current standard assessments for acute and chronic toxicity (Brack et al., 2009). The application of a broader range of effects-based assessment tools in screening could lead to a more accurate assessment of the ecological risk, and potentially better environmental management, mitigation, and remediation strategies.

Ultimately the decision of how we regulate and manage chemicals in the environment, and what level of environmental impact we consider acceptable/unacceptable, is a societal one. The problem of managing contamination in the marine environment is complex, uncertain, and multi-scale, and affects multiple stakeholders. Where pollutants are associated with commercial interests the impetus for change will be driven by both economic and social agendas. Green economics, and the triple bottom-line approach (TBL—an accounting framework that incorporates social, environmental, and financial measures of performance) or the three Ps triangle (people, planet, profit) (Elkington, 1999), are playing an increasingly important role in incorporating societal views. Stakeholders are becoming more informed, and as a result investors are seeking economic, environmental, and social sustainability from industry, government, and all potential polluters (Slaper and Hall, 2011).

Chemicals are essential components of modern life, but, given the critical role of the marine environment in delivering ES, it is clear that the complex effects of most chemical pollutants (Maltby, 2013), and the relative importance of combined stressors (Elliott et al., 2007; Apitz, 2013) must be better understood and managed. There is a need to develop indicators and predictors of ecosystem change across different spatial scales (local to global), and to establish protection goals (and in some cases restoration targets) with respect to specific ecosystem services (Munns et al., 2009; Maltby, 2013).

Assessment and management approaches need to be developed that take into account the entire life cycle of pollutants (including the original contaminant, any by-products, and relevant process interactions). These are the challenges to be faced in order to accurately evaluate and manage the risks associated with chemical pollutants in marine ecosystems into the future.

References

Abarnou, A. & Miossec, L. (1992). Chlorinated waters discharged to the marine environment chemistry and environmental impact—an overview. *Science of the Total Environment*, **126**(1–2), 173–197.

Agus, E., Voutchkov, N., & Sedlak, D.L. (2009). Disinfection by-products and their potential impact on the quality of water produced by desalination systems: a literature review. *Desalination*, **237**(1), 214–237.

Allen, Y., Scott, A.P., Matthiessen, P., et al. (1999). Survey of oestrogenic activity in United Kingdom estuarine and coastal waters and its effects on gonadal development of the flounder *Platichthys flesus*. *Environmental Toxicology and Chemistry*, **18**(8), 1791–1800.

Alzieu, C., Heral, M., Thibaud, Y., et al. (1981). Influence des Peintures Antisalissures a Base d'Organostanniques sur la Calcification de la Coquille de l'Huitre *Crassostrea gigas*. *Revue des travaux de l'Institut des pêches maritimes*, **45**(2), 101–116.

Alzieu, C.L., Sanjuan, J., Deltreil, J.P., & Borel, M. (1986). Tin contamination in Arcachon Bay: effects on oyster shell anomalies. *Marine Pollution Bulletin*, **17**(11), 494–498.

AMSA (2012). *How Australia Responds to Oil and Chemical Spills in the Marine Environment*. Australian Maritime Safety Authority (AMSA), Canberra ACT Australia.

Anon (2011). Editorial: the dose makes the poison. *Nature Nanotechnology*, **6**, 329.

Anon (2012). Editorial: join the dialogue. *Nature Nanotechnology*, **7**, 545.

ANZECC (1997). *Code of Practice for Antifouling and In-water Hull Cleaning and Maintenance*. Maritime Safety Authority of New Zealand, Australian and New Zealand Environment and Conservation Council, Australian Transport Council, and Australian Maritime Safety Authority, Canberra.

ANZECC (2000). *Australian and New Zealand Guidelines for Fresh and Marine Water Quality*. Australian and New Zealand Environment and Conservation Council and Agriculture and Resource Management Council of Australia and New Zealand, Canberra.

Apitz, S.E. (2011). Integrated risk assessments for the management of contaminated sediments in estuaries and coastal systems. In: Wolanski, E. & McLusky, D.S. (eds), *Treatise on Estuarine and Coastal Science*, Vol. 4, pp. 311–338. Academic Press, Waltham, MA.

Apitz, S.E. (2013). Ecosystem services and environmental decision making: seeking order in complexity. *Integrated Environmental Assessment and Management*, **9**(2), 414–430.

Apitz, S.E. & Agius, S. (2013). Anatomy of a decision: potential regulatory outcomes from changes to chemistry protocols in the Canadian disposal at Sea Program. *Marine Pollution Bulletin*, **69**, 76–90.

Apitz, S.E. & Black, K. (2010). *Research and Support for Developing a UK Strategy for Managing Contaminated Sediments: An Analysis of the Project Findings*. Partrac and SEA Environmental Decisions Ltd; Report to DEFRA, Glasgow, Scotland, 100 pp.

Apitz, S.E., Elliot, M., Fountain, M., & Galloway, T. (2006). European environmental management: moving to an ecosystem approach. *Integrated Environmental Assessment and Management*, **2**, 80–86.

Arcand-Hoy, L.D. & Benson, W.H. (1998). Fish reproduction: an ecologically relevant indicator of endocrine disruption. *Environmental Toxicology and Chemistry*, **17**(1), 49–57.

Arfsten, D.P., Schaeffer, D.J., & Mulveny, D.C. (1996). The effects of near ultraviolet radiation on the toxic effects of polycyclic aromatic hydrocarbons in animals and plants: a review. *Ecotoxicology and Environmental Safety*, **33**(1), 1–24.

Aricò, S. (2008). Advances in concepts and methods for the marine environment: implications for policy. *Cell Biology and Toxicology*, **24**(6), 475–481.

Atkins, J.P., Burdon, D., Elliott, M., & Gregory, A.J. (2011). Management of the marine environment: integrating ecosystem services and societal benefits with the DPSIR framework in a systems approach. *Marine Pollution Bulletin*, **62**, 215–226.

Barker, D.E., Khan, R.A., & Hooper, R. (1994). Bioindicators of stress in winter flounder *Pleuronectes americanus*, captured adjacent to a pulp and paper mill in St. George's Bay, Newfoundland. *Canadian Journal of Fisheries and Aquatic Sciences*, **51**, 2203–2209.

Barnes, D.K.A. (2002). Biodiversity: invasions by marine life on plastic debris. *Nature*, **416**, 808–809.

Barnes, D.K., Galgani, F., Thompson, R.C., & Barlaz, M. (2009). Accumulation and fragmentation of plastic debris in global environments. *Philosophical Transactions of the Royal Society Series B*, **364**(1526), 1985–1998.

Batley, G. (1995). Heavy metals and tributyltin in Australian coastal and estuarine waters. In: Zann, L.P. & Sutton, P.C. (eds) *State of the Marine Environment Report for Australia: Pollution—Technical Annex 2*. Department of the Environment and Water Resources, Canberra.

Batley, G. & Maher, W.A. (2001). The development and application of ANZECC and ARMCANZ sediment quality guidelines. *Australasian Journal of Ecotoxicology*, **7**, 81–92.

Beiras, R., Dura´n, I., Parra, S., et al. (2012). Linking chemical contamination to biological effects in coastal pollution monitoring. *Ecotoxicology*, **21**, 9–17.

Boa, V.W., Leung, K.M.Y., Kwok, K.W.H., et al. (2008). Synergistic toxic effects of zinc pyrithione and copper to three marine species: implications on setting appropriate water quality criteria. *Marine Pollution Bulletin*, **57**(6–12), 616–623.

Bodkin, J.L., Esler, D., Rice, S.D., et al. (2014). The effects of spilled oil on coastal ecosystems: lessons from the Exxon Valdez spill. In: Maslo, B. & Lockwood, J.L. (eds), *Coastal Conservation*. Conservation Biology Series No. 19, pp. 311–346. Cambridge University Press, Cambridge, UK.

Bortone, S.A. & Cody, R.P. (1999). Morphological masculinization in poeciliid females from a paper mill effluent receiving tributary of the St. Johns River, Florida, USA. *Bulletin of Environmental Contamination and Toxicology*, **63**(2), 150–156.

Boyle, E.A., Lee, J.-M., Echegoyen, Y., et al. (2014). Anthropogenic lead emissions in the ocean: the evolving global experiment. *Oceanography*, **27**(1), 69–75.

Brack, W., Apitz, S.E., Borchardt, D., et al. (2009). Toward a holistic and risk-based management of European river basins. *Integrated Environmental Assessment and Management*, **5**(1), 5–10.

Brack, W., Klamer, H.J.C., Lo´pez de Alda, M., & Barcelo´, D. (2007). Effect-directed analysis of key toxicants in European river basins: a review. *Environmental Science and Pollution. Research International*, **14**, 30–38.

Brand, L.E., Sunda, S.G., & Guillard, R.R.L. (1986). Reduction of marine phytoplankton reproduction rates by copper and cadmium. *Journal of Experimental Marine Biology and Ecology*, **96**, 225–250.

Bridges, T.S., Ells, S., Hayes, D., et al. (2008). *The Four Rs of Environmental Dredging: Resuspension, Release, Residual, and Risk*. U.S. Army Engineer Research and Development Center, Dredging Operations and Environmental Research Program, Vicksburg, MS.

Brodin, T., Fick, J., Jonsson, M., & Klaminder, J. (2013). Dilute concentrations of a psychiatric drug alter behavior of fish from natural populations. *Science*, **339**, 814–815.

Brown, J.E., Hosseini, A., Børretzen, P., & Thørring, H. (2006). Development of a methodology for assessing the environmental impact of radioactivity in Northern Marine environments. *Marine Pollution Bulletin*, **52**, 1127–1137.

Cajaraville, M.P., Bebianno, M.J., Blasco, J., et al. (2000). The use of biomarkers to assess the impact of pollution in coastal environments of the Iberian Peninsula: a practical approach. *Science of the Total Environment*, **247**(2), 295–311.

Carson, R. (1994). *Silent Spring*. Houghton Mifflin, Boston.

CCME (Canadian Council of Ministers of the Environment) (1999). *Canadian Water Quality Guidelines*. Prepared by the Task Force on Water Quality Guidelines of the Canadian Council of Ministers of the Environment, Eco-Health Branch, Ottawa, Ontario, Canada.

Chapman, P.M., McDonald, B.G., & Lawrence, G.S. (2002). Weight-of-evidence issues and frameworks for sediment quality (and other) assessments. *Human and Ecological Risk Assessment*, **8**(7),1489–1515.

Christiansen, L.B., Pedersen, K.L., Pedersen, S.N., et al. (2000). In vivo comparison of xenoestrogens using rainbow trout vitellogenin induction as a screening system. *Environmental Toxicology and Chemistry*, **19**(7), 1867–1874.

Cleuvers, M. (2004). Mixture toxicity of the anti-inflammatory drugs diclofenac, ibuprofen, naproxen, and acetylsalicylic acid. *Ecotoxicology and Environmental Safety*, **59**(3), 309–315.

Cole, M., Lindeque, P., Halsband, C., & Galloway, T.S. (2011). Microplastics as contaminants in the marine environment: a review. *Marine Pollution Bulletin*, **62**(12), 2588–2597.

Collier, T.K., Johnson, L.L., Stehr, C.M., et al. (1998). A comprehensive assessment of the impacts of contaminants on fish from an urban waterway. *Marine Environmental Research*, **49**(1–5), 243–247.

Dafforn, K.A., Lewis, J.A., & Johnston, E.L. (2011). Antifouling strategies: history and regulation, ecological impacts and mitigation. *Marine Pollution Bulletin*, **62**(3), 453–465.

Daughton, C.G. (2002). Environmental stewardship and drugs as pollutants. *Lancet*, **360**(9339), 1035–1036.

Daughton, C.G. & Ternes, T.A. (1999). Pharmaceuticals and personal care products in the environment: agents of subtle change? *Environmental Health Perspectives*, **107**(Suppl. 6), 907.

Davies, I.M. & Vethaak, A.D. (2012). *Integrated Marine Environmental Monitoring of Chemicals and Their Effects*. ICES Cooperative Research Report No. 315, p. 277. ICES, Copenhagen.

De Lange, H.J., Noordoven, W., Murk, A.J., et al. (2006). Behavioural responses of *Gammarus pulex* (Crustacea, Amphipoda) to low concentrations of pharmaceuticals. *Aquatic Toxicology*, **78**, 209–216.

De Mora, S.J., Stewart, C., & Phillips, D. (1995). Sources and rate of degradation of Tri(nbutyl) tin in marine sediments near Auckland, New Zealand. *Marine Pollution Bulletin*, **30**, 50–57.

Di Poi, C., Darmaillacq, A.S., Dickel, L., et al. (2013). Effects of perinatal exposure to waterborne fluoxetine on memory processing in the cuttlefish (*Sepia officinalis*). *Aquatic Toxicology*, **132**, 84–91.

Dodds, F. (ed.) (2014). *Earth Summit 2002: A New Deal*. Routledge, London.

Drury, B., Scott, J., Rosi-Marshall, E.J., & Kelly, J.J. (2013). Triclosan exposure increases triclosan resistance and influences taxonomic composition of benthic bacterial communities. *Environmental Science and Technology*, **47**, 8923–8930.

EEA (2005). *Conceptual Framework: How We Reason*. European Environment Agency, Copenhagen, Denmark.

Eisenbud, M. & Gesell, T.F. (1997). *Environmental Radioactivity from Natural, Industrial & Military Sources*. Academic Press, Boston.

Ekstrom, E.B., Morel, F.M., & Benoit, J.M. (2003). Mercury methylation independent of the acetyl-coenzyme A pathway in sulfate-reducing bacteria. *Applied and Environmental Microbiology*, **69**(9), 5414–5422.

Elkington, J. (1999). *Cannibals with Forks—The Triple Bottom Line of the 21st Century*. Capstone Publishing Ltd, Oxford.

Elliott, M. (2002). The role of the DPSIR approach and conceptual models in marine environmental management: an example for offshore wind power. *Marine Pollution Bulletin*, **44**, 3–7.

Elliott, M., Burdon, D., Hemingway, K.L., & Apitz, S.E. (2007). Estuarine, coastal and marine habitat and ecosystem restoration: confusing management and science—a revision of concepts. *Estuarine, Coastal and Shelf Science*, **74**, 349–366.

Evans, S.M. & Nicholson, G.J. (2000). The use of imposex to assess tributyltin contamination in coastal waters and open seas. *Science of the Total Environment*, **258**, 73–80.

FAO—Food and Agriculture Organization of the United Nations (2002). *International Code of Conduct on the Distribution and Use of Pesticides*. Food and Agriculture Organization of the United Nations, Rome.

Farrington, J.W. (2013). Oil pollution in the marine environment I: inputs, big spills, small spills, and dribbles. *Environment: Science and Policy for Sustainable Development*, **55**, 3–13.

Finnie, A. (2006). Improved estimates of environmental copper release rates from antifouling products. *Biofouling*, **22**, 279–291.

Fitzgerald, W.F., Lamborg, C.H., & Hammerschmidt, C.R. (2007). Marine biogeochemical cycling of mercury. *Chemical Reviews*, **107**(2), 641–662.

Flegal, A.R., Brown, C.L., Squire, S., et al. (2007). Spatial and temporal variations in silver contamination and toxicity in San Francisco Bay. *Environmental Research*, **105**(1), 34.

Folmar, L.C., Denslow, N.D., Rao, V., et al. (1996). Vitellogenin induction and reduced serum testosterone concentrations in feral male carp (*Cyprinus carpio*) captured near a major metropolitan sewage treatment plant. *Environmental Health Perspectives*, **104**(10), 1096.

Fong, P.P. & Ford, A.T. (2014). The biological effects of antidepressants on the molluscs and crustaceans: a review. *Aquatic Toxicology*, **151**, 4–13.

Fong, P.P. & Hoy, C.M. (2012). Antidepressants (venlafaxine and citalopram) cause foot detachment from the substrate in freshwater snails at environmentally relevant concentrations. *Marine and Freshwater Behaviour and Physiology*, **45**(2), 145–153.

Fong, P.P. & Molnar, N. (2013). Antidepressants cause foot detachment from substrate in five species of marine snail. *Marine Environmental Research*, **84**, 24–30.

Fossi, M.C., Casini, S., Ancora, S., et al. (2001). Do endocrine disrupting chemicals threaten Mediterranean swordfish? Preliminary results of vitellogenin and Zona radiata proteins in *Xiphias gladius*. *Marine Environmental Research*, **52**, 447–483.

Fossi, M.C., Casini, S., Marsili, L., et al. (2002). Biomarkers for endocrine disruptors in three species of Mediterranean large pelagic fish. *Marine Environmental Research*, **54**(3), 667–671.

Franzellitti, S., Buratti, S., Valbonesi, P., & Fabbri, E. (2013). The mode of action (MOA) approach reveals interactive effects of environmental pharmaceuticals on (*Mytilus galloprovincialis*). *Aquatic Toxicology*, **140**, 249–256.

Galante-Oliveira, S., Oliveira, I., Jonkers, N., et al. (2009). Imposex levels and tributyltin pollution in Ria de Aveiro (NW Portugal) between 1997 and 2007: evaluation of legislation effectiveness. *Journal of Environmental Monitoring*, **11**, 1405–1411.

GESAMP (1993). *Impact of Oil and Related Chemicals and Wastes on the Marine Environment*. GESAMP Report 50. Joint Group of Experts on the Scientific Aspects of Marine Environmental Protection, London.

Gibbs, P.E., Bryan, G.W., & Pascoe, P.L. (1991). TBT-induced imposex in the dogwhelk Nucella lapillus: geographical uniformity of the response and effects. *Marine Environmental Research*, **32**, 79–87.

Gillan, D.C., Baeyens, W., Bechara, R., et al. (2012). Links between bacterial communities in marine sediments and trace metal geochemistry as measured by *in situ* DET/DGT approaches. *Marine Pollution Bulletin*, **64**(2), 353–362.

Gottschalk, F., Sonderer, T., Scholz, R.W., & Nowack, B. (2009). Modeled environmental concentrations of engineered nanomaterials (TiO_2, ZnO, Ag, CNT, Fullerenes) for different regions. *Environmental Science and Technology*, **43**, 9216–9222.

Gregory, M.R. (2009). Environmental implications of plastic debris in marine settings: entanglement, ingestion, smothering, hangers-on, hitch-hiking and alien invasions. *Philosophical Transactions of the Royal Society Series B*, **364**, 2013–2025.

Guler, Y. & Ford, A.T. (2010). Anti-depressants make amphipods see the light. *Aquatic Toxicology*, **99**(3), 397–404.

Halling-Sørensen, B., Nors Nielsen, S., Lanzky, P.F., et al. (1998). Occurrence, fate and effects of pharmaceutical substances in the environment—a review. *Chemosphere*, **36**(2), 357–393.

Haynes, D. & Johnson, J.E. (2000). Organochlorine, heavy metal and polyaromatic hydrocarbon pollutant concentrations in the Great Barrier Reef (Australia) environment: a review. *Marine Pollution Bulletin*, **41**(7), 267–278.

Hellou, J., Lebeuf, M., & Rudic, M. (2013). Review on DDT and metabolites in birds and mammals of aquatic ecosystems. *Environmental Reviews*, **21**(1), 53–69.

Hoffmann, L., Breitbarth, E., Boyd, P.W., & Hunter, K.A. (2012). Influence of ocean warming and acidification on trace metal biogeochemistry. *Marine Ecology Progress Series*, **470**, 191–205.

Hood, E., Fellman, J., Spencer, R.G.M., et al. (2009). Glaciers as a source of ancient and labile organic matter to the marine environment. *Nature*, **462**, 1044–1048.

Humborg, C., Ittekkot, V., Cociascu, A., & Bodungen, B. (1997). Effect of Danube River Dam on Black Sea biogeochemistry and ecosystem structure. *Nature*, **386**, 385–388.

Hvistendahl, M. (2008). China's Three Gorges Dam: An Environmental Catastrophe? *Scientific American*. March, 2008.

IAEA (International Atomic Energy Agency) (2011). *Coordinated Research Programme on Radioactive Particles*. Report by international advisory committee. IAEA Technical document. IAEA, Vienna.

ICES (2000). *Biological Assessment of Toxicity of Marine Dredged Materials*. Report of the ICES Working Group on Biological Effects of Contaminants, Nantes, France, 27–31 March 2000, ch. 7, ICES CM 2000/E: Ref.: ACME. ICES, Copenhagen.

IMO (2002). *International Convention on the Control of Harmful Anti-fouling Systems on Ships*. International Maritime Organization, London.

IMO (2009). *Guidance for the Development of Action Lists and Action Levels for Dredged Material*. London Convention, London Protocol, International Maritime Organization, London.

Isidori, M., Lavorgna, M., Nardelli, A., et al. (2005). Ecotoxicity of naproxen and its phototransformation products. *Science of the Total Environment*, **348**(1), 93–101.

Islam, M.S. & Tanaka, M. (2004). Impacts of pollution on coastal and marine ecosystems including coastal and marine fisheries and approach for management: a review and synthesis. *Marine Pollution Bulletin*, **48**, 624–649.

Janssen, P.A.H., Lambert, J.G.D., Vethaak, A.D., & Goos, H.J.Th. (1997). Environmental pollution caused elevated concentrations of oestradiol and vitellogenin in the female flounder, *Platichthys flesus* (L.). *Aquatic Toxicology*, **39**, 195–214.

Jobling, S., Casey, D., Rodgers-Gray, T., et al. (2003). Comparative responses of molluscs and fish to environmental estrogens and an estrogenic effluent. *Aquatic Toxicology*, **65**(2), 205–220.

Jobling, S., Sheahan, D., Osborne, J.A., et al. (1996). Inhibition of testicular growth in rainbow trout (*Oncorhynchus mykiss*) exposed to estrogenic alkylphenolic chemicals. *Environmental Toxicology and Chemistry*, **15**(2), 194–202.

Klaine, S.J., Koelmans, A.A., Horne, N., et al. (2012). Paradigms to assess the environmental impact of manufactured nanomaterials. *Environmental Toxicology and Chemistry*, **31**, 3–14.

Konar, B. & Stephenson, M. D. (1995). Gradients of subsurface water toxicity to oyster larvae in bays and harbors in California and their relation to mussel watch bioaccumulation data. *Chemosphere*, **30**(1), 165–172.

Kukulka, T., Proskurowski, G., Morét-Ferguson, S., et al. (2012). The effect of wind mixing on the vertical distribution of buoyant plastic debris. *Geophysical Research Letters*, **39**(7), 1–6.

Kumar, K., Guptar, S.C., Chander, Y., & Singh, A.K. (2005). Antibiotic use in agriculture and its impact on the terrestrial environment. *Advances in Agronomy*, **87**, 1–54.

Kümmerer, K. (2009a). The presence of pharmaceuticals in the environment due to human use—present knowledge and future challenges. *Journal of Environmental Management*, **90**(8), 2354–2366.

Kümmerer, K. (2009b). Antibiotics in the aquatic environment–a review–part II. *Chemosphere*, **75**(4), 435–441.

Kümmerer, K. (2010). Pharmaceuticals in the environment. *Annual Review of Environmental Resources*, **35**, 57–75.

Lamborg, C.H., Hammerschmidt, C.R., Bowman, K.L., et al. (2014). A global ocean inventory of anthropogenic mercury based on water column measurements. *Nature*, **512**, 65–68.

Langmann, B., Zaksek, K., Hort, M., & Duggen, S. (2010). Volcanic ash as fertiliser for the surface ocean. *Atmospheric Chemistry and Physics*, **10**(8), 3891–3899.

Leblanc, J., Couillard, C.M., & Brethes, J.F. (1997). Modifications of the reproductive period in mummichog (*Fundulus heteroclitus*) living downstream from a bleached kraft pulp mill in the Miramichi Estuary, New Brunswick, Canada. *Canadian Journal of Fisheries and Aquatic Sciences*, **54**, 2564–2573.

Lee, C., Lee, Y., Schmidt, C., et al. (2007). Oxidation of N-nitrosodimethylamine (NDMA) with ozone and chlorine dioxide: kinetics and effect on NDMA formation potential. *Environmental Science and Technology*, **41**, 2056–2063.

Luengen, A.C., Friedman, C.S., Raimondi, P.T., & Flegal, A.R. (2004). Evaluation of mussel immune responses as indicators of contamination in San Francisco Bay. *Marine Environmental Research*, **57**(3), 197–212.

Lye, C.M., Frid, C.L.J., Gill, M.E., et al. (1999). Estrogenic alkylphenols in fish tissues, sediments, and waters from the U.K. Tyne and Tees estuaries. *Environmental Science and Technology*, **33**, 1009–1014.

Maldonado, M.T., Hughes, M.P., Rue, E.L., & Wells, M.L. (2002). The effect of Fe and Cu on growth and domoic acid production by *Pseudo-nitzschia* multiseries and *Pseudo-nitzschia australis*. *Limnology and Oceanography*, **47**(2), 515–526.

Malins, D.C., McCain, B.B., Landahl, J.T., et al. (1988). Neoplastic and other diseases in fish in relation to toxic chemicals: an overview. *Aquatic Toxicology*, **11**, 43–67.

Maltby, L. (2013). Ecosystem services and the protection, restoration, and management of ecosystems exposed to chemical stressors. *Environmental Toxicology and Chemistry*, **32**(5), 974–983.

Mann, E.L., Ahlgren, N., Moffett, J.W., & Chisholm, S.W. (2002). Copper toxicity and cyanobacteria ecology in the Sargasso Sea. *Limnology and Oceanography*, **47**(4), 976–988.

Marine Strategy Framework Directive (2008). Directive 2008/56/EC of the European Parliament and of the Council of 17 June 2008 establishing a framework for community action in the field of marine environmental policy (Marine Strategy Framework Directive). *Official Journal of the European Union*, **164**, 19–40.

Maruya, K.A., Dodder, N.G., Tang, C.L., et al. (2014). Which coastal and marine environmental contaminants are truly emerging? *Environmental Science and Pollution Research*, **22**(3), 1644–1652.

Maskaoui, K. & Zhou, J.L. (2010). Colloids as a sink for certain pharmaceuticals in the aquatic environment. *Environmental Science and Pollution Research*, **17**(4), 898–907.

Mearns, A.J., Reish, D.J., Oshida, P.S., et al. (2008). Effects of pollution on marine organisms. *Water Environment Research*, **80**, 1918–1979.

Méndez-Arriaga, F., Esplugas, S., & Giménez, J. (2008). Photocatalytic degradation of non-steroidal anti-inflammatory drugs with TiO_2 and simulated solar irradiation. *Water Research*, **42**(3), 585–594.

Milliman, J.D. (1997). Blessed dams or damned dams? *Nature*, **386**, 325–327.

Moffett, J.W., Brand, L.E., Croot, P.L., & Barbeau, K.A. (1997). Cu speciation and cyanobacterial distribution in harbors subject to anthropogenic Cu inputs. *Limnology and Oceanography*, **42**(5), 789–799.

Money, C., Braungardt, C.B., Jha, A.N., et al. (2011). Metal speciation and toxicity of Tamar Estuary water to larvae of the Pacific oyster *Crassostrea gigas*. *Marine Environmental Research*, **72**(1), 3–12.

Monteiro, P.R.R., Reis-Henriques, M.A., & Coimbra, J. (2000). Polycyclic aromatic hydrocarbons inhibit in vitro ovarian steroidogenesis in the founder (*Platichthys flesus* L.). *Aquatic Toxicology*, **48**, 549–559.

Morrisey, D.J., Underwood, A.J., & Howitt, L. (1996). Effects of copper on the faunas of marine soft-sediments: an experimental field study. *Marine Biology*, **125**(1), 199–213.

Moss, A., Brodie, J., & Furnas, M. (2005). Water quality guidelines for the Great Barrier Reef World Heritage Area: a basis for development and preliminary values. *Marine Pollution Bulletin*, **51**, 76–88.

MPMMG (Marine Pollution Monitoring Management Group) (1998). *National Monitoring Programme Survey of the Quality of UK Coastal waters*. Marine Pollution Monitoring Management Group, Aberdeen.

Munns, W.R., Jr, Helm, R.C., Adams, W.J., et al. (2009). Translating ecological risk to ecosystem service loss. *Integrated Environmental Assessment and Management*, **5**, 500–514.

Nakari, T. & Erkomaa, K. (2003). Effects of phytosterols on zebrafish reproduction in multigeneration test. *Environmental Pollution*, **123**(2), 267–273.

Naylor, C., Davison, W., Motelica-Heino, M., et al. (2004). Simultaneous release of sulfide with Fe, Mn, Ni and Zn in marine harbour sediments measured using a combined metal/sulfide DGT probe. *Science of the Total Environment*, **328**, 275–286.

Nowack, B. & Bucheli, T. (2007). Occurrence, behavior and effects of nanoparticles in the environment. *Environmental Pollution*, **150**, 5–22.

NRC (National Research Council) (2003). *Oil in the Sea III: Inputs, Fates, and Effects*. National Academies Press, Washington, DC.

NRC (National Research Council) (2013). *An Ecosystem Services Approach to Assessing the Impacts of the Deepwater Horizon Oil Spill in the Gulf of Mexico*. National Academies Press, Washington, DC.

Oaks, J.L., Gilbert, M., Virani, M.Z., et al. (2004). Diclofenac residues as the cause of vulture population decline in Pakistan. *Nature*, **427**(6975), 630–633.

Oberdörster, E. & Cheek, A.O. (2001). Gender benders at the beach: endocrine disruption in marine and estuarine organisms. *Environmental Toxicology and Chemistry*, **20**(1), 23–36.

Ohlauson, C., Eriksson, K.M., & Blanck, H. (2012). Short-term effects of medetomidine on photosynthesis and protein synthesis in periphyton, epipsammon and plankton communities in relation to predicted environmental concentrations. *Biofouling*, **28**(5), 491.

OSPAR Commission (2010). Quality status report 2010. *Rapport*, **497**, 2010.

Paine, R.T., Ruesink, J.L., Sun, A., et al. (1996). Trouble on oiled waters: lessons from the Exxon Valdez oil spill. *Annual Review of Ecology and Systematics*, **27**, 197–235.

Paytan, A., Mackey, K.R., Chen, Y., et al. (2009). Toxicity of atmospheric aerosols on marine phytoplankton. *Proceedings of the National Academy of Sciences of the USA*, **106**(12), 4601–4605.

Penela-Arenaz, M., Bellas, J., & Vázquez, E. (2009). Chapter five: effects of the Prestige oil spill on the biota of NW Spain: 5 years of learning. *Advances in Marine Biology*, **56**, 365–396.

Peng, S.H., Wang, W.X., Li, X., & Yen, Y. (2004). Metal partitioning in river sediments measured by sequential extraction and biomimetic approaches. *Chemosphere*, **57**, 839–851.

Pentreath, R.J. (2002). Radiation protection of people and the environment: developing a common approach. *Journal of Radiological Protection*, **22**, 45–56.

Peters, K., Bundschuh, M., & Schäfer, R.B. (2013). Review on the effects of toxicants on freshwater ecosystem functions. *Environmental Pollution*, **180**, 324–329.

Peterson, C.H. (2001). The 'Exxon Valdez' oil spill in Alaska: acute, indirect and chronic effects on the ecosystem. *Advances in Marine Biology*, **39**, 1–103.

Peterson, C.H., Rice, S.D., Short, J.W., et al. (2003). Long-term ecosystem response to the Exxon Valdez oil spill. *Science*, **302**(5653), 2082–2086.

PIANC (2006a). *Environmental Risk Assessment of Dredging and Disposal Operations*. International Navigation Association, Brussels, Belgium.

PIANC (2006b). *Biological Assessment Guidance for Dredged Material*. Report of WG 8. International Navigation Association, Brussels, Belgium.

PIANC (2009). *Dredging Management Practices for the Environment—A structured Selection Approach*. International Navigation Association, Brussels, Belgium.

Piatt, J.F., Lensink, C.J., Butler, W., et al. (1990). Immediate impact of the 'Exxon Valdez' oil spill on marine birds. *Auk*, **107**(2), 387–397.

Pisa, L.W., Amaral-Rogers, V., Belzunces, L.P., et al. (2014). Effects of neonicotinoids and fipronil on non-target invertebrates. *Environmental Science and Pollution Research*, **22**(1), 68–102.

Porte, C., Janer, G., Lorusso, L.C., et al. (2006). Endocrine disruptors in marine organisms: approaches and perspectives. *Comparative Biochemistry and Physiology Part C*, **143**(3), 303–315.

Preston, A. & Jefferies, D.F. (1967). The assessment of the principal public radiation exposure from, and the resulting control of, discharges of aqueous radioactive waste from the United Kingdom Atomic Energy Authority factory at Windscale, Cumberland. *Health Physics*, **13**(5), 477–485.

Preston, A. & Jefferies, D.F. (1969). The ICRP critical group concept in relation to the Windscale sea discharges. *Health Physics*, **16**(1), 33–46.

Pulliainen, K., Korhonen, L., Kankaanranta, L., & Maki, K. (1992). Non-spawning burbot on the northern coast of Bothnian Bay. *Ambio*, **21**(2), 170–175.

Rajinder, P., Bandral, R.S., & Zhang, W.J. (2009). Integrated pest management: a global overview of history, programs and adoption. In: Rajinder, P. & Dhawan, A. (eds), *Integrated Pest Management: Innovation-Development Process*, Vol. 1, pp. 1–50. Springer Netherlands, Dordrecht.

Rogers, S.I. & Greenaway, B. (2005). A UK perspective on the development of marine ecosystem indicators. *Marine Pollution Bulletin*, **50**, 9–19.

Rubin, B.S. (2011). Bisphenol A: an endocrine disruptor with widespread exposure and multiple effects. *Journal of Steroid Biochemistry and Molecular Biology*, **127**(1), 27–34.

Ruoppolo, V., Woehler, E.J., Morgan, K., & Clumpner, C.J. (2012). Wildlife and oil in the Antarctic: a recipe for cold disaster. *Polar Record*, **49**, 97–109.

Salbu, B. (2007). Speciation of radionuclides—analytical challenges within environmental impact and risk assessments. *Journal of Environmental Radioactivity*, **96**, 47–53.

Salbu, B. (2011). Radionuclides released to the environment following nuclear events. *Integrated Environmental Assessment and Management*, **7**(3), 362–364.

Salbu, B., Nikitin, A.I., Strand, P., et al. (1997). Radioactive contamination from dumped nuclear waste in the Kara Sea—results from joint Russia-Norwegian expeditions in 1992–1994. *Science of the Total Environment*, **202**, 185–198.

Schmidt, A.S., Bruun, M.S., Dalsgaard, I., & Larsen, J.L. (2001). Incidence, distribution, and spread of tetracycline resistance determinants and integron-associated antibiotic resistance genes among motile aeromonads from a fish farming environment. *Applied and Environmental Microbiology*, **67**(12), 5675–5682.

Schultz, M.M., Painter, M.M., Bartell, S.E., et al. (2011). Selective uptake and biological consequences of environmentally relevant antidepressant pharmaceutical exposures on male fathead minnows. *Aquatic Toxicology*, **104**(1), 38–47.

Sinoir, M., Butler, E.C., Bowie, A.R., et al. (2012). Zinc marine biogeochemistry in seawater: a review. *Marine and Freshwater Research*, **63**(7), 644–657.

Slaper, T.F. & Hall, T.J. (2011). The triple bottom line: what is it and how does it work? *Indianan Business Review*, **Spring 2011**, 4–8.

Smith, C.R. & Baco, A.R. (2003). Ecology of whale falls at the deep-sea floor. *Oceanography and Marine Biology*, **41**, 311–354.

Smith, D.R., Stephenson, M.D., & Flegal, A.R. (1986). Trace metals in mussels transplanted to San Francisco Bay. *Environmental Toxicology and Chemistry*, **5**(2), 129–138.

Smith, P., Hiney, M., & Samuelsen, O. (1994). Bacterial resistance to antimicrobial agents used in fish farming: a critical evaluation of method and meaning. *Annual Review of Fish Diseases*, **4**, 273–313.

Stauber, J.L. & Florence, T.M. (1987). Mechanism of toxicity of ionic copper and copper complexes to algae. *Marine Biology*, **94**(4), 511–519.

Steinberg, C.E.W., Stürzenbaumb, S.R., & Menzel, R. (2008). Genes and environment—striking the fine balance between sophisticated biomonitoring and true functional environmental genomics. *Science of the Total Environment*, **400**, 142–161.

Stewart, M., Ahrens, M., & Olsen, G. (2009). *Field Analysis of Chemicals of Emerging Environmental Concern in Auckland's Aquatic Sediments*. Auckland Regional Council. Technical Report No.021 March 2009. ISSN 1179–0512 (Online).

Thomas, K.V. & Hilton, M.J. (2004). The occurrence of selected human pharmaceutical compounds in UK estuaries. *Marine Pollution Bulletin*, **49**(5), 436–444.

Thomas, K.V., McHugh, M., & Waldock, M. (2002). Antifouling paint booster biocides in UK coastal waters: inputs, occurrence, and environmental fate. *Science of the Total Environment*, **298**, 117–127.

Thompson, R.C. (2006). Plastic debris in the marine environment: consequences and solutions. In: Krause, J.C., Nordheim, H., & Bräger, S. (eds), *Marine Nature Conservation in Europe*, pp. 107–115. Federal Agency for Nature Conservation, Stralsund, Germany.

Thompson, R.C., Moore, C.J., vom Saal, F.S., & Swan, S.H. (2009). Plastics, the environment and human health: current consensus and future trends. *Philosophical Transactions of the Royal Society Series B*, **364**, 2153–2166.

Ullrich, S.M., Tanton, T.W., & Abdrashitova, S.A. (2001). Mercury in the aquatic environment: a review of factors affecting methylation. *Critical Reviews in Environmental Science and Technology*, **31**(3), 241–293.

US Department of Health and Human Services (1998). *Toxicological Profile for Total Petroleum Hydrocarbons (TPH)*. Agency for Toxic Substances and Disease Registry, Atlanta, Georgia, USA.

USEPA (2003). *Ambient Aquatic Life Water Quality Criteria for Tributyltin*. US Environmental Protection Agency, Office of Water, Washington, DC.

Van Boxtel, A.L., Kamstra, J.H., Fluitsma, D.M., & Legler, J. (2010). Dithiocarbamates are teratogenic to developing zebrafish through inhibition of lysyl oxidase activity. *Toxicology and Applied Pharmacology*, **244**, 156–161.

Van Dam, J.W., Negri, A.P., Mueller, J.F., & Uthicke, S. (2012). Symbiont-specific responses in foraminifera to the herbicide Diuron. *Marine Pollution Bulletin*, **65**, 373–383.

Vethaak, A.D. & ap Rheinhallt, T. (1992). Fish disease as a monitor for marine pollution: the case of the North Sea. *Reviews in Fish Biology and Fisheries*, **2**, 1–32.

Vethaak, D. & Legler, J. (2013). Endocrine disruption in wildlife: background, effects, and implications. In: Mathiessen, P. (ed.), *Endocrine Disrupters: Hazard Testing and Assessment Methods*, pp. 7–58. Wiley-Blackwell, Hoboken, NJ.

Votier, S.C., Hatchwell, B.J., Beckerman, A., et al. (2005). Oil pollution and climate have wide-scale impacts on seabird demographics. *Ecology Letters*, **8**(11), 1157–1164.

Wagner, W.E. (1995). The science charade in toxic risk regulation. *Columbia Law Review*, **95**(7), 1613–1723.

Weigel, S., Berger, U., Jensen, E., et al. (2004). Determination of selected pharmaceuticals and caffeine in sewage and seawater from Tromsø/Norway with emphasis on ibuprofen and its metabolites. *Chemosphere*, **56**(6), 583–592.

Wenning, R.J., Batley, G.E., Ingersoll, C.G., & Moore, D.W. (eds) (2005). *Use of Sediment Quality Guidelines and Related Tools for the Assessment of Contaminated Sediments*. Setac Press, Brussels, Belgium.

WHO (World Health Organization) (1998). *Environmental Health Criteria 202: Polycyclic Aromatic Hydrocarbons—Selected Non-heterocyclic*. World Health Organization, Geneva, Switzerland.

Widdicombe, S. & Somerfield, P.J. (2012). Marine biodiversity: its past development, present status, and future threats. In: Solan, M., Aspden, R.J., & Paterson, D.M. (eds), *Marine Biodiversity and Ecosystem Functioning: Frameworks, Methodologies, and Integration*, pp. 1–15. Oxford University Press, Oxford.

Wong, S.W., Leung, K.M., & Djurišić, A.B. (2013). A comprehensive review on the aquatic toxicity of engineered nanomaterials. *Reviews in Nanoscience and Nanotechnology*, **2**, 79–105.

Wu, J. & Boyle, E.A. (1997). Lead in the western North Atlantic Ocean: completed response to leaded gasoline phaseout. *Geochimica et Cosmochimica Acta*, **61**(15), 3279–3283.

Yang, C.Z., Yaniger, S.I., Jordan, V.C., et al. (2011). Most plastic products release estrogenic chemicals: a potential health problem that can be solved. *Environmental Health Perspectives*, **119**(7), 989.

Yan Ting, C., Teo, S.L.M., Wai, L., & Chai, C.L.L. (2014). Searching for 'environmentally-benign' antifouling biocides. *International Journal of Molecular Sciences*, **15**(6), 9255–9284.

Zarrelli, A., DellaGreca, M., Iesce, M.R., et al. (2014). Ecotoxicological evaluation of caffeine and its derivatives from a simulated chlorination step. *Science of the Total Environment*, **470**, 453–458.

Zhang, W.J., Jiang, F.B., & Ou, J.F. (2011). Global pesticide consumption and pollution: with China as a focus. *Proceedings of the International Academy of Ecology and Environmental Sciences*, **1**(2), 125–144.

Zuccato, E., Castiglioni, S., Fanelli, R., et al. (2006). Pharmaceuticals in the environment in Italy: causes, occurrence, effects and control. *Environmental Science and Pollution Research*, **13**(1), 15–21.

CHAPTER 14

Importance of species interactions in moderating altered levels of reactive nitrogen

Martin Solan

14.1 Introduction

Human activities, primarily the use of fertilizers in food production and the burning of fossil fuel for energy, are altering the nitrogen cycle because they are increasing the amount of reactive nitrogen (N_r; Table 14.1) that is generated and emitted at a global scale (Galloway et al., 2013), from a pre-industrial level of ~ 44 Tg N yr^{-1} (Vitousek et al., 2013) to a present day estimate of ~ 210 Tg N yr^{-1} (Galloway et al., 2004, 2013). In ecosystems that are N-limited, anthropogenic N_r can bolster primary production and have a stimulatory effect on the food web as a whole, but, once the availability of N_r exceeds the assimilative capacity of the system, the natural balance between N_2 fixation and denitrification becomes disrupted (Canfield et al., 2010), leading to N_r accumulation and adverse impacts on human health, biodiversity, ecosystem services, and climate change (Galloway et al., 2003; Duce et al., 2008; Doney, 2010; Erisman et al., 2013). While acknowledging that there are regional variations in the production and distribution of anthropogenic N_r (Galloway et al., 2008; Howarth, 2008; Liu et al., 2013), and in the expression of any associated ecosystem effects, the pooled outcome of these system responses is now a major source of concern to society, because, ultimately, it may threaten the integrity of Earth-system functioning and human well-being (Gruber and Galloway, 2008; Rockström et al., 2009). Consequently, understanding the threshold beyond which anthropogenic change in N_r supply will transpose ecosystems outside the 'safe operating space' for humanity is fast becoming a prominent research goal (Kaiser, 2001) and is increasingly attracting interest from those tasked with managing the environment or developing environmental policy (Austin et al., 2013).

How altered N_r availability and/or supply affects natural systems has received a significant amount of attention (summarized in, for example, the treatises of Capone et al., 2008 and Sutton et al., 2011), particularly in terrestrial soils and aquatic planktonic systems. For the most part, emphasis has been placed firmly on:

(i) establishment of the chemistry, microbial, and genetic basis for biogeochemical transformations and pathways (Strous et al., 2006; Thamdrup, 2012; Kartal et al., 2013; Isobe and Ohte, 2014; Hu and Ribbe, 2015);

(ii) determination of elemental budgets and dynamics, either globally (Galloway et al., 2008; Vitousek et al., 2013), or for specific circumstances (Capone and Hutchins, 2013), regions (Karl et al., 1997; Liu et al., 2013; Kim et al., 2014), or habitats (e.g. open oceans, Zehr and Kudela, 2011; marine sediments, Devol, 2015; wetlands, Zhou et al., 2014);

(iii) summarizing the occurrence, extent, and type of major ecosystem responses to exemplar alterations in N_r and/or other nutrients (e.g. algal blooms, Lyons et al., 2014; hypoxia/anoxia, Diaz and Rosenberg, 2008); and/or

(iv) parameterizing models that allow exploration of past and future scenarios of N_r budgets (Swaney et al., 2008; Krishnamurthy et al., 2009).

Hence, investigators have tended to focus either on elucidating the mechanistic minutiae of microbial processes and biogeochemical pathways or, conversely, on establishing the net consequences of altered N_r at the very broadest ecosystem level.

Stressors in the Marine Environment. Edited by Martin Solan and Nia M. Whiteley
© Oxford University Press 2016. Published in 2016 by Oxford University Press.

Table 14.1 Summary of the main groupings of nitrogen (N). Non-reactive N is nitrogen gas. Reactive N (N_r = 'fixed N') includes all biologically, photochemically, and radiatively active forms of N.

Non-reactive N	Reactive N (N_r)		
	Inorganic forms		Organic compounds
	Reduced	Oxidized	
Dinitrogen [N_2] (~78% atmosphere)	Ammonia [NH_3]	Nitrogen oxide [NO_x]	Urea [CH_4N_2O]
	Ammonium [NH_4^+]	Nitric acid [HNO_3]	Amines
		Nitrous oxide [N_2O]	Proteins
		Nitrate [NO_3^-]	Nucleic acids

The study of natural and anthropogenic gradients of N_r (and other nutrients) have been instrumental in building understanding of the pervasive effects of altered levels of N_r across a variety of ecosystems, and have much to offer as a means of establishing the sequential changes typical of this type of forcing (Fukami and Wardle, 2005; Godbold and Solan, 2009). Several notable reviews summarize the generic effects of N_r alleviation and N_r addition in marine systems (Rabalais, 2002; Howarth and Marino, 2006; Purvaja et al., 2008; Voss et al., 2013), but it is difficult to extrapolate findings beyond specific case studies (Smyth et al., 2015) because the magnitude and form of any observed effects are known to vary with seasonal timing, geographical location, and environmental context (Voss et al., 2013). Indeed, establishing the cause–effect relationships between altered levels of N_r and observed ecosystem responses is not as straightforward as it may seem; there have been many instances in the literature where major ecological changes have been linked to altered levels of N_r loading, but most of these have not separated the effects that can be directly attributed to the availability of N_r from those that can be attributed to concomitant changes in other components of the system. Such linked effects can be important, yet a significant proportion of studies do not take the necessary holistic view needed to explore the potential of alternative mechanisms. Increased turbidity (Pedersen et al., 2014) and decreased oxygen availability in the water column following an algal bloom that has been fuelled by changes in nutrient loading, for example, can alter nutrient uptake rates and the stoichiometric balance of nutrient ratios that are important determinants of algal and plant growth (Bracken et al., 2015). This can lead to a different growth response to that which otherwise would have been observed if the effects of N_r alone were expressed, and points to fundamentally different mechanisms.

Similar interpretative difficulties result from the tight coupling between the nitrogen cycle and other nutrient cycles, particularly phosphorus, silica, and iron, raising questions about whether the separation of the independent effects of N_r are scientifically useful, ecologically relevant, or even possible to achieve (Ward, 2012). The key problem is that an approximate match between field (or experimental) data and theory or expectation (based on intuitive deduction, which can be erroneous, e.g. Govers et al., 2014) has been assumed to be sufficient to ascribe causation, without explicit validation or testing of the underpinning mechanisms that have been ascribed. Thus, a key challenge for ecologists lies in determining when alterations in N_r are the primary determinant of observed ecosystem responses and, if they are, how to partition out the effects of multiple dynamic variables. Nevertheless, placing debate aside (Benton et al., 2007; Stewart et al., 2013), it is clear that there can be great benefit in understanding how changes to the stocks and flows of N_r may affect individuals, species, and species–environment interactions that, in turn, may either exacerbate or buffer the wider ecosystem effects of N_r forcing. Indeed, a cursory evaluation of the literature suggests that there is much to be done to understand how the interplay between species interactions and environmental context moderate the effects of N_r forcing. Hence, the objective of this short chapter is to stimulate discussion and provide a starting point for researchers that are interested in designing experiments that seek insights about how N_r effects are expressed and mediated by natural communities. The chapter starts with a brief summary of the nitrogen cycle to familiarize and orientate the reader with the various forms of N and the main transformative pathways that constitute the nitrogen cycle. The intention is not to provide an exhaustive review of all the known effects that alterations to N_r may have on the biotic components of marine ecosystems, although broad ecosystem responses commonly associated with these are acknowledged. Instead, consideration is given to how interspecific and species–environment interactions may alter, or be altered by, the effects of N_r forcing. As other recent reviews have considered the role of nitrogen toxicology (Camargo and Alonso, 2006), trophic biology (Clements et al., 2009), and physiological competence (Young and Berges, Chapter 6, this volume) in eliciting species responses or influencing species interactions, these will not be included. Instead, the emphasis of this chapter is to explore the view that the expression of N_r effects is not simply a function of the type, magnitude, and timing of N_r forcing, but also a product of the way in which species interact with one another and the environment to mediate those effects.

14.2 A brief summary of the nitrogen cycle

Nitrogen is an essential element for the growth and functioning of all organisms on Earth, but the most abundant form of nitrogen exists in an unreactive molecular form (dinitrogen, N_2) that is unavailable to most organisms. It is only when the very strong triple bond that holds the two N atoms together is broken, either through exposure to high temperature (Sano et al., 2001) or through biological nitrogen fixation performed by diazotrophs (mostly bacteria, but some archaea; Sohm et al., 2011; Vitousek et al., 2013), that N_2 is reduced (catalysed by the nitrogenase enzyme) to two molecules of highly reactive ammonia (NH_3):

$$N_2 + 8H^+ + 8e^- + 16ATP \rightarrow 2\,NH_3 + H_2 + 16ADP + 16\,P_i$$

This process requires a significant amount of energy, requiring the metabolism of at least 16 ATP (adenosine triphosphate) molecules and the provision of eight electrons per molecule of fixed N_2 (Hoffman et al., 2014). NH_3 is a much more convenient source of nitrogen because it requires a comparatively lower amount of metabolic energy to be utilized and, consequently, it is readily assimilated into amino acids and proteins, primarily during prokaryotic organismal growth (= organic nitrogen, N_{org}). Most eukaryotic organisms, however, are unable to access nitrogen in this way because they lack nitrogenase enzymes. Instead, they rely on recycled N_{org} that is made available when organisms die, largely as NH_4^+ following the mineralization (= ammonification) of organic matter by heterotrophic microbes. Hence, NH_3 (ammonia) or NH_4^+ (ammonium), and subsequently their derivatives, are typically the first forms of usable nitrogen for most prokaryotic and eukaryotic organisms. The fate of NH_3 and NH_4^+ is largely controlled by whether transformation into various oxidation states by specific groups of diazotrophs to generate energy and/or foster growth takes place under aerobic (nitrification, Fig. 14.1) or anaerobic (denitrification, nitrate reduction, anammox; Fig. 14.1) conditions (Canfield et al., 2010). While nitrogen inputs, ultimately, are the most important determinant of N_r concentration, how much N_r is processed, stored (buried), returned to the atmosphere ('lost'), or relocated elsewhere is dependent on which of the

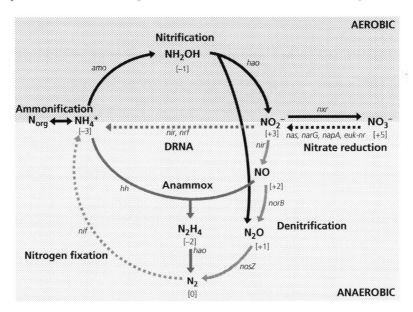

Figure 14.1 Summary of the nitrogen cycle indicating the principal transformation pathways (arrows: dotted light grey, nitrogen fixation; solid black, nitrification; solid light grey, denitrification; dotted black, nitrate reduction; solid dark grey, anammox; and solid black double arrow, ammonification) and how they relate to the presence (aerobic, lightly stippled) or absence (anaerobic, densely stippled) of oxygen. Enzymes (encoded by genes, abbreviated in italics) that catalyse nitrogen transformation include various nitrate reductases (*nas, euk-nr, narG, napA*), nitrite reductases (*nir, nrf*), nitric oxide reductase (*norB*), nitrous oxide reductase (*nosZ*), nitrogenase (*nif*), ammonium monooxygenase (*amo*), hydroxylamine oxidoreductase (*hao*), nitrite oxidoreductase (*nxr*), and hydrazine hydrolase (*hh*). The oxidation state of nitrogen is indicated in square parentheses. Formula: N_{org}, organic nitrogen; NH_4^+, ammonium; NH_2OH, hydroxylamine; NO_2^-, nitrite; NO_3^-, nitrate; NO, nitric oxide; N_2O, nitrous oxide; N_2H_4, hydrazine; and N_2, dinitrogen. Figure modified from Canfield et al. (2010).

pathways within the nitrogen cycle predominate. It is not within the scope of the present chapter to provide a detailed examination and summary of each of these pathways, or the specifics about their dynamics. Instead, the following is intended to sufficiently acquaint the reader with enough of an overview to appreciate the relevance of the ensuing discussion.

Nitrification (solid black arrows, Fig. 14.1) is the pathway where, in the presence of oxygen (even in hypoxic conditions; Abell et al., 2011), NH_4^+ is oxidized by autotrophic microbes, first to NH_2OH (hydroxylamine) and subsequently to NO_2^- (nitrite) and NO_3^- (nitrate), in order to facilitate the fixation of inorganic carbon (CO_2) during photosynthetic growth. Both ammonia oxidizing bacteria (AOB) and ammonia oxidizing archaea (AOA) are involved, although there is some debate regarding their relative contributions to nitrification in natural environments (Prosser and Nicol, 2008). Nevertheless, a by-product of this process is the production of the greenhouse gas N_2O (nitrous oxide), which can be regionally significant (Seitzinger and Kroeze, 1998), even in circumstances when nitrification contributes less to N_2O production, as a fraction of N turnover, than denitrification (Meyer et al., 2008).

In the absence of oxygen, the nitrate formed during nitrification is the preferred electron acceptor (respiratory substrate) for denitrifying bacteria and archaea. Nitrate reduction can occur by dissimilatory nitrate reduction (dotted black and dotted dark grey arrows, Fig. 14.1) to produce ammonia (NH_4^+) or, more frequently, by denitrification (solid light grey arrows, Fig. 14.1) in which nitrate (NO_3^-) is sequentially reduced to nitrite (NO_2^-), followed by nitric oxide (NO), and then nitrous oxide (N_2O), before being returned to the system (often referred to as 'lost') as biologically unavailable dinitrogen gas (N_2). However, this is not the only means by which NH_4^+ can be oxidized, as NH_4^+ oxidation can also be coupled to NO_2^- reduction, with hydrazine (N_2H_4) as an intermediary substrate, by a specialized group of bacteria (Planktomycetes, known mostly from their role in wastewater treatment) to facilitate chemoautotrophic growth (i.e. anaerobic ammonium oxidation or 'anammox'; solid dark grey arrows, Fig. 14.1). Unlike denitrification, the anammox pathway does not produce N_2O, and may be responsible for up to 40% of nitrogen loss from freshwater and terrestrial soil systems and ~50% of nitrogen loss from marine systems (Dalsgaard et al., 2005; Hong et al., 2009; Hu et al., 2011; Devol, 2015). Hence, in unperturbed systems, denitrification and anammox largely offset the fixation and aerobic oxidization (nitrification) of NH_4^+ (and its derivatives) such that the net balance of these pathways does not lead to excessive N_r storage within the system. This is most effective when the aerobic conditions for nitrification occur in close proximity to the anaerobic conditions required for denitrification/anammox, such as in bioturbated sediments (Bertics and Ziebis, 2009) and other interfaces (haloclines, Santoro, 2010) or within the stratified water column of oxygen minimum zones (De Brabandere et al., 2014), where a substantial portion of the N load can be removed to the atmosphere as dinitrogen gas (Seitzinger, 1988).

This view of a reasonably balanced nitrogen cycle was more or less globally valid until the late 1800s (Holtgrieve et al., 2011; for chronology of anthropogenic mediated change, see Galloway et al., 2013; for an extended view throughout the Holocene, see McLauchlan et al., 2013), when human understanding of the nitrogen cycle provided opportunity to generate and apply various forms of N_r to natural systems. Human production of food (i.e. in particular the widespread cultivation of legumes and rice that promote the conversion of N_2 to N_{org} through biological nitrogen fixation) and energy (i.e. combustion of fossil fuels that convert atmospheric N_2 and geological stores of N to NO_x) are the dominant drivers of this change, compounded by the widespread adoption of the Haber–Bosch process (converts N_2 to NH_3) that sustains fertilizer production and other industrial activities (Galloway et al., 2003). Consequently, in recent decades anthropogenic N_r production has become greater than that generated naturally (Ward, 2012), with significant enrichment occurring in localized, largely terrestrial, regions around the world (particularly North America, Europe, India, and SE China). In practice, however, as estuaries and coastal margins are often adjacent to significant natural sources of N_r (terrestrial soil, forests, grasslands, and wetlands) and tend to be close to human populations (which enhance agricultural practices, organic pollution, and fossil fuel combustion), they often serve as N_r sinks as the assimilative capacity of the system is challenged by local inputs. Regional watersheds provide a substantive conduit for N to be transported from the terrestrial point of origin (~29% of the total applied to land is transported in this way, = 43 Tg N yr^{-1}) to adjacent, or otherwise linked, coastal and open marine habitats. When this occurs, N_r tends to quickly accumulate because denitrification rates are lowered under excessive loading and dissimilatory nitrate reduction to ammonium (DNRA) rates are higher than, for example, in terrestrial settings (Seitzinger et al., 2006; Purvaja et al., 2008; Santoro, 2010; Seitzinger et al., 2010; Howarth et al., 2012; Billen et al., 2013). Future population growth, economic development, and the 'westernization' of diets mean that further increases in N_r are anticipated (Seitzinger et al., 2002; Bouwman et al., 2013), with associated increases in the likelihood of algal blooms and hypoxia in coastal regions (Kim et al., 2011) and any linked areas of open ocean (Kim et al., 2014).

14.3 Effects of reactive nitrogen on biodiversity and biological interactions

A lot of the information in the literature about the effects of N_r reports aspects of growth, tolerance, and recovery potential for individual species, or summarizes observed changes in community structure following nutrient enrichment that have led to varying degrees of eutrophication (i.e. excessive nutrient richness leading to oxygen depletion, reviewed in Diaz and Rosenberg, 1995). As noted earlier, most investigators report responses associated with the effects of multiple nutrients (typically N and P) such that, for the present purpose of distinguishing the effects of N_r from other variables, any inferences about the relative role of N_r are likely to be confounded, or at least ambiguous, if not elusive. An open empirical question remains as to whether inferences made about the role of nutrient loading per se are equally valid when considering the effects of single nutrients. Nevertheless, leaving interpretative difficulties aside, it would seem a useful exercise to extract possible mechanistic processes worthy of further investigation. In the following sections, focused mainly on marine macroalgae and invertebrates, consideration is given as to how changes in environmental conditions, resource availability, or other opportunities (e.g. competitor/predator release) following nitrogen adjustment may change some functionally important aspects of species behaviour and/or species interactions.

14.3.1 Alteration of food-web structure and biodiversity

Nitrogen is typically regarded as the limiting nutrient for the growth of primary producers (macrophytes, algae, and cyanobacteria) in marine waters, although this view has widened in recent decades (Howarth and Marino, 2006) to recognize the importance of temporal and spatial variation in N_r supply/bioavailability (Peierls and Paerl, 1997) and the mediating role of other macronutrients, most notably phosphorus, silica, and iron (Elser et al., 2007; Harpole et al., 2011). In waters

where the N:P ratio is low (~16:1, estuaries and coasts), N_r addition tends to have a stimulatory effect on primary production, because greater amounts of N and P are recycled as oxygen levels become depleted and disrupt nitrification. However, in waters where the N:P ratio is high (~100:1, open ocean), the availability of P, Si, and/or Fe becomes more important because the supply of these elements in relation to N_r (less terrestrial run-off, less organic matter available, and no significant atmospheric source) becomes increasingly important (Downing, 1997). For instance, long-term studies in the Black Sea have shown that decreased N_r and a lower N:P stoichiometric ratio are preconditions for coccolithophore domination, while increased concentrations of N_r and P near and above the Redfield stoichiometry (i.e. C:N:P = 106:16:1) encourage the development of diatoms (Silkin et al., 2014). It is also evident that nutrient addition can result in increased abundance of ephemeral and filamentous green algae (Worm and Lotze, 2006) and/or large perennial brown algae (kelps and fucoids; Agatsuma et al., 2014) in many intertidal and subtidal environments, often resulting in the dominance of a limited number of algal species that, in turn, affect subsequent algal recruitment and growth (Alestra and Schiel, 2014) and associated invertebrate communities (Raffaelli et al., 1998). There are, of course, many exceptions to these commonalities that can be explained by specific circumstances (Hessing-Lewis et al., 2015), but it is clear that the addition of either N or P generally leads to enhanced productivity across a range of marine habitats, and that a simultaneous increase in both nutrients leads to an even greater growth response than that achieved under single constituent nutrient enrichment (Elser et al., 2000, 2007; Harpole et al., 2011). However, the extent to which the full ecological consequences of altering N_r and other nutrient levels are expressed is dependent on seasonal timing (e.g. Piedras and Odebrecht, 2012; Whalen et al., 2013) and factors other than water/habitat quality, in particular the occurrence of, and interactions with, additional local stressors (Hondorp et al., 2010).

Despite such a strong correlation between the availability of nitrogen and the growth of phytoplankton (Micheli, 1999), it is important to emphasize that corresponding positive effects at higher trophic levels do not necessarily follow because the efficiency of production across each trophic level boundary can be significantly lowered by nutrient addition (Luong et al., 2014). In contrast to small inputs of N_r that augment biomass across multiple trophic levels, large or sustained inputs of N_r commonly evoke mortality of invertebrates and fish (Bishop et al., 2006). In the worst case, nutrient addition leads to eutrophication and low oxygen levels that can impact the entire food web by shifting ecosystems from supporting higher taxa to microbe-dominated communities (Diaz and Rosenberg, 2008). Indeed, the sequential changes in community structure that accompany a progressive change in organic matter and inorganic nutrient addition are well known and do not need rehearsing here; suffice to say that there is a high degree of conformity (i.e. unperturbed community ⇔ transitory ⇔ pioneer ⇔ opportunistic ⇔ microbial), albeit phenomenological, across a range of different habitats (e.g. seaweeds, Schramm, 1999; sediment-dwelling invertebrates, Pearson and Rosenberg, 1978; whole system overview, Cloern, 2001, Rabalais, 2002). It is important to recognize, however, that the effects of organic matter input are not necessarily synonymous with those of inorganic nutrient input, and can lead to quite different ecosystem level effects. In a field experiment on a sheltered intertidal shore, for example, Fitch and Crowe (2012) showed that changes in organic matter had a greater impact on the total abundance and diversity of sediment-dwelling invertebrates than that of inorganic nutrients. Further, changes in total abundance (largely driven by the opportunistic polychaete *Capitella* sp.) were found primarily to be in response to organic matter, but were further influenced by the addition of inorganic nutrients to the system. The mechanistic explanation for this synergistic effect is not clear, but it may be that the impact of nutrient addition on species assemblage structure may be dependent on subtle differences in nutrient bioavailability and how readily, and in what form, associated impacts cascade through the food web (Fitch and Crowe, 2011).

Depending on the timing, duration, and severity of nutrient input, the most common changes in assemblage structure, irrespective of habitat type, tend to relate to alterations in abundance, biomass, and dominance (e.g. Shin et al., 2006). Variations in species richness tend to be associated with substantive instances of altered nutrient loading, and are less likely to be observed under subtle modifications to the nutrient regime. More often than not, the numeric or biomass responses of individual species are interpreted as a product of their physiological tolerance/sensitivity to associated changes in environmental conditions (reviewed in Diaz and Rosenberg, 1995). However, changes in the composition and density of genera can also reflect other attributes of the system that are also directly, or indirectly, responding to changes in nutrient load and availability. Using an *in situ* continuous inorganic fertilizer addition experiment in intertidal

mudflats, for example, Ferreira et al. (2015) demonstrated that compensatory growth of either diatoms (low level enrichment) or cyanobacteria (high level enrichment) tends to occur and, as the microphytobenthos are a primary food source, concurrent changes in the trophic structure of the nematode community can take place. These compositional changes coincide with dramatic decreases in macrofaunal abundance, richness, and evenness, and an increase in dominance, that can only be partly be explained by physical and chemical disturbance associated with fertilizer enrichment; it is likely that the shift to a low palatable microphytobenthic community also negatively affects the macrofaunal community (Botter-Carvalho et al., 2014). In other cases, there is no evidence of such cascading indirect effects. Daudi et al. (2012) showed that epiphytic meiofauna and diatoms on seagrass can be negatively affected by nutrient addition, while corresponding benthic meiofauna and other biofilm characteristics remain unaffected. Hence, the effects of nutrient addition are not necessarily direct or universally expressed across different trophic levels within an assemblage; rather, they also depend on secondary effects linked to nutrient addition mediated through complex species interactions and responsive patterns of behaviour.

14.3.2 Herbivory

While it is clear that the addition of N or P, alone or in combination, generally leads to enhanced productivity (Elser et al., 2000, 2007; Harpole et al., 2011), herbivory of autotrophic biomass by a range of invertebrate and fish species can remove a significant amount of production across a wide range of environmental conditions. However, the relative role of nutrient loading and herbivory in determining various ecosystem properties varies across latitudes and primary producer identity, and with the inherent productivity of ecosystems (Burkepile and Hay, 2006). As a general rule, herbivores exert a stronger effect relative to nutrient enrichment in low productivity systems, while the reverse is true in highly productive systems, but these patterns are complicated by tremendous spatial and seasonal variation in the level and type of grazing (Lee et al., 2015; Sakihara et al., 2015), which, in turn, can be influenced by biotic (e.g. predator presence, Stoner et al., 2014) and abiotic context. Indeed, even within a single location, the interaction between nutrients and herbivory can vary between different groups of autotrophs. McClanahan et al. (2003), for example, found that brown frondose algal cover was negatively influenced by both herbivory and nutrients, whereas red frondose algal cover was negatively affected by herbivory but was unaffected by nutrients. In addition, turf algae performed best under conditions of either low herbivory and high nutrients or high herbivory and low nutrients, whereas frondose brown algae performed best under low herbivory and low nutrient conditions and was inhibited by high nutrient concentrations.

Changes in the balance of grazing and autotrophic growth under altered nutrient regimes affect the type, quantity, and quality of food resource made available to primary consumers (Pascal and Fleeger, 2013). This is important because even subtle changes in nutrient availability can lead to changes in the quality of the food resource (Bi et al., 2014), which can have significant effects on the condition and long-term fitness of consumer species (Schoo et al., 2014). Recent work combining outdoor mesocosm experiments with carbon budget modelling and ecological network analysis indicates that immediately following the onset of nutrient enrichment, herbivory is promoted in zooplankton communities, replacing the previous reliance on detritus-derived N_r with a direct source of N_{org} (Luong et al., 2014). Importantly, the diversity of the grazing community may not be the single most important determinant of resource use as observed responses tend to reflect species-specific traits associated with algal consumption and processing (Godbold et al., 2009b), as well as preferences associated with the nitrogen composition of the food resource (Nordhaus et al., 2011) and the palatability or otherwise (e.g. toxin production, Selander et al., 2008, Liu et al., 2010, Pistocchi et al., 2011; metal accumulation, Lee and Wang, 2001) of the grazed autotrophic community. Studies in fish communities show that, in addition to the nutritional value of the food resource, the physiological processes involved in nutrient extraction may also be important, but to date these have received little attention (Clements et al., 2009; Goldenberg and Erzini, 2014).

In addition to reducing autotrophic biomass through consumption, it is important to recognize that mobile grazing species can also enhance autotrophic biomass via non-consumptive facilitative mechanisms (Bracken et al., 2014). For example, the physical process of grazing can remove secondary epiphytic growth to restore maximal photosynthetic capacity (Parnoja et al., 2014), while faecal pellet production can enhance nutrient availability (Williams and Carpenter, 1988) and the active consumption of neighbouring species can reduce exposure to interspecific competition for light and/or space. Similarly, signs of habitat degradation during organic enrichment may directly alter, either

negatively or positively, the abundance and behaviour of autotrophs (Dyson et al., 2007) and of key invertebrate (Bulling et al., 2008; Godbold et al., 2011) and fish grazers, such that the relocation of mobile consumers away from the area of enrichment can cascade habitat degradation to other areas by elevating consumer density in areas that act as disturbance refugia (Lenihan et al., 2001). Many of these effects are often overlooked, yet the evidence to date suggests that they can be particularly influential in determining how altered nutrient budgets are expressed in natural systems.

14.3.3 Predation

The number of studies that have considered predation within the context of altered nutrient regimes in marine systems is limited. Of those that have, there is a tendency to predominantly focus on testing ecological theory rather than exploring the mechanistic basis of nutrient–predation interactions. A particular focus has been trophic cascades and the relative roles of bottom-up (nutrient enrichment) versus top-down (predation) forcing in shaping natural communities (e.g. Hughes et al., 2013). From these data, it can be seen that the presence of predators tends to increase autotrophic productivity via non-consumptive effects (e.g. threat avoidance) on the grazing community (Borer et al., 2006; Spivak et al., 2007; Premo and Tyler, 2013). This can have functional consequences, such as altering species–environment interactions (Armitage and Fong, 2006; Maire et al., 2010) and reducing the presence of prey populations (Posey et al., 2006), but very little is known about the wider functional importance of such behaviour.

Predation can also be affected by temporal and spatial context. Seasonal shifts in nutrient supply alter pelagic productivity, especially in upwelling regions, and can facilitate a change in the foraging patterns of predators, which switch, for example, from benthic to pelagic dwelling prey following planktonic bloom periods (Hamaoka et al., 2014). Similarly, increased canopy density and complexity due to epiphytic algal growth in seagrass beds provides protection from predators for prey species and can alter fish predation success rates (Williams et al., 2002). When rates of predation avoidance are high, it can contribute to the maintenance of high genetic diversity within the prey population (Strom and Bright, 2009). The vulnerability of a species to predation can also be influenced by recent diet history and, particularly in planktonic pelagic systems, can reduce predation risk for certain organisms (Strom and Bright, 2009). Conversely, when nutrient additions are sufficient to cause the deterioration of habitat, normal behaviour patterns can be compromised (e.g. emergence of infaunal sediment invertebrates in periods of hypoxia; Nilsson and Rosenberg, 1994) and raise the likelihood of predation (Long et al., 2008; Ortiz-Santaliestra et al., 2010).

14.3.4 Facilitation and competition

Anthropogenic alterations to the stocks and flows of N_r, and concomitant changes in other associated ecosystem properties, provide considerable opportunity for positive (facilitative) and negative (competitive) interspecific interactions to establish, but the dynamics, processes, and mechanisms affecting these interactions are inadequately understood. The input of new nitrogen by N_2-fixing cyanobacteria, for example, will not only benefit the diazotrophs themselves but may also positively affect other phototrophs and, importantly, other chemotrophic bacteria by releasing substantial amounts of fixed nitrogen and organic carbon (Brauer et al., 2015). Similarly, the secretion of allelopathic substances by macroalgae can have positive effects on growth and nutrient uptake in other macroalgae and microphytobenthos (Xu et al., 2013). At larger scales, many tropical systems are witnessing phase shifts from coral reef to algal dominated systems that are linked to the input of excess nitrogen. Under these circumstances, both facilitative and competitive interactions are common, and often occur simultaneously. For example, Easson et al. (2014) investigated tropical algal–sponge interactions and found that algal abrasion and shading negatively affected a sponge and its cyanobacterial symbionts, while, at the same time, elevated nutrients associated with enrichment enhanced the condition of the algae, which, in turn, may make it more palatable to herbivores (Longo and Hay, 2015). Extending this further, the predisposition of a location to invasive species could depend on the competitive ability of a native species, which can be affected by relevant nitrogen concentrations (e.g. rate of photoaccumulation, and hence growth, in seaweeds increases with greater nutrient loading; Hamel and Smith, 2011).

14.3.5 Anthropogenic nitrogen and other stressors

One of the difficulties with the N_r literature base is that, more often than not, multiple stressors are evoked and the effects of N_r alone are difficult to distinguish. This is because, even though N_r is highly likely to be a major contributory factor to, for example, eutrophication

leading to hypoxia and algal blooms, other parameters are arguably likely to be equally (or at least significantly) important in determining the ecological outcome (Howarth and Marino, 2006). For some of the other stressors addressed in this volume, the reverse problem is true; experiments investigating the effects of a particular stressor that can be isolated from other factors tend not to consider the role of other stressors simultaneously, so the challenge for experimentalists is to add complexity, rather than remove it, in a meaningful way. Either way, the interest lies in quantifying the effects of multiple stressors. In most cases, however, the interest lies in determining how two or more stressors interact with one another to affect an additional process (e.g. NO_3^- limitation × UV-B exposure × ocean acidification effects on diatom photosynthetic performance; Li et al., 2015) or ecosystem function (e.g. nutrient flux or primary productivity), rather than seeking how changes to N_r and/or the inclusion of other stressors alter fundamental mechanisms within the N_r cycle. This distinction is subtle, but important, because the two approaches answer fundamentally different sets of questions.

There are examples emerging in the literature of both additive and non-additive effects of interacting multiple stressors (including N_r), but the number of experiments is low and the focus tends to be on N_r with appropriate climatic variables (e.g. temperature, ocean acidification, UV-B, and/or salinity) in pelagic (Li et al., 2015), saltmarsh, or intertidal settings (Ryan and Boyer, 2012). Interestingly, interactions with other properties of the food web (e.g. species richness, abundance, biomass, evenness, functional groupings, and trophic ecology), floral or faunal condition (immunological, physiological or structural variation, niche envelope positioning, age class, and sex), pollutants (heavy metals, hydrocarbons, and organic matter), physical stress (hydrodynamics, immersion or emersion exposure, abrasion, desiccation, heat or cold shocks, hydrogen sulphide, and low oxygen), and/or many of the other stressors featured within this book are either rare or have not been addressed. Similarly, aspects of variation in the magnitude, timing, frequency, and geographical location have been largely ignored to date, as has the influence of human attitudes, cultural values and alternative management and governance regimes.

There are good reasons to incorporate various aspects of the nitrogen cycle into factorial experiments. The ratio of N_r to other nutrients is influential in processes such as photosynthesis, while the concentration and form of N_r may or may not be bioavailable, toxic, or inhibitive to species important in mediating transformative pathways (particularly microbes, but also invertebrates, algae and plants, and vertebrates). There is also merit in manipulating the microbial community to determine which groups are functionally the most important, and/or which groups can acclimatize or adapt to novel conditions. Other parameters could be used to alter the fate of N_{org} (e.g. aerobic versus anaerobic conditions to alter the relative role of nitrification; use of inhibitors, Gilbertson et al., 2012; or alteration of environmental conditions to favour direct over coupled N removal using increased salinity regimes, Hines et al., 2015). The question is not about whether multiple stressors should be incorporated, rather it should be about whether N_r is included and, if so, whether it should be considered as a response or explanatory variable. To date, it has only really been considered as the latter.

14.4 A comment on future research directions

How much N_r is likely to matter for a range of ecosystem properties and the future will depend, at least in part, on whether species operate then in the same way as they do now to influence, and be influenced by, various components of N_r. In compiling this chapter, a surprising difficulty was gathering quantitative information that unambiguously confirmed or rejected specific mechanistic processes or hypotheses relating to how species related to specific parts of the nitrogen cycle. Only a minority of cases consider N_r in isolation, or within a factorial framework that allows additive versus interactive relationships to be identified, and many contributions proffer conclusions based on little more than phenomenological or correlative evidence. Thus:

Recommendation 1. Place greater emphasis on the use of highly controlled experiments to aid understanding of the interplay and the interdependencies within and between various abiotic and biotic explanatory variables that influence ecosystem level effects of N_r.

Rigorous testing of candidate mechanistic processes and theory using carefully designed experiments is not, of course, a panacea in formulating a definitive understanding of the fundamental principles underpinning nitrogen forcing. Such efforts will need to be underpinned by theory and supported by field observation, broadening the evidence base to include the effects of multiple seasons, habitat types, and trophic levels, as well as contrasting socio-ecological settings. In doing so, it will be important to recognize and

embrace the way in which N_r forcing is expressed in natural systems (unequal distribution, pulsed versus press forcing, sequential versus simultaneous forcing, etc.). Hence, a second research priority should be:

Recommendation 2. Investigate the effects of periodic, stochastic, sequential, and/or cyclical variations in the nitrogen cycle from directional change imposed by both short- and long-term N_r forcing.

A logical extension to this is to consider the long-term legacy of localized changes in ecosystem properties. The predominate research focus to date has been on the immediate and negative consequences of N_r addition, perhaps at the expense of contemplating initial gains (i.e. stimulatory effects) or identifying where thresholds of resilience and/or processes of recovery during N_r alleviation may lie. This could include, for instance, establishing generality about when, and under what circumstances, ecosystems change state, or whether or not they have the bidirectional capacity to withstand (acclimate) or adapt to altered levels of N_r. Therefore:

Recommendation 3. Consider the mediating role of resilience and the adaptive capacity of species and ecosystems in determining N_r-driven regime shifts, rebound trajectories, and recovery potential.

These three recommendations, in essence, serve an overarching goal to identify the mechanisms underpinning the variability in how species respond to and effect N_r. Estimates of the functional consequences associated with altered levels of anthropogenic nitrogen that incorporate the error associated with such variation will provide improved levels of certainty when determining the consequences of future global change. This next step will be necessary, especially given the multiple pressures marine systems experience, if we are to effectively manage the functioning and longevity of our estuarine and coastal heritage.

References

Abell, G.C.J., Banks, J., Ross, D.J., et al. (2011). Effects of estuarine sediment hypoxia on nitrogen fluxes and ammonia oxidizer gene transcription. *FEMS Microbial Ecology*, **75**, 111–122.

Agatsuma, Y., Endo, H., Yoshida, S., et al. (2014). Enhancement of *Saccharina* kelp production by nutrient supply in the Sea of Japan off Southwestern Hokkaido, Japan. *Journal of Applied Phycology*, **26**, 1845–1852.

Alestra, T. & Schiel, D.R. (2014). Effects of opportunistic algae on the early life history of a habitat-forming fucoid: influence of temperature, nutrient enrichment and grazing pressure. *Marine Ecology Progress Series*, **508**, 105–115.

Armitage, A.R. & Fong, P. (2006). Predation and physical disturbance by crabs reduce the relative impacts of nutrients in a tidal mudflat. *Marine Ecology Progress Series*, **313**, 205–213.

Austin, A.T., Bustamante, M.M.C., Nardoto, G.B., et al. (2013). *Science*, **340**, 149.

Benton, T.G., Solan, M., Travis, J.M.J., & Sait, S.M. (2007). Microcosm experiments can inform global ecological problems. *Trends in Ecology and Evolution*, **22**, 516–521.

Bertics, V.J. & Ziebis, W. (2009). Biodiversity of benthic microbial communities in bioturbated coastal sediments is controlled by geochemical microniches. *ISME Journal*, **3**, 1269–1285.

Bi, R., Arndt, C., & Sommer, U. (2014). Linking elements to biochemical: effects of nutrient supply ratios and growth rates on fatty acid composition of phytoplankton species. *Journal of Phycology*, **50**, 117–130.

Billen, G., Garnier, J., & Lassaletta, L. (2013). The nitrogen cascade from agricultural soils to the sea: modelling nitrogen transfers at regional watershed and global scales. *Philosophical Transactions of the Royal Society Series B*, **368**, 20130123.

Bishop, M.J., Kelaher, B.P., Smith, M.P., et al. (2006). Ratio-dependent response of a temperate Australian estuarine system to sustained nitrogen loading. *Oecologia*, **149**, 701–708.

Borer, E.T., Halpern, B.S., & Seabloom, E.W. (2006). Asymmetry in community regulation: effects of predators and productivity. *Ecology*, **87**, 2813–2820.

Botter-Carvalho, M.L., Carvalho, P.V.V.C., Valenca, A.P.M.C., et al. (2014). Estuarine macrofauna responses to continuous *in situ* nutrient addition on a tropical mudflat. *Marine Pollution Bulletin*, **83**, 214–223.

Bouwman, A.F., Beusen, A.H.W., Griffioen, J., et al. (2013). Global trends and uncertainties in terrestrial denitrification and N$_2$O emissions. *Philosophical Transactions of the Royal Society Series B*, **368**, 20130112.

Bracken, M.E.S., Dolecal, R.E., & Long, J.D. (2014). Community context mediates the top-down vs. bottom-up effects of grazers on rocky shores. *Ecology*, **95**, 1458–1463.

Bracken, M.E.S., Hillebrand, H., Borer, E.T., et al. (2015). Signatures of nutrient limitation and co-limitation: responses of autotroph internal nutrient concentrations to nitrogen and phosphorus additions. *Oikos*, **124**, 113–121.

Brauer, V.S., Stomp, M., Bouvier, T., et al. (2015). Competition and facilitation between the marine nitrogen-fixing cyanobacterium *Cyanothece* and its associated bacterial community. *Frontiers in Microbiology*, **5**, 795.

Bulling, M.T., Solan, M, Dyson, K.E., et al. (2008). Species effects on ecosystem processes are modified by faunal responses to habitat quality. *Oecologia*, **158**, 511–520.

Burkepile, D.E. & Hay, M.E. (2006). Herbivore versus nutrient control of marine primary producers: context dependent effects. *Ecology*, **87**, 3128–3139.

Camargo, J.A. & Alonso, Á. (2006). Ecological and toxicological effects of inorganic nitrogen pollution in aquatic ecosystems: a global assessment. *Environmental International*, **32**, 831–849.

Canfield, D.E., Glazer, A.N., & Falkowski, P.G. (2010). The evolution and future of Earth's nitrogen cycle. *Science*, **330**, 192–196.

Capone, D.G., Bronk, D.A., Mulholland, M.R., & Carpenter, E.J. (2008). *Nitrogen in the Marine Environment*. Elsevier, Amsterdam.

Capone, D.G. & Hutchins, D.A. (2013). Microbial biogeochemistry of coastal upwelling regimes in a changing ocean. *Nature Geoscience*, **6**, 711–717.

Clements, K.D., Raubenheimer, D., & Choat, J.H. (2009). Nutritional ecology of marine herbivorous fishes: ten years on. *Functional Ecology*, **23**, 79–92.

Cloern, J.E. (2001). Our evolving conceptual model of the coastal eutrophication problem. *Marine Ecology Progress Series*, **210**, 223–253.

Dalsgaard, T., Thamdrup, B., & Canfield, D.E. (2005). Anaerobic ammonium oxidation (anammox) in the marine environment. *Research in Microbiology*, **156**, 457–464.

Daudi, L.N., Lugomela, C., Uko, J.N., et al. (2012). Effect of nutrient enrichment on seagrass associated meiofauna in Tanzania. *Marine Environmental Research*, **82**, 49–58.

De Brabandere, L., Canfield, D.E., Dalsgarrd, T., et al. (2014). Vertical partitioning of nitrogen-loss processes across the oxic-anoxic interface of an oceanic oxygen minimum zone. *Environmental Microbiology*, **16**, 3041–3054.

Devol, A.H. (2015). Denitrification, anammox, and N$_2$ production in marine sediments. *Annual Review of Marine Science*, **7**, 403–423.

Diaz, R.J. & Rosenberg, R. (1995). Marine benthic hypoxia: a review of its ecological effects and the behavioural responses of benthic macrofauna. *Oceanography and Marine Biology*, **33**, 245–303.

Diaz, R.J. & Rosenberg, R. (2008). Spreading dead zones and consequences for marine ecosystems. *Science*, **321**, 926–929.

Doney, S.C. (2010). The growing human footprint on coastal and open ocean biogeochemistry. *Science*, **328**, 1512–1516.

Downing, J.A. (1997). Marine nitrogen: phosphorus stoichiometry and the global N:P cycle. *Biogeochemistry*, **37**, 237–252.

Duce, R.A., LaRoche, J., Altieri, K., et al. (2008). Impacts of atmospheric anthropogenic nitrogen on the open ocean. *Science*, **320**, 893–897.

Dyson, K.E., Bulling, M.T., Solan, M., et al. (2007). Influence of macrofaunal assemblages and environmental heterogeneity on microphytobenthic production in experimental systems. *Proceedings of the Royal Society of London Series B*, **274**, 2547–2554.

Easson, C.G., Slattery, D.M., Baker, D.M. & Gochfeld, D.J. (2014). Complex ecological associations: competition and facilitation in a sponge-algal interaction.. *Marine Ecology Progress Series*, **507**, 153–167.

Elser, J.J., Bracken, M.E.S., Cleland, E.E., et al. (2007). Global analysis of nitrogen and phosphorus limitation of primary producers in freshwater, marine and terrestrial ecosystems. *Ecology Letters*, **10**, 1135–1142.

Elser, J.J., Sterner, R.W., Gorokhova, E., et al. (2000). Biological stoichiometry from genes to ecosystems. *Ecology Letters*, **3**, 540–550.

Erisman, J.W., Galloway, J.N., Seitzinger, S., et al. (2013). Consequences of human modification of the global nitrogen cycle. *Philosophical Transactions of the Royal Society Series B*, **368**, 20130116.

Ferreira, R.C., Nascimento-Junior, A.B., Santos, P.J.P., et al. (2015). Responses of estuarine nematodes to an increase in nutrient supply: an *in situ* continuous addition experiment. *Marine Pollution Bulletin*, **90**, 115–120.

Fitch, J.E. & Crowe, T.P. (2011). Combined effects of temperature, inorganic nutrients and organic matter on ecosystem processes in intertidal sediments. *Journal of Experimental Marine Biology and Ecology*, **400**, 257–263.

Fitch, J.E. & Crowe, T.P. (2012). Combined effects of inorganic nutrients and organic enrichment on intertidal benthic macrofauna: an experimental approach. *Marine Ecology Progress Series*, **461**, 59–70.

Fukami, T. & Wardle, D.A. (2005). Long-term ecological dynamics: reciprocal insights from natural and anthropogenic gradients. *Proceedings of the Royal Society of London Series B*, **272**, 2105–2115.

Galloway, J.N., Aber, J.D., Erisman, J.W., et al. (2003). The nitrogen cascade. *Bioscience*, **53**, 341–356.

Galloway, J.N., Dentener, F.J., Capone, D.B., et al. (2004). Nitrogen cycles: past, present and future. *Biogeochemistry*, **70**, 153–226.

Galloway, J.N., Leach, A.M., Bleeker, A., et al. (2013). A chronology of human understanding of the nitrogen cycle. *Philosophical Transactions of the Royal Society Series B*, **368**, 20130120.

Galloway, J.N., Townsend, A.R., Erisman, J.W., et al. (2008). Transformation of the nitrogen cycle: recent trends, questions, and potential solutions. *Science*, **320**, 889–892.

Gilbertson, W.W., Solan, M., & Prosser, J.I. (2012). Differential effects of microorganism–invertebrate interactions on benthic nitrogen cycling. *FEMS Microbiology Ecology*, **82**, 11–22.

Godbold, J.A., Bulling, M.T., & Solan, M. (2011). Habitat structure mediates biodiversity effects on ecosystem properties. *Proceedings of the Royal Society of London Series B*, **278**, 2510–2518.

Godbold, J.A. & Solan, M. (2009). Relative importance of biodiversity and the abiotic environment in mediating an ecosystem process. *Marine Ecology Progress Series*, **396**, 281–290.

Godbold, J.A., Solan, M., & Killham, K. (2009b). Consumer and resource diversity effects on marine macroalgal decomposition. *Oikos*, **118**, 77–86.

Goldenberg, S.U. & Erzini, K. (2014). Seagrass feeding choices and digestive strategies of the herbivorous fish *Sarpa salpa*. *Journal of Fish Biology*, **84**, 1474–1489.

Govers, L.L., Pieck, T., Bouma, T.J., et al. (2014). Seagrasses are negatively affected by organic matter loading and *Arenicola marina* activity in a laboratory experiment. *Oecologia*, **175**, 677–685.

Gruber, N. & Galloway, J.N. (2008). An Earth system perspective of the global nitrogen cycle. *Nature*, **451**, 293–296.

Hamaoka, H., Kaneda, A., Okuda, N., et al. (2014). Upwelling-like bottom intrusion enhances the pelagic-benthic coupling by a fish predator in a coastal food web. *Aquatic Ecology*, **48**, 63–71.

Hamel, K. & Smith, C.M. (2011). Effects of nitrogen on rate of photoacclimation by invasive and native species of Gracilaria (Rhodophyta). *Journal of Phycology*, **47**, S65–S66.

Harpole, W.S., Ngai, J.T., Cleland, E.E., et al. (2011). Nutrient co-limitation of primary producer communities. *Ecology Letters*, **14**, 852–862.

Hessing-Lewis, M.L., Hacker, S.D., Menge, B.A., et al. (2015). Are large macroalgal blooms necessarily bad? Nutrient impacts on seagrass in upwelling-influenced estuaries. *Ecological Applications*, **25**, 1330–1347.

Hines, D.E., Lisa, J.A., Sonmg, B., et al. (2015). Estimating the effects of seawater intrusion on an estuarine nitrogen cycle by comparative network analysis. *Marine Ecology Progress Series*, **524**, 137–154.

Hoffman, B.M., Lukoyanov, D., Yang, Z.-Y., et al. (2014). Mechanism of nitrogen fixation by nitrogenase: the next stage. *Chemical Reviews*, **114**, 4041–4062.

Holtgrieve, G.W., Schindler, D.E., Hobbs, W.O., et al. (2011). A coherent signature of anthropogenic nitrogen deposition to remote watersheds of the Northern hemisphere. *Science*, **334**, 1545–1548.

Hondorp, D.W., Breitburg, D.L., & Davias, L.A. (2010). Eutrophication and Fisheries: separating the effects of nitrogen loads and hypoxia on the pelagic-to-demersal ratio and other measures of landings composition. *Marine and Coastal Fisheries*, **2**, 339–361.

Hong, Y., Li, M., & Gu, J. (2009). Bacterial anaerobic ammonia oxidation (Anammox) in the marine nitrogen cycle—a review. *Weishengwu Xuebao*, **49**, 281–286.

Howarth, R.W. (2008). Coastal nitrogen pollution: a review of sources and trends globally and regionally. *Harmful Algae*, **8**, 14–20.

Howarth, R.W. & Marino, R. (2006). Nitrogen as the limiting nutrient for eutrophication in coastal marine ecosystems: evolving views over three decades. *Limnology and Oceanography*, **51**, 364–376.

Howarth, R.W., Swaney, D., Billen, G., et al. (2012). Nitrogen fluxes from the landscape are controlled by net anthropogenic nitrogen inputs and by climate. *Frontiers in Ecology and the Environment*, **10**, 37–43.

Hu, B.L., Shen, L.D., Xu, X.Y., et al. (2011). Anaerobic ammonium oxidation (anammox) in different natural ecosystems. *Biochemical Society Transactions*, **39**, 1811–1816.

Hu, Y.L. & Ribbe, M.W. (2015). Nitrogenase and homologs. *Journal of Biological Inorganic Chemistry*, **20**, 435–445.

Hughes, B.B., Eby, R., van Dyke, V., et al. (2013). Recovery of a top predator mediates negative eutrophic effects on seagrass. *Proceedings of the National Academy of Sciences of the USA*, **110**, 15313–15318.

Isobe, K. & Ohte, N. (2014). Ecological perspectives on microbes involved in N-cycling. *Microbes and Environments*, **29**, 4–16.

Kaiser, J. (2001). The other global pollutant: nitrogen proves tough to curb. *Science*, **294**, 1268–1269.

Karl, D., Letelier, R., Tupas, L., et al. (1997). The role of nitrogen fixation in biogeochemical cycling in the subtropical North Pacific Ocean. *Nature*, **388**, 533–538.

Kartal, B., de Almeida, N.M., Maalcke, W.J., et al. (2013). How to make a living from anaerobic ammonium oxidation. *FEMS Microbiological Reviews*, **37**, 428–461.

Kim, I.-N., Lee, K., Gruber, N., et al. (2014). Increasing anthropogenic nitrogen in the North Pacific Ocean. *Science*, **346**, 1102–1106.

Kim, T.-W., Lee, K., Najjar, R.G., et al. (2011). Increasing N abundance in the Northwestern Pacific Ocean due to atmospheric Nitrogen deposition. *Science*, **334**, 505–509.

Krishnamurthy, A., Moore, J.K., Mahowald, N., et al. (2009). Impacts of increasing anthropogenic soluble iron and nitrogen deposition on ocean biogeochemistry. *Global Biogeochemical Cycles*, **23**, GB3016.

Lee, C.L., Huang, Y.H., Chung, C.Y., et al. (2015). Herbivory in multi-species, tropical seagrass beds. *Marine Ecology Progress Series*, **525**, 65–80.

Lee, W.Y. & Wang, W.X. (2001). Metal accumulation in the green macroalga *Ulva fasciata*: effects of nitrate, ammonium and phosphate. *Science of the Total Environment*, **278**, 11–22.

Lenihan, H.S., Peterson, C.H., Byers, J.E., et al. (2001). Cascading of habitat degradation: oyster reefs invaded by refugee fishes escaping stress. *Ecological Applications*, **11**, 764–982.

Li, W., Gao, K., & Beardall, J. (2015). Nitrate limitation and ocean acidification interact with UV-B to reduce photosynthetic performance in the diatom *Phaeodactylum tricornutum*. *Biogeosciences*, **12**, 2383–2393.

Liu, W., Huang, L., Yang, W., et al. (2010). Effect of some nutrients on the growth and haemolytic toxins production of *Chattonella marina*. *Asian Journal of Ecotoxicology*, **5**, 394–401.

Liu, X., Zhang, Y., Wenxuan, H., et al. (2013). Enhanced nitrogen deposition over China. *Nature*, **494**, 459–462.

Long, W.C., Brylawski, B.J., & Seitz, R.D. (2008). Behavioral effects of low dissolved oxygen on the bivalve *Macoma balthica*. *Journal of Experimental Marine Biology and Ecology*, **359**, 34–39.

Longo, G.O. & Hay, M.E. (2015). Does seaweed-coral competition make seaweeds more palatable? *Coral Reefs*, **34**, 87–96.

Luong, A.D., De Laender, F., Olsen, Y., et al. (2014). Inferring time-variable effects of nutrient enrichment on marine ecosystems using inverse modelling and ecological network analysis. *Science of the Total Environment*, **493**, 708–718.

Lyons, D.A., Arvanitidis, C., Blight, A.J., et al. (2014). Macroalgal blooms alter community structure and primary productivity in marine ecosystems. *Global Change Biology*, **20**, 2712–2724.

Maire, O., Merchant, J.N., Bulling, M., et al. (2010). Indirect effects of non-lethal predation on bivalve activity and sediment reworking. *Journal of Experimental Marine Biology and Ecology*, **395**, 30–36.

McClanahan, T.R., Sala, E., Stickels, P.A., et al. (2003). Interaction between nutrients and herbivory in controlling algal communities and coral condition on Glover's Reef, Belize. *Marine Ecology Progress Series*, **261**, 135–147.

McLauchlan, K.K., Williams, J.J., Craine, J.M., et al. (2013). Changes in global nitrogen cycling during the Holocene epoch. *Nature*, **495**, 352–357.

Meyer, R.L., Allen, D.E., & Schmidt, S. (2008). Nitrification and denitrification as sources of sediment nitrous oxide production: a microsensor approach. *Marine Chemistry*, **110**, 68–76.

Micheli, F. (1999). Eutrophication, fisheries, and consumer-resource dynamics in marine pelagic ecosystems. *Science*, **285**, 1396–1399.

Nilsson, H.C. & Rosenberg, R. (1994). Hypoxic response of two benthic communities. *Marine Ecology Progress Series*, **115**, 209–217.

Nordhaus, I., Salewski, T., & Jennerjahn, T.C. (2011). Food preferences of mangrove crabs related to leaf nitrogen compounds in the Segara Anakan Lagoon, Java, Indonesia. *Journal of Sea Research*, **65**, 414–426.

Ortiz-Santaliestra, M.E., Fernandez-Beneitez, M.J., Marco, A., et al. (2010). Influence of ammonium nitrate on larval anti-predatory responses of two amphibian species. *Aquatic Toxicology*, **99**, 198–204.

Parnoja, M., Kotta, J., & Orav-Kotta, H. (2014). Effect of short-term elevated nutrients and mesoherbivore grazing on photosynthesis of macroalgal communities. *Proceedings of the Estonian Academy of Sciences*, **63**, 93–103.

Pascal, P.Y. & Fleeger, J.W. (2013). Diverse dietary responses by saltmarsh consumers to chronic nutrient enrichment. *Estuaries and Coasts*, **36**, 1115–1134.

Pearson, T.H. & Rosenberg, R. (1978). Macrobenthic succession in relation to organic enrichment and pollution of the marine environment. *Oceanography and Marine Biology*, **16**, 229–311.

Pedersen, T.M., Sand-Jensen, K., & Markager, S. (2014). Optical changes in a eutrophic estuary during reduced nutrient loadings. *Estuaries and Coasts*, **37**, 880–892.

Peierls, B.L. & Paerl, H.W. (1997). Bioavailability of atmospheric organic nitrogen deposition to coastal phytoplankton. *Limnology and Oceanography*, **42**, 1819–1823.

Piedras, F.R. & Odebrecht, C. (2012). The response of surf-zone phytoplankton to nutrient enrichment (Cassino Beach, Brazil). *Journal of Experimental Marine Biology and Ecology*, **432**, 156–161.

Pistocchi, R., Pezzolesi, L., Guerrini, F., et al. (2011). A review on the effects of environmental conditions on growth and toxin production of *Ostreopsis ovate*. *Toxicon*, **57**, 421–428.

Posey, M.H., Alphin, T.D., & Cahoon, L. (2006). Benthic community responses to nutrient enrichment and predator exclusion: influence of background nutrient concentrations and interactive effects. *Journal of Experimental Marine Biology and Ecology*, **330**, 105–118.

Premo, K.M. & Tyler, A.C. (2013). Threat of predation alters the ability of benthic invertebrates to modify sediment biogeochemistry and benthic microalgal abundance. *Marine Ecology Progress Series*, **494**, 29–39.

Prosser, J.I. & Nicol, G.W. (2008). Relative contributions of archaea and bacteria to aerobic ammonia oxidation in the environment. *Environmental Microbiology*, **10**, 2391–2941.

Purvaja, R., Ramesh, R., Ray, A.K., et al. (2008). Nitrogen cycling: a review of the processes, transformations and fluxes in coastal ecosystems. *Current Science*, **94**, 1419–1438.

Rabalais, N.N. (2002). Nitrogen in aquatic ecosystems. *Ambio*, **31**, 102–112.

Raffaelli, D.G., Raven, J.A., & Poole, L.J. (1998). Ecological impact of green macroalgal blooms. *Oceanography and Marine Biology*, **36**, 97–125.

Rockström, J., Steffen, W., Noone, K., et al. (2009). A safe operating space for humanity. *Nature*, **461**, 472–475.

Ryan, A.B. & Boyer, K.E. (2012). Nitrogen further promotes a dominant salt marsh plant in an increasingly saline environment. *Journal of Plant Ecology*, **5**, 429–441.

Sakihara, T.S., Dudley, B.D., MacKenzie, R.A., & Beets, J.P. (2015). Endemic grazers control benthic microalgal growth in a eutrophic tropical brackish ecosystem. *Marine Ecology Progress Series*, **519**, 29–45.

Sano, Y., Takahata, N., Nishio, Y., et al. (2001). Volcanic flux of nitrogen from the Earth. *Chemical Geology*, **171**, 263–271.

Santoro, A.E. (2010). Microbial nitrogen cycling at the saltwater-freshwater interface. *Hydrogeology Journal*, **18**, 187–202.

Schoo, K.L., Aberle, N., Malzahn, A.M., et al. (2014). The reaction of European lobster larvae (*Homarus gammarus*) to different quality food: effects of ontogenetic shifts and pre-feeding history. *Oecologia*, **174**, 581–594.

Schramm, W. (1999). Factors influencing seaweed responses to eutrophication: some results from EU-project EUMAC. *Journal of Applied Phycology*, **11**, 69–78.

Seitzinger, S.P. (1988). Denitrification in freshwater and coastal marine ecosystems—ecological and geochemical significance. *Limnology and Oceanography*, **33**, 702–724.

Seitzinger, S.P., Harrison, J.A., Bohlke, J.K., et al. (2006). Denitrification across landscapes and waterscapes: a synthesis. *Ecological Applications*, **16**, 2064–2090.

Seitzinger, S.P. and Kroeze, C. (1998). Global distribution of nitrous oxide production and N inputs in freshwater and coastal marine ecosystems. *Global Biogeochemical Cycles*, **12**, 93–113.

Seitzinger, S.P., Kroeze, C., Bouwman, A.F., et al. (2002). Global patterns of dissolved inorganic and particulate nitrogen inputs to coastal systems: recent conditions and future projections. *Estuaries*, **25**, 640–655.

Seitzinger, S.P., Mayorga, E., Bouwman, A.F., et al. (2010). Global river nutrient export: a scenario analysis of past and future trends. *Global Biogeochemical Cycles*, 24, GB0A08.

Selander, E., Cervin, G., & Pavia, H. (2008). Effects of nitrate and phosphate on grazer-induced toxin production in *Alexandrium minutum*. *Limnology and Oceanography*, **53**, 523–530.

Shin, P.K.S., Cheung, C.K.C., & Cheung, S.G. (2006). Effects of nitrogen and sulphide on macroinfaunal community: a microcosm study. *Marine Pollution Bulletin*, **52**, 1333–1339.

Silkin, V.A., Pautova, L.A., Pakhomova, S.V., et al. (2014). Environmental control on phytoplankton community structure in the NE Black Sea. *Journal of Experimental Marine Biology and Ecology*, **461**, 267–274.

Smyth, A.R., Plehler, M.F., & Grabowski, J.H. (2015). Habitat context influences nitrogen removal by restored oyster reefs. *Journal of Applied Ecology*, **52**, 716–725.

Sohm, J.A., Webb, E.A., & Capone, D.G. (2011). Emerging patterns of marine nitrogen fixation. *Nature Reviews Microbiology*, **9**, 499–508.

Spivak, A.C., Canuel, E.A., Duffy, J.E., et al. (2007). Top-down and bottom-up controls on sediment organic matter composition in an experimental seagrass ecosystem. *Limnology and Oceanography*, **52**, 2595–2607.

Stewart, R.I.A., Dossena, M., Bohan, D.A., et al. (2013). Mesocosm experiments as a tool for ecological climate-change research. *Advances in Ecological Research*, **48**, 71–181.

Stoner, E.W., Yeager, L.A., Sweatman, J.L., et al. (2014). Modification of a seagrass community by benthic jellyfish blooms and nutrient enrichment. *Journal of Experimental Marine Biology and Ecology*, **461**, 185–192.

Strom, S.L. & Bright, K.J. (2009). Inter-strain differences in nitrogen use by the coccolithophore *Emiliania huxleyi*, and consequences for predation by a planktonic ciliate. *Harmful Algae*, **8**, 811–816.

Strous, M., Pelletier, E., Mangenot, S., et al. (2006). Deciphering the evolution and metabolism of an anammox bacterium from a community genome. *Nature*, **440**, 790–794.

Sutton, M.A., Howard, C.M., Erisman, J.W., et al. (2011). *The European Nitrogen Assessment: Sources, Effects and Policy Perspectives*. Cambridge University Press, Cambridge, UK.

Swaney, D.P., Scavia, D., Howarth, R.W., et al. (2008). Estuarine classification and response to nitrogen loading: insights from simple ecological models. *Estuarine Coastal and Shelf Science*, **77**, 253–263.

Thamdrup, B. (2012). New pathways and processes in the global nitrogen cycle. *Annual Review of Ecology, Evolution, and Systematics*, **43**, 407–428.

Vitousek, P.M., Menge, D.N.L., Reed, S.C., et al. (2013). Biological nitrogen fixation: rates, patterns and ecological controls in terrestrial ecosystems. *Philosophical Transactions of the Royal Society Series B*, **368**, 20130119.

Voss, M., Bange, H.W., Dippner, J.W., et al. (2013). The marine nitrogen cycle: recent discoveries, uncertainties, and the potential relevance of climate change. *Philosophical Transactions of the Royal Society Series B*, **368**, 20130121.

Ward, B.B. (2012). The global nitrogen cycle. In: Knoll, A.H., Canfield, D.E., & Konhauser, K.O. (eds), *Fundamentals of Geomicrobiology*, pp. 36–48. Wiley-Blackwell, Chichester, UK.

Whalen, M.A., Duffy, J.E., & Grace, J.B. (2013). Temporal shifts in top-down vs. bottom-up control of epiphytic algae in a seagrass ecosystem. *Ecology*, **94**, 510–520.

Williams, B.S., Hughes, J.E., & Hunter-Thomson, K. (2002). Influence of epiphytic algal coverage on fish predation rates in simulated Eelgrass habitats. *Biological Bulletin*, **203**, 248–249.

Williams, S.L. & Carpenter, R.C. (1988). Nitrogen-limited primary productivity of coral reef algal turfs: potential contribution of ammonium excreted by *Diadema antillarum*. *Marine Ecology Progress Series*, **47**, 145–152.

Worm, B. & Lotze, H.K. (2006). Effects of eutrophication, grazing, and algal blooms on rocky shores. *Limnology and Oceanography*, **51**, 569–579.

Xu, D., Li, F., Gao, Z.Q., et al. (2013). Facilitative interactions between the green-tide macroalga *Monostroma arctium* and the red macroalga *Porphyra yezoensis*. *Journal of Experimental Marine Biology and Ecology*, **444**, 8–15.

Young, E.B. & Berges, J.A. (2016). Nitrogen stress in the marine environment: from scarcity to surfeit. In: Solan, M. & Whiteley, N. (eds), *Stressors in the Marine Environment: Physiological and Ecological Responses, Societal Implications*, pp. 93–116. Oxford University Press, Oxford, UK. [Chapter 6, this volume.]

Zehr, J.P. & Kudela, R.M. (2011). Nitrogen cycle of the open ocean: from genes to ecosystems. *Annual Review of Marine Science*, **3**, 197–225.

Zhou, N.Q., Zhao, S., & Shen, X.P. (2014). Nitrogen cycle in the hyporheic zone of natural wetlands. *Chinese Science Bulletin*, **59**, 2945–2956.

CHAPTER 15

Ecological impacts of ultraviolet-B radiation on marine ecosystems

Sébastien Moreau, Francesca Vidussi, Gustavo Ferreyra, and Behzad Mostajir

15.1 Introduction

Ultraviolet B radiation (UVBR, 280–320 nm), the most biologically damaging portion of the solar spectrum reaching the Earth, received considerable scientific attention after the discovery of the spring stratospheric 'ozone hole' in the late 1970s over Antarctica. Recently, similar low ozone conditions were observed over the Arctic and occasionally at lower latitudes. Furthermore, expected increases in ocean acidification, surface water temperatures, and modifications in the structure of the water column due to global change exacerbated general concerns about the potential impact that such changes may have on the structure of marine food webs. In this chapter, we review the effects of ultraviolet-B radiation (UVBR) on various marine ecosystems. We start by providing a description of factors that influence the UVBR intensity, including latitude, season, stratospheric ozone layer thickness, and penetration within the water column. Then, we depict the effects of UVBR on the food webs of some important marine ecosystems such as polar oceans, coastal waters, fronts and upwellings, oceanic gyres, and benthic ecosystems. Finally, we investigate the potential interactions of enhanced UVBR along with other climate change stressors such as global warming and ocean acidification.

15.1.1 UVBR penetration through the atmosphere: the role of stratospheric ozone

When solar energy reaches the surface of the atmosphere (Fig. 15.1), it is mainly distributed between ultraviolet C radiation (UVCR, 200–280 nm), UVBR, ultraviolet A radiation (UVAR, 320–400 nm), visible light (or photosynthetically available radiation, PAR, 400–700 nm), and infrared radiation (IR, > 700 nm). Ultraviolet radiation (UVCR + UVBR + UVAR) only represents ~8% of this incoming solar energy. During its passage through the atmosphere, solar energy is scattered and absorbed, which reduces its intensity by 35% (Fig. 15.1) and modifies its spectral distribution depending on atmospheric gases concentration (e.g. ozone, O_3, which absorbs UVBR and UVCR), aerosols concentration, and solar zenith angle, hence daytime, season, and latitude. As a result, the proportion of solar radiation that reaches the surface of the Earth decreases with increasing latitude (i.e. towards the poles) mainly because of a longer path length to the Earth's surface which increases the chance for light scattering and light absorption by gases and other particles (for a review of light penetration within the atmosphere, see Whitehead et al., 2000). UVCR never reaches the lower half of the stratosphere. Therefore, within the rest of this section, we will focus on the role of stratospheric ozone in the penetration of UVBR.

The ozone molecules have a great potential to absorb UVBR and the thickness of the stratospheric ozone layer will mainly determine the amount of UVBR that reaches the surface of the oceans. Within the stratosphere, short wavelengths (< 240 nm) react with oxygen (O_2) molecules to form ozone (O_3):

$$O_2 + h\nu \rightarrow O + O \qquad (15.1)$$

$$O_2 + O \rightarrow O_3 \qquad (15.2)$$

where h is the Planck constant, υ is the wave frequency, and $h\upsilon$ represents the energy of radiation. The inverse reaction, the destruction of ozone into single oxygen

Stressors in the Marine Environment. Edited by Martin Solan and Nia M. Whiteley
© Oxford University Press 2016. Published in 2016 by Oxford University Press.

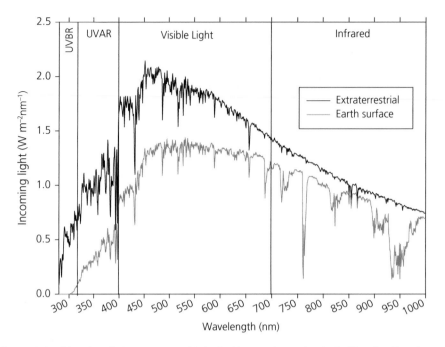

Figure 15.1 Characteristics of the solar radiation spectrum outside the Earth's atmosphere and at the Earth's surface. The solar spectrum is only given between 280 and 1000 nm as UVCR (100–280 nm) does not penetrate through the troposphere.

molecules, may also occur naturally, through recombination with O:

$$O_3 + O \rightarrow 2O_2 \qquad (15.3)$$

However, this recombination is enhanced by catalytic cycles involving a free radical X combined with O:

$$O_3 + XO \rightarrow 2O_2 + X \qquad (15.4)$$

where the free radical X may be NO, HO, Cl, I, or Br. Among these radicals, the anthropogenic chlorofluorocarbons (CFCs) have the greatest potential for ozone destruction. This recombination process, therefore, decreases the total amount of O_3 within the stratosphere and may increase the penetration of UVBR into the lower atmosphere.

Because of the rise in CFC concentrations within the atmosphere during the last century, and their accumulation within the stratosphere, the stratospheric ozone (O_3) concentration has decreased dramatically over the Southern Ocean and moderately at higher latitudes. This is especially true over the Antarctic, where a seasonal ozone minimum period is observed every year from August to November (Fig. 15.2; Salby et al., 2012). This phenomenon, named an 'ozone hole', results from the formation of the stratospheric polar

vortex during the austral winter within which CFCs accumulate (Rowland, 2006). With the increase in light reaching the atmosphere over Antarctica in early spring, catalytic reactions begin. The ozone hole is considered to be present when O_3 < 220 Dobson Units (DU, with 1 DU being equivalent to an ozone layer 0.01 mm thick at 0 °C and 1 atm)—'normal' stratospheric ozone thickness over Antarctica is ~ 340 DU. During the ozone hole period, the stratospheric ozone concentration over Antarctica may be reduced by as much as 50% (Casiccia et al., 2008). A similar phenomenon is sometimes observed in the Arctic, at the beginning of spring, with stratospheric ozone concentrations being reduced by 45% of their normal values (e.g. Fioletov et al., 1997) and this event seems to be becoming more intense and comparable to the Antarctic ozone hole (Manney et al., 2011). In the Arctic, climate change may induce stronger and more frequent ozone holes due to cooler stratospheric temperatures (Rex et al., 2004), while there is evidence of links between stratospheric and tropospheric temperatures and ozone loss over the Antarctic (Turner and Overland, 2009). Other stratospheric ozone loss events have been recorded at lower latitudes, which may have significant impacts on UVBR reaching the Earth's surface (Stolarski et al.,

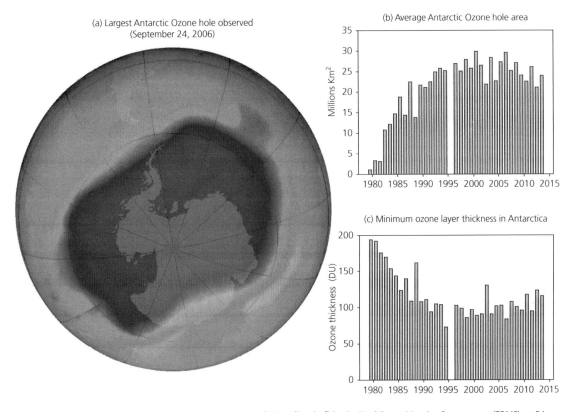

(a) Largest Antarctic Ozone hole observed
(September 24, 2006)

(b) Average Antarctic Ozone hole area

(c) Minimum ozone layer thickness in Antarctica

Figure 15.2 (a) The largest ozone hole observed over Antarctica (29.9 million km²) by the Total Ozone Mapping Spectrometer (TOMS) on 24 September 2006. Dark area in centre refers to low ozone concentrations while paler area surrounding it refers to higher total ozone column. Source: http://www.nasa.gov/vision/earth/lookingatearth/ozone_record.html; Annual records of (b) the average ozone hole area and (c) the average minimum stratospheric ozone thickness. Source: http://ozonewatch.gsfc.nasa.gov/.

1992; Reinsel et al., 2005). These ozone loss events may be related to the export of ozone poor stratospheric air masses from polar regions (Lee et al., 2002; Ajtić et al., 2004; Hadjinicolaou and Pyle, 2004). On the contrary, the transport of low latitudes stratospheric air masses, which are relatively ozone rich, contributes to ozone reconstruction at higher latitudes (Harris et al., 2008).

During these low stratospheric ozone concentration events, the proportion of UVBR that reaches the surface of the Earth with respect to other wavelengths, and more particularly to UVAR and PAR, rises (Frederick and Lubin, 1994). These changes modify the balance between the potentially positive effects of PAR and UVAR (although UVAR may also have detrimental effects) and the potentially negative effects of UVBR on living organisms (Davidson, 1998). On average, the UVBR only represents 0.8% of the total solar radiation that reaches the surface of the Earth but is responsible for half of the photochemical effects observed in

marine environments (Whitehead et al., 2000). Clouds, aerosols, and greenhouse gases may also modify the spectrum of the radiation that reaches the surface of the Earth (Balis et al., 2004; Chubarova, 2006, 2009). Because these factors are influenced by climate change and show great spatial variability, they may have a long-term impact on the UVBR that reaches the Earth's surface and, eventually, marine ecosystems.

In the early 1980s, the international community ratified the Montreal Protocol on Substances that Deplete the Ozone Layer, which aimed at a drastic reduction of CFC emissions. The objective was a reconstruction of the stratospheric ozone layer by 2050 (IPCC, 2005). Even if the Montreal protocol has avoided large-scale UVBR depletion, new challenges concerning UVBR are arising in relation to climate change. In fact, climate change and the release of other ozone-destroying substances such as N_2O from heavily fertilized soils interfere with the Montreal protocol objective (Bouwman,

1998; Rex et al., 2004), and, according to model predictions, one or two more decades may be necessary for significant ozone recovery over polar regions (Knudsen et al., 2004; Newman et al., 2006). In addition, models considering both ozone recovery and climate trends predict that ultraviolet radiation (UVR; i.e. including all types) will increase in the tropics by 2100, while it will decrease in polar regions (Bais et al., 2011). Actually, the most recent observations show that the 2011 Arctic ozone hole was particularly large (Manney et al., 2011) while, according to NASA, the 2014 Antarctic ozone hole covers a larger than average area compared to the last decade (http://ozonewatch.gsfc. nasa.gov/).

15.1.2 PAR and UVR penetration in the water column

When radiation strikes the ocean's surface, it is partly reflected towards the atmosphere to a percentage that depends on the angle of incoming light (which depends on the time of the day and latitude), the nature of radiation (direct or diffuse), and on sea surface roughness (see Table 1.1 of Whitehead et al., 2000 for percentage reflectance of incoming radiation on a smooth water surface). Once radiation has entered the seawater, its intensity decreases with depth due to scattering and absorption by water molecules and dissolved and particulate (living and non-living) matter present in the water column (Wozniak and Dera, 2007). The spectral irradiance attenuation at the depth z within the water column ($E(z, \lambda)$) is calculated from the Beer–Lambert law as:

$$E(z,\lambda) = E\left(0^-,\lambda\right) * e^{-K_d(\lambda)*z} \qquad (15.5)$$

with $K_d(\lambda)$ being the diffuse attenuation coefficient at a wavelength λ and $E(0^-, \lambda)$ the irradiance just below the water surface. Radiation wavelengths are differently absorbed and scattered in the water column. Coloured dissolved organic matter (CDOM) is one of the main absorbing substances that attenuate UVR penetration in seawater (Diaz et al., 2000) while phytoplankton and suspended non-living particles (detrital and inorganic) also significantly attenuate UVR penetration in coastal and estuarine waters (Hargreaves, 2003). The absorption of radiation by CDOM in seawater increases with decreasing wavelength (Blough et al., 1993); hence UVBR is attenuated at lower depths than UVAR and PAR. The distribution of CDOM, which has a higher concentration in coastal than in open waters, will, therefore, have a great impact on UVBR penetration (Tedetti and Sempéré, 2006).

When studying the effect of UVBR in marine systems, the depth at which the effects of UVBR cease is defined as the photoactive depth ($Z_{10\%}$) and corresponds to the depth of ~10% penetration of the incident UVBR (Neale et al., 2003), although UVBR effects may occur deeper. In terms of visible light (i.e. PAR), the euphotic zone (Z_{eu}) is defined as the maximum depth where autotrophic organisms can use PAR for photosynthesis and mainly corresponds to the depth of 1% penetration of the incident light. Collecting published radiation penetration data obtained from all over the world's oceans, Tedetti and Sempéré (2006) showed that the photoactive depth of UVBR can be as deep as 17, 11.5, 5, and 6.7 m in, respectively, oceanic, Antarctic, Arctic, and coastal waters.

15.1.3 UVBR effects at the cell/organism levels

UVBR is recognized to have significant impacts on marine organisms ranging from viruses, bacteria, phytoplankton, zooplankton, and fish eggs, larvae, and adults to macroalgae, sea grasses, corals, crustaceans, echinoderms, molluscs, etc. (Llabrés et al., 2012). These impacts include DNA damage (Häder and Sinha, 2005) and photoinhibition, the inhibition of primary production by radiation (e.g. Fritz et al., 2008). By forming cyclobutane pyrimidine dimers (CPD), UVBR may lead to DNA destruction (Häder and Sinha, 2005). Meanwhile, DNA may be repaired by several processes such as photoreactivation which requires UVAR and PAR and by dark repair processes (Sinha and Häder, 2002; Häder and Sinha, 2005). In turn, marine organisms may also avoid harmful UVBR by actively or passively migrating to deeper waters (Neale et al., 2003; Hylander and Hansson, 2010) and/or use screening compounds such as mycosporine-like amino acids (MAAs; Helbling et al., 1996). These physiological and behavioural effects are the focus of Chapter 7 and will only be considered here with respect to their role in the ecology of marine systems.

UVBR effects on marine organisms are species specific. These effects (which have been described thoroughly in the literature; e.g. de Mora et al., 2000; Sinha and Häder, 2002; Helbling and Zagarese, 2003) may alter or change the competition for resources or the trophic interactions between marine organisms and have significant effects on marine ecosystem functioning, a topic that has been less studied. In this chapter, we will review the potential effects of UVBR on the food webs of several major marine ecosystems: the polar oceans, coastal waters, fronts and upwellings, the oceanic gyres, and finally benthic ecosystems, including macroalgae, grazers such as urchins, and coral reefs.

15.2 UVBR effects on the food webs of different marine ecosystems

As UVBR effects mainly concern organisms living near the surface of the oceans, we will mainly focus on the role of UVBR on the ecology of the pelagic ecosystem (excepted in Section 15.2.4 dedicated to the effects of UVBR on benthic ecosystems). Different types of plankton food webs are described in the literature. Communities where bacteria are being grazed on by heterotrophic flagellates and ciliates in a closed system correspond to what Azam et al. (1983) defined as the microbial loop. Communities dominated by heterotrophs (bacteria, flagellates, and ciliates) and by small autotrophs (cyanobacteria and small flagellates) are generally low in productivity and correspond to what Legendre and Rassoulzadegan (1995) defined as the microbial food web. In this type of food web, primary production is generally based on ammonium (i.e. a form of production that relies on remineralization; Dugdale and Goering, 1967; Eppley and Peterson, 1979). In addition, the ratio of gross primary production to respiration (GPP:R) is generally < 1, with low export of organic carbon to higher trophic levels (Legendre and Rassoulzadegan, 1995) and towards deep waters (Michaels and Silver, 1988; Olli et al., 2002). Both the microbial loop and the microbial food web are encountered in oligotrophic waters (i.e. where nutrients are limiting) and towards the end of the planktonic succession in other areas. Communities composed of heterotrophic bacteria, flagellates, ciliates, and both small (< 5 µm) and medium (> 5 µm) phytoplankton cells are defined as multivorous food webs, while planktonic food webs dominated by microphytoplankton (e.g. diatoms > 20 µm) being grazed on by mesozooplankton (e.g. copepods and krill) are defined as herbivorous food webs by Cushing (1989). In these two types of planktonic food webs, primary production is usually based on nitrate (i.e. new production; Dugdale and Goering, 1967; Eppley and Peterson, 1979). In addition, these food webs generally have a GPP:R ratio ≥ 1 and have a greater potential for carbon export to higher trophic levels and towards deep waters (Pollard et al., 2009). These two types of food web are generally encountered in mesotrophic to eutrophic waters (i.e. where nutrients are not limiting; Fenchel, 2008). Fig. 15.3a represents schematically the

trophic continuum from the herbivorous to the multivorous and microbial food webs, ending by the microbial loop.

15.2.1 Effects of UVBR on the food web components of the polar oceans

The largest proportional amount of UV to reach the Earth's surface are recorded at the equator and at high-altitude land sites. However, because of ozone loss, abnormal high UVBR doses are recorded at high latitudes where organisms are more sensitive and less adapted to relatively high UVBR doses. Because ozone depletion has been the strongest over Antarctica (McKenzie et al., 2007), the effects of UVBR on marine organisms have been the focus of intensive research in this region (Karentz and Bosch, 2001).

To discuss the effects of UVBR on polar oceans, we will focus here on the seasonally ice-covered regions which are some of the most productive waters of these oceans (Carmack and Wassmann, 2006; Ducklow et al., 2006). In these areas, biomass accumulation usually starts in the austral spring (around October–December) in Antarctica and in summer (around June–September) in the Arctic, along with a receding ice-edge (Ferreyra et al., 2004; Wassmann, 2006; Smith et al., 2008), although the timing of phytoplankton blooms is changing due to climate change both in the Arctic Ocean (Arrigo and Van Dijken, 2011) and in some regions of the Southern Ocean (e.g. the Bellingshausen Sea, Moreau et al., 2015). The release of fresh water by sea-ice melt confines phytoplankton within a highly illuminated and nutrient-rich shallow mixed layer (~ 20 m; Vernet et al., 2008). These conditions favour the development of large blooms. Relatively high incident UVBR is observed at the sea surface during sea-ice retreat, which corresponds to the period of ozone depletion in Antarctica (Moreau et al., 2015). Z_{ph} of UVBR can be as deep as 12 m in Antarctic waters (Tedetti and Sempéré, 2006). Hence, the confinement of ice-edge phytoplankton blooms within shallow waters exposes planktonic communities to relatively high UVBR. In comparison, in the Arctic the photoactive depth for UVBR is shallower (i.e. 5 m, Tedetti and Sempéré, 2006) but Arctic phytoplankton may be more sensitive to UVBR than Antarctic phytoplankton, possibly because they are less adapted to high UVBR intensities (Helbling and Villafañe, 2002). UVBR may have negative effects on the ice-edge blooming phytoplankton by affecting DNA (Buma et al., 2001a) and primary production rates (Neale et al., 1998) although there is some debate regarding to what extent

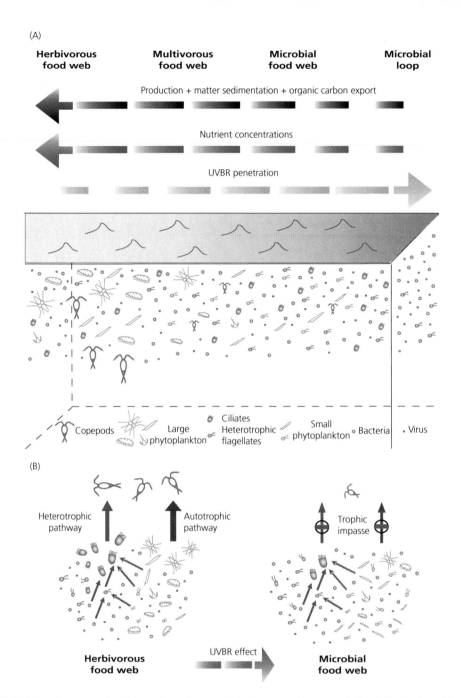

Figure 15.3 (A) Schematic representation of the continuum between the herbivorous, multivorous, and microbial food webs and the microbial loop. The UVBR penetration is usually higher when the microbial food web or the microbial loop are dominant because of the lower dissolved and particulate matter present within the water column due to oligotrophic conditions. On the other hand, the herbivorous food web is usually more present when nutrients concentrations are higher. In addition, the production of the system, the sedimentation of organic matter, and the export of organic carbon to higher trophic levels are higher when the system is dominated by the herbivorous food web. In the herbivorous food web, large copepods graze on large phytoplankton cells, so that predators are mostly metazoans. In the multivorous food web, both copepods (metazoans) and ciliates (protozoans) graze on both large and small prey. In the microbial food web, all components are microorganisms and small phytoplankton and heterotrophic bacteria are predated by protozoans (ciliates and flagellates). Finally, in the microbial loop all components are heterotrophic microorganisms (bacteria, flagellates, and ciliates). It should be noted that viruses are present in all types of food webs. (B) Based on the results of Mostajir et al. (1999) and Ferreyra et al. (2006), UVBR seem to drive planktonic communities from being dominated by a herbivorous food web to being dominated by a microbial food web and, therefore, towards less productive systems with less food transfer to higher trophic levels. The solid arrows represent 'predation'. [PLATE 12]

UVBR may inhibit primary production in the Southern Ocean. For example, Arrigo et al. (2003) estimated that UVR photoinhibition reduced primary production by 13% in the surface waters of the Southern Ocean in 1992 but that photoinhibition was less than 0.25% when integrated over the daily light cycle and to the bottom of the euphotic layer. UVBR may also affect phytoplankton quality as a food resource (in terms of lipid quantity and quality) to higher trophic levels (Hessen et al., 2012) and, therefore, have consequences for polar marine food webs. It should be noted that sea ice-edge blooms contribute to up to 60% of the annual autotrophic production in some polar seas (e.g. Sakshaug, 2004) and are an essential source of energy for the rest of the ecosystem. Large zooplankton rely on it for their growth (Leu et al., 2011) and higher trophic levels (e.g. fish, birds, and marine mammals) rely on large zooplankton for their development (Steen et al., 2007; Nicol et al., 2008; Trivelpiece et al., 2011).

Before ice edge phytoplankton blooms take place, ice algae may also be at threat due to high incident UVBR reaching polar ecosystems. Although sea ice greatly attenuates the penetration of UVBR into the water column (Cockell and Córdoba-Jabonero, 2004; Lesser et al., 2004), marine organisms thriving within or just below sea ice may be impacted by UVBR. For example, photoinhibition of sea ice algae by UVBR was measured as up to 23% in Antarctica during October (Schofield et al., 1995), while Ryan et al. (2012) discriminated between more UVBR-tolerant ice algae present in sea-ice brine inclusions and less UVBR-tolerant bottom ice algae. UVBR stress on ice algae may become a major problem for polar ecosystems as many organisms rely on them for their survival (i.e. ice algae account for an average of 10% of the annual primary production of ice covered regions; Deal et al., 2011). More particularly, large Arctic copepods such as *Calanus glacialis*, which represent 80% of the zooplankton biomass in the Arctic (Blachowiak-Samolyk et al., 2008), may have timed their seasonal migration, foraging, and reproduction period with ice algae blooms rather than sea ice-edge water column blooms (Søreide et al., 2010; Leu et al., 2011). In turn, a key component of the Antarctic ecosystem, the krill *Euphausia superba*, also relies on ice algae for its survival during winter and early spring (Flores et al., 2012). In addition, as for pelagic algae, UVBR may reduce the nutritional quality of sea-ice algae by reducing their content in polyunsaturated fatty acids (Leu et al., 2006, 2010) which are essential to higher trophic levels in lipid-based polar ecosystems' food webs (Müller-Navarra, 2008).

Besides these negative cascading trophic effects, UVBR may also have a direct impact on higher trophic level organisms through exposure to high UVBR levels when these organisms are close to the surface. Antarctic krill DNA, which is rich in thymine residues, may be particularly sensitive to UVBR stress (Jarman et al., 1999). In fact, Naganobu et al. (1999) observed a connection between krill populations' variability in the western Antarctic Peninsula and ozone level fluctuations from 1977 to 1997. However, through grazing, krill may acquire photoprotective pigments (Newman et al., 2000) that can partially protect them from ambient UVBR. For polar fishes and zooplankton, the larvae stage, which are planktonic and constrained near the sea surface, may be particularly affected (Browman et al., 2000; Dethlefsen et al., 2001; Dahms et al., 2011).

In addition, the UVBR effects for different components of the food web presented here for polar ecosystems may be enhanced or reduced by trophic interactions, pointing to the need to study the effects of UVBR stress on communities comprising all trophic levels from virus to, at least, zooplankton (Mostajir et al., 2000). Incubations of Antarctic marine phytoplankton showed that chlorophyll *a* (Chl-*a*) concentration, cell abundance, and carbon biomass do not differ between normal and enhanced UVBR treatments (Davidson and Marchant, 1994; Davidson et al., 1996). However, these authors observed important changes in species composition towards UVBR-tolerant species such as *Phaeocystis antarctica* with possible consequences for the food web. In terms of community structure, Davidson and Belbin (2002) noticed a decrease in Antarctic phytoplankton cell size, concentration, and biomass and a change in dominant species under UVBR stress. These authors noticed that small microzooplankton concentrations were higher under the high UVBR treatments. They suggested that phytoplankton cell death resulted in an increase in the release of dissolved organic matter on which bacteria may have fed. During the same experiment, Davidson and van der Heijden (2000) observed that bacterial biomass was not impacted by UVBR. This suggests that microzooplankton exerted a high grazing pressure on bacteria, which may also explain the high concentration of microzooplankton observed during this study. The authors concluded that UVBR stress may modify the biomass and composition of the microbial communities towards smaller cells and more heterotrophs (i.e. the microbial food web) and, therefore, towards less food availability for higher trophic levels.

Regarding the effect of UVBR on the structure of plankton communities, Wängberg et al. (2008) observed that, in an Arctic Fjord on the west coast of Spitsbergen (Svalbard), UVBR reduced the overall biomass of the plankton community and favoured the development of choanoflagellates over nanoflagellates. The authors pointed that UVBR reduced the capacity of the system to allocate photosynthetically produced organic carbon to the rest of the food web or that grazing pressure was enhanced under high UVBR stresses. Wickhman and Carstens (1998) studied the effects of UVBR on Arctic planktonic communities near the coast of Greenland using seawater from a sea-ice meltwater pond. They observed no effects of UVBR on bacteria, heterotrophic flagellates, and autotrophs, and only species-specific UVBR effects on microzooplankton. These results highlight the potential acclimation of upper sea-ice communities to high levels of UVBR (Ryan et al., 2012). Hence, in polar regions, we may expect negative individual UVBR effects to cascade throughout the food web with, for example, lower food quality of ice algae having a negative impact on *Calanus glacialis* and, as a consequence, on higher trophic levels (e.g. whales), although there is no experimental evidence of UVBR effects on the entire food web.

15.2.2 UVBR effects on the food web components of meso- to eutrophic coastal waters, fronts, and upwellings

Coastal waters, oceanic fronts, and upwellings generally range from mesotrophic to eutrophic conditions. Coastal waters comprise shorelines, estuaries, and shelf waters, and have long been seen as the most productive areas of the world's oceans (Mann and Lazier, 1991). Fronts (e.g. the Polar Front) are coastal and oceanic features that result from the encounter of two water masses of different origins and distinct properties. This encounter leads to a relatively narrow transition region where properties (e.g. pressure and density) follow steep gradients (Cushman-Roisin and Beckers, 2011). Geostrophic adjustment leads to an intense flow in the direction of the front. Upwellings are a particular type of oceanic front: they are wind-driven phenomena that trigger the displacement of a deeper, dense, cold, and usually nutrient-rich water mass to the surface, replacing the warmer and usually nutrient-depleted surface water mass (Stewart, 2008). Upwellings may occur coastally, associated with eddies or topography, and in large oceanic areas such as the equatorial upwelling. Fronts and upwellings are major sites of primary production (Sokolov and Rintoul, 2007; Landry et al., 2012) and a large part of the world fisheries catches are made in these regions (Jennings et al., 2001).

These productive waters are generally characterized by a high biomass and, for coastal waters, high concentrations of particulate and dissolved matter of terrigenous origin, reducing the penetration of UVBR in the water column (Diaz et al., 2000). The first consequence of this low penetration of UVBR is the photoprotection of marine organisms. However, within these systems, the absorption of UVBR breaks down dissolved organic matter (DOM) and CDOM into low molecular weight, labile carbon compounds and nutrients (Kieber, 2000; Mopper and Kieber, 2002; Aarnos et al., 2012). This photodegradation of DOM and CDOM reduces their ability to absorb UVBR and may lead to increased negative UVBR effects on marine organisms. On the contrary, it can stimulate bacterial activity (Abboudi et al., 2008), especially in the presence of inorganic nutrients (Tedetti et al., 2009).

In a mesocosm experiment performed in the St Lawrence Estuary, Mostajir et al. (1999) observed the effects of UVBR on a coastal planktonic community. The increased UVBR resulted in an important decrease of ciliates (66%) and large phytoplankton (63%), together with an increase in heterotrophic flagellates (300%), bacteria (50%), and small phytoplankton (40%). Mostajir et al. (1999) proposed that the decrease in abundance of ciliates led to the development of their prey: bacteria, small phytoplankton, and heterotrophic flagellates. In the same experiment, Chatila et al. (1999) observed a respective decrease and increase in the bacterivory of ciliates and of heterotrophic flagellates. Mostajir et al. (1999) concluded that, under UVBR stress, the marine ecosystem was more likely to develop towards a microbial food web with a predominance of small organisms (i.e. bacteria, picoplankton, and heterotrophic flagellates), which may have large consequences on carbon fluxes towards higher trophic levels and deep oceanic waters. In another experiment in the St Lawrence Estuary, Ferreyra et al. (2006) observed a decrease in ciliates biomass (70–80%) under high UVBR stress in coincidence with an increase in bacterial abundance, although phytoplankton biomass was not affected by UVBR. Ferreyra et al. (2006) also concluded that UVBR stress led the community towards a microbial food web. In a mesocosm experiment performed off the coast of Sweden, Wängberg et al. (2001) failed to find significant effects of UVBR on the total biomass of phytoplankton and bacteria. However, UVBR had marked effects on carbon allocation, size distribution of primary production, and phytoplankton species composition. Wängberg et al. (1998) also showed an increase in bacterial predators' abundances and a decrease in Chl-*a* due to UVBR in

Swedish waters. In conclusion, it seems that microbial food webs are more resilient to UVBR relative to herbivorous food webs and that UVBR drives the structure of food webs towards less productive microbial food webs. It turns out that higher trophic levels may be affected directly by UVBR (e.g. fish larvae; Sucré et al., 2012) or indirectly by the presence of less productive microbial food webs under UVBR stress. This can have significant impacts on fisheries in these regions.

15.2.3 UVBR effects on the food web components of oligotrophic systems such as oceanic gyres

Oligotrophic areas encompass ~40% of the Earth's surface area (Field et al., 1998) and, hence, constitute an important part of global primary production (Viviani et al., 2011). On the contrary to coastal and frontal waters, oligotrophic areas such as oceanic gyres are characterized by a deep penetration of radiation, and hence UVBR, within the water column, due to the low biomass and particulate matter concentrations generally observed in these systems. In this section, we mainly refer to subtropical gyres. In these waters, picophytoplankton (*Prochlorococcus*, *Synechococcus*, and picoeukaryotes) accounts for the major part of the plankton biomass and primary production (Agawin et al., 2000). These organisms are adapted to low nutrient concentrations, although *Synechococcus* and picoeukaryotes require higher nutrient concentrations than *Prochlorococcus* (Vaulot et al., 1996). Bacteria also play a major role both in terms of DOM recycling and as prey, thus transferring energy to higher trophic levels through microzooplankton grazing. Hence, the microbial loop or the microbial food web are usually dominant in these oligotrophic waters.

Picophytoplankton can be greatly affected by UVBR (Häder et al., 2003). For example, Llabrés and Agustí (2006) showed that ambient UVR levels could significantly induce picophytoplankton mortality in the clear oligotrophic waters of the Atlantic Ocean. In addition, Agustí and Llabrés (2007) and Llabrés et al. (2010) observed that *Synechococcus* and picoeukaryotes were more resistant to UVBR than *Prochlorococcus*. The smaller cell size of *Prochlorococcus* could be responsible for its higher UVBR sensitivity because larger phytoplankton cells synthesize more photoprotective pigments and have a longer pathlength for UVBR to reach significant cell targets (Buma et al., 2001b; Helbling et al., 2001). It is also possible that *Synechococcus* and picoeukaryotes have better photoprotection

and/or repair systems than *Prochlorococcus*. Llabrés et al. (2010) suggested that UVBR is a primary driver of the structure of the microbial community in clear, oligotrophic oceanic areas. Because *Prochlorococcus* represents a major part of the microbial biomass in oligotrophic gyres, its sensitivity to UVBR is an issue. However, because UVBR effects at the food web level have not been studied for such ecosystems, more experimental studies involving both microcosms and mesocosms should be performed to determine UVBR effects on the microbial community.

Nutrient limitation is a major driver of the community structure in oligotrophic environments. Nutrient limitation may increase the sensitivity of phytoplankton to UVBR stress (Bouchard et al., 2008). Firstly, nitrogen limitation may reduce the algal cells' capacity to produce MAAs and/or enzymes responsible for photoprotective antioxidants that scavenge reactive oxygen species (ROS) formed either under enhanced intracellular UVBR stress (Beardall et al., 2009) or extracellularly by photochemical reactions between UVBR, CDOM, and DOC (Kieber, 2000; Mopper and Kieber, 2002). Moreover, nitrogen limitation reduces the repair cycle of the protein D1 in photosystem II, increasing the effects of UVBR on photosynthesis (Bouchard et al., 2006, 2008). Phosphorus limitation may also increase phytoplankton sensitivity to UVBR stress (Aubriot et al., 2004) and limit repair processes which are energy dependent and thus need adenosine triphosphate (ATP; Beardall et al., 2009). Although large areas of the world's oceans, such as high nutrient–low chlorophyll (HNLC) regions, are iron (Fe) limited, almost no studies have investigated the potential effects of Fe limitation on UVBR effects on phytoplankton. Only one study reported that the diatom *Chaetoceros brevis* from the Southern Ocean was less sensitive to UVBR stress under Fe limitation due to elevated concentrations of ROS scavengers in Fe-limited cells (van de Poll et al., 2005). Therefore, enhanced UVBR may have large consequences on the food web functioning of oligotrophic oceanic areas.

15.2.4 Effects of UVBR on benthic and coral reefs food web components

In this section, we refer to benthic ecosystems worldwide, with focuses on shallow benthic ecosystems including macroalgae canopies, sea grass meadows, and coral reefs. According to Wahl et al. (2004), UVBR effects on benthic organisms should mainly concern shallow depths (because of the relatively low penetration of UVBR within seawater) and sessile organisms that do not have the capacity to escape exposure to this radiation. Among benthic organisms, juveniles are particularly at threat and more sensitive to UVBR because they have a higher metabolism, they are less pigmented, and have no or thinner shells compared to adults (Wiencke et al., 2000). In addition, differential UVBR sensitivity may exist among benthic organisms' larvae as some recruits may come from shallow or deep waters and possess or lack MAAs (Adams and Shick, 2001; Nahon et al., 2009). Wahl et al. (2004) observed a modification and a reduction of benthic organism diversity and biomass under UVBR stress in 10 different coastal regions of the world's oceans, and this effect mainly concerned the early succession stages of the benthic community. In agreement with these results, Nozais et al. (1999) observed negative UVBR effects on the young stages of harpacticoid copepods. Other studies found no strong negative UVBR effects on benthic communities, regardless of communities' ages (e.g. Fricke et al., 2011), showing the high variability in the response of benthic organisms and ecosystems to UVBR stress. Once the benthic community is developed, organisms have mechanisms to avoid UVBR stress (Molis and Wahl, 2009). These mechanisms range from the synthesis (Cockell and Knowland, 1999) or absorption by feeding (Nahon et al., 2012) of photoprotective molecules to photoprotection by the development of a canopy and self-shading mechanisms (Wahl et al., 2004).

Some key benthic organisms are very sensitive to UVBR, such as the early growing stages of macroalgae and sea urchins (Wiencke et al., 2000; Adams and Shick, 2001). Macroalgae are a major source of biomass in rocky shores and continental shelves ecosystems and serve as nurseries for larval and juvenile marine organisms. However, macroalgae are sensitive to UVBR and there exists a range of sensitivity among species and habitats (see Häder et al., 2007 for a review). Deeper macroalgae and younger specimens (zoospores, zygotes, gametes, and young germlings) are generally more sensitive to UVBR than intertidal macroalgae and adults (e.g. Bischof et al., 2002a, b; Roleda et al., 2005). Contrarily to subtidal species, intertidal macroalgae can produce UV-absorbing compounds. In addition, red algae produce more MAAs than brown or green algae (Xu and Gao, 2007; Lee and Shiu, 2009) and polar macroalgae are more sensitive to UVBR than their equatorial counterparts (van de Poll et al., 2003). Sea grass meadows, which cover large coastal areas and largely contribute to the production of biomass, are also sensitive to UVBR, as their photosynthetic activity decreases when exposed to UVBR (Figueroa et al., 2002). Interestingly, epiphytes growing on sea grasses strongly absorb UVBR and partly protect sea grasses from the detrimental UVBR effects (Brandt and Koch, 2003).

The sensitivity of sea urchins' eggs and larvae to UVBR stress may also have further consequences on benthic ecosystems because of the important role that they play in controlling algal assemblages (e.g. Humphries et al., 2014). UVBR may also negatively affect benthic organisms through the transfer of UVBR-altered organic detritus (i.e. with a lower nutritional value) from the upper water column as highlighted by Nahon et al. (2011); organic matter settlement is an essential source of food for benthic communities (Quijón et al., 2008). Roux et al. (2002) also showed that UVBR stress had trophic effects on a microbenthic community of the St Lawrence Estuary, with negative effects on diatoms' photosynthesis but positive effects on grazers. Therefore, it seems that UVBR plays mostly a role in shaping the structure of benthic communities through a direct impact of UVBR on organisms, more particularly during early development stages, or through trophic interactions, by the transfer of UVBR-altered organic detritus from the upper water column.

Among benthic ecosystems, coral reef ecosystems develop in shallow waters at tropical latitudes (Fig. 15.4), and hence are exposed to naturally high levels of UVBR. In addition, coral reefs are generally found within oligotrophic waters which allow a deep penetration of UVBR (i.e. > 20 m; Shick et al., 1996; Banaszak et al., 1998), although coastal reefs may be significantly influenced by terrigenous particulate and dissolved materials and CDOM inputs as well as upwelling events (Zepp et al., 2008), which would limit the penetration of UVBR within seawater. For a thorough review of direct UVBR effects on corals and coral reef organisms, we refer our readers to the excellent review of Banaszak and Lesser (2009). In terms of ecosystem function, corals live in symbiosis with dinoflagellates (genus *Symbiodinium*) that perform primary production and benefit corals (Trench, 1979). In turn, the endosymbionts live under the protection of their hosts, which also provide them with nutrients (Lesser, 2004). UVBR may have negative effects on these fragile coral-based ecosystems. For example, UVBR may reduce the photosynthesis of symbionts (Ferrier-Pagès et al., 2007) with negative effects on the rest of the food web. In addition, enhanced UVBR in the future may also have negative effects on coral reefs through cascading negative trophic effects if keystone species such as the sea urchin *Diadema antillarum* decline. Indeed, *D. antillarum* plays an important role in controlling algal

Figure 15.4 Elkhorn coral on a fringing reef on the coast of Jamaica.

development in the Caribbean's corals reefs. In the 1980s, 80% of the coral cover disappeared in the Caribbean's reefs because of a decline in the *D. antillarum* population (Carpenter, 1988), although other factors may also have been responsible for this coral reef decline (e.g. hurricanes, eutrophication, thermal stress, diseases, or sedimentation; Hughes, 1994). Finally, as discussed in Section 15.3.1, high UVBR, together with elevated temperature, may also disrupt the coral–dinoflagellates symbiosis, leading to coral bleaching with negative consequences on coral reef ecosystems (Lesser, 2004).

15.3 UVBR impacts and other major global change processes

Many other anthropogenic stressors can simultaneously occur along with UVBR stress, acting synergistically or antagonistically with it on marine ecosystems. Of these potential stressors, major global change issues (i.e. rising temperature and ocean acidification due to increasing CO_2) are of particular interest, and their interactions with enhanced UVBR will be the focus of the following two sections.

15.3.1 Simultaneous effects of UVBR and rising temperatures on marine ecosystems

One of the main outcomes of global change is increasing air temperature and the consequent increase of water temperatures. In fact, perspective and review articles focusing on the effects of UVBR call attention to new rising challenges concerning the evolution of surface UVBR in a changing climate (Williamson et al., 2014). Biological metabolic reactions are influenced positively by temperature and it is predicted that higher water temperatures will lead to higher primary production and organisms' growth rates, but should also enhance photorepair mechanisms (Bouchard et al., 2006). A number of laboratory-based studies have assessed the combined effects of rising temperatures and UVBR. Halac et al. (2010) observed lower UVBR-induced photoinhibition in two diatom species when they were exposed to a 5 °C increase in water temperature, and Sobrino and Neale (2007) observed that UVR-induced sensitivity of the diatom *Thalassiosira pseudonana* was highly affected by temperature. Roos and Vincent (1998) found that the UVR-induced inhibition of growth of Antarctic cyanobacteria increased linearly with decreasing temperature. However, these responses are highly species specific as other studies reported stronger UVBR effects under high temperatures due to the production of superoxide radicals (e.g. Lesser, 1996).

In addition, indirect interactions between global warming and enhanced UVBR stress may occur. For example, in polar regions, global change has already led to large decreases in the extent and duration of sea ice. This is especially true for the Arctic where sea ice has decreased dramatically in the last decades (Lindsay and Zhang, 2005; Stroeve et al., 2007) and continues to decline. This seems not to be the case in the Southern Ocean, where sea ice extent has only decreased in the waters west of the Antarctic Peninsula (Stammerjohn et al., 2008; Parkinson and Cavalieri, 2012). However, future projections of climate change predict that sea-ice cover will also decrease in other Antarctic seas in the next decades (Christensen et al., 2007). This decrease in the extent and duration of sea ice has led to an increase in primary production and the overall plankton biomass in the Arctic Ocean (Arrigo et al., 2008) and is predicted to have the same effect in the Southern Ocean in the future (Arrigo and Thomas, 2004; Sarmiento et al., 2004) due to longer growth seasons. However, sea-ice retreat may happen earlier in spring in the Southern Ocean and could potentially expose planktonic communities to enhanced UVBR stress at the peak of the ozone depletion as described by Moreau et al. (2010) for the western Antarctic Peninsula region.

Rising temperatures may also enhance water column stratification, exposing planktonic communities to higher UVBR stress (Sarmiento et al., 2004; Häder et al., 2007) and reducing primary productivity (Behrenfeld et al., 2006). Vertical mixing plays an essential role in photorepair processes by reducing the exposure of planktonic cells to high surface UVBR intensities (Neale et al., 2003). For example, a decrease in phytoplankton abundance and Chl-*a* concentration across the world's oceans over the last decades was noticed by Gregg and Conkwright (2002) and Gregg et al. (2003), while Behrenfeld et al. (2006) and Boyce et al. (2010) noticed a decrease in primary production in the world's oceans over, respectively, the past decade and the past century. The authors suggest that this decrease in phytoplankton biomass was due to enhanced stratification caused by global change. It is possible that the concomitant increased exposition to UVBR stress also played a role on this phytoplankton biomass decrease. In this sense, Halpern et al. (2008) suggested that UVBR stress ranks second, after increasing temperature, in

terms of anthropogenically induced impacts on surface ocean ecosystems. Hence, UVBR may also have played direct or indirect roles on the significant losses in marine biota that have been observed in the last decades (corals: Gardner et al., 2003; predatory fish: Myers and Worm, 2003; seagrasses: Waycott et al., 2009).

Enhanced stratification will also reduce nutrient availability and therefore potentially limit primary production but also repair processes (Behrenfeld et al., 2006). The expected slowdown of the Atlantic meridional overturning circulation, part of the global thermohaline circulation (Gregory et al., 2005), which may result from global change, may further reduce the supply of nutrients to oceanic surface waters. Lower nutrient concentrations should also favour smaller cells (Raven, 1998) such as *Prochlorococcus* spp. or coccolithophores (e.g. *Emiliania huxleyi*) which are more sensitive to UVBR than larger cells. This enhanced stratification and lower nutrient hypothesis is true for open-water systems whereas coastal systems are likely to see more nutrient supply due to increased frequency of storms and hence upwelling events (Bakun, 1990).

Global warming may also lead to changes in the composition and structure of planktonic communities and hence modify their resistance to UVBR stress. Because of regional warming and the observed decrease in sea-ice cover, Montes-Hugo et al. (2008) have already noticed a reduction in the size of phytoplankton together with a general decrease in Chl-*a* in the waters west of the Antarctic Peninsula. Changes in the planktonic community structure have also been noticed in the Arctic Ocean, where Li et al. (2009) also observed a reduction in the size of phytoplankton as the Arctic Ocean receives more heat and freshwater due to regional warming. In addition, as proposed by Riebesell et al. (2009), bacteria have higher Q_{10} values (Pomeroy and Wiebe, 2001) than phytoplankton and may benefit more from an increase in water temperature along with global warming. Changes have also been observed at higher trophic levels (i.e. zooplankton) with a gradual borealization (i.e. a displacement towards higher latitudes) of the copepod species dominating the northern Atlantic and Arctic Oceans (from south to north: *Calanus finmarchicus*, *Calanus hyperboreus*, and *C. glacialis*) with negative trophic consequences (i.e. reduced nutritional quality; Falk-Petersen et al., 2007; Søreide et al., 2010). Overall, a modification of the marine food webs towards microbial food webs may have significant consequences in terms of carbon transfer to higher trophic levels and export to the deep ocean. However, Vidussi et al. (2011) found no synergistic effects of increased

UVBR and temperature on the planktonic community of a productive coastal Mediterranean site because of the overall low effect of UVBR on this community, suggesting that such communities are resistant to UVBR and can have complex responses to these stressors.

Finally, global warming may also act synergistically or antagonistically with increased UVBR on benthic communities. For example, Fredersdorf et al. (2009) observed that higher temperatures augmented the damage that UVBR caused to the early life stages of the Arctic kelp *Alaria esculenta*. In addition, changing sea-ice and snow cover due to increasing temperature can drive polar shallow sea-bed ecosystems to abrupt transformations, as shown in Antarctica by Clark et al. (2013). Coral bleaching (i.e. a paling or whitening of corals), which corresponds to the breakdown of the symbiotic relationship between corals and dinoflagellates with possible dramatic consequences on coral reef ecosystems may be enhanced by increasing temperature (Lesser, 2004), together with enhanced UVBR stress as shown by Drohan et al. (2005). In fact, UVBR may itself cause coral bleaching under prolonged calm water conditions known as doldrums, although this has only been shown once, by Gleason and Wellington (1993). Coral bleaching is threatening coral reefs, as in 1998 when a single warming event related to El Niño was responsible for a loss of 16% of the world's corals (Wilkinson, 2000).

15.3.2 Simultaneous effects of enhanced UVBR and CO_2 on marine ecosystems

The world's oceans have absorbed around half of the anthropogenically produced carbon since the beginning of the industrial revolution (Sabine et al., 2004). In the future, atmospheric CO_2 is further predicted to increase up to 700 to 1000 ppm by 2100 with a coincident increase in oceanic pCO_2 and a reduction of pH for oceanic waters by 0.3 to 0.5 pH units (Arrigo and Van Dijken, 2007; Riebesell et al., 2009).

This is a major concern for the oceans as elevated CO_2 concentrations and, therefore, ocean acidification may have both detrimental and positive effects on the physiology of marine organisms (Sobrino et al., 2008; Gao et al., 2009). This particularly concerns calcifying organisms such as coccolithophores, calcified macroalgae, foraminifera, corals, molluscs, crustaceans, and echinoderms (Fabry et al., 2008). Ocean acidification may also change the composition of phytoplankton communities away from diatoms, as observed in the South China Sea by Gao et al. (2012), or deteriorate food quality with a reduction in fatty acids (FAs)

content and the ratio of polyunsaturated FAs to saturated FAs as observed by Rossoll et al. (2012) for algae grown under elevated CO_2 concentrations. On the contrary, Riebesell et al. (2007) observed enhanced primary production and carbon consumption by phytoplankton under elevated CO_2.

Gao et al. (2009) showed that ocean acidification enhanced UVBR sensitivity and reduced photosynthesis and coccolith thickness in the widespread coccolithophore, *Emiliania huxleyi*, and Gao et al. (2012) showed that the interaction of ocean acidification and enhanced UVBR decreased photosynthesis and growth of marine primary producers. Although UVBR and elevated CO_2 may both have detrimental effects on marine phytoplankton and may act synergistically on marine organisms, their metabolic targets are different (Beardall et al., 2009) such that their individual effects do not necessarily add. Beardall et al. (2002) observed that the combination of UVBR and elevated CO_2 stresses led to a higher carbon uptake by the marine chlorophyte *Dunaliella tertiolecta* due to the high CO_2 concentration effects on the CO_2 concentrating mechanisms (CCMs are used by algae to concentrate inorganic CO_2 to perform photosynthesis). However, this high carbon uptake was assimilated at a lower rate due to negative UVBR effects on photosystem II activity (i.e. inhibiting the synthesis of protein D1, thus slowing down the photosynthetic electron transport chain; Bouchard et al., 2006), leading to a higher internal cell CO_2 pool. It has been suggested that the interaction between UVBR and elevated CO_2 stresses is species dependent (Beardall and Raven, 2004; Sobrino et al., 2005) and may lead to a modification of the phytoplankton composition as observed by Gao et al. (2012). However, as a whole, it seems that the interaction between UVBR and elevated CO_2 stresses will lead to a higher UVBR sensitivity of marine organisms (Beardall et al., 2009), which is a concern in the context of global change.

15.4 Conclusion and perspectives

UVBR has deleterious effects on the food web components of all marine ecosystems. Almost all the pelagic systems studied in this chapter seem to respond more or less similarly to increased UVBR levels, with a change in the structure of the ecosystem. Polar (Davidson and Belbin, 2002; Wängberg et al., 2008) and temperate (Burkhardt and Riebesell, 1997; Mostajir et al., 1999; Ferreyra et al., 2006) herbivorous food webs tend to evolve towards microbial food webs and, therefore, towards less productive systems with less food availability for higher trophic levels. In addition, global warming also tends to drive planktonic systems towards the dominance of smaller cells (Montes-Hugo et al., 2008; Li et al., 2009; Moreau et al., 2014) leading to the establishment of microbial food webs. One may, therefore, argue that stresses on the planktonic communities of marine systems tend to change the structure of these systems towards less productive planktonic communities and engender a trophic impasse for matter (Fig. 15.3b). It should be noted, however, that significant UVBR effects were not always observed for the components of pelagic food webs (Wängberg et al., 2001; Vidussi et al., 2011; Lionard et al., 2012) and that UVBR may have greater effects on the DNA of small phytoplankton than on larger phytoplankton (Buma et al., 2001b; Helbling et al., 2001), probably because larger phytoplankton synthesize more photoprotective pigments and because the path through the cytoplasm before UVBR reaches significant cell targets is longer (Helbling et al., 1996).

UVBR also has deleterious effects on benthic communities. These effects are strongest for early life stages and subtidal species. Key benthic organisms such as macroalgae, sea grasses and sea urchins may be negatively impacted by UVBR with possible consequences for the rest of the ecosystems (Wiencke et al., 2000; Adams and Shick, 2001). Among the benthos, intertidal organisms seem to be best adapted to UVBR as well as to other stresses (elevated temperature, desiccation, etc.) while corals are particularly threatened by UVBR stress, with possible consequences for whole coral reef ecosystems. Overall, UVBR seems to shape the structure of benthic organisms during early growing stages.

The main differences in the responses of all the marine ecosystems seem to reside in the timing of observations or experimentations, the duration and dose of UVBR that reaches the water column, and the ratio of UVBR/UVAR + PAR. As described in this chapter, the intensity of UVBR that reaches the water column and the UVBR/UVAR + PAR ratios depend on the overlying atmospheric ozone concentration, the presence of clouds and aerosols, sea surface roughness, and the optical properties of the water column. The timing and duration of exposure to UVBR depend mainly on vertical mixing, which can reduce the exposure of planktonic cells to surface high UVBR intensities (Neale et al., 2003) but has no impact on benthic organisms.

In this chapter, we mainly considered the effects of UVBR on marine ecosystems. UVAR, however, may also play an important role in these environments. Firstly, although UVBR photons contain more energy than UVAR ones, UVAR comprises a much larger band

of UVR than UVBR and may induce a stronger biological response than UVBR only (Karentz and Bosch, 2001). For example, UVAR can lead to more bacterial death than UVBR (Helbling et al., 1995) and UVAR and PAR are responsible for some photoinhibition in Antarctic phytoplankton (Holm-Hansen et al., 1997). Although UVAR participates in photorepair processes, UVAR induces photochemical reactions both in seawater and in cellular fluids that result in the production of ROS (Karentz and Bosch, 2001). Hence, the effects of UVAR and PAR on marine ecosystems should also be considered, especially in the context of future global warming and the expected increase in water column stratification, which may increase the degree of exposure of marine ecosystems to the whole irradiance spectra (PAR, UVAR, and UVBR).

In the future, meta-analyses should be used to address the effects of UVBR on all marine biota. Using this type of analysis for UVBR effects on marine biota (from viruses to fish), Llabrés et al. (2012) pointed that southern hemisphere organisms are more resistant to UVBR stress than their northern hemisphere counterparts. They also observed that the resistance to UVBR among marine organisms of all marine systems seems to have increased in 1979–2009, since the beginning of the ozone layer thinning, suggesting that UVBR has acted as a selective driver among marine organisms over the last three decades, more particularly over the Southern Ocean, which has suffered the most ozone layer thinning. With the increase in the frequency and strength of ozone depletion events in the Arctic (Manney et al., 2011), together with the dramatic decrease in the Arctic Ocean sea ice extent and duration (Stroeve et al., 2007), it is probable that such selection will also occur among Arctic marine organisms. On the other hand, it is uncertain how the stratospheric ozone layer will evolve at mid and low latitudes in the next decades and thus how marine ecosystems will cope with UVBR stress.

References

Aarnos, H., Ylostalo, P., & Vahatalo, A.V. (2012). Seasonal phototransformation of dissolved organic matter to ammonium, dissolved inorganic carbon, and labile substrates supporting bacterial biomass across the Baltic Sea. *Journal of Geophysical Research*, **117**, G01004.

Abboudi, M., Jeffrey, W.H., Ghiglione, J.F., et al. (2008). Effects of photochemical transformations of dissolved organic matter on bacterial metabolism and diversity in three contrasting coastal sites in the Northwestern Mediterranean Sea during summer. *Microbial Ecology*, **55**, 344–357.

Adams, N.L. & Shick, J.M. (2001). Mycosporine-like amino acids prevent UVB-induced abnormalities during early development of the green sea urchin *Strongylocentrotus droebachiensis*. *Marine Biology*, **138**, 267–280.

Agawin, N.S.R., Duarte, C.M., & Agustí, S. (2000). Nutrient and temperature control of the contribution of picoplankton to phytoplankton biomass and production. *Limnology and Oceanography*, **45**, 591–600.

Agustí, S. & Llabrés, M. (2007). Solar radiation-induced mortality of marine pico-phytoplankton in the oligotrophic ocean. *Photochemistry and Photobiology*, **83**, 793–801.

Ajtić, J., Connor, B.J., Lawrence, B.N., et al. (2004). Dilution of the Antarctic ozone hole into southern midlatitudes, 1998–2000. *Journal of Geophysical Research D*, **109**(D17107), 17101–17109.

Arrigo, K.R., Lubin, D., van Dijken, G.L., et al. (2003). Impact of a deep ozone hole on Southern Ocean primary production. *Journal of Geophysical Research C*, **108**, 23–21.

Arrigo, K.R. & Thomas, D.N. (2004). Large scale importance of sea ice biology in the Southern Ocean. *Antarctic Science*, **16**, 471–486.

Arrigo, K.R. & van Dijken, G.L. (2007). Interannual variation in air-sea CO_2 flux in the Ross Sea, Antarctica: a model analysis. *Journal of Geophysical Research C*, **112**(C3), C03020.

Arrigo, K.R. & van Dijken, G.L. (2011). Secular trends in Arctic Ocean net primary production. *Journal of Geophysical Research C*, **116**(C9), CO9011.

Arrigo, K.R., van Dijken, G., & Pabi, S. (2008). Impact of a shrinking Arctic ice cover on marine primary production. *Geophysical Research Letters*, **35**, 1–6.

Aubriot, L., Conde, D., Bonilla, S., & Sommaruga, R. (2004). Phosphate uptake behavior of natural phytoplankton during exposure to solar ultraviolet radiation in a shallow coastal lagoon. *Marine Biology*, **144**, 623–631.

Azam, F., Fenchel, T., Field, J.G., et al. (1983). The ecological role of water-column microbes in the sea. *Marine Ecology Progress Series*, **10**, 257–263.

Bais, A.F., Tourpali, K., Kazantzidis, A., et al. (2011). Projections of UV radiation changes in the 21st century: impact of ozone recovery and cloud effects. *Atmospheric Chemistry and Physics*, **11**, 7533–7545.

Bakun, A. (1990). Global climate change and intensification of coastal ocean upwelling. *Science*, **247**, 198–201.

Balis, D.S., Amiridis, V., Zerefos, C., et al. (2004). Study of the effect of different type of aerosols on UV-B radiation from measurements during EARLINET. *Atmospheric Chemistry and Physics*, **4**, 307–321.

Banaszak, A.T. & Lesser, M.P. (2009). Effects of solar ultraviolet radiation on coral reef organisms. *Photochemical and Photobiological Sciences*, **8**, 1276–1294.

Banaszak, A.T., Lesser, M.P., Kuffner, I.B., & Ondrusek, M. (1998). Relationship between ultraviolet (UV) radiation and mycosporine-like amino acids (MAAs) in marine organisms. *Bulletin of Marine Science*, **63**, 617–628.

Beardall, J., Heraud, P., Roberts, S., et al. (2002). Effects of UV-B radiation on inorganic carbon acquisition by the marine microalga *Dunaliella tertiolecta* (Chlorophyceae). *Phycologia*, **41**, 268–272.

Beardall, J. & Raven, J.A. (2004). The potential effects of global climate change on microalgal photosynthesis, growth and ecology. *Phycologia*, **43**, 26–40.

Beardall, J., Sobrino, C., & Stojkovic, S. (2009). Interactions between the impacts of ultraviolet radiation, elevated CO_2, and nutrient limitation on marine primary producers. *Photochemical and Photobiological Sciences*, **8**, 1257–1265.

Behrenfeld, M.J., O'Malley, R.T., Siegel, D.A., et al. (2006). Climate-driven trends in contemporary ocean productivity. *Nature*, **444**, 752–755.

Bischof, K., Hanelt, D., Aguilera, J., et al. (2002a). Seasonal variation in ecophysiological patterns in macroalgae from an Arctic fjord: I. Sensitivity of photosynthesis to ultraviolet radiation. *Marine Biology Research*, **140**, 1097–1106.

Bischof, K., Peralta, G., Kraebs, G., et al. (2002b). Effects of solar UV-B radiation on canopy structure of Ulva communities from Southern Spain. *Journal of Experimental Botany*, **53**, 2411–2421.

Blachowiak-Samolyk, K., Søreide, J.E., Kwasniewski, S., et al. (2008). Hydrodynamic control of mesozooplankton abundance and biomass in northern Svalbard waters (79–81° N). *Deep Sea Research. Part II*, **55**, 2210–2224.

Blough, N.V., Zafiriou, O.C., & Bonilla, J. (1993). Optical absorption spectra of waters from the Orinoco River outflow: terrestrial input of colored organic matter to the Caribbean. *Journal of Geophysical Research*, **98**, 2271–2278.

Bouchard, J.N., Longhi, M.L., Roy, S., et al. (2008). Interaction of nitrogen status and UVB sensitivity in a temperate phytoplankton assemblage. *Journal of Experimental Marine Biology and Ecology*, **359**, 67–76.

Bouchard, J.N., Roy, S., & Campbell, D.A. (2006). UVB effects on the photosystem II-D1 protein of phytoplankton and natural phytoplankton communities. *Photochemistry and Photobiology*, **82**, 936–951.

Bouwman, A.F. (1998). Nitrogen oxides and tropical agriculture. *Nature*, **392**, 866–867.

Boyce, D.G., Lewis, M.R., & Worm, B. (2010). Global phytoplankton decline over the past century. *Nature*, **466**, 591–596.

Brandt, L.A. & Koch, E.W. (2003). Periphyton as a UV-B filter on seagrass leaves: a result of different transmittance in the UV-B and PAR ranges. *Aquatic Botany*, **76**, 317–327.

Browman, H.I., Rodriguez, C.A., Bélaud, F., et al. (2000). Impact of ultraviolet radiation on marine crustacean zooplankton and ichthyoplankton: a synthesis of results from the estuary and Gulf of St. Lawrence, Canada. *Marine Ecology Progress Series*, **199**, 293–311.

Buma, A.G.J., de Boer, M.K., & Boelen, P. (2001a). Depth distributions of DNA damage in Antarctic marine phyto- and bacterioplankton exposed to summertime UV radiation. *Journal of Phycology*, **37**, 200–208.

Buma, A.G.J., Helbling, E.W., de Boer, M.K., & Villafañe, V.E. (2001b). Patterns of DNA damage and photoinhibition in temperate South-Atlantic picophytoplankton exposed to solar ultraviolet radiation. *Journal of Photochemistry and Photobiology B*, **62**, 9–18.

Burkhardt, S. & Riebesell, U. (1997). CO_2 availability affects elemental composition (C:N:P) of the marine diatom *Skeletonema costatum*. *Marine Ecology Progress Series*, **155**, 67–76.

Carmack, E. & Wassmann, P. (2006). Food webs and physical-biological coupling on pan-Arctic shelves: unifying concepts and comprehensive perspectives. *Progress in Oceanography*, **71**, 446–477.

Carpenter, R.C. (1988). Mass mortality of a Caribbean sea urchin: immediate effects on community metabolism and other herbivores. *Proceedings of the National Academy of Sciences of the USA*, **85**, 511–514.

Casiccia, C., Zamorano, F., & Hernandez, A. (2008). Erythemal irradiance at the Magellan's region and Antarctic ozone hole 1999–2005. *Atmosfera*, **21**, 1–12.

Chatila, K., Dermers, S., Mostajir, B., et al. (1999). Bacterivory of a natural heterotrophic protozoan community exposed to different intensities of ultraviolet-B radiation. *Aquatic Microbial Ecology*, **20**, 59–74.

Christensen, J.H., Hewitson, B., Busuioc, A., et al. (2007). Regional climate projections. In: Solomon, S., Qin, D., Manning, M., et al. (eds), *Climate Change 2007: The Physical Science Basis. Contribution of Working Group I to the Fourth Assessment Report of the Intergovernmental Panel on Climate Change*, pp. 847–940. Cambridge University Press, Cambridge, United Kingdom and New York, NY, USA.

Chubarova, N.E. (2006). Role of tropospheric gases in the absorption of UV radiation. *Doklady Earth Sciences*, **407**, 294–297.

Chubarova, N.Y. (2009). Seasonal distribution of aerosol properties over Europe and their impact on UV irradiance. *Atmospheric Measurement Techniques*, **2**, 593–608.

Clark, G.F., Stark, J.S., Johnston, E.L., et al. (2013). Light-driven tipping points in polar ecosystems. *Global Change Biology*, **19**, 3749–3761.

Cockell, C.S. & Córdoba-Jabonero, C. (2004). Coupling of climate change and biotic UV exposure through changing snow–ice covers in terrestrial habitats. *Photochemistry and Photobiology*, **79**, 26–31.

Cockell, C.S. & Knowland, J. (1999). Ultraviolet radiation screening compounds. *Biological Reviews of the Cambridge Philosophical Society*, **74**, 311–345.

Cushing, D.H. (1989). A difference in structure between ecosystems in strongly stratified waters and in those that are only weakly stratified. *Journal of Plankton Research*, **11**, 1–13.

Cushman-Roisin, B. & Beckers, J.-M. (2011). *Introduction to Geophysical Fluid Dynamics*. Prentice-Hall, Waltham.

Dahms, H.U., Dobretsov, S., & Lee, J.S. (2011). Effects of UV radiation on marine ectotherms in polar regions. *Comparative Biochemistry and Physiology Part C*, **153**, 363–371.

Davidson, A. & Belbin, L. (2002). Exposure of natural Antarctic marine microbial assemblages to ambient UV radiation: effects on the marine microbial community. *Aquatic Microbial Ecology*, **27**, 159–174.

Davidson, A.T. (1998). The impact of UVB radiation on marine plankton. *Mutation Research*, **422**, 119–129.

Davidson, A.T. & Marchant, H.J. (1994). The impact of ultraviolet radiation on phaeocystis and selected species of Antarctic marine diatoms. In: Weiler, C.S. & Penhale, P.A. (eds), *Ultraviolet Radiation in Antarctica: Measurement and Biological Effects*. Antarctic Research Series, pp. 160–187. American Geophysical Union, Washington, DC.

Davidson, A.T., Marchant, H.J., & De La Mare, W.K. (1996). Natural UVB exposure changes the species composition of Antarctic phytoplankton in mixed culture. *Aquatic Microbial Ecology*, **10**, 299–305.

Davidson, A.T. & van der Heijden, A. (2000). Exposure of natural Antarctic marine microbial assemblages to ambient UV radiation: effects on bacterioplankton. *Aquatic Microbial Ecology*, **21**, 257–264.

Deal, C., Jin, M., Elliott, S., et al. (2011). Large-scale modeling of primary production and ice algal biomass within arctic sea ice in 1992. *Journal of Geophysical Research C*, **116**, C07004.

de Mora, S., Demers, S., & Vernet, M. (2000). *The Effects of UV Radiation in the Marine Environment*. Cambridge University Press, Cambridge, UK.

Dethlefsen, V., von Westernhagen, H., Tüg, H., et al. (2001). Influence of solar ultraviolet-B on pelagic fish embryos: osmolality, mortality and viable hatch. *Helgoland Marine Research*, **55**, 45–55.

Diaz, S.B., Morrow, J.H., & Booth, C.R. (2000). UV physics and optics. In: de Mora, S., Demers, S., & Vernet, M. (eds), *The Effects of UV Radiation in the Marine Environment*, pp. 35–71. Cambridge University Press, Cambridge, UK.

Drohan, A.F., Thoney, D.A., & Baker, A.C. (2005). Synergistic effect of high temperature and ultraviolet-B radiation on the gorgonian *Eunicea tourneforti* (Octocorallia: Alcyonacea: Plexauridae). *Bulletin of Marine Science*, **77**, 257–266.

Ducklow, H.W., Fraser, W., Karl, D.M., et al. (2006). Water-column processes in the West Antarctic Peninsula and the Ross Sea: interannual variations and foodweb structure. *Deep Sea Research Part II*, **53**, 834–852.

Dugdale, R.C. & Goering, J.J. (1967). Uptake of new and regenerated forms of nitrogen in primary productivity. *Limnology and Oceanography*, **12**, 196–206.

Eppley, R.W. & Peterson, B.J. (1979). Particulate organic matter flux and planktonic new production in the deep ocean. *Nature*, **282**, 677–680.

Fabry, V.J., Seibel, B.A., Feely, R.A., & Orr, J.C. (2008). Impacts of ocean acidification on marine fauna and ecosystem processes. *ICES Journal of Marine Science*, **65**, 414–432.

Falk-Petersen, S., Pavlov, V., Timofeev, S., & Sargent, J.R. (2007). Climate variability and possible effects on arctic food chains: the role of *Calanus*. *Arctic Alpine Ecosystems and People in a Changing Environment*, **2**, 147–166.

Fenchel, T. (2008). The microbial loop—25 years later. *Journal of Experimental Marine Biology and Ecology*, **366**, 99–103.

Ferreyra, G., Schloss, I., & Demers, S. (2004). Rôle de la glace saisonnière dans la dynamique de l'écosystème marin de l'Antarctique: impact potentiel du changement climatique global. *VertigO – La revue en sciences de l'environnement.*, **5**, 1–11.

Ferreyra, G.A., Mostajir, B., Schloss, I.R., et al, (2006). Ultraviolet-B radiation effects on the structure and function of lower trophic levels of the marine planktonic food web. *Photochemistry and Photobiology*, **82**, 887–897.

Ferrier-Pagès, C., Richard, C., Forcioli, D., et al. (2007). Effects of temperature a. d UV radiation increases on the photosynthetic efficiency in four scleractinian coral species. *Biological Bulletin*, **213**, 76–87.

Field, C.B., Behrenfeld, M.J., Randerson, J.T., & Falkowski, P. (1998). Primary production of the biosphere: integrating terrestrial and oceanic components. *Science*, **281**, 237–240.

Figueroa, F.L., Jiménez, C., Viñegla, B., et al. (2002). Effects of solar UV radiation on photosynthesis of the marine angiosperm *Posidonia oceanica* from southern Spain. *Marine Ecology Progress Series*, 230, 59–70.

Fioletov, V.E., Kerr, J.B., Wardle, D.I., et al. (1997). Long-term ozone decline over the Canadian Arctic to early 1997 from ground-based and balloon observations. *Geophysical Research Letters*, **24**, 2705–2708.

Flores, H., Atkinson, A., Kawaguchi, S., et al. (2012). Impact of climate change on Antarctic krill. *Marine Ecology Progress Series*, 458, 1–19.

Frederick, J.E. & Lubin, D. (1994). Solar ultraviolet radiation at Palmer Station, Antarctica. In: Weiler, C.S. & Penhale, P.A. (eds), *Ultraviolet Radiation in Antarctica. Measurements and Biological Effects.* Antarctic Research Series, pp. 43–52. American Geophysical Union, Washington, DC.

Fredersdorf, J., Müller, R., Becker, S., et al. (2009). Interactive effects of radiation, temperature and salinity on different life history stages of the Arctic kelp *Alaria esculenta* (Phaeophyceae). *Oecologia*, **160**, 483–492.

Fricke, A., Molis, M., Wiencke, C., et al. (2011). Effects of UV radiation on the structure of Arctic macrobenthic communities. *Polar Biology*, **34**, 995–1009.

Fritz, J.J., Neale, P.J., Davis, R.F., & Peloquin, J.A. (2008). Response of Antarctic phytoplankton to solar UVR exposure: inhibition and recovery of photosynthesis in coastal and pelagic assemblages. *Marine Ecology Progress Series*, **365**, 1–16.

Gao, K., Ruan, Z., Villafañe, V.E., et al. (2009). Ocean acidification exacerbates the effect of UV radiation on the calcifying phytoplankter *Emiliania huxleyi*. *Limnology and Oceanography*, **54**, 1855–1862.

Gao, K., Xu, J., Gao, G., et al. (2012). Rising CO_2 and increased light exposure synergistically reduce marine primary productivity. *Nature Climate Change*, **2**, 519–523.

Gardner, T.A., Côté, I.M., Gill, J.A., et al. (2003). Long-term region-wide declines in Caribbean corals. *Science*, **301**, 958–960.

Gleason, D.F. & Wellington, G.M. (1993). Ultraviolet radiation and coral bleaching. *Nature*, **365**, 836–838.

Gregg, W.W., Conkright, M.E., Ginoux, P., et al. (2003). Ocean primary production and climate: global decadal changes. *Geophysical Research Letters*, **30**, 3–1 to 3–4.

Gregg, W.W. & Conkwright, M.E. (2002). Decadal changes in global ocean chlorophyll. *Geophysical Research Letters*, **29**, 20–21.

Gregory, J.M., Dixon, K.W., Stouffer, R.J., et al. (2005). A model intercomparison of changes in the Atlantic thermohaline circulation in response to increasing atmospheric CO 2 concentration. *Geophysical Research Letters*, **32**, 1–5.

Häder, D.-P., Kumar, H.D., Smith, R.C., & Worrest, R.C. (2003). Aquatic ecosystems: effects of solar ultraviolet radiation and interactions with other climatic change factors. *Photochemical and Photobiological Sciences*, **2**, 39–50.

Häder, D.P., Kumar, H.D., Smith, R.C., & Worrest, R.C. (2007). Effects of solar UV radiation on aquatic ecosystems and interactions with climate change. *Photochemical and Photobiological Sciences*, **6**, 267–285.

Häder, D.P. & Sinha, R.P. (2005). Solar ultraviolet radiation-induced DNA damage in aquatic organisms: potential environmental impact. *Mutation Research*, **571**, 221–233.

Hadjinicolaou, P. & Pyle, J.A. (2004). The impact of Arctic ozone depletion in Northern middle latitudes: interannual variability and dynamical control. *Journal of Atmospheric Chemistry*, **47**, 25–43.

Halac, S.R., Villafañe, V.E., & Helbling, E.W. (2010). Temperature benefits the photosynthetic performance of the diatoms *Chaetoceros gracilis* and *Thalassiosira weissflogii* when exposed to UVR. *Journal of Photochemistry and Photobiology B*, **101**, 196–205.

Halpern, B.S., Walbridge, S., Selkoe, K.A., et al. (2008). A global map of human impact on marine ecosystems. *Science*, **319**, 948–952.

Hargreaves, B.R. (2003). Water column optics and penetration of UVR. In: Helbling, E.W. & Zagarese, H.E. (eds), *UV Effects in Aquatic Organisms and Ecosystems*, pp. 59–105. Royal Society of Chemistry, Cambridge, UK.

Harris, N.R.P., Kyrä, E., Staehelin, J., et al. (2008). Ozone trends at northern mid- and high latitudes—a European perspective. *Annales Geophysicae*, **26**, 1207–1220.

Helbling, E.W., Buma, A.G.J., de Boer, M.K., & Villafañe, V.E. (2001). *In situ* impact of solar ultraviolet radiation on photosynthesis and DNA in temperate marine phytoplankton. *Marine Ecology Progress Series*, **211**, 43–49.

Helbling, E.W., Chalker, B.E., Dunlap, W.C., et al. (1996). Photoacclimation of Antarctic marine diatoms to solar ultraviolet radiation. *Journal of Experimental Marine Biology and Ecology*, **204**, 85–101.

Helbling, E.W., Marguet, E.R., Villafane, V.E., & Holm-Hansen, O. (1995). Bacterioplankton viability in Antarctic waters as affected by solar ultraviolet radiation. *Marine Ecology Progress Series*, **126**, 293–298.

Helbling, E.W. & Villafañe, V.E. (2002). UV radiation effects on phytoplankton primary production: a comparison between Arctic and Antarctic marine ecosystems. In: Hessen, D. (ed.), *UV Radiation and Arctic Ecosystems*. Ecological Studies, pp. 203–226. Springer, Berlin, Heidelberg.

Helbling, E.W. & Zagarese, H. (2003). *UV Effects in Aquatic Organisms and Ecosystems*. Royal Society of Chemistry, Cambridge.

Hessen, D.O., Frigstad, H., Færøvig, P.J., et al. (2012). UV radiation and its effects on P-uptake in arctic diatoms. *Journal of Experimental Marine Biology and Ecology*, **411**, 45–51.

Holm-Hansen, O., Villafañe, V., & Helbling, E.W. (1997). Effects of solar ultraviolet radiation on primary production in Antarctic waters. In: Battaglia, B., Valencia, J., & Walton, D.W.H. (eds), *Antarctic Communities: Species, Structure and Survival*, pp. 375–380. Cambridge University Press, Cambridge.

Hughes, T.P. (1994). Catastrophes, phase shifts, and large-scale degradation of a Caribbean coral reef. *Science*, **265**, 1547–1551.

Humphries, A.T., McClanahan, T.R., & McQuaid, C.D. (2014). Differential impacts of coral reef herbivores on algal succession in Kenya. *Marine Ecology Progress Series*, **504**, 119–132.

Hylander, S. & Hansson, L.A. (2010). Vertical migration mitigates UV effects on zooplankton community composition. *Journal of Plankton Research*, **32**, 971–980.

IPCC (2005). *Safeguarding the Ozone Layer and the Global Climate System: Issues Related to Hydrofluorocarbons and Perfluorocarbons*. IPCC/TEAP Special report: Summary for policymakers. Intergovernmental Panel on Climate Change, Geneva.

Jarman, S., Elliott, N., Nicol, S., et al. (1999). The base composition of the krill genome and its potential susceptibility to damage by UV-B. *Antarctic Science*, **11**, 23–26.

Jennings, S., Kaiser, M.J., & Reynolds, J.D. (2001). *Marine Fisheries Ecology*. Blackwell Science Ltd, Oxford, UK.

Karentz, D. & Bosch, I. (2001). Influence of ozone-related increases in ultraviolet radiation on Antarctic marine organisms. *American Zoologist*, **41**, 3–16.

Kieber, D.J. (2000). Photochemical production of biological substrates. In: Mora, S.D., Demers, S., & Vernet, M. (eds), *The Effects of UV Radiation in the Marine Environment*, pp. 130–148. Cambridge University Press, Cambridge.

Knudsen, B.M., Harris, N.R.P., Andersen, S.B., et al. (2004). Extrapolating future Arctic ozone losses. *Atmospheric Chemistry and Physics*, **4**, 1849–1856.

Landry, M.R., Ohman, M.D., Goericke, R., et al. (2012). Pelagic community responses to a deep-water front in the California Current Ecosystem: overview of the A-Front Study. *Journal of Plankton Research*, **34**, 739–748.

Lee, A.M., Jones, R.L., Kilbane-Dawe, I., & Pyle, J.A. (2002). Diagnosing ozone loss in the extratropical lower stratosphere. *Journal of Geophysical Research D*, **107**, ACH 3–1–ACH 3–12.

Lee, T.M. & Shiu, C.T. (2009). Implications of mycosporine-like amino acid and antioxidant defenses in UV-B radiation tolerance for the algae species *Ptercladiella capillacea* and *Gelidium amansii*. *Marine Environmental Research*, **67**, 8–16.

Legendre, L. & Rassoulzadegan, F. (1995). Plankton and nutrient dynamics in marine waters. *Ophelia*, **41**, 153–172.

Lesser, M.P. (1996). Acclimation of phytoplankton to UV-B radiation: oxidative stress and photoinhibition of photo-synthesis are not prevented by UV-absorbing compounds in the dinoflagellate *Prorocentrum micans*. *Marine Ecology Progress Series*, **132**, 287–297.

Lesser, M.P. (2004). Experimental biology of coral reef ecosystems. *Journal of Experimental Marine Biology and Ecology*, **300**, 217–252.

Lesser, M.P., Lamare, D., & Barker, M.F. (2004). Transmission of ultraviolet radiation through the Antarctic annual sea ice and its biological effects on sea urchin embryos. *Limnology and Oceanography*, **49**, 1957–1963.

Leu, E., Søreide, J.E., Hessen, D.O., et al. (2011). Consequences of changing sea-ice cover for primary and secondary producers in the European Arctic shelf seas: timing, quantity, and quality. *Progress in Oceanography*, **90**, 18–32.

Leu, E., Wängberg, S.Å., Wulff, A., et al. (2006). Effects of changes in ambient PAR and UV radiation on the nutritional quality of an Arctic diatom (*Thalassiosira antarctica* var. *borealis*). *Journal of Experimental Marine Biology and Ecology*, **337**, 65–81.

Leu, E., Wiktor, J., Søreide, J.E., et al. (2010). Increased irradiance reduces food quality of sea ice algae. *Marine Ecology Progress Series*, **411**, 49–60.

Li, W.K.W., McLaughlin, F.A., Lovejoy, C., & Carmack, E.C. (2009). Smallest algae thrive as the Arctic Ocean freshens. *Science*, **326**, 539.

Lindsay, R.W. & Zhang, J. (2005). The thinning of Arctic sea ice, 1988–2003: have we passed a tipping point? *Journal of Climate*, **18**, 4879–4894.

Lionard, M., Roy, S., Tremblay-Létourneau, M., & Ferreyra, G.A. (2012). Combined effects of increased UV-B and temperature on the pigment-determined marine phytoplankton community of the St. Lawrence Estuary. *Marine Ecology Progress Series*, **445**, 219–234.

Llabrés, M. & Agustí, S. (2006). Picophytoplankton cell death induced by UV radiation: evidence for oceanic Atlantic communities. *Limnology and Oceanography*, **51**, 21–29.

Llabrés, M., Agustí, S., Alonso-Laita, P., & Herndl, G.J. (2010). *Synechococcus* and *Prochlorococcus* cell death induced by UV radiation and the penetration of lethal UVR in the Mediterranean Sea. *Marine Ecology Progress Series*, **399**, 27–37.

Llabrés, M., Agustí, S., Fernández, M., et al. (2012). Impact of elevated UVB radiation on marine biota: a meta-analysis. *Global Ecology and Biogeography*, **22**(1), 131–144.

Mann, K.H. & Lazier, J.R.N. (1991). *Dynamics of Marine Ecosystems: Biological-Physical Interactions in the Oceans*. Blackwell Scientific Publications, Cambridge, MA.

Manney, G.L., Santee, M.L., Rex, M., et al. (2011). Unprecedented Arctic ozone loss in 2011. *Nature*, **478**, 469–475.

McKenzie, R.L., Aucamp, P.J., Bais, A.F., et al. (2007). Changes in biologically-active ultraviolet radiation reaching the Earth's surface. *Photochemical and Photobiological Sciences*, **6**, 218–231.

Michaels, A.F. & Silver, M.W. (1988). Primary production, sinking fluxes and the microbial food web. *Deep Sea Research Part I*, **35**, 473–490.

Molis, M. & Wahl, M. (2009). Comparison of the impacts of consumers, ambient UV, and future UVB irradiance on mid-latitudinal macroepibenthic assemblages. *Global Change Biology*, **15**, 1833–1845.

Montes-Hugo, M.A., Vernet, M., Martinson, D., et al. (2008). Variability on phytoplankton size structure in the western Antarctic Peninsula (1997–2006). *Deep Sea Research Part II*, **55**, 2106–2117.

Mopper, K. & Kieber, D.J. (2002). Photochemistry and the cycling of carbon, sulfur, nitrogen and phosphorus. In: Hansell, D.A. & Carlson, C. (eds), *Biogeochemistry of Marine Dissolved Organic Matter*, pp. 455–507. Academic Press, New York.

Moreau, S., Mostajir, B., Almandoz, G.O., et al. (2014). Effects of enhanced temperature and ultraviolet B radiation on a natural plankton community of the Beagle Channel (southern Argentina): a mesocosm study. *Aquatic Microbial Ecology*, **72**, 155–173.

Moreau, S., Mostajir, B., Bélanger, S., et al. (2015). Climate change enhances primary production in the western Antarctic Peninsula. *Global Change Biology*, **21**(6), 2191–2205.

Moreau, S., G. A. Ferreyra, B. Mercier, K. Lemarchand, M. Lionard, S. Roy, B. Mostajir, S. Roy, B. v. Hardenberg and S. Demers (2010). Variability of the microbial community in the Western Antarctic Peninsula from late fall to spring during a low-ice cover year. *Polar Biology* **33**(12): 1599–1614

Mostajir, B., Demers, S., De Mora, S., et al. (1999). Experimental test of the effect of ultraviolet-B radiation in a planktonic community. *Limnology and Oceanography*, **44**, 586–596.

Mostajir, B., Demers, S., de Mora, S., et al. (2000). Implications of UV radiation for the food web structure and consequences on carbon flow. In: de Mora, S., Demers, S., & Vernet, M. (eds), *The Effects of UV Radiation in the Marine Environment*, pp. 310–320. Cambridge University Press, Cambridge.

Müller-Navarra, D.C. (2008). Food web paradigms: the biochemical view on trophic interactions. *International Review of Hydrobiology*, **93**, 489–505.

Myers, R.A. & Worm, B. (2003). Rapid worldwide depletion of predatory fish communities. *Nature*, **423**, 280–283.

Naganobu, M., Kutsuwada, K., Sasai, Y., et al. (1999). Relationships between Antarctic krill (*Euphausia superba*) variability and westerly fluctuations and ozone depletion in the Antarctic Peninsula area. *Journal of Geophysical Research C*, **104**, 20651–20665.

Nahon, S., Castro Porras, V.A., Pruski, A.M., & Charles, F. (2009). Sensitivity to UV radiation in early life stages of the Mediterranean sea urchin *Sphaerechinus granularis* (Lamarck). *Science of the Total Environment*, **407**, 1892–1900.

Nahon, S., Nozais, C., Delamare-Deboutteville, J., et al. (2012). Trophic relationships and UV-absorbing compounds in a Mediterranean medio-littoral rocky shore community. *Journal of Experimental Marine Biology and Ecology*, **424–425**, 59–65.

Nahon, S., Pruski, A.M., Duchêne, J.C., et al. (2011). Can UV radiation affect benthic deposit-feeders through biochemical alteration of food resources? An experimental study with juveniles of the benthic polychaete *Eupolymnia nebulosa*. *Marine Environmental Research*, **71**, 266–274.

Neale, P.J., Cullen, J.J., & Davis, R.F. (1998). Inhibition of marine photosynthesis by ultraviolet radiation: variable sensitivity of pytoplankton in the Weddell-Scotia Confluence during the austral spring. *Limnology and Oceanography*, **43**, 433–448.

Neale, P.J., Helbling, E.W., & Zagarese, H.E. (2003). Modulation of UVR exposure and effects by vertical mixing and advection. In: Helbling, E.W. & Zagarese, H. (eds), *UV Effects in Aquatic Organisms and Ecosystems*, pp. 107–134. Royal Society of Chemistry, Cambridge, UK.

Newman, P.A., Nash, E.R., Kawa, S.R., et al. (2006). When will the Antarctic ozone hole recover? *Geophysical Research Letters*, 33, L12814.

Newman, S.J., Dunlap, W.C., Nicol, S., & Ritz, D. (2000). Antarctic krill (*Euphausia superba*) acquire a UV-absorbing mycosporine-like amino acid from dietary algae. *Journal of Experimental Marine Biology and Ecology*, **255**, 93–110.

Nicol, S., Worby, A., & Leaper, R. (2008). Changes in the Antarctic sea ice ecosystem: potential effects on krill and baleen whales. *Marine and Freshwater Research*, **59**, 361–382.

Nozais, C., Desrosiers, G., Gosselin, M., et al. (1999). Effects of ambient UVB radiation in a meiobenthic community of a tidal mudflat. *Marine Ecology Progress Series*, **189**, 149–158.

Olli, K., Wexels Riser, C., Wassmann, P., et al. (2002). Seasonal variation in vertical flux of biogenic matter in the marginal ice zone and the central Barents Sea. *Journal of Marine Systems*, **38**, 189–204.

Parkinson, C.L. & Cavalieri, D.J. (2012). Antarctic sea ice variability and trends, 1979–2010. *Cryosphere*, **6**, 871–880.

Pollard, R.T., Salter, I., Sanders, R.J., et al. (2009). Southern Ocean deep-water carbon export enhanced by natural iron fertilization. *Nature*, **457**, 577–580.

Pomeroy, L.R. & Wiebe, W.J. (2001). Temperature and substrates as interactive limiting factors for marine heterotrophic bacteria. *Aquatic Microbial Ecology*, **23**, 187–204.

Quijón, P.A., Kelly, M.C., & Snelgrove, P.V.R. (2008). The role of sinking phytodetritus in structuring shallow-water benthic communities. *Journal of Experimental Marine Biology and Ecology*, **366**, 134–145.

Raven, J.A. (1998). The twelfth Tansley lecture. Small is beautiful: the picophytoplankton . *Functional Ecology*, **12**, 503–513.

Reinsel, G.C., Miller, A.J., Weatherhead, E.C., et al. (2005). Trend analysis of total ozone data for turnaround and dynamical contributions. *Journal of Geophysical Research*, **110**(D16), D16306.

Rex, M., Salawitch, R.J., von der Gathen, P., et al. (2004). Arctic ozone loss and climate change. *Geophysical Research Letters*, **31**, L04116.

Riebesell, U., Rtzinger, A.K., & Oschlies, A. (2009). Sensitivities of marine carbon fluxes to ocean change. *Proceedings of the National Academy of Sciences of the USA*, **106**, 20602–20609.

Riebesell, U., Schulz, K.G., Bellerby, R.G.J., et al. (2007). Enhanced biological carbon consumption in a high CO_2 ocean. *Nature*, **450**, 545–548.

Roleda, M., Wiencke, C., Hanelt, D., et al. (2005). Sensitivity of Laminariales zoospores from Helgoland (North Sea) to ultraviolet and photosynthetically active radiation: implications for depth distribution and seasonal reproduction. *Plant, Cell and Environment*, **28**, 466–479.

Roos, J.C. & Vincent, W.F. (1998). Temperature dependence of UV radiation effects on Antarctic cyanobacteria. *Journal of Phycology*, **34**, 118–125.

Rossoll, D., Bermúdez, R., Hauss, H., et al. (2012). Ocean acidification-induced food quality deterioration constrains trophic transfer. *PLoS ONE*, **7**(4), e34737.

Roux, R., Gosselin, M., Desrosiers, G., & Nozais, C. (2002). Effects of reduced UV radiation on a microbenthic community during a microcosm experiment. *Marine Ecology Progress Series*, **225**, 29–43.

Rowland, F.S. (2006). Stratospheric ozone depletion. *Philosophical Transactions of the Royal Society Series B*, **361**, 769–790.

Ryan, K.G., McMinn, A., Hegseth, E.N., & Davy, S.K. (2012). The effects of ultraviolet-B radiation on Antarctic sea-ice algae. *Journal of Phycology*, **48**, 74–84.

Sabine, C.L., Feely, R.A., Gruber, et al. (2004). The oceanic sink for anthropogenic CO_2. *Science*, **305**, 367–371.

Sakshaug, E. (2004). Primary and secondary production in the Arctic Seas. In: Stein, R. & Macdonald, R.W. (eds), *The Organic Carbon Cycle in the Arctic Ocean*, pp. 57–81. Springer, Berlin.

Salby, M.L., Titova, E.A., & Deschamps, L. (2012). Changes of the Antarctic ozone hole: controlling mechanisms, seasonal predictability, and evolution. *Journal of Geophysical Research D*, **117**(D10), D10111.

Sarmiento, J.L., Slater, R., Barber, R., et al. (2004). Response of ocean ecosystems to climate warming. *Global Biogeochemical Cycles*, **18**, GB3003.

Schofield, O., Kroon, B.M.A., & Prézelin, B.B. (1995). Impact of ultraviolet-B radiation on photosystem II activity and its relationship to the inhibition of carbon fixation rates for Antarctic ice algae communities. *Journal of Phycology*, **31**, 703–715.

Shick, J.M., Lesser, M.P., & Jokiel, P.L. (1996). Effects of ultraviolet radiation on corals and other coral reef organisms. *Global Change Biology*, **2**, 527–545.

Sinha, R.P. & Häder, D.-P. (2002). UV-induced DNA damage and repair: a review. *Photochemical and Photobiological Sciences*, **1**, 225–236.

Smith, R.C., Martinson, D.G., Stammerjohn, S.E., et al. (2008). Bellingshausen and western Antarctic Peninsula region: pigment biomass and sea-ice spatial/temporal distributions and interannual variability. *Deep Sea Research Part II*, **55**, 1949–1963.

Sobrino, C. & Neale, P.J. (2007). Short-term and long-term effects of temperature on photosynthesis in the diatom *Thalassiosira pseudonana* under UVR exposures. *Journal of Phycology*, **43**, 426–436.

Sobrino, C., Neale, P.J., & Lubián, L.M. (2005). Interaction of UV radiation and inorganic carbon supply in the inhibition of photosynthesis: spectral and temporal responses of two marine picoplankters. *Photochemistry and Photobiology*, **81**, 384–393.

Sobrino, C., Ward, M.L., & Neale, P.J. (2008). Acclimation to elevated carbon dioxide and ultraviolet radiation in the diatom *Thalassiosira pseudonana*: effects on growth, photosynthesis, and spectral sensitivity of photoinhibition. *Limnology and Oceanography*, **53**, 494–505.

Sokolov, S. & Rintoul, S.R. (2007). On the relationship between fronts of the Antarctic Circumpolar Current and surface chlorophyll concentrations in the Southern Ocean. *Journal of Geophysical Research C*, **112**(C7), C07030.

Søreide, J.E., Leu, E.V.A., Berge, J., et al. (2010). Timing of blooms, algal food quality and *Calanus glacialis* reproduction and growth in a changing Arctic. *Global Change Biology*, **16**, 3154–3163.

Stammerjohn, S.E., Martinson, D.G., Smith, R.C., & Iannuzzi, R.A. (2008). Sea ice in the western Antarctic Peninsula region: spatio-temporal variability from ecological and climate change perspectives. *Deep Sea Research Part II*, **55**, 2041–2058.

Steen, H., Vogedes, D., Broms, F., et al. (2007). Little auks (*Alle alle*) breeding in a High Arctic fjord system: bimodal foraging strategies as a response to poor food quality? *Polar Research*, **26**, 118–125.

Stewart, R. (2008). *Introduction to Physical Oceanography*. Texas A&M University. http://oceanworld.tamu.edu/resources/ocng_textbook/PDF_files/book_pdf_files.html

Stolarski, R., Bojkov, R., Bishop, L., et al. (1992). Measured trends in stratospheric ozone. *Science*, **256**, 342–349.

Stroeve, J., Holland, M.M., Meier, W., et al. (2007). Arctic sea ice decline: faster than forecast. *Geophysical Research Letters*, **34**(9), L09501.

Sucré, E., Vidussi, F., Mostajir, B., et al. (2012). Impact of ultraviolet-B radiation on planktonic fish larvae: alteration of the osmoregulatory function. *Aquatic Toxicology*, **109**, 194–201.

Tedetti, M., Joux, F., Charrière, B., et al. (2009). Contrasting effects of solar radiation and nitrates on the bioavailability of dissolved organic matter to marine bacteria. *Journal of Photochemistry and Photobiology A*, **201**, 243–247.

Tedetti, M. & Sempéré, R. (2006). Penetration of ultraviolet radiation in the marine environment. A review. *Photochemistry and Photobiology*, **82**, 389–397.

Trench, R.K. (1979). The cell biology of plant–animal symbiosis. *Annual Review of Plant Physiology*, **30**, 485–531.

Trivelpiece, W.Z., Hinke, J.T., Miller, A.K., et al. (2011). Variability in krill biomass links harvesting and climate warming to penguin population changes in Antarctica. *Proceedings of the National Academy of Sciences of the USA*, **108**, 7625–7628.

Turner, J. & Overland, J. (2009). Contrasting climate change in the two polar regions. *Polar Research*, **28**, 146–164.

van de Poll, W.H., Bischof, K., Buma, A.G.J., & Breeman, A.M. (2003). Habitat related variation in UV tolerance of tropical marine red macrophytes is not temperature dependent. *Physiologia Plantarum*, **118**, 74–83.

van de Poll, W.H., van Leeuwe, M.A., Roggeveld, J., & Buma, A.G.J. (2005). Nutrient limitation and high irradiance acclimation reduce par and UV-induced viability loss in the Antarctic diatom *Chaetoceros brevis* (Bacillariophyceae). *Journal of Phycology*, **41**, 840–850.

Vaulot, D., LeBot, N., Marie, D., & Fukai, E. (1996). Effect of phosphorus on the *Synechococcus* cell cycle in surface Mediterranean waters during summer. *Applied and Environmental Microbiology*, **62**, 2527–2533.

Vernet, M., Martinson, D., Iannuzzi, R., et al. (2008). Primary production within the sea-ice zone west of the Antarctic Peninsula: I—Sea ice, summer mixed layer, and irradiance. *Deep Sea Research Part II*, **55**, 2068–2085.

Vidussi, F., Mostajir, B., Fouilland, E., et al. (2011). Effects of experimental warming and increased ultraviolet B radiation on the Mediterranean plankton food web. *Limnology and Oceanography*, **56**, 206–218.

Viviani, D.A., Björkman, K.M., Karl, D.M., & Church, M.J. (2011). Plankton metabolism in surface waters of the tropical and subtropical Pacific Ocean. *Aquatic Microbial Ecology*, **62**, 1–12.

Wahl, M., Molis, M., Davis, A., et al. (2004). UV effects that come and go: a global comparison of marine benthic community level impacts. *Global Change Biology*, **10**, 1962–1972.

Wängberg, S.Å., Andreasson, K.I.M., Gustavson, K., et al. (2008). UV-B effects on microplankton communities in Kongsfjord, Svalbard—a mesocosm experiment. *Journal of Experimental Marine Biology and Ecology*, **365**, 156–163.

Wängberg, S.A., Selmer, J.S., & Gustavson, K. (1998). Effects of UV-B radiation on carbon and nutrient dynamics in marine plankton communities. *Journal of Photochemistry and Photobiology B*, **45**, 19–24.

Wängberg, S.Å., Wulff, A., Nilsson, C., & Stagell, U. (2001). Impact of UV-B radiation on microalgae and bacteria: a mesocosm study with computer modulated UV-B radiation addition. *Aquatic Microbial Ecology*, **25**, 75–86.

Wassmann, P. (2006). Structure and function of contemporary food webs on Arctic shelves: an introduction. *Progress in Oceanography*, **71**, 123–128.

Waycott, M., Duarte, C.M., Carruthers, T.J.B., et al. (2009). Accelerating loss of seagrasses across the globe threatens coastal ecosystems. *Proceedings of the National Academy of Sciences of the USA*, **106**, 12377–12381.

Whitehead, R.F., de Mora, S., & Demers, S. (2000). Enhanced UV radiation—a new problem for the marine environment. In: de Mora, S., Demers, S., & Vernet, M. (eds), *The Effects of UV Radiation in the Marine Environment*, pp. 1–34. Cambridge University Press, Cambridge.

Wickham, S. & Carstens, M. (1998). Effects of ultraviolet-B radiation on two arctic microbial food webs. *Aquatic Microbial Ecology*, **16**, 163–171.

Wiencke, C., Gómez, I., Pakker, H., et al. (2000). Impact of UV-radiation on viability, photosynthetic characteristics and DNA of brown algal zoospores: implications for depth zonation. *Marine Ecology Progress Series*, **197**, 217–229.

Wilkinson, C. (2000). *Status of Coral Reefs of the World: 2000*. Global Coral Reef Monitoring Network, AIMS, Townsville, Australia.

Williamson, C.E., Zepp, R.G., Lucas, R.M., et al. (2014). Solar ultraviolet radiation in a changing climate. *Nature Climate Change*, **4**, 434–441.

Wozniak, B. & Dera, J. (2007). *Light Absorption in Sea Water*. Springer, New York.

Xu, J. & Gao, K.S. (2007). Growth, pigments, UV-absorbing compounds and agar yield of the economic red seaweed *Gracilaria lemaneiformis* (Rhodophyta) grown at different depths in the coastal waters of the South China Sea. *Journal of Applied Phycology*, **20**, 681–686.

Zepp, R.G., Shank, G.C., Stabenau, E., et al. (2008). Spatial and temporal variability of solar ultraviolet exposure of coral assemblages in the Florida Keys: importance of colored dissolved organic matter. *Limnology and Oceanography*, **53**, 1909–1922.

CHAPTER 16

Ecological impacts of anthropogenic underwater noise

Jenni A. Stanley and Andrew G. Jeffs

16.1 Introduction

The underwater environment can be a very noisy place, with a variety of both natural and human activities that generate sound. Over the last few decades human-generated underwater sound has been increasing at dramatic rates and may have substantial ecological impacts/impacts on marine organisms. Anthropogenic sound in the underwater environment is often referred to as noise and is generated by a wide range of human activities, including shipping, hydrocarbon exploration, sea bed mining, coastal development, sonar sensing, electricity generation, fishing, research (e.g. airguns, sonars, telemetry, communication, and navigation), and recreational boating. For simplicity we will refer to this human-generated underwater sound as anthropogenic 'noise'. As such, anthropogenic noise in the marine environment has been increasing due to an increasing number and size of noise sources, especially from vessels used for international shipping. For example, in the North-eastern Pacific Ocean underwater sounds in the low frequency range of 30–50 Hz were measured to be 10–12 dB re 1 μPa @ 1 m higher in intensity in 2003–2004 than in 1964–1966 (note that the dB scale is logarithmic so that an increase of 3 dB is a doubling of sound power, and an increase of 10 dB is an order of magnitude increase in sound power) (McDonald et al., 2006). As well as increasing levels of background anthropogenic noise, the peak sound intensities from individual sound sources are also increasing (Hildebrand, 2009). For example, larger international vessels, such as supertankers, tend to produce higher peak sound intensities than smaller vessels due to the greater size of their engine capacity. In addition to the increasing quantity and intensity of anthropogenic noise sources, many of them

are located within coastal and continental shelf waters, and are often within the range of key habitats for marine organisms.

Over the last few decades it has become increasingly clear that natural sources of underwater sound play important roles in the ecology of many marine species, including settlement cues in larvae, as well as a useful sensory cue for short- and long-range orientation, communication, predator avoidance, and prey location and capture (Myrberg, 1997; Simpson et al., 2005; Montgomery et al., 2006; Radford et al., 2007; Fay, 2009; Stanley et al., 2012; Lillis et al., 2013). This knowledge has led to an increase in research on the effects of anthropogenic noise on the marine environment and its inhabitants, especially in relation to the use of natural sound sources and sound communication by marine organisms. The majority of the research efforts into the effects of anthropogenic noise in the underwater marine environment have been on the direct impacts on megafauna, such as marine mammals and certain commercially important species of fishes, such as cod and haddock (Engas et al., 1996; Madsen et al., 2006b; Weilgart, 2007; Popper and Hastings, 2009). However, some studies have uncovered a number of potential effects of anthropogenic noise on marine organisms, such as alteration of vocalization and/or behaviour, disruption of orientation, temporary and permanent hearing impairment, and acoustic masking (when a noise interferes with or obscures an acoustic signal) (Lesage et al., 1999; Finneran et al., 2002; Popper, 2003; Gotz et al., 2009; Popper and Hawkins, 2012; Holles et al., 2013). More recent studies have identified that a variety of marine invertebrates are also affected by anthropogenic noise. These studies have shown that noise can significantly delay development and settlement rates of larvae, cause body malformations, and increase settlement and growth rates of invasive

Stressors in the Marine Environment. Edited by Martin Solan and Nia M. Whiteley
© Oxford University Press 2016. Published in 2016 by Oxford University Press.

species in non-native habitats (Pine et al., 2012; Wilkens et al., 2012; Aguilar de Soto et al., 2013; Nedelec et al., 2014; Stanley et al., 2014). These results indicate that anthropogenic noise has a variety of impacts on a wide range of organisms in the marine environment, many of which are likely to interfere with processes involved in maintaining natural populations and may ultimately have impacts on marine ecosystems.

16.2 A primer on underwater sound

Some of the difficulties in progressing our understanding of the role of underwater sound in the environment comes from the limited ability of humans to directly perceive underwater sound using human hearing systems designed for sound carried in air. The greater density of water enables sound to travel five times more quickly in water than in air, and to travel for far greater distances due to low levels of attenuation of sound in water. The intensity of sound in both air and water is most commonly measured in decibels (dB), which is a relative measure of sound pressure within the medium and has a logarithmic scale. Therefore, different relative scales of sound pressure are used for sound in air versus water and is usually signified by reporting dB in relation to a reference pressure for the medium; i.e. airborne sound as dB re 20 µPa and waterborne as dB re 1 µPa. The intensity of underwater sound typically varies over time and to attempt to summarize this sound pressure levels are often reported as measures averaged over time as root-mean-square (RMS). For shorter duration sounds peak or peak-to-peak values are reported, which essentially represent the maximum intensity level of the sound. Intensities of measured sound sources can be reported as if they were measured at the source itself, or at a standard distance of 1 m from the source, or at another point away from the source where a measurement was taken, i.e. at the location of the receiver. Measurements of the intensity of underwater sound are usually undertaken within a limited range of sound frequencies, depending on the measurement apparatus, the characteristics of which should have been selected to match the source of the sound. Unfortunately, in many published reports of sound intensity the type of intensity measure and the range of frequencies covered by the measure is not fully reported, making comparisons among reports problematic. The vast majority of measurements of underwater sound intensity are taken as acoustic pressure. However, underwater sound energy has two physical components, consisting of pressure waves and particle motion, with the latter being much more challenging to use for measuring sound intensity. Greater research focus is needed for the particle motion component of anthropogenic sources of underwater sound as many marine organisms have hearing structures that rely on the detection of particle motion. Besides difficulties in some aspects of measuring the properties of underwater sound for research purposes, underwater sound is also difficult to manage in experimental settings. Underwater sound also has a great capacity to refract, scatter, and reflect when encountering changes in density, such as the sea surface and seafloor. This propensity of underwater sound creates major difficulties for experimental studies involving introducing underwater sound into enclosures, such as tanks holding aquatic animal subjects. Field research is also challenging with underwater sound because of difficulties containing sound, isolating natural from anthropogenic sounds, and making observations of organisms without introducing further anthropogenic sounds. For all these reasons, rigorous research on underwater sound has a degree of complexity that constrains progress towards improved understanding of the ecological impacts of anthropogenic underwater noise.

16.3 The importance of sound for marine organisms

Sound plays a significant role in marine ecosystems, being a critical sensory modality for many marine organisms that can be useful for both sensing the environment and communication (Tavolga, 1964; Richardson et al., 1991; Tyack, 1998). Underwater sound has some unique characteristics that make it a highly effective means for sending and receiving information in the marine environment (Rogers and Cox, 1988; Au and Hastings, 2008). Water is highly conductive of acoustic energy, with minimal absorption, especially of lower frequencies, allowing it to be transmitted over relatively long distances. Furthermore, the opacity of seawater, which greatly restricts the use of visual senses underwater, does not constrain the transmission of sound (Urick, 1983). Underwater sound sources produce omnidirectional sound that typically propagates in all directions, allowing it be distributed widely for reception (Rogers and Cox, 1988). Underwater sound radiating out from a source also generally retains its directional qualities so that with appropriate receptors the direction of the source can be determined from the sound (Rogers and Cox, 1988). This allows useful directional information to be conveyed to organisms over comparatively long distances, whether a predator warning call from a conspecific or determining the location of suitable distant settlement habitat for a pelagic larva. These useful characteristics of underwater sound have been shown to be used by a wide variety of marine animals for communication, orientation, settlement processes, and detection and interception of predators and prey (Hawkins and Myrberg, 1983; Montgomery et al., 2006; Miller, 2010; Stanley et al., 2010, 2012; Lillis et al., 2013).

Natural sources of sound in the marine environment are composed of a wide range of frequencies derived from a variety of sources, both abiotic and biotic in origin (Cato, 1992; Acosta et al., 1997). Abiotic sources of underwater sound are usually the result of the effects of wind, waves, and rain on the sea surface, which typically produces sound in the 100–1000 Hz frequency range with a peak in the spectra at around 500 Hz (Wenz, 1962). Other intermittent abiotic underwater sound sources include undersea volcanic activity, earthquakes, and lightning strikes (Urick, 1983). A considerable contribution to ambient underwater sound is biotic or biological in origin (Cato, 1992; Cato and McCauley, 2002). Biotic underwater sounds are produced both intentionally and unintentionally by a range of activities by marine animals. Intentional sounds include calling associated with reproductive displays and social interaction, territorial defence, and echolocation (Hawkins and Myrberg, 1983), while unintentional sound generation can be associated with feeding and movement. For example, the rasping of rock involved in the feeding of the sea urchin *Evechinus chloroticus* produces sound in the frequency range of 800–2800 Hz that is amplified by their ovoid calcareous skeleton (test), such that a population of urchins feeding on a reef becomes a major contributor to biotic underwater sound in some temperate marine environments (Radford et al., 2008). Underwater sounds from biological origins cover a large range of frequencies from < 100 Hz, such as social calls of the multiband butterfly fish (*Chaetodon multicinctus*), and up to 130 000 Hz in the echolocation signals of the bottlenose dolphin (*Tursiops truncatus*) (Tricas et al., 2006; Au and Hastings, 2008). These biological sounds can range from a few microseconds in duration to several hours, with some whale choruses lasting as long as 24 h (Au and Hastings, 2008).

Marine mammals, especially cetaceans (whales, dolphins, and porpoises), are one of the best known examples in the ocean of both producing and using underwater sound. Members of the cetaceans produce sounds that are among the highest source levels for an individual animal. For example, a sperm whale click has been measured at 233 dB re 1 μPa @ 1 m and covers a wide range of frequencies (10—130 000 Hz) (Møhl, 2001). Cetacean sounds are also an important contributor to the ambient underwater sound in some marine areas, such as localities within waters around Australia and the Pacific (Cato and McCauley, 2002). Cetaceans commonly have a very well-developed sensory system for utilization of sound for both orientation and communication (Madsen et al., 2006b). The large baleen whales (Mysticetes) produce calls generally no higher than 10 000 Hz, and some species, blue and fin whales, have been observed to produce infrasonic vocalizations in the 10–20 Hz frequency range, which are thought to be used for long distance communication (Cummings and Thompson, 1971; Payne and Webb, 1971; Watkins, 1981).

The toothed whales (Odontocetes) produce sound for communication during social interactions and for navigating, both of which are essential to their successful migration between habitats during their life cycle. Communication signals in Odontocetes are usually classified into two broad categories; whistles (continuous tonal sounds of varying frequency) and clicks (broadband) (Au and Hastings, 2008). The active use of sound for navigation or echolocation is especially well documented among the cetaceans. Echolocation is the process by which an animal produces acoustic signals and obtains a sense of its surroundings from the subsequent echoes it receives and interprets (Au and Hastings, 2008). Echolocation has been demonstrated in 13 species of Odontocetes (Richardson et al., 1995). Odontocetes use echolocation not only in navigation but also in prey location and capture. For example, sperm whales have been revealed to produce a series of regular clicks interspersed with rapid-click buzzes called 'creaks' that are used for localizing fast-moving prey such as squid (Miller et al., 2004).

It was recognized early on that soniferous fishes also make a substantial contribution to ambient underwater sound in some marine environments, with numerous species of fish producing a wide variety of sounds (i.e. pop, knock, chirp, drum, and growl types of calls), usually within the frequency range 50–4000 Hz (Kasumyan, 2008). Many of these sound-producing fish exhibit daily, lunar, seasonal, and spatial patterns in their sound production (Cato and McCauley, 2002). In the shallow waters of tropical regions, fish are regularly the major biotic component of ambient underwater sound. For example, in coral reef systems in northern Australia nocturnally emergent fish have been observed to increase ambient underwater sound levels by an average of 15 dB re 1 µPa @ 1 m in the 300–900 Hz frequency range. On occasions, usually during the summer new moon periods, fish chorusing has been measured to be up to 30 dB re 1 µPa @ 1 m above the normal ambient underwater sound levels (Cato and McCauley, 2002). Members of 50 families of fish are known to produce sound, and many of these use sound extensively in social interactions (Hawkins and Myrberg, 1983). These signals usually involve a series of complex tones in sequential order (Myrberg, 1997). For example, during nesting and mate attraction, the male gulf toadfish (*Opsanus beta*) produces a boat whistle advertisement call and an agonistic grunt call used in male–male competition and to attract females (Thorson and Fine, 2002). The grunt is a short-duration pulsatile call, while the boat whistle is a complex call usually consisting of up to three grunts to start, followed by long tonal 'boops', and finishing in up to three shorter 'boops'. Many species of damselfishes produce acoustic signals to mediate species recognition during courtship (Myrberg et al., 1978). Certain fish species are also known to use sound actively for orientation. For example, the hardhead catfish (*Arius felis*) can avoid objects at distances up to 5 cm when producing a knocking sound (Tavolga, 1976).

In addition to marine mammals and fishes, there is also a wide range of marine invertebrates that produce sound. One of the most ubiquitous biological sounds in coastal waters around the world is created by snapping shrimp (*Alpheus* spp. and *Synalpheus* spp.), with a measured peak in frequency between 2000 and 5000 Hz, with acoustic energy extending out to 200 000 Hz and peak sound levels as high as 210 dB re 1 µPa @ 1 m (Cato, 1992; Au and Kiara, 1998; Versluis et al., 2000). Snapping shrimp often live in high densities with burrows spaced 5–50 cm apart, and these animals are also highly territorial; therefore antagonistic interactions can be common between individuals. The acoustic pulse of the snap that is produced by the enlarged snapper claw can seriously injure other animals within close proximity and is used by some snapping shrimp to stun or kill prey (Schultz et al., 1998; Obermeier and Schmitz, 2003). Another major contributor to ambient underwater sound in reef habitats is the movements and feeding activity of sea urchins (Tait, 1962; Castle and Kibblewhite, 1975; Radford et al., 2008). The dominant frequencies produced by the urchins typically occur in the 800–2000 Hz frequency band, and the diurnal pattern in the sound intensity within this frequency range is due to the diurnal pattern of activity of the urchins (Radford et al., 2008).

As well as actively using sound for navigation, and social interactions, there are many marine organisms which passively use ambient underwater sound to gain information regarding their position in relation to natural sound sources. Many coastal reef species of fish and crustaceans have a dispersive pelagic larval phase of varying lengths which ends once a suitable habitat in which to settle is located. During this time these organisms rely on strong swimming and sensory abilities to guide them to the appropriate settlement habitat (Leis and Carson-Ewart, 1997; Leis and Carson-Ewart, 1999; Myrberg and Fuiman, 2002; Fisher et al., 2005). Underwater sound originating from shallow coastal reef is used by the larvae of some reef organisms, especially a wide variety of reef fishes, to orientate to a suitable settlement habitats from offshore (Stobutzki and Bellwood, 1997; Tolimieri et al., 2000; Jeffs et al., 2003, 2005; Leis and Lockett, 2005; Simpson et al., 2005; Montgomery et al., 2006; Radford et al., 2007). For example, several studies using underwater loud speakers broadcasting reef sound have been shown to attract more larvae from both temperate and tropical reef fish species than those without sound (Tolimieri et al., 2000; Simpson et al., 2004). Also a wide range of pelagic crustacean taxa, including copepods, ostracods, mysids, caridean shrimp, and gammarid and hyperiid amphipods, were found to be actively avoiding replayed coral reef sound, presumably due to the increasing risk of predation with closer proximity to a reef (Simpson et al., 2011). The larvae of some fish and crab species have also been found to be capable of localizing the direction of a sound source (Tolimieri et al., 2004; Leis and Lockett, 2005; Radford et al., 2007).

Ambient underwater sound in coastal habitats has also been shown to initiate physiological and morphological changes associated with settlement in several species of temperate and tropical decapod crustacean larvae (Stanley et al., 2010, 2011, 2012). The settlement stage larvae of some crab species have also shown the ability to initiate these physiological and morphological settlement responses according to habitat-related differences in underwater sound, indicating that they are highly sensitive to subtle differences in the composition of natural sources of underwater sound (Stanley et al., 2012).

16.4 Sources of anthropogenic noise in the ocean

Anthropogenic underwater noise has differing characteristics, such as intensity, frequency, duration, directionality, and temporality (Weilgart, 2007), which have varying potential to generate ecological effects in the marine environment.

A sound is often termed noise when any unwanted sound reaches a certain intensity threshold, which when surpassed is known to adversely affect the ecology or physiology of any exposed organism (Gotz et al., 2009). Noise also has a very specific acoustic definition, where it denotes a random waveform which contains broadband energy. Anthropogenic underwater sound often occurs in high peak intensities, covering a wide range of frequencies (1 Hz–200 kHz), and, due to the excellent propagation of sound underwater, it can travel very large distances from the source (Rogers and Cox, 1988). Among the various activities that create underwater noise, there is a great deal of variability in the temporal, spatial, and acoustic characteristics which can often result in generalizations when comparing the various sound sources. This variability makes it very difficult to generalize the impacts of underwater noise, and from the small amount we know, the impacts are wide reaching and vary depending on the taxa and type of impact (NRC, 2003; Gotz et al., 2009).

16.4.1 Vessels

There is substantial variability in the sound emitted by individual vessels due to characteristics such as vessel type and size (increasing size usually decreases frequency), method of propulsion, and the operation of additional machinery on board the vessel (Hildebrand, 2009). At low vessel speeds the sound produced by propellers passing through water and associated harmonics from dominant frequencies, propeller rotation rate, and other machinery-related dominant frequencies will often contribute to an acoustic signature of a vessel. With increasing speed, broadband noise generation becomes more apparent, as other sources become more important, such as noise from the flow of water over the hull and propeller cavitation. These will often mask the machinery-related frequencies which are more apparent at the lower speeds (NRC, 2003; Gotz et al., 2009).

16.4.1.1 Large vessels (greater than 100 m)

Among the many sources of anthropogenic sound in the marine environment, the major contributors are large steel-hulled vessels that are usually used for international trade and shipping coastal freight. This is because these vessels are numerous, widespread in their operations in all oceans around the world, and individually are significant sources of lower frequency

sound. For example, steel-hulled freight vessels usually emit relatively high intensity (180–190 dB re 1 μPa @ 1 m) noise at relatively low frequencies (highest intensity levels in the 5–500 Hz range but can extend up to 10 000 Hz) (Richardson et al., 1995; Arveson and Vendittis, 2000). This noise is mostly generated by propeller cavitation, propulsion machinery (engines and gears), large auxiliary engines (generators), and water turbulence around the hull (Ross, 1976). These low-frequency sounds generated from vessels radiate out in all directions for considerable distances (i.e. tens to hundreds of kilometres), before attenuating to the level of ambient background sound (Rogers and Cox, 1988; Au and Hastings, 2008). Consequently, the sound emitted by large vessels often dominates the low-frequency ambient underwater noise in many marine environments (Greene and Moore, 1995; McDonald et al., 2006).

Monitoring of ambient underwater sound on the west coast of North America over the past four decades has shown significant increases in anthropogenic sound (Andrew et al., 2002, 2011; McDonald et al., 2006). Ambient noise levels in the 30–50 Hz range were 10–12 dB re 1 μPa higher in 2003–2004 than in 1964–1966, averaging an increase of 2.5–3 dB re 1 μPa (doubling in intensity) per decade (McDonald et al., 2006). This increase in low-frequency background noise is largely attributed to the rapid growth in global ocean trade using large steel-hulled vessels. Between 1965 and 2003 the total gross tonnage (GT) of the world's shipping fleet quadrupled, and at the same time the number of commercial vessels in operation approximately doubled (Ross, 1993; McDonald et al., 2006; McKenna et al., 2012). The growing number and size of vessels is contributing to the increasing anthropogenic sound, not only along shipping channels in the open ocean, but also in areas with large aggregations of operating vessels, such as in ports and harbours (NRC, 2003). For example, a 125 m steel-hulled passenger and freight ferry at berth without its main propulsion engines operating was found to be emitting 126 dB re 1 μPa @ 1 m RMS at the source with dominant frequencies between 100 and 1000 Hz but reaching up to 11 000 Hz (Wilkens et al., 2012).

16.4.1.2 Medium to small vessels (no greater than 40 m in hull length)

Compared with large vessels, small to medium watercraft emit underwater sound that can often cover a wider range of frequencies. The characteristics of anthropogenic noise output among small to medium vessels varies substantially as it is highly dependent on speed and other features of the vessel, such as the nature of the propulsion unit (Richardson et al., 1995; Erbe, 2002). These vessels usually emit lower intensity (160–175 dB re 1 μPa) underwater sound at higher frequencies (highest intensity levels are usually above 1000 Hz) than larger vessels. The sound emitted by these vessels include dominant frequencies and their harmonics at the vibrational frequencies of propeller blades, engines and gearboxes, usually below 10 000 Hz, but with significant sound levels extending above 10 000 Hz as a result of propeller cavitation and harmonics. In the past it was assumed that, with the more rapidly attenuated higher frequencies and the near-shore operation of small vessels, their underwater noise emissions had limited impacts (Gotz et al., 2009). However, data from the National Marine Manufacturers' Association in the United States show the number of boats owned in the USA increased from 15.8 million in 1995 to nearly 17 million in 2001, suggesting that increasing number of acoustic sources could be leading to cumulative noise impacts (National Marine Manufacturers Association, 2002; Gotz et al., 2009). Small vessels may be less of a concern in offshore locations or in regards to overall increases in low frequency ambient noise; however, they can dominate many shallow coastal environments such as harbours, estuaries, and bays. These areas often support a high proportion of marine life and overlap with sensitive habitats, such as breeding, feeding and nursery grounds and migratory corridors, and also where a high number of species use underwater sound, for example, coral and temperate reefs and estuaries (Beck et al., 2001; Holles et al., 2013; Nedelec et al., 2014; Simpson et al., 2014) where impacts from the sound generated by small watercraft may be substantial.

16.4.2 Sonar

Sonar systems use the reflections from broadcast acoustic energy to characterize physical properties, locate objects, and communicate underwater. Sonar was one of the first intentionally introduced anthropogenic sounds in the marine environment on a large scale (Gotz et al., 2009). The wide range of sonar applications involves systems that vary in terms of specification (frequency) and deployment methods. Sonar systems have both military and civilian application and are broadly divided for convenience into low-frequency (below 1000 Hz), mid-frequency (1000–10 000 Hz), and high frequency (greater than 10 000 Hz) applications.

16.4.2.1 Civilian/commercial sonar

Generally, sonar systems used for civilian or commercial applications are designed for depth sounding, seabed characterization, detection of solid objects, and fish finding, and these types of sonar typically operate at the high frequency range (> 10 000 Hz). These largely operate at lower intensities than military sonar systems and seabed survey sonar can provide high spatial resolution over small areas with narrow beam patterns and shorter pulse length than military sonars. There are also civilian systems that operate in the mid-frequency range for surveys of low resolution and wide areas, and they are usually involved in remote sensing activities, such as bathymetry mapping of the seafloor. These systems typically operate with source levels of approximately 223–230 dB re 1 μPa @ 1 m (Gotz et al., 2009). Commercial depth sounders and fish finders often operate at high frequencies (> 10 000 Hz) with lower intensity output and have a narrow acoustic beam focused downward, or sometimes sideways for fish finders or for locating towed trawl nets, at high resolution. The use of sonar is very common among the millions of global recreational and commercial vessels. These vessels are most commonly used in the shallow coastal waters and use of the sonar systems occurs throughout the year, during both day and night.

16.4.2.2 Military sonar

Military active sonar systems more commonly use a very wide range of frequencies and operate at higher power levels than civilian sonars and are used typically for target detection, localization, classification, and interception, as well interference with sonar detection.

Low-frequency sonar is used for surveillance and to gather information over vast areas, if the right conditions exist, because low-frequency military sonar is capable of being effective over entire ocean basins (NRC, 2003). Mid-frequency military active sonar systems are usually more tactical systems used to locate and track underwater targets over tens of kilometres. Hull-mounted systems used by naval vessels, such as the AN/SQS-53C and -56 use acoustic pulses in the 1000–10 000 Hz range and operate at relatively high source levels of up to 235 dB re 1 μPa @ 1 m (Evans and England, 2001). High-frequency military sonar systems (above 10 000 Hz) are used in weapons (mines or torpedoes) or counter-weapons (anti-torpedo or mine systems), and these are designed to operate over hundreds of metres to a few kilometres (NRC, 2003; Gotz et al., 2009). High-frequency systems are also used in a wide range of modes, with different signal types, lengths, and strengths, but typically over a relatively narrow frequency range. The use of the full range of military active sonar systems is typically limited to specified military operational areas.

16.4.3 Seismic seabed profiling

Seismic profiling covers a wide variety of methods, all which use underwater sound sources to gather information about the geological structure beneath the seafloor. Seismic profiling methods are used most extensively by the oil and gas industry to locate and characterize new hydrocarbon reservoirs, as well as monitoring the depletion of already tapped reserves (Gotz et al., 2009). Seismic methods are also extensively used by governmental and research agencies, to gather data on the geological structure of subsurface rock up to several thousand metres below the Earth's surface. The technique normally relies on detecting and interpreting the reflections of intense pulses of directed sound waves emitted by seismic airguns located at the sea surface, typically in an array (12–48 airguns in each array). Seismic airguns emit low frequency sound pulses below 250 Hz, with the peak of the energy in the 30–50 Hz range; however, low levels of high-frequency sound are also produced and can extend up to approximately 100 000 Hz (Bain and Williams, 2006; Goold and Coates, 2006; Madsen et al., 2006a). The power of modern airguns has been increased due to the requirement for exploration at greater geological depths in search for hydrocarbons (Barlow and Gentry, 2004). The source level of an airgun array can reach up to 262 dB re 1 µPa @ 1 m (peak-to-peak), but measuring the exact source levels of an operational airgun array is very difficult due to the high intensity of the peak signal and destructive interference caused by the lateral separation of airguns in the array, as well as sound reflections from the sea surface (NRC, 2003; Gotz et al., 2009). Seismic airguns have previously been detected thousands of kilometres from their source (Nieukirk et al., 2004). For example, a hydrophone array used to record whale vocalizations moored near the mid-Atlantic Ridge (35–15° N, 50–33° W) frequently recorded the sounds from seismic airguns, particularly during the summer, from surveys conducted at locations over 3000 km away (Nieukirk et al., 2004). Large commercial cargo vessels travelling at over 11 km h^{-1} can be heard tens of kilometres away, while a leisure launch with an inboard engine is detectable within kilometres (Pine et al., 2013).

16.4.4 Other activities

16.4.4.1 Renewable energy

Due to the demand for renewable energy sources, over the last few decades there has been an increase in the construction and operation of sea-based renewable energy generation devices, such as wind and tidal turbines placed in the marine environment (Nedwell and Howell, 2004; Madsen et al., 2006b; Pine et al., in press). Recordings of the underwater sound emitted from operating wind turbines have shown that they generate sound in frequencies up to 2000 Hz, with peak sound pressure levels in the range of 109–153 dB re 1 µPa @ 1 m (Nedwell and Howell, 2004; Thomsen et al., 2006; Madsen et al., 2006b). There are very few reported field-based measurements of the sound emitted by an operating tidal turbine. However, field measurements of the underwater sound emitted from a SeaFlow-0.3 in midwater and an OpenHydro tidal turbine in 6 m of water showed that they produced continuous underwater noise that was composed of a large range of frequencies from 10 to 30 000 Hz (Parvin et al., 2005; Halvorsen et al., 2011) with a sound level of about 165–175 dB re 1 µPa @ 1 m under a maximum current flow rate of 3 m s^{-1} (Lloyd et al., 2011). Sound levels emitted from both wind turbines and tidal turbines are likely to vary depending on the operational design and local conditions (e.g. wind and current speed).

Not only does the operation of these sea-based renewable energy generators create noise in the marine environment, but there are also many noise sources associated with the construction and decommissioning of these devices. These activities include vessel movements, trenching, dredging, drilling, pile driving, and rock laying (Nedwell and Howell, 2004). One of the most significant activities during the installation of a windfarm, in terms of emission of anthropogenic noise, is the installation of foundations, with pile installation often using impact or vibro-pile hammers which can produce a wide range of peak source sound levels of between 189 and 260 dB re 1 µPa @ 1 m (Nedwell and Howell, 2004). Measurements of hopper dredges taken in shallow water, similar to techniques used in windfarm construction, have found peak sound source levels of up to 177 dB re 1 µPa @ 1 m over 80–200 Hz (Nedwell and Howell, 2004).

16.4.4.2 Acoustic deterrent devices

There are also a large number of smaller, comparatively localized anthropogenic noise sources in the marine environment. These include a number of different types of acoustic deterrent devices used to deter marine organisms from entering certain areas. Acoustic deterrent devices are often relatively high powered, with peak source levels of no higher than 195 dB re 1 µPa @ 1 m, and are generally used to deter small marine mammals, especially pinnipeds (seals and sea lions), dolphins, and porpoises, from a range of fishing and aquaculture equipment to reduce by-catch and crop losses (Reeves et al., 1996). Acoustic deterrent devices or pingers vary in their acoustic characteristics, dependent on their target animals as well as the design and manufacturer. Peak source levels can range from 132 dB re 1 µPa @ 1 m at 10 000 Hz (Airmar Technologies Group—gill net pinger) to 155 re 1 µPa @ 1 m with two partial frequency 5000–40 000 Hz and 30 000–160 000 Hz wideband sweeps (SaveWave—seal deterrent for fish farms). Acoustic harassment devices were originally developed to prevent pinniped predation on fish farms and range in peak source level from 178 dB re 1 µPa @ 1 m at 5000 Hz (sweeps of frequency range of 1800–3000 Hz) (Terecos Ltd) to 200 dB re 1 µPa @ 1 m at 25 000 Hz (pulses centred at five different frequencies, up to ~40 000 Hz) (Ferranti-Thomson) (Yurk and Trites, 2000; Reeves et al., 2001; Lepper et al., 2004). A variety of high-frequency sound generating devices are also widely used on the hulls of small vessels to deter hull fouling organisms—despite their wide use there is very little, or no, documented evidence of their efficiency or mode of action against fouling marine organisms.

16.4.4.3 Research activities

Advances in technology have enabled the increased use of acoustics during research in ocean science (Southall and Nowacek, 2009). Research uses a variety of different sound sources to investigate many properties of the ocean: for example, acoustic doppler current profilers for measuring the speed and direction of water currents, side-scan sonar used extensively for mapping of the seafloor and tracking organisms, and standard and multi-beam sonar for density estimation of plankton and fish and seafloor habitat determination. Ocean tomography studies also occur whereby researchers measure the physical properties of the ocean using very large sound sources with frequencies between 50 and 200 Hz with high source levels (165—220 dB re 1 µPa peak-to-peak). The most well known of these studies is the 'Heard Island Feasibility Test' to establish the limits of usable, long-range acoustic transmissions. Sound signals of ~ 57 Hz were broadcast in the SOFAR channel (175 m depth) at source levels up to 220 dB re 1 µPa for one hour each day (Munk et al., 1994). The broadcast signal was detected across ocean basins with received levels of up to 160 dB re 1 µPa @ 1 km.

16.5 Ecological impacts of underwater noise on marine systems

16.5.1 Extent of evidence for ecological impacts

There is a wide range of evidence to support the growing concern that anthropogenic noise is causing ecological impacts in both terrestrial and aquatic environments (Slabbekoorn et al., 2010; Kight et al., 2012; Morley et al., 2014). While anthropogenic underwater noise in the oceans has been shown to directly affect a variety of marine organisms in a number of ways, the overall importance of anthropogenic sound to the fitness of individuals, populations, and marine ecosystems as a whole is only beginning to be researched. While there are a number of studies which have found minimal or no effects on marine organisms from some sources of anthropogenic sound (Popper et al., 2005; Wysocki et al., 2007; Bruintjes and Radford, 2014), sufficient studies are now available to indicate the relative importance of various anthropogenic sound sources, the taxa they affect, and the types of effects they elicit.

Anthropogenic underwater noise has been demonstrated to cause physiological effects, such as temporary or permanent threshold shifts in hearing ability and damage to auditory systems, as well as causing the potential for ecological effects, such as the alteration in natural behaviour, reduction in communication ranges, reduction in foraging ability, prevention of predator avoidance, and in extreme cases complete habitat avoidance or death of individuals (McCauley et al., 2003; Gotz et al., 2009). To date, the focus of the research into the effects of noise in the marine environment has largely been on the direct or physiological impacts, such as triggering stress responses or impairment of cognitive processes, in high-profile vertebrate species, especially marine mammals and fishes (Popper, 2003; Popper and Hastings, 2009; Kastelein and Jennings, 2012; Voellmy et al., 2014a, b). This is largely due to the existing knowledge that these taxa are active users of underwater sound for echolocation, prey capture, and intraspecific communication. However, there is a much wider range of marine organisms that utilize sound in a variety of ways, such as mediating behaviour during larval settlement. If anthropogenic sound interferes with key biological processes involved in maintaining populations, such as reproduction, settlement, and recruitment behaviour, then there is the potential for more significant ecological impacts than just through the direct impacts on adult individuals.

Overall, there are now numerous examples of the physiological and behavioural functioning of individuals from a wide range of taxa being altered by the presence of anthropogenic underwater sound, some of which are documented in the following sections. However, extensive evidence on how a change in functioning of individuals translates into reducing their fitness, or the fitness of populations, or causes wider ecological impacts is somewhat lacking, but is likely to be significant based on the information available.

16.5.2 Swimming behavioural responses

Over the past few decades the marked growth in the scale of commercial shipping has greatly increased anthropogenic underwater noise, especially at low frequencies (10–150 Hz) (McDonald et al., 2006). This low-frequency noise has the potential to overlap with the frequencies produced by, used by, and within the hearing range of many marine organisms for which underwater sound is critical for maintaining normal behaviour and ecological functions (Gotz et al., 2009). The most well-known examples and most recognizable of these animals are the cetaceans. For example, a number of dolphin and whale species, when exposed to close approaches from vessels, have been observed to dramatically change their swimming behaviour by increasing swimming speeds, increasing angle between successive dives, spending more time below the sea surface, and decreasing inter-animal distances (Nowacek et al., 2001; Williams et al., 2002). Several species of baleen whales, including the grey, bowhead, blue, sei, minke, and fin whales, have shown active avoidance behaviour of the underwater noise produced during seismic surveys (Gotz et al., 2009). Sea turtles have also been observed to display alarm responses when approximately 2 km from an active seismic vessel, and demonstrate strong avoidance behaviour at 1 km from the seismic sounder (McCauley et al., 2000).

The avoidance of approaching large vessels has also been observed in certain fish species, such as the walleye pollock (*Theragra chalcogramma*) (De Robertis et al., 2010). Fish avoiding areas in which there is vessel noise could result in important feeding or breeding grounds being less used or even abandoned (Engas and Lokkeborg, 2002; Popper, 2003). This is also supported by observations of fewer fish and reduced catch rates, by as much as 40–80% less, in the vicinity of seismic surveys for a variety of fish species including herring, cod, haddock, rockfish, sand eel, and blue whiting (Engas et al., 1996; Slotte et al., 2004).

16.5.3 Vocalization behavioural responses

Alteration in vocalization due to a variety of anthropogenic sound sources has been commonly observed in a number of cetacean species. For example, the Cuvier's beaked whale (*Ziphius cavirostris*) has been observed to reduce the production of foraging sounds in the presence of passing cargo ships (Aguilar Soto et al., 2006), completely stop echolocating, and extend dive duration and the subsequent non-foraging interval when exposed to short pulses of active mid-range sonar signals (DeRuiter et al., 2013). Beluga whales in the St Lawrence River Estuary also decreased their call rate and shifted the frequency range of their calls from 3.6 kHz to up to 8.8 kHz when a commercial ferry was approaching (Lesage et al., 1999). It was also observed that when the ferry was less than 1 km from the whales they increased the repetition rate of specific calls. Alteration in acoustic activity and swimming behaviour has also been observed in several cetacean species during pile-driving activities during the construction of offshore wind farms, and also in the presence of military sonar (Tougaard et al., 2003). Acoustic activity decreases in the harbour porpoise (*Phocoena phocoena*) shortly after each pile-driving event and the response was observed up to 15 km away (Tougaard et al., 2003). It was also found that densities of porpoises in the area during the construction were significantly lower than those during baseline surveys. Long-finned pilot whales (*Globicephala melas*) changed the type of vocalization in response to the presence of active sonar, while humpbacks have been known to respond by lengthening their song cycles (Gotz et al., 2009). During a seismic survey off western Europe an array of hydrophones detected that a group of 205 fin whales ceased vocalization across an area of 10 000 square nautical miles and only recommenced once the survey was completed (Clark and Gagnon, 2006). Alterations in vocalization behaviour has also been observed in fishes in response to a variety of anthropogenic sound sources including vessels, naval sonar, pile driving, and seismic surveying (reviewed by Gotz et al., 2009). For example, during experiments in a nearshore Mediterranean marine reserve the mean pulse rate in the calls from the brown meagre (*Sciaena umbra*) increased during multiple passages by boats (Picciulin et al., 2012).

16.5.4 Reproductive behavioural responses

Underwater sounds are thought to serve a role in maintaining cohesion and synchronizing gamete release in breeding aggregations of certain fish species (Myrberg and Lugli, 2006). Underwater noise created by a fast approaching power boat has been reported to interrupt spawning in the fresh and brackish water roach (*Rutilus rutilus*) (Boussard, 1981). Exposure to the noise of a passing boat was also found to alter the social cohesion of, as well as reduce nest-building and the defence of offspring by, the cooperatively breeding cichlid fish *Neolamprologus pulcher* (Bruintjes and Radford, 2013).

16.5.5 Antipredator behavioural responses

The antipredator behaviour of the three-spined stickleback (*Gasterosteus aculeatus*) in response to visual predator stimulus was enhanced in the presence of ship noise, but for the European minnow (*Phoxinus phoxinus*) it made no difference (Voellmy et al., 2014b). Similar experiments with exposing juvenile European eels (*Anguilla anguilla*) to ship noise also found their antipredator behaviour was compromised (Simpson et al., 2014). The playback of the noise of a passing boat to nesting adult cichlid fish, *Neolamprologus pulcher*, decreased their defence of their eggs and fry from predators (Bruintjes and Radford, 2013). Replaying ship noise versus ambient noise to captive shore crabs made no difference to their ability to detect and respond to a simulated predatory attack; however, in the presence of ship noise the crabs were slower to return to their shelters (Wale et al., 2013a).

16.5.6 Physiological responses

Juvenile European eels subjected to ship noise were found to have diminished spatial performance and elevated ventilation and metabolic rates, which are consistent with increased physiological stress in fish (Simpson et al., 2014). Shore crabs exposed to ship noise were also found to have elevated oxygen consumption compared to crabs in ambient noise, which may indicate a physiological stress response to the anthropogenic sound (Wale et al., 2013b). Anthropogenic sound has been shown to increase physiological processes associated with the settlement and growth rates of a number of invertebrate biofouling species (Wilkens et al., 2012; Stanley et al., 2014). Within four weeks of experimental settlement panels being placed in the oceans there was an almost doubling of numbers of some fouling species that had settled on the panels exposed to vessel noise compared to the controls (Stanley et al., 2014). Additionally, several species also grew significantly larger in the presence of vessel noise. Therefore, anthropogenic noise, such as vessel noise, may also be having significant ecological effects by encouraging the translocation of species and subsequent alteration of marine ecosystems.

16.5.7 Masking of acoustic communication

More recently, there is emerging concern that anthropogenic noise in the marine environment is affecting the regular use of biologically relevant sounds via acoustic masking (Popper, 2003; Gotz et al., 2009). Masking by anthropogenic noise can have serious implications if it affects the ability of organisms to recognize or detect underwater signals of interest due to interference from extraneous sound sources, which may vary over time (Clark et al., 2009). Masking has the potential to lead to a reduction in the biological fitness of the organism attempting to receive or communicate via the masked underwater sounds of interest; however, the determination of the presence of masking as well the consequential effects of masking on the organism is somewhat challenging for aquatic animals. For example, it was found that small vessels travelling at 5 knots in 5–7 m of water reduced the communication range of bottlenose dolphins, *Tursiops* sp., within 50 m by 26%, and short-finned pilot whales, *Globicephala macrorhynchus*, could experience a reduction of 58% (Jensen et al., 2009). However, in both cases the impact on the fitness of these cetaceans experiencing masking remained uncertain.

One of the most common contexts in which fishes are known to produce and respond to sound is in courtship interactions and spawning aggregations (Myrberg et al., 1986; Saucier and Baltz, 1993; Aalbers, 2008). Masking of the biological sound associated with these behaviours has the potential to lead to a reduction in fitness as result of a lowered reproductive efficiency in soniferous fishes that use sound, or react to sound, during courtship, mating, and spawning (Boyle and Cox, 2009; Mann et al., 2009; Boyle and Tricas, 2010). These types of masking impacts from acoustic pollution have often been reported in terrestrial animals (Reijnen and Foppen, 1994; Sun and Narins, 2005; Kaiser and Hammers, 2009; Kaiser et al., 2011) and therefore warrant further study in order to improve our understanding of the impacts of masking in aquatic environments.

In the last few decades there has been increased concern and awareness about the ecological risks associated with increased anthropogenic noise in the coastal environment. Research has demonstrated that natural sources of underwater sound emanating from coastal habitats play a significant role in the larval settlement and recruitment processes of a wide range of coastal organisms, especially in reef habitats (Montgomery et al., 2006; Radford et al., 2011; Stanley et al., 2012). For coastal organisms, effective settlement choices are often critical for ensuring subsequent survival and recruitment into appropriate habitat conditions in which individuals can become established and recruit to the adult population (Leis and McCormick, 2002). It is for these reasons it is likely that anthropogenic sound sources in the vicinity of coastal reefs have the potential to mask natural underwater reef sound and interfere with recruitment processes. For instance, boat noise was shown to disrupt the natural positive directional response to reef sound in the larvae of coral reef fish (Holles et al., 2013). Another study has demonstrated that settlement behaviour and time-to-metamorphosis (TTM) in the megalopae of the estuarine crabs *Austrohelice crassa* and *Hemigrapsus crenulatus* was significantly increased (by 38–60%) when exposed to the noise of either tidal turbines or sea-based wind turbines (Pine et al., 2012).

16.5.8 Anthropogenic sound as additional stressor

There is growing consensus that multiple stressors acting in concert, or even synergistically, play an important role in driving what is often abrupt and catastrophic change in marine ecosystems (Mollmann et al., 2015). Climate change effects on the marine environment are highly likely to further exacerbate any stress effects from anthropogenic noise. Increasing temperature, greater dissolved carbon dioxide, lowered pH, and decreasing salinity in many oceans as a result of climate change look set to extend the range of propagation of underwater anthropogenic sound mostly through reduced absorption, especially at lower frequencies (Joseph and Chiu, 2010; Reeder and Chiu, 2010). The combined effects of these stressors warrants more attention than the current tendency for researchers to focus on the potential impacts of ocean change parameters in isolation (Harvey et al., 2013).

16.6 Conclusion

In reality, most of the evidence we have to date on the effects of anthropogenic underwater noise in the underwater marine environment involves either short-term behavioural or physiological modifications in individual species, whereas evidence for long-term population or ecological consequences is scant. However, on the basis of collective evidence it would seem likely that anthropogenic sound is likely to be interfering with some key biological processes in the marine environment, such as the settlement and recruitment of animals to coastal populations. Determining the ecological effects of anthropogenic sound in the marine environment is challenging due to the difficulties of conducting rigorous *in situ* experiments with sound, and the problem of detecting effects on a wide scale among extensive natural variation in population and ecosystem characteristics.

Here we have discussed the current state of knowledge in the field of anthropogenic sound effects, and highlighted the need for future research to move beyond describing short-term impacts on individuals to improving our understanding of chronic impacts of anthropogenic sound on populations and ecological processes. To this end, research in coastal habitats is likely to be most fruitful where low-frequency noise is largely continuous due to human activities and appears very likely to be affecting predator–prey interactions, breeding, and recruitment, which are fundamental processes for sustaining viable populations.

References

Aalbers, S.A. (2008). Seasonal, diel, and lunar spawning periodicities and associated sound production of white seabass (*Atractoscion nobilis*). *Fishery Bulletin*, **106**, 143–151.

Acosta, C.A., Matthews, T.R., & Butler, M.J. (1997). Temporal patterns and transport processes in recruitment of spiny lobster (*Panulirus argus*) postlarvae to south Florida. *Marine Biology*, **129**, 79–85.

Aguilar de Soto, N., Delorme, N., Atkins, J., et al. (2013). Anthropogenic noise causes body malformations and delays development in marine larvae. *Scientific Reports*, **3**, 2831.

Aguilar de Soto, N., Johnson, M., Madsen, P.T., et al. (2006). Does intense ship noise disrupt foraging in deep-diving Cuvier's beaked whales (*Ziphius cavirostris*). *Marine Mammal Science*, **22**, 690–699.

Andrew, R.K., Howe, B.M., & Mercer, J.A. (2011). Long-time trends in ship traffic noise for four sites off the North American West Coast. *Journal of the Acoustical Society of America*, **129**, 642–651.

Andrew, R.K., Howe, B.M., Mercer, J.A., & Dzieciuch, M.A. (2002). Ocean ambient sound: comparing the 1960s with the 1990s for a receiver off the California coast. *Acoustics Research Letters Online*, **3**, 65–70.

Arveson, P.T. & Vendittis, D.J. (2000). Radiated noise characteristics of a modern cargo ship. *Journal of the Acoustical Society of America*, **107**, 118–129.

Au, W.W.L. & Hastings, M.C. (2008). *Principles of Marine Bioacoustics*. Springer, New York.

Au, W.W.L. & Kiara, B. (1998). The acoustics of the snapping shrimp *Synalpheus parneomeris* in Kaneohe Bay. *Journal of the Acoustical Society of America*, **103**, 41–47.

Bain, D.E. & Williams, R. (2006). *Long-Range Effects of Airgun Noise on Marine Mammals: Responses as a Function of Received Sound Level and Distance*. IWC-SC/58E35. International Whaling Commission Scientific Committee, Cambridge, UK.

Barlow, J. & Gentry, R. (2004). *Report of the NOAA Workshop on Anthropogenic Sound and Marine Mammals, 19–20 February 2004*. U.S. Department of Commerce, National Oceanic and Atmospheric Administration.

Beck, M.W., Heck, K.L., Able, K.W., et al. (2001). The identification, conservation, and management of estuarine and marine nurseries for fish and invertebrates. *BioScience*, 51, 633–641.

Boussard, A. (1981). *The reactions of roach* (Rutilus rutilus) *and rudd* (Scardinius erythrophthalmus) *to noises produced by high speed boating*. Proceedings of the 2nd British Freshwater Fisheries Conference, Liverpool. pp. 188–200.

Boyle, K.S. & Cox, T.E. (2009). Courtship and spawning sounds in bird wrasse *Gomphosus varius* and saddle wrasse *Thalassoma duperrey*. *Journal of Fish Biology*, **75**, 2670–2681.

Boyle, K.S. & Tricas, T.C. (2010). Pulse sound generation, anterior swim bladder buckling and associated muscle activity in the pyramid butterflyfish, *Hemitaurichthys polylepis*. *Journal of Experimental Biology*, **213**, 3881–3893.

Bruintjes, R. & Radford, A.N. (2013). Context-dependent impacts of anthropogenic noise on individual and social behaviour in a cooperatively breeding fish. *Animal Behaviour*, **85**, 1343–1349.

Bruintjes, R. & Radford, A.N. (2014). Chronic playback of boat noise does not impact hatching success or post-hatching larval growth and survival in a cichlid fish. *PeerJ*, **2**, e594.

Castle, M.J. & Kibblewhite, A.C. (1975). The contribution of the sea urchin to ambient sea noise. *Journal of the Acoustical Society of America*, **58**, S122.

Cato, D.H. (1992). The biological contribution to the ambient noise in waters near Australia. *Acoustics Australia*, **20**, 76–80.

Cato, D.H. & McCauley, R.D. (2002). Australian research into ambient sea noise. *Acoustics Australia*, **30**, 13–20.

Clark, C.W., Ellison, W.T., Southall, B.L., et al. (2009). Acoustic masking in marine ecosystems: intuitions, analysis, and implication. *Marine Ecology Progress Series*, **395**, 201–222.

Clark, C.W. & Gagnon, G.C. (2006). *Considering the temporal and spatial scales of noise exposures from seismic surveys on baleen whales*. Paper SC/58/E9 presented to the IWC Scientific Committee. International Whaling Commission, Cambridge, UK.

Cummings, W.C. & Thompson, P.O. (1971). Underwater sounds from the blue whale, *Balaenoptera musculus*. *Journal of the Acoustical Society of America*, **50**, 1193–1198.

De Robertis, A., Wilson, C.D., Williamson, N.J., et al. (2010). Silent ships sometimes do encounter more fish. 1. Vessel comparisons during winter pollock surveys. *ICES Journal of Marine Science*, **67**, 985–995.

Deruiter, S.L., Southall, B.L., Calambokidis, J., et al. (2013). First direct measurements of behavioural responses by Cuvier's beaked whales to mid-frequency active sonar. *Biology Letters*, **9**.

Engas, A. & Lokkeborg, S. (2002). Effects of seismic shooting and vessel-generated noise on fish behaviour and catch rates. *Bioacoustics*, **12**, 131–135.

Engas, A., Lokkeborg, S., Ona, E., & Soldal, A.V. (1996). Effects of seismic shooting on local abundance and catch rates of cod (*Gadus morhua*) and haddock (*Melanogrammus aeglefinus*). *Canadian Journal of Fisheries and Aquatic Sciences*, **53**, 2238–2249.

Erbe, C. (2002). Underwater noise of whale-watching boats and potential effects on killer whales (*Orcinus orca*), based on an acoustic impact model. *Marine Mammal Science*, **18**, 394–418.

Evans, D.L. & England, G.R. (2001). *Joint Interim Report Bahamas Marine Mammal Stranding Event of 15–16 March 2000*. Joint report of US Department of Commerce and US Navy.

Fay, R. (2009). Soundscapes and the sense of hearing of fishes. *Integrative Zoology*, **4**, 26–32.

Finneran, J.J., Schlundt, C.E., Dear, R., et al. (2002). Temporary shift in masked hearing thresholds in odontocetes after exposure to single underwater impulses from a seismic watergun. *Journal of the Acoustical Society of America*, **111**, 2929–2940.

Fisher, R., Leis, J.M., Clark, D.L., & Wilson, S. K. (2005). Critical swimming speeds of late-stage coral reef fish larvae: variation within species, among species and between locations. *Marine Biology*, **147**, 1201–1212.

Goold, J.C. & Coates, R.F.W. (2006). *Near Source, High Frequency Air-Gun Signatures*. IWCSC/58/E30. International Whaling Commission Scientific Committee, Cambridge, UK.

Gotz, T., Hastie, G., Hatch, L.T., et al. (2009). *Overview of the Impacts of Anthropogenic Underwater Sound in the Marine Environment*. OSPAR Commission.

Greene, J. & Moore, S.E. (1995). Man-made noise. In: Richardson, J.W., Greene, J., Malme, C.I., & Thomson, D.H. (eds), *Marine Mammals and Noise*, pp. 101–158. Academic Press, San Diego.

Halvorsen, M., Carlson, T. & Copping, A. (2011). *Effects of Tidal Turbine Noise on Fish Hearing and Tissues. Environmental Effects of Marine and Hydrokinetic Energy*. Prepared for the U.S. Department of Energy under Contract DE-AC05–76RL01830. Pacific Northwest National Laboratory.

Harvey, B.P., Gwynn-Jones, D., & Moore, P.J. (2013). Meta-analysis reveals complex marine biological responses to the interactive effects of ocean acidification and warming. *Ecology and Evolution*, **3**, 1016–1030.

Hawkins, A.D. & Myrberg, A.A. Jr (1983). Hearing and sound communication underwater. In: Lewis, B. (ed.), *Bioacoustics: A Comparative Approach*, pp. 347–405. Academic Press, London.

Hildebrand, J.A. (2009). Anthropogenic and natural sources of ambient noise in the ocean. *Marine Ecology Progress Series*, **395**, 5–20.

Holles, S., Simpson, S.D., Radford, A.N., et al. (2013). Boat noise disrupts orientation behaviour in a coral reef fish. *Marine Ecology Progress Series*, **485**, 295–300.

Jeffs, A., Tolimieri, N., & Montgomery, J.C. (2003). Crabs on cue for the coast: the use of underwater sound for orientation by pelagic crab stages. *Marine and Freshwater Research*, **54**, 841–845.

Jeffs, A.G., Montgomery, J.C., & Tindle, C.T. (2005). How do spiny lobster post-larvae find the coast? *New Zealand Journal of Marine and Freshwater Research*, **39**, 605–617.

Jensen, F.H., Bejder, L., Wahlberg, M., et al. (2009). Vessel noise effects on delphinid communication. *Marine Ecology Progress Series*, **395**, 161–175.

Joseph, J.E. & Chiu, C.S. (2010). A computational assessment of the sensitivity of ambient noise level to ocean acidification. *Journal of the Acoustical Society of America*, **128**, EL144–EL149.

Kaiser, K. & Hammers, J.L. (2009). The effect of anthropogenic noise on male advertisement call rate in the neotropical treefrog, *Dendropsophus triangulum*. *Behaviour*, **146**, 1053–1069.

Kaiser, K., Scofield, D., Alloush, M., et al. (2011). When sounds collide: the effect of anthropogenic noise on a breeding assemblage of frogs in Belize, Central America. *Behaviour*, **148**, 215–232.

Kastelein, R. & Jennings, N. (2012). Impacts of anthropogenic sounds on Phocoena phocoena (harbor porpoise). In: Popper, A. N. & Hawkins, A. (eds), *Effects of Noise on Aquatic Life*, pp. 311–315. Springer, New York.

Kasumyan, A. (2008). Sounds and sound production in fishes. *Journal of Ichthyology*, **48**, 981–1030.

Kight, C.R., Saha, M.S., & Swaddle, J.P. (2012). Anthropogenic noise is associated with reductions in the productivity of breeding Eastern Bluebirds (*Sialia sialis*). *Ecological Applications*, **22**, 1989–1996.

Leis, J.M. & Carson-Ewart, B.M. (1997). *In situ* swimming speeds of the late pelagic larvae of some Indo-Pacific coral-reef fishes. *Marine Ecology Progress Series*, **159**, 165–174.

Leis, J.M. & Carson-Ewart, B.M. (1999). *In situ* swimming and settlement behaviour of larvae of an Indo-Pacific coral-reef fish, the coral trout *Plectropomus leopardus* (Pisces: Serranidae). *Marine Biology*, **134**, 51–64.

Leis, J.M. & Lockett, M.M. (2005). Localization of reef sounds by settlement-stage larvae of coral-reef fishes (Pomacentridae). *Bulletin of Marine Science*, **76**, 715–724.

Leis, J.M. & McCormick, M.I. (2002). The biology, behaviour and ecology of the pelagic, larval stage of coral reef fishes. In: Sale, P.F. (ed.), *Coral Reef Fishes: Dynamics and Diversity in a Complex Ecosystem*, pp. 171–199. Academic Press, San Diego.

Lepper, P.A., Turner, V.L.G., Goodson, A.D., & Black, K.D. (2004). *Source levels and spectra emitted by three commercial aquaculture anti-predation devices*. Seventh European Conference on Underwater Acoustics, ECUA. Delft, the Netherlands.

Lesage, V., Barrette, C., Kingsley, M.C.S., & Sjare, B. (1999). The effects of vessel noise on the vocal behavior of Belugas in the St. Lawrence river estuary, Canada. *Marine Mammal Science*, **15**, 65–84.

Lillis, A., Eggleston, D.B., & Bohnenstiehl, D.R. (2013). Oyster larvae settle in response to habitat-associated underwater sounds. *PLoS ONE*, **8**, e79337.

Lloyd, T.P., Turnock, S.R., & Humphrey, V.F. (2011). *Modelling techniques for underwater noise generated by tidal turbines in shallow water*. ASME 2011 30th International Conference on Ocean, Offshore and Arctic Engineering, pp. 777–785. American Society of Mechanical Engineers.

Madsen, P.T., Johnson, M., Miller, P.J.O., et al. (2006a). Quantitative measures of air-gun pulses recorded on sperm whales (*Physeter macrocephalus*) using acoustic tags during controlled exposure experiments. *Journal of the Acoustical Society of America*, **120**, 2366–2379.

Madsen, P.T., Wahlberg, M., Tougaard, J., et al. (2006b). Wind turbine underwater noise and marine mammals: implications of current knowledge and data needs. *Marine Ecology Progress Series*, **309**, 279–295.

Mann, D.A., Locascio, J.V., Coleman, F.C., & Koenig, C.C. (2009). Goliath grouper *Epinephelus itajara* sound production and movement patterns on aggregation sites. *Endangered Species Research*, **7**, 229–236.

McCauley, R., Fewtrell, J., Duncan, A., et al. (2000). Marine seismic surveys: a study of environmental implications. *APPEA Journal*, **2000**, 692–708.

McCauley, R.D., Fewtrell, J., & Popper, A.N. (2003). High intensity anthropogenic sound damages fish ears. *Journal of the Acoustical Society of America*, **113**, 638–642.

McDonald, M.A., Hildebrand, J.A., & Wiggins, S.M. (2006). Increases in deep ocean ambient noise in the Northeast Pacific west of San Nicolas Island, California. *Journal of the Acoustical Society of America*, **120**, 711–718.

McKenna, M.F., Ross, D., Wiggins, S.M., & Hildebrand, J.A. (2012). Underwater radiated noise from modern commercial ships. *Journal of the Acoustical Society of America*, **131**, 92–103.

Miller, L.A. (2010). Prey capture by harbor porpoises (*Phocoena phocoena*): a comparison between echolocators in the field and in captivity. *Journal of the Marine Acoustic Society of Japan*, **37**, 156–168.

Miller, P.J.O., Johnson, M.P., & Tyack, P.L. (2004). Sperm whale behaviour indicates the use of echolocation click buzzes 'creaks' in prey capture. *Proceedings of the Royal Society of London Series B*, **271**, 2239–2247.

Møhl, B. (2001). *Sperm whale acoustics: the clicking machine and its output*. 17th International Congress on Acoustics, Rome.

Mollmann, C., Folke, C., Edwards, M., & Conversi, A. (2015). Marine regime shifts around the globe: theory, drivers and impacts. *Philosophical Transactions of the Royal Society Series B*, **370**: 20130273, 1–12.

Montgomery, J.C., Jeffs, A., Simpson, S.D., et al. (2006). Sound as an orientation cue for the pelagic larvae of reef fishes and decapod crustaceans. *Advances in Marine Biology*, **51**, 143–196.

Morley, E.L., Jones, G., & Radford, A.N. (2014). The importance of invertebrates when considering the impacts of anthropogenic noise. *Proceedings of the Royal Society of London Series B*, **281**, 21032683.

Munk, W.H., Spindel, R.C., Baggeroer, A., & Birdsall, T.G. (1994). The Heard Island Feasibility Test. *Journal of the Acoustical Society of America*, **96**, 2330–2342.

Myrberg, A.A. (1997). Underwater sound: its relevance to behavioral functions among fishes and marine mammals. *Marine and Freshwater Behaviour and Physiology*, **29**, 3–21.

Myrberg, A.A. & Lugli, M. (2006). Reproductive behavior and acoustical interactions. In: Ladich, F., Collin, S.P., Moller, P., & Kapoor, B.G. (eds), *Communication in Fishes. Volume 1: Acoustical and Chemical Communication*, pp. 149–176. Science Publishers, Enfield, New Hampshire.

Myrberg, A.A., Mohler, M., & Catala, J.D. (1986). Sound production by males of a coral reef fish (*Pomacentrus partitus*)—its significance to females. *Animal Behaviour*, **34**, 913–923.

Myrberg Jr, A.A. (1997). Underwater sound: its relevance to behavioral functions among fishes and marine mammals. *Marine and Freshwater Behaviour and Physiology*, **29**, 3–21.

Myrberg Jr, A.A. & Fuiman, L.A. (2002). The sensory world of coral reef fishes. In: Sale, P.F. (ed.) *Coral Reef Fishes: Dynamics and Diversity in a Complex Ecosystem*, pp. 123–148. Academic Press, San Diego.

Myrberg Jr, A.A., Spanier, E., & Ha, S.J. (1978). Temporal patterning in acoustical communication. In: Reese, E.S. & Lighter, F. (eds), *Contrasts in Behavior*, pp. 137–179. Wiley and Sons, New York.

National Marine Manufacturers Association (2002). *NMMA Research Library* [Online]. http://www.nmma.org.

Nedelec, S.L., Radford, A.N., Simpson, S.D., et al. (2014). Anthropogenic noise playback impairs embryonic development and increases mortality in a marine invertebrate. *Scientific Reports*, **4**, 5891.

Nedwell, J. & Howell, D. (2004). *A Review of Offshore Windfarm Related Underwater Noise Sources*. Tech. Rep. 544R0308. Prepared by Subacoustech Ltd, Hampshire, UK.

Nieukirk, S.L., Stafford, K.M., & Fox, C.G. (2004). Low-frequency whale and seismic airgun sounds recorded in the mid-Atlantic Ocean. *Journal of the Acoustical Society of America*, **115**, 1832–1843.

Nowacek, S.M., Wells, R.S., & Solow, A.R. (2001). Short-term effects of boat traffic on bottlenose dolphins, *Tursiops truncatus*, in Sarasota Bay, Florida. *Marine Mammal Science*, **17**, 673–688.

NRC (National Research Council) (2003). *Ocean Noise and Marine Mammals*. National Academic Press, Washington, DC.

Obermeier, M. & Schmitz, B. (2003). Recognition of dominance in the big-clawed snapping shrimp (*Alpheus heterochaelis* Say 1818) part I: individual or group recognition? *Marine and Freshwater Behaviour and Physiology*, **36**, 1–16.

Parvin, S.J., Workman, R., Bourke, P., & Nedwell, J.R. (2005). *Assesment of Tidal Current Turbine Noise at Lybmouth Site and Predicted Impact of Underwater Noise in Strangford Lough*. Report 628 R 0102 628 R 0102. Subacoustech Ltd, Hampshire, UK.

Payne, R. & Webb, D. (1971). Orientation by means of long range acoustic signaling in baleen whales. *Annals of the New York Academy of Sciences*, **188**, 110–141.

Picciulin, M., Sebastianutto, L., Codarin, A., et al. (2012). Brown meagre vocalization rate increases during repetitive boat noise exposures: a possible case of vocal compensation. *Journal of the Acoustical Society of America*, **132**, 3118–3124.

Pine, M.K., Jeffs, A., & Radford, C.A. (2013). The torment beneath. *Planning Quarterly*, **188**, 15–19.

Pine, M. K., Jeffs, A., & Radford, C.A. (in press). The propagation of underwater sound from tidal turbines in shallow water. *Journal of Applied Ecology*.

Pine, M.K., Jeffs, A.G., & Radford, C.A. (2012). Turbine sound may influence the metamorphosis behaviour of estuarine crab megalopae. *PLoS ONE*, **7**, e51790.

Popper, A. & Hawkins, A.D. (eds) (2012). *The Effects of Noise on Aquatic Life*. Springer, New York.

Popper, A.N. (2003). Effects of anthropogenic sounds on fishes. *Fisheries*, **28**, 24–31.

Popper, A.N. & Hastings, M.C. (2009). The effects of anthropogenic sources of sound on fishes. *Journal of Fish Biology*, **75**, 455–489.

Popper, A.N., Smith, M.E., Cott, P.A., et al. (2005). Effects of exposure to seismic airgun use on hearing of three fish species. *Journal of the Acoustical Society of America*, **117**, 3958–3971.

Radford, C., Stanley, J., Simpson, S., & Jeffs, A. (2011). Juvenile coral reef fish use sound to locate habitats. *Coral Reefs*, **30**, 295–305.

Radford, C.A., Jeffs, A.G., & Montgomery, J.C. (2007). Directional swimming behavior by five species of crab postlarvae in response to reef sound. *Bulletin of Marine Science*, **80**, 369–378.

Radford, C.A., Jeffs, A.G., Tindle, C.T., & Montgomery, J.C. (2008). Resonating sea urchin skeletons create coastal choruses. *Marine Ecology Progress Series*, **362**, 37–43.

Reeder, D.B. & Chiu, C.S. (2010). Ocean acidification and its impact on ocean noise: phenomenology and analysis. *Journal of the Acoustical Society of America*, **128**, EL137–EL143.

Reeves, R.R., Hofman, R.J., Silber, G.K., & Wilkinson, D. (1996). *Acoustic deterrence of harmful marine mammal-fishery interactions*. Proceedings of a workshop held in Seattle Washington, p. 70.

Reeves, R.R., Read, A.J., & Notarbartolo Di Sciara, G. (2001). *Report of the Workshop on Interactions between Dolphins and Fisheries in the Mediterranean: Evaluation of Mitigation Alternatives*. Istituto Centrale per la Ricerca Scientifica e Tecnologica Applicata al Mare, Rome, Italy.

Reijnen, R. & Foppen, R. (1994). The effects of car traffic on breeding bird populations in woodland. 1. Evidence of reduced habitat quality for willow warblers (*Phylloscopus trochilus*) breeding close to a highway. *Journal of Applied Ecology*, **31**, 85–94.

Richardson, W.J., Green, C.R., Malme, C.I., et al. (1991). *Effects of Noise on Marine Mammals*. OCS Study MMS90-0093. LGL Rep. TA834-1. Report from LGL Ecol. Res. Assoc., Inc., Bran, Texas, for US Minerals Management Service, Atlantic Outer Continental Shelf Region, Herndon, VA. NTIS PB91-168914. 462 pp.

Richardson, W.J., Greene, C.R.J., Malme, C.I., & Thomson, D.H. (1995). *Marine Mammals and Noise*. Academic Press, San Diego.

Rogers, P.H. & Cox, M. (1988). Underwater sound as a biological stimulus. In: Atema, J., Fay, R.R., Popper, A.N., & Tavolga, W.N. (eds), *Sensory Biology of Aquatic Animals*, pp. 131–149. Springer, New York.

Ross, D. (1976). *Mechanics of Underwater Noise*. Pergamon Press, New York.

Ross, D. (1993). On ocean underwater ambient noise. *Institute of Acoustics Bulletin*, **18**, 5–8.

Saucier, M.H. & Baltz, D.M. (1993). Spawning site selection by spotted sea-trout, *Cynoscion nebulosus*, and black drum, *Pogonias cromis*, in Louisiana. *Environmental Biology of Fishes*, **36**, 257–272.

Schultz, S., Wuppermann, K., & Schmitz, B. (1998). Behavioural interactions of the snapping shrimp (*Alpheus heterochaelis*) with conspecifics and sympatric crabs (*Eurypanopeus depressus*). *Zoology*, **101**(Suppl. 1), 85.

Simpson, S.D., Meekan, M.G., McCauley, R.D., & Jeffs, A. (2004). Attraction of settlement-stage coral reef fishes to reef noise. *Marine Ecology Progress Series*, **276**, 263–268.

Simpson, S.D., Meekan, M.G., Montgomery, J.C., et al. (2005). Homeward sound. *Science*, **308**, 221.

Simpson, S.D., Purser, J., & Radford, A.N. (2014). Anthropogenic noise compromises antipredator behaviour in European eels. *Global Change Biology*, **21**(2), 586–593.

Simpson, S.D., Radford, A.N., Tickle, E.J., et al. (2011). Adaptive avoidance of reef noise. *PLoS ONE*, **6**, e16625.

Slabbekoorn, H., Bouton, N., Van Opzeeland, I., et al. (2010). A noisy spring: the impact of globally rising underwater sound levels on fish. *Trends in Ecology and Evolution*, **25**, 419–427.

Slotte, A., Hansen, K., Dalen, J., & Ona, E. (2004). Acoustic mapping of pelagic fish distribution and abundance in relation to a seismic shooting area off the Norwegian west coast. *Fisheries Research*, **67**, 143–150.

Southall, B.L. & Nowacek, D.P. (2009). Acoustics in marine ecology: innovation in technology expands the use of sound in ocean science. *Marine Ecology Progress Series*, **395**, 1–3.

Stanley, J.A., Radford, C.A., & Jeffs, A.G. (2010). Induction of settlement in crab megalopae by ambient underwater reef sound. *Behavioral Ecology*, **21**, 113–120.

Stanley, J.A., Radford, C.A., & Jeffs, A.G. (2011). Behavioural response thresholds in New Zealand crab megalopae to ambient underwater sound. *PLoS ONE*, **6**.

Stanley, J.A., Radford, C.A., & Jeffs, A.G. (2012). Location, location, location: finding a suitable home among the noise. *Proceedings of the Royal Society of London Series B*, **279**, 3622–3631.

Stanley, J.A., Wilkens, L.S., & Jeffs, A.G. (2014). Fouling in your own nest: vessel noise increases biofouling. *Biofouling*, **30**, 837–844.

Stobutzki, I.C. & Bellwood, D.R. (1997). Sustained swimming abilities of the late pelagic stages of coral reef fishes. *Marine Ecology Progress Series*, **149**, 35–41.

Sun, J.W.C. & Narins, P.M. (2005). Anthropogenic sounds differentially affect amphibian call rate. *Biological Conservation*, **121**, 419–427.

Tait, R. (1962). *The Evening Chorus: A Biological Noise Investigation*. Report No. 26. Naval Research Laboratory, HMNZ Dockyard. Auckland, New Zealand.

Tavolga, W.N. (1964). *Marine Bio-Acoustics*. Pergamon Press, New York.

Tavolga, W.N. (1976). Acoustic obstacle detection in the sea catfish (*Arius felis*). In: Schuijf, A. & Hawkins, A.D. (eds), *Sound Reception in Fish*, pp. 185–204. Elsevier, Amsterdam.

Thomsen, F., Ludemann, K., Kafemann, R., & Piper, W. (2006). *Effects of Offshore Wind Farm Noise on Marine Mammals and Fish*. Biola, Hamburg, Germany on behalf of COWRIE Ltd.

Thorson, R.F. & Fine, M.L. (2002). Acoustic competition in the gulf toadfish *Opsanus beta*: acoustic tagging. *Journal of the Acoustical Society of America*, **111**, 2302–2307.

Tolimieri, N., Haine, O., Jeffs, A., et al. (2004). Directional orientation of pomacentrid larvae to ambient reef sound. *Coral Reefs*, **23**, 184–191.

Tolimieri, N., Jeffs, A., & Montgomery, J.C. (2000). Ambient sound as a cue for navigation by the pelagic larvae of reef fishes. *Marine Ecology Progress Series*, **207**, 219–224.

Tougaard, J., Carstensen, J., Henriksen, O.H., et al. (2003). *Short-Term Effects of the Construction of Wind Turbines on Harbour Porpoises at Horns Reef*. Technical report to Techwise A/S. Hedeselskabet.

Tricas, T.C., Kajiura, S.M., & Kosaki, R.K. (2006). Acoustic communication in territorial butterflyfish: test of the sound production hypothesis. *Journal of Experimental Biology*, **209**, 4994–5004.

Tyack, P.L. (1998). *Acoustic Communication under the Sea*. Springer, Berlin.

Urick, R.J. (1983). *Principles of Underwater Sound*. McGraw-Hill, New York.

Versluis, M., Schmitz, B., Von Der Heydt, A., & Lohse, D. (2000). How Snapping Shrimp Snap: through Cavitating Bubbles. *Science*, **289**, 2114–2117.

Voellmy, I.K., Purser, J., Flynn, D., et al. (2014a). Acoustic noise reduces foraging success in two sympatric fish species via different mechanisms. *Animal Behaviour*, **89**, 191–198.

Voellmy, I.K., Purser, J., Simpson, S.D., & Radford, A.N. (2014b). Increased noise levels have different impacts on the anti-predator behaviour of two sympatric fish species. *PLoS ONE*, **9**, e102946.

Wale, M.A., Simpson, S.D., & Radford, A.N. (2013a). Noise negatively affects foraging and antipredator behaviour in shore crabs. *Animal Behaviour*, **86**, 111–118.

Wale, M.A., Simpson, S.D., & Radford, A.N. (2013b). Size-dependent physiological responses of shore crabs to single and repeated playback of ship noise. *Biology Letters*, **9**, 20121103.

Watkins, W.A. (1981). Activities and underwater sounds of fin whales. *Scientific Reports of the Whales Research Institute*, **33**, 83–117.

Weilgart, L.S. (2007). The impacts of anthropogenic ocean noise on cetaceans and implications for management. *Canadian Journal of Zoology*, **85**, 1091–1116.

Wenz, G.M. (1962). Acoustic ambient noise in ocean—spectra and sources. *Journal of the Acoustical Society of America*, **34**, 1936–1956.

Wilkens, S.L., Stanley, J.A., & Jeffs, A.G. (2012). Induction of settlement in mussel (*Perna canaliculus*) larvae by vessel noise. *Biofouling*, **28**, 65–72.

Williams, R., Trites, A.W., & Bain, D.E. (2002). Behavioural responses of killer whales (*Orcinus orca*) to whale-watching boats: opportunistic observations and experimental approaches. *Journal of Zoology*, **256**, 255–270.

Wysocki, L.E., Davidson III, J.W., Smith, M.E., et al. (2007). Effects of aquaculture production noise on hearing, growth, and disease resistance of rainbow trout *Oncorhynchus mykiss*. *Aquaculture*, **272**, 687–697.

Yurk, H. & Trites, A.W. (2000). Experimental attempts to reduce predation by harbour seals (*Phoca vitulina*) on out-migrating juvenile salmonids. *Transactions of the American Fisheries Society*, **129**, 1360–1366.

PART III

Societal Implications

Managing complex systems to enhance sustainability

Simon Willcock, Sarwar Hossain, and Guy M. Poppy

17.1 Introduction

It is widely recognized that global natural resources and processes must be managed sustainably to ensure the continuation of livelihoods. More recent developments demonstrate that this must be achieved in a fair and just manner, safeguarding people's most basic needs and satisfying human rights. With increasing populations, the window through which human needs can be met without exceeding the planet's 'environmental ceiling' narrows by the day. Through the utilization of research frameworks, we can use our understanding of smaller-scale social and ecological processes to estimate the emergent properties of global socio-ecological systems. The resulting holistic models enable us to explore the impacts of proposed management options, reducing the risk associated with large-scale changes in policy and practice; enabling society to make the necessary progression towards sustainable resource management more rapidly. Here we aim to shed light on the science required to develop evidence-based policy enabling the sustainable management of complex systems. Firstly, we introduce planetary boundaries, just social systems, and tipping points, discussing each concept separately as it is introduced. Following this, we use examples from freshwater and marine ecosystems to present existing frameworks and modelling techniques used to address complex transdisciplinary issues. Finally, we discuss how risks, often derived from our current lack of understanding, can be managed, enabling policy makers to address urgent global issues (e.g. climate change) as soon as possible.

17.2 The planetary boundaries concept

Rockström et al. (2009) provided the planetary boundary concept, quantifying and defining Earth's biophysical limits. Exceeding these limits by use of natural resources is expected to change the planet's systems to the extent that consequences become unpredictable and/or unmanageable. As such, this group of scientists, led by the Stockholm Resilience Centre, identified a 'safe operating space' using biophysical data from the Holocene (the last 11 000 years) as baseline markers.

Rockström et al. (2009) identified boundaries for seven biophysical changes:

- Climate change: The boundary value for CO_2 concentration (the proxy for global climate change) is set at 350 ppm, beyond which there is a risk of loss of some parts of the large polar ice sheets and impacts for ecosystems.
- Ocean acidification: As atmospheric CO_2 concentrations increase, so too does ocean acidity. Here, ocean acidification was measured by using the aragonite saturation state (Q_{arag}) which is declining due to rising acidity. Marine organisms are very sensitive to changes in ocean acidity, currently occurring at least 100 times faster than at any other time in the last 20 million years. Declining aragonite concentrations makes water more corrosive and poses a great threat to coral reefs. Rockström et al. (2009) suggest we must maintain at 80% or higher of the average global pre-industrial aragonite saturation state as the boundary for this process.
- Stratospheric ozone: A less than 5% reduction of stratospheric O_3 with respect to 1964–1980 has been set as the threshold level for stratospheric ozone depletion.
- Biogeochemical flows (nitrogen and phosphorus cycle): Anthropogenic nitrogen (N) and phosphorus (P) flows from regional to global scales could be a major threat to marine systems due to global-scale oceanic anoxic events and coastal eutrophication.

Stressors in the Marine Environment. Edited by Martin Solan and Nia M. Whiteley
© Oxford University Press 2016. Published in 2016 by Oxford University Press.

Thus, no more than 35 Mt N yr^{-1} and less than 10 times of anthropogenic P flow with respect to natural weathering of N flow have been set as the safe operating spaces.

- Global freshwater use: Alteration of inputs from freshwater ecosystems could substantially influence marine systems. Reduction in freshwater (e.g. rainfall run-off and river water flow) could affect freshwater–ocean mixing, increasing eutrophication in coastal ecosystems. Therefore, crossing the boundary of < 4000 km^3 yr^{-1} could result in collapse of terrestrial and freshwater ecosystems as well as major shifts in marine systems.

- Land system change: Current ice-free land used in agriculture is 12% of that available and the boundary is considered to be when land systems converted to crop land result in 15% of ice-free land being cropped. This aggregated global boundary is dependent on local-level food production and cropping intensity; in addition, this slow variable influences the other biophysical processes (e.g. climate change and biodiversity)

- Biological diversity: These above-planetary processes are negatively affecting biodiversity in both land and ocean systems. The background rate of extinction for marine organisms is 0.1–1 extinctions per million species per year (E/MYS). Current extinction rates are estimated to be between 100 and 1000-fold larger than the background rate; this exceeds the preliminary boundary of less than 10 E/MYS.

Rockström et al. (2009) did not identify the boundary for chemical pollution due to data deficiency. The scientific community lacks sufficient knowledge of the aggregation of global chemicals, as well as their atmospheric aerosol loading. Both of these processes are important for marine systems, where the benthic community have shown marked changes due to chemical pollution, for example dichlorodiphenyltrichloroethane (DDT; Schramm and Nienhuis (1996)), transported to marine systems through the atmosphere (Singer, 1970). Moreover, the aerosol loading can also alter the hydrological cycle which can ultimately affect the marine system in various ways (Ferek et al., 2000).

The planetary boundary concept has already received much attention among sustainability scientists supporting and critiquing it, as well as from governments/regulators. Carpenter and Bennett (2011) followed up the concept by using more specific theories to define the phosphorus boundaries for ecosystems. Though Rockström et al. (2009) concluded that the

ocean phosphorus loading has not yet transgressed the boundary for oceanic anoxic events, Carpenter and Bennett (2011) concluded that the freshwater phosphorus loading has already crossed the planetary standards, but that planetary boundaries for phosphorus flow from freshwater to sea are below the boundary level.

17.3 Linking planetary boundaries with social systems

A striking criticism of the planetary boundary concept is that it mostly ignores the aspects of social systems, the patterned series of interrelationships existing between individuals, groups, and institutions. Social systems drive us towards many of the boundaries, but also have the potential to hold keep us from crossing these thresholds. For example, Raworth (2012) argues that the planetary boundary concept ignores the inequality of access to natural resources and that drawing a global-scale boundary would not fulfil the goal of sustainable development. Half of global carbon emissions are emitted by only 11% of the world's population. More than one third of the global nitrogen budget is used by only 7% of the people. However, small changes can have great impacts; for example, it would only require a 1% increase in global supply to provide food for over one tenth of the world population currently facing hunger (Raworth, 2012).

Raworth (2012) revises the planetary boundaries concept to also include social systems such as access to food and water, gender equality, energy, income, education, social equity, jobs, voice, and health. The results can be presented as a 'doughnut' where sustainable development can take place; outer and inner circles represent the boundaries for the safe and just operating spaces respectively (Fig. 17.1). However, she does not identify the boundaries for social systems, which could be set as per the national, regional, or global scale norms, but highlights the possibility of increasing poverty if the planetary boundaries are exceeded. For example, Bangladesh, where around 40% of people are below the poverty line, is one of the most vulnerable countries to climate change and thus there is a high risk of poverty increases due to global climate change, for which they are not responsible.

Dearing et al. (2014) propose a framework to define the safe and just operating space on a regional scale using the planetary boundary and doughnut concepts. The researchers use time-series analysis to define safe operating space in two provinces of China, where

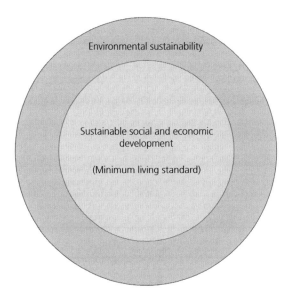

Figure 17.1 Safe and just operating space, where the outer circle represents the environmental boundary, beyond which there is a dangerous zone (undesirable condition) for humanity. The area between the two circles represents desirable environmental conditions, where poverty eradication can be achieved for sustainable economic and social development. To sustain minimum living standards for all people we have to ensure we stay within this safe operating space. Adapted from Raworth (2012).

numerous environmental variables (e.g. air and water quality, sediment regulation, etc.) and social variables (e.g. sanitation, food, health, education, etc.) can be considered. Though the planetary approach focused on a safe zone at/for the global scale and Dearing et al. demonstrate these approaches on a regional scale, the regional scale sustainability in the safe zone is mostly unexplored. Focusing on safe operating spaces at regional scales may increase tractability and buy-in from policy makers. Moreover, the indicators for regional scale changes could be varied, addressing the needs and limits of that specific system, as opposed to averaged global-scale planetary boundaries. The application of the planetary boundaries–social systems doughnut is difficult due to the complex nature of the systems involved, coupled with the needs to truly undertake interdisciplinary research, frequently spoken about but rarely achieved, as identified in the Future Earth Initiative (http://www.futureearth.org/). For example, the investigation of fish capture at a specific scale must take into account marine environmental issues at the global scale (e.g. climate change), as well as other local affects (e.g. pollution) and the identification

of 'winners' and 'losers' within the socio-economic system.

17.4 Tipping points

Both planetary boundaries and social systems can involve complex, non-linear relationships and may well include a wide range of tipping points. Tipping points are said to occur when a small change in a driving force results in a strongly non-linear response in structure, function, composition, or productivity, and potential large-scale shifts in state when a critical threshold, known as a 'catastrophic bifurcation', is passed (Lenton, 2011). Tipping points are typified by the large impacts of very small changes which require significant investment to reverse. Simply returning the driver of the change back to its previous levels may not be enough to recreate the former state due to internal positive feedback effects. Because of the small perturbations causing state changes and the inherent complexity of natural systems, the exact timing and spatial location of tipping points is difficult to predict; making setting thresholds for planetary boundaries and social systems is fraught with uncertainty. Here, we introduce tipping points and the cutting-edge research involved in their prediction, using examples from freshwater and marine systems.

An experiment with tipping points in a controlled marine environment has been provided by Drake and Griffen (2010). In a large replicated experiment, the authors slowly reduced the food available to populations of zooplankton (*Daphnia*), ultimately resulting in the populations crashing suddenly and dramatically. However, do these controlled conditions illustrate a true tipping point? In this instance, the removal of the perturbation via the provision of additional food sources would likely reverse the population crash for all but small populations.

Real-world examples of tipping points are more complex and often only well understood with hindsight. For decades, Caribbean coral reefs have been intensively researched, but few predicted the dramatic and sudden shift of the vast majority of reefs into an algal encrusted state (Nyström et al., 2000). The nutrient loading on the system gradually increased over several years, providing ideal conditions for algal growth. This went unnoticed by researchers as herbivorous fish suppressed algal populations, providing the system with the resilience to resist this perturbation. Overfishing severely impacted the ability of these fish to perform this function; however, the sea urchin *Diadema antillarum* increased in abundance, once again demonstrating the extraordinary complexity and re-silience contained within natural systems. In 1983, the dense *D. antillarum* populations were afflicted by a pathogen, reducing the grazing pressure on the algae, causing the reefs to rapidly become overgrown. Al-though the stress that ultimately resulted in the shift to an algal encrusted state gradually built over several years, it is thought a tipping point occurred as the final perturbation, an outbreak of disease, was relatively small and the phase-shift it brought about will require significant investment to reverse. Similar dramatic re-gime shifts are now documented in numerous marine and freshwater systems, including sudden synchron-ized population changes of numerous organisms in the open ocean (Reid et al., 1998), standing waters abruptly becoming overgrown by floating plants (Scheffer et al., 2003), and lakes shifting from clear to turbid (Carpen-ter et al., 1999). Examples are also evident in terrestrial

systems (e.g. Hirota et al. (2011)); however, these are typically less well understood.

While important knowledge has been gained from hindsight, there is a widespread need to develop an early warning mechanism for the detection of indica-tors of tipping points in order to avoid the negative consequences that often accompany them (Lenton, 2011). These concepts can be visualized as a ball in a well that gets shallower as the tipping point is neared (see Fig. 17.2). The wells in the system represent sta-ble attractors and the ball represents the state of the system. In Fig. 17.2, the right-hand side well becomes shallower over time as the result of a gradual forcing. The tipping point is finally reached as the ball shifts left and settles in the alternative stable state. This visu-alization is accompanied with several predictions al-lowing us to anticipate the point at which the ball will shift leftwards. Firstly, for a given perturbation, the

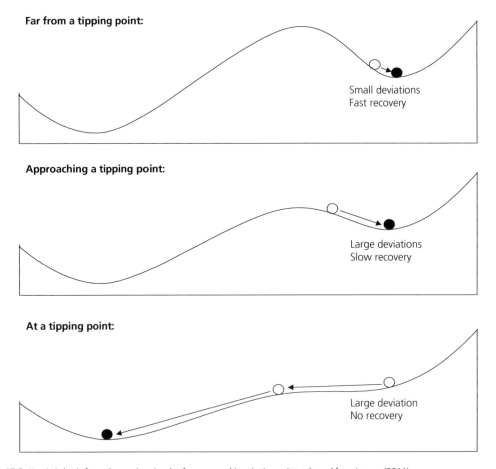

Far from a tipping point:

Small deviations
Fast recovery

Approaching a tipping point:

Large deviations
Slow recovery

At a tipping point:

Large deviation
No recovery

Figure 17.2 Heuristic basis for early warning signals of an approaching tipping point, adapted from Lenton (2011).

ball (representing the system) will undergo increasingly large oscillations around the base of the well (representing the stable state). Secondly, as the system approaches the tipping point, illustrated by the shallowing of the left-hand side barrier separating the wells, the ball can be expected to undergo greater deviations from the stable state in this direction, termed increasing 'skewness' (Lenton, 2011). Thirdly, as the well becomes shallower, the ball can be expected to roll back to the stable state much more slowly, indicating that the rates of change in the system decrease as the tipping point is approached. This phenomenon is termed 'critical slowing down'. By collecting observational data about the system we are able to detect all these characteristics in a system: increased variability (known as 'flickering'), increased skewness, and decreased rate of change (Lenton, 2011). The latter is often detected within time-series data as each point becomes more correlated with the point next to it as the rate of change slows. Thus, an increased autocorrelation in the natural fluctuations can also be considered as a signal that a tipping point is being approached.

Returning to the aforementioned experiment conducted by Drake and Griffen, as food availability declined and the zooplankton populations neared collapse, the populations displayed increases in the coefficients of variation, skewness, and autocorrelation as much as eight generations before extinction occurred (Drake and Griffen, 2010). Real-world examples are less forthcoming due to a lack of appropriate high-resolution time-series data, although evidence from diatom communities obtained via lake sediment cores in China show similar early warning signals prior to a critical transition (Wang et al., 2012). Further work is needed to convincingly demonstrate that these phenomena are common to most tipping points in most systems.

It must be noted, however, that the ability to predict some tipping points does not necessarily lead to the ability to prevent sudden shifts in system state. Firstly, some state shifts can be purely noise-induced. These changes do not have an explicit driver but can still be described as tipping points. A likely real-world example of such noise-driven changes are the abrupt warming events, known as Dansgaar–Oeschger events, that occurred during the last ice age (Ditlevsen and Johnsen, 2010). Secondly, real-world systems are complex and may possess a significant lag. Hence, by the time early warning signals are detected, systems may already be committed to undergoing a state shift. Thirdly, tipping points show a distinct scale effect. That is to say that, alternative states can exist in close proximity and, in spatially heterogeneous landscapes, state shifts are smoothed on larger scales. Thus, the detection of tipping points, and their associated early warning signals, is highly dependent on the spatial and temporal scale over which observations take place. For example, both clear and turbid states can coexist within the same lake (Carpenter et al., 1999). Finally, real-world systems are innately complex. Those systems dominated by anthropogenic actions are typically impacted by the decisions and actions of numerous individuals and not as the result of a single decision. Therefore, to anticipate tipping points in a system we must develop an understanding of ecosystem function over multiple temporal and spatial scales, as well as identifying the drivers and pressures affecting a system and how they result from the decisions taken by individuals, communities, governments, and nations. Assessments of this kind over large time frames and incorporating many nations are exceedingly challenging for the scientific community, highlighting a need for long-term funding sources and increasing communication and consultation with policy makers, a process that has already begun within many global science and political bodies (e.g. the Future Earth Initiative).

17.5 Policy support oriented research frameworks

Despite the complex nature of real-world systems, policy makers often desire simplified outputs before taking the necessary action which may prevent a tipping point being reached or a planetary boundary threshold being breached. The Driver-Pressure-State-Impact-Response framework (DPSIR) can be useful to 'bridge the gap' between real-world complexity and our ability to conceptualize systems and make decisions based on that. In this section, we introduce DPSIR, providing examples from marine ecosystems (see Box 17.1).

DPSIR can be traced back over three decades, to the Stress-Response framework developed by Statistics Canada in the late 1970s. The framework was further developed over time by, among others, the Organisation for Economic Co-operation and Development and the United Nations, before the European Environmental Agency published it in its final form in the mid-1990s (European Environmental Agency, 1995). Broadly, DPSIR characterizes D̲riving forces, including socio-economic and environmental variables, which exert P̲ressures on the system of interest. As a result of such pressures, the S̲tate of that system may undergo change. A shift in the state of a system, in turn, then

---

Box 17.1 The management of marine biodiversity using a DPSIR approach

Atkins et al. (2011) detail two case studies in which DPSIR has been applied in marine ecosystems. Here, we summarize one of those studies, the management of marine biodiversity at Flamborough Head, United Kingdom. Flamborough Head is designated as a European Marine Site for its diversity and importance for seabird life, but also has many other uses including fishing, recreation, waste disposal, and aggregate extraction. Given the multiple uses and users of this site, multiple interacting DPSIR cycles can be envisaged (see Fig. 17.3). In this case, the responses to the multiple cycles are dictated by an integrated management scheme, produced in 2000, and later revised in 2007. Many activities that take place at Flamborough Head place several pressures upon the system. For example, sewage treatment works discharge up to 2500 m^3 day^{-1} into the area, which receives over 56 000 visitors per year (Atkins et al., 2011). If management is not effective, these pressures could lead to changes in the state of this protected environment, such as an increase in the level of pollutants in the water and/or sediments. These, in turn, could lead to impacts such as reductions in seabird numbers or fewer tourists visiting the beaches. The Flamborough Head Management Plan involves the monitoring of key resources so state changes can be detected and management altered to ensure sustainable use. As well as reactive measures, precautionary measures were also established. For example, prohibited trawl zones and no take zones were established to preserve and maintain diversity. Through continual monitoring and assessment of the uses of Flamborough Head and the impact of these uses, this important marine area can be maintained in a sustainable state.

Impacts those individuals and communities that were either partially or wholly reliant on the system in its previous state. Adapting to such impacts may involve changes in life-history strategies and behaviour, termed Responses, which in turn alter the driving forces experienced by the system.

DPSIR is an extremely flexible framework and can be adapted to address a variety of questions for a wide range of ecosystems. The framework follows a sequential thought process but this is not unidirectional, allowing for the evaluation of policy via the assessment of how these responses alter drivers and impact the state of the system, as well as understanding the impact of drivers (Atkins et al., 2011). However, the sequential thinking underlying DPSIR has been raised as a criticism, with claims that the framework is only capable of representing linear cause-and-effect schemes. These criticisms are unjustified, confusing sequential thinking with linearity (Atkins et al., 2011). The cyclical DPSIR system can result in non-linear, large-scale changes if positive feedbacks reinforce current drivers, shifting the system to an alternative state. By contrast, negative feedbacks can ensure the cycle is stable, with the system oscillating around its stable state.

Other criticisms of DPSIR are more justified. In real-world examples, the relationships between the DPSIR categories may be synergistic. For example, numerous drivers could interact to cause a solitary pressure, altering a state but causing multiple impacts and responses (Atkins et al., 2011). DPSIR encourages the sequential analyses of each category which may potentially lead to a neglect of the connections between them. While this compartmentalized approach may cause difficulties, it can help to highlight knowledge gaps, suggesting areas of future research by emphasizing specific aspects of the systems that are data deficient or highly uncertain. Furthermore, while DPSIR is not scale-constrained, being applicable over all temporal and spatial scales, in practice each category of the framework is bounded to a set scale and the scale of each category need not overlap (Atkins et al., 2011). For example, the scale of the impact may be local, leading to local responses, but the main driver can be global (e.g. anthropogenic climate change). In reality, the demarcation of these boundaries is strongly influenced by the perspective of those performing the assessment. This user bias can create ambiguity, reducing comparability between studies.

Despite these difficulties, DPSIR offers a framework with enough flexibility to be applicable to a wide range of situations, while facilitating the communication and transparency between scientists and policy makers. DPSIR emphasizes the effect of anthropogenic actions by combining both socio-economic and biophysical systems, allowing for the development of the holistic evidence-based policy required to maintaining safe and just operating spaces. However, it is important to remember that DPSIR is simply one of numerous available frameworks which should be considered. Many frameworks are not mutually exclusive and there is often added benefit from combining multiple frameworks. For example, Atkins et al. (2011) combine 'The Ecosystem Approach', a strategy promoting conservation and sustainable use via integrated and equitable management of land, water, and living resources,

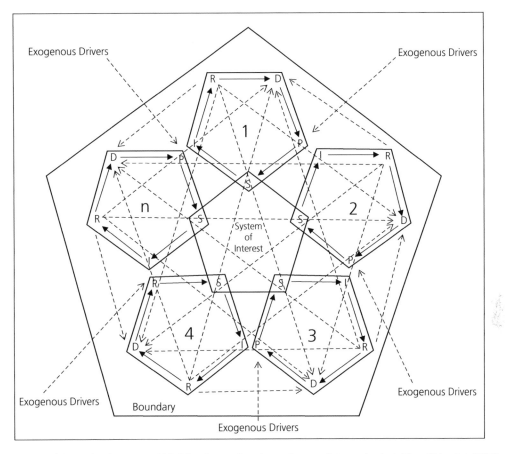

Figure 17.3 Example interactions between multiple Driver-Pressure-State-Impact-Response Framework, adapted from Atkins et al. (2011).

with DPSIR for marine systems (see Box 17.1 for further details). This combination allows a single DPSIR cycle (e.g. that related to a marine fishery) to be nested within numerous other DPSIR cycles representing other sectors (e.g. marine aggregates, energy generation, and aquaculture), incorporating linkages between all cycles and better reflecting the complexity of the real world (Fig. 17.3).

17.6 Research frameworks translated into interdisciplinary models

Using DPSIR provides guidance on linking biophysical and socio-economic data, but, typically, finding reliable and comprehensive data is an issue. While biophysical data are often based on several variables but a low number of observations, socio-economic data often focus on fewer variables but lack the necessary

spatial aggregation. Often, experiments cannot be conducted for ethical or logistical reasons. For example, trialling different management options of a marine fishery on which numerous rural poor are reliant could have grave impacts on their livelihoods, perhaps even resulting in fatalities if the trials are unsuccessful. These ethical dilemmas limit policy as well as science development. Given the associated risk, decision makers are often unwilling to develop and trial the radical policies needed to address complex and important issues, such as the detrimental impact of exceeding the planetary boundaries of biodiversity loss, climate change, and impacts on the nitrogen cycle. Hence, decision makers often opt for remaining with the status quo or react with small-scale tinkering of current policy. Linking models of biophysical and socio-economic processes, both spatially and temporally, can provide a testing ground for radical policies, reducing their associated risk.

Parker et al. (2008) outline current strategies to address the challenges of modelling complex systems. Firstly, despite the complexities of real-world systems, the models must be as simple as possible. Typically, this is achieved by identifying the minimum set of variables through which model uncertainty can be reduced to the desired level. Any further additions of processes or data will reduce model transparency without adding further insight to the questions at hand. Where possible, models should be constructed to share data inputs, reducing data requirements and assisting in linking different aspects of the system. Secondly, the highest possible aggregates relevant to the subject of interest must be selected. For example, models could incorporate individual decisions, but the overall actions of households or communities may be sufficient if these span the required variations in behaviour. Utilizing higher levels of aggregations can increase model tractability and assist in linkages between scales, without compromising the research question or biasing results.

The strategies to model complex systems can be applied across most categories of model. Here, we briefly introduce six broad categories of model (system models, empirical models, economic models, cellular-based models, agent-based models, and hybrid models) used to understand and link planetary boundaries and social systems.

Systems models (1) represent stocks, flows, and sinks of information and/or resources, through differential equations and structural ontologies. Socio-economic structures (e.g. governments, communities, and markets) interact at numerous spatial and temporal scales and cumulatively provide many of the factors on which an individual's decisions are based. Time is divided into discrete steps of varying lengths, allowing feedback loops to be included in the system. Furthermore, human and ecological interactions can be represented in these models, although such approaches require specific data sets that are often deficient and so poorly represent the apparent spatial variation (Baker, 1989; Parker et al., 2003).

Empirical models (2) differ from systems models in that they are not derived from an understanding of the underlying socio-economic pressures, but are often simple correlations between response variables and explanatory variables, developed using field data for a single time point. These models are readily applied in areas where detailed data are lacking. In many cases, these models fit the data well, explaining a large amount of the observed spatial variation. However, caution must be applied when interpreting these models, as correlations identified do not prove causation.

Economic models (3) can be broadly divided into two subcategories: microeconomic models that describe equilibrium patterns within a local context or regional economic models that encompass the flows of various resources and goods across regions. Due to over-simplification, neither microeconomic nor regional economic models capture the complex spatial and temporal patterns well (Anas et al., 1998).

Cellular models (4) include both cellular automata and Markov models, and operate over a network of similar cells. Cellular automata characterize the behaviour of the system using a set of deterministic or probabilistic rules to assign the state of a cell based on the state of its neighbours, although non-local neighbours (Takeyama and Couclelis, 1997) and networks can also be used (O'Sullivan, 2001). The system is fully homogenous, with each cell capable of being assigned all states and the same transition rules apply to each cell (Parker et al., 2003). Markov models adopt a similar approach, but cell states depend on temporally lagged transition rules. Thus, Markov models and cellular automata can be combined to model complex systems. However, again, this approach is limited by over-simplification, and thus may prove erroneous in predicting the effects of various social phenomena.

The **agent-based modelling (5)** approach takes a different perspective to the above techniques, focusing on individual decision making rather than landscape variation and transitions. Agent-based modelling represents the motivations behind decisions and identifies the external factors that influence these decisions through the use of autonomous agents. Each agent represents a social unit (e.g. an individual or household) which can act intelligently to achieve desirable goals (e.g. increased household income or food security). As a minimum, agent-based models can be driven using semi-quantitative data via Bayesian belief networks and used to test reactions to environmental shocks or policy decisions.

In reality, a wide variety of modelling approaches are used in combination to form **hybrid models (6)**. By using a combination of approaches many of the limitations discussed above can be overcome. For example, the limitations of economic models can be somewhat avoided by combining this approach with empirical models, using socio-economic theory to provide a sound basis by which remote sensing data are used to explain the variation observed (Nelson and Hellerstein, 1997; Pfaff, 1999). However, when linking models, it is important to consider the manner in which they are linked. Linkages can vary from single uni-directional linkages to fully coupled systems. We suggest that the complex nature of planetary boundaries and social systems can only be adequately envisaged through fully coupled systems. Thus, the bare minimum sufficient to provide advice for managing complex socio-ecological systems requires integrated two-way feedbacks between anthropogenic and environmental systems; for example, this feature is vital if trade-offs and tipping points are to be considered. Today, cutting-edge research strives to provide these complex, fully coupled models capable of addressing the global challenge of maintaining safe and just operating spaces. For example, ASSETS (Attaining Sustainable Services from Ecosystems through Trade-off Scenarios; http://espa-assets.org/) aims to explicitly quantify the linkages between natural resources and food security, using a modified holistic DPSIR approach. ASSETS fully couple cellular models describing natural resource processes (e.g. photosynthesis) with agent-based models which represent how individuals make decisions, aiming to influence policy decisions. Similarly, a related project (WISER: Which Ecosystem Service Models Best Capture the Needs of the Rural Poor?; http://www.espa.ac.uk/projects/ne-l001322–1) attempts to identify the minimum adequate degree of complexity of modelling required to enable the model's usefulness in a decision-making process using binary discriminator tests (described below).

To help ensure models are used appropriately by decision makers, the uncertainty contained within the models must be accurately portrayed. The Managing Uncertainty in Complex Models toolkit provides guidance to quantify model discrepancy (the accuracy of the model when compared to independent field data), the degree this changes by location, service, and spatial resolution (model uncertainty), and the sensitivity of the model outputs to parameter values (see http://www.mucm.ac.uk for more information). Finally, the impact of model discrepancies, model uncertainties, and model sensitivities on the usefulness of the complex model in a decision-making process can be evaluated using binary discriminator tests (Stow et al., 2009). Binary discriminator tests compare observations and model predictions with respect to a user-defined threshold value that is then adjusted to explore the limits of acceptable error for policy contexts. The amount of acceptable error can be defined in agreement with stakeholders. In simple terms, these tests score whether the models represent the observed distribution within a system using four possible outcomes: correctly positive (CP), correctly negative (CN), incorrectly positive (IP), and incorrectly negative (IN). Varying the threshold value builds non-parametric measures of the models' ability to predict the distribution of a given variable. The effect on decision making can be assessed by calculating the ratios of correct and incorrect predictive values for action taken above (CP/[CP+IP]) and below (CN/[CN+IN]) the variable thresholds. A predictive value ratio near 1 indicates that the model output is sufficiently accurate for policy decisions; a value near 0 indicates it is not. This simple output readily communicates the ability for the model to support evidence-based policy in a manner decision makers can easily understand.

17.7 Managing risk in spite of uncertainty

While models help to develop an understanding of the system, to demonstrate trade-offs and potential tipping points, and to provide a testing-ground for new practices and policies, they do not remove the risk associated with decision making. Here, we introduce the United States Environmental Protection Agency (US EPA) ecological risk assessment, and a school of thought placing the onus of risk evaluation on the proponent of potential risk-producing activities, known as the precautionary principle.

The US EPA began to develop ecological risk assessments in the 1980s, publishing guidelines in 1992, and later refining these to emphasize the need for stakeholder consultations in 1998 (Hope, 2006). In order to ensure the ecological risk assessment could be applied in many varying situations, the US EPA adopted a tiered approach. The first tier is generally a qualitative determination of potential risk, typically resulting from rapid surveys of ecological resources. This first tier is usually regarded as conservative because positive results, where risks are identified, indicate potential and not realized risks. Thus, positive results from tier 1 progress through to the remaining tiers, where further acquisition of more or better quality data can reduce uncertainty, clarifying the situation. The second tier, often termed a 'screening' or 'generic' assessment, develops and interprets data on ecological conditions, stressors, and predicted impacts. If sufficient data are available, numerical 'hazard quotients' can be produced to estimate risk. However, it must be noted that quotients such as this are not true estimates of risk. As with tier 1, the second tier is also biased towards conservatism but, if risk is still suggested, a third tier may be undertaken. The third tier fully characterizes the nature and extent of the risk, using a wide range of evidence developed from modelling and ecological surveys. It is during this final tier that cutting-edge risk assessments provide spatially explicit estimates of risk, using probabilistic assessments to quantify the potential impact of multiple stressors.

Within each tier, the US EPA risk assessment adopts a framework with three phases: problem formulation, analysis, and risk characterization (see Fig. 17.4). The problem formulation phase integrates the available information and develops an analysis plan for obtaining the data required for an assessment. During this phase, the environmental values to be protected, known as assessment endpoints, are defined and a conceptual model is developed, describing the theorized pathway of impact. It is at this stage that stakeholders and decision makers develop ownership of the risk assessment, being involved in a consultation defining the endpoints and levels of risk that are unacceptable. The evaluation of data concerning exposure and the effects of this exposure occurs during the analysis phase. Changes of state, e.g. altered temperature or chemical composition, must be measured, indicating the nature, temporal and spatial distribution, and amount of exposure at points of potential impact. Similarly, the impact on endpoints must be monitored and quantified, developing correlations and/or relationships between the magnitude and duration of exposure to the endpoint effects. Finally, the results of the analysis phase are integrated and risks described during the risk characterization phase. During this process, the parameters of exposure–response models are refined and the associated uncertainty should be indicated. These values should then be interpreted and risk described for communication to stakeholders. Given this information, the stakeholders or risk managers may determine a course of action, or further consultation may be necessary to build cost–benefit analyses of potential mitigating actions.

Risk assessments, such as that developed by the US EPA, help provide a quantitative estimate of risk, but how do we decide how much risk is acceptable for a given system? This decision is made all the more difficult when impacts are uncertain. For example, risk assessments using a probabilistic approach convey uncertainty more openly to stakeholders; however, this can present risk managers with ambiguity, delaying much needed remedial action. In the remainder of this chapter, we introduce the precautionary principle, placing the onus of risk evaluation on the proponent of potential risk-producing activities.

The precautionary principle has four central concepts: taking preventative actions in spite of uncertainty, shifting the emphasis of risk evaluation to proponents of activities, exploring the range of

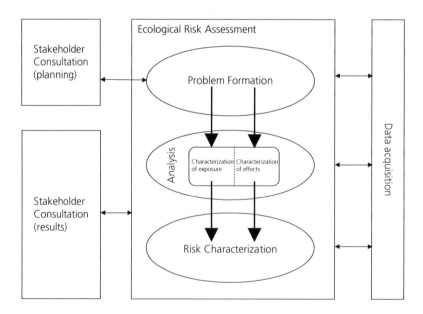

Figure 17.4 The US EPA framework for ecological risk assessment.

potential outcomes, and increasing public participation and ownership in the decision-making process (Kriebel et al., 2001). Central to the principle is the proposition that precautionary measures should be taken to reduce the potential threats of activities, even if some of the cause–effect relationships are not fully understood. This concept is not necessarily new to marine systems. For example, maintaining catch levels below the maximum standard yield can be regarded as precautionary. However, how do we decide what level of precaution is necessary given the potential risk? Lauck et al. (1998) provide the example of a fishery whereby stock estimates are only valid to within 30%, annual productivity varies over 50%, and that fishing mortality varies within 25% of the total allowable catch (TAC). Given this uncertainty, the TAC should be set at 28% of the mean estimated value to ensure sustainability and minimize risk. However, such a safety margin may be considered extreme and some in the industry may argue for the best-case scenario of TAC of over 200% of the mean estimate. One can imagine negotiations between conservationists and industry resulting in quotas similar to the original mean estimate; however, given the uncertainty, this value is extremely risky. In this example, the precautionary principle places the emphasis of proof of safety on industry.

While the protection afforded to the environment under the precautionary principle is immediately evident, a lively debate about its usefulness is underway (Kriebel et al., 2001). Critics suggest that current regulatory procedures are already precautionary, citing the conservative stance of risk assessments when faced with data deficiency. Furthermore, the precautionary principle could stifle innovation, requiring proof of safety before new technologies can be adopted and making precautionary decisions up to that point without adequate knowledge to formulate a robust scientific justification (Holm and Harris, 1999). For example, field trials of genetically modified crops would have been prevented under the precautionary principle; however, without these trials, the data illustrating the real risks of these crops would have never been produced. Finally, while in the case of industrial-scale fishing operations, the proof of safety required by the precautionary principle is probably affordable and appropriate; this cost would be prohibitive for subsistence fishers in developing nations (Aronson and Precht, 2006). However, given the dire consequences of exceeding planetary boundaries, perhaps the conservativeness afforded by the precautionary principle is necessary.

17.8 Conclusions

Some of the most pressing issues facing humanity today (e.g. climate change) are exceedingly complex and highly uncertain, yet require action. We have already breached three vital planetary thresholds and are yet to address important social issues, such as world hunger. Delaying decisions until socio-ecological systems and the linkages between them are better understood is not an option. Even small delays in reducing pressures on environmental systems may result in catastrophic changes if it allows ecosystems to reach tipping points, where their characteristics and functions fundamentally change. With current monitoring techniques, by the time we detect the possibility of a tipping point, if we pre-empt it at all, it may be too late to act as the system may already be committed to large shifts in state. Research frameworks (e.g. DPSIR) may enable us to better conceptualize complex, global socio-ecological systems using our understanding of smaller-scale social and ecological processes. Using such approaches, fully coupled hybrid models linking anthropogenic and environmental systems can be developed, enabling the radical policies that may be necessary to address humanity's most pressing issues to be tested. Despite recent scientific advances such as this providing new and insightful evidence, large uncertainties remain. Quantifying and communicating the uncertainties and risks associated with both undertaking potential actions and maintaining the status quo is necessary to enable decision makers to develop the evidence-based policy we need to sustainably manage Earth's natural resources.

Acknowledgements

The ideas and discussion developed in this paper were developed during the authors' participation in the ASSETS project. ASSETS is funded under the Ecosystem Services for Poverty Alleviation Programme (ESPA) project no. NE-J002267-1. The ESPA programme is funded by the Department for International Development (DFID), the Economic and Social Research Council (ESRC), and the Natural Environment Research Council (NERC), as part of the UK's Living with Environmental Change Programme (LWEC). In addition, SW acknowledges financial support from the ESPA-funded WISER project (NE/L001322/1) and SH was provided with financial support by a joint NERC/ESRC PhD studentship award, and the University of Southampton.

References

Anas, A., Arnott, R., & Small, K.A. (1998). Urban spatial structure. *Journal of Economic Literature*, **36**, 1426–1464.

Aronson, R. & Precht, W. (2006). Conservation, precaution, and Caribbean reefs. *Coral Reefs*, **25**, 441–450.

Atkins, J.P., Burdon, D., Elliott, M., & Gregory, A.J. (2011). Management of the marine environment: integrating ecosystem services and societal benefits with the DPSIR framework in a systems approach. *Marine Pollution Bulletin*, **62**, 215–226.

Baker, W. (1989). A review of models of landscape change. *Landscape Ecology*, **2**, 111–133.

Carpenter, S.R. & Bennett, E.M. (2011). Reconsideration of the planetary boundary for phosphorus. *Environmental Research Letters*, **6**, 014009.

Carpenter, S.R., Ludwig, D., & Brock, W.A. (1999). Management of eutrophication for lakes subject to potentially irreversible change. *Ecological Applications*, **9**, 751–771.

Dearing, J.A., Wang, R., Zhang, K., et al. (2014). Safe and just operating spaces for regional social-ecological systems. *Global Environmental Change*, **28**, 227–238.

Ditlevsen, P.D. & Johnsen, S.J. (2010). Tipping points: early warning and wishful thinking. *Geophysical Research Letters*, **37**, L19703.

Drake, J.M. & Griffen, B.D. (2010). Early warning signals of extinction in deteriorating environments. *Nature*, **467**, 456–459.

European Environmental Agency (1995). *Europe's Environment: The Dobris Assessment*. European Environmental Agency, London, UK.

Ferek, R.J., Garrett, T., Hobbs, P.V., et al. (2000). Drizzle suppression in ship tracks. *Journal of the Atmospheric Sciences*, **57**, 2707–2728.

Hirota, M., Holmgren, M., Van Nes, E.H., & Scheffer, M. (2011). Global resilience of tropical forest and savanna to critical transitions. *Science*, **334**, 232–235.

Holm, S. & Harris, J. (1999). Precautionary principle stifles discovery. *Nature*, **400**, 398.

Hope, B.K. (2006). An examination of ecological risk assessment and management practices. *Environment International*, **32**, 983–995.

Kriebel, D., Tickner, J., Epstein, P., et al. (2001). The precautionary principle in environmental science. *Environmental Health Perspectives*, **109**, 871.

Lauck, T., Clark, C.W., Mangel, M., & Munro, G.R. (1998). Implementing the precautionary principle in fisheries management through marine reserves. *Ecological Applications*, **8**, S72–S78.

Lenton, T.M. (2011). Early warning of climate tipping points. *Nature Climate Change*, **1**, 201–209.

Nelson, G.C. & Hellerstein, D. (1997). Do roads cause deforestation? Using satellite images in econometric analysis of land use. *American Journal of Agricultural Economics*, **79**, 80–88.

Nyström, M., Folke, C., & Moberg, F. (2000). Coral reef disturbance and resilience in a human-dominated environment. *Trends in Ecology and Evolution*, **15**, 413–417.

O'Sullivan, D. (2001). Graph-cellular automata: a generalised discrete urban and regional model. *Environment and Planning B*, **28**, 687–705.

Parker, D.C., Hessl, A., & Davis, S.C. (2008). Complexity, land-use modeling, and the human dimension: fundamental challenges for mapping unknown outcome spaces. *Geoforum*, **39**, 789–804.

Parker, D.C., Manson, S.M., Janssen, M.A., et al. (2003). Multi-agent systems for the simulation of land-use and land-cover change: a review. *Annals of the Association of American Geographers*, **93**, 314–337.

Pfaff, A.S.P. (1999). What drives deforestation in the Brazilian Amazon? Evidence from satellite and socioeconomic data. *Journal of Environmental Economics and Management*, **37**, 26–43.

Raworth, K. (2012). *A safe and just space for humanity: can we live within the doughnut?* Oxfam Discussion Paper. Oxfam, Oxford.

Reid, P.C., Edwards, M., Hunt, H.G., & Warner, A.J. (1998). Phytoplankton change in the North Atlantic. *Nature*, **391**, 546–546.

Rockström, J., Steffen, W., Noone, K., et al. (2009). Planetary boundaries: exploring the safe operating space for humanity. *Ecology and Society*, **14**, 32.

Scheffer, M., Szabó, S., Gragnani, A., et al. (2003). Floating plant dominance as a stable state. *Proceedings of the National Academy of Sciences of the USA*, **100**, 4040–4045.

Schramm, W. & Nienhuis, P.H. (1996). *Marine Benthic Vegetation: Recent Changes and the Effects of Eutrophication*. Springer, Berlin.

Singer, S.F. (1970). Global effects of environmental pollution. *Eos, Transactions American Geophysical Union*, **51**, 476–478.

Stow, C.A., Jolliff, J., McGillicuddy Jr, et al. (2009). Skill assessment for coupled biological/physical models of marine systems. *Journal of Marine Systems*, **76**, 4–15.

Takeyama, M. & Couclelis, H. (1997). Map dynamics: integrating cellular automata and GIS through Geo-Algebra. *International Journal of Geographical Information Science*, **11**, 73–91.

Wang, R., Dearing, J.A., Langdon, P.G., et al. (2012). Flickering gives early warning signals of a critical transition to a eutrophic lake state. *Nature*, **492**, 419–422.

CHAPTER 18

Using the Ecosystem Approach to manage multiple stressors in marine environments

Zoë Austin and Piran C.L. White

18.1 Introduction

There are many stressors caused by anthropogenic activity which are currently affecting the marine environment in a host of different ways. As demonstrated in previous chapters, stressors such as chemical pollution, ocean acidification, and nitrogen deposition invoke multi-level responses in marine ecosystems, from effects on the physiology of individual organisms to ecosystem-level impacts (Lenihan et al., 2003). Many of these responses can have detrimental impacts on the services that human society derives from the marine environment, such as food, coastal protection, and recreational activities. How best to manage these multiple stressors, considering the many sectors of human society that utilize the oceans and all that they provide, is a globally significant challenge.

The majority of the current management practices that seek to reduce the impact of marine stressors are likely to consider each activity in isolation, using a sector-by-sector approach. However, as most stressors are likely to accumulate and interact both spatially and temporally, managing each stressor in isolation will rarely meet environmental, social, or economic goals and needs. The Ecosystem Approach provides a framework whereby the approach is focused around maintaining the ecosystem services that support human well-being (Halpern et al., 2008a), rather than sector-by-sector objectives focusing on specific processes or benefits. In this chapter, we explore how using an Ecosystem Approach to understanding and managing multiple stressors in marine environments may offer many advantages over sector-by-sector approaches.

In Section 18.1.1 we discuss what we mean by an Ecosystem Approach and why ecosystem-based management can be applied to complex environmental management issues. In Section 18.2 we track the concept of ecosystem services and identify the main ecosystem services which are provided by the marine environment. In Section 18.3 we discuss some of the principal human-induced stressors and give examples of how they may impact upon various ecosystem services. We consider the need to manage multiple stressors, taking into account various factors, such as the accumulation and interaction of multiple stressors and ecosystem services in the marine environment. In Section 18.4 we re-emphasize how the Ecosystem Approach will be beneficial in the management of multiple stressors and present a framework for its application based on a set of key principles. These principles highlight, among other important factors, the need to consider ecosystem services goals, trade-offs, and assessment when using this approach. Section 18.5 then examines some tools which could be used in the application of the framework and includes some case studies to demonstrate how these tools have been used in previous studies. We finish (Section 18.6) with a discussion of the challenges that need to be faced in order to implement the Ecosystem Approach, but emphasizing that many of the principles we present here are achievable despite these challenges.

18.1.1 The Ecosystem Approach

The Ecosystem Approach is often seen as a way of making decisions in order to manage human activities sustainably. It recognizes that humans are part of the ecosystem and that our activities both affect the ecosystem and depend on the services we gain from it. It was first defined by the Convention on Biological

---

Box 18.1 Principles of the Ecosystem Approach, as identified by the Convention on Biological Diversity

1. The objectives of management of land, water, and living resources are a matter of societal choices.
2. Management should be decentralized to the lowest appropriate level.
3. Ecosystem managers should consider the effects (actual or potential) of their activities on adjacent and other ecosystems.
4. Recognizing potential gains from management, there is usually a need to understand and manage the ecosystem in an economic context. Any such ecosystem programme should: reduce those market distortions that adversely affect biological diversity; align incentives to promote biodiversity conservation and sustainable use; and, internalize costs and benefits in the given ecosystem to the extent feasible.
5. Conservation of ecosystem structure and functioning, in order to maintain ecosystem services, should be a priority target of the Ecosystem Approach.

6. Ecosystems must be managed within the limits of their functioning.
7. The Ecosystem Approach should be undertaken at the appropriate spatial and temporal scales.
8. Recognizing the varying temporal scales and lag-effects that characterize ecosystem processes, objectives for ecosystem management should be set for the long term.
9. Management must recognize that change is inevitable.
10. The Ecosystem Approach should seek the appropriate balance between, and integration of, conservation and use of biological diversity.
11. The Ecosystem Approach should consider all forms of relevant information, including scientific and indigenous and local knowledge, innovations, and practices.
12. The Ecosystem Approach should involve all relevant sectors of society and scientific disciplines.

Diversity (CBD) as 'a strategy for the integrated management of land, water, and living resources that promotes conservation and sustainable use in an equitable way' (CBD, 2000). The approach therefore requires an understanding of the way in which the ecological system functions while at the same time understanding the way that society manages and impacts on ecosystems (Atkins et al., 2011). Inclusion of all relevant stakeholders in the decision-making process is a key component of the ecosystem approach and is one of the 12 key principles (Box 18.1) that guide the approach as outlined by the CBD (CBD, 2000).

The Ecosystem Approach is now a term used in many management and policy documents across multiple sectors in both the terrestrial and marine environments. It is embedded in, for example, the European Commission's Marine Strategy Framework Directive, the EU Water Framework Directive, and the Common Fisheries Policy. In March 2013, the European Commission, European Parliament, and the European Fisheries Council agreed to an MSP Directive that requires member states to establish and implement Maritime Spatial Plans (MSPs) for their marine areas, according to a set of minimum common requirements that adhere to the Ecosystem Approach.

Outside of Europe, in the United States the approach is part of the National Policy for the Stewardship of the Ocean, Our Coasts, and the Great Lakes. It also forms part of the strategy for managing multiple uses of the Great Barrier Reef in Australia and Marine Protected Area (MPA) planning in New Zealand. The Ecosystem Approach has also been identified as a key policy objective in West Africa, Latin America, and the Caribbean for managing water resources, wetlands, and land (UNEP, 2012).

However, despite the importance of the approach to policy, implementation of the approach to the marine environment is still not widespread, although clear examples of where it has been applied do exist (see Pisces, 2012; UNEP, 2012).

18.2 Ecosystem services

The Ecosystem Approach is underpinned by the concept of ecosystem services, which originated in the academic literature in the 1980s (Dick et al., 2011) in a paper published by Ehrlich and Mooney (1983). Ehrlich and Mooney (1983) recognized the impact of human activity such as agriculture, deforestation, and fishing on ecosystems and the further implications it could have for various ecosystem services. Despite this, it was another 10–15 years before the concept was further developed, including the introduction of techniques to establish values for ecosystem services (Costanza et al., 1997). However, it was with the publication of the Millennium Ecosystem Assessment (MEA, 2005) that the ecosystem service concept was brought to the forefront of the research community, and since then, the number of publications on ecosystem services has continued to grow exponentially (Raffaelli and White, 2013).

The MEA defined ecosystems services as 'the benefits people obtain from ecosystems' and developed four service categories: provisioning, regulating, cultural and supporting services (that are necessary for the production of all other services) (MEA, 2005). The MEA definition conflates benefits with services, and there has been considerable debate over how ecosystem services should be classified, most recently over the validity of supporting services as a distinct category. Nevertheless, the MEA categorization is the most well known and widely applied to date (Fletcher et al., 2011). However, the MEA did not deliver the tools necessary to make the approach operational (Armsworth et al., 2007), which has led to other, alternative schemes and new definitions (Raffaelli and White, 2013). Here, we are using a scheme developed by the UK's National Ecosystem Assessment (UK NEA, 2011) from Fisher and Turner (2008). The UK NEA effectively combines these various typologies (Raffaelli and White, 2013), and, like the MEA, it recognizes four categories of ecosystem services: provisioning, regulating, cultural, and supporting.

18.2.1 Ecosystem services from the marine environment

Marine habitats provide a range of ecosystem services of significant value to society. According to the UK's National Ecosystem Assessment technical report, these include: 'food such as fish and shellfish; the reduction of climate stress by regulating carbon and other biogases; genetic resources for aquaculture; industrial inputs for blue biotechnology such as biocatalysts, natural medicines; fertilizer (seaweed); coastal protection; waste breakdown and detoxification leading to pollution control, waste removal, and waste degradation; disease and pest control; tourism, leisure, and recreation opportunities; a focus for engagement with the natural environment; physical and mental health benefits; and cultural heritage and learning experiences' (UKNEA, 2011). In addition, energy provision is also likely to be an increasingly important marine ecosystem service as the technology is developed to exploit wave power and biofuels from macro- and microalgae (Hossain et al., 2008). We now discuss these ecosystem services further based on their categorization within the framework (Table 18.1).

18.2.1.1 Provisioning services

The benefits that people derive from the marine provisioning services include fish and shellfish as foods derived either from wild harvest or aquaculture, algae

Table 18.1 Ecosystem services from the marine environment. UKNEA categorizations are used.

Service type	Ecosystem service or good	Example
Provisioning	Food	Wild fish and shellfish for harvesting
	Aquaculture	Fish farming
	Blue biotechnology	Provision of inputs such as biocatalysts
Regulating	Climate regulation	Regulation of carbon
	Flood, storm, and coastal protection	Protection from storms by coastal habitat
	Waste breakdown and detoxification	Pollution control and waste degradation
Supporting	Nutrient cycling	Carbon or nitrogen cycling
	Biologically mediated habitat	Habitat created by one organism that benefits others
Cultural	Education and research	Schools programmes and scientific research
	Leisure, tourism, and recreation	Sailing, fishing, and wildlife-watching tourism
	Health	Mental and physical health benefits of marine recreation
	Heritage	Aesthetic, inspirational, and cultural properties of the marine environment

and seaweed as inputs for pharmaceuticals, and biofuels (Table 18.1). A number of secondary services are also supported by the provision of fish. Globally, fisheries provide direct employment to around 200 million people (Botsford et al., 1997), and the fisheries and seafood service sector is beneficial to millions of other workers and consumers including hotels and restaurants.

18.2.1.2 Regulating services

Regulating services from the marine environment include waste breakdown and detoxification, climate regulation, and flood, storm, and coastal protection (Beaumont et al., 2007; Table 18.1). Waste materials such as industrial waste and human waste from sewerage systems are often discharged into the marine environment and therefore marine ecosystems are providing a waste breakdown and detoxification service to society (Wurl and Obbard, 2005). When waste levels are too high, however, for the marine environment to cope, pollution can occur. Climate regulation in the marine environment occurs due to biogeochemical processes, for example, marine organisms can regulate carbon fluxes by acting as a sink for carbon dioxide (Riebesell et al., 2007). Coastal protection can occur physically, due to coastal habitats such as mudflats and salt marshes that dissipate energy from storms and floods (Möller and Spencer, 2002).

18.2.1.3 Supporting services

Supporting services from marine environments include nutrient cycling and biologically mediated habitat (Table 18.1). Nutrient cycling takes place in many components of the marine environment, is essential for living marine organisms, and supports all other marine ecosystem services (Costanza et al., 1997). Biologically mediated habitat occurs when organisms provide habitats for other organisms. Cold water corals, for example, can form reef colonies that provide habitats for various fish species where they are often found in far greater densities than in the background environment (Roberts et al., 2006).

18.2.1.4 Cultural services

There are many cultural services that society derives from the marine environment (Costanza et al., 1997; Beaumont et al., 2007). The marine environment provides opportunities for education, research, leisure, tourism, and recreation (Table 18.1). Many people also report health benefits obtained from the marine environment, such as the physical benefits of engaging in angling, sailing, and other activities that take place at sea. In addition, these activities can lead to broader well-being and mental health benefits through reduction in stress and enhanced social interactions (Pretty et al., 2007).

18.2.1.5 Wild species diversity

Wild species diversity has not been included as a separate service in Table 18.1 as it can span many of the ecosystems service categories. Indeed, the diversity of living organisms underpins the functioning and resilience of all ecosystem services (Hails and Ormerod, 2013). There are three main ways by which biodiversity contributes to the ecosystem services framework: (i) through the underpinning of processes or supporting services; (ii) through the delivery of 'diversity' as a final ecosystem service in its own right; and (iii) in the delivery of specific species and landscapes which hold great cultural significance for society and can be valued as 'goods' (Mace et al., 2012).

18.3 Stressors to the marine environment and the impacts on ecosystem services

Human-induced stressors affecting the marine environment have been increasing as human use and demand from many sectors has intensified. As Section 18.2 has shown, the marine environment provides society with many ecosystem services and goods. However, the activities that we undertake in order to utilize these services and goods can themselves lead to impacts on the sustainability and continued provision of these services. Human activities associated with development, over-harvesting, and recreation in the marine environment can lead to a number of associated stressors that can have adverse impacts on a range of ecosystem services. The United Nations Environment Programme states that primary threats to the world's oceans include the 'Big Five' stressors (Nellemann et al., 2008). These stressors are (not in order of magnitude): (i) climate change; (ii) pollution (mainly coastal); (iii) fragmentation and habitat loss (e.g. from dredging, trawling, and the use of explosives in fishing on coral reefs); (iv) invasive species infestations; and (v) over-harvesting from fisheries (Nellemann et al., 2008).

The specific stressors discussed in previous chapters of this book such as ocean acidification, thermal stress, and chemical pollution can all be linked to one of these 'Big Five' anthropogenic impacts. But it is the impact of multiple stressors acting together and their combined management that is the focus of this chapter. Multiple, co-occurring stressors are common, especially when correlated with centres of human population (Crain et al., 2009). When mapping human impact on marine ecosystems, studies have found that no fewer than five threats overlap anywhere in the world and that the number of geographically overlapping stressors may in fact be much higher in certain areas (Halpern et al., 2008b). So, it is important that as well as understanding how stressors impact on ecosystem services, we also need to consider the cumulative and interactive effects of these multiple stressors when considering how best to intervene from an ecosystem services perspective.

18.3.1 The impact of human-induced stressors on marine ecosystem services

The impact of multiple human-induced stressors on the marine environment will have implications for the provision of a number of ecosystem services, either directly or indirectly via changes to ecosystem processes and functioning (Fig. 18.1). The delivery of many provisioning and regulating services in the marine environment is declining because of over-exploitation and other human-induced stressors and the resulting direct and indirect impacts. For example, wild fisheries are

declining due to over-fishing (Watson and Pauly, 2001; Myers and Worm, 2003), while trawling fishing methods also have an adverse effect on seabed life, which plays a key role in cycling nutrients crucial to ensuring the productivity of the oceans (Hiddink et al., 2006). Such impacts will also have cumulative and interactive effects on other ecosystem services. For example, a reduction in wild fisheries can have a knock-on effect for aquaculture (Atkins et al., 2011).

The complex interacting relationships between drivers, ecosystem processes, and ecosystem service provision are beyond the scope of this chapter and much is still the focus of ongoing research aimed at further defining these relationships. In the following subsections we therefore aim to discuss some examples of ecosystem impacts as a result of some key stressors and the potential cumulative impacts and interactions associated with multiple stressors. Fig. 18.1 is a summary of some of the examples discussed here.

18.3.1.1 Over-exploitation

The over-exploitation of marine resources, and the resulting impacts on the provision of ecosystem services, results from a number of indirect and direct drivers. Policy failures in terms of unsustainable catch rates/

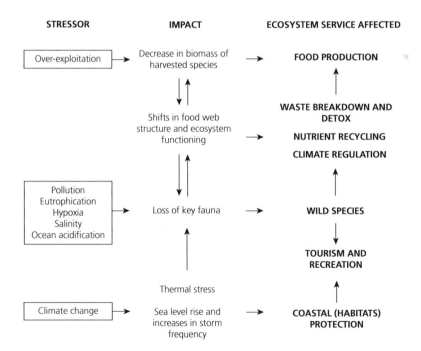

Figure 18.1 Impacts of human-induced stressors on marine ecosystem services. Examples are shown (not all stressors and ecosystem services are included). Arrows show direction of impact.

quotas and advances in fishing technology (e.g. trawling methods) have led to unprecedented harvest rates
achieved in some parts of the world (Daw and Gray,
2005; Bromley, 2009). While there are indications that,
for many of the formally assessed fisheries, harvest
rates have declined and recovery plans are in place, the
same is not true for unassessed fisheries. Many of these
fisheries, which represent > 80% of global catch, are still
in decline (Costello et al., 2012). This not only has implications for economic yields and the ecology of the
target species, but also for non-target species and the
wider marine ecosystem. When affected species play
keystone roles in the ecosystem, the knock-on implications for the ecosystem can be significant. For example,
the removal of Alaskan sea otters can lead to the replacement of kelp forests with urchin barrens as urchins
no longer have any key predators (Crain et al., 2009).
When the over-harvesting of species leads to shifts in
the food-web structure and ecosystem functioning, this
can then lead to further implications for other supporting and regulating ecosystem services (Fig. 18.1).

The most measurable impact of over-exploitation in
terms of ecosystem services however is the substantial impact on global marine food supply (Fig. 18.1). In
the UK, the provision of food from marine fisheries is
lower now than at any time in the last century, and the
weight of landings per unit of fishing power has declined by 94% in 118 years (Thurstan et al., 2010).

18.3.1.2 Pollution

Human activities can result in a large range of pollutants entering the marine environment, usually as a result of marine dumping or land-based run-off (Islam
and Tanaka, 2004; Crain et al., 2009). Persistent organic
pollutants (POPs) are of particular concern as they usually have long half-lives and persist in the marine environment (Wurl and Obbard, 2005). As they accumulate,
they biomagnify and become more concentrated, especially at higher trophic levels (Crain et al., 2009). These
compounds are now found in all parts of the ocean but
the effects have only been studied at a species level,
and the ecosystem-level impacts are not well understood. They can lead to mortality and sublethal effects
in a range of marine organisms, and potentially in humans that feed on them (Kelly et al., 2007). The knock-
on implications for ecosystem function are likely to
have an impact on a wide range of ecosystem services
where pollution levels are high (Fig. 18.1).

18.3.1.3 Eutrophication and hypoxia

Eutrophication is caused by excessive nitrogen enrichment in the water that can lead to algal blooms, algal

decomposition, and then oxygen depletion. Increased
nitrogen supply can be caused by human activities,
usually as a result of fossil fuel combustion or the use
of synthetic fertilizers (Crain et al., 2009). Eutrophication can impact upon many ecosystem services via
direct losses of key fauna and indirectly via shifts in
food web structure and functioning (Österblom et al.,
2007; Fig. 18.1). However, the greatest negative impacts resulting from eutrophication are usually from
the hypoxic conditions that result from the microbial
decomposition (Diaz, 2001; Diaz and Rosenberg, 2008;
Crain et al., 2009; Rabalais et al., 2009).

18.3.1.4 Ocean acidification

Ocean acidification is largely a consequence of increases in CO_2 levels in the atmosphere. As atmospheric levels increase, so does the amount of CO_2
absorbed by the ocean. This CO_2 lowers the pH of the
seawater by forming carbonic acid, therefore lowering
the concentration of dissolved carbonate ions (Billé
et al., 2013). This can have a wide range of biological
impacts on calcifying and photosynthesizing marine
organisms and indirect impacts on associated species
(Crain et al., 2009). Acquisition of calcium carbonate is
particularly essential to shellfish and reef ecosystems.
Calcification rates of reef-building corals have been declining in line with increasing CO_2 levels, with knock-
on implications for reef organisms dependent on the
reef structure (Crain et al., 2009). This will have an implication on the provision of ecosystem services such
as recreation and tourism as well as capture fisheries
and aquaculture (Rodrigues et al., 2013).

The other major stressor related to increased emissions of greenhouse gases is human-induced climate change and the impacts of this are discussed in
Section 18.3.1.5.

18.3.1.5 Climate change

The consequences of human-induced climate change
on ocean ecosystem services are mediated largely via
changes in sea temperature and sea-level rise coupled
with increases in storm frequency (Fig. 18.1). Range
shifts in fish and shellfish populations as a result in
thermal changes in the sea will have implications for
fisheries and therefore food production (Worm et al.,
2009; Fulton, 2011). But it is sea-level rise and the potential increase in storm frequency that can have a
large impact on coastal habitats such as sea grass, kelp
forests, and wetlands that currently provide regulating
services such as storm and coastal protection (Ruckelshaus et al., 2013). Many of these coastal habitats also
act as nurseries for key fauna important for commercial

fisheries and recreation and their destruction could lead to further implications for the provision of these ecosystem services (Ruckelshaus et al., 2013).

18.3.2 Cumulative impacts and interactions associated with multiple stressors

We have seen from the examples in Section 18.3.1 that a single stressor can have implications for multiple ecosystem services and that multiple stressors can therefore potentially impact upon a suite of ecosystem services. But these multiple stressors and ecosystem services are likely to interact in different ways and a consideration of such interactions will be necessary in an Ecosystem Approach to management.

Research that focuses on the cumulative and the possible interactive effects of multiple stressors is not as common as that on single stressor impacts. The research which has taken place has focused on multiple stressors and the impact on ecosystem function, usually at a small scale (e.g. O'Gorman et al., 2012; Godbold and Solan, 2013). Applying these insights to marine systems on a larger, policy-relevant scale presents many challenges. The cumulative impact depends not only on the number of stressors but also the severity of the different stressors and the vulnerability of different ecosystems to withstand these stressors (Orth et al., 2006; Crain et al., 2009). These cumulative impacts can be additive, synergistic, or indeed antagonistic, and often non-linear in nature (Crain et al., 2009). For example, nutrient loading can lead to increased productivity in fisheries when inputs are low, but when inputs reach a level to which they can cause hypoxic conditions this can lead to fishery collapse (Halpern et al., 2008a).

The impact of stressors on systems that have never previously experienced high levels of stress may be far lower than on similar ecosystems that have already been affected by multiple threats. For example, coral reefs in Australia have been shown to recover after recurrent cyclones while Jamaican reefs that were already experiencing over-fishing and outbreaks of disease were unable to recover from similar physical damage caused by repeated hurricanes (Halpern et al., 2008a). In addition, not all stressors are equal in terms of their impact on ecosystems and, indeed, some stressors have very different impacts depending on the different components of the marine habitats. For example, trawling can have devastating impacts on corals and seamounts, leading to almost total habitat destruction (Hall-Spencer et al., 2002; Halpern et al.,

2008b; Althaus et al., 2009), but the same activity on muddy benthic habitat will have less impact (Halpern et al., 2008a).

18.4 Using an ecosystem services approach to managing multiple stressors: a framework

The conventional sector-by-sector approach to managing stressors in the marine environment is largely ineffective at meeting the environmental, economic, and societal objectives of the many sectors that utilize the marine environment. This approach rarely takes into consideration many of the factors that we have outlined in previous sections such as: (i) interactions among activities; (ii) cumulative impacts of these activities both spatially and temporally; (iii) the process by which activities affect the delivery of ecosystem services; and (iv) inevitable trade-offs among activities (Halpern et al., 2008a). The Ecosystem Approach, however, may be more effective at meeting these challenges.

Here we present a generic framework for managing multiple marine stressors using an Ecosystem Approach largely focused around the management of ecosystem services (Fig. 18.2). By using ecosystem services as our point of focus, we can understand the range of impacts more clearly by focusing on changes to those marine services that society values highly and accounting for cumulative impacts through changes in these ecosystem services (Ruckelshaus et al., 2013). Firstly we present a series of principles that underpin the framework. We have chosen principles instead of stages as stages imply a linear progression, whereas in reality the management process will need to consider a number of activities or processes simultaneously, each of which inform one another. These principles are taken from common themes mentioned in the literature.

18.4.1 Engage stakeholders in a participatory process

Stakeholder involvement is assumed to take place throughout the framework. However, in order for this to take place, the relevant stakeholders and stakeholder groups need to be identified and methods of stakeholder engagement need to be decided. In the marine environment, stakeholders are likely to include at least several of the following groups: fisheries managers, government bodies, local or commercial fishermen, conservation groups, scientific researchers, recreational users, and other members of the general

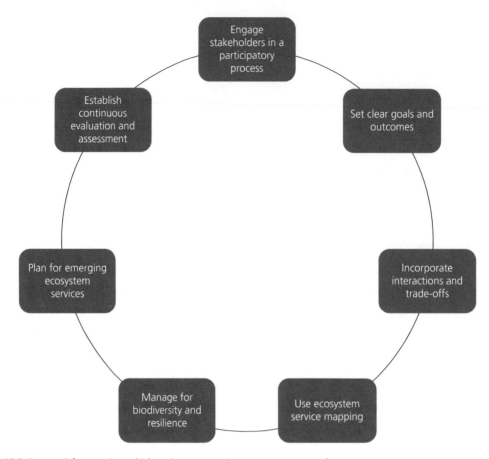

Figure 18.2 Framework for managing multiple marine stressors using an ecosystem approach.

public. There are many benefits of this multi-sector approach, not least to establish those ecosystem services that are important across a number of sectors, and how different services may interact due to differing demands from stakeholders. In addition, the participatory process itself helps to build trust between stakeholders and promote understanding of alternative viewpoints, therefore potentially leading to conflict reduction (Levin and Cross, 2004; Holt et al., 2011; Fazey et al., 2013).

18.4.2 Set clear goals and outcomes

For a collaborative management process focused on ecosystem services, the goals for delivery across the whole suite of ecosystem services need to be established at the outset (Halpern et al., 2008a). Within the stakeholder group, using a participatory process, key ecosystem services need to be identified and common

objectives need to be established. Where possible, management activities need to be planned to work towards these goals and the roles and responsibilities of each stakeholder within such a collaboration need to be explicit. If there are conflicting objectives regarding ecosystem service outcomes, then an ecosystem service trade-off plan (principle 3; see Section 18.4.3) may be an option. A system of monitoring, evaluation, and assessment needs to be in place to check if the objectives are being achieved as a result of the activities, and if not, where adjustment may be needed (see principle 7 (Section 18.4.7) for more on evaluation and assessment).

18.4.3 Incorporate interactions and trade-offs

If it is not possible for all interacting services to be maximized simultaneously, trade-offs among ecosystem services are required, and the stakeholder group must make decisions about their relative preferences for different services and how this affects management (Lester et al., 2013). Some management activities may support the production of some ecosystem services but may have detrimental impacts on others, as demonstrated in previous sections. Not all such interactions between ecosystem services are linear in nature and trade-offs can be convex or concave or form a number of other relationships (see Lester et al., 2013). Marine spatial planning (MSP) has been cited as a potential solution to balancing trade-offs, coordinating human activities and reducing impacts across sectors (White et al., 2012). Using this approach, decisions can be made as to which objectives and activities will have priority within each zone to protect selected ecosystem services (Halpern et al., 2008a).

18.4.4 Use ecosystem service mapping

The mapping of ecosystem services and benefits is a useful means of visualizing and communicating information on ecosystem services in an accessible way, in particular in relation to potential impacts as a result of management or environmental change (Daily and Matson, 2008; Burkhard et al., 2012). The framework therefore supports the use of ecosystem service mapping, not only in the assessment of process but also in terms of a tool for stakeholder engagement (Cowling et al., 2008; Tallis and Polasky, 2009). Ecosystem service mapping is a rapidly expanding area, but it poses significant challenges, both technically and practically, for both terrestrial and marine environments. These are explained further in Section 18.5.3.

18.4.5 Manage for biodiversity and resilience

The framework promotes the implementation of management activities to conserve biodiversity wherever possible, since this provides an insurance value by underpinning ecosystem resilience and health and the continued provision of ecosystem services. Marine diversity loss is increasingly impairing the oceans' capacity to provide food, maintain water quality, and recover from other stressors (Worm et al., 2006). Therefore, preserving diversity of species and ecosystems can be broadly beneficial to a wide range of important ecosystem processes and services (Folke et al., 2004; Palumbi et al., 2009). Because of this supporting role, managing for the maintenance of biological diversity has the potential to serve as a common goal to unite sectors and devise management options for restricting human stressors but alongside other

management objectives and trade-offs. While trade-offs will allow the continued sustainable use of certain ecosystem services (particularly provisioning and supporting), addressing multiple stressors for the purpose of biodiversity conservation will automatically support a broad range of ecological interactions and therefore multiple ecosystem service provision (Palumbi et al., 2009).

18.4.6 Plan for emerging ecosystem services

Emerging ecosystem services from the marine environment include wave energy and offshore aquaculture. Energy provision from the sea in particular is likely to be increasingly important over the coming years. Emerging ecosystem services, such as renewable energy production and recreational whale-watching, provide ideal opportunities for integrated planning prior to development (White et al., 2012; Lester et al., 2013). A key part of this framework is therefore to remain vigilant of such opportunities so that other principles in the framework can be practised from an early stage in the development process and include as many of the new and existing stakeholders as possible.

18.4.7 Establish continuous evaluation and assessment

Continuous evaluation and assessment of ecosystem service changes as a result of management interventions are needed in the framework in order to adapt management goals and objectives as necessary through time. However, to date, most applications of evaluation in conservation have been at the level of species and habitats, or in relation to projects or programmes, rather than at the level of ecosystem services (Raffaelli and White, 2013). This is a common problem for ecosystem service frameworks in both terrestrial and marine environments and yet assessment is essential for an adaptive ecosystems approach to management. Nonetheless, proxies or indicators can be used to monitor changes in ecosystem services (Boyd and Banzhaf, 2007; Layke, 2009) and the choice of indicator will depend on the ecosystem service being measured and the location of the habitat being monitored. Provisioning services (such as fish stocks) can be relatively simple to monitor when compared to other ecosystem services categories (Raffaelli and White, 2013). However, regulating and cultural services represent significant challenges for indicator development, and the application of ecosystem service indicators in practice to evaluate the state of marine ecosystems remains extremely limited (Feld et al., 2009).

18.5 Framework tools

The framework principles list a number of approaches to management, and these often require management tools. Some of these tools are well established and others are constantly being developed and adapted for use. In this section we discuss some of these tools in relation to three different parts of the framework: stakeholder engagement, ecosystem service trade-offs, and mapping ecosystem services. We also offer some case studies of where some of these tools have been used in studies of marine areas experiencing multiple stressors.

18.5.1 Stakeholder engagement tools

The increasingly complex and multifaceted nature of many environmental challenges means that participatory approaches have been increasingly incorporated into policy and decision-making processes (Reed, 2008). Different participatory approaches are needed to engage different stakeholder groups. Key stakeholders can be identified through stakeholder mapping (Pomeroy and Douvere, 2008), and techniques for

engagement should be chosen and adapted according to the purpose of the engagement, the audience, the scale of management, and the desired outcomes. The use of the correct engagement tool is important, since using a tool poorly suited to the situation can reduce the likelihood of a successful outcome (Lynam et al., 2007). Engagement tools include scenario-based approaches, which can be particularly useful in bringing together diverse stakeholders to consider future options for management (Tompkins et al., 2008), and participatory mapping, which can be used to establish how stakeholders use and value ecosystem services (Holt et al., 2011). Stakeholders can also be involved in participatory modelling exercises, for example using Bayesian belief network-based approaches (Metcalf et al., 2014). One of the most significant aspects of all these participatory approaches is their capacity to empower stakeholders through developing social learning and shared understanding, which can combine to deliver more generally favourable outcomes (Stringer et al., 2006). Much of the work that has been conducted on methods for stakeholder engagement in marine environments has focused on stakeholder engagement for the establishment of marine protected areas (MPAs). Box 18.2 provides a case study from Fox et al. (2013) which demonstrates how stakeholder engagement tools have been used for this purpose.

Box 18.2 Case study 1: Stakeholder engagement for marine protected areas in California.

California's Marine Life Protection Act (MLPA) led to the creation of a network of MPAs to meet ecosystem-based goals. In 2004, a partnership was launched to assist the state in implementing the MLPA, and a process was initiated to redesign California's existing MPAs. California's state waters were divided into five study regions, and in each study region individuals representing diverse interests were appointed to a Regional Stakeholder Group (RSG). The role of each RSG was to produce alternative MPA proposals, including proposed boundaries and regulations for an array of MPAs that together addressed the goals of the MLPA. These MPA proposals were forwarded to the MLPA Blue Ribbon Task Force (BRTF). The BRTF reviewed proposals from the RSG and forwarded a range of alternatives, including a preferred alternative, to the formal decision-making body, the California Fish and Game Commission. Each proposal was also reviewed by a Science Advisory Team (SAT) relative to science-based design guidelines. Through this process, the RSG-developed proposals had

significant influence on the MPA network designs ultimately delivered to the Commission for decision making and implementation.

The process included many best practices for successful stakeholder engagement. It allowed for a high level of stakeholder participation, incorporated active public engagement, and encouraged the pursuit of mutual benefits. The Initiative was also committed to self-evaluation and improvement, including widely available 'lessons learned' reports generated for each regional process. MLPA Initiative study regions were planned sequentially, which allowed for the incorporation of lessons learned through early monitoring in other areas and variations between study regions to fit locally specific requirements. The process also enabled the integration of input from state agencies into the stakeholder-driven MPA design process.

See Fox et al. (2013) for further details.

Taken from Fox et al. (2013).

18.5.2 Tools for investigating ecosystem service trade-offs

Quantitatively evaluating trade-offs among multiple ecosystem services may be necessary when there are multiple ecosystem services to manage for and not all interacting services can be maximized simultaneously. The analysis of trade-offs is therefore a useful tool for helping to inform the resolution of conflicts between stakeholders where there are multiple, conflicting objectives. Tools for assessing trade-offs between ecosystem services build on a number of approaches from the economics literature. Production function and expected damage approaches (Barbier, 2007) can help to inform a process-based understanding of the relationships among different ecosystem services. Economic approaches can also be used to quantify preferences for different ecosystem service outcomes, and in some cases attach a monetary value by determining how much participants would be 'willing to pay' for certain quantities of the service. The technique now most widely applied in this context is choice modelling, which was originally developed to determine consumer preferences for multi-attribute goods (Louviere and Woodworth, 1983). Choice modelling is especially well suited to the explicit consideration of trade-offs which stakeholders make between competing natural resource priorities (Breffle and Rowe, 2002; Milon and Scrogin, 2006). While performing such analyses requires specific expertise, it does demonstrate and help to clarify where there are common preferences for certain ecosystem services and how management efforts could be directed.

18.5.3 Tools for mapping ecosystem services

The mapping of ecosystem services is a rapidly growing technique, with no set methodology; organizations or stakeholder groups often adapt their technique based on the availability of data sets and expertise. However, this technique can be extremely important, not only in the setting of goals and the evaluation of trade-offs, but also as a way of engaging stakeholders and opening communication between groups.

Box 18.3 Case study 2: Application of marine InVEST to the West Coast of Vancouver Island, Canada.

The Integrated Valuation of Ecosystem Services and Trade-offs (InVEST) is a tool that maps, quantifies, and values the services provided by landscapes and seascapes. It is a set of spatially explicit computer-based models that focus on ecosystem services and provide outputs in both biophysical and monetary and nonmonetary value terms. It can be used to evaluate alternative management scenarios and can reveal relationships among multiple ecosystem services. InVEST was designed to be integrated with stakeholder engagement processes, and is therefore iterative and interactive.

The West Coast of Vancouver Island (WCVI) has multiple, often competing interests and stakeholders. These stakeholders have come together as The West Coast Aquatic Management Board (WCA). This is a public–private partnership with participation from four levels of government (federal, provincial, local, and First Nations) and diverse stakeholders, tasked with developing a marine spatial plan for the region. The marine area is under stress due to demands on provisioning, regulating, and cultural services. There are pressures from extractive, industrial, and commercial uses, traditional First Nations subsistence and ceremonial uses, recreation and tourism, and emerging ocean uses such as the extraction of wave energy. The goal of the WCA is to manage resources for the benefit of current and future generations of people and the natural systems on which they depend. WCA has partnered with the Natural Capital Project, a collaboration between Stanford University, the Nature Conservancy, World Wildlife Fund, and the University of Minnesota to explore how alternative spatial plans might affect a wide range of ecosystem services and to provide information about trade-offs among multiple key ecosystem services to governments and stakeholders in the region.

The use of InVEST has allowed the team to identify spatially explicit management scenarios that maintain a range of benefits (e.g. shellfish harvest and good water quality) while minimizing conflicts among damaging uses and sensitive habitats, such as eelgrass habitats. The use of InVEST in this way to generate quantitative outputs (maps), to document explicit connections between potentially competing activities, and to demonstrate ways in which these conflicts could be reduced, has helped to build mutual understanding among different stakeholder groups and to shape the development of future plans for the region.

See Guerry et al. (2012) for further details.

Taken from Guerry et al. (2012).

While there is no universally accepted methodology, there are a number of established tools that are being used to build maps for these purposes. In most cases, a version of a Geographic Information System (GIS) is used to construct maps using habitat or land-use data as proxies for ecosystem services, although not all projects will use this method. Ecosystem service mapping is currently more common for terrestrial environments than marine ones, and the range of approaches being used for terrestrial environments is exemplified by the UK Ecosystem Service Mapping Gateway (http://www.nerc-bess.net/ne-ess/). The relative paucity of ecosystem service mapping for marine systems is due to a number of challenges with mapping ecosystem services in a marine setting, which are discussed further in Section 18.6. However, an example of where an established tool has been applied to evaluate and map ecosystem services in the marine environment (Guerry et al., 2012) is provided in Box 18.3.

18.6 Challenges for the future

The Ecosystem Approach is particularly well suited for informing the management of multiple stressors in marine environments because of its emphasis on integrating ecosystem services and stakeholder groups. The Ecosystem Approach recognizes that different stressors, management sectors, and ecosystem services can interact and influence one another, and provides a means by which these different effects can be brought together in an integrative manner. Our framework outlines how such an approach might be implemented by focusing on seven core principles. However, these principles and the tools that are needed to implement them can also lead to challenges, especially in the marine environment. In the following two sections we discuss these challenges specifically in relation to our framework but also more generally.

18.6.1 Challenges of marine systems

Like many terrestrial systems, marine environments have a large range of stakeholders from tourists to fishers, policy makers, scientists, and the general public. Unlike many terrestrial environments currently, much of the marine environment is under common, international ownership, which presents significant challenges for governance (Ardron et al., 2008). In particular, it makes assigning responsibilities for managing large parts of the marine environment a major challenge. This is compounded by the fact that

the marine environment has no physical boundaries, which in terrestrial environments can provide a clear limit to certain activities or management responsibilities, and can also help to restrict the impact of certain management activities on neighbouring areas.

Marine systems also face data challenges. Because the marine environment represents a more physically challenging one for research compared with most terrestrial environments, there may be fewer data relating to some ecosystem services and therefore less knowledge regarding the system when compared with terrestrial areas. Moreover, there are challenges in making the data that exist accessible and understandable to the different stakeholder groups (Hiscock et al., 2003).

18.6.2 Framework-specific challenges

A key challenge regarding the implementation of the Ecosystem Approach is the continuous evaluation and assessment of ecosystem service changes in order to adapt management goals and objectives as necessary through time. Although an essential part of the process, monitoring and evaluation is often hindered by a lack of financial and human resources, but also by a lack of understanding of the importance of assessing change in the system. Furthermore, suitable metrics and indicators need to be developed at the scale that ecosystem services operate. Often, monitoring and evaluation for biodiversity conservation has focused on protecting stocks of either species or of biodiversity in key habitats (Bhaskar and Adams, 2009). As a consequence, indicators have tended to be focused on measures of these stocks. While important, this provides an incomplete picture of the status of ecosystem services, which are not likely to be related in a simple way to levels of biodiversity stock (Raffaelli and White, 2013). In most cases, we do not have a complete understanding of the relationship between these stocks of biodiversity and the ecosystem services they provide (Raffaelli and White, 2013).

Another key challenge is establishing ecosystem service mapping in the marine environment. We have stated several times in this chapter that this technique can be extremely useful, not only in the setting of goals and the evaluation of trade-offs, but also as a way of engaging stakeholders and opening communication between groups, and yet conducting the technique in the marine environment can present a number of challenges. Firstly, maps of habitat type and condition are more difficult and costly to obtain in marine environments than in terrestrial environments. This is because many of the habitat types (with the exception of coral reefs and kelp forests) are not visible on satellite imagery or other remote sensing technology (Guerry et al., 2012). Secondly, knowledge of the spatial distribution of uses and the impacts of those uses on the habitats is often not known or formally recorded. This is especially true in much of the marine environment which is managed as commons (Guerry et al., 2012). Despite these challenges, some areas of research are starting to make headway in this area and we have provided an example earlier, in Box 18.3.

18.7 Conclusion

The marine environment is currently facing unprecedented levels of anthropogenic stressors. The principal stressors are climate change, pollution, fragmentation and habitat loss, invasive species, and over-harvesting from fisheries. Many of these stressors act on the marine system at a variety of scales, from genetic through to the ecosystem. The effects of these stressors are frequently synergistic, with detrimental impacts across a range of ecosystem services that humans derive from the ocean, including fisheries, coastal protection, climate regulation, and recreation. Furthermore, the effects of some stressors may be exacerbated by failures of policy and governance. Management of these stressors to ensure the sustainable delivery of ecosystem services for present and future generations therefore represents a considerable challenge.

The Ecosystem Approach, whereby interventions are focused around maintaining functionality and a diversity of service provision from the ecosystem in an equitable way, provides many advantages over more traditional, single-sector-focused approaches. An ecosystem service focus provides a means of understanding the range of impacts of stressors more clearly by focusing on changes to those marine services that society values highly and accounting for cumulative impacts through changes in these ecosystem services.

For an ecosystem service-based approach to be effective, we have identified seven key principles that need to be adopted: (i) engage stakeholders in a participatory process; (ii) set clear goals and outcomes; (iii) incorporate ecosystem service interactions and trade-offs; (iv) use ecosystem service mapping; (v) manage for biodiversity and resilience; (vi) plan for emerging ecosystem services; and (vii) establish continuous evaluation and assessment. There are a number of stakeholder engagement tools that can facilitate an ecosystem service-based approach. Various mechanisms for enhancing participation, tools for illustrating and

analysing trade-offs between different ecosystem services, and ecosystem service mapping can all contribute towards more active engagement, help to inform debate around potential conflict situations, and build trust and understanding among different stakeholder groups.

Marine systems provide particular challenges for integrative management. The large extent of common, international ownership means that roles and responsibilities are frequently unclear or difficult to enforce. The challenging nature of the marine environment also means that data to inform monitoring and evaluation may be scarce, especially in more remote areas. Understanding of the relationships between biodiversity, ecosystem function, and ecosystem services at the ecosystem level remains incomplete, and system-level models may not be very robust, especially when used to make predictions for alternative futures. There are also challenges in making data and outputs from models available and understandable to all the different stakeholder groups. Perhaps the greatest challenge lies in establishing effective governance systems throughout the oceans to ensure that sustainable management can be practised and policed effectively.

Despite the many challenges, frameworks such as the one presented here offer an integrated management approach for addressing multiple stressors in marine environments. A focus on ecosystem services and the interactions and trade-offs that can occur, mediated via participatory approaches, provides a strong framework for an integrative approach to management that is essential to build a sustainable future for the oceans.

References

Althaus, F., Williams, A., Schlacher, T.A., et al. (2009). Impacts of bottom trawling on deep-coral ecosystems of seamounts are long-lasting. *Marine Ecology Progress Series*, **397**, 279–294.

Ardron, J., Gjerde, K., Pullen, S., & Tilot, V. (2008). Marine spatial planning in the high seas. *Marine Policy*, **32**, 832–839.

Armsworth, P.R., Chan, K.M.A., Daily, G.C., et al. (2007). Ecosystem service science and the way forward for conservation. *Conservation Biology*, **21**, 1383–1384.

Atkins, J.P., Burdon, D., Elliott, M., & Gregory, A.J. (2011). Management of the marine environment: integrating ecosystem services and societal benefits with the DPSIR framework in a systems approach. *Marine Pollution Bulletin*, **62**, 215–226.

Barbier, E.B. (2007). Valuing ecosystem services as productive inputs. *Economic Policy*, **22**, 177–229.

Beaumont, N.J., Austen, M.C., Atkins, J.P., et al. (2007). Identification, definition and quantification of goods and services provided by marine biodiversity: implications for the ecosystem approach. *Marine Pollution Bulletin*, **54**, 253–265.

Bhaskar, V. & Adams, W.M. (2009). Ecosystem services and conservation strategy: beware the silver bullet. *Conservation Letters*, **2**, 158–162.

Bille, R., Kelly, R., Biastoch, A., et al. (2013). Taking action against ocean acidification: a review of management and policy options. *Environmental Management*, **52**, 761–779.

Botsford, L.W., Castilla, J.C., & Peterson, C.H. (1997). The management of fisheries and marine ecosystems. *Science*, **277**, 509–515.

Boyd, J. & Banzhaf, S. (2007). What are ecosystem services? The need for standardized environmental accounting units. *Ecological Economics*, **63**, 616–626.

Breffle, W.S. & Rowe, R.D. (2002). Comparing choice question formats for evaluating natural resource tradeoffs. *Land Economics*, **78**, 298–314.

Bromley, D.W. (2009). Abdicating responsibility: the deceits of fisheries policy. *Fisheries*, **34**, 280–290.

Burkhard, B., Kroll, F., Nedkov, S., & Müller, F. (2012). Mapping ecosystem service supply, demand and budgets. *Ecological Indicators*, **21**, 17–29.

Convention on Biological Diversity (CBD) (2000). *The Conference of the Parties, at its Fifth Meeting, endorsed the description of the ecosystem approach and operational guidance and recommended the application of the principles and other guidance on the Ecosystem Approach (decision V/6).* http://www.cbd.int/ecosystem/

Costanza, R., d'Arge, R., deGroot, R., et al. (1997). The value of the world's ecosystem services and natural capital. *Nature*, **387**, 253–260.

Costello, C., Ovando, D., Hilborn, R., et al. (2012). Status and solutions for the world's unassessed fisheries. *Science*, **338**, 517–520.

Cowling, R.M., Egoh, B., Knight, A.T., et al. (2008). An operational model for mainstreaming ecosystem services for implementation. *Proceedings of the National Academy of Sciences of the USA*, **105**, 9483–9488.

Crain, C.M., Halpern, B.S., Beck, M.W., & Kappel, C.V. (2009). Understanding and managing human threats to the coastal marine environment. *Annals of the New York Academy of Sciences*, **1162**, 39–62.

Daily, G.C. & Matson, P.A. (2008). Ecosystem services: from theory to implementation. *Proceedings of the National Academy of Sciences of the USA*, **105**, 9455–9456.

Daw, T. & Gray, T. (2005). Fisheries science and sustainability in international policy: a study of failure in the European Union's Common Fisheries Policy. *Marine Policy*, **29**, 189–197.

Diaz, R.J. (2001). Overview of hypoxia around the world. *Journal of Environmental Quality*, **30**, 275–281.

Diaz, R.J. & Rosenberg, R. (2008). Spreading dead zones and consequences for marine ecosystems. *Science*, **321**, 926–929.

Dick, J. McP., Smith, R.I., & Scott, E.M. (2011). Ecosystem services and associated concepts. *Envirometrics*, **22**, 598–607.

Ehrlich P.R. & Mooney, H.A. (1983). Extinction, substitution, and ecosystem services. *Bioscience*, **33**, 248–254.

Fazey, I., Evely, A.C., Reed, M.S., et al. (2013). Knowledge exchange: a review and research agenda for environmental management. *Environmental Conservation*, **40**, 19–36.

Feld, C.K., Martins da Silva, P., Paulo Sousa, J., et al. (2009). Indicators of biodiversity and ecosystem services: a synthesis across ecosystems and spatial scales. *Oikos*, **118**, 1862–1871.

Fisher, B. & Turner, R.K. (2008). Ecosystem services: classification for valuation. *Biological Conservation*, **141**, 1167–1169.

Fletcher, S., Saunders, J., & Herbert, R.J.H. (2011). A review of the ecosystem services provided by broad-scale marine habitats in England's MPA network. *Journal of Coastal Research*, **64**, 378–383.

Folke, C., Carpenter, S., Walker, B., et al. (2004). Regime shifts, resilience, and biodiversity in ecosystem management. *Annual Review of Ecology, Evolution, and Systematics*, **35**, 557–581.

Fox, E., Poncelet, E., Connor, D., et al. (2013). Adapting stakeholder processes to region-specific challenges in marine protected area network planning. *Ocean and Coastal Management*, **74**, 24–33.

Fulton, E.A. (2011). Interesting times: winners, losers, and system shifts under climate change around Australia. *ICES Journal of Marine Science*, **68**, 1329–1342.

Godbold, J.A. & Solan, M. (2013). Long-term effects of warming and ocean acidification are modified by seasonal variation in species responses and environmental conditions. *Philosophical Transactions of the Royal Society Series B*, **368**, 20130186.

Guerry, A.D., Ruckelshaus, M.H., Arkema, K.K., et al. (2012). Modeling benefits from nature: using ecosystem services to inform coastal and marine spatial planning. *International Journal of Biodiversity Science, Ecosystem Services and Management*, **8**, 107–121.

Hails, R.S. & Ormerod, S.J. (2013). Ecological science for ecosystem services and the stewardship of natural capital. *Journal of Applied Ecology*, **50**, 807–811.

Hall–Spencer, J., Allain, V., & Fosså, J.H. (2002). Trawling damage to Northeast Atlantic ancient coral reefs. *Proceedings of the Royal Society of London Series B*, **269**, 507–511.

Halpern, B.S., McLeod, K.L., Rosenberg, A.A., & Crowder, L.B. (2008a). Managing for cumulative impacts in ecosystem-based management through ocean zoning. *Ocean and Coastal Management*, **51**, 203–211.

Halpern, B.S., Walbridge, S., Selkoe, K.A., et al. (2008b). A global map of human impact on marine ecosystems. *Science*, **319**, 948–952.

Hiddink, J.G., Jennings, S., Kaiser, M.J., et al. (2006). Cumulative impacts of seabed trawl disturbance on benthic biomass, production, and species richness in different habitats. *Canadian Journal of Fisheries and Aquatic Sciences*, **63**, 721–736.

Hiscock, K., Elliott, M., Laffoley, D., & Rogers, S. (2003). Data use and information creation: challenges for marine scientists and for managers. *Marine Pollution Bulletin*, **46**, 534–541.

Holt, A., Godbold, J.A., White, P.C.L., et al. (2011). Mismatches between legislative frameworks and benefits restrict the implementation of the Ecosystem Approach in coastal environments. *Marine Ecology Progress Series*, **434**, 213–228.

Hossain, A.B.M.S., Salleh, A., Boyce, A.N., et al. (2008). Biodiesel fuel production from algae as renewable energy. *American Journal of Biochemistry and Biotechnology*, **4**, 250–254.

Islam, S. & Tanaka, M. (2004). Impacts of pollution on coastal and marine ecosystems including coastal and marine fisheries and approach for management: a review and synthesis. *Marine Pollution Bulletin*, **48**, 624–649.

Kelly, B.C., Ikonomou, M.G., Blair, J.D., et al. (2007). Food web-specific biomagnification of persistent organic pollutants. *Science*, **317**, 236–239.

Layke, C. (2009). *Measuring Nature's Benefits: A Preliminary Roadmap for Improving Ecosystem Service Indicators*. World Resources Institute, Washington, DC.

Lenihan, H.S., Peterson, C.H., Kim, S.L., et al. (2003). Variation in marine benthic community composition allows discrimination of multiple stressors. *Marine Ecology Progress Series*, **261**, 63–73.

Lester, S.E., Costello, C., Halpern, B.S., et al. (2013). Evaluating tradeoffs among ecosystem services to inform marine spatial planning. *Marine Policy*, **38**, 80–89.

Levin, D.Z. & Cross, R. (2004). The strength of weak ties you can trust: the mediating role of trust in effective knowledge transfer. *Management Science*, **50**, 1477–1490.

Louviere, J.J. & Woodworth, G. (1983). Design and analysis of simulated consumer choice or allocation experiments: an approach based on aggregate data. *Journal of Marketing Research*, **20**, 350–367.

Lynam, T., de Jong, W., Sheil, D., et al. (2007). A review of tools for incorporating community knowledge, preferences, and values into decision making in natural resources management. *Ecology and Society*, **12**, 5.

Mace, G.M., Norris, K., & Fitter, A.H. (2012). Biodiversity and ecosystem services: a multilayered relationship. *Trends in Ecology and Evolution*, **27**, 19–26.

Metcalf, S.J., van Putten, E.I., Frusher, S.D., et al. (2014). Adaptation options for marine industries and coastal communities using community structure and dynamics. *Sustainability Science*, **9**, 247–261.

Millennium Ecosystem Assessment (2005). *Ecosystems and Human Well-being: Synthesis*. Island Press, Washington, DC.

Milon, J.W. & Scrogin, D. (2006). Latent preferences and valuation of wetland ecosystem restoration. *Ecological Economics*, **56**(2), 162–175.

Möller, I. & Spencer, T. (2002). Wave dissipation over macro-tidal saltmarshes: effects of marsh edge typology and vegetation change. *Journal of Coastal Research*, **36**, 506–521.

Myers, R.A. & Worm, B. (2003). Rapid worldwide depletion of predatory fish communities. *Nature*, **423**, 280–283.

Nellemann, C., Hain, S., & Alder, J. (eds) (2008). *In Dead Water – Merging of Climate Change with Pollution, Over-harvest, and Infestations in the World's Fishing Grounds*. United Nations Environment Programme, GRID-Arendal, Norway. http://www.grida.no

O'Gorman, E.J., Fitch, J.E., & Crowe, T.P. (2012). Multiple anthropogenic stressors and the structural properties of food webs. *Ecology*, **93**, 441–448.

Orth, R.J., Carruthers, T.J., Dennison, W.C., et al. (2006). A global crisis for seagrass ecosystems. *Bioscience*, **56**, 987–996.

Österblom, H., Hansson, S., Larsson, U., et al. (2007). Human-induced trophic cascades and ecological regime shifts in the Baltic Sea. *Ecosystems*, **10**, 877–889.

Palumbi, S.R., Sandifer, P.A., Allan, J.D., et al. (2009). Managing for ocean biodiversity to sustain marine ecosystem services. *Frontiers in Ecology and the Environment*, **7**, 204–211.

Pomeroy, R. & Douvere, F. (2008). The engagement of stakeholders in the marine spatial planning process. *Marine Policy*, **32**, 816–822.

PISCES (2012). *Towards sustainability in the Celtic Sea. A guide to implementing the ecosystem approach through the MSFD.* http://www.projectpisces.eu/guide/

Pretty, J., Peacock, J., Hine, R., et al. (2007). Green exercise in the UK countryside: effects on health and psychological well-being, and implications for policy and planning. *Journal of Environmental Planning and Management*, **50**, 211–231.

Rabalais, N.N., Turner, R.E., Díaz, R.J., & Justić, D. (2009). Global change and eutrophication of coastal waters. *ICES Journal of Marine Science*, **66**, 1528–1537.

Raffaelli, D.G. & White, P.C.L. (2013). Ecosystems and their services in a changing world: an ecological perspective. *Advances in Ecological Research*, **48**, 1–70.

Reed, M.S. (2008). Stakeholder participation for environmental management: a literature review. *Biological Conservation*, **141**, 2417–2431.

Riebesell, U., Schulz, K.G., Bellerby, R.G.J., et al. (2007). Enhanced biological carbon consumption in a high CO_2 ocean. *Nature*, **450**, 545–548.

Roberts, J.M., Wheeler, A.J., & Freiwald, A. (2006). Reefs of the deep: the biology and geology of cold-water coral ecosystems. *Science*, **312**, 543–547.

Rodrigues, L.C., van den Bergh, J.C.J.M., & Ghermandi, A. (2013). Socio-economic impacts of ocean acidification in the Mediterranean Sea. *Marine Policy*, **38**, 447–456.

Ruckelshaus, M., Doney, S.C., Galindo, H.M., et al. (2013). Securing ocean benefits for society in the face of climate change. *Marine Policy*, **40**, 154–159.

Stringer, L.C., Dougill, A.J., Fraser, E., et al. (2006). Unpacking 'participation' in the adaptive management of social–ecological systems: a critical review. *Ecology and Society*, **11**, 39.

Tallis, H. & Polasky, S. (2009). Mapping and valuing ecosystem services as an approach for conservation and natural-resource management. *Annals of the New York Academy of Sciences*, **1162**, 265–283.

Thurstan, R.H., Brockington, S., & Roberts, C.M. (2010). The effects of 118 years of industrial fishing on UK bottom trawl fisheries. *Nature Communications*, **1**, 15.

Tompkins, E.L., Few, R., & Brown, K. (2008). Scenario-based stakeholder engagement: incorporating stakeholders preferences into coastal planning for climate change. *Journal of Environmental Management*, **88**, 1580–1592.

UK National Ecosystem Assessment (2011). *The UK National Ecosystem Assessment: Technical Report*. UNEP-WCMC, Cambridge.

UNEP (United Nations Environment Programme) (2012). *Global Environment Outlook 5 (GEO-5)*. UNEP, Nairobi, Kenya.

Watson, R. & Pauly, D. (2001). Systematic distortions in world fisheries catch trends. *Nature*, **414**, 534–536.

White, C., Halpern, B.S., & Kappel, C.V. (2012). Ecosystem service tradeoff analysis reveals the value of marine spatial planning for multiple ocean uses. *Proceedings of the National Academy of Sciences of the USA*, **109**, 4696–4701.

Worm, B., Barbier, E.B., Beaumont, N., et al. (2006). Impacts of biodiversity loss on ocean ecosystem services. *Science*, **314**, 787–790.

Worm, B., Hilborn, R., Baum, J., et al. (2009). Rebuilding global fisheries. *Science*, **325**, 578–585.

Wurl, O. & Obbard, J. P. (2005). Organochlorine pesticides, polychlorinated biphenyls and polybrominated diphenyl ethers in Singapore's coastal marine sediments. *Chemosphere*, **58**, 925–933.

CHAPTER 19

Quantifying the economic consequences of multiple stressors on the marine environment

Nick Hanley

19.1 Introduction

In this chapter I explain the principles behind estimating economic values for changes in the quality of marine ecosystems. I consider four case studies where economists have combined with ecologists to actually estimate such values, for both marine and coastal systems. A number of cross-cutting themes are contained within the chapter, including (i) the kinds of data and information which economists need from scientists in order to come up with valuation estimates, and (ii) how economic values can vary between different ecosystem services, across beneficiaries and over space.

19.2 Valuing changes in ecosystem services and in biodiversity

Since the Millennium Ecosystem Assessment (2005), the language of ecosystem services (ES) has come to dominate scientific and policy discourse on protecting the natural environment, in exploring how humans derive well-being from this environment, and in thinking about how environmental change might effect future human well-being. Ecosystem services are conventionally grouped into supporting, regulating, provisioning, and cultural services. Table 19.1 gives examples of these service flows for a coastal wetland. This classification system dominates the ES scientific literature and public discourse, although it is rather less useful from an economic viewpoint since it does not provide guidance on which kinds of method are best suited to estimate economic values for a specific ES. In parallel to this ES debate, a scientific literature and policy discussion on the economic value of protecting and enhancing biodiversity has emerged, for

example through the series of TEEB reports (The Economics of Ecosystems and Biodiversity, see http://www.teebweb.org), and in collections of work such as Kontoleon et al. (2007) and Helm and Hepburn (2014).

A range of contributions in economics has discussed how best to map changes in the flows of ES and/or changes in biodiversity, due perhaps to changes in multiple pressures on ecosystems, to economic measures of well-being (Boyd and Banzhaf, 2007; Bateman et al., 2011). ES contribute to human well-being in both indirect and direct ways: indirectly through their contribution to the production of goods or services which people value, and directly through their contribution to utility. Supporting ES, such as nutrient cycling and biogenic habitat provision, contribute to human well-being via their role in maintaining the flow of provisioning, regulatory, and cultural services. These three service flows yield economic benefits when they improve the well-being of people, whether these impacts are priced by the market or not. For example, coastal waters provide fishermen with fish to catch and sell—a benefit valued by the market, since the process of buying and selling puts a price on these outputs. Coastal waters also provide recreational opportunities for dinghy sailors and wind-surfers, who derive utility (happiness, well-being, contentment) from being able to sail and surf, even though they pay no market price for this pleasure, since we do not pay an entry fee to use the sea. People do incur costs in undertaking these activities, such as travelling to the coast: as we will see, these expenditures are one source of information on the value of such activities.

Moreover, marine areas are home to a vast range of biodiversity which people derive utility from, whether it is a value from whale watching, seeing sea birds, or

Stressors in the Marine Environment. Edited by Martin Solan and Nia M. Whiteley
© Oxford University Press 2016. Published in 2016 by Oxford University Press.

Table 19.1 Examples of ecosystem service (ES) flows from a coastal wetland.

Type of ES	Examples of this ES type for a coastal wetland	Examples of valuation methods applicable to measuring economic values associated with these ecosystem services
Supporting	Nutrient cycling	Production functions (link to commercial fish catches)
Provisioning	Fish Timber (for mangroves)	Market prices
Regulating	Flood and storm protection Carbon sequestration and storage	Avoided cost; hedonic pricing; production functions (links to flood risks and storm damages)
Cultural	Recreational fishing and hunting Wildlife viewing (birds, fish, reptiles, etc.)	Stated preferences; travel cost models

just being aware of the existence of cold-water corals (Aanesen et al., 2014). Such biodiversity as a direct source of utility has an economic value even though there is typically no market to price this (Kontoleon et al., 2007). Moreover, biodiversity plays a role in maintaining the supply of a range of ecosystem services which themselves benefit people. Biodiversity thus has indirect economic values as well as a direct value.

The fact that markets fail to adequately reflect the economic value of many ES and almost all biodiversity is one example of what economists refer to as *market failure* (Hanley et al., 2007). Markets are missing for these environmental resources due to a lack of private property rights and the 'public good' nature of the benefits that derive from ecosystems and biodiversity. A public good is a commodity or service which has the characteristics of non-rivalness and non-excludability. Non-rivalness means that the number of people benefitting from such a good does not determine the benefit to any individual; thus, the fact that people in one village are protected from tsunamis by a mangrove forest does not reduce the protection available to another village. My utility from knowing that albatrosses are being conserved does not reduce your utility from knowing that they are being conserved. Non-excludability means that individuals cannot be excluded from benefitting from a good once it is provided. Thus, global agreements to reduce climate change confer benefits to all who are likely to suffer from sea level rise, whether they contribute to paying for reductions in emissions or not. Markets cannot work for non-rival and non-excludable goods, partly because there is an insufficient incentive for private agents (firms or individuals) to supply them at a socially desirable level. Governments must therefore intervene to increase the supply

of public goods (by setting regulations on pollution emissions, for instance).

However, the lack of a market price for marine biodiversity or marine ES does not signal a zero economic value—far from it. Economic value for any change in an ES or in some measure of biodiversity is measured in conceptually the same way as all other goods, whether priced by the market or not. One conceptual measure of economic values asks what is the most an individual would give up to have more of a desirable thing, or less of an undesirable thing. This is their maximum Willingness to Pay (WTP), determined both by peoples' preferences (what they want, and how much they want it relative to other goods) and by their ability to pay for it. Alternatively, we can measure the minimum compensation someone would accept to tolerate more of a bad thing (more pollution) or less of a good thing (access to a nice beach). This is their minimum Willingness to Accept Compensation, WTAC. Adding up measures of WTP or WTAC, expressed in dollars, euros, or any other currency, across all the people who appreciate the good gives a measure of social or aggregate value.

Thus, the aggregate value of improvements to coastal water quality resulting from tougher legislation on water pollution is determined by the maximum WTP (or minimum WTAC) of everyone within a country who is affected by such a change, whether as local residents living on the coast, beach visitors, or concerned citizens in faraway cities. The fact that environmental changes can be expressed in monetary units means that environmental benefits can be compared with the monetary costs of achieving such changes in environmental quality (e.g. the costs of reducing pollution from sewage works or farmland), or with the benefits of alternative policies (e.g. the benefits of reducing urban air pollution). Such a comparison of benefits and costs in the same units through the technique of cost–benefit analysis has been argued to be a useful guide to public policy making (Hanley and Barbier, 2009).

Since the early 1970s, economists have developed a range of methods for measuring the economic value of what are referred to as 'non-market goods'. Biodiversity and some ES are examples of non-market goods. These empirical methods may be divided into two groups:

- indirect, or production function, approaches, which define environmental values in terms of their role as inputs to the production of market-valued goods; and
- direct approaches, which measure the contribution of environmental goods directly to utility.

Within the category of direct approaches, we can further distinguish:

- revealed preference approaches such as travel cost models and hedonic pricing, which rely on measures of actual behaviour by people which is related to variations in environmental quality or access to environmental resources; and
- stated preference methods such as choice modelling and contingent valuation, which use constructed, hypothetical markets to measure willingness to pay or willingness to accept compensation for changes in environmental quality.

Full details on all of these approaches can be found in Hanley and Barbier (2009). Here, I simply provide a short summary, before reviewing a set of case studies which apply a subset of these approaches to coastal and marine contexts. These case studies are used partly to illustrate the spatial variability in economic values of ES and biodiversity conservation, and the ways in which these values vary across people.

In *production function approaches*, the researcher assembles data which enable the estimation of a model which relates changes in ecosystem condition to changes in the production and, perhaps, price of some marketed good. For example, a reduction in nutrient pollution entering the Baltic Sea could be related statistically to changes in submerged aquatic communities, and then to the effects on fish catches. Or, a reduction in the area of mangrove wetlands in the Gulf of Mexico could be related to changes in the growth rate and steady-state stock size of fish available to coastal fishermen (Barbier and Strand, 1998). Such an approach is sometimes referred to as the Barbier–Strand model.

For each extra hectare of wetlands lost, the economic cost is estimated by using the expected effect on fish catch and thus on profits to fishermen. This cost will depend on the property rights regime operating within the fishery. The data and knowledge that the economist needs to acquire from the scientist in such settings are clearly considerable: the method only works if there are models and/or data available which relate the change in ecosystem condition (here, the loss of mangrove wetland) into the variable which enters a function determining economic well-being (here, a catch equation for a coastal fishery).

In *revealed preference approaches* the researcher makes use of observations of actual behaviour by people to estimate direct effects of changes in ES and biodiversity on well-being. There are two main methods here that are relevant to marine and coastal settings. The *travel cost model* is focused on measuring cultural ES such as recreation, and how the economic value of such cultural ES change when ecological quality changes (Hynes et al., 2009; Chae et al., 2012). The Hauraki Gulf in New Zealand is intensively used for a range of recreational activities, including sailing, dolphin watching, and recreational sea fishing. The travel cost model estimates a statistical relationship between the costs of engaging in such recreational activities to users (focusing on travel costs from their home to the site) and the number of visits each person makes. Once such a relationship is estimated, then the value that an average person derives from a sailing trip, say, can be estimated as so many dollars per trip. To see the effect of changes in ecological quality on this value, observations are needed on how recreational visits (their number or location) vary with some observable measure of environmental quality (this could be an objective or subjective measure). The change in both the number of users at a particular location and the utility they get from each trip could then be estimated. However, this requires data sets and models which relate changes in environmental pressures (e.g. nutrient pollution) to changes in variables which drive recreational behaviour (e.g. expected fish catch or species distributions). Moreover, the method cannot measure the value that 'non-users' get from changes in marine biodiversity or cultural ES.

An alternative revealed preference method with some application to changes in coastal environmental quality is the *hedonic pricing method*. This involves the estimation of a statistical relationship between spatial variations in environmental quality and variations in house prices. For instance, one could look at the empirical relationship between water quality and house prices to estimate values for improving coastal water quality. Leggett and Bocksteal (2000) do this for water quality in the eastern USA, estimating a hedonic price function relating house prices to variations in two indicators of coastal water quality. Poor et al. (2007) is another example. One could also study how house prices vary with flooding probabilities, and relate this to spatial variations in coastal wetlands. However, for many coastal and marine ES, and especially for biodiversity, the hedonic pricing method is of very limited interest.

Stated preference methods are much more general in their applicability. Here, the researcher attempts to directly measure WTP or WTAC through the use of constructed or hypothetical markets, where a sample of individuals are asked how they would behave if a carefully specified market for a change in ES provision or biodiversity conservation existed. In contingent valuation, individuals are asked whether they would be WTP a specific amount (e.g. $20 per year) for a given change in an ES or a biodiversity indicator (Ressurreicao et al., 2011). For example, they could be asked whether they would pay this amount in higher local taxes if better pollution treatment led to an increase in coastal water quality from 'unsafe for swimming' to 'safe for swimming'. Research has shown that hypothetical WTP is closest to true WTP (which we cannot typically observe) if (i) the payment mechanism is non-voluntary and (ii) the survey is consequential for the individual, e.g. they believe that it will help guide policy decisions. However, it is still likely that true WTP is overstated on average. In choice experiments, individuals choose from a set of options over changes in ES or biodiversity described in terms of attributes and the levels these take. For example, a deep-sea conservation project could be described in terms of which ecosystem was targeted for protection from mining or other commercial activities (e.g. sea mounts or canyons), where it was located (within or outside national waters), how many species would be likely to be protected, how long the protection would occur for, etc. If one of the attributes describing the hypothetical choice sets is a price (for instance, the cost to taxpayers of setting up the protection system), then choices reveal respondents' economic values for each attribute included in the design (Hynes et al., 2013a; Jobsvogt et al., 2014).

Avoided cost methods generate value estimates by considering what costs society avoids by ES being maintained at a particular level. For instance, sea grass beds accumulate and store carbon (Luisetti et al., 2013). If x tonnes per km^2 per year are accumulated, the value of this could be measured by the costs avoided by

reducing carbon emissions in some other way (e.g. by investing in wind energy, or by improving public transport links to reduce car use). It is important that the lowest alternative cost avoided is used in such measurement. Another example relates to the flood defence function of coastal wetlands. The avoided costs of investing in conventional flood defences to achieve a given level of flood protection could be used as a measure of one of the benefits of increasing the area of coastal wetland. Economists need to know the effects of changes in ecosystem management on the ES flows that are valued before this method can be used. For example, if the area of sea grasses in a given location falls by 50%, what is the effect on carbon sequestration and storage, as well as on other ES such as sediment control and fisheries production? What is the spatial variability in these relationships, and how does the level of ES supply respond over time to a change in ecosystem pressures? The other main problems with the avoided costs method are: (i) it uses measures of cost to estimate the value of a benefit; (ii) for many ES, and for biodiversity in particular, there are no avoided costs that are relevant, so that the method is very limited in when it can be applied; and (iii) the actions avoided will typically generate a different set of benefits than the ES being valued. For instance, investing in wind energy has a range of impacts that are additional to their impacts on net carbon emissions, while creating new coastal wetlands has benefits in terms of other ES and biodiversity in addition to increasing flood protection.

Many of the methods described above are expensive and time-consuming to apply. Given the desirability of including monetary values of ES and biodiversity in policy-making and management, *value transfer* has emerged as an alternative method (Brander et al., 2012). Value transfer at best consists of the statistical analysis of a range of existing valuation studies for a given ecosystem, ES type, or biodiversity indicator. The aim is to generate a statistical function using meta analysis which explains 'enough' of the variation in measures of monetary value across studies and across sites. This function can then be used to predict values (e.g. in WTP per hectare or per km^2) for a given site where no original valuation function has been undertaken.

19.3 Some case studies

The purpose of this section is to show how the methods described in Section 19.2 have been applied to a range of ecosystems and a range of ES.

19.3.1 Valuing the deep sea

The deep sea (areas of greater than 200 m in depth) constitutes the largest assemblage and diversity of ecosystems or biomes on the planet (*Brandt et al, (2007)*). It is an area rich in biodiversity, although there is much uncertainty over the number of species present (Grassle and Maciolek, 1992). Pressures on this system are increasing globally, with mining for minerals, oil and gas exploration, cable laying, and deep water fishing all increasing (Barbier et al., 2014). Climate change, leading to ocean acidification and rising temperatures, is also an important future threat (Ramirez-Lodra et al., 2011). Armstrong et al. (2012) produced an initial listing of the

ES which the deep sea provides us with: Table 19.2 is taken from their paper. As may be seen, the deep sea provides a range of supporting, regulatory, provisioning, and cultural services. One problem with valuing many of these is the severe lack of evidence linking ecosystem function, service provision, and economic well-being. While the supporting services of the deep sea with respect to coastal systems are potentially highly economically valuable, we currently lack the knowledge basis to quantify the economic costs of declines in these supporting services through their impacts on ES flows in coastal waters, or with any provisioning services other than fish (Foley et al., 2010).

Table 19.2 Ecosystem services from the deep sea and state of evidence.

Services/ecosystems and habitats		Cold water corals	Open slopes and basins	Canyons	Sea-mounts	Chemo-synthetic	Water column	Sub-seabed
Supporting services	Nutrient cycling	?	+	?	?	+	+	0
	Habitat	+	+	+	+	+	+	0
	Resilience	?	?	?	?	?	?	0
	Primary production	?	?	?	?	+	+	0
	Biodiversity	+	+	+	+	+	+	?
	Water circulation and exchange	0	+	+	?	0	+	0
Provisioning services	Carbon capture and storage (artificial)	0	0	0	0	0	+	€
	Finfish, shellfish, marine mammals	+	+	+	+	+	€	0
	Energy: oil, gas, minerals	?	?	0	?	?	0	€
	Chemical compounds: industrial / pharmaceutical	+	?	?	?	+	?	?
	Waste disposal sites	0	+	+	0	0	0	+
Regulating services	Gas and climate regulation	0	?	+	0	+	+	+
	Waste absorption and detoxification	0	+	+	0	0	+	0
	Biological regulation	?	+	?	?	+	+	0
Cultural services	Educational	+	+	+	+	+	+	+
	Scientific	+	+	+	+	+	+	+
	Aesthetic	+	?	?	?	+	+	0
	Existence/bequest	+	?	?	?	?	+	?

Source: Armstrong et al. (2012).

Another problem relates specifically to cultural ES: the deep sea is not something most people can directly experience or which they have much knowledge about. Aanesen et al. (2014) show this with regard to perhaps the best-known aspect of the deep sea, namely cold water corals, in Norway.

However, a few stated preference studies have recently been published which investigate public willingness to pay to conserve deep sea habitats (Wattage et al., 2011). Using choice experiments, Aanesen et al. (2014) show a positive WTP on the part of Norwegians to increase the protection of cold water coral locations from fishing and the oil and gas industry. Jobstvogt et al. (2014) report the results of a choice experiment which quantified the willingness to pay of the Scottish general public for conservation of deep sea areas off the Scottish coastline. A split sample treatment was used to test whether respondents' WTP depended on which sectors (e.g. oil and gas, or fishing) were restricted in their activities as part of such conservation actions. The attributes and levels used in the choice experiment were:

- number of protected species (1000, 1300, or 1600);
- potential for the discovery of new medicinal products from deep sea species (unknown, high); and
- annual tax increases for Scottish residents to fund the scheme (0, 5, 10, 20, 30, 40, 60 pounds sterling).

An example of a choice card from their study is shown as Fig. 19.1. The utility from protecting deep sea biodiversity ('number of protected species') is an example of a non-use or existence value. Benefits from the potential discovery of new medicines in the future are an example of a future provisioning value.

Results showed that respondents valued increased protection of deep sea protection, and were WTP on average around £70–77 per year for the highest level of protection offered to them. There was a significantly higher WTP for protecting 1300 species than 1000, and 1600 rather than 1300. There were no statistically significant differences in WTP according to which sectors faced restrictions. WTP was higher for males, those who ate fish, and those who were members of an environmental NGO. This shows evidence of the economic value of biodiversity conservation varying across people according to their characteristics.

19.3.2 Values from reducing pathogen levels in coastal waters

The deep sea case study provides some evidence of the economic benefits of conserving marine biodiversity. Another example of such evidence is provided by Hynes et al. (2013a), who also apply the choice experiment method. They study the benefits of improvements

SCENARIO 1		Option A	Option B	Option C ['Business as usual']
New medicinal products (potential for the discovery of new medicinal products from deep-sea organisms		**Unknown** (potential for new medicinal products unknown)	**High potential** for new medicines (protect animals with potential for new medicinal products)	**Unknown** (potential for new medicinal products unknown)
Number of protected species (includes animals such as fish, starfish, corals, worms, lobsters, sponges, & anemones)		**1300 species** (300 more than 'business as usual')	**1600 species** (600 more than 'business as usual')	**1000 species** (base level)
Addional costs (per household per year)	**£**	£ 5	£ 60	£ 0
Your choice for scenario 1 (*please tick A, B or C*)		☐	☐	☐

Figure 19.1 Choice card from Jobstvogt et al. (2014).

to coastal waters around Ireland from the tightening of standards under the revised Bathing Waters Directive. Active users of a number of beach sites (e.g. surfers, swimmers, kayakers, and walkers) were surveyed. The choice experiment used the following attributes to describe potential improvements in coastal water quality from implementing the revised directive:

- benthic health;
- human health risks; and
- beach debris.

Each attribute is described in more detail below, and an example choice card is presented as Fig. 19.2.

19.3.2.1 Benthic health

Measures taken as part of complying with the revised directive will impact upon the 'health of the seas' through improvements at the benthic level. However, the concept of benthic health is not likely to be understandable to most members of the public, and so was related here to probable outcomes on vertebrate populations (birds, fish, and marine mammal species). Levels selected were:

- **No improvement** to the current situation, which will mean no changes to the numbers or chance of seeing fish, birds, and mammals.

- **Small improvement in benthic health,** which will mean that there will be more fish, birds, and mammals. This will mean that endangered species will be less likely to disappear from the seas around Ireland. However, respondents were told that it was unlikely that they would see more fish, birds, or mammals on a typical visit to the beach.
- **Large improvement in benthic health** means that there will be many more fish, birds, and mammals, resulting in an increased chance of seeing them on a typical visit to the beach.

19.3.2.2 Health risks

Health risk was included as a design attribute since bacteria concentrations in bathing waters are expected to be reduced under the new directive standards. These bacteria will always be present in marine systems; however, it is the level of untreated or poorly treated waste and agricultural run-off within the system which is most associated with increased risk of human infections from bathing in the sea. The levels of faecal coliforms under current standards, the future 'good' (current excellent) standards, future 'excellent' standards, and, as a point of reference, the levels allowed in swimming pools were identified to respondents. These were then related to the risk of a stomach

	Beach A	Beach B	Beach C
Benthic Health and Population	**Small Increase** More fish, mammals, and birds. Limited potential to notice the change in species numbers.	**Large Increase** More fish, mammals, and birds and an increased potential of seeing these species.	**No Improvement**
Health Risk (of stomach upsets and ear infections)	**5% Risk** —good water quality	**10% Risk** —no improvement	**10% Risk** —no improvement
Debris Management	**No Improvement**	**Collection and Prevention** Debris collected from beaches more regularly in addition to filtration and policing.	**No Improvement**
Additional cost of travelling to each beach	€0.60	€12.00	€0
Please tick the **one** option you prefer.	☐	☐	☐

Figure 19.2 Example choice card from Hynes et al. (2013a).

upset or ear infection, based upon dose response relationships. Levels selected for this attribute were:

- **10% risk**—no change to the current risk of a stomach upset or ear infection from bathing in the sea (current risk as assessed by the EU).
- **5% risk**—good water quality achieved with a somewhat reduced risk of stomach upsets and ear infections, although risks would still be present for vulnerable groups such as children.
- **Very little risk**—excellent water quality achieved with a larger reduction in the risk of stomach upsets and ear infections.

19.3.2.3 Debris management

Changes to sewage treatment and storm water collection made necessary by the directive will likely have effects on debris deposited in beaches. Previous work had shown such debris to be a significant determinant of choices of beach visits. Three levels were used in this choice experiment:

- **No change**—current levels of debris on beaches and in coastal waters will remain.
- **Prevention**—more filtration of storm water, more regular cleaning of filters, and better policing of fly tipping, which will all reduce the generation of new debris.
- **Collection and prevention**—debris collected from beaches more regularly in addition to filtration and policing.

Finally, in order to estimate measures of economic benefit (value) from changes in the environmental attributes listed above, a cost attribute was included in the design. Choices would then show how much people are willing to trade off improvements in an environmental attribute for a decrease in their income. The per visit travel cost to the individual of visiting a beach with a given set of characteristics was used as this cost attribute. Six levels of cost were selected, ranging from €0.90 to €16. Variations in travel costs were used as a means of estimating willingness to pay for changes in the environmental attributes.

Results showed that people were, on average, willing to pay for improvements in all of the attributes in the design. Factors such as personal income, nature of beach use, and level of education of the respondent all had a significant effect on the size of willingness to pay. Table 19.3, which is adapted from their paper, summarizes results for a scenario where all attributes are improved relative to the baseline. The table also shows how results vary according to which 'latent class' respondents are more likely to belong to. Again, this illustrates the way in which the economic value of water quality improvements varies across people. In particular, it shows differences in willingness to pay for risk reductions associated with water pollution according to what kinds of recreational activities people engage in.

19.3.3 Values from reductions in nutrient pollution in the Baltic Sea

Both of the case studies reported in Sections 19.3.1 and 19.3.2 make use of a stated preference method—choice experiments—to estimate the economic values for

Table 19.3 Values of an improvement in coastal water quality due to a reduction in pollution inflows in Ireland (€ per person per year).

Attribute changes	Business as usual	Policy change
Health of the Seabed (benthic health)	Small improvement	Large improvement
Health Risk	5%	virtually zero
Debris Management	Prevention	Collection and Prevention
Economic values (Willingness to pay)		
Conditional Logit	5.59*** (0.86)	
Latent Class 1	9.19*** (1.59)	
Latent Class 2	2.53*** (0 0.59)	

Notes: Figures in parenthesis indicate the values of the standard errors. ***indicates significant at 1%, **indicates significant at 5%, *indicates significant at 10%. The Conditional Logit model shows the mean values across all respondents, with no allowance for preference heterogeneity. Latent class 1 shows the equivalent mean value for people more likely to belong to this latent class, being those who undertake water sports such as surfing and swimming, which mean more immersion in water (and thus more exposure to health risks). Latent class 2 shows values for those who engage in activities with lower levels of immersion and thus risk, such as kayaking and kite-surfing.

environmental changes in marine and coastal waters. Revealed preference methods such as the travel cost model have also been extensively used. *Czajkowski et al (2015)* report results from a large, nine-country study of recreational use of the Baltic Sea, and the benefits which people would derive from improvements in water quality.

The Baltic Sea provides multiple ecosystem services to people from the nine countries which border it, while 85 million people live within its catchment area (Ahtiainen et al., 2014). However, the flow of these services is currently threatened by poor water quality in many areas of the Baltic, mainly due to excessive inflows of nutrients from agriculture and sewage (Ahtiainen et al., 2014). Previous work has applied mainly stated preference approaches to estimating the economic benefits of policies which reduce nutrient inflows to the Baltic at the level of individual nations (e.g. Eggert and Olsson, 2009).

Czajkowski et al. carried out a survey of 9127 individuals living in each of the nine Baltic nations. As part of the survey, people were asked (i) how many recreation trips they had made to the Baltic in the last 12 months, and (ii) how they rated water quality in that part of the Baltic on which their country had a shoreline. Using the information from (i), we related visits per year for each individual to the travel costs they incur per visit. These travel costs were calculated based on how far away each person lived from the closest section of the coastline. Travel distances were converted into costs using estimates of fuel costs and the value of leisure time (since time is a scarce resource). A simple travel cost model as in (1) was estimated:

$$r_i = f(p_i, z_i), \qquad (1)$$

where r_i is the number of trips taken by individual i to a given site during a given time period, p_i is the cost of access to the site, and z_i is a vector of individual characteristics that are believed to influence the number of trips an individual takes, such as their income, age, or family situation. Since r can only take on integer values, we used a negative binomial estimator. We also allowed for the large number of zero values that r could take (since not everyone in the sample had visited the Baltic in the previous 12 months).

The survey data showed that on average 45% of respondents had visited the Baltic in the last 12 months. This varied from 74% in Sweden to 13% in Russia (we only sampled in the two Russian regions on the Baltic coast). The mean number of trips per year among those who made at least one trip varied from 1.1 in

Poland to 6.9 in Sweden, while there was also considerable variation in the estimated cost per trip. Average rating of current water quality, on a scale from 1 (very bad) to 5 (very good), was highest for German respondents and lowest for Russian respondents. Estimation of (1) above allowed the calculation of the economic value of a single visit to the Baltic under current environmental conditions (note that some trips could be multi-day). This value ranged from 439 euros for Swedish respondents to 38 euros for Latvians. Total economic benefits attributable to visits to the Baltic Sea for recreation varied from 22.5 billion euros in Sweden to 0.238 billion euros in Lithuania. This enormous variation across countries reflects differences in: (i) the percentage of people within a country who currently visit the Baltic; (ii) the economic benefit per trip for these people; and (iii) the population level of each country. Differences also reflect the variation in household income across countries, through the link this has with the value of leisure time.

Finally, we used the model to simulate what the economic benefits of an improvement in water quality throughout the Baltic would be. This was done by using the model in (1) above, where we include water quality rating as one of the elements of **z**. This showed that an improvement in perceived water quality equal to a one-unit move up the Likert scale rating (e.g. from 3 to 4) in each country would produce an increase in economic benefits from recreation in the Baltic equal to between + 5% and + 25% of existing economic benefits across the nine countries, dependent on how important perceived water quality was to trips in that country. Across all nine Baltic states such an improvement would be worth around 3 billion euros *per year*. While this is based on the statistical link between recreational use and perceived water quality, some studies in the same area show a close match between variations in perceived and measured water quality in the Baltic (Artell et al., 2013).

19.3.4 Galway Bay

The case study in Section 19.3.3 showed how the value of improvements to the Baltic Sea, in terms of reducing nutrient pollution, varied to a very considerable degree across the nine countries which border the Baltic. Such spatial variation in economic values is a common finding in the economics literature, and can be found at most lower levels of spatial resolution. An example in a coastal setting is Hynes et al. (2013b). They investigate the use of value transfer to estimate values for

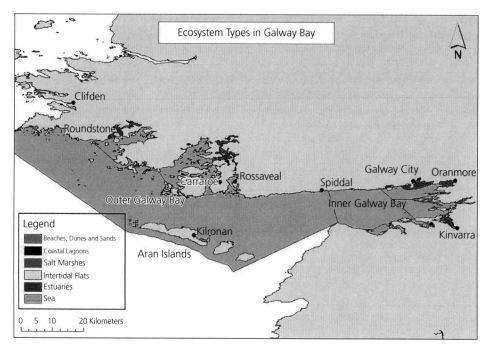

Figure 19.3 The Galway Bay coastal zone with associated ecosystems.

Source: Hynes et al. (2013b).

(i) different habitats and (ii) different ES flows from Galway Bay, Ireland (Fig. 19.3). Value transfer means taking economic value estimates from one or a set of existing studies, and then adjusting them in some manner to predict values at a new site for which no original study values exist (Johnston and Rosenberger, 2010). Hynes et al. compare transfer values which adjust simply for differences in income levels between countries, and those which also adjust for differences in cultural values as proxied by a set of indicators. The study area in Galway Bay consisted of 143 430 ha of aquatic habitats (such as salt marsh, rocky coastline, and inter-tidal mud flats) and 2840 ha of terrestrial habitats. Some 209 value estimates from 123 separate studies were used in the value transfer.

Tables 19.4–19.6 show both the split in economic values for the total flow of ES from Galway Bay in its current state by type of service (Table 19.5) and by type of habitat (Table 19.6). In each case, we present three values: one where the Galway Bay number is a value transfer from the 209 value estimates for other sites which only adjusts for differences in national income levels; one which only adjusts for differences in the cultural indices; and one which uses both adjustments.

This final set of 'both adjustments' figures is the preferred value estimate for Galway Bay. Table 19.4 gives the per-hectare transfer mean values which are used to produce these aggregate benefits by service or by habitat; the highest value is for coastal lagoons, associated with the mitigation of eutrophication problems. As may be seen, sea areas provide the highest aggregate benefit since such areas account for 95% of the study area, but provide low per-hectare values. Significant economic values are associated with ES flows from beaches and dunes, and from coastal lagoons and estuary areas. The highest valued single ES is eutrophication mitigation, with sediment retention and 'other' pollution control also being valuable regulatory services. Recreation and nature values sum together to around 60 million euros per year. While there is no suggestion that any of these benefit estimates are precise, they do provide an indication of the range of relative values across space and by type of service, which could be useful in targeting management measures. However, the authors point out that these valuation estimates for each ES were carried out independent of all other ES values, so that no account was taken of synergies or substitution effects.

Table 19.4 Per-hectare values for value of ecosystem services, Galway Bay.

	Beach, dunes	Salt marsh	Intertidal flats	Coastal lagoons	Estuaries	Open sea
Sediment regulation	22.7, 9	1.4, 3				
Biological regulation						0.03, 1
Pollution control	13.0, 7				0.8, 2	0.01, 3
Eutrophication mitigation		9.3, 6		58.7, 2	5.2, 16	0.7, 15
Recreation	22.2, 13	10.2, 10		2.7, 2	0.6, 6	0.1, 41
Aesthetic values	4.3, 2					0.1, 3
Nature values	3.4, 3	4.0, 4	1.4, 1	31.7, 3	1.5, 1	0.06, 11

In each cell, the first value is the mean value per hectare for this service supplied by a given habitat obtained via value transfer. The second value (after the comma) is the number of studies on which this value transfer is based. Blank cells show that no studies were available for this service/habitat combination. For more details, see Hynes et al. (2013b).

Table 19.5 Total values of service flows from Galway Bay by type of service (thousand euro per year).

	Adjusted for income differences	Adjusted for cultural differences	Both adjustments
Sediment regulation	15 265	25 111	16 060
Biological regulation	3845	4064	3735
Pollution control	12 713	11 645	13 536
Eutrophication mitigation	136 817	270 129	143 522
Recreation	34 533	51 685	35 939
Aesthetic values	15 266	15 294	15 530
Nature values	34 849	44 663	36 389

Source: Adapted from Hynes et al. (2013b).

Table 19.6 Total values of ecosystem service flows from Galway Bay by habitat (thousand euro per year).

	Area in Galway Bay (ha)	Adjusted for income differences	Adjusted for cultural differences	Both adjustments
Beaches and dunes	691	43 466	56 857	45 330
Salt marsh	279	6636	11 191	6966
Intertidal flats	1584	2194	2411	2252
Coastal lagoons	400	34 547	65 928	37 255
Estuaries	3976	30 158	34 496	31 958
Open sea	139 386	132 962	248 385	137 615
Seagrass and kelp beds	1622	3206	3203	3217

Source: Adapted from Hynes et al. (2013b).

19.4 Conclusions

As other chapters in this book show, coastal and marine ecosystems are under increasing pressure from a number of sources. These sources have interactive effects on the supply of ES, such that cumulative impacts are typically greater than the sum of individual impacts. The same stressors also produce effects on indicators of biodiversity. In this chapter, I have explained the basis by which ES and biodiversity have an economic value, even though many of these values are not recognized by markets. I then described briefly the range of methods which environmental economists have developed to estimate such non-market economic values. Finally, four case studies were presented, which show results from empirical applications of these valuation methods to different marine and coastal management problems.

What lessons can be drawn from this discussion? Firstly, that it is possible to provide estimates of the economic value of measures to protect or enhance marine biodiversity. However, there are considerable difficulties in performing such estimations, most importantly the lack of knowledge of ordinary people of the nature and importance of many marine species. Apart from some coastal species (e.g. dolphins and some sea birds) and iconic species (e.g. whales), many people lack any familiarity with such animals. While it is possible to still get them to express a monetary value for protecting the unfamiliar (as in Jobstvogt et al., 2014), the process is complicated by this lack of knowledge. Yet economists would argue that it is the values of ordinary people, rather than those of experts, which are most relevant to the economic assessment of policies or projects funded by or approved by the government on behalf of its citizens (Hanley, 2001). Techniques such as valuation workshops can be of assistance here (e.g. as in Colombo et al., 2013; Aaneson et al., 2014), since they allow people to learn about unfamiliar goods before being asked to value them (LaRiviere et al., 2014).

Secondly, a given environmental change which produces changes in ES supply and/or biodiversity can have outcomes in terms of economic values which differ greatly across space and across groups of citizens. This is seen most particularly in the Baltic Sea case study. Such spatial differences in economic values of environmental change are particularly relevant when the costs of producing such changes (e.g. of reducing nutrient loads going into the Baltic) also vary considerably with space. For example, if the costs of a nutrient management plan fall particularly heavily on countries or groups of stakeholders who receive a fairly low share of the benefits, then this makes it much harder to reach agreement on the need for such actions, and much more desirable to engage in cost-sharing (Hasler et al., 2014).

Thirdly, economists need very considerable model inputs and information from natural scientists if they are to conduct valuation exercises. This is true in any environmental setting, but the complexity of ecosystem linkages in coastal and marine areas can make this undertaking particularly difficult. This is especially true if production function approaches are used to estimate the indirect economic value of ES and/or biodiversity, since finding robust and generalizable empirical relationships which relate changes in ecosystem functions to changes in human well-being is very challenging. This perhaps explains the relative lack of production function applications to ecosystem services and biodiversity in marine and coastal settings, when compared to applications in terrestrial settings (Hanley and Barbier, 2009).

References

Aanesen, M., Armstrong, C., Czajkowski, M., et al. (2014). *Willingness to pay for unfamiliar public goods: preserving cold-water corals in Norway.* Paper to the 2014 BIOECON conference, Cambridge, UK.

Ahtiainen, H., Artell, J., Czajkowski, M., et al. (2014). Benefits of meeting nutrient reduction targets for the Baltic Sea—a contingent valuation study in the nine coastal states. *Journal of Environmental Economics and Policy*, **3**(3), 278–305.

Armstrong, C.W., Foley, N.S., Tinch, R., & van den Hove, S. (2012). Services from the deep: steps towards valuation of deep sea goods and services. *Ecosystem Services*, **2**, 2–13.

Artell, J., Ahtiainen, H., & Pouta, E. (2013). Subjective vs. objective measures in the valuation of water quality. *Journal of Environmental Management*, **130**, 288–296.

Barbier, E.B., Moreno-Mateos, D., Rogers, A.D., et al. (2014). Protect the deep sea. *Nature*, **505**, 475–477.

Barbier, E.B. & Strand, I. (1998). Valuing mangrove-fishery linkages. *Environmental and Resource Economics*, **12**, 151–166.

Bateman, I.J., Mace, G., Fezzi, C., et al. (2011). Economic analysis for ecosystem service assessments. *Environmental and Resource Economics*, **48**, 177–218.

Boyd, J., Banzhaf, S. (2007) "What are ecosystem services? The need for standardized environmental accounting units" *Ecological Economics*, 63 (2–3), pp. 616–626.

Brander, L., Wagtendonk, A., Hussain, S., et al. (2012). Ecosystem service values for mangroves in Southeast Asia. *Ecosystem Services*, **1**, 62–69.

Brandt, A., Gooday, A.J., Brandão, S.N., Brix, S., Brökeland, W., Cedhagen, T., Choudhury, M., Cornelius, N., Danis, B., De Mesel, I., Diaz, R.J., Gillan, D.C., Ebbe, B., Howe, J.A.,

Janussen, D., Kaiser, S., Linse, K., Malyutina, M., Pawlowski, J., Raupach, M., Vanreusel, A. "First insights into the biodiversity and biogeography of the Southern Ocean deep sea" (2007) Nature, 447 (7142), pp. 307–311.

Chae, D.-R., Wattage, P., & Pascoe, S. (2012). Recreational benefits from a marine protected area: a travel cost analysis of Lundy. *Tourism Management*, **33**(4), 971–977.

Colombo, S., Christie, M., & Hanley, N. (2013). What are the consequences of ignoring attributes in choice experiments? Implications for ecosystem service values. *Ecological Economics*, **96**, 25–35.

Czajkowski M, H. Ahtiainen, J. Artell, W. Budziński, B. Hasler, L. Hasselström, J. Meyerhoff, T. Nõmmann, D. Semeniene, T. Söderqvist, H. Tuhkanen, T. Lankia, A. Vanags, M. Zandersen, T. Żylicz and N. Hanley. "Valuing the commons: an international study on the recreational benefits of the Baltic Sea" *Journal of Environmental Management*, Volume 156, 1 June 2015, Pages 209–217.

Eggert, H. & Olsson, B. (2009). Valuing multi-attribute marine water quality. *Marine Policy*, **33**(2), 201–206.

Foley, N.S., Van Rensburg, T.M., & Armstrong, C.W. (2010).The ecological and economic value of cold-water coral ecosystems. *Ocean and Coastal Management*, **53**(7), 313–326.

Grassle, J.F. & Maciolek, N.J. (1992). Deep-sea species richness: regional and local diversity estimates from quantitative bottom samples. *American Naturalist*, **139**(2), 313–341.

Hanley, N. (2001). Cost-benefit analysis and environmental policy making. *Environment and Planning C*, **19**, 103–118.

Hanley, N. & Barbier, E.B. (2009). *Pricing Nature: Cost-Benefit Analysis and Environmental Policy*. Edward Elgar, Cheltenham.

Hanley, N., Shogren, J., & White, B. (2007). *Environmental Economics in Theory and Practice*, 2nd edition. Palgrave MacMillan,London.

Hasler, B., Smart, J.C.R., Fonnesbech-Wulff, A., et al. (2014). Hydro-economic modelling of cost-effective transboundary water quality management in the Baltic Sea. *Water Resources and Economics*, **5**, 1–23.

Helm, D. & Hepburn, C. (2014). *Nature in the Balance: The Economics of Biodiversity*. Oxford University Press,Oxford.

Hynes, S., Hanley, N., & O'Donoghue, C. (2009). Alternative treatments of the cost of time in recreational demand models: an application to whitewater kayaking in Ireland. *Journal of Environmental Management*, **90**(2), 1014–1021.

Hynes, S., Hanley, N., & Tinch, D. (2013a). Valuing improvements to coastal waters using choice experiments: an application to revisions of the EU Bathing Waters Directive. *Marine Policy*, **40**, 137–144.

Hynes, S., Norton, D., & Hanley, N. (2013b). Accounting for cultural dimensions when estimating the value of coastal zone ecosystem services using benefit transfer. *Environmental and Resource Economics*, **56**, 499–519.

Jobstvogt, N., Hanley, N., Hynes, S., et al. (2014). Twenty thousand sterling under the sea: estimating the value of protecting deep sea biodiversity. *Ecological Economics*, **97**, 10–19.

Johnston, R. & Rosenberger, R. (2010). Methods, trends and controversies in contemporary benefit transfer. *Journal of Economic Surveys*, **24**, 479–510.

Kontoleon, A., Pascual, U., & Swanson, T. (2007). *Biodiversity Economics*. Cambridge University Press, Cambridge, UK.

LaRiviere, J., Czajkowski, M., Hanley, N., et al. (2014). The value of familiarity: effects of knowledge and objective signals on willingness to pay for a public good. *Journal of Environmental Economics and Management*, **68**(2), 376–389.

Leggett, C.G. & Bockstael, N. (2000). Evidence on the effects of water quality on residential land prices. *Journal of Environmental Economics and Management*, **39**, 121–144.

Luisetti, T., Jackson, E., & Turner, R.K. (2013). Valuing the European 'coastal blue' carbon storage benefit. *Marine Pollution Bulletin*, **71**, 101–106.

Millennium Ecosystem Assessment (2005). *Ecosystems and Human Well-being: Synthesis*. Island Press, Washington, DC.

Poor, P.J., Pessagno, K.L., & Paul, R.W. (2007). Exploring the hedonic value of ambient water quality: a local watershed-based study. *Ecological Economics*, **60**, 797–806.

Ramirez-Llodra, E., Tyler, P., Baker, M.C., et al. (2011). Man and the last great wilderness: human impact on the deep sea. *PLoS ONE*, **6**(7), 1–25.

Ressurreincao, A., Gibbons, J., Ponce, T., et al. (2011). Economic valuation of species loss in the open sea. *Ecological Economics*, **70**(4), 729–739.

Wattage, P., Glenn, H., Mardle, S., et al. (2011). Economic value of conserving deep-sea corals in Irish waters: a choice experiment study on marine protected areas. *Fisheries Research*, **107**(1–3), 59–67.

Subject Index

Notes

Page numbers followed by f indicate a figure, t a table and b boxed material. *vs.* indicates a comparison or differential diagnosis

Abbreviations

OA – ocean acidification

UV-A – ultraviolet radiation type A

UV-B – ultraviolet radiation type B

UV-R – ultraviolet radiation

A

abiotic sources, noise 284

ABR (auditory brain response) 140–1

ABTs (Arrhenius break temperatures) 62

Acartia tonsa, low nitrogen stress 104–5

acid–base balance, carbon dioxide increases 40, 43, 46, 198

acid phosphatase (ACP), salinity stress 16

acorn barnacle (*Amphibalanus amphitrite*), OA effect on shell 42

acoustic characteristics 141

acoustic communication, masking by noise stress 291–2

acoustic deterrent devices 288–9

acoustic Doppler current profilers 289

acoustic impedance, air *vs.* water 137

active oxygen uptake, environmental hypoxia 28

N-acyl homoserine lactose (AHL) 108

adaptive immune responses 77

Adélie penguin (*Pygoscelis adeline*), noise stress 150–1

adenylate energy charge (AEC) 10

Adriatic Sea, as hypoxic environment 178

AEC (adenylate energy charge) 10

agent-based modelling, sustainability 308

agricultural chemicals as chemical pollutants 230–1

AHL (N-acyl homoserine lactose) 108

Alaria marginata (winged kelp) 124–5

alarm response, noise stress in seabirds 150

algae

  Dunaliella terticolecta 273

  Fucus vesiculosus (bladder wrack) 129

  harmful algal blooms 107

  Macrocystis pyrifera (giant kelp) 128

  Palmaria palmata (dulse) 129

  UV-R effects

    salinity stress with 129

    in sea ice 267

    thermal stress with 124–5

  see also macroalgae; microalgae

alkaline phosphatase (ALP), salinity stress 16

ALP (alkaline phosphatase), salinity stress 16

Alphaeus (snapping shrimp), noise production 285

alternative biocides 232

ammonia (NH$_3$)

  assimilation, chemical pollutants with UV-R 128

  nitrogen cycle 249

  toxicity in animals 105–6

ammonium (NH$_4^+$)

  from chemical fertilizers 230

  high nitrogen stress 99

  hypoxic environments 177

  mangrove 101–2

  pH effects on conversion 163

  saltmarsh 101

Amoco Cadiz 233

Amphibalanus amphitrite (acorn barnacle), OA effect on shell 42

anaerobic metabolism, environmental hypoxia effects 30–1

analysis phase, US-EPA risk management 310

anammox pathway 249, 249f, 250

androgenic activity of chemical pollutants 80

angiosperms

  Halodule wrightii (shoalweed) 102

  high nitrogen stress 102–3

  low nitrogen stress 101–2

  Ruppia maritima 102

  Spartina alterniflora (smooth cordgrass) 103

  Zostera marina (common eelgrass) 102

animals

  ammonia toxicity 105–6

  high nitrogen stress 105–6

  low nitrogen stress 103–5

Antarctic

  invertebrates, thermal stress 58

  OA effects 202

  ozone hole 262, 263f, 265

Anthropocene 175

anthropogenic effects

  freshwater on estuaries 166

  reactive nitrogen (N$_r$) 254–5

  salinity stress 169–70

  salt pans 169–70

  thermal stress 215

anti-androgenic chemicals, chemical pollutants 80

antibiotics 228f

antidepressant drugs 230

antifoulants 231–2, 234

definition 162
in estuaries 165
St Lawrence Estuary, food web effects
  of UV-B 268–9
storm drain outfalls, salinity
  reduction 170
strandings, mammal response to noise
  stress 148–50
stress
  definition 3, 95
  organism-specific measures 96
Strongylocentrotus droebachiensis (green
  sea urchin), UV-R effects 120
Stylocheilus striatus (sea hare),
  noise stress effects on
  development 147
sub-detection sonar 148
sulphate, hypoxia with 188
superoxide dismutase (SOD)
  chemical pollutants 74
  reactive oxygen species control 7
  thermal stress 59
supporting services, Ecosystem
  Approach 316
surveillance monitoring, chemical
  pollutants 239
sustainability 301–12
  chemical pollutant management 238
  interdisciplinary models 307–9
  planetary boundaries concept
    301–2
  policy support oriented research
    frameworks 305–7
  see also Driver-Pressure State
    Impact-Response
    frameworks (DPSIR)
  risk management 309–11
  tipping points 303–5
swim bladder, noise effects in fish 140
swimming behaviour, noise stress 290
Synalpheus (snapping shrimp), noise
  production 285
systems models, sustainability 308

T

T4 (thyroxine) 80
taurine 9
TBARS (thiobarbituric acid reactive
  substances) 75
TBT *see* tributyltin (TBT)
TCDD (digoxin) 86
temperate regions, OA effects 203–4
temperature coefficient (Q_{10}) 57
temporal context of predation 253–4
temporary threshold shifts (TTS) 135
  fish 140–1
  hearing system effects 138
  mammals 141–2
  see also noise stress

THC (total haemocyte count), salinity
  stress 16
thermal sensitivity, species
  variations 216–17
thermal stress 56–72
  anthropogenic impacts 215
  community responses 221–3,
    221f, 222t
    community structure 223
    habitat complexity 222–3
    microbial loop 221–2
    phytoplankton 221–2
  ecosystem effects 213–27
    biotic interactions 217
    food chains 219
    intertidal habitats 214
    localised processes 214–15
    organism-level impacts 216–17
    species distribution 217–19, 218b
  global spatial trends 213–16
  metabolic responses 58
  multiple stressors with 64–6
    chemical pollution 224
    ecosystem effects 223–4
    environmental hypoxia 64,
      65–6, 188
    nitrogen stress 108
    noise stress 292
    OA 44, 46, 64, 65
    oxygen depletion 224
    salinity stress 64, 66
    ultraviolet radiation *see* ultraviolet
      radiation (UV-R)
    UV-R *see* ultraviolet radiation
      (UV-R)
  physiological response 56, 57–8
    chlorophyll increase 205
    latitudinal variation 60–3, 66
    metabolic rate 61
    vertical zonation 63–4,
      215–16, 216f
  recent observations 215
  repair mechanisms 58–60
    oxidative stress protection 59–60
    protein denaturation 58–9
  temperature projections 215–16
  temporal scales 214
  thermal tolerance limits 57
  transgenerational phenotypic
    plasticity 56–7
thermal tolerance limits 57
  species variations 216–17
thermohaline circulation 213
thiobarbituric acid reactive substances
  (TBARS) 75
Three Gorges Dam (China) 235
thyroxine (T4) 80
tidal pools, salinity 168
tidal turbines 288

Tigriopus brevicornis, tidal pool
  ecosystems 168
time–series analyses, OA 195–6
tipping points, sustainability 303–5
tissue pathology, salinity stress 15–17
tocopherol (vitamin E) 74
toothed whales (Odontocetes), sound
  production 284
total haemocyte count (THC), salinity
  stress 16
trade-off investigations, ecosystem
  services 323
trade-offs, ecosystem services 320–1
transcription
  cellular homeostatic response 9–10
  salinity stress in immune
    function 17
transgenerational phenotypic
  plasticity, thermal stress 56–7
translation, cellular homeostatic
  response 9–10
tributyltin (TBT) 231–2
  gastropods 82
triclosan 230
trophic cascades, reactive nitrogen
  effects 253
tropical species
  OA effects 202–4
  oxidative stress in thermal stress 63
  species range expansion 219–20b
  temperature tolerance 60–1
TTS *see* temporary threshold shifts
  (TTS)
Tursiops truncatus (bottlenose dolphin)
  endocrine disruption by chemical
    pollutants 80–1
  noise stress
    acoustic communication
      masking 291
    barotrauma 145
    tests 142
  sound production 284
turtles, noise stress effects 141, 148
  barotrauma 145
  swimming behavior 290

U

Uccinum undatum, salinity stress
  effects 14
ultraviolet radiation (UV-R) 117–34
  atmospheric penetration 261, 262f
  biological weighting functions
    120–2, 121f
  changes in 117
  concentrations of dissolved organic
    matter 127
  increase causes 117
  multiple stressors with 122–9,
    123f, 124t

Printed and bound by CPI Group (UK) Ltd, Croydon, CR0 4YY